VOLUME FOUR HUNDRED AND EIGHTY-FOUR

Methods in ENZYMOLOGY

Constitutive Activity in Receptors and Other Proteins, Part A

METHODS IN ENZYMOLOGY

Editors-in-Chief

JOHN N. ABELSON AND MELVIN I. SIMON

Division of Biology
California Institute of Technology
Pasadena, California

Founding Editors

SIDNEY P. COLOWICK AND NATHAN O. KAPLAN

VOLUME FOUR HUNDRED AND EIGHTY-FOUR

Methods in ENZYMOLOGY

Constitutive Activity in Receptors and Other Proteins, Part A

EDITED BY

P. MICHAEL CONN
Divisions of Reproductive Sciences and Neuroscience (ONPRC)
Departments of Pharmacology and Physiology,
Cell and Developmental Biology, and Obstetrics and
Gynecology (OHSU)
Beaverton, OR, USA

AMSTERDAM • BOSTON • HEIDELBERG • LONDON
NEW YORK • OXFORD • PARIS • SAN DIEGO
SAN FRANCISCO • SINGAPORE • SYDNEY • TOKYO
Academic Press is an imprint of Elsevier

Academic Press is an imprint of Elsevier
525 B Street, Suite 1900, San Diego, CA 92101-4495, USA
30 Corporate Drive, Suite 400, Burlington, MA 01803, USA
32 Jamestown Road, London NW1 7BY, UK

First edition 2010

Copyright © 2010, Elsevier Inc. All Rights Reserved.

No part of this publication may be reproduced, stored in a retrieval system or transmitted in any form or by any means electronic, mechanical, photocopying, recording or otherwise without the prior written permission of the publisher

Permissions may be sought directly from Elsevier's Science & Technology Rights Department in Oxford, UK: phone (+44) (0) 1865 843830; fax (+44) (0) 1865 853333; email: permissions@elsevier.com. Alternatively you can submit your request online by visiting the Elsevier web site at http://elsevier.com/locate/permissions, and selecting *Obtaining permission to use Elsevier material*

Notice
No responsibility is assumed by the publisher for any injury and/or damage to persons or property as a matter of products liability, negligence or otherwise, or from any use or operation of any methods, products, instructions or ideas contained in the material herein. Because of rapid advances in the medical sciences, in particular, independent verification of diagnoses and drug dosages should be made

For information on all Academic Press publications
visit our website at elsevierdirect.com

ISBN: 978-0-12-381298-8
ISSN: 0076-6879

Printed and bound in United States of America
10 11 12   10 9 8 7 6 5 4 3 2 1

Working together to grow
libraries in developing countries

www.elsevier.com | www.bookaid.org | www.sabre.org

ELSEVIER   BOOK AID International   Sabre Foundation

Contents

Contributors — xv
Preface — xxv
Volumes in Series — xvii

Section I. Identification and Measurement of Constitutive Activity — 1

1. Constitutive Activity at the Cannabinoid CB_1 Receptor and Behavioral Responses — 3
Katherine E. Hanlon and Todd W. Vanderah

1. Introduction — 4
2. Modifying CB_1 Activity — 6
3. Behavioral Models — 12
References — 26

2. Detecting Constitutive Activity and Protean Agonism at Cannabinoid-2 Receptor — 31
Massimiliano Beltramo, Rossella Brusa, Isabella Mancini, and Paola Scandroglio

1. Introduction — 32
2. General Considerations — 34
3. Evaluation of Constitutive Activity Using GTPγS Assay — 35
4. Evaluation of Constitutive Activity Using cAMP Assay — 39
5. Evaluation of Constitutive Activity Using RT-CES — 42
6. Evaluation of Protean Agonism with cAMP Assay — 46
7. Comparison of the Methods — 48
References — 49

3. Modulation of the Constitutive Activity of the Ghrelin Receptor by Use of Pharmacological Tools and Mutagenesis — 53
Jacek Mokrosiński and Birgitte Holst

1. Introduction — 54
2. The Ghrelin Receptor and Its Constitutive Activity — 55
3. Structural Basis of Constitutive Activity — 57

4. Residues Responsible for the Inverse Agonism and Efficacy Swap	61
5. Physiological Relevance of Constitutive Activity	65
6. Experimental Procedures	66
References	69

4. Assessment of Constitutive Activity and Internalization of GPR54 (KISS1-R) — 75

Macarena Pampillo and Andy V. Babwah

1. Introduction	76
2. Materials	77
3. Methods	79
Acknowledgment	91
References	92

5. Assessment of Constitutive Activity in E-Type Prostanoid Receptors — 95

Hiromichi Fujino, Toshihiko Murayama, and John W. Regan

1. Introduction	96
2. Assays Used to Assess EP3 Receptor Constitutive Activity	97
3. Assays Used to Assess EP4 Receptor Constitutive Activity	105
References	107

6. α_{1D}-Adrenergic Receptors: Constitutive Activity and Reduced Expression at the Plasma Membrane — 109

J. Adolfo García-Sáinz, M. Teresa Romero-Ávila, and Luz del Carmen Medina

1. Introduction	110
2. Methods	111
3. Constitutive Activity	114
4. Plasma Membrane α_{1D}-Adrenergic Receptors	117
5. Possible Physiological Implications	119
Acknowledgments	121
References	121

7. Constitutive Activity of the Histamine H_1 Receptor — 127

Saskia Nijmeijer, Rob Leurs, and Henry F. Vischer

1. Introduction	128
2. Methods to Study Constitutive H_1R Signaling	133
3. Constitutive Activity as Tool to Elucidate Receptor Activation and Crosstalk	142

	4. Conclusion	144
	Acknowledgments	144
	References	145

8. Constitutive Activity of Somatostatin Receptor Subtypes — 149
Anat Ben-Shlomo, Kolja Wawrowsky, and Shlomo Melmed

1. Introduction — 150
2. Choosing Cells to be Studied — 151
3. Modifying SSTR Density — 152
4. Summary — 163
References — 163

9. Assessment of Homologous Internalization of Constitutively Active N111G Mutant of AT_1 Receptor — 165
Mohiuddin Ahmed Bhuiyan and Takafumi Nagatomo

1. Introduction — 166
2. Preparation of Receptor Plasmid and Protein — 167
3. Radioligand Binding Assay — 168
4. Inositol Phosphate Accumulation Assay — 170
5. Internalization Assay — 171
6. Western Blot Analysis — 173
7. Data Analysis — 174
8. Concluding Remark — 175
References — 175

10. Methods to Detect Cell Surface Expression and Constitutive Activity of GPR6 — 179
Balakrishna M. Prasad, Bettye Hollins, and Nevin A. Lambert

1. Introduction — 180
2. GPR6 is Expressed in Intracellular Compartments — 180
3. Comparison of the Cell Surface Protein Detection Methods — 187
4. Comparison of the Constitutive Gs-Activity Detection Methods — 192
5. Conclusions — 194
References — 194

11. β_3-Adrenoceptor Agonists and (Antagonists as) Inverse Agonists: History, Perspective, Constitutive Activity, and Stereospecific Binding — 197
Maria Grazia Perrone and Antonio Scilimati

1. Introduction — 198
2. β_3-Adrenoceptor — 201

3. Methodologies	214
Acknowledgments	224
References	225

12. Constitutive Activity of the Lutropin Receptor and Its Allosteric Modulation by Receptor Heterodimerization — 231

Deborah L. Segaloff

1. Introduction	232
2. General Principles for Quantifying Receptor Activation	235
3. Modifying Cell Surface Expression Levels of Recombinant hLHR and Mutants Thereof	239
4. Quantifying Cell Surface hLHR Expression	243
5. Quantifying cAMP Production in Cells Expressing the hLHR	245
6. Experimental Strategies for Characterizing the Attenuating Effects of a Signaling Inactive hLHR on a Coexpressed wt or CAM hLHR	246
Acknowledgment	250
References	250

13. Assessing Constitutive Activity of Extracellular Calcium-Sensing Receptors *In Vitro* and in Bone — 253

Wenhan Chang, Melita Dvorak, and Dolores Shoback

1. Introduction	254
2. Materials and Methods	258
3. Results	262
4. Conclusions	263
Acknowledgments	265
References	265

14. Constitutive Activity of Neural Melanocortin Receptors — 267

Ya-Xiong Tao, Hui Huang, Zhi-Qiang Wang, Fan Yang, Jessica N. Williams, and Gregory V. Nikiforovich

1. Introduction	268
2. Signaling Assay for the Neural Melanocortin Receptors	268
3. Naturally Occurring Constitutively Active MC4R Mutants	270
4. Inverse Agonism of AgRP at the MC3R	273
5. Computational Modeling of the Constitutively Active MC4R Mutants	273
Acknowledgments	277
References	277

15. Measurement of Constitutive Activity of BMP Type I Receptors 281

David J. J. de Gorter, Maarten van Dinther, and Peter ten Dijke

1. Introduction 281
2. Determining ALK2 Constitutive Activity 284
3. Determining the Effects of ALK2 Constitutive Activity on Osteoblast Differentiation 289
4. Concluding Remarks 291
Acknowledgments 292
References 292

16. Probing the Constitutive Activity Among Dopamine D1 and D5 Receptors and Their Mutants 295

Bianca Plouffe, Jean-Philippe D'Aoust, Vincent Laquerre, Binhui Liang, and Mario Tiberi

1. Introduction 296
2. Design of Genetically Modified D1-Like Receptor Constructs and Cloning Strategy 299
3. Transfection of D1R and D5R Expression Constructs in HEK293 Cells 301
4. Radioligand-Binding Assays 303
5. Whole Cell cAMP Assays 305
6. Results Validating Experimental Approaches 313
7. Concluding Remarks 324
Acknowledgments 326
References 327

17. Identification of Gain-of-Function Variants of the Human Prolactin Receptor 329

Vincent Goffin, Roman L. Bogorad, and Philippe Touraine

1. Introduction 330
2. Experimental Procedures 333
3. Identification of Constitutive Activity: Results and Discussion 345
4. Conclusions 353
Acknowledgments 353
References 353

18. Investigations of Activated ACVR1/ALK2, a Bone Morphogenetic Protein Type I Receptor, That Causes Fibrodysplasia Ossificans Progressiva 357

Frederick S. Kaplan, Petra Seemann, Julia Haupt, Meiqi Xu, Vitali Y. Lounev, Mary Mullins, and Eileen M. Shore

1.	Introduction	358
2.	Patient Methodologies	359
3.	Cellular Methodologies	362
4.	Tissue Methodologies	366
5.	*In Vivo* Methodologies	369
	Acknowledgments	371
	References	371

19. Identification and Evaluation of Constitutively Active Thyroid Stimulating Hormone Receptor Mutations — 375

Joaquin Lado-Abeal, Leah R. Quisenberry, and Isabel Castro-Piedras

1.	Introduction	376
2.	*TSHR* Gene Mutational Screening	382
3.	Determination of TSHR Constitutive Activity *In Vitro*	386
4.	Measurement of TSHR Expression at Cell Surface by Flow Cytometry Analysis	388
5.	TSH–TSHR Binding Assays	389
6.	TSHR Phosphorylation Analysis	391
	Acknowledgments	392
	References	392

20. Assessment of Constitutive Activity of a G Protein-Coupled Receptor, Cpr2, in *Cryptococcus neoformans* by Heterologous and Homologous Methods — 397

Chaoyang Xue, Yina Wang, and Yen-Ping Hsueh

1.	Introduction of Receptors and Constitutive Receptors	398
2.	Identification of Cpr2 as a Natural Occurring Constitutively Active Receptor	399
3.	Additional Constitutively Active Receptors Identified in Fungi	409
	Acknowledgments	409
	References	409

21. *In Vitro* and *In Vivo* Assessment of Mu Opioid Receptor Constitutive Activity — 413

Edward J. Bilsky, Denise Giuvelis, Melissa D. Osborn, Christina M. Dersch, Heng Xu, and Richard B. Rothman

1.	Introduction	414
2.	Measuring Opioid Receptor Constitutive Activity *In Vitro*	415
3.	cAMP Quantification Assay in CHO Cells Expressing Cloned Opioid Receptors	425

4.	*In Vivo* Assessment of Antagonist Potency in Opioid Naïve Subjects	430
5.	*In Vivo* Assessment of Antagonist Potency to Precipitate Withdrawal	435
6.	Summary	440
	Acknowledgments	441
	References	441

22. Constitutively Active μ-Opioid Receptors 445

Mark Connor and John Traynor

1.	Introduction	446
2.	Methods for Measuring Constitutive Activity	447
3.	Conclusions	465
	Acknowledgments	465
	References	466

23. Protein Kinase CK2 Is a Constitutively Active Enzyme that Promotes Cell Survival: Strategies to Identify CK2 Substrates and Manipulate Its Activity in Mammalian Cells 471

Jacob P. Turowec, James S. Duncan, Ashley C. French, Laszlo Gyenis, Nicole A. St. Denis, Greg Vilk, and David W. Litchfield

1.	Introduction	472
2.	Purification of CK2 for *In Vitro* Studies	474
3.	Assays for CK2 Activity	481
4.	Modulation of CK2 in Mammalian Cells	485
5.	Conclusions	491
	Acknowledgments	491
	References	492

24. Assessment of CK2 Constitutive Activity in Cancer Cells 495

Maria Ruzzene, Giovanni Di Maira, Kendra Tosoni, and Lorenzo A. Pinna

1.	Introduction	496
2.	Assay of CK2 in Crude Biological Samples	499
3.	In-Cell Assay of Endogenous CK2 Activity	504
4.	Identification/Validation of *In Vivo* CK2 Targets with Specific Inhibitors	506
	Acknowledgments	511
	References	511

25. Structural Basis of the Constitutive Activity of Protein Kinase CK2 515

Birgitte B. Olsen, Barbara Guerra, Karsten Niefind, and Olaf-Georg Issinger

1. Introduction 516
2. A Constitutively Active CK2α Structure and Its Stabilizing Elements 516
3. Analyzing the Constitutive Activity of Protein Kinase CK2 521
Acknowledgments 528
References 528

26. Measuring the Constitutive Activation of c-Jun N-terminal Kinase Isoforms 531

Ryan T. Nitta, Shawn S. Badal, and Albert J. Wong

1. Introduction 532
2. Important Reagents for Studying JNK Activity 534
3. Protein Expression and Purification of JNK Proteins and c-JUN 535
4. Measuring the Autophosphorylation Ability of the JNK Isoforms 538
5. Determining the Kinase Activity of the JNK Isoforms 540
6. Monitoring the Formation of JNK Homodimers 540
7. Measuring Nuclear Translocation of JNK Protein 541
8. Future Directions 544
Acknowledgments 546
References 546

27. Measurement of Constitutive MAPK and PI3K/AKT Signaling Activity in Human Cancer Cell Lines 549

Kim H. T. Paraiso, Kaisa Van Der Kooi, Jane L. Messina, and Keiran S. M. Smalley

1. Introduction 550
2. Maintaining Melanoma Cell Lines 552
3. Western Blotting 553
4. Phospho-Flow Cytometry 559
5. Immunofluorescence 562
6. Conclusions 565
Acknowledgments 566
References 566

28. Constitutive Activity of GPR40/FFA1: Intrinsic or Assay Dependent? 569
Leigh A. Stoddart and Graeme Milligan

1. Introduction 570
2. Measuring FFA1-Mediated Calcium Mobilization 575
3. Measuring Direct Activation of G Proteins via FFA1 581
References 587

29. Constitutive Activity of TRP Channels: Methods for Measuring the Activity and Its Outcome 591
Shaya Lev and Baruch Minke

1. Introduction 592
2. TRP Channels and Cellular Degeneration 594
3. Constitutive TRP Channel Activity Which Does Not Lead to Cellular Degeneration 595
4. Constitutive TRP Channel Activity Which Leads to Cellular Degeneration 604
Acknowledgments 609
References 609

30. Measurement of Orexin (Hypocretin) and Substance P Effects on Constitutively Active Inward Rectifier K^+ Channels in Brain Neurons 613
Yasuko Nakajima and Shigehiro Nakajima

1. Introduction 614
2. Dissociated Culture of Cholinergic Neurons in the Basal Forebrain 615
3. Effects of Orexin (Hypocretin) and Substance P on Constitutively Active Inward Rectifier K^+ (KirNB) Channels 620
4. Signal Transduction of Substance P and Orexin Effects on KirNB Channels 625
Acknowledgments 628
References 629

31. Characterization of G Protein-Coupled Receptor Kinase 4 and Measuring Its Constitutive Activity *In Vivo* 631
Bradley T. Andresen

1. Introduction 632
2. Selection of Cells/Model Systems to Study GRK4 Function 634
3. Generation of Kinase Dead-GRK4 637
4. Functional Characterization of GRK4 Constitutive Activity 639

5. Agonist-Mediated GRK4 Activity	648
6. Summary	649
Acknowledgments	649
References	650

32. Voltage-Clamp-Based Methods for the Detection of Constitutively Active Acetylcholine-Gated $I_{K,ACh}$ Channels in the Diseased Heart — 653

Niels Voigt, Samy Makary, Stanley Nattel, and Dobromir Dobrev

1. Introduction	654
2. Recording of Constitutive $I_{K,ACh}$ Using Patch-Clamp Techniques	655
3. Conclusions and Perspective	671
Acknowledgments	673
References	673

33. Assaying WAVE and WASH Complex Constitutive Activities Toward the Arp2/3 Complex — 677

Emmanuel Derivery and Alexis Gautreau

1. Introduction	678
2. Establishment of Stable Cell Lines Expressing Tagged WAVE and WASH Complexes	680
3. Large-Scale Purification of WAVE and WASH Complexes	682
4. Aggregation Analysis of WAVE and WASH Multiprotein Complexes	684
5. Activity Measurements Using Pyrene Actin Polymerization Assays	687
6. Detection of an Endogenous Activity of the WAVE Complex	691
7. Concluding Remarks	693
Acknowledgments	693
References	694

Author Index	*697*
Subject Index	*735*

Contributors

Bradley T. Andresen
Department of Internal Medicine, Division of Endocrinology, and Department of Medical Pharmacology and Physiology, University of Missouri; Research Scientist, Harry S Truman VAMC, Columbia, Missouri, USA

Andy V. Babwah
The Children's Health Research Institute; Lawson Health Research Institute, London, Ontario, Canada; Department of Obstetrics and Gynecology; and Department of Physiology and Pharmacology, The University of Western Ontario, London, Ontario, Canada

Shawn S. Badal
Department of Neurosurgery, Cancer Biology Program, Stanford University Medical Center, Stanford, California, USA

Massimiliano Beltramo[1]
Schering-Plough Research Institute, Milan, Italy

Anat Ben-Shlomo
Pituitary Center, Department of Medicine, Cedars Sinai Medical Center, David Geffen School of Medicine at UCLA, Los Angeles, California, USA

Mohiuddin Ahmed Bhuiyan
Department of Pharmacy, The University of Asia Pacific, Dhaka, Bangladesh, and Department of Pharmacology, Faculty of Pharmaceutical Sciences, Niigata University of Pharmacy and Applied Life Sciences, Niigatashi, Japan

Edward J. Bilsky
Department of Biomedical Sciences, College of Osteopathic Medicine, University of New England, Biddeford, Maine, USA

Roman L. Bogorad[2]
Inserm, Unit 845, Research Center Growth and Signaling, Team "PRL/GH Pathophysiology," University Paris Descartes, Faculty of Medicine, Paris, France

---

[1] Present address: Institut National de la Recherche Agronomique (INRA), Physiologie de la Reproduction et de Comportements, UMR0085, Nouzilly, France
[2] Current address: The David H. Koch Institute for Integrative Cancer Research, Massachusetts Institute of Technology, Cambridge, Massachusetts, USA

Rossella Brusa
Schering-Plough Research Institute, Milan, and Novartis, Origgio, Varese, Italy

Isabel Castro-Piedras
Department of Internal Medicine, Tech University Health Sciences Center-SOM, Lubbock, Texas, USA, and UETeM, Department of Medicine, University of Santiago de Compostela-SOM, Santiago de Compostela, Spain

Wenhan Chang
Endocrine Research Unit, Department of Veterans Affairs Medical Center, Department of Medicine, University of California, San Francisco, California, USA

Mark Connor
Australian School of Advanced Medicine, Macquarie University, New South Wales, Australia

Jean-Philippe D'Aoust
Ottawa Hospital Research Institute (Neurosciences), Departments of Medicine/Cellular and Molecular Medicine/Psychiatry, University of Ottawa, Ontario, Canada

David J. J. de Gorter
Department of Molecular Cell Biology and Centre for Biomedical Genetics, Leiden University Medical Center, RC Leiden, The Netherlands

Kaisa Van Der Kooi
Department of Pathology and Cell Biology, University of South Florida College of Medicine, Tampa, Florida, USA

Emmanuel Derivery
Laboratoire d'Enzymologie et de Biochimie Structurales, Gif sur Yvette, France

Christina M. Dersch
Clinical Psychopharmacology Section, IRP/NIDA/NIH, Baltimore, Maryland, USA

Dobromir Dobrev
Division of Experimental Cardiology, Department of Internal Medicine I—Cardiology, Angiology, Pneumology, Intensive Care and Hemostaseology, Medical Faculty Mannheim, University of Heidelberg, Theodor-Kutzer-Ufer, Mannheim, Germany

James S. Duncan
Department of Biochemistry, Schulich School of Medicine & Dentistry, University of Western Ontario, London, Ontario, Canada

Melita Dvorak
Endocrine Research Unit, Department of Veterans Affairs Medical Center, Department of Medicine, University of California, San Francisco, California, USA

Ashley C. French
Department of Biochemistry, Schulich School of Medicine & Dentistry, University of Western Ontario, London, Ontario, Canada

Hiromichi Fujino
Laboratory of Chemical Pharmacology, Graduate School of Pharmaceutical Sciences, Chiba University, Chiba, Japan

J. Adolfo García-Sáinz
Departamento de Biología Celular y Desarrollo, Instituto de Fisiología Celular, Universidad Nacional Autónoma de México, México

Alexis Gautreau
Laboratoire d'Enzymologie et de Biochimie Structurales, Gif sur Yvette, France

Denise Giuvelis
Department of Biomedical Sciences, College of Osteopathic Medicine, University of New England, Biddeford, Maine, USA

Vincent Goffin
Inserm, Unit 845, Research Center Growth and Signaling, Team "PRL/GH Pathophysiology," University Paris Descartes, Faculty of Medicine, Paris, France

Maria Grazia Perrone
Department of Medicinal Chemistry, University of Bari "A. Moro", Bari, Italy

Barbara Guerra
Department of Biochemistry and Molecular Biology, University of Southern Denmark, Odense, Denmark

Laszlo Gyenis
Department of Biochemistry, Schulich School of Medicine & Dentistry, University of Western Ontario, London, Ontario, Canada

Katherine E. Hanlon
Department of Pharmacology, University of Arizona, Tucson, Arizona, USA

Julia Haupt
Berlin Brandenburg Center for Regenerative Therapies (BCRT), Charité-Universitätsmedizin Berlin, Berlin, Germany

Bettye Hollins
Department of Pharmacology and Toxicology, Medical College of Georgia, Augusta, Georgia, USA

Birgitte Holst
Department of Neuroscience and Pharmacology, University of Copenhagen, Copenhagen N, Denmark

Yen-Ping Hsueh
Division of Biology, California Institute of Technology, Pasadena, California, USA

Hui Huang
Department of Anatomy, Physiology and Pharmacology, College of Veterinary Medicine, Auburn University, Auburn, Alabama, USA

Olaf-Georg Issinger
Department of Biochemistry and Molecular Biology, University of Southern Denmark, Odense, Denmark

Frederick S. Kaplan
Department of Orthopaedic Surgery, and Department of Medicine; The Center for Research in FOP and Related Disorders, The University of Pennsylvania School of Medicine, Philadelphia, USA

Joaquin Lado-Abeal
Department of Internal Medicine, Tech University Health Sciences Center-SOM, and Department of Cell Biology and Biochemistry, Texas Tech University Health Sciences Center-SOM, Lubbock, Texas, USA; UETeM, Department of Medicine, University of Santiago de Compostela-SOM, Santiago de Compostela, Spain

Nevin A. Lambert
Department of Pharmacology and Toxicology, Medical College of Georgia, Augusta, Georgia, USA

Vincent Laquerre
Ottawa Hospital Research Institute (Neurosciences), Departments of Medicine/Cellular and Molecular Medicine/Psychiatry, University of Ottawa, Ontario, Canada

Rob Leurs
Leiden/Amsterdam Center for Drug Research, VU University Amsterdam, Amsterdam, The Netherlands

Shaya Lev
Department of Medical Neurobiology and the Kühne Minerva Center for Studies of Visual Transduction, Institute of Medical Research Israel-Canada (IMRIC), The Edmond & Lily Safra Centre for Brain Sciences (ELSC), The Hebrew University, Jerusalem, Israel

Binhui Liang
Ottawa Hospital Research Institute (Neurosciences), Departments of Medicine/Cellular and Molecular Medicine/Psychiatry, University of Ottawa, Ontario, Canada

David W. Litchfield
Department of Biochemistry, Schulich School of Medicine & Dentistry, University of Western Ontario, London, Ontario, Canada

Vitali Y. Lounev
Department of Orthopaedic Surgery, and The Center for Research in FOP and Related Disorders, The University of Pennsylvania School of Medicine, Philadelphia, USA

Samy Makary
Research Center, Montreal Heart Institute, Montreal, Quebec, Canada

Isabella Mancini
Schering-Plough Research Institute, Milan, and OPIS s.r.l., Palazzo Aliprandi, Desio, Italy

Luz del Carmen Medina
Departamento de Biología de la Reproducción, División de CBS, Universidad Autónoma Metropolitana-Iztapalapa, México

Giovanni Di Maira
Department of Biological Chemistry, and VIMM (Venetian Institute of Molecular Medicine), University of Padova, Padova, Italy

Shlomo Melmed
Pituitary Center, Department of Medicine, Cedars Sinai Medical Center, David Geffen School of Medicine at UCLA, Los Angeles, California, USA

Jane L. Messina
Department of Pathology and Cell Biology, University of South Florida College of Medicine, Tampa, Florida, USA

Graeme Milligan
Molecular Pharmacology Group, College of Medical, Veterinary and Life Sciences, University of Glasgow, Glasgow, United Kingdom

Baruch Minke
Department of Medical Neurobiology and the Kühne Minerva Center for Studies of Visual Transduction, Institute of Medical Research Israel-Canada (IMRIC), The Edmond & Lily Safra Centre for Brain Sciences (ELSC), The Hebrew University, Jerusalem, Israel

Jacek Mokrosiński
Department of Neuroscience and Pharmacology, University of Copenhagen, Copenhagen N, Denmark

Mary Mullins
Department of Cell and Developmental Biology, The University of Pennsylvania School of Medicine, Philadelphia, USA

Toshihiko Murayama
Laboratory of Chemical Pharmacology, Graduate School of Pharmaceutical Sciences, Chiba University, Chiba, Japan

Takafumi Nagatomo
Department of Pharmacology, Faculty of Pharmaceutical Sciences, Niigata University of Pharmacy and Applied Life Sciences, Niigatashi, Japan

Shigehiro Nakajima
Department of Pharmacology, College of Medicine, University of Illinois at Chicago, Chicago, Illinois, USA

Yasuko Nakajima
Department of Anatomy and Cell Biology, College of Medicine, University of Illinois at Chicago, Chicago, Illinois, USA

Stanley Nattel
Research Center, Montreal Heart Institute, Montreal, Quebec, Canada

Karsten Niefind
Department of Chemistry, Institute of Biochemistry, University of Cologne, Cologne, Germany

Saskia Nijmeijer
Leiden/Amsterdam Center for Drug Research, VU University Amsterdam, Amsterdam, The Netherlands

Gregory V. Nikiforovich
MolLife Design LLC, St. Louis, Missouri, USA

Ryan T. Nitta
Department of Neurosurgery, Cancer Biology Program, Stanford University Medical Center, Stanford, California, USA

Birgitte B. Olsen
Department of Biochemistry and Molecular Biology, University of Southern Denmark, Odense, Denmark

Melissa D. Osborn
Department of Biomedical Sciences, College of Osteopathic Medicine, University of New England, Biddeford, Maine, USA

Macarena Pampillo
The Children's Health Research Institute; Lawson Health Research Institute, London, Ontario, Canada; and Department of Obstetrics and Gynecology, The University of Western Ontario, London, Ontario, Canada

Kim H. T. Paraiso
Department of Molecular Oncology, and Department of Cutaneous Oncology, The Moffitt Cancer Center and Research Institute, Tampa, Florida, USA

Lorenzo A. Pinna
Department of Biological Chemistry, and VIMM (Venetian Institute of Molecular Medicine), University of Padova, Padova, Italy

Bianca Plouffe
Ottawa Hospital Research Institute (Neurosciences), Departments of Medicine/Cellular and Molecular Medicine/Psychiatry, University of Ottawa, Ontario, Canada

Balakrishna M. Prasad
Department of Clinical Investigation, Eisenhower Army Medical Center, Georgia, USA

Leah R. Quisenberry
Department of Cell Biology and Biochemistry, Texas Tech University Health Sciences Center-SOM, Lubbock, Texas, USA

John W. Regan
Department of Pharmacology and Toxicology, College of Pharmacy, The University of Arizona, Arizona, USA

M. Teresa Romero-Ávila
Departamento de Biología Celular y Desarrollo, Instituto de Fisiología Celular, Universidad Nacional Autónoma de México, México

Richard B. Rothman
Clinical Psychopharmacology Section, IRP/NIDA/NIH, Baltimore, Maryland, USA

Maria Ruzzene
Department of Biological Chemistry, and VIMM (Venetian Institute of Molecular Medicine), University of Padova, Padova, Italy

Paola Scandroglio
Schering-Plough Research Institute, Milan, and Imagine s.r.l., Gallarate, Italy

Antonio Scilimati
Department of Medicinal Chemistry, University of Bari "A. Moro", Bari, Italy

Petra Seemann
Berlin Brandenburg Center for Regenerative Therapies (BCRT), Charité-Universitätsmedizin Berlin, Berlin, Germany

Deborah L. Segaloff
Department of Physiology and Biophysics, The University of Iowa Roy J. and Lucille A. Carver College of Medicine, Iowa City, Iowa, USA

Dolores Shoback
Endocrine Research Unit, Department of Veterans Affairs Medical Center, Department of Medicine, University of California, San Francisco, California, USA

Eileen M. Shore
Department of Orthopaedic Surgery, and Department of Genetics; The Center for Research in FOP and Related Disorders, The University of Pennsylvania School of Medicine, Philadelphia, USA

Keiran S. M. Smalley
Department of Molecular Oncology, and Department of Cutaneous Oncology, The Moffitt Cancer Center and Research Institute, Tampa, Florida, USA

Nicole A. St. Denis
Department of Biochemistry, Schulich School of Medicine & Dentistry, University of Western Ontario, London, Ontario, Canada

Leigh A. Stoddart
Institute of Cell Signalling, School of Biomedical Science, Medical School, University of Nottingham, Nottingham, United Kingdom

Ya-Xiong Tao
Department of Anatomy, Physiology and Pharmacology, College of Veterinary Medicine, Auburn University, Auburn, Alabama, USA

Peter ten Dijke
Department of Molecular Cell Biology and Centre for Biomedical Genetics, Leiden University Medical Center, RC Leiden, The Netherlands

Mario Tiberi
Ottawa Hospital Research Institute (Neurosciences), Departments of Medicine/Cellular and Molecular Medicine/Psychiatry, University of Ottawa, Ontario, Canada

Philippe Touraine
Inserm, Unit 845, Research Center Growth and Signaling, Team "PRL/GH Pathophysiology," University Paris Descartes, Faculty of Medicine, and Assistance Publique—Hôpitaux de Paris, Department of Endocrinology and Reproductive Medicine, GH Pitié Salpêtrière, Paris, France

Kendra Tosoni
Department of Biological Chemistry, and VIMM (Venetian Institute of Molecular Medicine), University of Padova, Padova, Italy

John Traynor
Department of Pharmacology, Substance Abuse Research Centre, University of Michigan Medical School, Ann Arbor, Michigan, USA

Jacob P. Turowec
Department of Biochemistry, Schulich School of Medicine & Dentistry, University of Western Ontario, London, Ontario, Canada

Maarten van Dinther
Department of Molecular Cell Biology and Centre for Biomedical Genetics, Leiden University Medical Center, RC Leiden, The Netherlands

Todd W. Vanderah
Department of Pharmacology, University of Arizona, Tucson, Arizona, USA

Greg Vilk
Department of Biochemistry, Schulich School of Medicine & Dentistry, University of Western Ontario, London, Ontario, Canada

Henry F. Vischer
Leiden/Amsterdam Center for Drug Research, VU University Amsterdam, Amsterdam, The Netherlands

Niels Voigt
Division of Experimental Cardiology, Department of Internal Medicine I—Cardiology, Angiology, Pneumology, Intensive Care and Hemostaseology, Medical Faculty Mannheim, University of Heidelberg, Theodor-Kutzer-Ufer, Mannheim, Germany

Yina Wang
Public Health Research Institute, University of Medicine and Dentistry of New Jersey, Newark, New Jersey, USA

Zhi-Qiang Wang
Department of Anatomy, Physiology and Pharmacology, College of Veterinary Medicine, Auburn University, Auburn, Alabama, USA, and College of Veterinary Medicine, Yangzhou University, Jiangsu, People's Republic of China

Kolja Wawrowsky
Pituitary Center, Department of Medicine, Cedars Sinai Medical Center, David Geffen School of Medicine at UCLA, Los Angeles, California, USA

Jessica N. Williams
Department of Anatomy, Physiology and Pharmacology, College of Veterinary Medicine, Auburn University, Auburn, Alabama, USA

Albert J. Wong
Department of Neurosurgery, Cancer Biology Program, Stanford University Medical Center, and Cancer Biology Program, Stanford University School of Medicine, Stanford, California, USA

Heng Xu
Clinical Psychopharmacology Section, IRP/NIDA/NIH, Baltimore, Maryland, USA

Meiqi Xu
Department of Orthopaedic Surgery, and The Center for Research in FOP and Related Disorders, The University of Pennsylvania School of Medicine, Philadelphia, USA

Chaoyang Xue
Public Health Research Institute, and Department of Microbiology and Molecular Genetics, University of Medicine and Dentistry of New Jersey, Newark, New Jersey, USA

Fan Yang
Department of Anatomy, Physiology and Pharmacology, College of Veterinary Medicine, Auburn University, Auburn, Alabama, USA

Preface

The observation that mutant (and sometimes wild type) receptors, ion channels, and enzymes exist in states of constitutive activation has provided insight into the etiology of disease and the mechanism of protein function in ligand recognition, effector coupling, ion conductance, and catalysis. The observation that constitutive activity is a surprisingly common event has supported the view that biologically active molecules exist in both inactive and active states. Moreover, the observation that many drugs already at market are actually inverse agonists (agents that inhibit constitutive activity) makes understanding constitutive activity important for therapeutic drug design.

This volume provides descriptions of methods used to assess the mechanism of protein function and pharmacological tools for and methodological approaches to the analysis of constitutive activity. The authors explain how these methods are able to provide important biological insights.

Authors were selected based on research contributions in the area about which they have written and based on their ability to describe their methodological contribution in a clear and reproducible way. They have been encouraged to make use of graphics, comparisons with other methods, and to provide tricks and approaches not revealed in prior publications that make it possible to adapt methods to other systems.

The editor expresses appreciation to the contributors for providing their contributions in a timely fashion, to the senior editors for guidance, and to the staff at Academic Press for the helpful input.

June, 2010
P. Michael Conn

Methods in Enzymology

Volume I. Preparation and Assay of Enzymes
Edited by Sidney P. Colowick and Nathan O. Kaplan

Volume II. Preparation and Assay of Enzymes
Edited by Sidney P. Colowick and Nathan O. Kaplan

Volume III. Preparation and Assay of Substrates
Edited by Sidney P. Colowick and Nathan O. Kaplan

Volume IV. Special Techniques for the Enzymologist
Edited by Sidney P. Colowick and Nathan O. Kaplan

Volume V. Preparation and Assay of Enzymes
Edited by Sidney P. Colowick and Nathan O. Kaplan

Volume VI. Preparation and Assay of Enzymes *(Continued)*
Preparation and Assay of Substrates
Special Techniques
Edited by Sidney P. Colowick and Nathan O. Kaplan

Volume VII. Cumulative Subject Index
Edited by Sidney P. Colowick and Nathan O. Kaplan

Volume VIII. Complex Carbohydrates
Edited by Elizabeth F. Neufeld and Victor Ginsburg

Volume IX. Carbohydrate Metabolism
Edited by Willis A. Wood

Volume X. Oxidation and Phosphorylation
Edited by Ronald W. Estabrook and Maynard E. Pullman

Volume XI. Enzyme Structure
Edited by C. H. W. Hirs

Volume XII. Nucleic Acids (Parts A and B)
Edited by Lawrence Grossman and Kivie Moldave

Volume XIII. Citric Acid Cycle
Edited by J. M. Lowenstein

Volume XIV. Lipids
Edited by J. M. Lowenstein

Volume XV. Steroids and Terpenoids
Edited by Raymond B. Clayton

VOLUME XVI. Fast Reactions
Edited by KENNETH KUSTIN

VOLUME XVII. Metabolism of Amino Acids and Amines (Parts A and B)
Edited by HERBERT TABOR AND CELIA WHITE TABOR

VOLUME XVIII. Vitamins and Coenzymes (Parts A, B, and C)
Edited by DONALD B. MCCORMICK AND LEMUEL D. WRIGHT

VOLUME XIX. Proteolytic Enzymes
Edited by GERTRUDE E. PERLMANN AND LASZLO LORAND

VOLUME XX. Nucleic Acids and Protein Synthesis (Part C)
Edited by KIVIE MOLDAVE AND LAWRENCE GROSSMAN

VOLUME XXI. Nucleic Acids (Part D)
Edited by LAWRENCE GROSSMAN AND KIVIE MOLDAVE

VOLUME XXII. Enzyme Purification and Related Techniques
Edited by WILLIAM B. JAKOBY

VOLUME XXIII. Photosynthesis (Part A)
Edited by ANTHONY SAN PIETRO

VOLUME XXIV. Photosynthesis and Nitrogen Fixation (Part B)
Edited by ANTHONY SAN PIETRO

VOLUME XXV. Enzyme Structure (Part B)
Edited by C. H. W. HIRS AND SERGE N. TIMASHEFF

VOLUME XXVI. Enzyme Structure (Part C)
Edited by C. H. W. HIRS AND SERGE N. TIMASHEFF

VOLUME XXVII. Enzyme Structure (Part D)
Edited by C. H. W. HIRS AND SERGE N. TIMASHEFF

VOLUME XXVIII. Complex Carbohydrates (Part B)
Edited by VICTOR GINSBURG

VOLUME XXIX. Nucleic Acids and Protein Synthesis (Part E)
Edited by LAWRENCE GROSSMAN AND KIVIE MOLDAVE

VOLUME XXX. Nucleic Acids and Protein Synthesis (Part F)
Edited by KIVIE MOLDAVE AND LAWRENCE GROSSMAN

VOLUME XXXI. Biomembranes (Part A)
Edited by SIDNEY FLEISCHER AND LESTER PACKER

VOLUME XXXII. Biomembranes (Part B)
Edited by SIDNEY FLEISCHER AND LESTER PACKER

VOLUME XXXIII. Cumulative Subject Index Volumes I-XXX
Edited by MARTHA G. DENNIS AND EDWARD A. DENNIS

VOLUME XXXIV. Affinity Techniques (Enzyme Purification: Part B)
Edited by WILLIAM B. JAKOBY AND MEIR WILCHEK

VOLUME XXXV. Lipids (Part B)
Edited by JOHN M. LOWENSTEIN

VOLUME XXXVI. Hormone Action (Part A: Steroid Hormones)
Edited by BERT W. O'MALLEY AND JOEL G. HARDMAN

VOLUME XXXVII. Hormone Action (Part B: Peptide Hormones)
Edited by BERT W. O'MALLEY AND JOEL G. HARDMAN

VOLUME XXXVIII. Hormone Action (Part C: Cyclic Nucleotides)
Edited by JOEL G. HARDMAN AND BERT W. O'MALLEY

VOLUME XXXIX. Hormone Action (Part D: Isolated Cells, Tissues, and Organ Systems)
Edited by JOEL G. HARDMAN AND BERT W. O'MALLEY

VOLUME XL. Hormone Action (Part E: Nuclear Structure and Function)
Edited by BERT W. O'MALLEY AND JOEL G. HARDMAN

VOLUME XLI. Carbohydrate Metabolism (Part B)
Edited by W. A. WOOD

VOLUME XLII. Carbohydrate Metabolism (Part C)
Edited by W. A. WOOD

VOLUME XLIII. Antibiotics
Edited by JOHN H. HASH

VOLUME XLIV. Immobilized Enzymes
Edited by KLAUS MOSBACH

VOLUME XLV. Proteolytic Enzymes (Part B)
Edited by LASZLO LORAND

VOLUME XLVI. Affinity Labeling
Edited by WILLIAM B. JAKOBY AND MEIR WILCHEK

VOLUME XLVII. Enzyme Structure (Part E)
Edited by C. H. W. HIRS AND SERGE N. TIMASHEFF

VOLUME XLVIII. Enzyme Structure (Part F)
Edited by C. H. W. HIRS AND SERGE N. TIMASHEFF

VOLUME XLIX. Enzyme Structure (Part G)
Edited by C. H. W. HIRS AND SERGE N. TIMASHEFF

VOLUME L. Complex Carbohydrates (Part C)
Edited by VICTOR GINSBURG

VOLUME LI. Purine and Pyrimidine Nucleotide Metabolism
Edited by PATRICIA A. HOFFEE AND MARY ELLEN JONES

VOLUME LII. Biomembranes (Part C: Biological Oxidations)
Edited by SIDNEY FLEISCHER AND LESTER PACKER

VOLUME LIII. Biomembranes (Part D: Biological Oxidations)
Edited by SIDNEY FLEISCHER AND LESTER PACKER

VOLUME LIV. Biomembranes (Part E: Biological Oxidations)
Edited by SIDNEY FLEISCHER AND LESTER PACKER

VOLUME LV. Biomembranes (Part F: Bioenergetics)
Edited by SIDNEY FLEISCHER AND LESTER PACKER

VOLUME LVI. Biomembranes (Part G: Bioenergetics)
Edited by SIDNEY FLEISCHER AND LESTER PACKER

VOLUME LVII. Bioluminescence and Chemiluminescence
Edited by MARLENE A. DELUCA

VOLUME LVIII. Cell Culture
Edited by WILLIAM B. JAKOBY AND IRA PASTAN

VOLUME LIX. Nucleic Acids and Protein Synthesis (Part G)
Edited by KIVIE MOLDAVE AND LAWRENCE GROSSMAN

VOLUME LX. Nucleic Acids and Protein Synthesis (Part H)
Edited by KIVIE MOLDAVE AND LAWRENCE GROSSMAN

VOLUME 61. Enzyme Structure (Part H)
Edited by C. H. W. HIRS AND SERGE N. TIMASHEFF

VOLUME 62. Vitamins and Coenzymes (Part D)
Edited by DONALD B. MCCORMICK AND LEMUEL D. WRIGHT

VOLUME 63. Enzyme Kinetics and Mechanism (Part A: Initial Rate and Inhibitor Methods)
Edited by DANIEL L. PURICH

VOLUME 64. Enzyme Kinetics and Mechanism
(Part B: Isotopic Probes and Complex Enzyme Systems)
Edited by DANIEL L. PURICH

VOLUME 65. Nucleic Acids (Part I)
Edited by LAWRENCE GROSSMAN AND KIVIE MOLDAVE

VOLUME 66. Vitamins and Coenzymes (Part E)
Edited by DONALD B. MCCORMICK AND LEMUEL D. WRIGHT

VOLUME 67. Vitamins and Coenzymes (Part F)
Edited by DONALD B. MCCORMICK AND LEMUEL D. WRIGHT

VOLUME 68. Recombinant DNA
Edited by RAY WU

VOLUME 69. Photosynthesis and Nitrogen Fixation (Part C)
Edited by ANTHONY SAN PIETRO

VOLUME 70. Immunochemical Techniques (Part A)
Edited by HELEN VAN VUNAKIS AND JOHN J. LANGONE

VOLUME 71. Lipids (Part C)
Edited by JOHN M. LOWENSTEIN

VOLUME 72. Lipids (Part D)
Edited by JOHN M. LOWENSTEIN

VOLUME 73. Immunochemical Techniques (Part B)
Edited by JOHN J. LANGONE AND HELEN VAN VUNAKIS

VOLUME 74. Immunochemical Techniques (Part C)
Edited by JOHN J. LANGONE AND HELEN VAN VUNAKIS

VOLUME 75. Cumulative Subject Index Volumes XXXI, XXXII, XXXIV–LX
Edited by EDWARD A. DENNIS AND MARTHA G. DENNIS

VOLUME 76. Hemoglobins
Edited by ERALDO ANTONINI, LUIGI ROSSI-BERNARDI, AND EMILIA CHIANCONE

VOLUME 77. Detoxication and Drug Metabolism
Edited by WILLIAM B. JAKOBY

VOLUME 78. Interferons (Part A)
Edited by SIDNEY PESTKA

VOLUME 79. Interferons (Part B)
Edited by SIDNEY PESTKA

VOLUME 80. Proteolytic Enzymes (Part C)
Edited by LASZLO LORAND

VOLUME 81. Biomembranes (Part H: Visual Pigments and Purple Membranes, I)
Edited by LESTER PACKER

VOLUME 82. Structural and Contractile Proteins (Part A: Extracellular Matrix)
Edited by LEON W. CUNNINGHAM AND DIXIE W. FREDERIKSEN

VOLUME 83. Complex Carbohydrates (Part D)
Edited by VICTOR GINSBURG

VOLUME 84. Immunochemical Techniques (Part D: Selected Immunoassays)
Edited by JOHN J. LANGONE AND HELEN VAN VUNAKIS

VOLUME 85. Structural and Contractile Proteins (Part B: The Contractile Apparatus and the Cytoskeleton)
Edited by DIXIE W. FREDERIKSEN AND LEON W. CUNNINGHAM

VOLUME 86. Prostaglandins and Arachidonate Metabolites
Edited by WILLIAM E. M. LANDS AND WILLIAM L. SMITH

VOLUME 87. Enzyme Kinetics and Mechanism (Part C: Intermediates, Stereo-chemistry, and Rate Studies)
Edited by DANIEL L. PURICH

VOLUME 88. Biomembranes (Part I: Visual Pigments and Purple Membranes, II)
Edited by LESTER PACKER

VOLUME 89. Carbohydrate Metabolism (Part D)
Edited by WILLIS A. WOOD

VOLUME 90. Carbohydrate Metabolism (Part E)
Edited by WILLIS A. WOOD

VOLUME 91. Enzyme Structure (Part I)
Edited by C. H. W. HIRS AND SERGE N. TIMASHEFF

VOLUME 92. Immunochemical Techniques (Part E: Monoclonal Antibodies and General Immunoassay Methods)
Edited by JOHN J. LANGONE AND HELEN VAN VUNAKIS

VOLUME 93. Immunochemical Techniques (Part F: Conventional Antibodies, Fc Receptors, and Cytotoxicity)
Edited by JOHN J. LANGONE AND HELEN VAN VUNAKIS

VOLUME 94. Polyamines
Edited by HERBERT TABOR AND CELIA WHITE TABOR

VOLUME 95. Cumulative Subject Index Volumes 61–74, 76–80
Edited by EDWARD A. DENNIS AND MARTHA G. DENNIS

VOLUME 96. Biomembranes [Part J: Membrane Biogenesis: Assembly and Targeting (General Methods; Eukaryotes)]
Edited by SIDNEY FLEISCHER AND BECCA FLEISCHER

VOLUME 97. Biomembranes [Part K: Membrane Biogenesis: Assembly and Targeting (Prokaryotes, Mitochondria, and Chloroplasts)]
Edited by SIDNEY FLEISCHER AND BECCA FLEISCHER

VOLUME 98. Biomembranes (Part L: Membrane Biogenesis: Processing and Recycling)
Edited by SIDNEY FLEISCHER AND BECCA FLEISCHER

VOLUME 99. Hormone Action (Part F: Protein Kinases)
Edited by JACKIE D. CORBIN AND JOEL G. HARDMAN

VOLUME 100. Recombinant DNA (Part B)
Edited by RAY WU, LAWRENCE GROSSMAN, AND KIVIE MOLDAVE

VOLUME 101. Recombinant DNA (Part C)
Edited by RAY WU, LAWRENCE GROSSMAN, AND KIVIE MOLDAVE

VOLUME 102. Hormone Action (Part G: Calmodulin and Calcium-Binding Proteins)
Edited by ANTHONY R. MEANS AND BERT W. O'MALLEY

VOLUME 103. Hormone Action (Part H: Neuroendocrine Peptides)
Edited by P. MICHAEL CONN

VOLUME 104. Enzyme Purification and Related Techniques (Part C)
Edited by WILLIAM B. JAKOBY

VOLUME 105. Oxygen Radicals in Biological Systems
Edited by LESTER PACKER

VOLUME 106. Posttranslational Modifications (Part A)
Edited by FINN WOLD AND KIVIE MOLDAVE

VOLUME 107. Posttranslational Modifications (Part B)
Edited by FINN WOLD AND KIVIE MOLDAVE

VOLUME 108. Immunochemical Techniques (Part G: Separation and Characterization of Lymphoid Cells)
Edited by GIOVANNI DI SABATO, JOHN J. LANGONE, AND HELEN VAN VUNAKIS

VOLUME 109. Hormone Action (Part I: Peptide Hormones)
Edited by LUTZ BIRNBAUMER AND BERT W. O'MALLEY

VOLUME 110. Steroids and Isoprenoids (Part A)
Edited by JOHN H. LAW AND HANS C. RILLING

VOLUME 111. Steroids and Isoprenoids (Part B)
Edited by JOHN H. LAW AND HANS C. RILLING

VOLUME 112. Drug and Enzyme Targeting (Part A)
Edited by KENNETH J. WIDDER AND RALPH GREEN

VOLUME 113. Glutamate, Glutamine, Glutathione, and Related Compounds
Edited by ALTON MEISTER

VOLUME 114. Diffraction Methods for Biological Macromolecules (Part A)
Edited by HAROLD W. WYCKOFF, C. H. W. HIRS, AND SERGE N. TIMASHEFF

VOLUME 115. Diffraction Methods for Biological Macromolecules (Part B)
Edited by HAROLD W. WYCKOFF, C. H. W. HIRS, AND SERGE N. TIMASHEFF

VOLUME 116. Immunochemical Techniques
(Part H: Effectors and Mediators of Lymphoid Cell Functions)
Edited by GIOVANNI DI SABATO, JOHN J. LANGONE, AND HELEN VAN VUNAKIS

VOLUME 117. Enzyme Structure (Part J)
Edited by C. H. W. HIRS AND SERGE N. TIMASHEFF

VOLUME 118. Plant Molecular Biology
Edited by ARTHUR WEISSBACH AND HERBERT WEISSBACH

VOLUME 119. Interferons (Part C)
Edited by SIDNEY PESTKA

VOLUME 120. Cumulative Subject Index Volumes 81–94, 96–101

VOLUME 121. Immunochemical Techniques (Part I: Hybridoma Technology and Monoclonal Antibodies)
Edited by JOHN J. LANGONE AND HELEN VAN VUNAKIS

VOLUME 122. Vitamins and Coenzymes (Part G)
Edited by FRANK CHYTIL AND DONALD B. MCCORMICK

VOLUME 123. Vitamins and Coenzymes (Part H)
Edited by FRANK CHYTIL AND DONALD B. MCCORMICK

VOLUME 124. Hormone Action (Part J: Neuroendocrine Peptides)
Edited by P. MICHAEL CONN

VOLUME 125. Biomembranes (Part M: Transport in Bacteria, Mitochondria, and Chloroplasts: General Approaches and Transport Systems)
Edited by SIDNEY FLEISCHER AND BECCA FLEISCHER

VOLUME 126. Biomembranes (Part N: Transport in Bacteria, Mitochondria, and Chloroplasts: Protonmotive Force)
Edited by SIDNEY FLEISCHER AND BECCA FLEISCHER

VOLUME 127. Biomembranes (Part O: Protons and Water: Structure and Translocation)
Edited by LESTER PACKER

VOLUME 128. Plasma Lipoproteins (Part A: Preparation, Structure, and Molecular Biology)
Edited by JERE P. SEGREST AND JOHN J. ALBERS

VOLUME 129. Plasma Lipoproteins (Part B: Characterization, Cell Biology, and Metabolism)
Edited by JOHN J. ALBERS AND JERE P. SEGREST

VOLUME 130. Enzyme Structure (Part K)
Edited by C. H. W. HIRS AND SERGE N. TIMASHEFF

VOLUME 131. Enzyme Structure (Part L)
Edited by C. H. W. HIRS AND SERGE N. TIMASHEFF

VOLUME 132. Immunochemical Techniques (Part J: Phagocytosis and Cell-Mediated Cytotoxicity)
Edited by GIOVANNI DI SABATO AND JOHANNES EVERSE

VOLUME 133. Bioluminescence and Chemiluminescence (Part B)
Edited by MARLENE DELUCA AND WILLIAM D. MCELROY

VOLUME 134. Structural and Contractile Proteins (Part C: The Contractile Apparatus and the Cytoskeleton)
Edited by RICHARD B. VALLEE

VOLUME 135. Immobilized Enzymes and Cells (Part B)
Edited by KLAUS MOSBACH

VOLUME 136. Immobilized Enzymes and Cells (Part C)
Edited by KLAUS MOSBACH

VOLUME 137. Immobilized Enzymes and Cells (Part D)
Edited by KLAUS MOSBACH

VOLUME 138. Complex Carbohydrates (Part E)
Edited by VICTOR GINSBURG

VOLUME 139. Cellular Regulators (Part A: Calcium- and Calmodulin-Binding Proteins)
Edited by ANTHONY R. MEANS AND P. MICHAEL CONN

VOLUME 140. Cumulative Subject Index Volumes 102–119, 121–134

VOLUME 141. Cellular Regulators (Part B: Calcium and Lipids)
Edited by P. MICHAEL CONN AND ANTHONY R. MEANS

VOLUME 142. Metabolism of Aromatic Amino Acids and Amines
Edited by SEYMOUR KAUFMAN

VOLUME 143. Sulfur and Sulfur Amino Acids
Edited by WILLIAM B. JAKOBY AND OWEN GRIFFITH

VOLUME 144. Structural and Contractile Proteins (Part D: Extracellular Matrix)
Edited by LEON W. CUNNINGHAM

VOLUME 145. Structural and Contractile Proteins (Part E: Extracellular Matrix)
Edited by LEON W. CUNNINGHAM

VOLUME 146. Peptide Growth Factors (Part A)
Edited by DAVID BARNES AND DAVID A. SIRBASKU

VOLUME 147. Peptide Growth Factors (Part B)
Edited by DAVID BARNES AND DAVID A. SIRBASKU

VOLUME 148. Plant Cell Membranes
Edited by LESTER PACKER AND ROLAND DOUCE

VOLUME 149. Drug and Enzyme Targeting (Part B)
Edited by RALPH GREEN AND KENNETH J. WIDDER

VOLUME 150. Immunochemical Techniques (Part K: *In Vitro* Models of B and T Cell Functions and Lymphoid Cell Receptors)
Edited by GIOVANNI DI SABATO

VOLUME 151. Molecular Genetics of Mammalian Cells
Edited by MICHAEL M. GOTTESMAN

VOLUME 152. Guide to Molecular Cloning Techniques
Edited by SHELBY L. BERGER AND ALAN R. KIMMEL

VOLUME 153. Recombinant DNA (Part D)
Edited by RAY WU AND LAWRENCE GROSSMAN

VOLUME 154. Recombinant DNA (Part E)
Edited by RAY WU AND LAWRENCE GROSSMAN

VOLUME 155. Recombinant DNA (Part F)
Edited by RAY WU

VOLUME 156. Biomembranes (Part P: ATP-Driven Pumps and Related Transport: The Na, K-Pump)
Edited by SIDNEY FLEISCHER AND BECCA FLEISCHER

VOLUME 157. Biomembranes (Part Q: ATP-Driven Pumps and Related Transport: Calcium, Proton, and Potassium Pumps)
Edited by SIDNEY FLEISCHER AND BECCA FLEISCHER

VOLUME 158. Metalloproteins (Part A)
Edited by JAMES F. RIORDAN AND BERT L. VALLEE

VOLUME 159. Initiation and Termination of Cyclic Nucleotide Action
Edited by JACKIE D. CORBIN AND ROGER A. JOHNSON

VOLUME 160. Biomass (Part A: Cellulose and Hemicellulose)
Edited by WILLIS A. WOOD AND SCOTT T. KELLOGG

VOLUME 161. Biomass (Part B: Lignin, Pectin, and Chitin)
Edited by WILLIS A. WOOD AND SCOTT T. KELLOGG

VOLUME 162. Immunochemical Techniques (Part L: Chemotaxis and Inflammation)
Edited by GIOVANNI DI SABATO

VOLUME 163. Immunochemical Techniques (Part M: Chemotaxis and Inflammation)
Edited by GIOVANNI DI SABATO

VOLUME 164. Ribosomes
Edited by HARRY F. NOLLER, JR., AND KIVIE MOLDAVE

VOLUME 165. Microbial Toxins: Tools for Enzymology
Edited by SIDNEY HARSHMAN

VOLUME 166. Branched-Chain Amino Acids
Edited by ROBERT HARRIS AND JOHN R. SOKATCH

VOLUME 167. Cyanobacteria
Edited by LESTER PACKER AND ALEXANDER N. GLAZER

VOLUME 168. Hormone Action (Part K: Neuroendocrine Peptides)
Edited by P. MICHAEL CONN

VOLUME 169. Platelets: Receptors, Adhesion, Secretion (Part A)
Edited by JACEK HAWIGER

VOLUME 170. Nucleosomes
Edited by PAUL M. WASSARMAN AND ROGER D. KORNBERG

VOLUME 171. Biomembranes (Part R: Transport Theory: Cells and Model Membranes)
Edited by SIDNEY FLEISCHER AND BECCA FLEISCHER

VOLUME 172. Biomembranes (Part S: Transport: Membrane Isolation and Characterization)
Edited by SIDNEY FLEISCHER AND BECCA FLEISCHER

VOLUME 173. Biomembranes [Part T: Cellular and Subcellular Transport: Eukaryotic (Nonepithelial) Cells]
Edited by SIDNEY FLEISCHER AND BECCA FLEISCHER

VOLUME 174. Biomembranes [Part U: Cellular and Subcellular Transport: Eukaryotic (Nonepithelial) Cells]
Edited by SIDNEY FLEISCHER AND BECCA FLEISCHER

VOLUME 175. Cumulative Subject Index Volumes 135–139, 141–167

VOLUME 176. Nuclear Magnetic Resonance (Part A: Spectral Techniques and Dynamics)
Edited by NORMAN J. OPPENHEIMER AND THOMAS L. JAMES

VOLUME 177. Nuclear Magnetic Resonance (Part B: Structure and Mechanism)
Edited by NORMAN J. OPPENHEIMER AND THOMAS L. JAMES

VOLUME 178. Antibodies, Antigens, and Molecular Mimicry
Edited by JOHN J. LANGONE

VOLUME 179. Complex Carbohydrates (Part F)
Edited by VICTOR GINSBURG

VOLUME 180. RNA Processing (Part A: General Methods)
Edited by JAMES E. DAHLBERG AND JOHN N. ABELSON

VOLUME 181. RNA Processing (Part B: Specific Methods)
Edited by JAMES E. DAHLBERG AND JOHN N. ABELSON

VOLUME 182. Guide to Protein Purification
Edited by MURRAY P. DEUTSCHER

VOLUME 183. Molecular Evolution: Computer Analysis of Protein and Nucleic Acid Sequences
Edited by RUSSELL F. DOOLITTLE

VOLUME 184. Avidin-Biotin Technology
Edited by MEIR WILCHEK AND EDWARD A. BAYER

VOLUME 185. Gene Expression Technology
Edited by DAVID V. GOEDDEL

VOLUME 186. Oxygen Radicals in Biological Systems (Part B: Oxygen Radicals and Antioxidants)
Edited by LESTER PACKER AND ALEXANDER N. GLAZER

VOLUME 187. Arachidonate Related Lipid Mediators
Edited by ROBERT C. MURPHY AND FRANK A. FITZPATRICK

VOLUME 188. Hydrocarbons and Methylotrophy
Edited by MARY E. LIDSTROM

VOLUME 189. Retinoids (Part A: Molecular and Metabolic Aspects)
Edited by LESTER PACKER

VOLUME 190. Retinoids (Part B: Cell Differentiation and Clinical Applications)
Edited by LESTER PACKER

VOLUME 191. Biomembranes (Part V: Cellular and Subcellular Transport: Epithelial Cells)
Edited by SIDNEY FLEISCHER AND BECCA FLEISCHER

VOLUME 192. Biomembranes (Part W: Cellular and Subcellular Transport: Epithelial Cells)
Edited by SIDNEY FLEISCHER AND BECCA FLEISCHER

VOLUME 193. Mass Spectrometry
Edited by JAMES A. MCCLOSKEY

VOLUME 194. Guide to Yeast Genetics and Molecular Biology
Edited by CHRISTINE GUTHRIE AND GERALD R. FINK

VOLUME 195. Adenylyl Cyclase, G Proteins, and Guanylyl Cyclase
Edited by ROGER A. JOHNSON AND JACKIE D. CORBIN

VOLUME 196. Molecular Motors and the Cytoskeleton
Edited by RICHARD B. VALLEE

VOLUME 197. Phospholipases
Edited by EDWARD A. DENNIS

VOLUME 198. Peptide Growth Factors (Part C)
Edited by DAVID BARNES, J. P. MATHER, AND GORDON H. SATO

VOLUME 199. Cumulative Subject Index Volumes 168–174, 176–194

VOLUME 200. Protein Phosphorylation (Part A: Protein Kinases: Assays, Purification, Antibodies, Functional Analysis, Cloning, and Expression)
Edited by TONY HUNTER AND BARTHOLOMEW M. SEFTON

VOLUME 201. Protein Phosphorylation (Part B: Analysis of Protein Phosphorylation, Protein Kinase Inhibitors, and Protein Phosphatases)
Edited by TONY HUNTER AND BARTHOLOMEW M. SEFTON

VOLUME 202. Molecular Design and Modeling: Concepts and Applications (Part A: Proteins, Peptides, and Enzymes)
Edited by JOHN J. LANGONE

VOLUME 203. Molecular Design and Modeling: Concepts and Applications (Part B: Antibodies and Antigens, Nucleic Acids, Polysaccharides, and Drugs)
Edited by JOHN J. LANGONE

VOLUME 204. Bacterial Genetic Systems
Edited by JEFFREY H. MILLER

VOLUME 205. Metallobiochemistry (Part B: Metallothionein and Related Molecules)
Edited by JAMES F. RIORDAN AND BERT L. VALLEE

VOLUME 206. Cytochrome P450
Edited by MICHAEL R. WATERMAN AND ERIC F. JOHNSON

VOLUME 207. Ion Channels
Edited by BERNARDO RUDY AND LINDA E. IVERSON

VOLUME 208. Protein–DNA Interactions
Edited by ROBERT T. SAUER

VOLUME 209. Phospholipid Biosynthesis
Edited by EDWARD A. DENNIS AND DENNIS E. VANCE

VOLUME 210. Numerical Computer Methods
Edited by LUDWIG BRAND AND MICHAEL L. JOHNSON

VOLUME 211. DNA Structures (Part A: Synthesis and Physical Analysis of DNA)
Edited by DAVID M. J. LILLEY AND JAMES E. DAHLBERG

VOLUME 212. DNA Structures (Part B: Chemical and Electrophoretic Analysis of DNA)
Edited by DAVID M. J. LILLEY AND JAMES E. DAHLBERG

VOLUME 213. Carotenoids (Part A: Chemistry, Separation, Quantitation, and Antioxidation)
Edited by LESTER PACKER

VOLUME 214. Carotenoids (Part B: Metabolism, Genetics, and Biosynthesis)
Edited by LESTER PACKER

VOLUME 215. Platelets: Receptors, Adhesion, Secretion (Part B)
Edited by JACEK J. HAWIGER

VOLUME 216. Recombinant DNA (Part G)
Edited by RAY WU

VOLUME 217. Recombinant DNA (Part H)
Edited by RAY WU

VOLUME 218. Recombinant DNA (Part I)
Edited by RAY WU

VOLUME 219. Reconstitution of Intracellular Transport
Edited by JAMES E. ROTHMAN

VOLUME 220. Membrane Fusion Techniques (Part A)
Edited by NEJAT DÜZGÜNEŞ

VOLUME 221. Membrane Fusion Techniques (Part B)
Edited by NEJAT DÜZGÜNEŞ

VOLUME 222. Proteolytic Enzymes in Coagulation, Fibrinolysis, and Complement Activation (Part A: Mammalian Blood Coagulation Factors and Inhibitors)
Edited by LASZLO LORAND AND KENNETH G. MANN

VOLUME 223. Proteolytic Enzymes in Coagulation, Fibrinolysis, and Complement Activation (Part B: Complement Activation, Fibrinolysis, and Nonmammalian Blood Coagulation Factors)
Edited by LASZLO LORAND AND KENNETH G. MANN

VOLUME 224. Molecular Evolution: Producing the Biochemical Data
Edited by ELIZABETH ANNE ZIMMER, THOMAS J. WHITE, REBECCA L. CANN, AND ALLAN C. WILSON

VOLUME 225. Guide to Techniques in Mouse Development
Edited by PAUL M. WASSARMAN AND MELVIN L. DEPAMPHILIS

VOLUME 226. Metallobiochemistry (Part C: Spectroscopic and Physical Methods for Probing Metal Ion Environments in Metalloenzymes and Metalloproteins)
Edited by JAMES F. RIORDAN AND BERT L. VALLEE

VOLUME 227. Metallobiochemistry (Part D: Physical and Spectroscopic Methods for Probing Metal Ion Environments in Metalloproteins)
Edited by JAMES F. RIORDAN AND BERT L. VALLEE

VOLUME 228. Aqueous Two-Phase Systems
Edited by HARRY WALTER AND GÖTE JOHANSSON

VOLUME 229. Cumulative Subject Index Volumes 195–198, 200–227

VOLUME 230. Guide to Techniques in Glycobiology
Edited by WILLIAM J. LENNARZ AND GERALD W. HART

VOLUME 231. Hemoglobins (Part B: Biochemical and Analytical Methods)
Edited by JOHANNES EVERSE, KIM D. VANDEGRIFF, AND ROBERT M. WINSLOW

VOLUME 232. Hemoglobins (Part C: Biophysical Methods)
Edited by JOHANNES EVERSE, KIM D. VANDEGRIFF, AND ROBERT M. WINSLOW

VOLUME 233. Oxygen Radicals in Biological Systems (Part C)
Edited by LESTER PACKER

VOLUME 234. Oxygen Radicals in Biological Systems (Part D)
Edited by LESTER PACKER

VOLUME 235. Bacterial Pathogenesis (Part A: Identification and Regulation of Virulence Factors)
Edited by VIRGINIA L. CLARK AND PATRIK M. BAVOIL

VOLUME 236. Bacterial Pathogenesis (Part B: Integration of Pathogenic Bacteria with Host Cells)
Edited by VIRGINIA L. CLARK AND PATRIK M. BAVOIL

VOLUME 237. Heterotrimeric G Proteins
Edited by RAVI IYENGAR

VOLUME 238. Heterotrimeric G-Protein Effectors
Edited by RAVI IYENGAR

VOLUME 239. Nuclear Magnetic Resonance (Part C)
Edited by THOMAS L. JAMES AND NORMAN J. OPPENHEIMER

VOLUME 240. Numerical Computer Methods (Part B)
Edited by MICHAEL L. JOHNSON AND LUDWIG BRAND

VOLUME 241. Retroviral Proteases
Edited by LAWRENCE C. KUO AND JULES A. SHAFER

VOLUME 242. Neoglycoconjugates (Part A)
Edited by Y. C. LEE AND REIKO T. LEE

VOLUME 243. Inorganic Microbial Sulfur Metabolism
Edited by HARRY D. PECK, JR., AND JEAN LEGALL

VOLUME 244. Proteolytic Enzymes: Serine and Cysteine Peptidases
Edited by ALAN J. BARRETT

VOLUME 245. Extracellular Matrix Components
Edited by E. RUOSLAHTI AND E. ENGVALL

VOLUME 246. Biochemical Spectroscopy
Edited by KENNETH SAUER

VOLUME 247. Neoglycoconjugates (Part B: Biomedical Applications)
Edited by Y. C. LEE AND REIKO T. LEE

VOLUME 248. Proteolytic Enzymes: Aspartic and Metallo Peptidases
Edited by ALAN J. BARRETT

VOLUME 249. Enzyme Kinetics and Mechanism (Part D: Developments in Enzyme Dynamics)
Edited by DANIEL L. PURICH

VOLUME 250. Lipid Modifications of Proteins
Edited by PATRICK J. CASEY AND JANICE E. BUSS

VOLUME 251. Biothiols (Part A: Monothiols and Dithiols, Protein Thiols, and Thiyl Radicals)
Edited by LESTER PACKER

VOLUME 252. Biothiols (Part B: Glutathione and Thioredoxin; Thiols in Signal Transduction and Gene Regulation)
Edited by LESTER PACKER

VOLUME 253. Adhesion of Microbial Pathogens
Edited by RON J. DOYLE AND ITZHAK OFEK

VOLUME 254. Oncogene Techniques
Edited by PETER K. VOGT AND INDER M. VERMA

VOLUME 255. Small GTPases and Their Regulators (Part A: Ras Family)
Edited by W. E. BALCH, CHANNING J. DER, AND ALAN HALL

VOLUME 256. Small GTPases and Their Regulators (Part B: Rho Family)
Edited by W. E. BALCH, CHANNING J. DER, AND ALAN HALL

VOLUME 257. Small GTPases and Their Regulators (Part C: Proteins Involved in Transport)
Edited by W. E. BALCH, CHANNING J. DER, AND ALAN HALL

VOLUME 258. Redox-Active Amino Acids in Biology
Edited by JUDITH P. KLINMAN

VOLUME 259. Energetics of Biological Macromolecules
Edited by MICHAEL L. JOHNSON AND GARY K. ACKERS

VOLUME 260. Mitochondrial Biogenesis and Genetics (Part A)
Edited by GIUSEPPE M. ATTARDI AND ANNE CHOMYN

VOLUME 261. Nuclear Magnetic Resonance and Nucleic Acids
Edited by THOMAS L. JAMES

VOLUME 262. DNA Replication
Edited by JUDITH L. CAMPBELL

VOLUME 263. Plasma Lipoproteins (Part C: Quantitation)
Edited by WILLIAM A. BRADLEY, SANDRA H. GIANTURCO, AND JERE P. SEGREST

VOLUME 264. Mitochondrial Biogenesis and Genetics (Part B)
Edited by GIUSEPPE M. ATTARDI AND ANNE CHOMYN

VOLUME 265. Cumulative Subject Index Volumes 228, 230–262

VOLUME 266. Computer Methods for Macromolecular Sequence Analysis
Edited by RUSSELL F. DOOLITTLE

VOLUME 267. Combinatorial Chemistry
Edited by JOHN N. ABELSON

VOLUME 268. Nitric Oxide (Part A: Sources and Detection of NO; NO Synthase)
Edited by LESTER PACKER

VOLUME 269. Nitric Oxide (Part B: Physiological and Pathological Processes)
Edited by LESTER PACKER

VOLUME 270. High Resolution Separation and Analysis of Biological Macromolecules (Part A: Fundamentals)
Edited by BARRY L. KARGER AND WILLIAM S. HANCOCK

VOLUME 271. High Resolution Separation and Analysis of Biological Macromolecules (Part B: Applications)
Edited by BARRY L. KARGER AND WILLIAM S. HANCOCK

VOLUME 272. Cytochrome P450 (Part B)
Edited by ERIC F. JOHNSON AND MICHAEL R. WATERMAN

VOLUME 273. RNA Polymerase and Associated Factors (Part A)
Edited by SANKAR ADHYA

VOLUME 274. RNA Polymerase and Associated Factors (Part B)
Edited by SANKAR ADHYA

VOLUME 275. Viral Polymerases and Related Proteins
Edited by LAWRENCE C. KUO, DAVID B. OLSEN, AND STEVEN S. CARROLL

VOLUME 276. Macromolecular Crystallography (Part A)
Edited by CHARLES W. CARTER, JR., AND ROBERT M. SWEET

VOLUME 277. Macromolecular Crystallography (Part B)
Edited by CHARLES W. CARTER, JR., AND ROBERT M. SWEET

VOLUME 278. Fluorescence Spectroscopy
Edited by LUDWIG BRAND AND MICHAEL L. JOHNSON

VOLUME 279. Vitamins and Coenzymes (Part I)
Edited by DONALD B. MCCORMICK, JOHN W. SUTTIE, AND CONRAD WAGNER

VOLUME 280. Vitamins and Coenzymes (Part J)
Edited by DONALD B. MCCORMICK, JOHN W. SUTTIE, AND CONRAD WAGNER

VOLUME 281. Vitamins and Coenzymes (Part K)
Edited by DONALD B. MCCORMICK, JOHN W. SUTTIE, AND CONRAD WAGNER

VOLUME 282. Vitamins and Coenzymes (Part L)
Edited by DONALD B. MCCORMICK, JOHN W. SUTTIE, AND CONRAD WAGNER

VOLUME 283. Cell Cycle Control
Edited by WILLIAM G. DUNPHY

VOLUME 284. Lipases (Part A: Biotechnology)
Edited by BYRON RUBIN AND EDWARD A. DENNIS

VOLUME 285. Cumulative Subject Index Volumes 263, 264, 266–284, 286–289

VOLUME 286. Lipases (Part B: Enzyme Characterization and Utilization)
Edited by BYRON RUBIN AND EDWARD A. DENNIS

VOLUME 287. Chemokines
Edited by RICHARD HORUK

VOLUME 288. Chemokine Receptors
Edited by RICHARD HORUK

VOLUME 289. Solid Phase Peptide Synthesis
Edited by GREGG B. FIELDS

VOLUME 290. Molecular Chaperones
Edited by GEORGE H. LORIMER AND THOMAS BALDWIN

VOLUME 291. Caged Compounds
Edited by GERARD MARRIOTT

VOLUME 292. ABC Transporters: Biochemical, Cellular, and Molecular Aspects
Edited by SURESH V. AMBUDKAR AND MICHAEL M. GOTTESMAN

VOLUME 293. Ion Channels (Part B)
Edited by P. MICHAEL CONN

VOLUME 294. Ion Channels (Part C)
Edited by P. MICHAEL CONN

VOLUME 295. Energetics of Biological Macromolecules (Part B)
Edited by GARY K. ACKERS AND MICHAEL L. JOHNSON

VOLUME 296. Neurotransmitter Transporters
Edited by SUSAN G. AMARA

VOLUME 297. Photosynthesis: Molecular Biology of Energy Capture
Edited by LEE MCINTOSH

VOLUME 298. Molecular Motors and the Cytoskeleton (Part B)
Edited by RICHARD B. VALLEE

VOLUME 299. Oxidants and Antioxidants (Part A)
Edited by LESTER PACKER

VOLUME 300. Oxidants and Antioxidants (Part B)
Edited by LESTER PACKER

VOLUME 301. Nitric Oxide: Biological and Antioxidant Activities (Part C)
Edited by LESTER PACKER

VOLUME 302. Green Fluorescent Protein
Edited by P. MICHAEL CONN

VOLUME 303. cDNA Preparation and Display
Edited by SHERMAN M. WEISSMAN

VOLUME 304. Chromatin
Edited by PAUL M. WASSARMAN AND ALAN P. WOLFFE

VOLUME 305. Bioluminescence and Chemiluminescence (Part C)
Edited by THOMAS O. BALDWIN AND MIRIAM M. ZIEGLER

VOLUME 306. Expression of Recombinant Genes in Eukaryotic Systems
Edited by JOSEPH C. GLORIOSO AND MARTIN C. SCHMIDT

VOLUME 307. Confocal Microscopy
Edited by P. MICHAEL CONN

VOLUME 308. Enzyme Kinetics and Mechanism (Part E: Energetics of Enzyme Catalysis)
Edited by DANIEL L. PURICH AND VERN L. SCHRAMM

VOLUME 309. Amyloid, Prions, and Other Protein Aggregates
Edited by RONALD WETZEL

VOLUME 310. Biofilms
Edited by RON J. DOYLE

VOLUME 311. Sphingolipid Metabolism and Cell Signaling (Part A)
Edited by ALFRED H. MERRILL, JR., AND YUSUF A. HANNUN

VOLUME 312. Sphingolipid Metabolism and Cell Signaling (Part B)
Edited by ALFRED H. MERRILL, JR., AND YUSUF A. HANNUN

VOLUME 313. Antisense Technology
(Part A: General Methods, Methods of Delivery, and RNA Studies)
Edited by M. IAN PHILLIPS

VOLUME 314. Antisense Technology (Part B: Applications)
Edited by M. IAN PHILLIPS

VOLUME 315. Vertebrate Phototransduction and the Visual Cycle (Part A)
Edited by KRZYSZTOF PALCZEWSKI

VOLUME 316. Vertebrate Phototransduction and the Visual Cycle (Part B)
Edited by KRZYSZTOF PALCZEWSKI

VOLUME 317. RNA–Ligand Interactions (Part A: Structural Biology Methods)
Edited by DANIEL W. CELANDER AND JOHN N. ABELSON

VOLUME 318. RNA–Ligand Interactions (Part B: Molecular Biology Methods)
Edited by DANIEL W. CELANDER AND JOHN N. ABELSON

VOLUME 319. Singlet Oxygen, UV-A, and Ozone
Edited by LESTER PACKER AND HELMUT SIES

VOLUME 320. Cumulative Subject Index Volumes 290–319

VOLUME 321. Numerical Computer Methods (Part C)
Edited by MICHAEL L. JOHNSON AND LUDWIG BRAND

VOLUME 322. Apoptosis
Edited by JOHN C. REED

VOLUME 323. Energetics of Biological Macromolecules (Part C)
Edited by MICHAEL L. JOHNSON AND GARY K. ACKERS

VOLUME 324. Branched-Chain Amino Acids (Part B)
Edited by ROBERT A. HARRIS AND JOHN R. SOKATCH

VOLUME 325. Regulators and Effectors of Small GTPases
(Part D: Rho Family)
Edited by W. E. BALCH, CHANNING J. DER, AND ALAN HALL

VOLUME 326. Applications of Chimeric Genes and Hybrid Proteins
(Part A: Gene Expression and Protein Purification)
Edited by JEREMY THORNER, SCOTT D. EMR, AND JOHN N. ABELSON

VOLUME 327. Applications of Chimeric Genes and Hybrid Proteins
(Part B: Cell Biology and Physiology)
Edited by JEREMY THORNER, SCOTT D. EMR, AND JOHN N. ABELSON

VOLUME 328. Applications of Chimeric Genes and Hybrid Proteins (Part C: Protein–Protein Interactions and Genomics)
Edited by JEREMY THORNER, SCOTT D. EMR, AND JOHN N. ABELSON

VOLUME 329. Regulators and Effectors of Small GTPases (Part E: GTPases Involved in Vesicular Traffic)
Edited by W. E. BALCH, CHANNING J. DER, AND ALAN HALL

VOLUME 330. Hyperthermophilic Enzymes (Part A)
Edited by MICHAEL W. W. ADAMS AND ROBERT M. KELLY

VOLUME 331. Hyperthermophilic Enzymes (Part B)
Edited by MICHAEL W. W. ADAMS AND ROBERT M. KELLY

VOLUME 332. Regulators and Effectors of Small GTPases (Part F: Ras Family I)
Edited by W. E. BALCH, CHANNING J. DER, AND ALAN HALL

VOLUME 333. Regulators and Effectors of Small GTPases (Part G: Ras Family II)
Edited by W. E. BALCH, CHANNING J. DER, AND ALAN HALL

VOLUME 334. Hyperthermophilic Enzymes (Part C)
Edited by MICHAEL W. W. ADAMS AND ROBERT M. KELLY

VOLUME 335. Flavonoids and Other Polyphenols
Edited by LESTER PACKER

VOLUME 336. Microbial Growth in Biofilms (Part A: Developmental and Molecular Biological Aspects)
Edited by RON J. DOYLE

VOLUME 337. Microbial Growth in Biofilms (Part B: Special Environments and Physicochemical Aspects)
Edited by RON J. DOYLE

VOLUME 338. Nuclear Magnetic Resonance of Biological Macromolecules (Part A)
Edited by THOMAS L. JAMES, VOLKER DÖTSCH, AND ULI SCHMITZ

VOLUME 339. Nuclear Magnetic Resonance of Biological Macromolecules (Part B)
Edited by THOMAS L. JAMES, VOLKER DÖTSCH, AND ULI SCHMITZ

VOLUME 340. Drug–Nucleic Acid Interactions
Edited by JONATHAN B. CHAIRES AND MICHAEL J. WARING

VOLUME 341. Ribonucleases (Part A)
Edited by ALLEN W. NICHOLSON

VOLUME 342. Ribonucleases (Part B)
Edited by ALLEN W. NICHOLSON

VOLUME 343. G Protein Pathways (Part A: Receptors)
Edited by RAVI IYENGAR AND JOHN D. HILDEBRANDT

VOLUME 344. G Protein Pathways (Part B: G Proteins and Their Regulators)
Edited by RAVI IYENGAR AND JOHN D. HILDEBRANDT

VOLUME 345. G Protein Pathways (Part C: Effector Mechanisms)
Edited by RAVI IYENGAR AND JOHN D. HILDEBRANDT

VOLUME 346. Gene Therapy Methods
Edited by M. IAN PHILLIPS

VOLUME 347. Protein Sensors and Reactive Oxygen Species (Part A: Selenoproteins and Thioredoxin)
Edited by HELMUT SIES AND LESTER PACKER

VOLUME 348. Protein Sensors and Reactive Oxygen Species (Part B: Thiol Enzymes and Proteins)
Edited by HELMUT SIES AND LESTER PACKER

VOLUME 349. Superoxide Dismutase
Edited by LESTER PACKER

VOLUME 350. Guide to Yeast Genetics and Molecular and Cell Biology (Part B)
Edited by CHRISTINE GUTHRIE AND GERALD R. FINK

VOLUME 351. Guide to Yeast Genetics and Molecular and Cell Biology (Part C)
Edited by CHRISTINE GUTHRIE AND GERALD R. FINK

VOLUME 352. Redox Cell Biology and Genetics (Part A)
Edited by CHANDAN K. SEN AND LESTER PACKER

VOLUME 353. Redox Cell Biology and Genetics (Part B)
Edited by CHANDAN K. SEN AND LESTER PACKER

VOLUME 354. Enzyme Kinetics and Mechanisms (Part F: Detection and Characterization of Enzyme Reaction Intermediates)
Edited by DANIEL L. PURICH

VOLUME 355. Cumulative Subject Index Volumes 321–354

VOLUME 356. Laser Capture Microscopy and Microdissection
Edited by P. MICHAEL CONN

VOLUME 357. Cytochrome P450, Part C
Edited by ERIC F. JOHNSON AND MICHAEL R. WATERMAN

VOLUME 358. Bacterial Pathogenesis (Part C: Identification, Regulation, and Function of Virulence Factors)
Edited by VIRGINIA L. CLARK AND PATRIK M. BAVOIL

VOLUME 359. Nitric Oxide (Part D)
Edited by ENRIQUE CADENAS AND LESTER PACKER

VOLUME 360. Biophotonics (Part A)
Edited by GERARD MARRIOTT AND IAN PARKER

VOLUME 361. Biophotonics (Part B)
Edited by GERARD MARRIOTT AND IAN PARKER

VOLUME 362. Recognition of Carbohydrates in Biological Systems (Part A)
Edited by YUAN C. LEE AND REIKO T. LEE

VOLUME 363. Recognition of Carbohydrates in Biological Systems (Part B)
Edited by YUAN C. LEE AND REIKO T. LEE

VOLUME 364. Nuclear Receptors
Edited by DAVID W. RUSSELL AND DAVID J. MANGELSDORF

VOLUME 365. Differentiation of Embryonic Stem Cells
Edited by PAUL M. WASSAUMAN AND GORDON M. KELLER

VOLUME 366. Protein Phosphatases
Edited by SUSANNE KLUMPP AND JOSEF KRIEGLSTEIN

VOLUME 367. Liposomes (Part A)
Edited by NEJAT DÜZGÜNEŞ

VOLUME 368. Macromolecular Crystallography (Part C)
Edited by CHARLES W. CARTER, JR., AND ROBERT M. SWEET

VOLUME 369. Combinational Chemistry (Part B)
Edited by GUILLERMO A. MORALES AND BARRY A. BUNIN

VOLUME 370. RNA Polymerases and Associated Factors (Part C)
Edited by SANKAR L. ADHYA AND SUSAN GARGES

VOLUME 371. RNA Polymerases and Associated Factors (Part D)
Edited by SANKAR L. ADHYA AND SUSAN GARGES

VOLUME 372. Liposomes (Part B)
Edited by NEJAT DÜZGÜNEŞ

VOLUME 373. Liposomes (Part C)
Edited by NEJAT DÜZGÜNEŞ

VOLUME 374. Macromolecular Crystallography (Part D)
Edited by CHARLES W. CARTER, JR., AND ROBERT W. SWEET

VOLUME 375. Chromatin and Chromatin Remodeling Enzymes (Part A)
Edited by C. DAVID ALLIS AND CARL WU

VOLUME 376. Chromatin and Chromatin Remodeling Enzymes (Part B)
Edited by C. DAVID ALLIS AND CARL WU

VOLUME 377. Chromatin and Chromatin Remodeling Enzymes (Part C)
Edited by C. DAVID ALLIS AND CARL WU

VOLUME 378. Quinones and Quinone Enzymes (Part A)
Edited by HELMUT SIES AND LESTER PACKER

VOLUME 379. Energetics of Biological Macromolecules (Part D)
Edited by JO M. HOLT, MICHAEL L. JOHNSON, AND GARY K. ACKERS

VOLUME 380. Energetics of Biological Macromolecules (Part E)
Edited by JO M. HOLT, MICHAEL L. JOHNSON, AND GARY K. ACKERS

VOLUME 381. Oxygen Sensing
Edited by CHANDAN K. SEN AND GREGG L. SEMENZA

VOLUME 382. Quinones and Quinone Enzymes (Part B)
Edited by HELMUT SIES AND LESTER PACKER

VOLUME 383. Numerical Computer Methods (Part D)
Edited by LUDWIG BRAND AND MICHAEL L. JOHNSON

VOLUME 384. Numerical Computer Methods (Part E)
Edited by LUDWIG BRAND AND MICHAEL L. JOHNSON

VOLUME 385. Imaging in Biological Research (Part A)
Edited by P. MICHAEL CONN

VOLUME 386. Imaging in Biological Research (Part B)
Edited by P. MICHAEL CONN

VOLUME 387. Liposomes (Part D)
Edited by NEJAT DÜZGÜNEŞ

VOLUME 388. Protein Engineering
Edited by DAN E. ROBERTSON AND JOSEPH P. NOEL

VOLUME 389. Regulators of G-Protein Signaling (Part A)
Edited by DAVID P. SIDEROVSKI

VOLUME 390. Regulators of G-Protein Signaling (Part B)
Edited by DAVID P. SIDEROVSKI

VOLUME 391. Liposomes (Part E)
Edited by NEJAT DÜZGÜNEŞ

VOLUME 392. RNA Interference
Edited by ENGELKE ROSSI

VOLUME 393. Circadian Rhythms
Edited by MICHAEL W. YOUNG

VOLUME 394. Nuclear Magnetic Resonance of Biological Macromolecules (Part C)
Edited by THOMAS L. JAMES

VOLUME 395. Producing the Biochemical Data (Part B)
Edited by ELIZABETH A. ZIMMER AND ERIC H. ROALSON

VOLUME 396. Nitric Oxide (Part E)
Edited by LESTER PACKER AND ENRIQUE CADENAS

VOLUME 397. Environmental Microbiology
Edited by JARED R. LEADBETTER

VOLUME 398. Ubiquitin and Protein Degradation (Part A)
Edited by RAYMOND J. DESHAIES

VOLUME 399. Ubiquitin and Protein Degradation (Part B)
Edited by RAYMOND J. DESHAIES

VOLUME 400. Phase II Conjugation Enzymes and Transport Systems
Edited by HELMUT SIES AND LESTER PACKER

VOLUME 401. Glutathione Transferases and Gamma Glutamyl Transpeptidases
Edited by HELMUT SIES AND LESTER PACKER

VOLUME 402. Biological Mass Spectrometry
Edited by A. L. BURLINGAME

VOLUME 403. GTPases Regulating Membrane Targeting and Fusion
Edited by WILLIAM E. BALCH, CHANNING J. DER, AND ALAN HALL

VOLUME 404. GTPases Regulating Membrane Dynamics
Edited by WILLIAM E. BALCH, CHANNING J. DER, AND ALAN HALL

VOLUME 405. Mass Spectrometry: Modified Proteins and Glycoconjugates
Edited by A. L. BURLINGAME

VOLUME 406. Regulators and Effectors of Small GTPases: Rho Family
Edited by WILLIAM E. BALCH, CHANNING J. DER, AND ALAN HALL

VOLUME 407. Regulators and Effectors of Small GTPases: Ras Family
Edited by WILLIAM E. BALCH, CHANNING J. DER, AND ALAN HALL

VOLUME 408. DNA Repair (Part A)
Edited by JUDITH L. CAMPBELL AND PAUL MODRICH

VOLUME 409. DNA Repair (Part B)
Edited by JUDITH L. CAMPBELL AND PAUL MODRICH

VOLUME 410. DNA Microarrays (Part A: Array Platforms and Web-Bench Protocols)
Edited by ALAN KIMMEL AND BRIAN OLIVER

VOLUME 411. DNA Microarrays (Part B: Databases and Statistics)
Edited by ALAN KIMMEL AND BRIAN OLIVER

VOLUME 412. Amyloid, Prions, and Other Protein Aggregates (Part B)
Edited by INDU KHETERPAL AND RONALD WETZEL

VOLUME 413. Amyloid, Prions, and Other Protein Aggregates (Part C)
Edited by INDU KHETERPAL AND RONALD WETZEL

VOLUME 414. Measuring Biological Responses with Automated Microscopy
Edited by JAMES INGLESE

VOLUME 415. Glycobiology
Edited by MINORU FUKUDA

VOLUME 416. Glycomics
Edited by MINORU FUKUDA

VOLUME 417. Functional Glycomics
Edited by MINORU FUKUDA

VOLUME 418. Embryonic Stem Cells
Edited by IRINA KLIMANSKAYA AND ROBERT LANZA

VOLUME 419. Adult Stem Cells
Edited by IRINA KLIMANSKAYA AND ROBERT LANZA

VOLUME 420. Stem Cell Tools and Other Experimental Protocols
Edited by IRINA KLIMANSKAYA AND ROBERT LANZA

VOLUME 421. Advanced Bacterial Genetics: Use of Transposons and Phage for Genomic Engineering
Edited by KELLY T. HUGHES

VOLUME 422. Two-Component Signaling Systems, Part A
Edited by MELVIN I. SIMON, BRIAN R. CRANE, AND ALEXANDRINE CRANE

VOLUME 423. Two-Component Signaling Systems, Part B
Edited by MELVIN I. SIMON, BRIAN R. CRANE, AND ALEXANDRINE CRANE

VOLUME 424. RNA Editing
Edited by JONATHA M. GOTT

VOLUME 425. RNA Modification
Edited by JONATHA M. GOTT

VOLUME 426. Integrins
Edited by DAVID CHERESH

VOLUME 427. MicroRNA Methods
Edited by JOHN J. ROSSI

VOLUME 428. Osmosensing and Osmosignaling
Edited by HELMUT SIES AND DIETER HAUSSINGER

VOLUME 429. Translation Initiation: Extract Systems and Molecular Genetics
Edited by JON LORSCH

VOLUME 430. Translation Initiation: Reconstituted Systems and Biophysical Methods
Edited by JON LORSCH

VOLUME 431. Translation Initiation: Cell Biology, High-Throughput and Chemical-Based Approaches
Edited by JON LORSCH

VOLUME 432. Lipidomics and Bioactive Lipids: Mass-Spectrometry–Based Lipid Analysis
Edited by H. ALEX BROWN

VOLUME 433. Lipidomics and Bioactive Lipids: Specialized Analytical Methods and Lipids in Disease
Edited by H. ALEX BROWN

VOLUME 434. Lipidomics and Bioactive Lipids: Lipids and Cell Signaling
Edited by H. ALEX BROWN

VOLUME 435. Oxygen Biology and Hypoxia
Edited by HELMUT SIES AND BERNHARD BRÜNE

VOLUME 436. Globins and Other Nitric Oxide-Reactive Protiens (Part A)
Edited by ROBERT K. POOLE

VOLUME 437. Globins and Other Nitric Oxide-Reactive Protiens (Part B)
Edited by ROBERT K. POOLE

VOLUME 438. Small GTPases in Disease (Part A)
Edited by WILLIAM E. BALCH, CHANNING J. DER, AND ALAN HALL

VOLUME 439. Small GTPases in Disease (Part B)
Edited by WILLIAM E. BALCH, CHANNING J. DER, AND ALAN HALL

VOLUME 440. Nitric Oxide, Part F Oxidative and Nitrosative Stress in Redox Regulation of Cell Signaling
Edited by ENRIQUE CADENAS AND LESTER PACKER

VOLUME 441. Nitric Oxide, Part G Oxidative and Nitrosative Stress in Redox Regulation of Cell Signaling
Edited by ENRIQUE CADENAS AND LESTER PACKER

VOLUME 442. Programmed Cell Death, General Principles for Studying Cell Death (Part A)
Edited by ROYA KHOSRAVI-FAR, ZAHRA ZAKERI, RICHARD A. LOCKSHIN, AND MAURO PIACENTINI

VOLUME 443. Angiogenesis: *In Vitro* Systems
Edited by DAVID A. CHERESH

VOLUME 444. Angiogenesis: *In Vivo* Systems (Part A)
Edited by DAVID A. CHERESH

VOLUME 445. Angiogenesis: *In Vivo* Systems (Part B)
Edited by DAVID A. CHERESH

VOLUME 446. Programmed Cell Death, The Biology and Therapeutic Implications of Cell Death (Part B)
Edited by ROYA KHOSRAVI-FAR, ZAHRA ZAKERI, RICHARD A. LOCKSHIN, AND MAURO PIACENTINI

VOLUME 447. RNA Turnover in Bacteria, Archaea and Organelles
Edited by LYNNE E. MAQUAT AND CECILIA M. ARRAIANO

VOLUME 448. RNA Turnover in Eukaryotes: Nucleases, Pathways and Analysis of mRNA Decay
Edited by LYNNE E. MAQUAT AND MEGERDITCH KILEDJIAN

VOLUME 449. RNA Turnover in Eukaryotes: Analysis of Specialized and Quality Control RNA Decay Pathways
Edited by LYNNE E. MAQUAT AND MEGERDITCH KILEDJIAN

VOLUME 450. Fluorescence Spectroscopy
Edited by LUDWIG BRAND AND MICHAEL L. JOHNSON

VOLUME 451. Autophagy: Lower Eukaryotes and Non-Mammalian Systems (Part A)
Edited by DANIEL J. KLIONSKY

VOLUME 452. Autophagy in Mammalian Systems (Part B)
Edited by DANIEL J. KLIONSKY

VOLUME 453. Autophagy in Disease and Clinical Applications (Part C)
Edited by DANIEL J. KLIONSKY

VOLUME 454. Computer Methods (Part A)
Edited by MICHAEL L. JOHNSON AND LUDWIG BRAND

VOLUME 455. Biothermodynamics (Part A)
Edited by MICHAEL L. JOHNSON, JO M. HOLT, AND GARY K. ACKERS (RETIRED)

VOLUME 456. Mitochondrial Function, Part A: Mitochondrial Electron Transport Complexes and Reactive Oxygen Species
Edited by WILLIAM S. ALLISON AND IMMO E. SCHEFFLER

VOLUME 457. Mitochondrial Function, Part B: Mitochondrial Protein Kinases, Protein Phosphatases and Mitochondrial Diseases
Edited by WILLIAM S. ALLISON AND ANNE N. MURPHY

VOLUME 458. Complex Enzymes in Microbial Natural Product Biosynthesis, Part A: Overview Articles and Peptides
Edited by DAVID A. HOPWOOD

VOLUME 459. Complex Enzymes in Microbial Natural Product Biosynthesis, Part B: Polyketides, Aminocoumarins and Carbohydrates
Edited by DAVID A. HOPWOOD

VOLUME 460. Chemokines, Part A
Edited by TRACY M. HANDEL AND DAMON J. HAMEL

VOLUME 461. Chemokines, Part B
Edited by TRACY M. HANDEL AND DAMON J. HAMEL

VOLUME 462. Non-Natural Amino Acids
Edited by TOM W. MUIR AND JOHN N. ABELSON

VOLUME 463. Guide to Protein Purification, 2nd Edition
Edited by RICHARD R. BURGESS AND MURRAY P. DEUTSCHER

VOLUME 464. Liposomes, Part F
Edited by NEJAT DÜZGÜNEŞ

VOLUME 465. Liposomes, Part G
Edited by NEJAT DÜZGÜNEŞ

VOLUME 466. Biothermodynamics, Part B
Edited by MICHAEL L. JOHNSON, GARY K. ACKERS, AND JO M. HOLT

VOLUME 467. Computer Methods Part B
Edited by MICHAEL L. JOHNSON AND LUDWIG BRAND

VOLUME 468. Biophysical, Chemical, and Functional Probes of RNA Structure, Interactions and Folding: Part A
Edited by DANIEL HERSCHLAG

VOLUME 469. Biophysical, Chemical, and Functional Probes of RNA Structure, Interactions and Folding: Part B
Edited by DANIEL HERSCHLAG

VOLUME 470. Guide to Yeast Genetics: Functional Genomics, Proteomics, and Other Systems Analysis, 2nd Edition
Edited by GERALD FINK, JONATHAN WEISSMAN, AND CHRISTINE GUTHRIE

VOLUME 471. Two-Component Signaling Systems, Part C
Edited by MELVIN I. SIMON, BRIAN R. CRANE, AND ALEXANDRINE CRANE

VOLUME 472. Single Molecule Tools, Part A: Fluorescence Based Approaches
Edited by NILS G. WALTER

VOLUME 473. Thiol Redox Transitions in Cell Signaling, Part A Chemistry and Biochemistry of Low Molecular Weight and Protein Thiols
Edited by ENRIQUE CADENAS AND LESTER PACKER

VOLUME 474. Thiol Redox Transitions in Cell Signaling, Part B Cellular Localization and Signaling
Edited by ENRIQUE CADENAS AND LESTER PACKER

VOLUME 475. Single Molecule Tools, Part B: Super-Resolution, Particle Tracking, Multiparameter, and Force Based Methods
Edited by NILS G. WALTER

VOLUME 476. Guide to Techniques in Mouse Development, Part A Mice, Embryos, and Cells, 2nd Edition
Edited by PAUL M. WASSARMAN AND PHILIPPE M. SORIANO

VOLUME 477. Guide to Techniques in Mouse Development, Part B Mouse Molecular Genetics, 2nd Edition
Edited by PAUL M. WASSARMAN AND PHILIPPE M. SORIANO

VOLUME 478. Glycomics
Edited by MINORU FUKUDA

VOLUME 479. Functional Glycomics
Edited by MINORU FUKUDA

VOLUME 480. Glycobiology
Edited by MINORU FUKUDA

VOLUME 481. Cryo-EM, Part A: Sample Preparation and Data Collection
Edited by GRANT J. JENSEN

VOLUME 482. Cryo-EM, Part B: 3-D Reconstruction
Edited by GRANT J. JENSEN

VOLUME 483. Cryo-EM, Part C: Analyses, Interpretation, and Case Studies
Edited by GRANT J. JENSEN

VOLUME 484. Constitutive Activity in Receptors and Other Proteins, Part A
Edited by P. MICHAEL CONN

SECTION ONE

IDENTIFICATION AND MEASUREMENT OF CONSTITUTIVE ACTIVITY

CHAPTER ONE

Constitutive Activity at the Cannabinoid CB_1 Receptor and Behavioral Responses

Katherine E. Hanlon *and* Todd W. Vanderah

Contents

1. Introduction	4
2. Modifying CB_1 Activity	6
2.1. Pharmacological manipulation	7
3. Behavioral Models	12
3.1. Pain models	12
3.2. Other behavioral models	23
References	26

Abstract

The cannabinoid receptor type 1, found mainly on cells of the central and peripheral nervous system, is a major component of the endogenous cannabinoid system. Constitutive and endogenous activity at cannabinoid receptor type 1 regulates a diverse subset of biological processes including appetite, mood, motor function, learning and memory, and pain. The complexity of cannabinoid receptor type 1 activity is not limited to the constitutive activity of the receptor: promiscuity of ligands associated with and the capability of this receptor to instigate G protein sequestration also complicates the activity of cannabinoid receptor type 1. The therapeutic use of cannabinoid receptor type 1 agonists is still a heavily debated topic, making research on the mechanisms underlying the potential benefits and risks of cannabinoid use more vital than ever. Elucidation of these mechanisms and the quest for agonists and antagonists with greater specificity will allow a greater control of the side effects and risks involved in utilizing cannabinoids as therapeutic agents. In this chapter, we review a small subset of techniques used in the pharmacological application of and the behavioral effects of molecules acting at the paradoxical cannabinoid receptor type 1.

Department of Pharmacology, University of Arizona, Tucson, Arizona, USA

1. Introduction

The endocannabinoid system (ECS) is a complex and far reaching network comprised of the endogenous cannabinoids N-arachidonoylethanolamine (anandamide or AEA) and 2-arachidonyl glycerol (2-AG), cannabinoid hydrolyzing enzymes fatty acid amide hydrolase (FAAH), and monoacylglyceride lipase (MAGL), the endogenous cannabinoid receptors: type 1 (CB_1) and type 2 (CB_2); and the recently characterized G-protein-coupled receptor (GPCR) GPR55 (Alpini and DeMorrow, 2009; Pertwee, 2001a). Endogenous cannabinoid ligands and many synthetic cannabimimetic compounds also display activity at a number of other receptor classes, including transient receptor potential cation channel, subfamily V, member 1 (TRPV1) channels, and the peroxisome proliferator-activated receptors (PPARs) of the nucleus (De Petrocellis and Di Marzo, 2010; Di Marzo et al., 1998; Hermann et al., 2003; Patwardhan et al., 2006; Zygmunt et al., 1999). The promiscuity of cannabinoids is due in part to their lipophilic properties and contributes greatly to the complexity of the ECS.

Another major factor complicating the ECS is the abnormal activity of the CB_1 receptor. The CB_1 receptor is not only constitutively active, but it has the ability to sequester $G_{i/o}$ proteins (Vasquez and Lewis, 1999)—either preventing other receptors from signaling at all, or increasing the signaling of other receptors through G_s or G_q, ultimately forcing a nonnative response. Review of the literature reveals some degree of discrepancy with the claim of constitutive activity since it can be difficult to distinguish between endogenous ligand activity and constitutive activity of a receptor in an intact biological system (Seifert and Wenzel-Seifert, 2002; Turu and Hunyady, 2010; Turu et al., 2007); however, it is generally accepted that the CB_1 GPCR does behave constitutively in most systems (Bouaboula et al., 1997; Canals and Milligan, 2008; Fioravanti et al., 2008; Nie and Lewis, 2001; Pertwee, 2005). Howlett et al. (1998) described a peptide sequence in the juxtamembrane region of the proximal C-terminal tail of the CB_1 receptor that can activate G proteins. Nie and Lewis (2001) further enumerated the underlying mechanism of CB_1 constitutive activity by defining another region of the C-terminal tail, distal to the G-protein-activating region, acting as a regulatory sequence to prevent activation of G proteins. Truncation of this distal portion of the CB_1 C-terminus increases the constitutive activity of the receptor (Nie and Lewis, 2001). A third site within the CB_1 receptor also contributes to constitutive activity—a sequence in the second transmembrane domain stabilizes the receptor in one of two states: the inactive RG_{GDP} or constitutively active $R\star G_{GTP}$ conformation. Interestingly, alteration of this stabilization region also inhibits the CB_1 receptor's ability to sequester $G_{i/o}$ proteins (Howlett et al., 1998; Mukhopadhyay et al., 1999).

Pharmacological manipulation of the CB_1 receptor is significantly impacted by the constitutive activity of the receptor. Long after the introduction of the so-called selective antagonists of CB_1, including AM251 and SR141716A (Rimonabant©), it was revealed that many if not all are actually inverse agonists (MacLennan et al., 1998; Pertwee, 2005). The search for a neutral antagonist of CB_1 has been extensive, and only recently have compounds begun to surface claiming to be neutral antagonists—most notably NESS 0327 (Ruiu et al., 2003). Further characterization is still needed to determine the true activity of these compounds.

CB_1 receptors are widely distributed in the CNS and to a lesser extent in the periphery (Pertwee, 1999). CB_1 receptors are found in the gastrointestinal tract, mainly on cells of the enteric nervous system, as well as on neurons of the peripheral nervous system (Pertwee, 2001b). CB_1 receptors display a limited expression on immune cells as well (Maresz et al., 2007). The CB_1 receptor couples primarily to $G_{i/o}$ and is found in the highest density in GABAergic neurons, though CB_1 has also been shown to act via $G_{i/o}$ in excitatory neurons (Howlett et al., 2002). This results in inhibition of neurotransmitter release via a blockage of voltage-dependent Ca^{2+} channels and activation of inwardly rectifying K^+ channels. In addition, CB_1 affects the intracellular focal adhesion kinase cascade, the mitogen-activated protein kinase (MAPK) cascade, phosphatidylinositol 3-kinase (PI3K) pathway, and modulates production of nitric oxide through nNOS (Bouaboula et al., 1997; Carney et al., 2009; Howlett et al., 2002; Turu and Hunyady, 2010). CB_1 has been implicated in a number of different biological processes including, but not limited to, appetite/satiation, mood, learning and memory, pain processing, and inflammation (Abush and Akirav, 2009; Herkenham et al., 1990; Howlett et al., 2002; Pertwee, 2001b; Richardson et al., 1998a,b; Zhou and Shearman, 2004).

One of the most famous characteristics of exogenous cannabinoids is their ability to stimulate appetite. In 1992, Marinol© (dronabinol), a product composed of purified $(-)$-$trans$-D^9-tetrahydrocannabidiol (THC), was approved as an appetite stimulant in AIDS patients; for review on cannabinoids and patents see Galal et al. (2009). Cannabinoid agonists also make excellent antiemetics. Marinol© (dronabinol) was approved by the FDA in 1985 for nausea and vomiting caused by chemotherapeutic agents (Galal et al., 2009). Other studies have shown that cannabinoid agonists are effective in the relief of opioid-induced emesis (Simoneau et al., 2001), as well as emesis induced by radiotherapy (Stewart, 1990). The antiemetic effect of CB_1 agonists can be attributed to the CB_1 receptors present in the dorsal–vagal complex of the brain stem (Van Sickle et al., 2001), as well as CB_1 receptors present within the gastrointestinal tract; CB_1 activation also reduces electrically stimulated contractions of the small intestine by inhibiting evoked acetylcholine release (Gifford and Ashby, 1996; Izzo and Coutts, 2005).

Gastrointestinal inflammation may also be regulated by the ECS. Inflammatory bowel disease is a persistent state of inflammation within the gastrointestinal tract that comprises two main diseases—ulcerative colitis and Crohn's disease. Both conditions are treated as autoimmune disorders, though the status of ulcerative colitis as such is still debated (Kunos and Pacher, 2004). CB_1 activation within the enteric nervous system reduces acid production and reduces production of the proinflammatory cytokines TNFα and IL-1β (Engel et al., 2008). Further, $CB_1^{-/-}$ mice have increased susceptibility to the well-established *in vivo* model of IBD-like symptoms—trinitrobenzene sulfonic acid (TNBS)-induced colitis (Engel et al., 2010).

Endogenous CB_1 involvement in learning and memory processes has been well established (for review see Moreira and Lutz, 2008). Several studies implicate CB_1 agonists as inhibitors of memory, plasticity, and long-term potentiation while CB_1 antagonists improve these activities in hippocampus-mediated tasks (Abush and Akirav, 2009). Amygdala-dependent processes, including emotional memory, however, do not tend to follow the same etiology (Chhatwal and Ressler, 2007), further complicating the physiological role of CB_1.

Finally, one of the most notable central roles of the ECS lies in pain processing. CB_1 receptors are heavily colocalized with μ-opioid receptors (MOPs) (Pickel et al., 2004; Rodriguez et al., 2001; Salio et al., 2001) and the two receptor classes have been shown to heterodimerize (Rios et al., 2006; Schoffelmeer et al., 2006). Several groups have shown cross regulation by ligands at these two receptors (Fioravanti et al., 2008); Canals and Milligan (2008) demonstrated that the function of MOPs is regulated by constitutive activity at the CB_1 receptor. In addition, a substantial number of studies have shown the analgesic properties of CB_1 agonists (for review see Pertwee, 2001a) along with the upregulation of CB_1 receptors at the spinal cord level in models of neuropathic pain (Walczak et al., 2006; Wang et al., 2007). While the involvement of CB_1 in ascending pain pathways has been well established, the CB_1 role in descending pain modulation pathways is still being elucidated (Wilson et al., 2008). One is hard pressed to find a more diverse and interesting endogenous system than the ECS. In this chapter, we explore methods employed in the study of the CB_1 receptor and CB_1 selective compounds.

2. Modifying CB_1 Activity

While a great number of viable genetic manipulations exist and aid in either the silencing or the overexpression of a given receptor or protein, in this section we will focus on pharmacological manipulations. Pharmacological approaches are much simpler to carry out *in vivo* than gene targeted approaches due to their time- and dose-relationships without long-term compensation found in knockout animals. It should be mentioned that CB_1 knockout mice and rats are commercially available and have been instrumental in the study of endogenous CB_1 activity.

2.1. Pharmacological manipulation

When selecting a drug delivery method a number of factors must be considered, including animal size, solubility of the drug being administered, volume being administered, half-life of the drug, and the drug's ability to cross the blood–brain barrier (Scheld, 1989). Given the lipophilicity of compounds acting at the cannabinoid receptors, solubility is an issue that demands attention in the design of an effective vehicle; see Table 1.1 for a brief summary. In this section, we will focus on drug delivery methods in mice and rats, though many of these practices can be applied to a number of different species.

2.1.1. Vehicle preparation

While dimethyl sulfoxide (DMSO) and dimethyl formamide (DMF) seem to be the universal solvents as far as cannabinoid agonists and inverse agonists are concerned, they are by no means infallible. DMSO is less toxic than DMF (El, 1996); however, preparations of 100% DMSO are not recommended as it is neurotoxic at levels as low as 0.3 mL/kg (Junior et al., 2008). To avoid neurotoxicity the organic solvent must be diluted; this can be done relatively easily with a 0.9% saline solution. Typical vehicle dilutions are 10% DMSO with a volume of Tween 80 that ranges from 3% of the final volume for spinal preparations to 10% of the final volume for parenteral injection; the solution is then brought to volume with saline. The addition of Tween 80 is critical—diluting the concentration of DMSO will cause the compound to fall out of solution: adding a small volume of Tween 80 helps homogenize the mixture.

2.1.2. Routes of drug delivery

2.1.2.1. Intrathecal catheter
The placement of an intrathecal catheter in order to deliver drugs by the spinal route both acutely and chronically was first reported by (Yaksh and Rudy, 1976). Rats should be anesthetized by either injecting a mixture of ketamine/xylazine (100 mg/kg, i.p.), or isoflurane (inhaled, 5% in O_2 for induction and maintained at 2.5% in O_2 at 2 L/min). Shave from the dorsal nape of the neck to between the ears and clean the area with a povidone–iodine topical antiseptic followed by 70% ethanol in diH_2O. Anesthetized rats should be placed in a stereotaxic head holder by lightly gripping the skull between thumb and forefinger (thumb on top of the skull and forefinger beneath chin); support the rat's body weight with the remaining fingers and palm wrapping lightly around the animal's ribcage. The stereotaxic head holder should have one arm fixed in

Table 1.1 A brief summary of cannabinoid solubility

Commercial name	Activity	K_i at human receptor		Solubility		
		CB_1	CB_2	Ethanol	DMSO	DMF
AM251[a]	Inverse agonist	7.5 nM	2290 nM	×	×	×
AM630[b]	Inverse agonist	5.2 μM	31.2 nM		×	×
CAY10508[c]	Inverse agonist	243 nM	Unavailable		×	×
NESS 0327[d]	Neutral antagonist	0.35 pM	21 nM	×	×	×
(S)-SLV 319[e]	Inverse agonist	7.8 nM	>1000 nM	×	×	×
SR141716[f]	Inverse agonist	1.8 nM	>1000 nM	×	×	×
SR144528[g]	Inverse agonist	400 nM	0.6 nM	×	×	×
2-Arachidonyl Glycerol[h]	Agonist	472 nM	1400 nM		×	×
AM1241[i]	Agonist	280 nM	2 nM	×	×	×
Anandamide (AEA)[j]	Agonist	52 nM	90 nM	×	×	×
(±)-CP 47,497[k]	Agonist	2.2 nM	Unavailable	×	×	×
(±)-CP 55,940[l]	Agonist	0.58 nM	0.69 nM	×	×	×
HU-210[m]	Agonist	0.06 nM	.52 nM	×	×	×
HU-308[n]	Agonist	>1000 nM	22.7 nM	×	×	×
JWH 015[o]	Agonist	383 nM	13.8 nM	×	×	
JWH 018[p]	Agonist	9.0 nM	2.94 nM	×	×	×
O-2545[q]	Agonist	1.5 nM	0.32 nM	×	×	×
WIN 55,212-2[r]	Agonist	3.3 nM	62.3 nM	×	×	×

[a] Lan et al. (1999).
[b] Ross et al. (1999).
[c] Muccioli et al. (2006).
[d] Ruiu et al. (2003).
[e] Lange et al. (2004).
[f] Rinaldi-Carmona et al. (1994).
[g] Rinaldi-Carmona et al. (1998).
[h] Mechoulam et al. (1995).
[i] Malan et al. (2001).
[j] Devane et al. (1992).
[k] Huffman et al. (2008).
[l] Wiley et al. (1995).
[m] Howlett et al. (1990).
[n] Hanus et al. (1999).
[o] Pertwee (1999).
[p] Aung et al. (2000).
[q] Martin et al. (2006).
[r] Felder et al. (1995).

place. Which arm remains fixed is up to the user; in general a right handed person holding the animal with the right hand will find it most comfortable to tighten the left arm of the stereotaxic device and vice versa. Gently maneuver the rat onto the fixed arm by inserting the arm into the rat's ear. Still holding the rat steadily, slide the free arm into the rat's other ear and tighten. Slide the nose piece back until the rat's head is bent downward such that the top of the nose is perpendicular to the bench. Using a number 10 blade, make a skin incision approximately 2 cm in length down the midline beginning slightly higher than ear level. Using the same number 10 blade and holding the scalpel nearly horizontal, make an incision through the exposed muscle down the midline by resting the blade on the midline and applying a gentle rocking pressure. Expose the cisterna magna by inserting a retractor and opening the muscle incision just enough to place a cotton swab inside of the opening. Using a number 11 blade at a 90° angle, very gently make a shallow 2 mm incision in the cistern magna. It is extremely important to only allow the blade to puncture the membrane; a deeper incision will damage the brain stem and cause paralysis. Drain the cerebrospinal fluid using cotton swabs and gently massaging the spine. In general a 300 g rat will require two cotton swabs. Widen the opening in the cisterna magna slightly using the number 11 blade again at a 90° angle but this time perpendicular to the original incision to yield a diamond shaped opening. Insert the catheter (PE: 10, 8 mm) down the length of the spinal column such that it terminates in the lumbar region of the spinal cord. At this stage it is important to avoid nicking the spinal cord with the catheter to avoid paralysis: holding the catheter horizontally, insert the tubing into the opening while holding the tail of the animal with the opposite hand to maintain a straight and rigid spinal column. While inserting the catheter, attempt to keep the catheter at the dorsal side of the column away from the spinal cord, which should now be lying on the ventral surface of the column. If any resistance is felt during insertion of the catheter, stop, retract the catheter, turn slightly and try again. Once the catheter has been inserted it can be stitched into place deep in the muscle with 3-0 silk and the skin closed above it via 3-0 silk suture or staples. Allow the animals to recover 5–7 days postsurgery before any pharmacological manipulations are made.

For intrathecal drug administration to the lumbar section of the spinal cord, all drugs should be injected in a 5 µL volume followed by a 9 µL saline flush using a 25 mL Hamilton syringe fitted with tubing connected to an injection cannula (C3131I) to allow the animals to move freely within their home cages during injection. Saline (9 µL) followed by a 1 µL of air then 5 µL of drug should be drawn up into the tubing. The 1 µL of air is important to observe while injecting to assure that the intrathecal catheter is not blocked. If the catheter is blocked, the 1 µL of air will compress alarming the experimenter that the drug may not be moving to the site of action. The 9 µL of saline is important to push the drug entirely through the intrathecal catheter and into the site of action. If drug is to be administered to other areas of the spinal cord the volume of saline should be adjusted (i.e., less saline to thoracic and more to sacral).

2.1.2.2. Lumbar puncture While the intrathecal catheter is an excellent method of spinal delivery in rats, it is a cumbersome procedure in mice. Alternatively a simple spinal administration can be made via lumbar puncture in the oft significantly smaller mouse without requiring anesthesia. For lumbar puncture, inject a volume of 5 μL using a 30 gauge 0.5 in. needle fitted to a 10 μL Hamilton syringe. Holding the mouse firmly by the pelvic girdle, insert the needle at a 20° angle into one side of the L5 or L6 region in between the spinous and transverse processes. Decrease the angle to approximately 10° and insert 0.5 cm into the vertebral column. Expel the volume slowly (over a period of at least 5 s) and rotate the needle on withdrawal to ensure delivery of the entire volume. No recovery period is required.

2.1.2.3. Oral gavage Oral gavage can be performed in mice or rats when oral dosing is required and the researcher is unable to dissolve the compound into the animal's food or water. Additionally, oral gavage allows for a greater control of the dosing schedule and volume. A maximum of 10 mL/kg may be administered via oral gavage and may be repeated up to three times daily. Round tip feeding tubes or ball tip feeding needles may be used: 14–18 gauge for rats or 20–24 gauge for mice; length should be determined by measuring from mouth to most caudal rib. Anesthesia is not required for oral gavage; however, the animal will need to be restrained. This condition makes oral gavage administration easiest to perform on mice, as they can be held by the scruff, pinching the skin over the shoulders such that the forelimbs extend to either side of the body. Holding the animal horizontally parallel to the work surface, straighten head so that no bends in the esophagus occur. Insert the gavage tube or needle over the tongue through the pharynx in one motion; if any resistance is felt withdraw and try again, do not attempt to push through the obstruction, as this is a clear sign of entering into the lungs. The researcher should observe a natural gag and swallow reflex so that the animal swallows the feeding tube or needle. After administering the compound, observe the animal for 5 min, watching for difficulty in breathing or distress.

2.1.2.4. Parenteral injections in mice The most effective parenteral injections into a mouse include, subcutaneous, intraperitoneal, and intravenous. Concentrations should be based upon a 10 mL/kg injection volume. Intramuscular injections are discouraged due to the small amount of muscle mass compared to injection volume and the potential for muscle destruction.

Subcutaneous (s.c.): insert a 30 gauge, 0.5 in. needle fitted to a 1 cc syringe parallel to the body of the animal into the base of the fold formed when pinching the skin constituting the scruff and gently pulling away from the mouse's body. In general, no more than 1 mL may be administered subcutaneously to an adult mouse. If a greater volume is required, repeat the procedure at multiple sites sufficiently separated such that the sites are not overlapping.

Intraperitoneal (i.p.): insert a 30 gauge 0.5 in. needle fitted to a 1 cc syringe at a 45° angle, positioned so that the tip points towards the midline. The mouse should be lightly restrained by hand, pinching the skin at the scruff of the neck between thumb and forefinger and securing the tail to the researchers palm with the pinky finger. Invert the animal prior to injecting. An adult mouse can withstand up to a 2 mL total volume of injection with i.p. administration.

Intravenous (i.v.): place the animal into a restraining device (available through VWR) such that the tail protrudes from the device. Hold the tail under warm water for 10 s (do not allow the water temperature to exceed 55 °C) to dilate the tail vein. Insert a 30 gauge 0.5 in. needle attached to a 1 mL syringe into the tail vein at an angle of approximately 20°, confirm placement by observation of blood in the tip of the syringe prior to injection. Intravenous injection volume in an adult mouse should not exceed 0.5 mL, and should be administered slowly (over a period of at least 10 s). If an out-pocketing of the tail skin is observed at the injection site the researcher should stop, remove the needle and try again.

2.1.2.5. Parenteral injections in rats
Parenteral injections in rats are similar to those in mice with the exception of concentration and volume. Parenteral injections in rats should have drug concentrations based on a 1 mL/kg volume. In addition, intramuscular (i.m.) injections in rats can be extremely useful: i.m. injections are not as time consuming as i.v, and the compound reaches the circulatory system more quickly than s.c. or i.p injections.

Subcutaneous (s.c.): insert a 25 or 30 gauge 0.5 in. needle fitted to a 1 cc syringe parallel to the body of the animal into the base of the fold formed when pinching skin on the dorsal thoracic area and gently pulling the skin away from the rat's body. In general, no more than 1 mL may be administered subcutaneously to an adult rat. If a greater volume is required, repeat the procedure at multiple sites sufficiently separated such that the sites are not overlapping.

Intramuscular (i.m.): insert a 25 or 30 gauge 0.5 in. needle fitted to a 1 cc syringe at a 20–30° angle into either the hamstring on the ventral side of the femur or the quadriceps on the dorsal side of the femur, aligning the needle length-wise to the direction of muscle fiber extension. A maximum volume of 0.25 mL should be injected i.m. in a rat.

Intraperitoneal (i.p.): insert a 25 or 30 gauge 0.5 in. needle fitted to a 1 cc syringe at a 45° angle, positioned so that the tip points towards the midline. The rat should be restrained by hand, wrapping the thumb and forefinger around the shoulder area (aligning the rat's spine to the soft area of the researchers hand where the forefinger and thumb meet) such that the animal's forelimbs are crossed in front and the weight of the animal is supported by the remaining fingers and palm surrounding the ribcage firmly, but gently. Invert the animal prior to injecting. An adult rat can withstand up to a 5 mL total volume of injection with i.p. administration.

Intravenous (i.v.): place the animal into a restraining device (available through VWR) such that the tail protrudes from the device. Hold the tail under warm water for 10 s (do not allow the water temperature to exceed 55 °C) to dilate the tail vein. Insert a 30 gauge 0.5 in. needle attached to a 1 mL syringe into the tail vein at an angle of approximately 20°, confirm placement by observation of blood in the tip of the syringe prior to injection. Intravenous injection volume in an adult rat should not exceed 1 mL, and should be administered slowly (over a period of at least 10 s). If an out-pocketing of the tail skin is observed at the injection site the researcher should stop, remove the needle and try again.

3. Behavioral Models

Because the scope of the authors experience lies primarily in the pain field, this section will be heavily focused on pain models and the associated behavioral testing. Other models relevant to the CB_1 receptor are also briefly described, including multiple sclerosis, emesis, and some models used in learning and memory studies. The CB_1 receptor has been described as the most abundant receptor in the brain (Howlett, 2005); its study is an ever expanding field and as such the scope of pharmacological manipulation is not limited to the methods described here.

3.1. Pain models

3.1.1. Neuropathic pain models

During the late twentieth century several pain models were developed that enabled the study of neuropathic pain. While the models use primarily mice and rats, the symptoms closely mimic those of humans with conditions such as causalgia, diabetic neuropathies, and trauma-induced neuropathies (Bennett and Xie, 1988). Initially, the models primarily focused on the transection of a peripheral nerve; often used was the sciatic nerve due to its size, location, and relative ease of manipulation. The constraints of these models lie in the inability to measure pain caused by a stimulus, and instead relied on spontaneous pain behaviors since all sensory information distal to the transection site was no longer transmitted (Decosterd and Woolf, 2000). Until the realization of the advantage that constriction models, either partial or full, offered, it was not possible to distinguish between allodynia and hyperalgesia, let alone study these two phenomena separately.

Described below are three widely used methods for inducing neuropathic pain via nerve ligation. In general, these models are usually employed in rats bearing intrathecal catheters; however, these can be modified for use in a mouse population as described below with exception to the spinal nerve ligation (SNL). It is recommended that rats be given a recovery period of at least 5–7 days between procedures; either the intrathecal catheter implantation or the nerve ligation may be performed first, however, due to the higher risk of paralysis with the catheter implantation it is in general recommended to first implant the intrathecal catheter.

3.1.1.1. Chronic constriction injury of the sciatic nerve (CCI)

In 1988, Bennett and Xie (Bennett and Xie, 1988) described the groundbreaking method of constricting the sciatic nerve. The model was initially described for use in rats; however, it can be modified and applied to mice by altering the length of incision and reducing the number of loose ligations. The procedure is relatively simple and provides a means of measuring both allodynia and hyperalgesia while simultaneously retaining the ability to study spontaneous pain. Spontaneous symptoms normally present in the form of self mutilation to the toes or foot of the affected limb. Because of this, the timeline in which this model is used may be limited to include acute and short chronic studies.

Induce anesthesia in a 250–300 g rat with 5% isoflurane in O_2, maintain the animal at 2.5% isoflurane for the duration of the procedure (or induce anesthesia in a 20 g mouse with 2.5% isoflurane in O_2 and maintain the mouse at 1.5% isoflurane). Shave or use depilatory cream to remove fur from the thigh of the animal, cleaning the area with a povidone–iodine topical antiseptic followed by 70% ethanol in diH_2O. Make an incision through the biceps femoris to expose the sciatic nerve. Separate a portion (7 mm in rats, 3 mm in mice) of the sciatic from surrounding tissue proximal to the trifurcation. Loop 5-0 suture silk around the freed section of nerve and loosely ligate such that the nerve appears to be barely constricted. Insert four ligations 1 mm apart in a rat or two ligations 1 mm apart in a mouse. Close the muscle incision with 3-0 suture silk; the skin may be closed with either surgical staples or 3-0 suture silk. Allow the animal a minimum of 5–7 days recovery prior to pharmacological manipulation. Appropriate controls for this model should include a sham surgery group in which the same surgery is performed without ligation of the aforementioned nerve.

3.1.1.2. L5/L6 spinal nerve ligation (SNL)

Full ligation of the L_5 and L_6 spinal nerves provides an excellent model of neuropathic pain. Onset is quick, the model induces both hyperalgesia and allodynia as well as spontaneous pain in the affected limb, and the model is ideal for both acute and chronic studies as the effects last upward of 60 days after the procedure. The procedure described here is based on that developed by Kim and Chung (1992).

Prior to beginning the SNL an operating board should be purchased or constructed. Constructing a Chung board can be done easily with a piece of 1/4 in. pressboard cut to approximately 18 in. lengthwise and 6 in. wide. Cut three strips of heavy grade tubing (at least a 1/4 in. thick for strength, but still malleable by hand) to approximately 15 in. and tightly tie widthwise around the board to form three taught loops. Cut the excess tubing after a tight knot has been fashioned. Affix thick ($\sim$1/4 in.) 1 in. diameter rubber bands to the tubing (one band per tubing loop) by looping the band around the tubing and pulling through itself. Arrange the board such that the long side sits parallel to the edge of the work surface and arrange the bands so that the two outer bands sit at the bottom (closest to the researcher's torso) and the middle band is at the top. The bands should be approximately 1 in. apart. Fold a bench underpad and wrap around the board so that the entire board is covered but the rubber bands protrude either side. Cut and sterilize three pieces of 18-gauge wire to approximately 3 in. in length. Bend each wire into a hook on one end and a loop on the other that can be affixed to each of the rubber bands on the board (see Fig. 1.1 for a visual guide). The purpose of the Chung board is to provide a firm stable surface that will double as nonimposing retractors to hold back muscle surrounding the incision site without blocking the operators view. Alternatively, instead of purchasing or constructing a board, long handle retractors can be used during the SNL. We do not recommend this, however, because the use of retractors without the Chung board does not immobilize the animal. The small movements induced by the operator's manipulation of the rat make locating, isolating, and ligating the L_5/L_6 spinal nerves very difficult. Use of the Chung board prevents the rat's body from shifting even slightly. The surgical procedure described below induces injury to the left hind paw. The procedure can be modified to affect the right hind paw as well.

Induce anesthesia in a 250–300 g rat with 5% isoflurane in O_2, maintain the animal at 2.5% isoflurane for the duration of the procedure. Shave or use depilatory cream to remove fur in a 1 in. wide and 2 in. long section up the midline centering on the lumbar region of the spinal cord. Clean the area with a povidone–iodine topical antiseptic followed by 70% ethanol in diH_2O. Lay the anesthetized rat in a prone position and make an incision slightly (about 5 mm) to the left of the spinal column from L_4 to S_2. Position the hooks protruding from the Chung board to hold the incision open wide enough to insert a cotton swab. Remove excess blood from the incision with cotton swabs and carefully remove the muscle from the dorsal aspect of the exposed pelvic bone with forceps. Using a small rongeur, remove the transverse process of the L_6 vertebra. Ensure that no bone fragments remain by checking the surrounding area with forceps. This is an extremely important step required to avoid damaging the spinal cord during recovery.

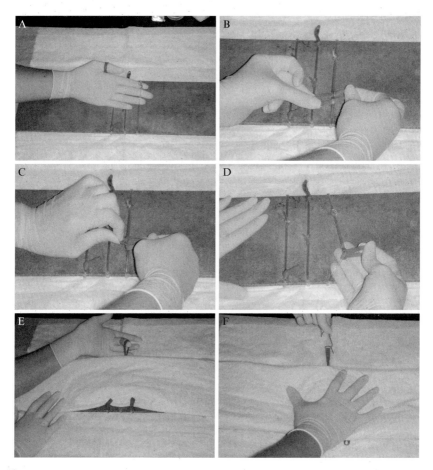

Figure 1.1 Setting up an SNL operating board. (A) Wrap tubing around short edge of board and tie snugly to form three loops. (B–D) Tie one rubber band around each tubing loop. (E) When wrapping board with bench pad, fold so that rubber bands are still reachable. (F) Form hooks from sterile wire with closed loop ends that can be attached to each of the rubber bands.

Slide a ball tipped (preferably glass) probe under the anterior edge of the S_1 transverse process. Lightly scraping the ventral surface of the process with the ball of the probe, slowly draw the probe back out until it catches on a slightly elastic thread extending diagonally out from the midline across the S_1 vertebra. This is the L6 spinal nerve. Carefully hook the nerve across the ball of the probe and draw it out into the space created by removing the transverse process of L_6. Carefully slide the loop of a piece of 5-0 suture silk 3 in. long and fitted with a slipknot around the ball of the probe. Loop the suture ligand around the nerve by grabbing one of the loose ends with a pair of forceps and pulling through the slipknot to completely encircle the nerve. Tighten until half to two-thirds of the diameter has been constricted. Repeat the same ligation to the L_5 spinal nerve which is easily locatable by sliding the probe with the handle extending vertically and the hook end perpendicular to the spinal column under the L_6 spinous process and pulling out the L_4/L_5 bundle. Do not injure L_4 as this will significantly impair motor function, causing paralysis in the left hind limb. L_5 is distinguishable from L_4 by recognizing that L_5 will appear to sit on top of L_4 and will have a slightly smaller diameter. Separate L_4 and L_5 by sliding the process between the two spinal nerves. Once both ligation sutures are in place, remove the wire hooks and close the muscle using 1 or 2 sutures with 3-0 silk. Staple or suture the skin incision. Allow the ligated animal to recover for a minimum of 7 days prior to pharmacological manipulation. Appropriate controls for this model should include a sham surgery group in which the same surgery is performed without ligation of the aforementioned nerves.

3.1.1.3. Spared nerve injury (SNI)

In order to minimize the degree of variability of results between animals and researchers that exists with the CCI as well as eliminate the possibility of inflammation affecting the L_4 spinal nerve introduced with the L_5/L_6 SNL, the spared nerve injury (SNI) was introduced in 2000 by Decosterd and Woolf (Decosterd and Woolf, 2000). The SNI is similar to the CCI except that the ligations are made distal to the trifurcation of the sciatic nerve in the common peroneal and tibial nerves (not affecting the sural nerve), only one ligation per nerve is performed, and an axotomy 2–4 mm above the distal stump is also performed on the common peroneal and tibial nerves. Decosterd and Woolf report symptoms of allodynia, hyperalgesia, and spontaneous pain that persist for greater than 100 days.

Induce anesthesia in a 200–250 g rat with 5% isoflurane in O_2, maintain the animal at 2.5% isoflurane for the duration of the procedure (or induce anesthesia in a 20 g mouse with 2.5% isoflurane in O_2 and maintain the mouse at 1.5% isoflurane). Shave or use depilatory cream to remove fur from the thigh of the animal, cleaning the area with a povidone–iodine topical antiseptic followed by 70% ethanol in diH_2O. Make an incision through the biceps femoris to expose the sciatic nerve and trifurcation. Separate a portion of the trifurcated common peroneal and tibial nerves from surrounding tissue. Distinction can be made between the sural, common peroneal, and tibial nerves by denoting the branching location from the sciatic nerve. The first (most proximal) to branch off will be the sural nerve, followed by the simultaneous bifurcation into the common peroneal and tibial nerves. Loop 5-0 suture silk around the freed section of nerve and ligate tightly such that the nerve appears to be constricted by about 1/2 to 2/3 of the diameter. Distal to the ligation, 2–4 mm proximal to the nerve stump, section the nerve. Close the muscle incision with 3-0 suture silk; the skin may be closed with either surgical staples or 3-0 suture silk. Allow the animal a minimum of 5–7 days recovery prior to pharmacological manipulation. Appropriate controls for this model should include a sham surgery group in which the same surgery is performed without ligation or axotomy of the aforementioned nerves.

3.1.2. Inflammatory pain models

Inflammatory pain models are a highly diverse set of models, ranging from models mimicking human conditions such as multiple sclerosis or arthritis to paw/intra-articular injections of irritant solutions aimed at isolating the mechanism of inflammatory pain. In this section, we describe some commonly used models targeting inflammatory pain mechanism.

3.1.2.1. Formalin A major breakthrough in analgesic study occurred in 1977 when Dubuisson and Dennis introduced the formalin test (Dubuisson and Dennis, 1977): subcutaneous injection of dilute formalin induced continuous rather than transient pain, and no restraint of the animal is required, avoiding potential confounding factors that may arise from restraint.

Allow a 250–300 g rat to acclimate to the testing environment for a period of 30 min for rats and 90 min for mice. In general, with the formalin test flinching and guarding are measured and should be carried out in a plastic or plexiglass enclosure of 30 cm^3. Dilute sterile formalin to 2–5% by volume in 0.9% saline (5% formalin will produce robust flinching and should be the maximum concentration used to avoid serious tissue damage). Using a 1 mL syringe equipped with a 30 gauge 0.5 in. needle, inject 0.05 mL subcutaneously into the dorsal aspect of either hind-paw. For ease of injection a leather workers glove can be used to gently restrain the rat by placing the glove on the work surface and allowing the rat to climb inside of the glove until the most caudal rib area is no longer visible. Gently hold the rat inside of the glove such that the animal cannot back out of the glove but being careful not to restrict breathing. By applying gentle pressure with the palm of the hand, the back limbs should be visible and planted securely on the work surface. Gently insert the needle subcutaneously to the dorsal aspect and discharge the formalin solution. Effects should be noticeable immediately and allow testing periods of at least 1 h; appropriate controls for this type of study should include a group injected with the vehicle saline.

3.1.2.2. Complete freund's adjuvant (CFA) CFA is a suspension of dessicated mycobacterium in paraffin oil and mannide monooleate that induces inflammation, tissue necrosis, and ulceration. It can be used subcutaneously in the paw, or intraperitonealy in mice and rats. A period of 24 h is required for onset after injection. Due to the rate of tissue necrosis, it is recommended to euthanize CFA injected animals within 1 week.

Intraperitoneal injections of CFA in rats should have a maximum volume of 0.5 mL or 0.25 mL in mice. For s.c. injections in the plantar aspect of a hind paw, use a CFA volume of 0.2 mL in rats and 0.1 mL in mice. Use a 30 gauge 0.5 in. needle on a 1 mL syringe. For ease of injection, a leather workers glove can be used to gently restrain the animal by placing the glove on the work surface and allowing the animal to climb inside of the glove until the most caudal rib area is no longer visible. Gently hold the animal inside of the glove such that the animal cannot back out of the glove but being careful not to restrict breathing. By applying gentle pressure with the palm of the hand the back limbs should be visible and planted securely on the work surface. Gently extend one of the hind legs until the plantar aspect of the paw is facing upwards away from the work surface. Insert the needle subcutaneously to the plantar aspect and discharge the CFA solution. Gentle pressure should be applied to the injection site for 10 s to prevent leakage. Appropriate controls for this model should include a group injected with incomplete Freund's adjuvant (mannide monooleate in paraffin oil). *Note*: the use of CFA will result in necrosis of the surrounding tissue over time. These areas of necrosis should be avoided for behavioral testing as they will give false positives of analgesic drugs.

3.1.2.3. Carrageenan Carrageenan is a seaweed extract that induces local inflammation in approximately 4 h when injected s.c. to the dorsal aspect of a hind paw in rats or mice. The inflammatory pain may persist for upwards of 12 hours. Use a 3% lambda-carrageenan in diH2O. Lambda-carrageenan is recommended due to its solubility in cool water. An injection volume of 0.2 mL in rats and 0.1 mL in mice is recommended and should be done with a 30 gauge 0.5 in. needle on a 1 mL syringe. For ease of injection, a leather workers glove can be used to gently restrain the animal by placing the glove on the work surface and allowing the animal to climb inside of the glove until the most caudal rib area is no longer visible. Gently hold the animal inside of the glove such that the animal cannot back out of the glove but being careful not to restrict breathing. By applying gentle pressure with the palm of the hand the back limbs should be visible and planted securely on the work surface. Insert the needle subcutaneously to the dorsal aspect and discharge the lambda-carrageenan solution.

3.1.3. Cancer pain
The prototypical model of cancer pain was developed by the Mantyh group in 1999 (Schwei *et al.*, 1999). It is a highly relevant model—it is estimated that up to 50% of cancer patients report pain as the first symptom of the disease. It is particularly relevant to cancers that have a high rate of metastasis to bone, and as such can be used with bone cancer cells, breast cancer cells, prostate cancer cells, lung cancer cells, or any other cancer line with a high rate of metastasis to bone. This model requires the use of a small drill (dental or dremel drill) and a faxitron imaging system to ensure proper placement, and consists of the intramedullary implantation of tumor cells into a mouse

femur. Aggressiveness of the tumor cell line used will determine the timeline of the experiment and must be determined empirically.

Mice should be anesthetized by injecting a mixture of ketamine/xylazine (100 mg/kg, i.p.); concentration of the solution should be based on a final volume of 10 mL/kg. Shave or use depilatory cream to remove fur from the right thigh of the animal, cleaning the area with a povidone–iodine topical antiseptic followed by 70% ethanol in diH_2O. Place the anesthetized mouse in a supine position and using a number 10 blade make a skin incision to expose the right distal femur and proximal tibia. Holding the scalpel at a 5° angle, section the patellar tendon just distal to the patella and gently move the patella proximal by about 3 mm to expose the condyles of the distal femur. Between the condyles, drill a hole with the bit of the drill extending in the same direction as the cortical shaft of the femur. Insert a cannulation needle (C31311) (Fig 1.2) through the hole into the intramedullary space and obtain a faxitron image to ensure that the needle lies completely within the intramedullary space without any other bone or tissue damage. Remove the cannulation needle and with a 10 μL Hamilton syringe inject a volume of 5 μL of either control culture medium or a population of tumor cells suspended in culture medium. Seal the hole with bone cement and apply 1 suture from the section of patellar tendon still attached to the patella to the portion attached to the tibia to replace the patella in the original location. Ensure that the full range of motion is retained. The skin incision may be closed with either 3-0 suture silk or surgical staples. Allow the animal a recovery period of at least 7 days, this period will also allow the tumor to establish.

3.1.4. Quantification of pain-related behaviors

3.1.4.1. Rotarod The rotarod test is not a measure of pain behavior, but rather a measure of the amount of sedation or motor impairment imposed upon an animal undergoing an experimental therapy. Prior to inducing a

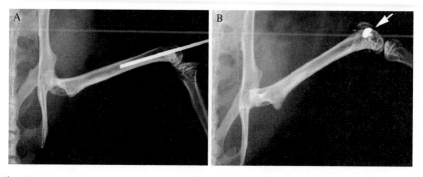

Figure 1.2 Needle placement (A) into the intramedullary space equidistant from the ends of the femur was verified prior to injection of 66.1 cells (10^5 in 5 μl). Dental amalgam (B) is used to seal the cells inside the bone (arrow).

particular pain state, uninjured animals should be subjected to the rotarod test with sufficient variety in dosages to construct a dose–response curve. The rotarod device (Columbus Instruments International, Columbus, OH) has interchangeable rods in order to accommodate rodents of differing size. Naïve animals should be trained to ambulate on the device for a period of 180 s by simply placing a group of up to four animals on the rod and turning the device on after a few minutes. If a rat falls to the lower chamber prior to 180 s or a mouse prior to 60 s, pick them up and put them back on the rod. In order to avoid fatigue, training should occur in increments no greater than 10 min in length for rats and 5 min in length for mice. It should take one training session to train a rat or mouse to ambulate on the device. The timecourse and testing intervals after drug administration will have to be empirically determined depending on the composition of the drug and the site of administration. A general rule of thumb for spinal administration is testing intervals of 15 min over a timecourse of 60 min. Measure the time a treated animal is able to ambulate on the device and compare between treatment groups with a cut off time of 180 s.

3.1.4.2. Tail flick The tail flick model is a measure of acute nociception and does not require use of an injury model. A water bath maintained at 52 °C is used to immerse the distal 2/3 of a rat or mouse's tail. The latency until withdrawal (rapid flick) should be measured prior to drug treatment (baseline) and compared among treatment groups. Typically a 10 s cut off time is utilized in order to avoid damage to the tail. Temperatures can be increased or decreased along with cut off times; however, temperatures should not increase over 55 °C as this temperature will cause tissue damage if tails remain in the water over 10 s.

$$\% \text{ Activity} = 100 \times (\text{tail withdrawal latency post drug} \\ - \text{baseline tail withdrawal latency}) / \\ (\text{cut off tail withdrawal latency} \\ - \text{baseline tail withdrawal latency}) \quad (1)$$

3.1.4.3. Von Frey Filament Up-Down Method The Von Frey up-down method (Dixon, 1980) is a measure of tactile allodynia and can be used with the injury models above with exception to the carrageenan model of inflammatory pain. This is because the inflammation induced by the carrageenan model is so significant on the plantar surface of the hind paw that mechanical nociceptors may not properly identify stimulus. Prior to inducing an injury model, all animals should be subjected to a pre-injury baseline test of paw withdrawal threshold. After induction of injury and the appropriate recovery period, the animal should be subjected to a postinjury baseline test. Within group comparisons will be made to this value with the Dixon nonparametric statistical test. The Von Frey up-down method can be employed in mice and rats, though the caliber of filaments used in mice is generally lower.

The calibration range used in adult mice is 2.44–4.56 (corresponding to a logarithmic range of force from 0.03 to 2.34 g); adult rats are subjected to a filament range from 3.61 to 5.18 (corresponding to a range of force from 0.41 to 15 g).

Place the animals in suspended wire mesh cages and allow to acclimate for an appropriate period of time (30 min for rats and 90 min for mice). Beginning with the middle calibration fiber, for example in a mouse this would be the 3.61 fiber, apply the tip of the filament to the plantar aspect of the affected paw for a duration of 3 s. If no response is received (the paw is not withdrawn) apply the next higher weight filament. If a response is observed, apply the next lower weight filament. Proceed until either the highest or lowest caliber filament has been exhausted or observing three "switches" in the direction of calibration. A switch in direction can be summarized as follows: if a mouse gives a positive response at the 3.61 filament and a positive response at the 3.08 filament then the direction of calibration is currently downward since the researcher is proceeding downward in filament stiffness. If the mouse then gives no response to the 2.44 filament then the direction of calibration has "switched," causing the researcher to then proceed upward until a response is received. The Dixon statistic should be calculated for each pattern of calibration using Flashcalc (University of Arizona) or a similar program. The Dixon statistic can then be averaged for each group and plotted against time points, or % activity can be calculated as follows for comparison between groups:

$$\% \text{ Activity} = 100 \times (\text{withdrawal threshold postdrug} \\ - \text{postinjury baseline threshold})/ \\ (\text{preinjury baseline threshold} \\ - \text{postinjury baseline threshold}) \quad (2)$$

Since it is not plausible to induce mechanical allodynia in an uninjured animal (it would require a caliber filament that would support the weight of the paw), no calculation exists for sham animals with the Von Frey up-down method.

3.1.4.4. Hargreaves method
The Hargreaves method (Hargreaves *et al.*, 1988) measures thermal hypersensitivity, a form of hyperalgesia. This method can be used with all of the pain models described above. A mobile

radiant heat source (Ugo Basile, Italy) is required for use of this method. Preinjury and postinjury baselines should be taken for comparison between and within groups.

Allow animals to acclimate to a plexiglass enclosure on a glass plate. Place the radiant heat source beneath the glass plate at a location corresponding to the plantar surface of the affected hind limb. During the preinjury baseline test, a baseline heat intensity must be established for each individual animal; this baseline intensity will be maintained for the duration of the experiment. The baseline intensity is identified as the heat intensity at which the animal is able to withstand the heat source for a given latency (i.e., the heat source can be adjusted for baselines within 10–15 s or 20–25 s depending on the desired outcome). Establishing baseline intensities that comply with a range will minimize standard error and standard deviations within groups. Once a baseline range is established, the intensity should be recorded for later application. When applying different heat intensities to identify the baseline intensity, it is essential to give the animal a rest period of at least 1 min between each intensity level to avoid tissue damage. A cut off latency for thermal testing will be 32 s. A thermometer or calibrating probe should be available to detect ramp heating and confirm temperature values for publication. For comparison to baseline or between groups, percent activity of the experimental animal groups can be calculated using Eq. (2) above, and sham, or control activities can be calculated as follows:

$$\% \text{ Activity} = 100 \times (\text{withdrawal latency postdrug} \\ -\text{postsham baseline latency})/(\text{cut off latency (32 s)} \quad (3) \\ -\text{postsham baseline latency})$$

3.1.4.5. Spontaneous pain Spontaneous pain is commonly measured by observing flinching and guarding. Allow animal to acclimate to a plastic or plexiglass enclosure. Establish a postinjury baseline (prior to treatment) by counting the number of flinches of the affected limb in a 2-min period and count the number of seconds the animal spends guarding the affected limb in a 2-min period. A flinch is defined as a fast jerking motion of the affected paw and the number of flinches will be recorded on a five-channel counter for each minute and totaled. Guarding stance is that in which the animal retracts the affected limb up against the torso and does not allow the limb to bear weight with recording performed using a stop watch. Comparisons can be made between groups (Lozano-Ondoua et al., 2010).

3.1.4.6. Movement-evoked pain Movement-evoked pain is measured in the same manner as spontaneous pain; however, it is only measured at a time in which the animal is moving throughout the enclosure and is most often associated with a bone cancer model of pain. Movement evoked pains evaluate the severity of pain the mouse experienced during normal ambulation. Animals will be placed in an empty housing pan and limping and guarding behavior of the right leg (injured or inoculated with cancer cells) will be observed for 2 min. As each animal walks across the empty pan, the use of the afflicted hind limb will be rated with the following scale: 0, no use of hind limb at all; 1, partial nonuse; 2, limp and guard; 3, limp; and 4, normal use (Lozano-Ondoua et al., 2010).

3.2. Other behavioral models

3.2.1. Emesis

Briefly, ferrets of approximately 1.5 kg should be acclimated in plexiglass enclosures for a period of 30 min. Anesthesia should be induced with 4% halothane in O_2 only until loss of righting occurs (less than 1 min). The dorsal nape of the neck should be shaven and cleaned with a povidone–iodine topical antiseptic followed by 70% ethanol in diH_2O. Make a subcutaneous injection of either control vehicle or experimental compound; volume of injection should not exceed 0.5 mL. The injection should be performed with a 1 cc syringe equipped with a 30 gauge 0.5 in. disposable needle. Observe the animal for a period of approximately 30 min, recording retching and vomiting. A retch is defined as any contraction of the abdomen without expulsion, while vomiting is any oral expulsion episode (Simoneau et al., 2001). Ferrets may be tested in this manner multiple times; however, a washout period is required. The length of the washout period is dependent on the half life of the compound administered; however, with opiates and cannabinoids, a washout period of 3 days is generally acceptable.

3.2.2. Multiple sclerosis

Multiple sclerosis (MS) is a debilitating disease of the central nervous system (CNS) that involves inflammation and demyelination (Rossi et al., 2010). To date there are two methods of inducing MS like symptoms in vivo: experimental autoimmune encephalomyelitis (EAE) and viral-induced demyelination models (Kubajewska and Constantinescu, 2010). Here, we will briefly describe EAE since it is an extensively studied and widely accepted model (Mokhtarian and Griffin, 1984). EAE can be induced by s.c. injection on the plantar surface of a hindlimb of either rats or mice. Hundred micrograms of myelin basic protein should be dissolved in 50 μL of 0.9% saline emulsified in CFA. The injection should be performed in the same manner described above for the CFA model. After the inoculation, animals should be scored for symptoms of EAE based on the following scale: 0, no neurological symptoms; 1, limp tail [stage E1, usually observed on day 10 postinjection (PI)]; 2, hindlimb weakness (stage E2, day 11 PI); 3, hindlimb paralysis (stage E3, day 12–13 PI); 4, quadriplegia (stage E4; day 14–15 PI); 5, moribund (past day 16 PI).

3.2.3. Learning and memory

There are a multitude of models for the study of learning and memory (for review see Ouerstreet and Russell (2010)); these models can be classified as hippocampus dependent and include tasks such as spatial learning and social recognition; or as amygdala-dependent- encompassing emotional memory. Traditionally emotional learning and memory tasks have been based on fear conditioning—due to a disputed range of emotion in animals and inability to positively identify "good" or "happy" emotional response in rodents, it has proven quite difficult to design positive emotional memory models. Here we briefly describe a few well developed models that can utilize cannulation of either the hippocampus or the basolateral amygdala, or systemic administration as methods of drug delivery.

3.2.3.1. Radial maze Developed by Iwasaki *et al.* (Iwasaki *et al.*, 1992), the radial maze comprises of a central platform and eight radial arms extending from the platform. The maze may either be elevated and open sided, or enclosed. An enclosed format will specify evaluation to spatial cognition and provide less stress to the animal. Food should be placed at the terminus of each arm prior to beginning. Whether naive animal training should be done prior to administration of the test compound will be determined by the scope and aims of the study. Place a rat (age and weight of the animal will also be determined by the aims of the experiment) on the central platform. Count the number of correct and incorrect choices and the time required for the rat to complete the task. A cut-off of 10 min should be established regardless of whether the animal has completed the task. A correct choice is defined as traversing an arm that has not previously been visited, yielding the food pellet; an incorrect choice is defined as traversing an arm that has already been visited. If the researcher chooses a format in which maze training occurs prior to compound administration, then animals that make at least seven correct choices and either 1 or 0 errors in the first eight decisions of three consecutive tests should be considered as having acquired spatial cognition.

3.2.3.2. Social recognition The social recognition model developed by Thor and Holloway (Thor *et al.*, 1982) and modified by Dantzer *et al.* (Dantzer *et al.*, 1987) deals with an adult male rat's ability to recognize a juvenile rat based on the amount of time the adult male spends investigating the juvenile on subsequent meetings. The model consists of two phases: a training phase and an experimental phase. The experimental phase should be comprised of four conditions (facilitation, nonfacilitation, interference, and noninterference), which each animal will experience in a random order on the testing day. The facilitation condition is described as placement of the juvenile in the adult's home cage for 5 min, withdrawal of the juvenile for 5 min, replacement of the juvenile into the adult's cage for an additional 5 min, withdrawal of the juvenile for 120 min, and final replacement of the juvenile into the adult's cage for 5 min. The nonfacilitation condition is the same as the facilitation condition but without the second 5-min "facilitatory" visit. The interference condition is similar to the facilitation condition; however, the second 5-min visit is done with a second, different juvenile and time of withdrawal between the interference juvenile and the final visit of the original juvenile is reduced to 30 min. The noninterference condition is the same as the nonfacilitatory condition but the withdrawal time is reduced to 30 min.

Prior to conducting the experimental phase, the training phase should be carried out by allowing the rats to acclimate for a period of 2 h in a dark room with a background of white noise. The room should be equipped with an infrared spotlight (only on during testing) and a video camera or other visual recording device. Introduce a juvenile into an adult cage for a period of 5 min, then withdraw the juvenile for varying lengths of time. A second 5-min visit with either the same or a different juvenile should be made after the withdrawal period. On the 1st and 2nd training days, the withdrawal period should be 5 min; on the 3rd and 4th days the period should be increased to 30 min; and finally to 120 min on the 5th and 6th days. On testing days, record the final visit and score—time positive for investigation includes nosing, sniffing, and grooming of the juvenile as well as pawing and close following. Adult males that display aggressive behavior towards the juveniles during the training phase should not be advanced to the experimental phase. Comparisons between treatment groups can be made based on the average time spent investigating a juvenile; this model assumes that a lower amount of time spent investigating indicates greater social recognition of the juvenile.

3.2.3.3. Morris water maze

A pool should be constructed (1–2 m in diameter), and filled with 26 °C water. Opaque water can be created by the addition of a small amount of either powdered or liquid milk. An escape platform can be constructed by filing large diameter (~ 9 cm) tubing cut to a height that would render the upper platform just beneath water level. Fill with rocks and seal both ends to create a base and platform. The escape platform should be placed in the pool. Four points around the perimeter of the pool should be designated as N, S, E, W; allowing the pool to be divided into four quadrants. Placing the hidden platform in a set location allows the researcher to train rats to escape from the water onto the hidden platform in a relatively short period of time. The model can be modified simply by moving or adding additional platforms, or changing water visibility or the color of the platform (for complete review see Morris (1984)).

REFERENCES

Abush, H., and Akirav, I. (2009). Cannabinoids modulate hippocampal memory and plasticity. *Hippocampus.*

Alpini, G., and DeMorrow, S. (2009). Changes in the endocannabinoid system may give insight into new and effective treatments for cancer. *Vitam. Horm.* **81,** 469–485.

Aung, M. M., Griffin, G., Huffman, J. W., Wu, M., Keel, C., Yang, B., Showalter, V. M., Abood, M. E., and Martin, B. R. (2000). Influence of the N-1 alkyl chain length of cannabimimetic indoles upon CB(1) and CB(2) receptor binding. *Drug Alcohol Depend.* **60,** 133–140.

Bennett, G. J., and Xie, Y. K. (1988). A peripheral mononeuropathy in rat that produces disorders of pain sensation like those seen in man. *Pain* **33,** 87–107.

Bouaboula, M., Perrachon, S., Milligan, L., Canat, X., Rinaldi-Carmona, M., Portier, M., Barth, F., Calandra, B., Pecceu, F., Lupker, J., Maffrand, J. P., Le, F. G., *et al.* (1997). A selective inverse agonist for central cannabinoid receptor inhibits mitogen-activated protein kinase activation stimulated by insulin or insulin-like growth factor 1. Evidence for a new model of receptor/ligand interactions. *J. Biol. Chem.* **272,** 22330–22339.

Canals, M., and Milligan, G. (2008). Constitutive activity of the cannabinoid CB1 receptor regulates the function of co-expressed Mu opioid receptors. *J. Biol. Chem.* **283,** 11424–11434.

Carney, S. T., Lloyd, M. L., MacKinnon, S. E., Newton, D. C., Jones, J. D., Howlett, A. C., and Norford, D. C. (2009). Cannabinoid regulation of nitric oxide synthase I (nNOS) in neuronal cells. *J. Neuroimmune Pharmacol.* **4,** 338–349.

Chhatwal, J. P., and Ressler, K. J. (2007). Modulation of fear and anxiety by the endogenous cannabinoid system. *CNS Spectr.* **12,** 211–220.

Dantzer, R., Bluthe, R. M., Koob, G. F., and Le, M. M. (1987). Modulation of social memory in male rats by neurohypophyseal peptides. *Psychopharmacology (Berl.)* **91,** 363–368.

De Petrocellis, L., and Di Marzo, V. (2010). Non-CB1, non-CB2 receptors for endocannabinoids, plant cannabinoids, and synthetic cannabimimetics: Focus on G-protein-coupled receptors and transient receptor potential channels. *J. Neuroimmune Pharmacol.* **5,** 103–121.

Decosterd, I., and Woolf, C. J. (2000). Spared nerve injury: An animal model of persistent peripheral neuropathic pain. *Pain* **87,** 149–158.

Devane, W. A., Hanus, L., Breuer, A., Pertwee, R. G., Stevenson, L. A., Griffin, G., Gibson, D., Mandelbaum, A., Etinger, A., and Mechoulam, R. (1992). Isolation and structure of a brain constituent that binds to the cannabinoid receptor. *Science* **258,** 1946–1949.

Di Marzo, V., Bisogno, T., Melck, D., Ross, R., Brockie, H., Stevenson, L., Pertwee, R., and De, P. L. (1998). Interactions between synthetic vanilloids and the endogenous cannabinoid system. *FEBS Lett.* **436,** 449–454.

Dixon, W. J. (1980). Efficient analysis of experimental observations. *Annu. Rev. Pharmacol. Toxicol.* **20,** 441–462.

Dubuisson, D., and Dennis, S. G. (1977). The formalin test: A quantitative study of the analgesic effects of morphine, meperidine, and brain stem stimulation in rats and cats. *Pain* **4,** 161–174.

El, J. A. (1996). Toxic effects of organic solvents on the growth of Chlorella vulgaris and Selenastrum capricornutum. *Bull. Environ. Contam. Toxicol.* **57,** 191–198.

Engel, M. A., Kellermann, C. A., Rau, T., Burnat, G., Hahn, E. G., and Konturek, P. C. (2008). Ulcerative colitis in AKR mice is attenuated by intraperitoneally administered anandamide. *J. Physiol. Pharmacol.* **59,** 673–689.

Engel, M. A., Kellermann, C. A., Burnat, G., Hahn, E. G., Rau, T., and Konturek, P. C. (2010). Mice lacking cannabinoid CB1-, CB2-receptors or both receptors show increased susceptibility to trinitrobenzene sulfonic acid (TNBS)-induced colitis. *J. Physiol. Pharmacol.* **61,** 89–97.

Felder, C. C., Joyce, K. E., Briley, E. M., Mansouri, J., Mackie, K., Blond, O., Lai, Y., Ma, A. L., and Mitchell, R. L. (1995). Comparison of the pharmacology and signal transduction of the human cannabinoid CB1 and CB2 receptors. *Mol. Pharmacol.* **48,** 443–450.

Fioravanti, B., De, F. M., Stucky, C. L., Medler, K. A., Luo, M. C., Gardell, L. R., Ibrahim, M., Malan, T. P. Jr. Yamamura, H. I., Ossipov, M. H., King, T., Lai, J., et al. (2008). Constitutive activity at the cannabinoid CB1 receptor is required for behavioral response to noxious chemical stimulation of TRPV1: Antinociceptive actions of CB1 inverse agonists. *J. Neurosci.* **28,** 11593–11602.

Galal, A. M., Slade, D., Gul, W., El-Alfy, A. T., Ferreira, D., and Elsohly, M. A. (2009). Naturally occurring and related synthetic cannabinoids and their potential therapeutic applications. *Recent Pat. CNS Drug Discov.* **4,** 112–136.

Gifford, A. N., and Ashby, C. R., Jr. (1996). Electrically evoked acetylcholine release from hippocampal slices is inhibited by the cannabinoid receptor agonist, WIN 55212-2, and is potentiated by the cannabinoid antagonist, SR 141716A. *J. Pharmacol. Exp. Ther.* **277,** 1431–1436.

Hanus, L., Breuer, A., Tchilibon, S., Shiloah, S., Goldenberg, D., Horowitz, M., Pertwee, R. G., Ross, R. A., Mechoulam, R., and Fride, E. (1999). HU-308: A specific agonist for CB(2), a peripheral cannabinoid receptor. *Proc. Natl. Acad. Sci. USA* **96,** 14228–14233.

Hargreaves, K., Dubner, R., Brown, F., Flores, C., and Joris, J. (1988). A new and sensitive method for measuring thermal nociception in cutaneous hyperalgesia. *Pain* **32,** 77–88.

Herkenham, M., Lynn, A. B., Little, M. D., Johnson, M. R., Melvin, L. S., de Costa, B. R., and Rice, K. C. (1990). Cannabinoid receptor localization in brain. *Proc. Natl. Acad. Sci. USA* **87,** 1932–1936.

Hermann, H., De, P. L., Bisogno, T., Schiano, M. A., Lutz, B., and Di, M. V. (2003). Dual effect of cannabinoid CB1 receptor stimulation on a vanilloid VR1 receptor-mediated response. *Cell. Mol. Life Sci.* **60,** 607–616.

Howlett, A. C. (2005). Cannabinoid receptor signaling. *Handb. Exp. Pharmacol.* **168,** 53–79.

Howlett, A. C., Champion, T. M., Wilken, G. H., and Mechoulam, R. (1990). Stereochemical effects of 11-OH-delta 8-tetrahydrocannabinol-dimethylheptyl to inhibit adenylate cyclase and bind to the cannabinoid receptor. *Neuropharmacology* **29,** 161–165.

Howlett, A. C., Song, C., Berglund, B. A., Wilken, G. H., and Pigg, J. J. (1998). Characterization of CB1 cannabinoid receptors using receptor peptide fragments and site-directed antibodies. *Mol. Pharmacol.* **53,** 504–510.

Howlett, A. C., Barth, F., Bonner, T. I., Cabral, G., Casellas, P., Devane, W. A., Felder, C. C., Herkenham, M., Mackie, K., Martin, B. R., Mechoulam, R., and Pertwee, R. G. (2002). International Union of Pharmacology. XXVII. Classification of cannabinoid receptors. *Pharmacol. Rev.* **54,** 161–202.

Huffman, J. W., Thompson, A. L., Wiley, J. L., and Martin, B. R. (2008). Synthesis and pharmacology of 1-deoxy analogs of CP-47, 497 and CP-55, 940. *Bioorg. Med. Chem.* **16,** 322–335.

Iwasaki, K., Matsumoto, Y., and Fujiwara, M. (1992). Effect of nebracetam on the disruption of spatial cognition in rats. *Jpn. J. Pharmacol.* **58,** 117–126.

Izzo, A. A., and Coutts, A. A. (2005). Cannabinoids and the digestive tract. *Handb. Exp. Pharmacol.* **168,** 573–598.

Junior, A. M., Arrais, C. A., Saboya, R., Velasques, R. D., Junqueira, P. L., and Dulley, F. L. (2008). Neurotoxicity associated with dimethylsulfoxide-preserved hematopoietic progenitor cell infusion. *Bone Marrow Transplant.* **41,** 95–96.

Kim, S. H., and Chung, J. M. (1992). An experimental model for peripheral neuropathy produced by segmental spinal nerve ligation in the rat. *Pain* **50,** 355–363.

Kubajewska, I., and Constantinescu, C. S. (2010). Cannabinoids and experimental models of multiple sclerosis. *Immunobiology* **215**(8), 647–657.
Kunos, G., and Pacher, P. (2004). Cannabinoids cool the intestine. *Nat. Med.* **10**, 678–679.
Lan, R., Liu, Q., Fan, P., Lin, S., Fernando, S. R., McCallion, D., Pertwee, R., and Makriyannis, A. (1999). Structure-activity relationships of pyrazole derivatives as cannabinoid receptor antagonists. *J. Med. Chem.* **42**, 769–776.
Lange, J. H., Coolen, H. K., van Stuivenberg, H. H., Dijksman, J. A., Herremans, A. H., Ronken, E., Keizer, H. G., Tipker, K., McCreary, A. C., Veerman, W., Wals, H. C., Stork, B., et al. (2004). Synthesis, biological properties, and molecular modeling investigations of novel 3, 4-diarylpyrazolines as potent and selective CB(1) cannabinoid receptor antagonists. *J. Med. Chem.* **47**, 627–643.
Lozano-Ondoua, A. N., Wright, C., Vardanyan, A., King, T., Largent-Milnes, T. M., Nelson, M., Jimenez-Andrade, J. M., Mantyh, P. W., and Vanderah, T. W. (2010). A cannabinoid 2 receptor agonist attenuates bone cancer-induced pain and bone loss. *Life Sci.* **86**, 646–653.
MacLennan, S. J., Reynen, P. H., Kwan, J., and Bonhaus, D. W. (1998). Evidence for inverse agonism of SR141716A at human recombinant cannabinoid CB1 and CB2 receptors. *Br. J. Pharmacol.* **124**, 619–622.
Malan, T. P., Jr., Ibrahim, M. M., Deng, H., Liu, Q., Mata, H. P., Vanderah, T., Porreca, F., and Makriyannis, A. (2001). CB2 cannabinoid receptor-mediated peripheral antinociception. *Pain* **93**, 239–245.
Maresz, K., Pryce, G., Ponomarev, E. D., Marsicano, G., Croxford, J. L., Shriver, L. P., Ledent, C., Cheng, X., Carrier, E. J., Mann, M. K., Giovannoni, G., Pertwee, R. G., et al. (2007). Direct suppression of CNS autoimmune inflammation via the cannabinoid receptor CB1 on neurons and CB2 on autoreactive T cells. *Nat. Med.* **13**, 492–497.
Martin, B. R., Wiley, J. L., Beletskaya, I., Sim-Selley, L. J., Smith, F. L., Dewey, W. L., Cottney, J., Adams, J., Baker, J., Hill, D., Saha, B., Zerkowski, J., et al. (2006). Pharmacological characterization of novel water-soluble cannabinoids. *J. Pharmacol. Exp. Ther.* **318**, 1230–1239.
Mechoulam, R., Ben-Shabat, S., Hanus, L., Ligumsky, M., Kaminski, N. E., Schatz, A. R., Gopher, A., Almog, S., Martin, B. R., and Compton, D. R. (1995). Identification of an endogenous 2-monoglyceride, present in canine gut, that binds to cannabinoid receptors. *Biochem. Pharmacol.* **50**, 83–90.
Mokhtarian, F., and Griffin, D. E. (1984). The role of mast cells in virus-induced inflammation in the murine central nervous system. *Cell. Immunol.* **86**, 491–500.
Moreira, F. A., and Lutz, B. (2008). The endocannabinoid system: Emotion, learning and addiction. *Addict. Biol.* **13**, 196–212.
Morris, R. (1984). Developments of a water-maze procedure for studying spatial learning in the rat. *J. Neurosci. Methods* **11**, 47–60.
Muccioli, G. G., Wouters, J., Charlier, C., Scriba, G. K., Pizza, T., Di, P. P., De, M. P., Poppitz, W., Poupaert, J. H., and Lambert, D. M. (2006). Synthesis and activity of 1, 3, 5-triphenylimidazolidine-2, 4-diones and 1, 3, 5-triphenyl-2-thioxoimidazolidin-4-ones: Characterization of new CB1 cannabinoid receptor inverse agonists/antagonists. *J. Med. Chem.* **49**, 872–882.
Mukhopadhyay, S., Cowsik, S. M., Lynn, A. M., Welsh, W. J., and Howlett, A. C. (1999). Regulation of Gi by the CB1 cannabinoid receptor C-terminal juxtamembrane region: Structural requirements determined by peptide analysis. *Biochemistry* **38**, 3447–3455.
Nie, J., and Lewis, D. L. (2001). Structural domains of the CB1 cannabinoid receptor that contribute to constitutive activity and G-protein sequestration. *J. Neurosci.* **21**, 8758–8764.
Overstreet, D., and Russell, R. W. (2010). Animal models of memory disorders. Animal Models in Psychiatry II. Vol. 19. Finders University of South Australia, Bedford Park, S.A., Australia, pp. 315–368.

Patwardhan, A. M., Jeske, N. A., Price, T. J., Gamper, N., Akopian, A. N., and Hargreaves, K. M. (2006). The cannabinoid WIN 55, 212-2 inhibits transient receptor potential vanilloid 1 (TRPV1) and evokes peripheral antihyperalgesia via calcineurin. *Proc. Natl. Acad. Sci. USA* **103**, 11393–11398.

Pertwee, R. G. (1999). Pharmacology of cannabinoid receptor ligands. *Curr. Med. Chem.* **6**, 635–664.

Pertwee, R. G. (2001a). Cannabinoid receptors and pain. *Prog. Neurobiol.* **63**, 569–611.

Pertwee, R. G. (2001b). Cannabinoids and the gastrointestinal tract. *Gut* **48**, 859–867.

Pertwee, R. G. (2005). Inverse agonism and neutral antagonism at cannabinoid CB1 receptors. *Life Sci.* **76**, 1307–1324.

Pickel, V. M., Chan, J., Kash, T. L., Rodriguez, J. J., and Mackie, K. (2004). Compartment-specific localization of cannabinoid 1 (CB1) and mu-opioid receptors in rat nucleus accumbens. *Neuroscience* **127**, 101–112.

Richardson, J. D., Aanonsen, L., and Hargreaves, K. M. (1998a). Antihyperalgesic effects of spinal cannabinoids. *Eur. J. Pharmacol.* **345**, 145–153.

Richardson, J. D., Kilo, S., and Hargreaves, K. M. (1998b). Cannabinoids reduce hyperalgesia and inflammation via interaction with peripheral CB1 receptors. *Pain* **75**, 111–119.

Rinaldi-Carmona, M., Barth, F., Heaulme, M., Shire, D., Calandra, B., Congy, C., Martinez, S., Maruani, J., Neliat, G., and Caput, D. (1994). SR141716A, a potent and selective antagonist of the brain cannabinoid receptor. *FEBS Lett.* **350**, 240–244.

Rinaldi-Carmona, M., Barth, F., Millan, J., Derocq, J. M., Casellas, P., Congy, C., Oustric, D., Sarran, M., Bouaboula, M., Calandra, B., Portier, M., Shire, D., *et al.* (1998). SR 144528, the first potent and selective antagonist of the CB2 cannabinoid receptor. *J. Pharmacol. Exp. Ther.* **284**, 644–650.

Rios, C., Gomes, I., and Devi, L. A. (2006). mu opioid and CB1 cannabinoid receptor interactions: Reciprocal inhibition of receptor signaling and neuritogenesis. *Br. J. Pharmacol.* **148**, 387–395.

Rodriguez, J. J., Mackie, K., and Pickel, V. M. (2001). Ultrastructural localization of the CB1 cannabinoid receptor in mu-opioid receptor patches of the rat Caudate putamen nucleus. *J. Neurosci.* **21**, 823–833.

Ross, R. A., Brockie, H. C., Stevenson, L. A., Murphy, V. L., Templeton, F., Makriyannis, A., and Pertwee, R. G. (1999). Agonist-inverse agonist characterization at CB1 and CB2 cannabinoid receptors of L759633, L759656, and AM630. *Br. J. Pharmacol.* **126**, 665–672.

Rossi, S., Bernardi, G., and Centonze, D. (2010). The endocannabinoid system in the inflammatory and neurodegenerative processes of multiple sclerosis and of amyotrophic lateral sclerosis. *Exp. Neurol.* **224**, 92–102.

Ruiu, S., Pinna, G. A., Marchese, G., Mussinu, J. M., Saba, P., Tambaro, S., Casti, P., Vargiu, R., and Pani, L. (2003). Synthesis and characterization of NESS 0327: A novel putative antagonist of the CB1 cannabinoid receptor. *J. Pharmacol. Exp. Ther.* **306**, 363–370.

Salio, C., Fischer, J., Franzoni, M. F., Mackie, K., Kaneko, T., and Conrath, M. (2001). CB1-cannabinoid and mu-opioid receptor co-localization on postsynaptic target in the rat dorsal horn. *NeuroReport* **12**, 3689–3692.

Scheld, W. M. (1989). Drug delivery to the central nervous system: General principles and relevance to therapy for infections of the central nervous system. *Rev. Infect. Dis.* **11** (Suppl. 7), S1669–S1690.

Schoffelmeer, A. N., Hogenboom, F., Wardeh, G., and De Vries, T. J. (2006). Interactions between CB1 cannabinoid and mu opioid receptors mediating inhibition of neurotransmitter release in rat nucleus accumbens core. *Neuropharmacology* **51**, 773–781.

Schwei, M. J., Honore, P., Rogers, S. D., Salak-Johnson, J. L., Finke, M. P., Ramnaraine, M. L., Clohisy, D. R., and Mantyh, P. W. (1999). Neurochemical and

cellular reorganization of the spinal cord in a murine model of bone cancer pain. *J. Neurosci.* **19,** 10886–10897.
Seifert, R., and Wenzel-Seifert, K. (2002). Constitutive activity of G-protein-coupled receptors: Cause of disease and common property of wild-type receptors. *Naunyn Schmiedebergs Arch. Pharmacol.* **366,** 381–416.
Simoneau, I. I., Hamza, M. S., Mata, H. P., Siegel, E. M., Vanderah, T. W., Porreca, F., Makriyannis, A., and Malan, T. P. Jr. (2001). The cannabinoid agonist WIN55 212–2 suppresses opioid-induced emesis in ferrets. *Anesthesiology* **94,** 882–887.
Stewart, D. J. (1990). Cancer therapy, vomiting, and antiemetics. *Can. J. Physiol. Pharmacol.* **68,** 304–313.
Thor, D. H., Wainwright, K. L., and Holloway, W. R. (1982). Persistence of attention to a novel conspecific: Some developmental variables in laboratory rats. *Dev. Psychobiol.* **15,** 1–8.
Turu, G., and Hunyady, L. (2010). Signal transduction of the CB1 cannabinoid receptor. *J. Mol. Endocrinol.* **44,** 75–85.
Turu, G., Simon, A., Gyombolai, P., Szidonya, L., Bagdy, G., Lenkei, Z., and Hunyady, L. (2007). The role of diacylglycerol lipase in constitutive and angiotensin AT1 receptor-stimulated cannabinoid CB1 receptor activity. *J. Biol. Chem.* **282,** 7753–7757.
Van Sickle, M. D., Oland, L. D., Ho, W., Hillard, C. J., Mackie, K., Davison, J. S., and Sharkey, K. A. (2001). Cannabinoids inhibit emesis through CB1 receptors in the brainstem of the ferret. *Gastroenterology* **121,** 767–774.
Vasquez, C., and Lewis, D. L. (1999). The CB1 cannabinoid receptor can sequester G-proteins, making them unavailable to couple to other receptors. *J. Neurosci.* **19,** 9271–9280.
Walczak, J. S., Pichette, V., Leblond, F., Desbiens, K., and Beaulieu, P. (2006). Characterization of chronic constriction of the saphenous nerve, a model of neuropathic pain in mice showing rapid molecular and electrophysiological changes. *J. Neurosci. Res.* **83,** 1310–1322.
Wang, S., Lim, G., Mao, J., Sung, B., Yang, L., and Mao, J. (2007). Central glucocorticoid receptors regulate the upregulation of spinal cannabinoid-1 receptors after peripheral nerve injury in rats. *Pain* **131,** 96–105.
Wiley, J. L., Barrett, R. L., Lowe, J., Balster, R. L., and Martin, B. R. (1995). Discriminative stimulus effects of CP 55, 940 and structurally dissimilar cannabinoids in rats. *Neuropharmacology* **34,** 669–676.
Wilson, A. R., Maher, L., and Morgan, M. M. (2008). Repeated cannabinoid injections into the rat periaqueductal gray enhance subsequent morphine antinociception. *Neuropharmacology* **55,** 1219–1225.
Yaksh, T. L., and Rudy, T. A. (1976). Chronic catheterization of the spinal subarachnoid space. *Physiol. Behav.* **17,** 1031–1036.
Zhou, D., and Shearman, L. P. (2004). Voluntary exercise augments acute effects of CB1-receptor inverse agonist on body weight loss in obese and lean mice. *Pharmacol. Biochem. Behav.* **77,** 117–125.
Zygmunt, P. M., Petersson, J., Andersson, D. A., Chuang, H., Sorgard, M., Di, M. V., Julius, D., and Hogestatt, E. D. (1999). Vanilloid receptors on sensory nerves mediate the vasodilator action of anandamide. *Nature* **400,** 452–457.

CHAPTER TWO

Detecting Constitutive Activity and Protean Agonism at Cannabinoid-2 Receptor

Massimiliano Beltramo,[*,1] Rossella Brusa,[*,†] Isabella Mancini,[*,‡] and Paola Scandroglio[*,§]

Contents

1. Introduction	32
2. General Considerations	34
3. Evaluation of Constitutive Activity Using GTPγS Assay	35
3.1. Instrumentations	35
3.2. Reagents and reagent preparation	36
3.3. Disposable	36
3.4. Preparation of membranes	37
3.5. GTPγS assay	37
3.6. Data analysis	38
4. Evaluation of Constitutive Activity Using cAMP Assay	39
4.1. Instrumentations	40
4.2. Reagents and reagent preparation	40
4.3. Disposable	41
4.4. cAMP assay	41
4.5. Data analysis	41
5. Evaluation of Constitutive Activity Using RT-CES	42
5.1. Instrumentations	42
5.2. Reagents and reagent preparation	42
5.3. Disposable	43
5.4. Cell impedance assay	43
5.5. Data analysis	45
6. Evaluation of Protean Agonism with cAMP Assay	46
7. Comparison of the Methods	48
References	49

[*] Schering-Plough Research Institute, Milan, Italy
[†] Novartis, Origgio, Varese, Italy
[‡] OPIS s.r.l., Palazzo Aliprandi, Desio, Italy
[§] Imagine s.r.l., Gallarate, Italy
[1] Present address: Institut National de la Recherche Agronomique (INRA), Physiologie de la Reproduction et de Comportements, UMR0085, Nouzilly, France

Abstract

Since the cannabinoid system is involved in regulating several physiological functions such as locomotor activity, cognition, nociception, food intake, and inflammatory reaction, it has been the subject of intense study. Research on the pharmacology of this system has enormously progressed in the last 20 years. One intriguing aspect that emerged from this research is that cannabinoid receptors (CBs) express a high level of constitutive activity. Investigation on this particular aspect of receptor pharmacology has largely focused on CB1, the CB subtype highly expressed in several brain regions. More recently, research on constitutive activity on the other CB subtype, CB2, was stimulated by the increasing interest on its potential as target for the treatment of various pathologies (e.g., pain and inflammation). There are several possible implications of constitutive activity on the therapeutic action of both agonists and antagonists, and consequently, it is important to have valuable methods to study this aspect of CB2 pharmacology. In the present chapter, we describe three methods to study constitutive activity at CB2: two classical methods relying on the detection of changes in cAMP level and GTPγS binding and a new one based on cell impedance measurement. In addition, we also included a section on detection of protean agonism, which is an interesting pharmacological phenomenon strictly linked to constitutive activity.

1. INTRODUCTION

In the 1980s, studies on benzodiazepine and β2-adrenergic receptors' pharmacology were the first ones to suggest the existence of constitutive activity at G-protein-coupled receptors (GPCR; Braestrup *et al.*, 1982; Cerione *et al.*, 1984). Nonetheless, for its own nature, constitutive activity could be quite elusive to study, and the existence of an endogenous ligand tone may have confounding effects on the interpretation of results. Hence, initially there were concerns that constitutive activity might actually represent the result of competition between antagonist and endogenous ligand. Different types of results supported the existence of *bona fide* constitutive activity, for example, increasing receptor concentration produced a corresponding increase of constitutive activity (Samama *et al.*, 1993). However, the more compelling evidence for constitutive activity could be obtained using true antagonists (Milligan and Bond, 1997). In the absence of endogenous agonists, true antagonists will have no effect *per se*, but they will block the effects of both agonists and inverse agonists. Despite the initial doubts by now constitutive activity at GPCR is considered to be the rule more than the exception (Kenakin, 2005).

Between GPCR, the cannabinoid receptors (CBs) are amongst those express a high level of constitutive activity. CBs come in two subtypes, CB1 and CB2. A third receptor (GPR55) has been proposed to belong to the CB group (Baker *et al.*, 2006). However, no general consensus has been so far achieved about its classification as a CB (Petitet *et al.*, 2006; Ross, 2009).

The synthesis of the first selective antagonists for CB1 and CB2, SR141716A and SR144528, respectively, represented a significant progress in the cannabinoid field (Rinaldi-Carmona *et al.*, 1995, 1998) and allowed the discovery of constitutive activity at CBs. Initial evidence of constitutive activity at CB2 was obtained using a cAMP assay and by incubating recombinant CHO cell line expressing CB2 with SR144528. Application of SR144528 not only reversed CP55940 agonistic effect, but also increased forskolin response above baseline (Portier *et al.*, 1999). A similar further augmentation of forskolin-induced cAMP level in CB2 transfected cells was also observed after pertussis toxin treatment (PTX), confirming an involvement of Gi protein. Consistent with a CB2-mediated event this effect was absent in nontransfected cells (Portier *et al.*, 1999). In the following years, additional evidence produced using various inverse agonists (AM630, JTE907, BML190, and SCH414319) supported the hypothesis that CB2 is constitutively active in various recombinant systems (Iwamura *et al.*, 2001; Lunn *et al.*, 2008; New and Wong, 2003; Ross *et al.*, 1999).

Expression of constitutive activity depends not only on receptor type but could also be influenced by cell line, receptor concentration, and receptor splice variant or single nucleotide polymorphism (Hasegawa *et al.*, 1996; Lefkowitz *et al.*, 1993; Samama *et al.*, 1993). In the case of CB2, it has been recently demonstrated that the H316Y and Q63R/H316Y polymorphic receptors exhibit higher constitutive activity than the CB2 wild-type receptor (Carrasquer *et al.*, 2010).

Classical methods used to unmask CB2 constitutive activity are based on GTPγS and cAMP assays. More recently, the development of label-free technologies made available a new way to reveal receptor constitutive activity (Peters and Scott, 2009; Scandroglio *et al.*, 2010). Label-free technologies have the advantage of performing real-time measurement of cellular response to receptor activation or inhibition. In addition, they do not require the use of prior pharmacological intervention such as forskolin stimulation in cAMP assay. The use of label-free technologies is not broadly diffused yet, but they substantially improve our capacity to detect constitutive activity. Of particular interest is their possible application to evaluate primary cell natively expressing the receptor of interest.

Another concept strictly linked to constitutive activity, that of protean agonist, was developed on a theoretical basis and then supported by experimental evidence (Kenakin, 2001). Protean agonists can act either as agonists, antagonists or inverse agonists, depending on the level of constitutive activity of the receptor and on the intrinsic activity of the ligand. The existence of protean agonists has been demonstrated at few receptors (e.g., H3, secretin receptor, and a-2-adrenoceptor; Ganguli *et al.*, 1998; Gbahou *et al.*, 2003; Jansson *et al.*, 1998; Pauwels *et al.*, 2002) including CB2

(Mancini *et al.*, 2009; Yao *et al.*, 2006). Protean agonists represent interesting pharmacological tools offering a novel opportunity to modulate signal transmission and may have interesting therapeutic applications. For example, they would set the level of receptor stimulation to a constant state without silencing it completely, as it would happen with an inverse agonist. Setting a constant level of receptor constitutive activity could represent a valid treatment for pathologies caused by abnormally high receptor constitutive activity (Parfitt *et al.*, 1996; Parma *et al.*, 1993).

2. General Considerations

Genetically engineered systems are the classical choice to study constitutive activity because their normally high level of recombinant receptor expression increases constitutive activity. Hence, to study constitutive activity, we generated our own CB2-expressing cell line. CHO cells were stably transfected with CB2 by using Lipofectamine. For transfection, a pcDNA 3.1 expression vector (Gibco/Invitrogen) containing the coding sequence of either human CB2 (Accession # AY242132) or rat CB2 (Accession # NM_020543), and the geneticin resistance gene were used.

The existence of an endogenous cannabinoid tone has been reported in several systems and may have confounding effects on the interpretation of results aimed at studying constitutive activity. To circumvent this issue, particular attention should be paid to the possible presence of endocannabinoids in the culture media, either produced by the cells themselves or as contaminants in the reagents, for example, in the serum. To evaluate the presence of constitutive activity, it is necessary to test inverse agonists and compare their effect with that of agonists and neutral antagonists. In the case of CB2, WIN55212-3 was described as a neutral antagonist (Savinainen *et al.*, 2005). However, no further studies confirmed this finding and in-house preliminary data do not seem to support the hypothesis that WIN55121-3 is a neutral CB2 antagonist (unpublished observation). Thus, the use of this compound as a standard to assess CB2 constitutive activity would require a further characterization to clearly establish its pharmacological profile.

On the other hand, the existence of CB2 protean agonists has been reported (Mancini *et al.*, 2009; Yao *et al.*, 2006). Consequently, it is extremely important to check the experimental setting using *bona fide* full agonists such as JWH133 and CP55940. The use of a partial agonist (e.g., WIN55212) as a standard should be avoided.

To successfully apply the methods described below, basic cell culture skills and a fully equipped biosafety level 1, or higher, cell culture facility are required.

Remind to take all the necessary safety precautions when manipulating radioactive material.

3. EVALUATION OF CONSTITUTIVE ACTIVITY USING GTPγS ASSAY

Using this method, it is possible to show inverse agonism as reduction of basal GTPγS binding. As test compounds we used the well-known CB2 inverse agonists, AM630 and SR144528. These compounds reduce the basal GTPγS binding by 23.5 ± 9.0% and 24.8 ± 12.3%, respectively (Fig. 2.1), unmasking CB2 constitutive activity.

3.1. Instrumentations

To measure radioactivity on a filter plate, we used the Microbeta Trilux counter (Perkin-Elmer).

The other instrumentation needed is a T25 basic Ultra Turrax (Ika-Werke).

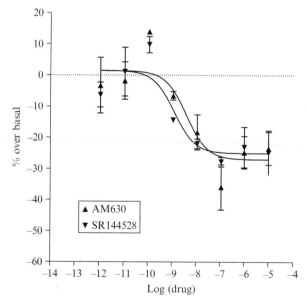

Figure 2.1 Detection of constitutive activity with GTPγS assay. Concentration–activity curve of two inverse agonists: AM630 and SR144528. Both inverse agonists reduce basal GTPγS binding showing the existence of constitutive activity at CB2. Basal activity is set at 0 (dotted line).

3.2. Reagents and reagent preparation

Cell growth medium contains F-12 Nutrient Mixture (HAM) (Gibco/Invitrogen) supplemented with 10% fetal bovine serum (FBS) heat inactivated (Gibco/Invitrogen) and 100 µg/ml Penicillin–Streptomycin (Gibco/Invitrogen). In our cell line, clone selection is maintained by addition to the medium of 600 µg/ml of G418 (Geneticin®, Gibco/Invitrogen). To harvest and expand the cells, we use D-PBS without Ca^{2+} and Mg^{2+} (Gibco/Invitrogen) and Trypsin–EDTA (Gibco/Invitrogen). Protein concentration in cell membranes is measured by BCA™ Protein Assay Kit (Pierce). To evaluate ligand activity, we use [^{35}S]GTPγS (Perkin Elmer) original stock (12387 nM) diluted 1:20 in 50 mM Tris–HCl (pH 7.4) and nonradioactive GTPγS (Sigma-Aldrich). AM630 and SR144528 are from Tocris.

Basic buffer contains 50 mM Tris–HCl, pH 7.4, 0.1% BSA, 100 mM NaCl, 3 mM $MgCl_2$, 0.2 mM EGTA, 50 µM GDP, and 0.5% DMSO (Sigma-Aldrich). The concentrations reported above are the final concentrations. Taking into consideration that during experimental procedure the buffer is diluted five times, all these reagents are prepared 5×.

Membranes (prepared as reported in Section 3.4) are dissolved at the final desired concentration in 50 mM Tris–HCl (pH 7.4) (obtained from the stock solution 1 M Tris–HCl, pH 7.2). To get better results and eliminate possible clots, it is advisable to pass the membranes through needles of different diameters, as reported in the Section 3.4.

Wash buffer (50 mM Tris–HCl, pH 7.4) is prepared from 1 mM Tris–HCl supplemented with 0.1% BSA, final concentration in the buffer, and is used to wet and wash filter plates.

Prepare test compounds in 10 mM stocks solution (100% DMSO). Dilute compound stock solution 1:10 in 50 mM Tris–HCl and DMSO in order to have 1 mM compound concentration in 50% DMSO. Then, dilute 1 mM solution to 50 µM using 50 mM Tris–HCl (2.5% DMSO final concentration). This is the first point of concentration–activity curves tested in the assay. Even in this case, both compounds and DMSO concentration should be prepared 5×. All the following concentrations are obtained by performing the dilution directly in the plate in 50 mM Tris–HCl supplemented with 2.5% DMSO.

3.3. Disposable

Filtering AcroWell™ Filter Plate (Sigma-Aldrich, Milan, Italy), low binding 96-well plates (Corning, NY, USA), for initial compound dilution prelubricated eppendorf (Corning, NY, USA), and low binding tips (Axigen).

3.4. Preparation of membranes

Prepare CB2-transfected CHO cells into T175 flasks (10 flasks of cells will be required to obtain a reasonable amount of membranes). When cells are 75–85% confluent, harvest them by first rinsing with Dulbecco's PBS without Ca^{2+} and Mg^{2+} and then dislodge the cells using Trypsin–EDTA (2 ml/flask). Add to flask F-12 (Ham) media supplemented with 10% FBS (10 ml/flask), collect cell suspension into tubes, and spin in a table top centrifuge at 2000 rpm for 7 min. Decant supernatants and resuspend pellets in media supplemented with 10% FBS and pull together. Spin down cells again at 2000 rpm for 7 min, remove media, fast freeze pellet, and store at −80 °C. Keep pellets frozen for at least 24 h before preparing membranes. To prepare membranes, resuspend pellets in 20-times their volume (we assume that cell density is 1 g/ml, and for 1 ml of starting material, we add 19 ml of buffer) of ice-cold membrane buffer (50 mM Tris–HCl, pH 7.4, 5 mM MgCl$_2$, 2.5 mM EDTA, 1 tablet/50 ml phosphatases inhibitors (Roche)) and homogenize with a T25 basic ULTRA TURRAX speed #1. Then, centrifuge cell suspension homogenate at 40,000×g for 20 min, at 4 °C. Repeat this step twice and then a third time using ice-cold membrane buffer. Discard supernatants, pool pellets, and resuspend in 1 ml ice-cold 50 mM Tris–HCl, pH 7.4. Homogenize again in a homogenizer pot (40 passes) and then using a sterile syringe (Microlane 3: 25GA 5/8″). Aliquot cell membranes and store at −80 °C. Perform protein determination using the BCA assay according to the manufacturer's instructions.

Before carrying out [^{35}S]GTPγS assay, it is important to perform a preliminary set of optimization experiments to establish the most appropriate membrane quantity to use in the assay for obtaining the best signal-to-noise ratio. We suggest testing the following membranes amounts: 2, 5, 10, 20, 50, and 100 μg under the stimulation of either WIN55212-2, CP55940, or HU210. In our experimental setting, the best stimulation over basal ratio is obtained using 5 μg. When membranes are prepared from recombinant cell lines, receptor expression is usually reasonably high to get good signal despite low membrane quantity. If high membrane quantity is required, you should keep in mind that a bigger pore diameter in filter plates may be necessary to avoid clogging the filters.

3.5. GTPγS assay

To evaluate specific GTPγS assay binding, three different experimental conditions are required in order to measure respectively basal activity, nonspecific binding, and stimulated binding. To obtain these data, prepare experimental plate according to the format reported in Table 2.1.

Table 2.1 GTPγS assay preparation

Reagents	Basal (μl)	NSB (nonspecific binding) (μl)	Stimulated (μl)
Basic buffer supplemented with DMSO (5×)	40	40	40
50 mM Tris–HCl, pH 7.4	40	0	0
GTPγS in 50 mM Tris–HCl, pH 7.4	0	40	0
Test compound	0	0	40
[^{35}S]GTPγS (5×)	40	40	40
Membranes	80	80	80

Prepare the assay on ice using low binding, white, round bottom 96-well plates. Basic buffer and membranes should be kept on ice up to the addition of radioligand.

For each data point, use 5 μg (80 μl/well) of CB2 transfected cell membranes prepared in Tris–HCl 50 mM. Prepare [^{35}S]GTPγS (1250 Ci/mmol) in Tris–HCl 50 mM and use at the final concentration of 0.1 nM (remind to prepare the solution 5×, see Table 2.1). Calculate each time the decay fraction from calibration date (see information supplied with [^{35}S]GTPγS). Final volume of radioligand to be added to all wells (basal, nonspecific binding, and stimulated) should be 40 μl. Distribute membranes in low-binding 96-well plates (Corning) and incubate for 60 min at 30 °C in buffer containing 50 mM Tris–HCl, 3 mM MgCl$_2$, 0.2 mM EGTA, 100 mM NaCl, 0.1% BSA, 5 mM GDP, 0.5% DMSO, 0.1 nM [^{35}S]GTPγS (Perkin-Elmer) and test compounds, in our example, AM630 or SR144528, at a concentration ranging: 10^{-12}–10^{-5} M.

Prewet filter plate using ice cold wash buffer prepared as reported above and filter with a vacuum system. Stop the assay by transferring the plate on ice. Then, take an aliquot of 150 μl of assay mixture (total volume 200 μl) and transfer to filter plates, filter the plates, and wash three times using 300 μl/well of wash buffer each time. Dry filters in the hood (1 h) under the air flow of a hair drier to speed the process. When the filters in the plate are completely dried, remove filter protection, add 25 μl/well of scintillation liquid, cover the plate with an aluminum foil, and gently shake for 1 h. Count radioactivity using the [^{35}S]GTPγS specific protocol.

3.6. Data analysis

Perform data analysis with GraphPad Prism 4 or analogous software, allowing sigmoidal dose–response curve fitting and calculation of EC$_{50}$ values. To obtain specific binding from both stimulated and basal samples, subtract nonspecific binding to each experimental data. To calculate the % of stimulation or inhibition over basal, apply the following formula: % over basal = (specific binding (from stimulated samples) × 100)/specific binding (from basal)) − 100. Negative values indicate an inhibition of basal activity.

4. EVALUATION OF CONSTITUTIVE ACTIVITY USING cAMP ASSAY

In the cAMP assay for Gi coupled receptor, such as the CB2, the most appropriate way to assess constitutive activity is to measure the increase of cAMP level induced by inverse agonists over basal. However, for technical reasons, this is quite difficult, and the presence of constitutive activity is normally revealed by an increase of cAMP level over that induced by forskolin. In our case, incubation of the cell with SR144528 or AM630 triggers a concentration-dependent augmentation (up to twofold) of cAMP production when compared to that obtained using forskolin alone (Fig. 2.2).

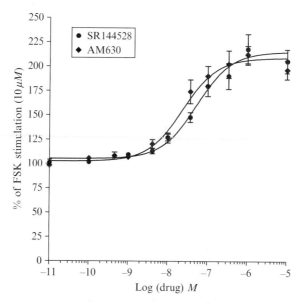

Figure 2.2 Detection of constitutive activity with cAMP assay. Concentration–activity curve of two inverse agonists (AM630 and SR144528) are reported in the plot. Both compounds significantly increase forskolin-stimulated cAMP levels indicating the existence of constitutive activity in this cell line (CHO expressing human CB2).

4.1. Instrumentations

To seed cells into plates, we use the Multidrop apparatus from MTX Lab System. Delivery of solubilised compounds into cell culture plates is performed using Thermo Scientific Matrix PlateMate 2×2 automatic pipettor (Thermo Scientific). Cell seeding and compound delivery could also be performed manually by using electronic pipettes, but this approach is time consuming, less accurate, and when handling a large number of samples, may cause ergonomic problems.

Chemiluminescence quantitation was performed with Victor 3 plate reader (Perkin Elmer Life and Analytical Sciences). This plate reader is now out of commerce but other counter such as the TopCount or the new generation plate readers from Perkin Elmer Life and Analytical Sciences like the Victor X are suited for the purpose.

4.2. Reagents and reagent preparation

To detach cell for replating use the cell dissociation buffer enzyme-free from Gibco/Invitrogen. Forskolin and IBMX are from Sigma. Test compounds are diluted in DMSO Hybry-Max (Sigma) and Hanks Balanced Salt Solution (HBSS, Gibco/Invitrogen). D-PBS w/o calcium, magnesium (Gibco/Invitrogen). HitHunter^{TM} Assay Kit cAMP II chemiluminescent Assay EFC (DiscoverX). Test compounds: JWH133, CP 55940, and AM630 (Tocris), R(+) AM1241 and L768242 were synthesized in-house. To be noticed that commercially available AM1241 is normally a racemate of R and S enantiomers.

Forskolin stock: Reconstitute 1× 10 mg vial in 2.44 ml of DMSO to obtain a 10 mM stock concentration. Prepare aliquot into microtubes and store at room temperature. Aliquots stored at room temperature have been found to be stable for many months.

Compound preparation: Prepare 10 points dilution starting at 1 µM as highest concentration. For compound dilution use nonbinding-surface 96-well plates. Prepare agonists as 7.5× working stock in HBSS buffer containing 3% DMSO as follows:

A. 10 mM in DMSO stock (prepared fresh on the day of experiment).
B. 1:10 from A (1 mM in DMSO).
C. 1:10 from B (0.1 mM in DMSO).
D. 1:13.3 from C 7.5 µM (7.5×) in HBSS buffer, final concentration 1 µM, 1% DMSO.
E. 1:3.33 from D 2.25 µM (7.5×) in 0.5% DMSO–HBSS buffer, final concentration 300 nM, 0.4% DMSO.
F. 1:3 serial dilution in HBSS buffer containing 3% DMSO, final concentration 0.4% DMSO.

HBSS buffer is prepared adding 20 mM Hepes buffer to HBSS.

4.3. Disposable

Polystyrene, white, 384 wells, assay microplates (Matrix); 96-well nonbinding-surface plates, white, flat bottom (Corning); low binding tips (Axigen).

4.4. cAMP assay

The following procedure was set up using CHO cell clone stably expressing CB2. The use of a different cell line could require adjustment of the method.

Cell preparation: Grow cells to 90% confluence in T175 or T125 ml flask. Decant medium and wash cells with D-PBS without calcium and magnesium. Incubate cells for 10 min at 37 °C, 5% CO_2, with cell dissociation buffer enzyme-free. Stop reaction by adding an excess (10 folds of medium, v/v) and detach cells by gently shaking the flask, harvest with a pipette, remove an aliquot, and count cells. Pellet the remaining cells by spinning in a table top centrifuge (1000 rpm, 8 min at room temperature) and resuspend in culture media at 3×10^6 cells/ml. Seed cells using a Multidrop into a 96-well white plate (30,000 cells/well in 100 µl of medium) or a 384-well white plate (15,000 cells/well in 50 µl of medium). Use cells for experiments 24 h after seeding.

After 24 h, discharge medium from the plates and, in order to inhibit phosphodiesterase, add with Multidrop 20 µl (96-well plates) or 10 µl (384-well plates) of PBS containing 0.5 mM IBMX. Then add, using the Thermo Scientific Matrix PlateMate 2×2, 5 µl/well (96-well plates) or 2.5 µl/well (384-well plates) of the agonist/inverse agonist at the desired concentration and 5 µl/well (96-well plates) or 2.5 µl/well (384-well plates) of 10 µM forskolin. Incubate cells for 30 min at 37 °C and 5% CO_2. Add, using the Thermo Scientific Matrix PlateMate 2×2, 80 µl/well (96-well plates) or 40 µl/well (384-well plates) of Hit Hunter cAMP II Assay EFC Chemiluscent Detection kit (40 or 20 µl/well of enzyme donor and substrate and 40 or 20 µl of enzyme acceptor and lysis buffer/anti-cAMP antibody). Spin 15 s in a table top centrifuge at 1000 rpm. Incubate cells at room temperature in the dark, in order to preserve the photosensitive reagents of the kit, for at least 4 h or overnight. Read the luminescent signal with Victor 3 plate reader at 1 s/well.

4.5. Data analysis

Perform data analysis with GraphPad Prism 4 or analogous software, allowing sigmoidal dose–response curve fitting and calculation of EC_{50} values. We express data as percent responses of the 10 µM forskolin-stimulated cAMP level. Counts per second (cps) relative to 10 µM forskolin correspond to 100% and the cps relative to the basal cAMP level corresponds to 0%.

5. Evaluation of Constitutive Activity Using RT-CES

Label-free technologies are based on the use of biosensors capable of sensing interaction between molecules. The label-free assay that we developed to evaluate constitutive activity at the CB2 is based on the RT-CES from ACEA Biosciences. This system relays on variation of cell impedance to detect receptor activity (Xi et al., 2008; Yu et al., 2006).

For each different cell line used it is extremely important to perform initial experiments of assay optimization to set appropriate conditions. This is because cell impedance assay is extremely sensitive to minimal cellular variations.

5.1. Instrumentations

To measure cell impedance, the RT-CES 96× System previously supplied by ACEA Biosciences, Inc. (CA, USA) and now supplied by Roche Applied Science under the name of xCELLigence, is employed. The system includes the following components:

- 1× W200 RT-CES Analyzer
- 1× 96× Device Station
- 1× Software Package and Computer

The device station (DS) must be hosted in a cell incubator for continuous cell recording. To avoid contamination and repeated transfer of the DS, it is convenient to dedicate an incubator to this system and to let the DS permanently inside the incubator. Incubators from Heraeus model Hera Cell, from Thermo Electron Corporation model Hepa Class 100, from RS Biotech model Galaxy S, from Forma Scientific model Steri-cult 200 all deserve the purpose.

For cell seeding the same instrumentation indicated in Section 4.1 (Multidrop and Matrix PlateMate 2×2 automatic pipettor) is used.

5.2. Reagents and reagent preparation

Dilute 10 mM compounds' stock solutions (100% DMSO) 1:10 in HBSS supplemented with 50% DMSO in order to have 1 mM compounds final concentration (50% DMSO). Then perform a second dilution from 1 mM to 30 µM in HBSS (1.6% DMSO). The first point of concentration–response curves (10 or 3 µM depending on the compound tested) is prepared in a plate performing a 1:3 or 1:10 dilution in HBSS supplemented with 1.6% DMSO. All the following points were obtained performing the dilution directly in the plate in HBSS supplemented with 1.6% DMSO. Prepare compound and DMSO concentration 10× (transfer 20 µl from the compound dilution plate to the experimental plate with the cells containing 180 µl of HBSS).

5.3. Disposable

To perform cell impedance measurement specific plates called E-Plate^{TM} (ACEA Biosciences) have been designed. In an E-Plate^{TM} a sensor constituted of a gold microelectrodes array has been incorporated on the bottom of each individual well so that cells plated in each well can be monitored over time. E-Plates^{TM} are supplied on different formats, for this method we used the 96-well format. These plates are sold as single-use disposable device. Low binding tips are from Axigen.

5.4. Cell impedance assay

Prepare cells using the same procedure described in Section 4.4.

Optimization of experimental conditions: One day before the experiment, add 50 µl of medium to each well of an empty E-Plate^{TM} and measure impedance background value. Following background measurement, add on each E-Plate^{TM} well 50 µl of media containing the cell suspension prepared as described above. Incubate the cells at room temperature for 1 h, then place them on the DS, hosted in an incubator at 37 °C with 5% CO_2, and monitor impedance constantly every 2 min for the first 2 h and every 30 min for the remaining time. Once seeded on the plate, cells start to adhere, grow, and duplicate. The interaction of the cells with microelectrodes in a well leads to changes in impedance over time which is reported by the software as changes in Cell Index (CI, see Section 5.5 for explanation on CI). As a consequence, cell impedance is increasing until these processes come to a plateau phase (Fig. 2.3). Experiment must be performed only after impedance has reached a plateau and has stabilized; otherwise, results will be biased by impedance changes related to cell adherence, growth, and division. The time necessary to reach a plateau is variable from different cell lines and on the same cell line transfected with different receptors. Hence incubation time necessary to reach a plateau has to be established for each cell clone. In our case the plateau was reached at about 24 h after seeding (Fig. 2.3).

There are two other important parameters to be optimized before performing the assay: number of cells seeded per well and DMSO concentration. The right amount of cells to be added on the well should be determined in preliminary experiments using different cell concentrations. In our experience changes in seeding density could affect signal-to-noise ratio. An example of this effect is shown in Fig. 2.4 where the effects of

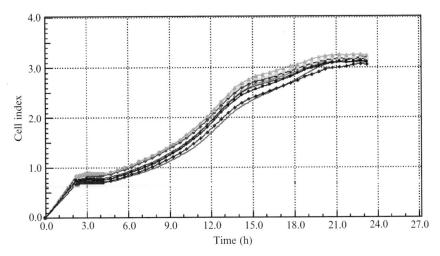

Figure 2.3 Time course of impedance variations during cell adhesion and growth. Cells were seeded at time zero and variation in cell impedance monitored over a period of 24 h. The cell index, an index related to cell impedance, is increased for several hours before reaching a plateau at around 24 h after cell seeding. Different curves represent different wells and each dot represents a recording point. Reported curves are representative of results obtained in several experiments. (See Color Insert.)

different seeding density (10,000, 20,000, 30,000, and 50,000 cell/well) on impedance variation following application of an agonist (JWH133) are reported. Low seeding densities result in a shallower curve, whereas higher densities (30,000 and 50,000) show a better signal-to-noise ratio. A good signal-to-noise ratio will facilitate constitutive activity detection.

The second parameter to be carefully checked for is the concentration of organic solvent used to dilute the test compound. One of the solvent most widely used in cellular pharmacology is DMSO. This solvent permits to properly dissolve the majority of small molecules of pharmacological interest. Based on these considerations we have set up our assay using DMSO, but it is important to stress that paramount attention should be paid at its final concentration. High DMSO concentrations have a dramatic effect on cell impedance. To establish which concentration does not alter significantly cell impedance it is necessary to perform a concentration–activity curve using different DMSO concentration. For this purpose discharge cell media, replace it with HBSS containing different DMSO's concentration and monitor impedance for the foreseen duration of experiment. In our experiment the final concentration of DMSO was kept at $\leq 0.1\%$.

Agonist and inverse agonist assays: To evaluate constitutive activity, decant culture media and replace with 180 μl of HBSS and record for 20 min. Then add 20 μl of test compound (agonists such as JWH133 or CP55940, and

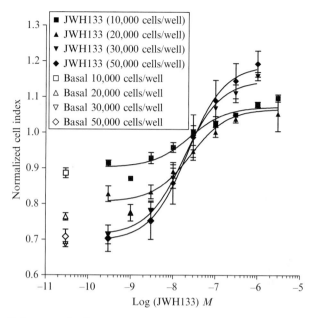

Figure 2.4 Influence of cell density on impedance variation. Effect of four different seeding cell densities (10,000, 20,000, 30,000, and 50,000) on cell impedance are reported in the graph. It could be observed that initial cell density clearly affects cell response to stimulation with a CB2 agonist (JWH133). Cell density of 30,000 and 50,000 per well gives the best signal-to-noise ratio. Reported curves are representative of results obtained in several experiments.

inverse agonists such as AM630 or SR144528) at different concentrations, using a PlateMate™ 2×2 (Matrix) and record impedance every 70 s until the end of the experiment. Activation of the receptor by agonists or blockade of constitutive activity by inverse agonists leads to impedance changes of opposite sign. In this case agonists increase cell impedance whereas inverse agonists decrease it (Fig. 2.5).

5.5. Data analysis

Perform data analysis with GraphPad Prism 4 or analogous software, allowing sigmoidal dose–response curve fitting and calculation of EC_{50} values. The results should be expressed by Normalized Cell Index (NCI), which is the ratio of the CI before and after the addition of the compounds. CI is a dimensionless number that tracks and compares changes in electrode impedance that occurs in the presence and absence of cells in the wells. CI is equal to the resistance in wells with cells minus the background resistance (in the same wells without cells), divided by 15 (Ω).

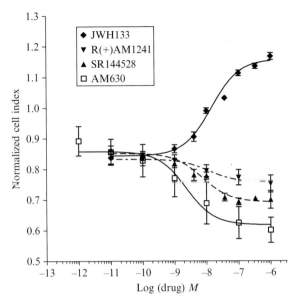

Figure 2.5 Detection of constitutive activity by recording cell impedance. In CHO cell transfected with rat CB2, the full agonist JWH133 increases cell impedance, whereas SR144528 and AM630, two inverse agonists, decrease it. In this particular setting, the protean agonist R(+)AM1241 decreases cell impedance acting as an inverse agonist.

6. Evaluation of Protean Agonism with cAMP Assay

As mentioned previously the phenomena of protean agonist is indissolubly linked to that of constitutive activity. To unveil the protean agonist behavior of molecules it is necessary to carry out parallel experiments in which constitutive activity is either present or abrogated. In order to do so we developed a cAMP-based assay in which a pretreatment with the inverse agonist AM630 is performed. Such pretreatment allows a reduction of constitutive activity as it has been previously shown (Mancini et al., 2009). In principle such treatment should enrich the inactive state of the receptor consequently reducing constitutive activity. Protean agonists will significantly modify their behavior after pretreatment. R(+)AM1241 and L768242 that in our normal experimental condition act as neutral agonists or inverse agonists, following AM630 pretreatment reveal an agonistic activity (Fig. 2.6).

It is important to note that after pretreatment and before performing the experiment, it is necessary to execute an extensive washing in order to

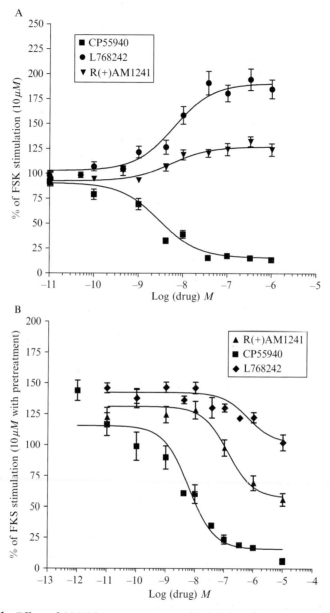

Figure 2.6 Effect of AM630 pretreatment on the activity of protean agonist. In the absence of treatment, R(+)AM1241 and L768242 behave as inverse agonists in the cAMP assay. For comparison, the activity of the full agonist CP55940 is shown. (A) After 24-h AM630 pretreatment and extensive wash, both compounds change their behavior and act as agonists. (B) Conversely, no difference is observed in the concentration–activity curves of CP55940 in the presence or absence of pretreatment, confirming that no residual AM630 antagonistic activity is present in the preparation.

eliminate from the media the inverse agonist. In order to verify washing effectiveness, it is recommended to evaluate full agonist efficacy. A reduction of full agonist efficacy will indicate an inadequate washing with residual inverse agonist present at receptor binding site (Fig. 2.6). To carry out this control experiment, we suggest the use of CP55940 as a full agonist and to follow the method mentioned above for cAMP assay (Section 4.4).

Abrogation of constitutive activity: To abolish the constitutive activity of CB2, seed cells into 96- or 384-well plates for the cAMP assay in the presence of 10 μM AM630 directly diluted from 10 mM stock in complete F-12 medium and incubate for 24 h at 37 °C and 5% CO_2. Discharge the medium, add 100 μl/well of complete F-12 medium, incubate for 10 min at 37 °C and 5% CO_2, and repeat six times for a total of 1 h wash. Now you can proceed at the evaluation of the activity of your test compound applying the method described in Section 4.4.

7. Comparison of the Methods

We presented here three methods that are suitable to study constitutive activity at the CB2 in recombinant cell lines. The three methods measure different aspects of the signal transduction system triggered by

Table 2.2 Advantages and disadvantages of the methods to evaluate constitutive activity described in this chapter

Method	Advantage	Disadvantage
GTPγS	– Assess an event very close to initial receptor–ligand interaction – No prestimulation required	– Use of radioactivity – Use of a rapidly decaying reagent: [^{35}S] GTPγS – End point method
cAMP	– Easy to apply – High throughput – Availability of detection kit	– Preliminary pharmacological manipulation required (stimulation with FSK) – Difficulty in detecting constitutive activity in nonstimulated cells – End point method
Impedance	– Real-time measurement performed on living cells – No prestimulation required	– High sensitivity to organic solvents – Elevated cost of E-plate

CB2 activation. The GTPγS is recording a step that is very close to the initial activation of the receptor: the exchange of the GDP of the $G_i\alpha$ protein with a GTP soon after ligand–receptor interaction. This exchange is at the base of the dissociation of the $G_i\alpha$ subunit from the $G\beta\gamma$ subunits. The second method is recording a downstream event of the amplification cascade: the decrease in cAMP synthesis that follows adenylate cyclase inhibition by $G_i\alpha$. The last method is associated with the detection of modification of cell shape and adhesion. CB2 ligand binding is triggering a series of intracellular events, not completely characterized yet but possibly including modification of RhoA-GTP, Rac-GTP, and PKB/Atk singaling, that induce cytoskeletal rearrangement and cell mobility (Kurihara et al., 2006). Thus, this last method is potentially measuring constitutive activity at multiple CB2 second messenger pathways.

All the three methods have pros and cons, and we summarize the major ones in Table 2.2.

REFERENCES

Baker, D., Pryce, G., Davies, W. L., and Hiley, C. R. (2006). In silico patent searching reveals a new cannabinoid receptor. *Trends Pharmacol. Sci.* **27,** 1–4.

Braestrup, C., Schmiechen, R., Neef, G., Nielsen, M., and Petersen, E. N. (1982). Interaction of convulsive ligands with benzodiazepine receptors. *Science* **216,** 1241–1243.

Carrasquer, A., Nebane, N. M., Williams, W. M., and Song, Z. H. (2010). Functional consequences of nonsynonymous single nucleotide polymorphisms in the CB2 cannabinoid receptor. *Pharmacogenet. Genomics* **20**(3), 157–166.

Cerione, R. A., Codina, J., Benovic, J. L., Lefkowitz, R. J., Birnbaumer, L., and Caron, M. G. (1984). The mammalian beta$_2$-adrenergic receptor: Reconstitution of functional interactions between pure receptor and pure stimulatory nucleotide binding protein of the adenylate cyclase system. *Biochemistry* **23,** 4519–4525.

Ganguli, S. C., Park, C. G., Holtmann, M. H., Hadac, E. M., Kenakin, T. P., and Miller, L. J. (1998). Protean effects of a natural peptide agonist of the G proteincoupled secretin receptor demonstrated by receptor mutagenesis. *J. Pharmacol. Exp. Ther.* **286,** 593–598.

Gbahou, F., Rouleau, A., Morisset, S., Parmentier, R., Crochet, S., Lin, J. S., Ligneau, X., Tardivel-Lacombe, J., Stark, H., Schunack, W., Ganellin, C. R., Schwartz, J. C., et al. (2003). Protean agonism at histamine H3 receptors in vitro and in vivo. *Proc. Natl. Acad. Sci. USA* **19,** 11086–11091.

Hasegawa, H., Negishi, M., and Ichikawa, A. (1996). Two isoforms of the prostaglandin E receptor EP3 subtype different in agonist-independent constitutive activity. *J. Biol. Chem.* **271**(4), 1857–1860.

Iwamura, H., Suzuki, H., Ueda, Y., Kaya, T., and Inaba, T. (2001). In vitro and in vivo pharmacological characterization of JTE-907, a novel selective ligand for cannabinoid CB2 receptor. *J. Pharmacol. Exp. Ther.* **296,** 420–425.

Jansson, C. C., Kukkonen, J. P., Nasman, J., Huifang, G., Wurster, S., Virtanen, R., Savola, J. M., Cockcroft, V., and Akerman, K. E. (1998). Protean agonism at alpha2A-adrenoceptors. *Mol. Pharmacol.* **53,** 963–968.

Kenakin, T. (2001). Inverse, protean, and ligand-selective agonism: Matters of receptor conformation. *FASEB J.* **15,** 598–611.

Kenakin, T. (2005). The physiological significance of constitutive receptor activity. *Trends Pharmacol. Sci.* **26**, 603–605.

Kurihara, R., Tohyama, Y., Matsusaka, S., Naruse, H., Kinoshita, E., Tsujioka, T., Katsumata, Y., and Hirohei, Y. (2006). Effects of peripheral cannabinoid receptor ligands on motility and polarization in neutrophil-like HL60 cells and human neutrophils. *J. Biol. Chem.* **281**, 12908–12918.

Lefkowitz, R. J., Cotecchia, S., Samama, P., and Costa, T. (1993). Constitutive activity of receptors coupled to guanine nucleotide regulatory proteins. *Trends Pharmacol. Sci.* **14**(8), 303–307.

Lunn, C. A., Reich, E.-P., Fine, J. S., Lavey, B., Kozlowski, J. A., Hipkin, R. W., Lundell, D. J., and Bober, L. (2008). Biology and therapeutic potential of cannabinoid CB2 receptor inverse agonists. *Br. J. Pharmacol.* **153**, 226–239.

Mancini, I., Brusa, R., Quadrato, G., Foglia, C., Scandroglio, P., Silverman, L. S., Tulahian, D., Reggiani, A., and Deluramo, M. (2009). Constitutive activity of cannabinoid receptor 2 plays an essential role in the protean agonism of (+)AM1241 and L768242. *Br. J. Pharmacol.* **158**, 382–391.

Milligan, G., and Bond, R. A. (1997). Inverse agonism and the regulation of receptor number. *Trends Pharmacol. Sci.* **18**, 468–474.

New, D. C., and Wong, Y. H. (2003). BML-190 and AM251 act as inverse agonists at the human cannabinoid CB2 receptor: Signalling via cAMP and inositol phosphates. *FEBS Lett.* **536**, 157–160.

Parfitt, A. M., Schipani, E., Rao, D. S., Kupin, W., Han, Z. H., and Jüppner, H. (1996). Hypercalcemia due to constitutive activity of the parathyroid hormone (PTH)/PTH-related peptide receptor: Comparison with primary hyperparathyroidism. *J. Clin. Endocrinol. Metab.* **81**(10), 3584–3588.

Parma, J., Duprez, L., Van Sande, J., Cochaux, P., Gervy, C., Mockel, J., Dumont, J., and Vassart, G. (1993). Somatic mutations in the thyrotropin receptor gene cause hyperfunctioning thyroid adenomas. *Nature* **365**(6447), 649–651.

Pauwels, P. J., Rauly, I., Wurch, T., and Colpaert, F. C. (2002). Evidence for protean agonism of RX 831003 at alpha 2A-adrenoceptors by co-expression with different G alpha protein subunits. *Neuropharmacology* **42**, 855–863.

Peters, M. F., and Scott, C. W. (2009). Evaluating cellular impedance assays for detection of GPCR pleiotropic signaling and functional selectivity. *J. Biomol. Screen.* **14**(3), 246–255.

Petitet, F., Donlan, M., and Michel, A. (2006). GPR55 as a new cannabinoid receptor: Still a long way to prove it. *Chem. Biol. Drug Des.* **67**, 252–253.

Portier, M., Rinaldi-Carmona, M., Pecceu, F., Combes, T., Poinot-Chazel, C., Calandra, B., Brath, F., Le Fur, G., and Casellas, P. (1999). SR144528, an antagonist for the peripheral cannabinoid receptor that behaves as an inverse agonist. *J. Pharmacol. Exp. Ther.* **288**, 582–589.

Rinaldi-Carmona, M., Barth, F., Héaulme, M., Alonso, R., Shire, D., Congy, C., Soubrié, P., Brelière, J. C., and Le Fur, G. (1995). Biochemical and pharmacological characterisation of SR141716A, the first potent and selective brain cannabinoid receptor antagonist. *Life Sci.* **56**(23–24), 1941–1947.

Rinaldi-Carmona, M., Barth, F., Millan, J., Derocq, J. M., Casellas, P., Congy, C., Oustric, D., Sarran, M., Bouaboula, M., Calandra, B., Portier, M., Shire, D., *et al.* (1998). SR144528, the first potent and selective antagonist of the CB2 cannabinoid receptor. *J. Pharmacol. Exp. Ther.* **284**, 644–650.

Ross, R. (2009). The enigmatic pharmacology of GPR55. *Trends Pharmacol. Sci.* **30**(3), 156–163.

Ross, R. A., Brockie, H. C., Stevenson, L. A., Murphy, V. L., Templeton, F., Makriyannis, A., and Pertwee, R. G. (1999). Agonist-inverse agonist characterization at CB1 and CB2 cannabinoid receptors of L759633, L759656 and AM630. *Br. J. Pharmacol.* **126,** 665–672.

Samama, P., Cotecchia, S., Costa, T., and Lefkowitz, R. J. (1993). A mutation-induced activated state of the beta 2-adrenergic receptor. Extending the ternary complex model. *J. Biol. Chem.* **268**(7), 4625–4636.

Savinainen, J. R., Kokkola, T., Salo, O. M. H., Poso, A., Järvinen, T., and Laitinen, J. T. (2005). Identification of WIN55212-3 as a competitive neutral antagonist of the human cannabinoid CB2 receptor. *Br. J. Pharmacol.* **145,** 636–645.

Scandroglio, P., Brusa, R., Lozza, G., Mancini, I., Petrò, R., Reggiani, A., and Beltramo, M. (2010). Evaluation of cannabinoid receptor 2 and metabotropic glutamate receptor 1 functional responses using a cell impedance based technology. *J. Biomol. Screen.* First published on September 1, 2010 as doi:10.1177/1087057110375615.

Xi, B., Yu, N., Wang, X., Xu, X., and Abassi, Y. A. (2008). The application of cell-based label-free technology in drug discovery. *Biotechnol. J.* **3**(4), 484–495.

Yao, B. B., Mukherjee, S., Fan, Y., Garrison, T. R., Daza, A. V., Grayson, G. K., Hooker, B. A., Dart, M. J., Sullivan, J. P., and Meyer, M. D. (2006). In vitro pharmacological characterization of AM1241: A protean agonist at the cannabinoid CB2 receptor? *Br. J. Pharmacol.* **149,** 145–154.

Yu, N., Atienza, J. M., Bernard, J., Blanc, S., Zhu, J., Wang, X., Xu, X., and Abassi, Y. A. (2006). Real-time monitoring of morphological changes in living cells by electronic cell sensor arrays: An approach to study G protein-coupled receptors. *Anal. Chem.* **78**(1), 35–43.

CHAPTER THREE

MODULATION OF THE CONSTITUTIVE ACTIVITY OF THE GHRELIN RECEPTOR BY USE OF PHARMACOLOGICAL TOOLS AND MUTAGENESIS

Jacek Mokrosiński *and* Birgitte Holst

Contents

1. Introduction	54
2. The Ghrelin Receptor and Its Constitutive Activity	55
3. Structural Basis of Constitutive Activity	57
4. Residues Responsible for the Inverse Agonism and Efficacy Swap	61
5. Physiological Relevance of Constitutive Activity	63
6. Experimental Procedures	65
6.1. Inositol phosphate turnover assay	67
6.2. SRE reporter gene assay	68
6.3. ELISA assay	68
References	69

Abstract

Ghrelin and its receptor are important regulators of metabolic functions, including appetite, energy expenditure, fat accumulation, and growth hormone (GH) secretion. The ghrelin receptor is characterized by an ability to signal even without any ligand present with approximately 50% of the maximally ghrelin-induced efficacy—a feature that may have important physiological implications.

The high basal signaling can be modulated either by administration of specific ligands or by engineering of mutations in the receptor structure. [D-Arg^1, D-Phe^5, D-Trp^{7,9}, Leu^{11}]-substance P was the first inverse agonist to be identified for the ghrelin receptor, and this peptide has been used as a starting point for identification of the structural requirements for inverse agonist properties in the ligand. The receptor binding core motif was identified as D-Trp-Phe-D-Trp-Leu-Leu, and elongation of this peptide in the amino-terminal end determined the efficacy. Attachment of a positively charged amino acid was responsible for full inverse agonism, whereas an alanin converted the peptide

Department of Neuroscience and Pharmacology, University of Copenhagen, Copenhagen N, Denmark into a partial agonist. Importantly, by use of mutational mapping of the residues critical for the modified D-Trp-Phe-D-Trp-Leu-Leu peptides, it was found that space-generating mutations in the deeper part of the receptor improved inverse agonism, whereas similar mutations located in the more extracellular part improved agonism.

Modulation of the basal signaling by mutations in the receptor structure is primarily obtained by substitutions in an aromatic cluster that keep TMs VI and VII in close proximity to TM III and thus stabilize the active conformation. Also, substitution of a Phe in TM V is crucial for the high basal activity of the receptor as this residue serves as a partner for Trp VI:13 in the active conformation.

It is suggested that inverse agonist and antagonist against the ghrelin receptor provide an interesting possibility in the development of drugs for treatment of obesity and diabetes and that improved structural understanding of the receptor function facilitates the drug development.

1. INTRODUCTION

The ghrelin receptor was initially cloned and described as the growth hormone secretagogue (GHS) receptor (Howard *et al.*, 1996) and was subsequently found to be a G-protein coupled receptor with the characteristic seven transmembrane spanning domains (7TM receptor) (Schwartz *et al.*, 2006). The GHSs were discovered during the seventies by Bowers and coworkers as small peptide fragments derived from endorphins with abilities to induce growth hormone (GH) secretion through a mechanism different from the GH releasing hormone receptor (Bowers *et al.*, 1984). In the following decade, the pharmaceutical industry developed both peptide and nonpeptide compounds in an attempt to obtain oral-active drugs with the capacity to induce GH secretion. Some of these compounds were tested in clinical trials for indications such as GH deficiency and bone fracture but they never made it to the market presumably due to lack of efficacy (Bach *et al.*, 2004; Murphy *et al.*, 2001; Patchett *et al.*, 1995).

In 1999, Kangawa and coworkers identified the endogenous ligand, ghrelin, for this highly interesting receptor (Kojima *et al.*, 1999). Today, the term GHS receptor is considered outdated, and the International Union of Pharmacology Committee on Receptor Nomenclature and Drug Classification has officially named it the ghrelin receptor (Davenport *et al.*, 2005). Ghrelin is a 28 amino acid long peptide that differs from all other known peptide hormones by an octanoylation. This fatty acid modification is essential both for the binding and activation of the receptor and for the pharmacokinetic properties (Kojima *et al.*, 2001). Ghrelin shows a high degree of conservation among the mammalian species, and in particular, the receptor binding amino-terminal part is highly conserved among vertebrates, from fish, through amphibians to mammals (Kojima and Kangawa, 2005).

The preprohormone for ghrelin that contains 117 amino acids is mainly expressed in the fundus part of the stomach, but a lower level of expression is observed in other parts of the upper GI tract. In patients and animal models of gastrectomy, the plasma level of ghrelin will initially decrease, however after a period of time, it will rise to the original level due to compensatory secretion in the duodenum and other parts of the GI tract (Ariyasu *et al.*, 2001). In the maturation process, preproghrelin is modified by ghrelin O-acylotransferase (GOAT), a membrane-enzyme catalyzing acylation of the serine residue in the 3rd position of the mature peptide with most often eight-carbon fatty acid. This recently discovered enzyme facilitates the physiological action of ghrelin and constitutes a major regulatory mechanism for the ghrelin receptor system (Kirchner *et al.*, 2009; Yang *et al.*, 2008). In recent years, it has been reported that des-acyl ghrelin might also have physiological function, though its specific receptor has not yet been described (Goodyear *et al.*, 2010; Granata *et al.*, 2007; Zhang *et al.*, 2008).

Ghrelin was named after its GH releasing (relin) properties but it was soon discovered that it also has other important functions (Tschop *et al.*, 2000). In particular the activation of the ghrelin receptor in the hypothalamus, and mainly in the arcuate nucleus of the hypothalamus, has attracted most focus over the last decade and indicates that ghrelin is responsible for increased activity in NPY/AgRP neurones leading to an increased appetite and decreased energy expenditure (Kojima *et al.*, 2001). The orexigenic effect of ghrelin has also been demonstrated to involve activation of AMP-activated protein kinase (AMPK) and a subsequent inactivation of the enzymatic steps in the fatty acid biosynthesis in the ventromedial hypothalamus (Lopez *et al.*, 2008). Furthermore, chronic administration of ghrelin has been shown to induce fat accumulation independent of any increase in food intake, in particular mediated through a central regulation of the sympathetic nervous system (Theander-Carrillo *et al.*, 2006). More recently ghrelin has been shown to modulate the activity of the dopaminergic neurons of the ventral tegmental area (VTA) and substantia niagra and has been shown to be involved in reward seeking behavior such as alcohol abuse, cocaine, or intake of palatable food (Abizaid *et al.*, 2006; Diano *et al.*, 2006; Jerlhag *et al.*, 2009; Perello *et al.*, 2010).

2. THE GHRELIN RECEPTOR AND ITS CONSTITUTIVE ACTIVITY

The ghrelin receptor belongs to the large family A of 7TM receptors. Together with six of the most closely related receptors it forms a ghrelin receptor family among which the motilin receptor (GPR38) shows the highest sequence homology (Fig. 3.1; Holst *et al.*, 2004). The gene encoding

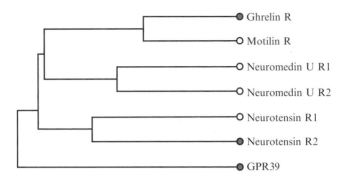

Figure 3.1 The phylogenic tree of the ghrelin receptor subfamily. The degree of protein sequence similarity is indicated by the order and proximity between the receptors. The red dots indicate the three receptors which have been demonstrated to display a high degree of constitutive signaling activity including the ghrelin receptor, NT-R2 and GPR39. (For interpretation of the references to color in this figure legend, the reader is referred to the Web version of this chapter.)

for the ghrelin receptor is located in locus 3q26.31 and consists of three exons separated with two introns. There are two splicing variants described; the full length ghrelin receptor known as the ghrelin receptor 1a composed of 366 amino acids, that forms seven alfahelical transmembrane domains, and a truncated version called the ghrelin receptor 1b which is composed of 289 amino acids and forms only the first five transmembrane domains. The molecular weights of 1a and 1b splicing variants of ghrelin receptor are around 41 kDa and 32 kDa, respectively (Korbonits et al., 2004).

Signal transduction from the extracellular environment via 7TM receptors in general requires a conformational change from an inactive (R) to an active (R*) state. Most often this process is facilitated by the binding of an agonist leading to rearrangement of the transmembrane segments and, subsequent activation of the intracellular signaling cascades (Gether et al., 1995). Certain 7TM receptors are stabilized in an active conformation without any ligand present. The ability to propagate the intracellular signal in the absence of agonist is commonly known as constitutive activity (Smit et al., 2007).

Constitutive activity is described for receptors from all major 7TM receptor super-families. The level of ligand-independent activation can be expressed as percent of the maximal agonist induced stimulation of the receptor and for most receptors the level of constitutive activity is very low. Moreover, constitutive activity observed for a specific receptor can also be dependent on the signaling pathway investigated and varies between both the pathways and the employed assay used to monitor the signal transduction. The ghrelin receptor shows strong, ligand-independent signaling via $G\alpha_{q/11}$ protein which activates phospholipase C (PLC), an enzyme cleaving phosphatidylinositol 4,5-bisphosphate (PIP2) into diacyl glycerol (DAG) and inositol 1,4,5-trisphosphate (IP3) (Holst et al., 2003).

The latter is, subsequently, released to the cytoplasm where it serves as a secondary messenger responsible for the release of Ca^{2+} ions causing further activation of intracellular signaling pathways. IP3 is very quickly degraded by dephosphorylation, which is important for tight regulation of the response. By use of Li^+ ions, as an inhibitor of inositol monophosphatase, intracellular accumulation of inositol monophosphate (IP) monitored in cell based assays described in the experimental section enables investigation of the receptor activation through $G\alpha_{q/11}$ pathway (Einat et al., 1998).

Constitutive activity of the ghrelin receptor is also detected by the cAMP response element (CRE) reporter gene assay (Holst et al., 2004). Activation of CRE binding protein (CREB) was initially described as a process caused by protein kinase A (PKA) which is controlled by another secondary messenger system, namely cAMP. Intracellular concentrations of cAMP are controlled by both $G\alpha_s$ and $G\alpha_i$ subunits. Alternatively, CREB can be phosphorylated in the $G\alpha_q$ pathway by downstream kinases such as Ca^{2+}/calmoduline-dependent kinase IV and protein kinase C (PKC) (Matthews et al., 1994; Singh et al., 2001). Another reporter gene assay, serum response element (SRE) luciferase assay can be employed to investigate constitutive activity of ghrelin receptor. Ghrelin receptor-induced SRE activity may partly be transduced via the $G\alpha_{12/13}$–Rho pathway (Holst et al., 2004; unpublished observations). Moreover, the ghrelin receptor is able to activate the ERK1/2 MAP kinase pathway but only in a ligand dependent manner as almost no basal ERK1/2 phosphorylation can be detected (Holst et al., 2004).

The ghrelin receptor 1b splice variant was shown to act in a dominant-negative manner for the ghrelin receptor. Co-expression of this truncated variant of ghrelin receptor together with the full length variant attenuated the constitutive signaling. This is possibly due to formation of heterodimerisation of both splice variants of the ghrelin receptor which leads to translocation of the ghrelin receptor from plasma membrane to cell nucleus (Leung et al., 2007).

The methods to investigate constitutive activity of ghrelin receptor are carefully described in the experimental section of this review. In principle, the constitutive activity is dependent on the level of receptor expression. Thus, for constitutively active receptors increase in the basal signaling efficiency is measured for cells with increase in of cell surface expression level of the receptor of interest. Furthermore, basal activity of the receptor can be investigated by use of inverse agonists if such are known.

3. Structural Basis of Constitutive Activity

Despite high level of sequence diversity, 7TM receptors are believed to share a common structural organization of the membrane spanning domains. For the family A of the 7TM receptors this organization is to a large degree constrained by the presence of highly conserved residues and clusters, characteristic for each TM domain, such as DRY motive at the intracellular end of TM III, proline residues in TMs II, IV, V, VI, and VII forming kinks of the α-helices, NPxxY motive at intracellular poll of TM VII, and others (Mirzadegan et al., 2003). These similarities in the structural organization together with common property of signal transduction through G proteins suggest the existence of a general activation mechanism shared by all 7TM receptors (Ballesteros et al., 2001). The ligands for the family A of 7TM receptors are highly diverse in their chemical structure, spanning from ions and small molecules like monoamines and large peptides such as glycoproteins and chemokines. The ligand binds to the specific orthosteric site located, most frequently on the extracellular part, such as in between extracellular part of the transmembrane domains, or within the extracellular loops and N-terminal domain, and leads to stabilization of the receptor either in the active or inactive conformation (Yao et al., 2009).

One possible model of the molecular mechanism of 7TM receptor activation is described as "The Global Toggle Switch Model" where activation is proposed to result from an inward movement of extracellular ends of TMs VI and VII toward TM III concomitant with a movement of the intracellular part of the transmembrane segments in the opposite direction (Elling et al., 2006; Schwartz et al., 2006). The conserved prolines in TMs VI and VII serve as pivots for a vertical seesaw movement of these two transmembrane segments, leading to closure of the ligand binding pocket at the extracellular side while opening a binding pocket for the G-protein at the intracellular part of the receptor (Scheerer et al., 2008).

The intracellular movements of TM VI away from TM III were initially investigated by biophysical methods including distance determination by use of fluorescence probes and spin labels. These experimental data has been confirmed by recent crystallographic studies of opsin in complex with the main opsin-binding fragment of the Gα protein (Ahuja et al., 2009; Gether, 2000; Hubbell et al., 2003; Scheerer et al., 2008). The movement of the extracellular part is based both on experimental data where activating metal ion sites have been introduced between TMs III, VI, and VII and computation docking data of small molecule agonist into the novel crystal structures of β-adrenergic receptors where an inward movement of in particular TM VI is required in accordance with "The Global Toggle Switch Model" to fit into the binding site (Elling et al., 2006; Nygaard et al., 2009).

It has been proposed that the inactive and active conformations of 7TM receptors are stabilized by interactions between the previously mentioned conserved residues of the receptors, called microswitches (Nygaard et al., 2009). One of the major constrains in the inactive conformation of the monoamine receptors is an ionic lock, a salt bridge, between the cytoplasmic part of TMs III and VI formed by Arg III:26, in the conserved DRY motive, and Glu VI:-05 at the intracellular end of TM VI. Comparison of rhodopsin and β2-adrenergic receptor structures, followed by molecular dynamic simulations of the latter, revealed that the ionic lock does not necessarily need to be formed to stabilize the inactive receptor conformation and other inactive receptor conformations may be relevant for other receptors (Dror *et al.*, 2009; Vanni *et al.*, 2009). In the inactive conformation the Arg III:26 also interacts with neighboring acidic residue, that is, Asp/Glu III:25 (Nygaard *et al.*, 2009). Mutation-induced constitutive activity can be caused by various naturally occurring genetic variants as well as achieved by *in vitro* engineered substitutions (Samama *et al.*, 1993; Seifert and Wenzel-Seifert, 2002). Increase in basal signaling seems to be caused by removal of stabilizing structural constrains that facilitates the inactive-to-active conformation transition (Gether *et al.*, 1997; Javitch *et al.*, 1997; Rasmussen *et al.*, 1999).

Agonists are capable of stabilizing the receptor in various active conformations that may differ in potency, efficacy as well as signaling pathway specificity. Identification of kinetically distinguishable conformational stages enlightens differences between full and partial agonists' mode of action (Swaminath *et al.*, 2004, 2005). Computational validation of the well described β2adrenergic receptor agonists as well as inverse agonists and antagonists shows direct link between chemical structure and pharmacological properties of the ligands, and receptor conformational changes occurring upon their binding (Graaf de and Rognan, 2008; Katritch *et al.*, 2009). These observations are also confirmed by available NMR spectroscopy data (Bokoch *et al.*, 2010).

The ghrelin receptor in addition to two other members of the subfamily, that is, the neurotensin receptor 2 and GPR39, shows high level of basal signaling. Structural and functional analysis of these structurally related receptors indicate that the constitutive activity is related to the presence of an aromatic cluster on the inner face of the extracellular pools of TMs VI and VII formed by Phe VI:16, Phe VII:06, and Phe VII:09, found in close spatial proximity (Fig. 3.2; Holst *et al.*, 2004). Formation of this cluster seems to be crucial for ghrelin receptor signaling. The level of constitutive activity is regulated by the size and hydrophobicity of the side chain of the residue in position VI:16. Pulling the extracellular ends of TMs VI and VII toward each together with an inward movement toward TM III is believed to stabilize the active conformation of the receptor. Gln III:05 and Arg VI:20 residues, which are right above described aromatic cluster, may form a polar interaction across main ligand binding pocket and it is suggested that this interaction contributes to the constitutive signaling of ghrelin receptor (Fig. 3.2; Holst *et al.*, 2004). An aromatic lock important for ghrelin receptor constitutive signaling is formed by the conserved Trp VI:13 and Phe V:13 residues (Fig. 3.2; Holst *et al.*, 2010). The Trp VI:13 is part of the CWxP motif in the middle of TM VI and is suggested to serve as a rotameric microswitch in 7TM receptors activation process (Ahuja and

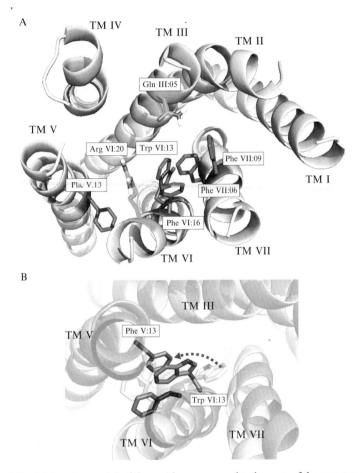

Figure 3.2 Molecular model of the residues proposed to be part of the structural basis for the high constitutive activity in the ghrelin receptor. (*Panel A*) Molecular model of the ghrelin receptor built over the inactive structure of the β2-adrenergic receptor. The seven helical bundles are displayed without the loops as viewed from the extracellular side. Only the residues on the inner faces of TMs III, VI, and VII, which in mutational analysis have been identified to be involved in the constitutive activity are shown. The residues forming the aromatic cluster in between TMs VI and VII are indicated *in purple* and ionic/hydrophilic interaction in TMs III and VI are indicated in *green*. (*Panel B*) Presumed interaction between Trp VI:13 and Phe V:13 seen in Monte Carlo simulations of the activation process where inactive and active conformations are shown in *white* and *blue ribbons*, respectively. (See Color Insert.)

Smith, 2009; Nygaard *et al.*, 2009; Shi *et al.*, 2002). Mutational analysis of Trp VI:13 showed elimination of constitutive activity as well as significant decrease in the agonist-induced signaling stimulation upon Ala and His substitution (24% and 49% of wild-type ghrelin receptor basal activity, respectively). Moreover, no change in ligand binding properties was noticed. Ala substitution of Phe V:13, the proposed interaction partner for Trp VI:13, impaired both basal and ligand-induced receptor activities. Hence, it is suggested that aromatic staking between Trp VI:13 and Phe V:13 stabilizes active conformation and formation of this lock plays an important role in transmembrane domains movements in activation mechanism not only for the ghrelin receptor but also for other family A 7TM receptors (Holst et al., 2010) (Fig. 3.2).

4. Residues Responsible for the Inverse Agonism and Efficacy Swap

In contrast to agonists, that stabilize the active conformation and induce increased intracellular signal transduction, ligands that stabilize the inactive conformation and suppress the basal signaling are known as inverse agonists (Costa and Cotecchia, 2005). [D-Arg^1, D-Phe^5, D-Trp^{7,9}, Leu^{11}]-substance P, which previously was reported to be a low potent ghrelin receptor antagonist, is able to inhibit the constitutive activity of the ghrelin receptor with high potency (Hansen et al., 1999; Holst et al., 2003). In order to dissect the core motif, responsible for the inverse agonist properties of the [D-Arg^1, D-Phe^5, D-Trp^{7,9}, Leu^{11}]-substance P peptide, N-terminally truncation was performed and revealed that the heptapeptide, fQwFwLL (for the peptides sequence single-letter amino acid code is used where capital letters stand for L-amino acids and small letters denote their D-isomers), was sufficient to be a potent and efficacious inverse agonist. Systematic Ala substitutions of each residue indicated that both Gln^6 and Leu^{10} side chains are of minor significance for the ligand binding and function, but truncation of C-terminal Leu^{11} resulted in loss of inverse agonism and decrease of binding affinity. Further N-terminal truncations and substitutions analysis revealed the pentapeptide, wFwLL, to be the active core peptide for the inverse agonism (Holst et al., 2006).

The inositol phosphate turnover for wFwLL peptide shows a biphasic dose–response curve. Partial agonism is observed for low concentrations of the core pentapeptide, followed by inverse agonism for increasing concentrations. Extension of this core peptide with positively charged Arg or Lys at the N-terminal end, rescue the high potent and efficient inverse agonism. Introduction of the small aliphatic Ala at the N-terminal end of wFwLL peptide, surprisingly, revealed partial agonist properties as measured by both IP accumulation and SRE reporter gene assays. Finally, when a negatively charged Asp residue was introduced instead at the N-terminal end, the core peptide lost both its high potency partial agonist and inverse agonist properties and became a neutral antagonist (Holst et al., 2006, 2007).

Mutational mapping of the ghrelin receptor ligand binding pocket revealed that the main binding crevice for its endogenous agonist is located on inner sites of TMs III, VI, and VII. Ligand binding occurring between the opposing faces of these transmembrane domains, which probably pulls them toward each other, support movement of the transmembrane segments leading to stabilization of the active conformation as described in "The Global Toggle Switch Model". In contrast, the inverse agonist peptides are dependent on residues spread all over the ligand binding pocket from Asp in position 20 of TM II across the main binding crevice between TMs III, VI, and VII to the extracellular part of TMs IV and V. Additionally, inverse agonism of [D-Arg^1, D-Phe^5, D-Trp^{7,9}, Leu^{11}]-substance P can be improved by space-generating Ala substitutions of residues in TMs IV and V located deeper in the binding pocket, that is, Ser IV:16, Val V:08, and Phe V:12 (Holst et al., 2006).

The N-terminally, Ala- and Lys-modified wFwLL peptide which display agonism and inverse agonism, respectively, seemed to be affected by mutations of the same residues as described for both [D-Arg^1, D-Phe^5, D-Trp^{7,9}, Leu^{11}]-substance P and its heptapeptide derivate, fQwFwLL. However, several substitutions showed differences in signaling effect for AwFwLL compared to KwFwLL. Val V:08 Ala and Phe V:12 Ala mutations abolish the agonism of the Ala peptide, while no effect on inverse agonism of the Lys peptide was observed. Two residues selectively involved in interaction with agonist and inverse agonist were identified at the interface between the extracellular ends of TMs IV and V. Ala substitution of Ser IV:16 resulted in increase of KwFwLL inverse agonist potency, while almost no effect on AwFwLL agonism was observed. In contrast, increase in AwFwLL agonist potency with no effect on KwFwLL inverse agonism was noticed for another space-generating mutation of Met V:05 to Ala (Table 3.1; Holst et al., 2007).

Substitutions of two residues located in TM III exactly one helical turn apart from each other, that is, Phe III:04 and Ser III:08 were found to induce swap in efficacy between agonism and inverse agonism of N-terminally modified wFwLL core peptide. Ala substitution of Phe III:08 converted the AwFwLL agonist into a potent inverse agonist, whereas the inverse agonist KwFwLL was turned into an agonist when analyzed on the Phe III:04 Ser mutation. The effect of substitution of Ile IV:20 into Ala was similar to the Phe III:04 Ser mutation as efficacy swap of KwFwLL, from inverse agonism into agonism, was observed. Additionally, this mutation improved AwFwLL efficacy from partial to almost full agonism (Figure 3.3).

It is concluded that inverse agonism was increased or efficacy swap from agonism into inverse agonism was caused by mutations of residues located relatively deep in the ligand binding pocket, for example, Ser III:08, Ser IV:16, Val V:08, and Phe V:12. In contrast, substitutions which bias the ligands toward agonism, that is, Phe III:04, Ile IV:20, and Met V:05 are located, more extracellularly (Table 3.1). These observations suggest that inactive conformation is stabilized by ligand binding in a deeper part of binding pocket, whereas receptor activation requires ligand binding in the upper, more extracellular site (Holst et al., 2007).

Table 3.1 Summary of the ghrelin receptor single-residue substitutions effects on inverse agonist [D-Arg^1, D-Phe^5, D-Trp^{7,9}, Leu^{11}]-substance P and KwFwLL as well as partial agonist AwFwLL agonist signaling phenotype

Phenotype	[D-Arg^1, D-Phe^5, D-Trp^{7,9}, Leu^{11}]-substance P	KwFwLL	AwFwLL
WT	Inverse agonist	Inverse agonist	Partial agonist
Phe III:04 Ser	Inverse agonist (decreased potency)	Partial agonist	Partial agonist
Ser III:08 Ala	Inverse agonist (improved potency)	Inverse agonist	Inverse agonist
Ser IV:16 Ala	Inverse agonist (improved potency)	Inverse agonist (improved potency)	Partial agonist
Ile IV:20 Ala	Inverse agonist (decreased potency)	Partial agonist	Full agonist (decreased potency)
Met V:05 Ala	Inverse agonist (improved potency)	Inverse agonist	Partial agonist (improved potency)
Val V:08 Ala	Inverse agonist (improved potency)	Inverse agonist	Neutral antagonist
Phe V:12 Ala	Inverse agonist (improved potency)	Inverse agonist	Neutral antagonist

5. PHYSIOLOGICAL RELEVANCE OF CONSTITUTIVE ACTIVITY

The *in vivo* function of the ligand-independent signaling observed for the ghrelin receptor has until recently been rather speculative and based on subtle facts such as the observation that the ghrelin receptor expression is highly regulated. Since the ghrelin receptor signals without any ligand present in a constitutive active manner the expression level of the receptor will be important for the activity. Physiological conditions associated with increased ghrelin receptor expression include fasting, leptin deficiency, and streptozotocin treatment (Hewson et al., 2002; Nogueiras et al., 2004; Petersen et al., 2009). Presumably the increased activity in the ghrelin receptor system mediated by the receptor expression contributes to increased appetite and feeding behavior.

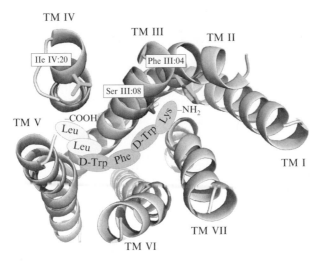

Figure 3.3 Molecular model of the ghrelin receptor α-helical transmembrane domains viewed from the extracellular side. The residues identified to drive the efficacy swap from agonism toward inverse agonism as well as from inverse agonism toward agonism are shown *in red* and *in green* sticks, respectively. A schematic model of the KwFwLL inverse agonist has been "docked" into the model in a configuration where the N-terminal molecular switch region of the ligand is placed in proximity to the efficacy switch region of the receptor—positions III:04 and III:08. It is proposed that the alpha NH_2-group of the N-terminal residue (Lys in this case Ala in the agonist version) makes a interaction with the region in between TM III and TM II and that the central aromatic cluster makes key interactions with the aromatic cluster presented at the interface between TMs VI and VII (including a potential with Arg VI:20) and that the C-terminal double Leu sequence is located in the pocket between TMs III, IV, and V. (See Color Insert.)

More direct evidence to prove the physiological function of the ligand independent signaling requires pharmacological or genetic ablation of the high basal signaling; approaches that both have been applied for the ghrelin receptor. The inverse agonist [D-Arg^1, D-Phe^5, D-Trp^{7,9}, Leu^{11}]-substance P, which has previously been demonstrated to inhibit the basal signaling of the ghrelin receptor *in vitro* with high potency and efficacy, has been used as a pharmacological tool. This peptide constitutes a valuable tool because it acts as a selective inverse agonist without antagonist function in a wide concentration range. The appropriate concentration was obtained by administration directly into the ventricular system of the brain through an osmotic minipump. Thus, only the high basal signaling was decreased, whereas the ghrelin-induced stimulation was unaffected. It was shown that the decreased level of constitutive activity gave rise to decreased food intake, body weight, and gene expression of neuropeptide Y (NPY) and uncoupling protein 2 (UCP2) in the hypothalamus (Petersen *et al.*, 2009).

A human mutation in the ghrelin receptor gene, where the constitutive activity is selectively eliminated without any effect on the ghrelin induced stimulation, has been observed and provides unique information about the *in vivo* consequences of the high basal signaling (Pantel *et al.*, 2006). This mutation is located in the first exon and leads to Ala to Glu substitution in position 204 located within the second extracellular loop (Figure 3.4). Occurrence of this mutation has been reported in two nonrelated patients, one of which being heterozygous and diagnosed to have isolated GH deficiency, whereas the other was homozygous and suffered from idiopathic short stature. Importantly, the mutation segregated with the pathologic condition. The probands as well as their families' members who carried the mutation shared similar phenotype of significantly reduced height, but not to same extent in all analyzed cases. It can be suggested that loss of constitutive activity of ghrelin receptor has a physiological effect in lower GH system which might be the reason for observed short stature phenotype (Holst and Schwartz, 2006; Pantel *et al.*, 2006).

The present data suggest that the ligand independent signaling from the ghrelin receptor is important for physiological functions including both linear growth and food intake—albeit not concomitantly. The data obtained from human mutations indicate that the basal signaling from the ghrelin receptor is important for linear growth before puberty whereas acute elimination of the signaling in adult life obtained by administration of an inverse agonist, affects the feeding behavior. It is possible that the most efficient inhibition of the ghrelin system would be obtained through the combined use of an inverse agonist, to decrease the basal tone of the system set by the constitutive activity of the ghrelin receptor, and an antagonist, to decrease the meal-related fluctuations in plasma ghrelin (Holst and Schwartz, 2004).

6. Experimental Procedures

Investigation of constitutive activity requires transient transfection of cells with variable receptor expression or stably transfected cell lines in which the cell surface expression levels can be regulated by use of specific agents like tetracycline-regulated expression system T-REx^{TM} (Invitrogen). The protocol for two cell-based assays suitable for measurement of ghrelin receptor signaling together with transient transfection protocols optimal for each assay is described below. Additionally, we provide a protocol for enzyme-linked immunosorbent assay (ELISA), providing an easy and efficient method to monitor cell surface expression of the ghrelin receptor.

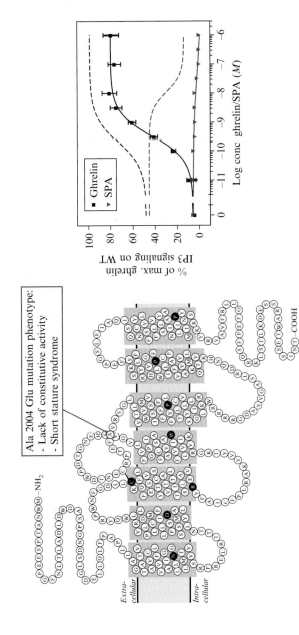

Figure 3.4 Serpentine model of the ghrelin receptor with Ala 204 residue located in the extracellular loop 2 indicated with *red circle*. The naturally occurring Ala 204 Glu mutation selectively eliminates the constitutive signaling of the ghrelin receptor and is associated with short stature. Dose-response curves for ghrelin (*blue*) and [D-Arg[1], D-Phe[5], D-Trp[7,9], Leu[11]]-substance P (*red*) on the Ala 204 Glu mutant (*solid lines*) compared with results obtained for the wild type receptor (*dotted lines*) indicate loss of the basal signaling of ghrelin receptor transduced via $G\alpha_{q/11}$ protein measured as inositol phosphates turnover. (For interpretation of the references to color in this figure legend, the reader is referred to the Web version of this chapter.)

6.1. Inositol phosphate turnover assay

COS-7 cells are seeded out at the density of 3×10^6 cells per 75 cm^2 flask 1 day prior to the transfection. Right before transfection cells, the culture medium is aspirated and half of the standard volume is added. For the receptor gene-dosing experiment we recommend use of a range of cDNA amount from 0 up to 30 µg. The appropriate amount of receptor-containing plasmid or "empty" vector plasmid, as a negative control, is mixed with 30 µL 2 mM CaCl$_2$ and TE buffer (10 mM Tris–HCl, 1 mM EDTA, pH 7.5) to a total volume 240 µL. The mixture is added drop-wise to another tube containing 240 µL 2× HBS buffer (280 mM NaCl, 50 mM HEPES, 1.5 mM HPO$_4$, pH 7.2) and left for 45 min at room temperature for DNA precipitation. Afterward, the precipitate is added carefully to the cells followed by addition of 150 µL 2 mg/mL chloroquine in PBS solution. The cells are incubated for 5 h at normal culture conditions and the transfection is stopped by change of the DNA-containing medium with 10 mL of fresh cell culture medium.

The day after transfection, the cells are harvested with PBS–EDTA, spun down, and resuspended in cell culture medium. The cells are counted and dissolved to the density 3.33×10^5 cells/mL and [^{3}H]-*myo*-inositol is added at concentration 5 µCi/mL. The cells are seeded out in poly-D-lysine coated 24-well plates, 300 µL of cell suspension per well (final density 1×10^5 cells/well) and incubated overnight.

Prior to the assay procedure, the cells are washed twice with 0.5 mL/well of Hank's balanced salt solution (HBSS) and 0.5 mL HBSS supplemented with 10 mM LiCl is added. For accumulation the cells are incubated for 1 h 15 min at 37 °C. If agonist stimulation of the receptor is investigated, the agonist is added after 30 min of initial LiCl incubation followed by 45-min incubation. If inverse agonist inhibition of the basal activity is studied it is important to add the ligand before the LiCl is added. The reaction is stopped by aspiration of incubation buffer and addition of 1 mL ice cold 10 mM formic acid followed by incubation on ice for at least 45 min.

The [^{3}H]-labeled product of inositol phosphates turnover, inositol monophosphate is purified on anion-exchange resin, Dowex 1x8-200. The packed columns are prepared by adding 3 mL regeneration buffer (3 M ammonium formate, 100 mM formic acid) and double washing with 5 mL deionized water. Formic acid extracts are loaded on columns and washed twice with 5 mL GPI buffer (60 mM sodium formate, 5 mM sodium tetraborate decahydrate). Finally, the product is eluted into 20-mL scintillation vials with 3 mL elution buffer (1 M ammonium formate, 100 mM formic acid) and 10 mL scintillation cocktail is added. The closed vials require thorough shaking and the radiation from β-emitting nuclides is measured in liquid.

6.2. SRE reporter gene assay

In the reporter gene assay the cells are transiently co-transfected with the receptor-encoding plasmid and the inducible *cis*-reporter plasmids, that is, pSRELuc (PathDetect System, Stratagene) which contains the luciferase reporter gene under control of synthetic promoter that contains direct repeats of the transcription recognition sequences for SRE. Activation of the SRE signal transduction pathway results in increased luciferase expression. Measurement of luminescence emitted upon addition of luciferase-specific substrate allows for determination of luciferase gene expression, which then reflects SRE-driven receptor signaling.

The SRE reporter gene assay for the ghrelin receptor activity is performed on HEK293 cells seeded in white 96-well plates (3.0–3.5×10^5 cells/well). The cells are transiently transfected on the day after seeding with use of cationic liposome based reagent, that is, Lipofectamine 2000 (Invitrogen). Briefly, cell culture medium is aspirated and 50 μL of OptiMem medium (Gibco) is added to each well. Solutions of 0.6 μL Lipofectamine 2000^{TM} and 25 μL of OptiMem medium per well is prepared and incubated at room temperature for 5 min. In a separate tube the receptor cDNA in range from 0 to 1 ng is dissolved together with 50 ng pSRE-Luc plasmid per well in 25 μL of OptiMem medium. Afterward, the Lipofectamine 2000^{TM} solution is mixed together with the cDNA solution and incubated at room temperature for 20 min prior to addition to cells. The transfection is carried out in normal cell culture conditions and is terminated after 5 h when the transfection medium is exchanged with cell culture medium supplemented with 0.5% heat-inactivated serum without antibiotics and followed by overnight incubation at regular cell culture conditions. If inverse agonism is studied the ligand is added to the cell culture medium right after the transfection is stopped.

On the assay day agonist stimulation or a new dose of inverse agonist inhibition is added to the cells followed by incubation with the ligands for 5 h. Afterward, the cell culture medium is aspirated and the cells are washed with 150 μL PBS buffer containing Ca^{2+} and Mg^{2+}. To detect luciferase activity 100 μL PBS buffer is dispensed to each well followed by addition of 100 μL luciferase substrate, that is, steadylite plus (PerkinElmer). The measurement is carried out in a luminescence reader such as TopCount (PerkinElmer).

6.3. ELISA assay

In order to examine the receptor cell surface expression level an ELISA assay can be employed. Due to lack of specific antibodies against the ghrelin receptor we recommend use of an N-terminal epitope tagged receptor, for example FLAG tag, which is small, highly immunoreactive and seems not to interfere with the ghrelin receptor function. Epitope tagging facilitates use of easily available, highly specific anti-tag antibodies.

The ELISA assay is performed on cells transfected with the receptor in the same way as described for functional assays. The cells are seeded out in 96-well plates in number of $3.0–3.5 \times 10^5$ cells/well. On the day of the assay the cells are fixed with 150 μL 3.7% formaldehyde in PBS buffer followed by triple washing with 200 μL PBS buffer. Unspecific binding sites are saturated by 2-h incubation with 150 μL blocking buffer (3% nonfat dry milk in PBS). Incubation with the mouse anti-FLAG antibody (Sigma) is carried out at room temperature for at least 2 h or preferentially overnight at 4 °C. Recommended dilution of primary antibody is 1:1000 and it should be prepared in 100 μL the blocking buffer per well. Excess of unbound antibodies is removed by triple washing with 200 μL PBS buffer. Antimouse IgG enzyme-linked anybody, that is, goat anti-mouse horse radish peroxidise (HRP)-bound antibody (Sigma) is used in the following step (recommended dilution 1:1250 in 1.5% nonfat dry milk in PBS, 100 μL/well). Incubation with secondary antibody should last 1–2 h at room temperature and be followed by triple washing with PBS buffer. Use of 150 μL/well of HRP substrate, that is, tetramethyl benzidine (TMB, KeMenTech) results in color product generation. The enzymatic reaction is terminated by addition of 100 μL 0.2 M H_2SO_4 per well. Hundred microliters of the color reaction product is transferred into separate transparent 96-well plate and an absorbance at 450 nm is measured.

REFERENCES

Abizaid, A., Liu, Z. W., Andrews, Z. B., Shanabrough, M., Borok, E., Elsworth, J. D., Roth, R. H., Sleeman, M. W., Picciotto, M. R., Tschop, M. H., Gao, X. B., and Horvath, T. L. (2006). Ghrelin modulates the activity and synaptic input organization of midbrain dopamine neurons while promoting appetite. *J. Clin. Invest.* **116**, 3229–3239.

Ahuja, S., and Smith, S. O. (2009). Multiple switches in G protein-coupled receptor activation. *Trends Pharmacol. Sci.* **30**, 494–502.

Ahuja, S., Hornak, V., Yan, E. C., Syrett, N., Goncalves, J. A., Hirshfeld, A., Ziliox, M., Sakmar, T. P., Sheves, M., Reeves, P. J., Smith, S. O., and Eilers, M. (2009). Helix movement is coupled to displacement of the second extracellular loop in rhodopsin activation. *Nat. Struct. Mol. Biol.* **16**, 168–175.

Ariyasu, H., Takaya, K., Tagami, T., Ogawa, Y., Hosoda, K., Akamizu, T., Suda, M., Koh, T., Natsui, K., Toyooka, S., Shirakami, G., Usui, T., *et al.* (2001). Stomach is a major source of circulating ghrelin, and feeding state determines plasma ghrelin-like immunoreactivity levels in humans. *J. Clin. Endocrinol. Metab.* **86**, 4753–4758.

Bach, M. A., Rockwood, K., Zetterberg, C., Thamsborg, G., Hebert, R., Devogelaer, J. P., Christiansen, J. S., Rizzoli, R., Ochsner, J. L., Beisaw, N., Gluck, O., Yu, L., *et al.* (2004). The effects of MK-0677, an oral growth hormone secretagogue, in patients with hip fracture. *J. Am. Geriatr. Soc.* **52**, 516–523.

Ballesteros, J. A., Shi, L., and Javitch, J. A. (2001). Structural mimicry in G protein-coupled receptors: Implications of the high-resolution structure of rhodopsin for structure-function analysis of rhodopsin-like receptors. *Mol. Pharmacol.* **60**, 1–19.

Bokoch, M. P., Zou, Y., Rasmussen, S. G., Liu, C. W., Nygaard, R., Rosenbaum, D. M., Fung, J. J., Choi, H. J., Thian, F. S., Kobilka, T. S., Puglisi, J. D., Weis, W. I., et al. (2010). Ligand-specific regulation of the extracellular surface of a G-protein-coupled receptor. *Nature* **463**, 108–112.

Bowers, C. Y., Momany, F. A., Reynolds, G. A., and Hong, A. (1984). On the in vitro and in vivo activity of a new synthetic hexapeptide that acts on the pituitary to specifically release growth hormone. *Endocrinology* **114**, 1537–1545.

Costa, T., and Cotecchia, S. (2005). Historical review: Negative efficacy and the constitutive activity of G-protein-coupled receptors. *Trends Pharmacol. Sci.* **26**, 618–624.

Davenport, A. P., Bonner, T. I., Foord, S. M., Harmar, A. J., Neubig, R. R., Pin, J. P., Spedding, M., Kojima, M., and Kangawa, K. (2005). International Union of Pharmacology. LVI. Ghrelin receptor nomenclature, distribution, and function. *Pharmacol. Rev.* **57**, 541–546.

Diano, S., Farr, S. A., Benoit, S. C., McNay, E. C., da Silva, I., Horvath, B., Gaskin, F. S., Nonaka, N., Jaeger, L. B., Banks, W. A., Morley, J. E., Pinto, S., et al. (2006). Ghrelin controls hippocampal spine synapse density and memory performance. *Nat. Neurosci.* **9**, 381–388.

Dror, R. O., Arlow, D. H., Borhani, D. W., Jensen, M. O., Piana, S., and Shaw, D. E. (2009). Identification of two distinct inactive conformations of the beta2-adrenergic receptor reconciles structural and biochemical observations. *Proc. Natl. Acad. Sci. USA* **106**, 4689–4694.

Einat, H., Kofman, O., Itkin, O., Lewitan, R. J., and Belmaker, R. H. (1998). Augmentation of lithium's behavioral effect by inositol uptake inhibitors. *J. Neural Transm.* **105**, 31–38.

Elling, C. E., Frimurer, T. M., Gerlach, L. O., Jorgensen, R., Holst, B., and Schwartz, T. W. (2006). Metal ion site engineering indicates a global toggle switch model for seven-transmembrane receptor activation. *J. Biol. Chem.* **281**, 17337–17346.

Gether, U. (2000). Uncovering molecular mechanisms involved in activation of G protein-coupled receptors. *Endocr. Rev.* **21**, 90–113.

Gether, U., Lin, S., and Kobilka, B. K. (1995). Fluorescent labeling of purified beta 2 adrenergic receptor. Evidence for ligand-specific conformational changes. *J. Biol. Chem.* **270**, 28268–28275.

Gether, U., Ballesteros, J. A., Seifert, R., Sanders-Bush, E., Weinstein, H., and Kobilka, B. K. (1997). Structural instability of a constitutively active G protein-coupled receptor. Agonist-independent activation due to conformational flexibility. *J. Biol. Chem.* **272**, 2587–2590.

Goodyear, S., Arasaradnam, R. P., Quraishi, N., Mottershead, M., and Nwokolo, C. U. (2010). Acylated and des acyl ghrelin in human portal and systemic circulations. *Mol. Biol. Rep.* (Epub ahead of print).

Graaf de, C., and Rognan, D. (2008). Selective structure-based virtual screening for full and partial agonists of the beta2 adrenergic receptor. *J. Med. Chem.* **51**, 4978–4985.

Granata, R., Settanni, F., Biancone, L., Trovato, L., Nano, R., Bertuzzi, F., Destefanis, S., Annunziata, M., Martinetti, M., Catapano, F., Ghe, C., Isgaard, J., et al. (2007). Acylated and unacylated ghrelin promote proliferation and inhibit apoptosis of pancreatic beta-cells and human islets: Involvement of 3',5'-cyclic adenosine monophosphate/protein kinase A, extracellular signal-regulated kinase 1/2, and phosphatidyl inositol 3-Kinase/Akt signaling. *Endocrinology* **148**, 512–529.

Hansen, B. S., Raun, K., Nielsen, K. K., Johansen, P. B., Hansen, T. K., Peschke, B., Lau, J., Andersen, P. H., and Ankersen, M. (1999). Pharmacological characterisation of a new oral GH secretagogue, NN703. *Eur. J. Endocrinol.* **141**, 180–189.

Hewson, A. K., Tung, L. Y., Connell, D. W., Tookman, L., and Dickson, S. L. (2002). The rat arcuate nucleus integrates peripheral signals provided by leptin, insulin, and a ghrelin mimetic. *Diabetes* **51,** 3412–3419.

Holst, B., and Schwartz, T. W. (2004). Constitutive ghrelin receptor activity as a signaling set-point in appetite regulation. *Trends Pharmacol. Sci.* **25,** 113–117.

Holst, B., and Schwartz, T. W. (2006). Ghrelin receptor mutations—Too little height and too much hunger. *J. Clin. Invest.* **116,** 637–641.

Holst, B., Cygankiewicz, A., Jensen, T. H., Ankersen, M., and Schwartz, T. W. (2003). High constitutive signaling of the ghrelin receptor—Identification of a potent inverse agonist. *Mol. Endocrinol.* **17,** 2201–2210.

Holst, B., Holliday, N. D., Bach, A., Elling, C. E., Cox, H. M., and Schwartz, T. W. (2004). Common structural basis for constitutive activity of the ghrelin receptor family. *J. Biol. Chem.* **279,** 53806–53817.

Holst, B., Lang, M., Brandt, E., Bach, A., Howard, A., Frimurer, T. M., Beck-Sickinger, A., and Schwartz, T. W. (2006). Ghrelin receptor inverse agonists: Identification of an active peptide core and its interaction epitopes on the receptor. *Mol. Pharmacol.* **70,** 936–946.

Holst, B., Mokrosinski, J., Lang, M., Brandt, E., Nygaard, R., Frimurer, T. M., Beck-Sickinger, A. G., and Schwartz, T. W. (2007). Identification of an efficacy switch region in the ghrelin receptor responsible for interchange between agonism and inverse agonism. *J. Biol. Chem.* **282,** 15799–15811.

Holst, B., Nygaard, R., Valentin-Hansen, L., Bach, A., Engelstoft, M. S., Petersen, P. S., Frimurer, T. M., and Schwartz, T. W. (2010). A conserved aromatic lock for the tryptophan rotameric switch in TM-VI of seven-transmembrane receptors. *J. Biol. Chem.* **285,** 3973–3985.

Howard, A. D., Feighner, S. D., Cully, D. F., Arena, J. P., Liberator, P. A., Rosenblum, C. I., Hamelin, M., Hreniuk, D. L., Palyha, O. C., Anderson, J., Paress, P. S., Diaz, C., *et al.* (1996). A receptor in pituitary and hypothalamus that functions in growth hormone release. *Science* **273,** 974–977.

Hubbell, W. L., Altenbach, C., Hubbell, C. M., and Khorana, H. G. (2003). Rhodopsin structure, dynamics, and activation: A perspective from crystallography, site-directed spin labeling, sulfhydryl reactivity, and disulfide cross-linking. *Adv. Protein Chem.* **63,** 243–290.

Javitch, J. A., Fu, D., Liapakis, G., and Chen, J. (1997). Constitutive activation of the beta2 adrenergic receptor alters the orientation of its sixth membrane-spanning segment. *J. Biol. Chem.* **272,** 18546–18549.

Jerlhag, E., Egecioglu, E., Landgren, S., Salome, N., Heilig, M., Moechars, D., Datta, R., Perrissoud, D., Dickson, S. L., and Engel, J. A. (2009). Requirement of central ghrelin signaling for alcohol reward. *Proc. Natl. Acad. Sci. USA* **106,** 11318–11323.

Katritch, V., Reynolds, K. A., Cherezov, V., Hanson, M. A., Roth, C. B., Yeager, M., and Abagyan, R. (2009). Analysis of full and partial agonists binding to beta2-adrenergic receptor suggests a role of transmembrane helix V in agonist-specific conformational changes. *J. Mol. Recognit.* **22,** 307–318.

Kirchner, H., Gutierrez, J. A., Solenberg, P. J., Pfluger, P. T., Czyzyk, T. A., Willency, J. A., Schurmann, A., Joost, H. G., Jandacek, R. J., Hale, J. E., Heiman, M. L., and Tschop, M. H. (2009). GOAT links dietary lipids with the endocrine control of energy balance. *Nat. Med.* **15,** 741–745.

Kojima, M., and Kangawa, K. (2005). Ghrelin: Structure and function. *Physiol. Rev.* **85,** 495–522.

Kojima, M., Hosoda, H., Date, Y., Nakazato, M., Matsuo, H., and Kangawa, K. (1999). Ghrelin is a growth-hormone-releasing acylated peptide from stomach. *Nature* **402,** 656–660.

Kojima, M., Hosoda, H., Matsuo, H., and Kangawa, K. (2001). Ghrelin: Discovery of the natural endogenous ligand for the growth hormone secretagogue receptor. *Trends Endocrinol. Metab.* **12,** 118–122.

Korbonits, M., Goldstone, A. P., Gueorguiev, M., and Grossman, A. B. (2004). Ghrelin—A hormone with multiple functions. *Front. Neuroendocrinol.* **25,** 27–68.

Leung, P. K., Chow, K. B., Lau, P. N., Chu, K. M., Chan, C. B., Cheng, C. H., and Wise, H. (2007). The truncated ghrelin receptor polypeptide (GHS-R1b) acts as a dominant-negative mutant of the ghrelin receptor. *Cell. Signal.* **19,** 1011–1022.

Lopez, M., Lage, R., Saha, A. K., Perez-Tilve, D., Vazquez, M. J., Varela, L., Sangiao-Alvarellos, S., Tovar, S., Raghay, K., Rodriguez-Cuenca, S., Deoliveira, R. M., Castaneda, T., et al. (2008). Hypothalamic fatty acid metabolism mediates the orexigenic action of ghrelin. *Cell Metab.* **7,** 389–399.

Matthews, R. P., Guthrie, C. R., Wailes, L. M., Zhao, X., Means, A. R., and McKnight, G. S. (1994). Calcium/calmodulin-dependent protein kinase types II and IV differentially regulate CREB-dependent gene expression. *Mol. Cell. Biol.* **14,** 6107–6116.

Mirzadegan, T., Benko, G., Filipek, S., and Palczewski, K. (2003). Sequence analyses of G-protein-coupled receptors: Similarities to rhodopsin. *Biochemistry* **42,** 2759–2767.

Murphy, M. G., Weiss, S., McClung, M., Schnitzer, T., Cerchio, K., Connor, J., Krupa, D., and Gertz, B. J. (2001). Effect of alendronate and MK-677 (a growth hormone secretagogue), individually and in combination, on markers of bone turnover and bone mineral density in postmenopausal osteoporotic women. *J. Clin. Endocrinol. Metab.* **86,** 1116–1125.

Nogueiras, R., Tovar, S., Mitchell, S. E., Rayner, D. V., Archer, Z. A., Dieguez, C., and Williams, L. M. (2004). Regulation of growth hormone secretagogue receptor gene expression in the arcuate nuclei of the rat by leptin and ghrelin. *Diabetes* **53,** 2552–2558.

Nygaard, R., Frimurer, T. M., Holst, B., Rosenkilde, M. M., and Schwartz, T. W. (2009). Ligand binding and micro-switches in 7TM receptor structures. *Trends Pharmacol. Sci.* **30,** 249–259.

Pantel, J., Legendre, M., Cabrol, S., Hilal, L., Hajaji, Y., Morisset, S., Nivot, S., Vie-Luton, M. P., Grouselle, D., de Kerdanet, M., Kadiri, A., Epelbaum, J., et al. (2006). Loss of constitutive activity of the growth hormone secretagogue receptor in familial short stature. *J. Clin. Invest.* **116,** 760–768.

Patchett, A. A., Nargund, R. P., Tata, J. R., Chen, M. H., Barakat, K. J., Johnston, D. B., Cheng, K., Chan, W. W., Butler, B., Hickey, G., Jacks, T., Schleim, K., et al. (1995). Design and biological activities of L-163, 191 (MK-0677): A potent, orally active growth hormone secretagogue. *Proc. Natl. Acad. Sci. USA* **92,** 7001–7005.

Perello, M., Sakata, I., Birnbaum, S., Chuang, J. C., Osborne-Lawrence, S., Rovinsky, S. A., Woloszyn, J., Yanagisawa, M., Lutter, M., and Zigman, J. M. (2010). Ghrelin increases the rewarding value of high-fat diet in an orexin-dependent manner. *Biol. Psychiatry* **67,** 880–886.

Petersen, P. S., Woldbye, D. P., Madsen, A. N., Egerod, K. L., Jin, C., Lang, M., Rasmussen, M., Beck-Sickinger, A. G., and Holst, B. (2009). In vivo characterization of high Basal signaling from the ghrelin receptor. *Endocrinology* **150,** 4920–4930.

Rasmussen, S. G., Jensen, A. D., Liapakis, G., Ghanouni, P., Javitch, J. A., and Gether, U. (1999). Mutation of a highly conserved aspartic acid in the beta2 adrenergic receptor: Constitutive activation, structural instability, and conformational rearrangement of transmembrane segment 6. *Mol. Pharmacol.* **56,** 175–184.

Samama, P., Cotecchia, S., Costa, T., and Lefkowitz, R. J. (1993). A mutation-induced activated state of the beta 2-adrenergic receptor. Extending the ternary complex model. *J. Biol. Chem.* **268,** 4625–4636.

Scheerer, P., Park, J. H., Hildebrand, P. W., Kim, Y. J., Krauss, N., Choe, H. W., Hofmann, K. P., and Ernst, O. P. (2008). Crystal structure of opsin in its G-protein-interacting conformation. *Nature* **455**, 497–502.

Schwartz, T. W., Frimurer, T. M., Holst, B., Rosenkilde, M. M., and Elling, C. E. (2006). Molecular mechanism of 7TM receptor activation—A global toggle switch model. *Annu. Rev. Pharmacol. Toxicol.* **46**, 481–519.

Seifert, R., and Wenzel-Seifert, K. (2002). Constitutive activity of G-protein-coupled receptors: Cause of disease and common property of wild-type receptors. *Naunyn Schmiedebergs Arch. Pharmacol.* **366**, 381–416.

Shi, L., Liapakis, G., Xu, R., Guarnieri, F., Ballesteros, J. A., and Javitch, J. A. (2002). Beta2 adrenergic receptor activation. Modulation of the proline kink in transmembrane 6 by a rotamer toggle switch. *J. Biol. Chem.* **277**, 40989–40996.

Singh, L. P., Andy, J., Anyamale, V., Greene, K., Alexander, M., and Crook, E. D. (2001). Hexosamine-induced fibronectin protein synthesis in mesangial cells is associated with increases in cAMP responsive element binding (CREB) phosphorylation and nuclear CREB: The involvement of protein kinases A and C. *Diabetes* **50**, 2355–2362.

Smit, M. J., Vischer, H. F., Bakker, R. A., Jongejan, A., Timmerman, H., Pardo, L., and Leurs, R. (2007). Pharmacogenomic and structural analysis of constitutive G protein-coupled receptor activity. *Annu. Rev. Pharmacol. Toxicol.* **47**, 53–87.

Swaminath, G., Xiang, Y., Lee, T. W., Steenhuis, J., Parnot, C., and Kobilka, B. K. (2004). Sequential binding of agonists to the beta2 adrenoceptor. Kinetic evidence for intermediate conformational states. *J. Biol. Chem.* **279**, 686–691.

Swaminath, G., Deupi, X., Lee, T. W., Zhu, W., Thian, F. S., Kobilka, T. S., and Kobilka, B. (2005). Probing the beta2 adrenoceptor binding site with catechol reveals differences in binding and activation by agonists and partial agonists. *J. Biol. Chem.* **280**, 22165–22171.

Theander-Carrillo, C., Wiedmer, P., Cettour-Rose, P., Nogueiras, R., Perez-Tilve, D., Pfluger, P., Castaneda, T. R., Muzzin, P., Schurmann, A., Szanto, I., Tschop, M. H., and Rohner-Jeanrenaud, F. (2006). Ghrelin action in the brain controls adipocyte metabolism. *J. Clin. Invest.* **116**, 1983–1993.

Tschop, M., Smiley, D. L., and Heiman, M. L. (2000). Ghrelin induces adiposity in rodents. *Nature* **407**, 908–913.

Vanni, S., Neri, M., Tavernelli, I., and Rothlisberger, U. (2009). Observation of "ionic lock" formation in molecular dynamics simulations of wild-type beta 1 and beta 2 adrenergic receptors. *Biochemistry* **48**, 4789–4797.

Yang, J., Brown, M. S., Liang, G., Grishin, N. V., and Goldstein, J. L. (2008). Identification of the acyltransferase that octanoylates ghrelin, an appetite-stimulating peptide hormone. *Cell* **132**, 387–396.

Yao, X. J., Velez, R. G., Whorton, M. R., Rasmussen, S. G., Devree, B. T., Deupi, X., Sunahara, R. K., and Kobilka, B. (2009). The effect of ligand efficacy on the formation and stability of a GPCR-G protein complex. *Proc. Natl. Acad. Sci. USA* **106**, 9501–9506.

Zhang, W., Chai, B., Li, J. Y., Wang, H., and Mulholland, M. W. (2008). Effect of des-acyl ghrelin on adiposity and glucose metabolism. *Endocrinology* **149**, 4710–4716.

CHAPTER FOUR

Assessment of Constitutive Activity and Internalization of GPR54 (KISS1-R)

Macarena Pampillo[*,†,‡] and Andy V. Babwah[*,†,‡,§]

Contents

1. Introduction	76
2. Materials	77
2.1. Materials common to all assays	77
2.2. Materials for HEK 293 cell transfection only	78
2.3. Materials for immunofluorescence assay only	78
2.4. Materials for flow cytometry assay only	78
2.5. Materials for receptor radiolabeling assay only	79
2.6. Materials for inositol phosphate formation assay only	79
3. Methods	79
3.1. Common assays	79
3.2. Constitutive receptor internalization assay measured by immunofluorescence and confocal imaging	81
3.3. Constitutive receptor internalization assay measured by flow cytometry	84
3.4. Constitutive receptor internalization assay measured by indirect receptor radiolabeling	87
3.5. Assessing constitutive receptor activity by measuring inositol phosphate turnover in intact cells	88
Acknowledgment	91
References	92

Abstract

The kisspeptin/GPR54 signaling system positively regulates GnRH secretion, thereby acting as an important regulator of the hypothalamic–pituitary–gonadal axis. It also negatively regulates tumor metastases and placental trophoblast

[*] The Children's Health Research Institute, London, Ontario, Canada
[†] Lawson Health Research Institute, London, Ontario, Canada
[‡] Department of Obstetrics and Gynecology, The University of Western Ontario, London, Ontario, Canada
[§] Department of Physiology and Pharmacology, The University of Western Ontario, London, Ontario, Canada

Methods in Enzymology, Volume 484   © 2010 Elsevier Inc.
ISSN 0076-6879, DOI: 10.1016/S0076-6879(10)84004-2   All rights reserved.

invasion. GPR54 is a $G_{q/11}$-coupled GPCR and activation by kisspeptin stimulates PIP_2 hydrolysis and inositol phosphate (IP) formation, Ca^{2+} mobilization, arachidonic acid release, and ERK1/2 and p38 MAP kinase phosphorylation. Recently, we reported that GPR54 displays constitutive activity and internalization in the heterologous human embryonic kidney 293 cell system. Given the physiological and clinical importance of GPR54 as well as other GPCRs, we present assays for measuring constitutive receptor internalization and activity. Specifically, we describe the use of immunofluorescence coupled to confocal imaging, flow cytometry and indirect receptor radiolabeling to measure constitutive receptor internalization, and IP turnover in intact cells to measure constitutive activity. While we use the FLAG-tagged GPR54 molecule as an example to describe these assays, the assays can be applied to a wide range of GPCRs.

1. INTRODUCTION

In 2003, the neuropeptide kisspeptin and its cognate receptor, G protein-coupled receptor 54 (GPR54), were positioned at the forefront of the neuroendocrine control of the hypothalamic–pituitary–gonadal axis (de Roux et al., 2003; Seminara et al., 2003). Originally, kisspeptin was identified as a potent suppressor of tumor metastasis; however, loss-of-function mutations in GPR54, in both man and mouse, led researchers to identify an important role for the kisspeptin/GPR54 signaling system in the control of puberty and reproductive function (de Roux et al., 2003; Seminara et al., 2003).

GPR54 is a $G_{q/11}$-coupled GPCR and activation by kisspeptin stimulates PIP_2 hydrolysis and inositol phosphate (IP) formation, Ca^{2+} mobilization, arachidonic acid release, and ERK1/2 and p38 MAP kinase phosphorylation (Kotani et al., 2001). Recently, we described the roles of G protein-coupled receptor kinase-2 (GRK2) and β-arrestin-1 and -2 in the desensitization and internalization of GPR54 and during the course of that study determined that GPR54 displays constitutive activity and internalization in the heterologous human embryonic kidney (HEK) 293 cell system (Pampillo et al., 2009). However, confirmation of this constitutive behavior requires the use of an inverse GPR54 agonist, but this is not yet available. Nevertheless, we have suggested that this is true constitutive activity and internalization because we were unable to detect the expression of the *KISS1* gene in HEK 293 cells.

Since constitutive activity has been described for more than 60 wild-type GPCRs (Smit et al., 2007), it should not be surprising to find that GPR54 does indeed display constitutive activity. Further support of constitutive GPR54 activity comes from *in vivo* studies performed on the $Kiss1^{-/-}$ and $Gpr54^{-/-}$ animals (Lapatto et al., 2007). In this study, the authors note a phenotypic variability observed among *Kiss1* knockout female mice and suggested modest constitutive GPR54 activity as one likely explanation for it. In our study, we report that maximum basal GPR54 activity is approximately 5% of the maximum kisspeptin (Kp)-10-induced IP formation. If the observations by Lapatto et al. (2007) are explained by constitutive activity, it appears that this level of activity is physiologically relevant. In our study (Pampillo et al., 2009), we provided strong evidence that basal GPR54 activity is under molecular regulation by GRK2 and β-arrestin, suggesting that basal receptor activity appears to be of physiological importance.

The high basal internalization rate we reported for GPR54 was higher (Pampillo et al., 2009) than what we previously observed for the metabotropic glutamate receptor 1a (mGluR1a) at an early time point (~70% of cell surface GPR54 internalized receptor after 5 min compared to ~50% of mGluR1a) (Bhattacharya et al., 2004). mGluR1a is a receptor well known for its constitutive internalization in heterologous cell cultures and primary neurons (Bhattacharya et al., 2004; Fourgeaud et al., 2003; Pula et al., 2004; Sallese et al., 2000). However, after 15 min of internalization under basal conditions, approximately 80% of GPR54 had internalized, and this compared to about 65% for mGluR1a (Bhattacharya et al., 2004) and about 75% for the ghrelin receptor (Holliday et al., 2007). The reason for the high basal internalization rate for GPR54 is still unknown and remains to be assessed in other cell systems.

GPCRs are the largest class of cell surface receptors and the target of most pharmaceuticals. GPCRs regulate an array of biological processes from the perception of light to reproductive functions and any given GPCR may regulate multiple functions. For example, GPR54 regulates tumor metastases and reproductive functions. Given their immense physiological and clinical importance, understanding all aspects of GPCR behavior is important. Here, we provide the optimal conditions and procedures to evaluate the constitutive activity and internalization of any GPCR transiently expressed in HEK 293 cells. The protocols include the use of immunofluorescence and imaging techniques, flow cytometry, receptor radiolabeling, and IP formation assays. The techniques described use the FLAG-tagged GPR54 as an example.

2. MATERIALS

2.1. Materials common to all assays

Cells of interest: HEK 293 cells can be obtained from the American Type Culture Collection (ATCC), ATCC Number: CRL-1573;

Complete growth media for HEK 293 cells: minimum essential medium (MEM), with Earle's salts and L-glutamine (Invitrogen, Cat No. 11095) with 10% (v/v) FBS (Sigma, Cat No. F-1051), 1% (v/v) MEM nonessential amino acids solution (Invitrogen, Cat No. 11140, 10 mM), 1% (v/v) penicillin–streptomycin (Invitrogen, Cat No. 15140, 10,000 units of penicillin and 10,000 µg of streptomycin/ml);

Trypsin–EDTA solution (trypsin, 0.25%, EDTA 4Na 0.38 g/l, Invitrogen, Cat No. 25200);
PBS (137 mM NaCl, 2.7 mM KCl, 10 mM Na$_2$HPO$_4$, 1.76 mM KH$_2$PO$_4$, pH 7.4);
HBSS (1.2 mM KH$_2$PO$_4$, 5 mM NaHCO$_3$, 20 mM HEPES, 11 mM glucose, 116 mM NaCl, 4.7 mM KCl, 1.2 mM MgSO$_4$, 2.5 mM CaCl$_2$, pH 7.4);
Cell culture supplies: 10 cm plates, 12- and 24-well dishes (BD Falcon or Corning).

2.2. Materials for HEK 293 cell transfection only

2.5 M CaCl$_2$, sterile;
2× HEPES-buffered saline (2× HBS) solution (0.28 M NaCl, 0.05 M HEPES, 1.5 mM Na$_2$HPO$_4$, pH 7.05), sterile;
Water, sterile;
DNA of interest: DNA of high quality is recommended, these are easily prepared using commercial kits from companies such as Sigma or Qiagen.

2.3. Materials for immunofluorescence assay only

Glass coverslips, round, 18 mm diameter, #1 (VWR, Cat No. CA48380-046);
Glass slides (VWR, Cat No. CA48311-950);
95% ethanol;
Collagen (type I from calf skin, 0.1% solution in 0.1 M acetic acid, 0.5 mg/ml, Sigma, Cat No. C819);
Blocking and antibody solution: 3% BSA-HBSS, filter-sterilized;
Anti-FLAG polyclonal antibody developed in rabbit, (Sigma, Cat No. F-7425) 1 mg prot/ml;
Fixing and permeabilizing solution (4% formaldehyde–0.2% Triton X-100 in HBSS);
Alexa Fluor 568 goat anti-rabbit IgG (H + L) (Invitrogen/Molecular Probes, Cat No. A-11011), 2 mg/ml;
Hoechst 33258 (Invitrogen, Cat No. H3569, 10 mg/ml) 1:50,000 in HBSS;
Shandon Immu-Mount^{TM} (Thermo Scientific, Cat No. 9990402).

2.4. Materials for flow cytometry assay only

Monoclonal anti-FLAG$^{®}$ M2 antibody produced in mouse (Sigma, Cat No. F3165), 4.8 mg prot/ml;
Anti-mouse IgG (Fc specific)–FITC antibody produced in goat (Sigma, Cat No. F4143), 3.4 mg prot/ml;
5 mM EDTA–2% BSA-PBS;

8% formaldehyde-PBS;
Plastic tubes, 12 × 75 mm (VWR, Cat No. 60818-383).

2.5. Materials for receptor radiolabeling assay only

Blocking and antibody solution: 10% FBS-1% BSA-HBSS, filter-sterilized;
Monoclonal anti-FLAG® M2 antibody produced in mouse (Sigma, Cat No. F3165), 4.8 mg prot/ml;
[^{125}I]-Goat anti-mouse IgG, Perkin-Elmer, 3.7 mBq/100 µCi;
0.2 N NaOH;
Plastic tubes, 12 × 75 mm, 5 ml (VWR, Cat No. CA60819-295).

2.6. Materials for inositol phosphate formation assay only

Dulbecco's Modified Eagle Medium (D-MEM, with 4500 mg/l D-glucose, but no L-glutamine or sodium pyruvate, Invitrogen, Cat No. 11960);
myo-[^{3}H]-inositol (Perkin-Elmer, Cat No. NET 114A), 1 mCi/ml (37 MBq/ml), sterile;
HBSS–10 mM LiCl;
0.8 M perchloric acid;
Neutralization buffer: 0.72 M KOH, 0.6 M KHCO$_3$;
Plastic tubes, 12 × 75 mm (VWR, Cat No. 60818-361);
AG1-X8 resin, formate form, 200–400 mesh (Bio-Rad, Cat No. 140-1454);
Poly-Prep polypropylene Chromatography Columns (9 cm high, 2 ml bed volume, Bio-Rad, Cat No. 731-1550);
Acrylic rack to hold columns;
60 mM ammonium formate;
Elution buffer: 0.1 M formic acid/1 M ammonium formate;
Wheaton plastic racks to hold plastic vials (VWR, Cat No. 66022-525);
Plastic scintillation vials, 20 ml (VWR, Cat No. 66022-263);
Ecolite scintillation cocktail (MP Biomedicals, VWR, Cat No. CAIC88247505).

3. METHODS

3.1. Common assays

3.1.1. Cell preparation and transfection

1. On day 1, seed HEK 293 cells onto 10 cm petri dishes in 10 ml of complete media (Section 2.1) and incubate at 37 °C and 5% CO$_2$. Seed at a density that will generate about 75% cell confluence on day 2. These cells should not have been passaged extensively, as this can result in abnormal cellular morphology and/or behavior.

2. On day 2, when the cells have reached approximately 75% confluence, transfect cells with the DNA constructs being examined. To do this, add 5 μg of each DNA construct to sterile 1.5 ml plastic tubes containing 0.45 ml of sterile water.
3. Next, add 0.05 ml of 2.5 M $CaCl_2$, and immediately after that add dropwise 0.5 ml of 2× HBS to each tube. Mix the transfection reagents right away by gently pipetting up and down 2–3 times.
4. Add the solution dropwise over the entire surface of each petri dish and incubate the cells for 18 h at 37 °C and 5% CO_2.
5. On the morning of day 3, wash cells twice with warm 1× PBS, add fresh media, and return cells to the incubator to permit recovery for at least 6 h. Depending on the exogenous DNA being expressed, cells can look unhealthy and display high cell death on day 3. If this is observed, we suggest titrating the amount of DNA used for transfection. For some DNA constructs, we have found as little as 0.5 μg is sufficient for many downstream assays. Also, to further assist in the recovery of stressed cells, instead of washing cells on the morning of day 3 with PBS, use warm HBSS. It is normal to find that all cells will "retract" a little or even "ball-up" after this wash, however, they will eventually spread out again.
6. Depending on the downstream assay that cells are being used for, transfected cells will be trypsinized and reseeded (see Section 3.1.2) onto glass coverslips in 12-well dishes if being used for confocal imaging (Section 3.2); 12-well dishes if being used for flow cytometry (Section 3.3) and radiolabeling (Section 3.4); and 24-well dishes if being used for IP formation (Section 3.5).
7. All experiments are conducted 40–44 h posttransfection.

3.1.2. Trypsinization and reseeding

1. Cells in the 10 cm petri dish are rinsed once with 10 ml of warm 1× PBS and then incubated in 1 ml of prewarmed trypsin–EDTA solution for no more than 2–3 min. The trypsin is then rapidly neutralized by the addition of 9 ml warm complete growth media (Section 2.1).
2. Cells are removed and pelleted at 1100 rpm (200 rcf) for 3 min at room temperature (RT).
3. Cell number is determined by counting and an appropriate number of cells are reseeded into new dishes. The number of cells reseeded is determined by the size of the dish or well the cells will be reseeded into. Cell numbers are suggested in the following sections.

3.1.3. Serum starvation

Immediately prior to any experiment, cells are serum starved to remove serum components that could affect the outcome of experiments. This is done by rinsing cells once with warm HBSS, followed by their incubation in HBSS (10 ml HBSS for a 10 cm plate) for 30 min at 37 °C and 5% CO_2.

3.2. Constitutive receptor internalization assay measured by immunofluorescence and confocal imaging

In this section, we provide a detailed protocol on how to assess constitutive receptor internalization by immunofluorescence analysis and confocal imaging of HEK 293 cells transiently expressing an amino-terminally tagged-receptor construct. In the example provided below, we describe the immunofluorescence detection of a FLAG-tagged receptor such as FLAG-GPR54.

To design this experiment, begin by calculating the amount of treatments/groups that will be studied. When doing so, remember to include appropriate controls, particularly if this is the first time you are testing a certain antibody or cell line. We recommend a minimum of three controls: an autofluorescence control, a negative control using epitope-tagged receptor transfected cells to assess nonspecific secondary antibody binding and a negative control using epitope-tagged empty vector transfected cells to assess nonspecific primary and secondary antibody binding. For a typical internalization assay, we examine internalization at 0, 2.5, 5, 10, and 15 min at 37 °C. We recommend using a different 12-well plate for each time point [see Section 3.2 (1)] and we normally run our controls only on our 0 min plate.

In the second negative control mentioned above instead of incubating cells in the primary antibody, cells are incubated in the antibody solution only. Although it is recommended to replace the antibody in this control with isotype-matched antibody (purified IgM or IgG of the same subclass as the antibody being tested), this option can prove very expensive particularly when examining a wide range of primary and secondary antibodies. We found that using the antibody solution only is sufficient to determine the level of background and nonspecific binding, and this is a less expensive and more practical option. Autofluorescence is defined as the fluorescence emission observed when certain cell molecules are excited by UV or visible radiation of suitable wavelength. This emission is an intrinsic property of cells and the majority of it originates in the mitochondria and the lysosomes (Monici, 2005). Since every cell line will have a different level of autofluorescence, it is important to become familiar with it and establish a baseline before starting an imaging experiment.

In the first negative control, instead of incubating cells in the primary antibody, they are incubated in the antibody solution only (or isotype matched antibody). This control assesses the level of nonspecific binding of the secondary antibody employed in the assay. The negative control should not be imaged first, simply because there is always a level of nonspecific binding and most cell lines exhibit autofluorescence. Therefore, to determine the level of nonspecific background, image the "positive" sample first, and then use the same "imaging settings" to image the negative control. In the second negative control, epitope-tagged empty vector transfected cells are incubated with both primary and secondary antibodies.

1. [This step continues from Section 3.1.1 (5).] Approximately 24 h posttransfection, transfected cells must be reseeded from the 10 cm petri dish onto glass coverslips in 12-well dishes for staining and imaging. To do this, begin by preparing the coverslips that the cells will be reseeded onto. Carefully insert a coverslip into each well of a 12-well dish, add 1 ml of ethanol; let coverslips stand for at least 1 min and then aspirate. Next, rinse the coverslips twice with 1 ml of PBS to remove traces of ethanol.
2. To support the attachment and growth of HEK 293 cells, the glass coverslips are coated with type I collagen. To do this, add 1 ml of collagen per well and let it stand for at least 1 min. Next, remove traces of the acetic acid that the collagen is dissolved in by carefully rinsing the coverslips once with excess PBS.
3. Transfected cells in the 10 cm dish [Section 3.1.1 (5)] are trypsinized, counted (Section 3.1.2) and $1–1.5 \times 10^5$ cells reseeded in complete growth media onto each coverslip. When studying the localization of a receptor in a cell by fluorescence microscopy, it is best to image samples where the cells are not too confluent and where you can clearly distinguish the plasma membrane and the processes of most cells.
4. Approximately 16–20 h after culturing cells on coverslips, cells are ready for studying receptor internalization.
5. Begin by serum starving cells as described in Section 3.1.3.
6. During step 5, place an aliquot of HBSS and 3% BSA-HBSS on ice, also place an aliquot of HBSS at 37 °C. Unless otherwise stated, in the following steps, all solutions used must be ice-cold and parts of the experiment are conducted on ice. Remember, these are agonist-independent experiments and it is the increase in temperature that triggers the measurable biological response, in this case receptor internalization. This cannot be overemphasized, especially in the case of a receptor like GPR54 where constitutive internalization is extremely rapid at 37 °C.
7. After serum-starvation, place cells on ice and rinse twice with 1 ml HBSS; incubate further on ice in 0.5 ml 3% BSA-HBSS blocking solution for 30 min.
8. To detect the FLAG-tagged receptor, replace the blocking solution with 350 µl of primary antibody for 45–60 min on ice. This method allows you to label FLAG-tagged cell surface receptors present at the plasma membrane and to subsequently follow those molecules as they undergo constitutive internalization at 37 °C. We use a polyclonal anti-FLAG rabbit antibody diluted 1:500 in 3% BSA-HBSS. If you are testing an antibody for the first time, it must be titrated to determine

the antibody dilution that allows for the strongest specific signal with the lowest amount of background under your experimental conditions. Include the suggested controls when performing the titration (as described in Section 3.2). If you would like to reduce the volume of antibody used in this staining procedure, you can incubate the coverslips in 60 µl of antibody solution and perform the incubation overnight at 4 °C. To do this, pipette 60 µl of the antibody solution onto a piece of parafilm and place the coverslip on top of the antibody droplet thereby allowing full antibody contact with the cells. Assemble a "wet chamber" by placing the parafilm sheet on a stack of wet filter paper or paper towels inside an air tight box with a flat bottom. Place the box in the fridge overnight. The next day, return coverslips (with cells facing up) to 12-well dishes.

9. Remove the primary antibody by gently but carefully washing the cells three times with 1 ml of HBSS; incubate on ice in 1 ml HBSS.
10. Except for the 0 min plate which stays on ice, immediately transfer the other plates to 37 °C to permit internalization. To rapidly warm up the cells, we gently float the 12-well dishes, with the lids on, in a 37 °C water bath. If you have selected time points 5 min and shorter, we recommend quickly replacing the cold HBSS that cells are incubated in with warm HBSS before starting the internalization assay.
11. Once a plate has been at 37 °C for the amount of time selected, return it to the ice and immediately replace the warm HBSS with ice cold HBSS, immediately replace this HBSS with another aliquot of ice cold HBSS.
12. The cells are now ready for fixing and permeabilizing. To do this, incubate cells in 1 ml/well of the fixing and permeabilizing solution for 20 min at RT. We do not recommend using cold methanol for this procedure because this treatment does not fix cells, instead it precipitates the cellular proteins and protein–protein interactions can be affected (Wouters and Bastiaens, 2006). Permeabilizing with Triton X-100 will extract cellular membranes without disturbing protein–protein interactions and allow access of cell-impermeable fluorescent probes to intracellular or intraorganellar antigens. However, because of the nonselective nature of this detergent, proteins as well as lipids are extracted so it is important to determine the lowest percentage of detergent necessary to perform this step (Jamur and Oliver, 2009). While some protocols separate the fixation and the permeabilization as two different steps, we have found that combining them does not affect the results and saves us time. Once the cells have been fixed, the following steps are all done at RT.
13. Next, rinse cells three times with 1 ml of HBSS and incubate in 0.5 ml 3% BSA-HBSS for 30 min at RT.

14. Replace blocking buffer with 350 μl of secondary antibody diluted in 3% BSA-HBSS and incubate cells for 45–60 min at RT. We use a goat anti-rabbit Alexa Fluor 568 antibody at a dilution of 1:1200. Remember that the cells in the autofluorescence control well should be incubated with 3% BSA-HBSS. Since the secondary antibody is coupled to a fluorophore, it is important to protect the samples from light from this step onward. Remember, only cell surface receptors that were labeled on ice with the primary antibody will be detected here. This method will not detect FLAG-tagged receptors that were present in the cytoplasm prior to constitutive internalization at 37 °C.
15. Remove the secondary antibody by gently but carefully washing the cells three times with 1 ml of HBSS.
16. Counterstain nuclei with the fluorescent Hoechst dye at a dilution of 1:50,000 (in HBSS). Cells are incubated in 0.5 ml of Hoechst dye for 7 min at RT.
17. Remove excess Hoechst dye by washing the cells three times with HBSS.
18. Finally, mount the coverslips on slides using a drop of Shandon Immu-Mount^{TM}. This is a water-based mounting medium that does not exhibit background fluorescence nor does it cause quenching of fluorescent emission.
19. Let the samples air dry in the dark at RT overnight before imaging. All images (Fig. 4.1) presented in this study were acquired with a Fluoview 1000 laser scanning confocal microscope (Olympus Corp), using the x60 Plan Apochromat 1.42 oil objective. Studies were performed using multiple excitation (405, 559) and emission (band pass 425–475 and 575–675 nm for Hoechst and AlexaFluor 568, respectively) filter sets. Multicolor images were acquired in the sequential acquisition mode to avoid cross-excitation.

3.3. Constitutive receptor internalization assay measured by flow cytometry

Constitutive receptor internalization can be quantified by flow cytometry by measuring the fluorescently tagged receptors still present at the plasma membrane following receptor internalization at 37 °C. An essential control for this assay is nontransfected (NT) cells (included in your 0 min plate) and these are incubated with the same antibodies as your receptor-transfected cells. As previously described for the immunofluorescence protocol (Section 3.2), cells are reseeded into collagen-coated 12-well dishes 24 h after transfection; however, for this assay cells are seeded directly on the plastic well. Again, we examine internalization at 0, 2.5, 5, 10, and 15 min at 37 °C and we use different 12-well plates for different time points.

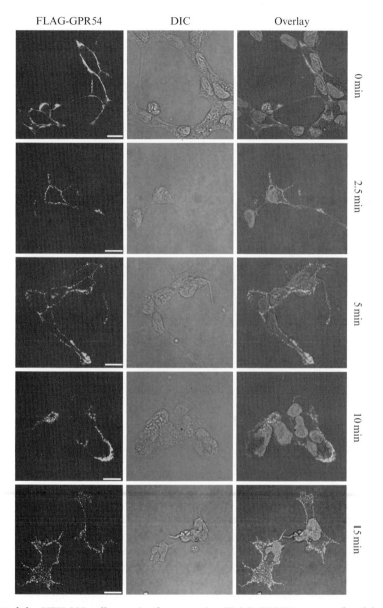

Figure 4.1 HEK 293 cells transiently expressing FLAG-GPR54 were surface labeled at 4 °C (0 min) using rabbit anti-FLAG antibody and then were incubated at 37 °C for 2.5, 5, 10, and 15 min. After fixation, surface GPR54 was detected by Alexa Fluor 568-conjugated anti-rabbit IgG. Images are representative of four independent experiments. Scale bars: 10 μm. DIC, differential interference contrast. (See Color Insert.)

Before starting this experiment, prepare the following: 5 mM EDTA–2% BSA-PBS and 8% formaldehyde-PBS.

1. Begin by serum starving HEK 293 cells transiently expressing the FLAG-tagged receptor as described in Section 3.1.3.
2. During step 1, place an aliquot of HBSS on ice; also place an aliquot of HBSS at 37 °C. Unless otherwise stated, in the following steps all solutions used must be ice-cold [see note in Section 3.2 (6)].
3. After serum starvation, place the cells on ice, rinse them once with 1 ml of HBSS and incubate them with 1 ml of HBSS for 10 min on ice.
4. To label cell surface receptors, incubate the cells with 350 µl/well of Monoclonal anti-FLAG® M2 antibody produced in mouse diluted in HBSS for 45 min on ice. In these internalization experiments, we use this antibody at a dilution of 1:500.
5. Remove the primary antibody by gently but carefully washing the cells three times with 1 ml of HBSS; incubate on ice in 1 ml HBSS. To allow the receptor molecules to internalize, warm the plates to 37 °C as stated in Section 3.2 (10) and (11).
6. Incubate the cells with 350 µl/well of anti-mouse IgG (Fc specific)-FITC antibody produced in goat for 45 min on ice. From this step onward care should be taken to protect the samples from exposure to the light.
7. After rinsing each well three times with 1 ml of HBSS, incubate the cells with 0.4 ml of 5 mM EDTA–2% BSA-PBS for 5 min. Both the presence of EDTA, which is a chelating agent, and the switch to PBS, which lacks calcium and magnesium, facilitate lifting the cells off the plastic. Gently pipet up and down to lift the cells and transfer the whole volume of each tube to plastic tubes containing 100 µl of 8% formaldehyde-PBS. Cover the tubes with parafilm and tin foil, and keep them at 4 °C until they are ready to be measured by flow cytometry.
8. Under the conditions detailed in this protocol, receptor molecules present at the plasma membrane are labeled with the primary antibody and then they are internalized during the incubation at 37 °C. Measuring the samples by flow cytometry will determine the amount of receptor molecules still present at the plasma membrane (and therefore available to be labeled with the FITC-conjugated secondary antibody) after the incubation at 37 °C. Therefore, receptor internalization is defined as the % loss of cell surface immunofluoresence (compared to cells kept on ice, that is, the 0 min plate).

Sample calculation:

$$100-[(MF\ RTC - MF\ NTC)/(MF\ RTC\ 0 - MF\ NTC) \times 100]$$
$$= \%\ internalization$$

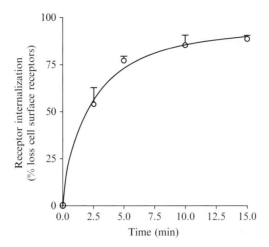

Figure 4.2 Time course for GPR54 basal or constitutive internalization. The data represent the mean ± SE of seven independent experiments. HEK 293 cells transiently expressing GPR54 were surface labeled at 4 °C using mouse anti-FLAG antibody. Cells were left untreated at 37 °C for the indicated times, receptor molecules remaining at the plasma membrane were labeled with anti-mouse FITC antibody and cells were fixed. Internalization was calculated as the percentage of loss of cell surface immunofluorescence over time and measured by flow cytometry.

where MF RTC, mean fluorescence of receptor transfected cells; MF NTC, mean fluorescence of NT cells; MF RTC 0, mean fluorescence of receptor transfected cells at 0 min.

Figure 4.2 shows the constitutive internalization of FLAG-GPR54 as measured by flow cytometry.

3.4. Constitutive receptor internalization assay measured by indirect receptor radiolabeling

Another quantitative method to study receptor internalization employs a secondary goat anti-mouse IgG antibody labeled with [^{125}I]. This protocol is modified after Pawson et al. (2008) and Hu et al. (2001). HEK 293 cells transiently expressing FLAG-tagged receptor are trypsinized, counted and 1–1.5 × 10^5 cells reseeded directly into 12-well plates as described earlier (Sections 3.1.1 and 3.1.2). Once again, remember to use different plates for the different time points selected in this assay and include the NT cells control on your 0 min plate (Section 3.2). We usually include two additional controls, one with NT cells and another one with receptor transfected cells, both of which are incubated with the antibody solution instead of with the primary antibody. These controls allow us to monitor nonspecific binding of the iodine-labeled secondary antibody. Before starting this experiment, you will need to prepare the following: an aliquot of warm HBSS and also ice-cold aliquots of the blocking and antibody solution (10% FBS-1% BSA-HBSS) and HBSS.

1. Begin by serum starving HEK 293 cells transiently expressing the FLAG-tagged receptor as described in Section 3.1.3. In the following steps, unless otherwise stated, the cells are kept on ice and all solutions used are ice-cold [see note in Section 3.2 (6)].
2. Put the plates on ice, rinse each well twice with 1 ml HBSS, and block nonspecific binding sites by incubating the cells with 0.5 ml of 10% FBS-1% BSA-HBSS for 30 min.
3. Replace blocking buffer with 350 µl/well of Monoclonal anti-FLAG® M2 antibody produced in mouse for 45–60 min.
4. Repeat steps 9–11 as described in Section 3.2.
5. Repeat the blocking step (described above in Section 3.4 (2)) on ice for 30 min.
6. Replace blocking solution in each well with 350 µl of [^{125}I]-Goat anti-mouse IgG for 1 h on ice (3 µCi/ml).
7. Next, wash cells four times with 1 ml of ice cold HBSS.
8. Add 250 µl of 0.2 N NaOH to each well and incubate the cells at RT for 15 min.
9. Finally, using a pipette tip, scrape each well and transfer the contents to plastic tubes and count the radioactivity in a gamma counter.

To quantify your results and determine the level of receptor internalization, use the formulas described in Section 3.3 (8). Figure 4.3 shows the constitutive internalization of FLAG-GPR54 as measured by indirect radiolabeling of FLAG-GPR54. The data obtained using this technique is similar to that using flow cytometry.

3.5. Assessing constitutive receptor activity by measuring inositol phosphate turnover in intact cells

A measure of constitutive activity in a GPCR coupled to Gq would be the formation of IP in the absence of agonist stimulation. To assess basal IP formation, we use a protocol modified after Shears (1997). HEK 293 cells, transiently expressing the receptor construct are used in this assay. Note, while we do not describe the following, coexpressing a second protein together with the receptor will allow you to determine the effect of the second protein on receptor constitutive activity. We recommend including a NT control, which can serve as an interassay control and also allows you to determine basal IP formation by endogenous receptors. For this assay, trypsinize, count and reseed the cells onto 24-well plates at a density of 10^5 cells/ml of complete growth media (as described earlier in Sections 3.1.1 and 3.1.2) 24 h after transfection.

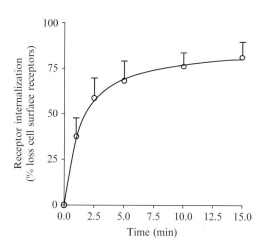

Figure 4.3 Time course for GPR54 basal or constitutive internalization. The data represent the mean ± SE of six independent experiments. HEK 293 cells transiently expressing GPR54 were surface labeled at 4 °C using mouse anti-FLAG antibody. Cells were left untreated at 37 °C for the indicated times, receptor molecules remaining at the plasma membrane were labeled with [^{125}I] anti-mouse antibody and cells were lysed and counted on a gamma counter. Internalization was calculated as the percentage of loss of cell surface activity over time.

The following day, before starting this experiment, prepare warm aliquots of HBSS and DMEM (no supplements are added) and prepare the inositol solution by diluting myo-[^{3}H]-inositol in the media to a concentration of 1 µCi inositol/ml of media.

1. Begin by washing the cells once with 1 ml of warm HBSS.
2. Next, incubate them overnight with 0.4 ml of the inositol solution at 37 °C for 16 h. Before going to the next step, prepare a solution of 10 mM LiCl–HBSS, and warm up aliquots of LiCl–HBSS and HBSS.
3. Wash the cells twice with 0.5 ml of warm HBSS and then incubate the cells in 0.5 ml of warm HBSS at 37 °C and 5% CO_2 for 1 h.
4. Wash the cells once with HBSS–10 mM LiCl and incubate them in 0.5 ml HBSS–10 M LiCl at 37 °C for the period of time selected. In our experiment, cells were incubated for 4 h to determine basal IP formation by GPR54. The phosphoinositides formed as a result of phospholipase C activation are rapidly metabolized, and the metabolites are dephosphorylated and some are hydrolyzed to free inositol. Since this reaction would represent a loss of products from this assay, the use of lithium is recommended to inhibit the phosphatases and reduce the metabolic flux from IP to free inositol (Shears, 1997; Skippen et al., 2006).

5. Without aspirating the HBSS–10 mM LiCl from the wells, stop the reaction by placing the cells on ice and adding 0.5 ml of 0.8 M perchloric acid/well to each well. Leave the cells undisturbed on ice for 30 min.
6. Gently transfer 0.8 ml of cell lysate (without removing cellular debris) into a plastic tube containing 0.4 ml of neutralization buffer.
7. Store the tubes on ice for at least 2 h or, preferably, overnight at 4 °C, to facilitate the precipitation of the poorly soluble potassium perchlorate. During this time, the tubes should be loosely covered to allow the release of CO_2. The addition of neutralization buffer to the tubes will ensure the neutralization of the acid-quenched samples and prevent the acid from reacting with the resin used for the anion exchange chromatography.
8. To perform the anion exchange chromatography, first calculate the amount of AG1-X8 resin (see Section 2.6) required for the experiment by taking into account that 0.4 g resin/column/4 ml water is required. Rinse the resin twice with distilled water by letting the resin settle before discarding the water each time and then take the resin to the required volume with distilled water. To load the resin onto the columns, first add some water to each column to prevent the formation of air pockets in the column support as the resin settles. Next, add 4 ml of the diluted resin into the column. In this assay, the columns are gravity-fed. Once the resin has settled in the columns, check that they all have the same amount of resin, as differences in the height column size will affect the results.
9. Next, load 0.9 ml of neutralized cell extract (from step 6) onto the resin bed immediately followed by a wash with 5 ml of water.
10. Repeat the washing step to remove free [^{3}H]-inositol. The free [^{3}H]-inositol must be completely removed, because it is present in such large quantities relative to the levels of [^{3}H]-IP.
11. Wash the columns twice with 5 ml of freshly prepared 60 mM ammonium formate to elute [3H] glycerophosphoinositol.
12. Finally, place the columns over 20 ml plastic vials, each already containing 15 ml of scintillation cocktail and elute phosphorylated inositols ([3H] InsP1, [3H] InsP2, [3H] InsP3 and [3H] InsP4) with 4 ml of freshly prepared elution buffer. Normally, [3H] InsP5 and [3H] InsP6 will not be eluted from the column under these conditions. We set up our columns on an acrylic rack and then place this rack on a plastic rack holding the 20 ml-plastic vials to elute our columns.
13. Remember to shake the vials before counting to mix the eluate with the scintillation cocktail and count the vials on a beta counter.
14. To normalize the total IPs produced in each well, transfer 50 µl of the neutralized cell extract (from step 6) to 20 ml plastic vials, add 5 ml of scintillation cocktail and count on a beta counter. The activity in the

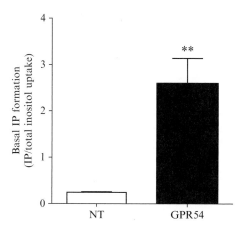

Figure 4.4 Basal IP formation in nontransfected (NT) versus FLAG-GPR54-expressing HEK 293 cells. Data are expressed as the fraction of the incorporated [^{3}H]-inositol converted to [^{3}H]-IP in the absence of agonist stimulation in NT HEK 293 cells and in cells expressing FLAG-GPR54. The data represent the mean ± SE for eight independent experiments. **$p < 0.001$ versus NT.

50 μl aliquot will depend on the amount of [^{3}H] inositol incorporated in the cells in that well, and thus serves as an indirect measure of the number of cells in that well. Although we initially seed the same amount of cells in each well, expression of some DNA constructs can reduce cell adhesion causing cell loss while conducting the assay. Therefore, it is important to normalize the data obtained from the anion exchange chromatography by counting the 50 μl aliquots of the neutralized cell extracts.

Sample calculation:

800 μl of extract + 400 μl of neutralization buffer = 1.2 ml of neutralized cell extract;
1.2 ml of neutralized cell extract: 900 μl to column and 50 μl to vials;
% conversion of [^{3}H]-inositol to [^{3}H]-IP;
=[(cpm from column) (1.2/0.9)]/[(cpm in 50 μl) (1.2/0.05)] × 100;

Figure 4.4 shows the constitutive activity of GPR54 as determined by measuring the agonist-independent formation of IP using the method described above.

ACKNOWLEDGMENT

Research reported in this study was supported by a grant from the Canadian Institutes of Health Research (CIHR) MOP 81383.

REFERENCES

Bhattacharya, M., Babwah, A. V., Godin, C., Anborgh, P. H., Dale, L. B., Poulter, M. O., and Ferguson, S. S. (2004). Ral and phospholipase D2-dependent pathway for constitutive metabotropic glutamate receptor endocytosis. *J. Neurosci.* **24,** 8752–8761.

de Roux, N., Genin, E., Carel, J. C., Matsuda, F., Chaussain, J. L., and Milgrom, E. (2003). Hypogonadotropic hypogonadism due to loss of function of the KiSS1-derived peptide receptor GPR54. *Proc. Natl. Acad. Sci. USA* **100,** 10972–10976.

Fourgeaud, L., Bessis, A. S., Rossignol, F., Pin, J. P., Olivo-Marin, J. C., and Hémar, A. (2003). The metabotropic glutamate receptor mGluR5 is endocytosed by a clathrin-independent pathway. *J. Biol. Chem.* **278,** 12222–12230.

Holliday, N. D., Holst, B., Rodionova, E. A., Schwartz, T. W., and Cox, H. M. (2007). Importance of constitutive activity and arrestin-independent mechanisms for intracellular trafficking of the ghrelin receptor. *Mol. Endocrinol.* **21,** 3100–3112.

Hu, W., Howard, M., and Lukacs, G. L. (2001). Multiple endocytic signals in the C-terminal tail of the cystic fibrosis transmembrane conductance regulator. *Biochem. J.* **354,** 561–572.

Jamur, M. C., and Oliver, C. (2009). Permeabilization of cell membranes immunocytochemical methods and protocols. *Methods Mol. Biol.* **588,** 63–66.

Kotani, M., Detheux, M., Vandenbogaerde, A., Communi, D., Vanderwinden, J. M., Le Poul, E., Brézillon, S., Tyldesley, R., Suarez-Huerta, N., Vandeput, F., Blanpain, C., Schiffmann, S. N., et al. (2001). The metastasis suppressor gene KiSS-1 encodes kisspeptins, the natural ligands of the orphan G protein-coupled receptor GPR54. *J. Biol. Chem.* **276,** 34631–34636.

Lapatto, R., Pallais, J. C., Zhang, D., Chan, Y. M., Mahan, A., Cerrato, F., Le, W. W., Hoffman, G. E., and Seminara, S. B. (2007). Kiss1−/− mice exhibit more variable hypogonadism than Gpr54−/− mice. *Endocrinology* **148,** 4927–4936.

Monici, M. (2005). Cell and tissue autofluorescence research and diagnostic applications. *Biotechnol. Annu. Rev.* **11,** 227–256.

Pampillo, M., Camuso, N., Taylor, J. E., Szereszewski, J. M., Ahow, M., Zajac, M., Millar, R. P., Bhattacharya, M., and Babwah, A. V. (2009). Molecular regulation of GPR54 activity by GRK-2 and β-arrestins. *Mol. Endocrinol.* **13,** 2060–2074.

Pawson, A. J., Faccenda, E., Maudsley, S., Lu, Z. L., Naor, Z., and Millar, R. P. (2008). Mammalian type I gonadotropin-releasing hormone receptors undergo slow, constitutive, agonist-independent internalization. *Endocrinology* **149,** 1415–1422.

Pula, G., Mundell, S. J., Roberts, P. J., and Kelly, E. (2004). Agonist-independent internalization of metabotropic glutamate receptor 1a is arrestin- and clathrin-dependent and is suppressed by receptor inverse agonists. *J. Neurochem.* **89,** 1009–1020.

Sallese, M., Salvatore, L., D'Urbano, E., Sala, G., Storto, M., Launey, T., Nicoletti, F., Knöpfel, T., and De Blasi, A. (2000). The G-protein-coupled receptor kinase GRK4 mediates homologous desensitization of metabotropic glutamate receptor 1. *FASEB J.* **14,** 2569–2580.

Seminara, S. B., Messager, S., Chatzidaki, E. E., Thresher, R. R., Acierno, J. S., Jr., Shagoury, J. K., Bo-Abbas, Y., Kuohung, W., Schwinof, K. M., Hendrick, A. G., Zahn, D., Dixon, J., et al. (2003). The GPR54 gene as a regulator of puberty. *N. Engl. J. Med.* **349,** 1614–1627.

Shears, S. B. (1997). Measurement of inositol phosphate turnover in intact cells and cell-free systems. *In* "Signalling by Inositides: A Practical Approach," (S. B. Shears and S. B. Shears, eds.), pp. 33–52. Oxford University Press, Oxford.

Skippen, A., Swigart, P., and Cockcroft, S. (2006). Measurement of phospholipase C by monitoring inositol phosphates using [3H]inositol-labeling protocols in permeabilized cells. *Methods Mol. Biol.* **312,** 183–193.

Smit, M. J., Vischer, H. F., Bakker, R. A., Jongejan, A., Timmerman, H., Pardo, L., and Leurs, R. (2007). Pharmacogenomic and structural analysis of constitutive G protein-coupled receptor activity. *Annu. Rev. Pharmacol. Toxicol.* **47,** 53–87.

Wouters, F. S., and Bastiaens, P. I. H. (2006). Imaging protein–protein interactions by fluorescence resonance energy transfer (FRET) microscopy. *Curr. Protoc. Neurosci.* 5.22.1–5.22.15.

CHAPTER FIVE

Assessment of Constitutive Activity in E-Type Prostanoid Receptors

Hiromichi Fujino,* Toshihiko Murayama,* and John W. Regan[†]

Contents

1. Introduction	96
2. Assays Used to Assess EP3 Receptor Constitutive Activity	97
2.1. cAMP assay	97
2.2. GTPase activity and GTPγS binding assays	99
2.3. GTPase activity assay	100
2.4. GTPγS binding assay	102
2.5. Use of the Gi inhibitor pertussis toxin	102
2.6. Determination of actin stress fiber formation by staining with TRITC-phalloidin	103
3. Assays Used to Assess EP4 Receptor Constitutive Activity	105
3.1. Tcf/β-Catenin luciferase reporter plasmid assay	105
References	107

Abstract

The potential for G-protein-coupled receptors (GPCRs) to show constitutive activity is emerging as one of the fundamental properties of GPCRs signal transduction. Indeed, of the four subtypes of E-type prostanoid (EP) receptors, the EP3 and EP4 subtypes show constitutive activity in addition to their innate ligand-dependent activation of signaling pathways. The constitutive activity of the EP3 and EP4 receptor subtypes was discovered during the initial characterizations of these receptors and may be important for setting the basal level of cellular tone in the given signaling pathway. This chapter introduces some of the methods that can be used to study the constitutive activity of the EP receptors.

---

* Laboratory of Chemical Pharmacology, Graduate School of Pharmaceutical Sciences, Chiba University, Chiba, Japan
[†] Department of Pharmacology and Toxicology, College of Pharmacy, The University of Arizona, Arizona, USA

1. INTRODUCTION

Prostanoids regulate a wide variety of physiological processes, including smooth muscle contraction, inflammation, pain, and fever. Five primary prostanoid metabolites are produced by various synthases following the initial actions of cyclooxygenase (COX) on arachidonic acid. These five metabolites are prostaglandin E_2 (PGE_2), PGD_2, $PGF_{2\alpha}$, PGI_2, and thromboxane A_2 (TXA_2) (Coleman et al., 1994). Prostanoids mediate their physiological effects by interacting with the prostanoid receptors, which are members of the rhodopsin-type subgroup of G-protein-coupled receptors (GPCRs). Pharmacologically, the receptors for these prostanoids are classified according to their endogenous metabolites. Therefore, PGE_2 activates E-type prostanoid (EP) receptors, PGD_2 activates D-type prostanoid (DP) receptors, $PGF_{2\alpha}$ activates F-type prostanoid (FP) receptors, PGI_2 activates I-type prostanoid (IP) receptors, and TXA_2 activates T-type prostanoid (TP) receptors (Coleman et al., 1994). The EP receptors have been further divided into four subtypes, EP1, EP2, EP3, and EP4, each the product of a separate gene (Regan, 2003; Sugimoto and Narumiya, 2007). EP1 receptors were traditionally thought to be coupled to the activation of phospholipase C via Gq, but have recently been shown to activate a phosphatidylinositol-3 kinase (PI3K) signaling pathway via additional coupling to Gi (Ji et al., 2010). Human EP3 receptors are subdivided into at least eight isoforms generated by alternative mRNA splicing of their carboxyl-terminal tails (Kotani et al., 1997). The predominant EP3 receptor signaling pathway is considered to be inhibition of adenylate cyclase via Gi. However, each EP3 receptor isoform has the ability to couple with additional G-proteins, including Gs, Gq, and $G_{12/13}$. EP2 and EP4 receptors were traditionally thought to couple with Gs to stimulate adenylate cyclase; however, it has since been found that the EP4 subtype has additional coupling to Gi and can activate PI3K signaling pathways (Fujino and Regan, 2006).

The mouse EP3α receptor isoform and later the EP3γ receptor isoform were found to have marked agonist-independent constitutive activity involving coupling to Gi, resulting in the inhibition of basal adenylate cyclase activity (Hasegawa et al., 1996; Negishi et al., 1996). A similar constitutive activity involving coupling to Gi was then found for the human $EP3_{III}$ and $EP3_{IV}$ receptor isoforms (Jin et al., 1997). Furthermore, the human $EP3_{III}$ and $EP3_{IV}$ receptor isoforms were found to be expressed intracellularly as well as on the cell surface, probably reflecting receptor internalization as a consequence of their constitutive activity (Bilson et al., 2004). In addition to constitutive coupling to Gi, the mouse EP3α receptor isoform also shows constitutive coupling to $G_{12/13}$ resulting in the agonist-independent activation of Rho and formation of actin stress fibers (Hasegawa et al., 1997).

So far, there is no correlation among the orthologous EP3 receptor isoforms in terms of their constitutive activity. Thus, the mouse EP3α and human EP3$_I$ isoforms are orthologs, as are the mouse EP3γ and human EP3$_{II}$; however, the two mouse isoforms, the EP3α and EP3γ, show constitutive activity, whereas, their human orthologs, the EP3$_I$ and EP3$_{II}$, do not. Interestingly, the two human isoforms that do show constitutive activity, the EP3$_{III}$ and EP3$_{IV}$, have relatively short C-tails like the constitutively active mouse EP3α and EP3γ isoforms; whereas the human EP3$_{III}$ and EP3$_{IV}$ isoforms and the mouse EP3β isoform have relatively longer C-tails and do not show constitutive activity. This has lead to a proposal that the longer C-tail of the EP3β isoform serves to inhibit constitutive activity. Indeed, truncated EP3β receptors showed constitutive activity in inverse relation to the length of their C-tail (Hizaki et al., 1997).

As noted above, the EP2 and EP4 receptor subtypes couple predominantly to Gs, although the EP4 receptor subtype has significant coupling to Gi and can activate PI3K signaling pathways (Fujino and Regan, 2006). Both of these receptor subtypes were found to stimulate T-cell factor (Tcf)/β-catenin signaling, an important pathway known to be involved in the development of colon cancer (Fujino and Regan, 2003). This stimulation of Tcf/β-catenin signaling involved the phosphorylation mediated inhibition of glycogen synthase kinase 3 (GSK-3), resulting in the stabilization of β-catenin expression and activation of Tcf/β-catenin transcriptional activity (Fujino et al., 2002). Notably, the EP4 receptor showed constitutive agonist-independent activation of Tcf/β-catenin signaling as compared with the EP2 receptor or control HEK cells, although both receptors were neutral with respect to their constitutive activation of cAMP formation. It was found that the stimulation of Tcf/β-catenin transcriptional activity by the EP2 receptor primarily involves Gs coupling and activation of protein kinase A (PKA); whereas, the stimulation of Tcf/β-catenin transcriptional activity by the EP4 receptor primarily involves Gi coupling and activation of PI3K. Constitutive agonist-independent activity of the EP4 receptor was also noted in HT-29 human colon cancer cells expressing recombinant human EP4 receptors (Hawcroft et al., 2007).

2. Assays Used to Assess EP3 Receptor Constitutive Activity

2.1. cAMP assay

The constitutive activity of Gi-coupled EP3 receptors was first identified in 1996 in a forskolin-stimulated cyclic adenosine monophosphate (cAMP) assay (Hasegawa et al., 1996). Forskolin, a naturally occurring diterpene, is a direct activator of adenylyl cyclase and is used to stimulate intracellular cAMP formation in the presence of isobutylmethylxanthine (IBMX), which inhibits phosphodiesterase and prevents the breakdown of cAMP (Fig. 5.1A).

In the variation of this assay described here, cAMP in the cell lysate is measured by competitive displacement of radiolabeled [^{3}H]cAMP binding to PKA. Bound [^{3}H]cAMP is separated from the free [^{3}H]cAMP by adsorption to powdered charcoal. Typically this assay is performed under

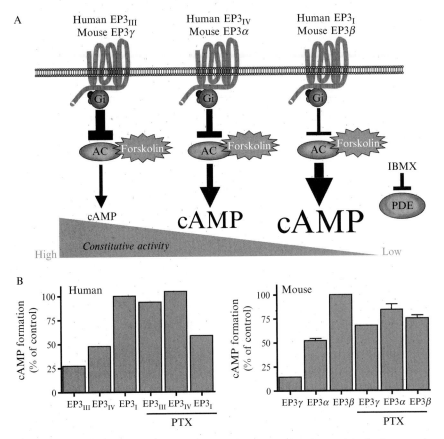

Figure 5.1 Forskolin-stimulated cAMP assay. (A) Graphic illustrating the increase in basal forskolin-stimulated cAMP formation with decreasing constitutive activity of the EP3 receptor isoforms. AC, adenylate cyclase; PDE, phosphodiesterase; IBMX, isobutylmethylxanthine. (B) Bar graphs showing the increase in basal forskolin-stimulated cAMP formation with decreasing constitutive activity of the EP3 receptor isoforms and the effect of pertussis toxin (PTX) pretreatment. Cells were treated with 20 μM (human) or 10 μM (mouse) forskolin for 10 min at 37 °C in the presence of the phosphodiesterase inhibitor IBMX. For PTX experiments, cells were pretreated with either 100 ng/ml PTX for 6 h (human) or 10 ng/ml PTX for 5 h (mouse) at 37 °C. Data are from Jin et al. (1997) and Negishi et al. (1996) and are regraphed and normalized to the forskolin-treated EP3$_I$ isoform (human) or EP3β isoform (mouse) as 100%.

basal conditions or following pretreatment of the cells with the Gi inhibitor, pertussis toxin (PTX). If there is significant Gi-coupled constitutive activity (e.g., human EP3$_{III}$), basal forskolin-stimulated cAMP formation will be less as compared with isoforms with less constitutive activity (e.g., human EP3$_I$, Fig. 5.1B). On the other hand, in cells that have been pretreated with PTX, isoforms with significant Gi-coupled constitutive activity (e.g., human EP3$_{III}$) will show an increase in basal forskolin-stimulated cAMP formation (Fig. 5.1B).

2.1.1. Materials

IBMX (product #I7018), forskolin (F6886), PKA from bovine heart (P5511), activated charcoal (C7606), and cAMP (A9501) are from Sigma-Aldrich (St. Louis, MO). [2,8-^{3}H]-cAMP 1 mCi/ml (37 MBq/ml) (NET-275) is from PerkinElmer (Boston, MA).

1. Cells are cultured in 10-cm plates and washed once with fresh culture medium such as Dulbecco's modified Eagles's medium, containing 0.1 mg/ml of IBMX for 20 min.
2. Cells are then treated with 3 μM forskolin for 10 min at 37 °C in culture medium containing IBMX, after which the medium is removed and the cells are placed on ice.
3. One milliliter of TE buffer (50 mM Tris–HCl, 4 mM EDTA (pH 7.5)) is added, and the cells are scraped off and transferred to microcentrifuge tubes.
4. The samples are boiled for 8 min, placed on ice, and centrifuged for 1 min at 14,000 rpm.
5. Fifty microliters of the supernatant is added to new tubes containing 50 μl of 9 pmol of [^{3}H]cAMP and 100 μl of 0.06 mg/ml PKA.
6. The mixture is vortexed and incubated on ice for 2 h, followed by the addition of 100 μl of TE buffer containing 2% bovine serum albumin (BSA) and 26 mg/ml of powdered charcoal.
7. After vortexing and centrifugation for 1 min at 14,000 rpm, 100 μl aliquots of the supernatant are removed for liquid scintillation counting
8. The amount of cAMP present is calculated from a standard curve prepared using nonradioactive cAMP (0–64 pmol).

2.2. GTPase activity and GTPγS binding assays

In addition to the forskolin-stimulated cAMP assay, the constitutive activity of Gi-coupled EP3 receptors can be assessed either by measuring the guanosine triphosphate (GTP) hydrolase (GTPase) activity or the GTP binding activity of Gi α-subunit. Upon formation of the active state of a GPCRs, whether it is through agonist binding or constitutive agonist-independent isomerization, there is an interaction with the corresponding heterotrimeric

G-protein resulting in the release of bound guanosine diphosphate (GDP) by the α-subunit, followed by the binding of GTP. This active state of the GTP bound α-subunit is terminated by the hydrolysis of GTP to GDP and release of inorganic phosphate (Pi). Thus, GPCRs constitutive activity can potentially be measured either by the binding of a radiolabeled hydrolysis-resistant analog of GTP known as [^{35}S]GTPγS, or by the hydrolysis of the terminal γ-phosphate of radiolabeled [γ-^{32}P]GTP (Fig. 5.2A).

2.2.1. Materials

AppNHp (AMPPNP), adenosin-5′-[(β,γ)-imido]triphosphate (NU407) is from Jena Bioscience (Jena, Germany). Activated charcoal (C7606) is from Sigma-Aldrich (St. Louis, MO). Norit A decolorizing carbon (102489) is from MP Biomedicals (Solon, OH). Shephadex G-50 (17-0042) is from GE Healthcare (Buckinghamshire, England). GTPγS, 10 mM, (20–176) is from Millipore (Temecula, CA). [γ-^{32}P]-GTP 10 mCi/ml (370 MBq/ml) (NEG004Z) and [^{35}S]-GTPγS 1 mCi/ml (37 MBq/ml) (NEG030X) are from PerkinElmer (Boston, MA).

2.3. GTPase activity assay

1. Cells are cultured in 10-cm plates and washed with 1× phosphate buffered saline (PBS).
2. Cells are scraped and homogenized in 50 mM Tris–HCl (pH 7.4) containing 10 mM MgCl$_2$, 1 mM EDTA, and 0.1 mM phenylmethylsulfonyl fluoride.
3. After centrifugation at 43,000× g for 20 min, the membrane pellet is washed, and suspended and incubated in 20 mM HEPES (pH 8.0), 10 mM MgCl$_2$, 1 mM EDTA, 160 mM NaCl, 0.2 mM AppNHp (AMPPNP), 1% BSA, 0.2 mM ascorbic acid, and 0.1 μM [γ-^{32}P]GTP at 30 °C for 30 min in a total volume of 100 μl.
4. The reaction is stopped by adding 10 μl of 50% trichloroacetic acid.
5. The mixture is centrifuged at 2,500 × g for 20–30 min at 4 °C.
6. The supernatant is added to 1 ml of ice-cold 5% charcoal or ice-cold 5% Norit A with 0.1% BSA in 20 mM sodium phosphate (pH 7.0).
7. [^{32}P]Pi release is measured using the supernatant obtained by centrifugation at 4,000 × g for 10 min (charcoal) or 2,000 × g for 5 min (Norit A) at 4 °C for liquid scintillation counting.
8. Alternatively, after step 5, the supernatant of the mixture is added to 4 ml of 1.25% ammonium molybdate in 1.2 M HCl containing 50 μM potassium phosphate followed by 5 ml of 2-methyl-2-propanol-benzene (1:1) and vortexed for 20 s.
9. [^{32}P]Pi release is measured using 2 ml of the supernatant for liquid scintillation counting.

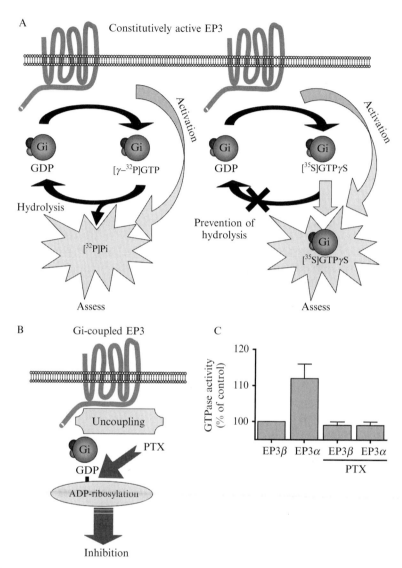

Figure 5.2 GTPase activity and GTPγS binding assays. (A) Graphic illustrating the functional basis leading to the generation of ^{32}Pi in the GTPase assay and the binding [^{35}S]GTPγS. (B) Graphic illustrating the uncoupling of Gi from the activated state of the EP3 receptor following ADP-ribosylation of Gi by pertussis toxin (PTX). (C) Bar graph showing the decrease in basal GTPase activity following the treatment of cells expressing the mouse EP3α and EP3β receptor isoforms with 10 ng/ml PTX for 6 h at 37 °C. Data are from Hasegawa et al. (1996) and are regraphed and normalized to the basal activity obtained from EP3β isoform (mouse) as 100%.

2.4. GTPγS binding assay

1. Cells are cultured in 10-cm plates and then washed with 1× PBS.
2. Cells are scraped and homogenized in 50 mM Tris–HCl (pH 7.4) containing 10 mM $MgCl_2$, 1 mM EDTA, and 0.1 mM phenylmethylsulfonyl fluoride.
3. After centrifugation at 43,000× g for 20 min, the membrane pellet is washed, suspended, and incubated in 20 mM HEPES (pH 8.0), 10 mM $MgCl_2$, 1 mM EDTA, 160 mM NaCl, 0.2 mM AppNHp (AMPPNP), 1% BSA, 0.2 mM ascorbic acid, and 0.1 μM [^{35}S]GTPγS at 30 °C for 30 min in a total volume of 100 μl.
4. The assay is stopped by adding 400 μl of a 0.5% cholate solution.
5. Free nucleotide is separated from nucleotide bound to G-protein by gel filtration on Sephadex G-50 (0.6 × 12.5 cm) that was preequilibrated and eluted for liquid scintillation counting with a buffer containing 20 mM Tris–HCl (pH 7.4), 100 mM NaCl, 25 mM $MgCl_2$, and 0.1% cholate.
6. Alternatively, after step 3, the mixture is centrifuged at 10,000× g for 5 min at 4 °C, or after step 3, the reaction is terminated by rapid filtration through a glass filter.
7. The pellet and/or filter are washed extensively with ice-cold buffer 50 mM Tris–HCl (pH 7.4) containing 10 mM $MgCl_2$, 1 mM EDTA, and 0.1 mM phenylmethylsulfonyl fluoride and dissolved in scintillation fluid for liquid scintillation counting.
8. Nonspecific binding is determined in the presence of 10 mM nonradioactive GTPγS.

2.5. Use of the Gi inhibitor pertussis toxin

PTX is a protein exotoxin from *Bordetella pertussis* that catalyzes the adenosine diphosphate(ADP)-ribosylation of the α-subunits of Gi, Go, and Gt, which thereby blocks the interaction of these G-proteins with the activated state of the GPCRs whether this activation is the result of agonist binding or constitutive activity. Blocking this receptor–G-protein interaction prevents the guanine nucleotide exchange reaction and leaves these G-proteins in their inactive GDP-bound state (Fig. 5.2B). Thus, treatment of cells with PTX will block the inhibition of adenylyl cyclase by constitutively active Gi-coupled GPCRs and will typically increase basal levels of intracellular cAMP. Similarly, in cells that have been pretreated with forskolin to stimulate adenylyl cyclase, PTX treatment will increase forskolin-stimulated cAMP formation by inhibiting the activity of constitutively active Gi-coupled GPCRs (Fig. 5.1B). Obviously by preventing the guanine nucleotide exchange reaction, treatment of cells with PTX will decrease Gi-mediated GTPase activity and GTP binding (Fig. 5.2C).

2.5.1. Materials
Opti-MEM (31985) is from Invitrogen (Carlsbad, CA). PTX (516560) is from Calbiochem-MERCK (Darmstadt, Germany).

1. Sixteen hours before the cAMP assay, GTPase assay and/or GTPγS binding assay, cells are switched to regular culture medium to serum free medium such as Opti-MEM.
2. Cells are then pretreated with either vehicle (water) or 100 ng/ml of PTX for 16 h at 37 °C.
3. Proceed to the cAMP assay, GTPase assay and/or GTPγS binding assay as described above.

2.6. Determination of actin stress fiber formation by staining with TRITC-phalloidin

As mentioned in Section 1, the mouse EP3α receptor isoform has been shown to have constitutive activity with respect to coupling to $G_{12/13}$ resulting in the activation of Rho and formation of actin stress fibers (Hasegawa et al., 1997). Actin stress fibers are a higher order form of actin filaments (F-actin), which in turn are linear polymers composed of actin monomers (G-actin). Together with intermediate filaments and microtubules, actin filaments are the main components of the cellular cytoskeleton. The activation of Rho signaling, which can be elicited by GPCRs coupled to $G_{12/13}$, is well known to promote the formation of actin stress fibers from F-actin. Phalloidin is a heptapeptide toxin from the mushroom, *Amantia phalloides*, that binds preferentially to F-actin (over G-actin) and fluorescent derivatives of phalloidin can be used to visualize cellular F-actin by fluorescence microscopy (Fig. 5.3A). In the protocol below, phalloidin conjugated to tetramethyl-rhodamine-isothiocyanate (TRITC-phalloidin) is used to stain cellular F-actin. Under basal conditions in HEK cells, and many other cells, most of the F-actin staining is cortical with diffuse and relatively disordered staining more centrally (Fig. 5.3B). Following the activation of Rho, however, the staining becomes highly organized and spread throughout the cell, which is characteristic of actin stress fiber formation (Fig. 5.3C). In cells expressing constitutively active GPCRs coupled to $G_{12/13}$, actin stress fibers may be present under basal conditions in the absence of agonist stimulation. This constitutive activity may be verified by pretreating the cells with C3 exoenzyme (C3 toxin) prior to labeling with TRITC-phalloidin. C3 toxin inactivates Rho and will block the effects of $G_{12/13}$-mediated constitutive activity resulting in the dissociation of the actin stress fibers.

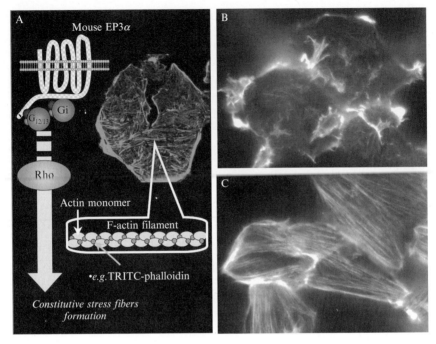

Figure 5.3 Determination of actin stress fiber formation by staining with TRITC-phalloidin. (A) Graphic illustrating the binding of TRITC-phalloidin to F-actin filaments and showing the constitutive activation of Rho signaling by the coupling of $G_{12/13}$ to the mouse EP3α receptor. (B) Fluorescence photomicrograph of quiescent HEK cells stained with TRITC-phalloidin. (C) Fluorescence photomicrograph of HEK cells stained with TRITC-phalloidin following induction of actin stress fiber formation.

2.6.1. Materials

TRITC-phalloidin (P1951) and p-phenylenediamine (P6001) are from Sigma–Aldrich (St. Louis, MO). C3 toxin (BML-G130) is from Enzo Life Science (Plymouth Meeting, PA) or can be purified from *Escherichia coli* transformed with a plasmid encoding a glutathione S-transferase/C3 toxin fusion protein (Pierce et al., 1999).

1. Cells are grown for 2 days on 22-mm round glass coverslips in 6-well plates under subconfluent conditions.
2. After washout with 1× PBS, they are fixed for 15 min in freshly made 4% paraformaldehyde in 1× PBS, quenched three times for 10 min in 0.1 M glycine, pH 7.4, and permeabilized for 15 min in 2× SSC (30 mM NaCl, 300 mM sodium citrate) containing 0.1% Triton X-100.
3. Cells are preincubated in blocking buffer (2× SSC, 0.05% Triton X-100, 2% goat serum, 1% BSA, 0.01% sodium azide) for 30 min.

4. Cells are then incubated with 0.1 units of TRITC-phalloidin in 50 µl of blocking buffer at room temperature for 1 h.
5. Cells are briefly washed in 1× PBS and mounted in media containing p-phenylenediamine.
6. Cells are viewed by fluorescence microscopy with a 60× oil objective using a Texas red isothiocyanate filter cube.
7. (Optional) Prior to step 2, pretreat cells for 48 h with 40 µg/ml C3 toxin.

3. Assays Used to Assess EP4 Receptor Constitutive Activity

3.1. Tcf/β-Catenin luciferase reporter plasmid assay

Constitutive activity of the EP4 receptor has been observed in stably transfected HEK cells following measurement of Tcf/β-catenin transcriptional activity using the TOPflash luciferase reporter plasmid (Fujino et al., 2002). The TOPflash reporter plasmid contains six Tcf binding sites upstream of a thymidine kinase minimal promotor and the gene encoding luciferase. It is typically used with FOPflash, which is the corresponding control plasmid in which the Tcf binding sites have been mutated. These plasmids were originally developed to study the Wnt signaling pathway in which Wnt, an extracellular morphogenic signaling molecule, interacts with a seven transmembrane receptor named frizzled to inhibit the activity of GSK-3 (Van de Wetering et al., 1997). GSK-3 is responsible for the phosphorylation of β-catenin, which targets β-catenin for ubiquitin-mediated degradation. Inhibition of GSK-3 prevents the phosphorylation of β-catenin allowing β-catenin to translocate to the nucleus where it interacts with members of the Tcf family of transcription factors to activate transcription (Fig. 5.4A). The EP2 and EP4 receptors have been found to cross talk with the Wnt signaling pathway at the level of GSK-3 by inhibiting its activity through phosphorylation, either by the activation of PKA and/or PI3K (Fujino et al., 2002). In the protocol described below, potential involvement of PI3K can be examined using either wortmannin or LY294002, which inhibit PI3K (Fig. 5.4B).

3.1.1. Materials
Opti-MEM (31985) is from Invitrogen (Carlsbad, CA). FuGENE-6 transfection reagent (11-815-091-001) is from Roche (Mannheim, Germany). TOPflash and FOPflash (TCF reporter plasmid kit, 17-285) are from Upstate-Millipore (Temecula, CA). Wortmannin (681675) and LY294002 (440202) are from Calbiochem-MERCK (Darmstadt, Germany). Luciferase Assay System (E1500) is from Promega (Madison, WI).

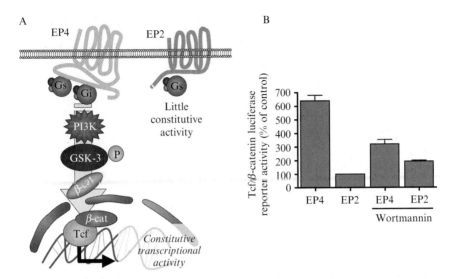

Figure 5.4 Tcf/β-catenin luciferase reporter plasmid assay. (A) Graphic illustrating the constitutive activation of Tcf/β-catenin transcriptional activity by the coupling of the human EP4 receptor to Gi and activation of a PI3K signaling pathway. PI3K, phosphatidylinositol 3-kinase; GSK-3, glycogen synthase kinase-3; β-cat, β-catenin; Tcf, T-cell factor. (B) Bar graphs showing the increase in basal Tcf/β-catenin luciferase reporter plasmid activity in HEK cells expressing the human EP4 receptor as compared with cells expressing the human EP2 receptor and the effect of pretreatment with 100 nM wortmannin for 1 h at 37 °C. Data are from Fujino et al. (2002) and are regraphed and normalized to the luciferase activity from EP2 receptor-expressing cells as 100%.

1. Cells are cultured with serum free medium such as Opti-MEM in 6-well plates and transiently transfected using a transfection reagent such as FuGENE-6 and 1.25 μg/well of either TOPflash or FOPflash.
2. To examine the potential involvement of PI3K, cells are pretreated with either vehicle (0.1% Me$_2$SO) or 100 nM wortmannin or 30 μM LY294002, for 1 h at 37 °C.
3. Cells are rapidly washed three times each with 1 ml/well of serum free medium such as Opti-MEM and then incubated for 16 h at 37 °C in 3 ml/well of medium containing antibiotics.
4. Cells are placed on ice and rinsed twice with ice-cold 1× PBS and extracts are prepared using the lysis buffer included in the Luciferase Assay System.
5. Luciferase activity in the extracts (using 1 μg of protein/sample) is measured using a luminometer and is corrected for background by subtracting FOPflash values from corresponding TOPflash values.

REFERENCES

Bilson, H. A., Mitchell, D. L., and Ashby, B. (2004). Human prostaglandin EP3 receptor isoforms show different agonist-induced internalization patterns. *FEBS Lett.* **572,** 271–275.
Coleman, R. A., Smith, W. L., and Narumiya, S. (1994). VIII. International Union of Pharmacology classification of prostanoid receptors: properties, distribution and structure of the receptors and their subtypes. *Pharmacol. Rev.* **46,** 205–229.
Fujino, H., and Regan, J. W. (2003). Prostanoid receptors and phosphatidylinositol 3-kinase: A pathway to cancer? *Trends Pharmacol. Sci.* **24,** 335–340.
Fujino, H., and Regan, J. W. (2006). EP4 prostanoid receptor coupling to a pertussis toxin-sensitive inhibitory G protein. *Mol. Pharmacol.* **69,** 5–10.
Fujino, H., West, K. A., and Regan, J. W. (2002). Phosphorylation of glycogen synthase kinase-3 and stimulation of T-cell factor signaling following activation of EP2 and EP4 prostanoid receptors by prostaglandin E_2. *J. Biol. Chem.* **277,** 2614–2619.
Hasegawa, H., Negishi, M., and Ichikawa, A. (1996). Two isoforms of the prostaglandin E receptor EP3 subtype different in agonist-independent constitutive activity. *J. Biol. Chem.* **271,** 1857–1860.
Hasegawa, H., Negishi, M., Katoh, H., and Ichikawa, A. (1997). Two isoforms of prostaglandin EP3 receptor exhibiting constitutive activity and agonist-dependent activity in Rho-mediated stress fiber formation. *Biochem. Biophys. Res. Commun.* **234,** 631–636.
Hawcroft, G., Ko, C. W. S., and Hull, M. A. (2007). Prostaglandin E_2-EP4 receptor signalling promotes tumorigenic behaviour of HT-29 human colorectal cancer cells. *Oncogene* **26,** 3006–3019.
Hizaki, H., Hasegawa, H., Katoh, H., Negishi, M., and Ichikawa, A. (1997). Functional role of carboxyl-terminal tail of prostaglandin EP3 receptor in Gi coupling. *FEBS Lett.* **414,** 323–326.
Ji, R., Chou, C. L., Xu, W., Chen, X. B., Woodward, D. F., and Regan, J. W. (2010). EP1 prostanoid receptor coupling to Gi/o upregulates the expression of hypoxia-inducible factor-1α through activation of a phosphoinositide-3 kinase signaling pathway. *Mol. Pharmacol.* **77,** 1025–1036.
Jin, J., Mao, G. F., and Ashby, B. (1997). Constitutive activity of human prostaglandin E receptor EP3 isoforms. *Br. J. Pharmacol.* **121,** 317–323.
Kotani, M., Tanaka, I., Ogawa, Y., Usui, T., Tamura, N., Mori, K., Narumiya, S., Yoshimi, T., and Nakao, K. (1997). Structural organization of the human prostaglandin EP3 receptor subtype gene (PTGER3). *Genomics* **40,** 425–434.
Negishi, M., Hasegawa, H., and Ichikawa, A. (1996). Prostaglandin E receptor EP3γ isoform, with mostly full constitutive Gi activity and agonist-dependent Gs activity. *FEBS Lett.* **386,** 165–168.
Pierce, K. L., Fujino, H., Srinivasan, D., and Regan, J. W. (1999). Activation of FP prostanoid receptor isoforms leads to Rho-mediated change in cell morphology and in the cell cytoskeleton. *J. Biol. Chem.* **274,** 35944–35949.
Regan, J. W. (2003). EP2 and EP4 prostanoid receptor signaling. *Life Sci.* **74,** 143–153.
Sugimoto, Y., and Narumiya, S. (2007). Prostanoid E receptors. *J. Biol. Chem.* **282,** 11613–11617.
Van de Wetering, M., Cavallo, R., Dooijes, D., van Beest, W., van Es, J., Loureiro, J., Ypma, A., Hursh, D., Jones, T., Bejsovec, A., Peifer, M., Mortin, M., *et al.* (1997). Armadillo coactivates transcription driven by the product of the drosophila segment polarity gene dTCF. *Cell* **88,** 789–799.

CHAPTER SIX

α_{1D}-ADRENERGIC RECEPTORS: CONSTITUTIVE ACTIVITY AND REDUCED EXPRESSION AT THE PLASMA MEMBRANE

J. Adolfo García-Sáinz,* M. Teresa Romero-Ávila,* and Luz del Carmen Medina[†]

Contents

1. Introduction	110
2. Methods	111
2.1. Cell culture and transfection	111
2.2. Radioligand binding assays and adrenoceptors photoaffinity labeling	112
2.3. Intracellular calcium concentration ([Ca^{2+}]$_i$)	113
3. Constitutive Activity	114
4. Plasma Membrane α_{1D}-Adrenergic Receptors	117
5. Possible Physiological Implications	119
Acknowledgments	121
References	121

Abstract

Adrenergic receptors are a heterogeneous family of the G protein-coupled receptors that mediate the actions of adrenaline and noradrenaline. Adrenergic receptors comprise three subfamilies (α_1, α_2, and β, with three members each) and the α_{1D}-adrenergic receptor is one of the members of the α_1 subfamily with some interesting traits. The α_{1D}-adrenergic receptor is difficult to express, seems predominantly located intracellularly, and exhibits constitutive activity. In this chapter, we will describe in detail the conditions and procedures used to determine changes in intracellular free calcium concentration which has been instrumental to define the constitutive activity of these receptors. Taking advantage of the fact that truncation of the first 79 amino acids of α_{1D}-adrenergic receptors markedly increased their membrane expression, we were able to

---

* Departamento de Biología Celular y Desarrollo, Instituto de Fisiología Celular, Universidad Nacional Autónoma de México, México
[†] Departamento de Biología de la Reproducción, División de CBS, Universidad Autónoma Metropolitana-Iztapalapa, México

show that constitutive activity is present in receptors truncated at the amino and carboxyl termini, which indicates that such domains are dispensable for this action. Constitutive activity could be observed in cells expressing either the rat or human α_{1D}-adrenergic receptor orthologs. Such constitutive activity has been observed in native rat arteries and we will discuss the possible functional implications that it might have in the regulation of blood pressure.

1. INTRODUCTION

The actions of the natural catecholamines, adrenaline, and noradrenaline (NA), and of those of synthetic ligands are mediated through nine G protein-coupled receptors coded by different genes. These adrenergic receptors or adrenoceptors (ARs) are grouped into three subfamilies (with three members each) on the basis of their pharmacological profiles, sequence similarities, and preferential signaling, that is, α_1-, α_2-, and β-ARs (Hieble et al., 1995). The α_1-AR subfamily is constituted by α_{1A}-, α_{1B}-, and α_{1D}-ARs. Mutants with constitutive activity of some of these receptors have been constructed and have shed important information on the process of receptor activation, their oncogenic potential, and are important tools to characterize pharmacological agents (Allen et al., 1991; Cotecchia, 2007; Kjelsberg et al., 1992; Rossier et al., 1999). A chapter of this volume is devoted to these ARs in general (Chapter 7, Vol. 485). In the present chapter, we will present some characteristics of α_{1D}-ARs that make them especially interesting and challenging, the main methods we have used in their study, how we found constitutive activity and finally, we will discuss the possible functional relevance that such constitutive activity might have in physiological contexts.

α_{1D}-ARs have been elusive (García-Sáinz and Villalobos-Molina, 2004). To define the correspondence of cloned receptors with those pharmacologically characterized in native tissues was very complicated due to the lack of selective agonists and antagonists and it is responsible for the odd nomenclature we currently use. At the present, BMY 7378 (8-[2-[4-(2-methoxyphenyl)-1-piperazinyl]ethyl]-8-azaspiro[4,5]decane-7,9-dione dihydrochloride) is the most useful selective antagonist, but the lack of pharmacological tools still persists. The rat and human orthologs of this receptor subtype have been studied more extensively than those of other species. Rat α_{1D}-ARs were initially cloned by Lomasney et al. (1991) and afterward identified as the α_{1D} subtype by Perez et al. (1991). The human ortholog was subsequently cloned and the gene located at chromosome 20 (Forray et al., 1994; Schwinn et al., 1995; Weinberg et al., 1994). The rat ortholog has 561 amino acids with a protein molecular weight of 59,364 Da whereas the human receptor has 572 amino acids and a protein molecular weight of 60,462 Da. Photoaffinity labeling of these receptor orthologs gave estimated M_r values of $\approx$ 70–80 kDa (García-Sáinz *et al.*, 2001, 2004; Schwinn and Kwatra, 1998), which is consistent with extensive *N*-glycosylation.

α_{1D}-ARs couple mainly to pertussis toxin-insensitive G proteins, likely of the Gq subfamily, which activate phosphoinositidase and consequently, phosphatidylinositol 4,5-bisphosphate hydrolysis, production of inositol 1,4,5-trisphosphate and diacylglycerol. These latter products are responsible of calcium signaling and protein kinase C activation that propagates the signal intracellularly (García-Sáinz and Villalobos-Molina, 2004). Other signaling pathways seem to participate also in some of the actions of these receptors (García-Sáinz and Villalobos-Molina, 2004; Kawanabe *et al.*, 2001; Perez *et al.*, 1993; Shinoura *et al.*, 2002).

There are other α_{1D}-AR interesting traits. First, these receptors are difficult to express in cellular model systems and they seem to be less efficient in their coupling to signaling pathways than the other members of the α_1-AR subfamily (Theroux *et al.*, 1996; Vázquez-Prado and García-Sáinz, 1996). Second, a predominant intracellular location has been detected (Chalothorn *et al.*, 2002; Hague *et al.*, 2004b; McCune *et al.*, 2000) as well as a low expression in most native tissues, that has greatly complicated their detection by radioligand binding studies (Yang *et al.*, 1997). It is likely that some of these characteristics might have a common structural/functional origin.

2. METHODS

In this section, we will briefly outline the basic methods used: (1) cell culture and transfection and (2) radioligand binding assays and ARs photo-affinity labeling. We will describe in more detail the conditions and procedures used to determine changes in intracellular free calcium concentration which has been instrumental to define the constitutive activity of these receptors.

2.1. Cell culture and transfection

The initial studies that suggested the possibility of constitutive activity of α_{1D}-ARs were performed in rat-1 fibroblast stably expressing the rat receptor kindly provided to us by Drs. L. Allen, R. J. Lefkowitz and M. G. Caron (Duke University Medical Center, Durham, NC, USA). Subsequent studies were performed using rat-1 fibroblast transfected with pcDNA, containing the cDNA of human α_{1D}-AR (generously provided to us by Dr. Marvin L. Bayne, Merck) (García-Sáinz *et al.*, 2004) or constructions in which the receptor amino terminus (amino acids 1–79 (ΔN)) or both the amino and carboxyl termini (amino acids 1–79 and 441–572 (ΔN–ΔC)) were deleted (Rodríguez-Pérez et al., 2009). Cells were transfected with these plasmids using Lipofectamine 2000, following the instructions of the manufacturer (Invitrogene-Life Technologies), and were cultured, for stable receptor expression, in a selection medium: glutamine-containing high-glucose Dulbecco's modified Eagle's medium supplemented with 10% fetal bovine serum, 600 μg/ml of the neomycin analog, G-418 sulfate, 100 μg/ml streptomycin, 100 U/ml penicillin, and 0.25 μg/ml amphotericin B at 37 °C under a 95% air/5% CO_2 atmosphere. Such high concentration of G-418 was required to select only clones with reasonable high receptor expression. In all cases several clones were isolated and were screened for α_{1D}-adrenergic responsiveness (NA-induced increase in intracellular calcium concentration) and receptor density (radioligand binding saturation isotherms). Clones were further cultured in the media described above, but with reduced G-418 (300 μg/ml). In spite of the presence of G-418 in the culture media some of the clones reduced receptor density and responsiveness after several passages.

2.2. Radioligand binding assays and adrenoceptors photoaffinity labeling

A crude membrane preparation was prepared as described (Vázquez-Prado et al., 2000). Binding studies were performed by incubating the radioligand (either [^{3}H]tamsulosin (56.3 Ci/mmol) or [^{3}H]prazosin (74.4 Ci/mmol)) with membranes (5–25 μg of protein) in a final volume of 0.25 ml of binding buffer (50 mM Tris, 10 mM $MgCl_2$; pH 7.5) for 60 min at 37 °C in a water bath shaker. Incubation was ended by addition of 5 ml of ice-cold buffer and filtration through GF/C filters using a Brandell harvester. Caution was exercised to avoid radioligand depletion and low concentrations of membrane protein were preferred when it was possible (García-Sáinz et al., 2004; Rodríguez-Pérez et al., 2009). Filters were washed twice, dried, and radioactivity was measured in a liquid scintillation counter. Nonspecific binding was determined in the presence of 10 μM BMY 7378; specific binding was > 90% of total binding at the K_D for tamsulosin and 75–80% for prazosin. Unfortunately, [^{3}H]tamsulosin, one of the "cleanest" radioligands we have used, is not commercially available (it was a generous gift from Yamonouchi Europe) but [^{3}H]prazosin gave similar Bmax data. The EBDA program from Biosoft-Elsevier (Cambridge, UK) was used to analyze radioligand binding saturation isotherms. For the photoaffinity labeling assays, membranes (usually 25 μg protein, but see below), were incubated in the dark with 6 nM [aryl-^{125}I]azido-prazosin and exposed to UV light as described (Vázquez-Prado et al., 2000). After this treatment, membranes were centrifuged, washed, solubilized, and subjected to electrophoresis in sodium dodecyl sulfate–polyacrylamide gels under reducing conditions

(García-Sáinz et al., 2004; Rodríguez-Pérez et al., 2009; Vázquez-Prado et al., 2000).

2.3. Intracellular calcium concentration ($[Ca^{2+}]_i$)

Cells were incubated overnight in Dulbecco's modified Eagle's medium without serum and antibiotics and were loaded with 5 μM Fura-2 AM in Krebs–Ringer-HEPES containing 10 mM glucose and 0.05% bovine serum albumin, pH 7.4 for 1 h at 37 °C. Cells were detached by gentle trypsinization; the amount of trypsin used and the time of incubation are empirically determined for each sample of enzyme. When cell detachment was not properly achieved, either the cells remained in clumps that results in light scattering with variations ("noise") in the signal amplitude and/or cell responsiveness decreased (likely due to over-trypsinized ("shaved") cell plasma membranes). After detachment, cells were gently washed with the same buffer containing albumin in order to remove any extracellular remains of the fluorescent indicator; cell dilution was usually $\approx 10^6$ cells per each 2 ml of buffer. Fluorescence measurements were carried out with an Aminco-Bowman Series 2 Spectrometer equipped with a water-jacketed cell holder to maintain temperature at 37 °C and a magnetic stirrer to preserve the cells in suspension. The excitation monochromator was set at 340 and 380 nm (chopper interval of 0.5 s), and the emission monochromator was set at 510 nm (Vázquez-Prado et al., 2000). Intracellular free calcium ($[Ca^{2+}]_i$) was calculated as described by Grynkiewicz et al. (1985) using the equation:

$$[Ca^{2+}]_i = K_d(F - F_{min}/F_{max} - F)$$

This equation required F (fluorescence), F_{min} (minimal fluorescence), and F_{max} (maximal fluorescence) to be determined at the same instrumental sensitivity, optical path length, and indicator concentration. To achieve this, cells were lysed by the addition of Triton X-100 (final concentration 0.2%), maximal fluorescence was recorded, and afterward calcium was chelated with an excess of EGTA (final concentration 10 mM) that allowed recording minimal fluorescence. The software provided by Aminco-Bowman allowed traces to be directly exported to the graphs. In Fig. 6.1, we present a representative tracing of fluorescence obtained with rat-1 fibroblasts expressing α_{1D}-ARs. It can be observed (upper panel) that the addition of 10 μM NA induced a rapid and robust increase in fluorescence than decreased slowly afterward. The tracing also shows the effect of cell lysis with Triton X-100 and calcium chelation with EGTA. In the lower panel, the determination of intracellular calcium is presented.

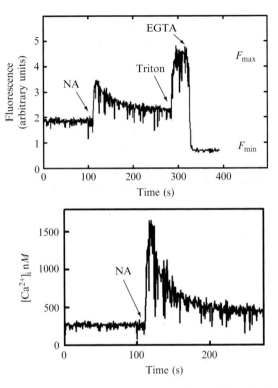

Figure 6.1 In the upper panel a representative tracing of the effect of 10 μM noradrenaline (NA) on intracellular calcium (in arbitrary units, fluorescence) in rat-1 fibroblasts expressing human amino-truncated (ΔN) α_{1D}-ARs and the procedure to obtain F_{max} and F_{min} are presented. In the lower panel the calibrated tracing is presented.

3. CONSTITUTIVE ACTIVITY

The observation of constitutive activity in rat wild-type α_{1D}-ARs expressed in rat-1 fibroblast came somehow as a surprise. We were studying the pharmacological actions of chloroethylclonidine (CEC). This agent was described as an irreversible α-AR agonist that induced phentolamine-sensitive contraction in rat aorta and increased blood pressure in pithed rats (Leclerc et al., 1980). Later studies indicated that CEC was an α_1-AR antagonist with some selectivity for the α_{1B} subtype (Han et al., 1987). We compared the effect of CEC in cell lines of rat-1 fibroblasts expressing each one of the cloned subtypes. Our results showed that CEC was able to decrease the effect of NA in cells expressing any of the three subtypes but that there were some differences, that is, it behaved as a classical antagonist in cells expressing α_{1B}-ARs, as a partial agonists in cells expressing the α_{1A}

subtype and in cells expressing α_{1D}-ARs as an "antagonist" that decreased the resting level of intracellular calcium (Villalobos-Molina et al., 1997b). It should be mentioned that determination of the real basal level of calcium is "tricky" and that values vary among cell preparations (usually between 100 and 200 nM); factors, difficult to control, contribute to this, including dye leaking, and cell lysis during the incubation (neither of which, obviously, has relationship with real "intracellular" calcium). Despite these difficulties in measuring differences in basal calcium levels, we continue our work observing that other antagonists such as phentolamine, 5-methyl urapidil (5-MU), and the α_{1D}-selective antagonist, BMY 7378, were all able to induce a similar effect in a concentration-dependent fashion. A very important additional finding was that the antagonist, WB4101, induced only a very small decrease as compared to the other agents and, even more interestingly, that it was able to block the effects of the other agents (García-Sáinz and Torres-Padilla, 1999). This allowed us to conclude unambiguously that rat wild-type α_{1D}-ARs exhibit constitutive activity and that therefore, the mentioned antagonists should be considered as inverse agonists; WB4101 could be considered as a classical antagonist or as an inverse partial agonist with very low intrinsic activity. Representative tracings showing the effects of BMY 7378 and WB4101 are shown in Fig. 6.2 together with a summary graph of the effects of these pharmacological agents. Soon afterward another group also observed that cells expressing wild-type rat α_{1D}-ARs showed high basal inositol phosphate production and extracellular signal regulated kinase (ERK) activity that could be reduced by prazosin (McCune et al., 2000).

We next explored the possibility that the wild-type human α_{1D}-AR ortholog might also exhibit constitutive activity. As mentioned already, α_{1D}-AR membrane expression is a complex process and most of the receptors remain intracellularly. Although we were able to express α_{1D}-ARs in rat-1 cells, membrane density was very low (García-Sáinz et al., 2004). In these cells, we observed α_{1D}-AR functional activity, receptor phosphorylation, and desensitization, but no constitutive activity (García-Sáinz et al., 2004). We attributed the absence of constitutive activity to the low membrane expression but the possibility that the human ortholog might not express this property could not be completely ruled out at that time. Taking advantage of the key observation that truncation of the first 79 amino acids of the amino terminus allows membrane expression (Pupo et al., 2003), we were able to obtain cells exhibiting high density human α_{1D}-ARs at the plasma membrane (Rodríguez-Pérez et al., 2009). In rat-1 fibroblasts expressing these amino terminus-truncated receptors or amino and carboxyl termini-truncated human α_{1D}-ARs, BMY 7378 was clearly able to decrease basal intracellular calcium, an action effectively blocked by WB4101 (Fig. 6.3; Rodríguez-Pérez et al., 2009). This clearly indicates that constitutive activity could be observed in cells expressing either rat or

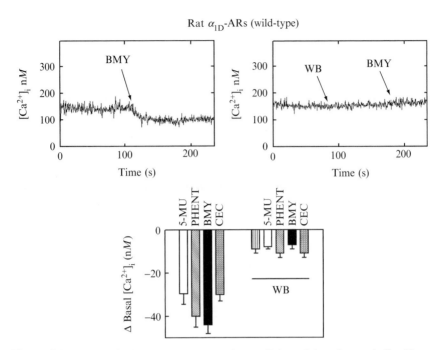

Figure 6.2 Effect of adrenergic agents on intracellular calcium in rat-1 fibroblasts expressing rat α_{1D}-ARs. In the upper panels representative tracings of the effects of BMY 7378 (BMY) and WB4101 (WB) on intracellular calcium are presented. In the lower panel the decrease in intracellular calcium induced by 5-methyl urapidil (5-MU), phentolamine (PHENT), BMY 7378 (BMY), and chloroethylclonidine (CEC) are plotted. Where indicated, WB4101 was added 2 min before the other agents. Plotted are the means and vertical lines represent the SEM of 3–4 experiments. Reprinted with permission (García-Sáinz and Torres-Padilla, 1999).

human α_{1D}-ARs orthologs at sufficient density. In addition, the data confirmed the finding that the amino terminus markedly reduces this receptor expression and further indicated that neither the amino nor the carboxyl termini were essential to observe α_{1D}-AR constitutive activity.

It is well known that activation of protein kinase C induces the phosphorylation of the three α_1-ARs and that such phosphorylation is associated to receptor desensitization (García-Sáinz et al., 2000). We have shown that in cells expressing rat or human α_{1D}-ARs, activation of protein kinase C with tetradecanoyl phorbol acetate (TPA) induced, as anticipated, a decrease in basal intracellular calcium (Fig. 6.4; García-Sáinz et al., 2001; Rodríguez-Pérez et al., 2009). This indicates that, in rat-1 fibroblasts expressing α_{1D}-ARs, receptor desensitization or inverse agonists induce phenomenologically the same response, that is, a decrease in intracellular calcium due to a reduction in the constitutive activity of these receptors.

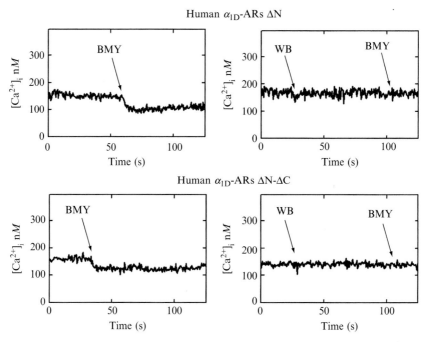

Figure 6.3 Representative tracings of the effects of BMY 7378 (BMY) and WB4101 (WB) on intracellular calcium in rat-1 fibroblasts expressing amino-truncated (ΔN) and amino- and carboxyl-truncated (ΔN-ΔC) human α_{1D}-ARs are presented. Reprinted (modified) with permission (Rodríguez-Pérez et al., 2009).

4. Plasma Membrane α_{1D}-Adrenergic Receptors

G protein-coupled receptors are usually located at the plasma membrane with a fraction of them in intracellular vesicles. These are newly formed receptors in their way to the external surface of the cell or receptors that have been internalized and are either recycled back to the plasma membrane or being processed for degradation. In the case of α_{1D}-ARs, the vast majority of the receptors seem to be in intracellular vesicles (Chalothorn et al., 2002; Hague et al., 2004a,b; McCune et al., 2000). It is well documented that receptor activation by agonists induces G protein-coupled receptor internalization. Therefore, it is possible that the constitutive activity could be favoring internalization. It has been observed that prazosin, which acts as inverse agonist, promotes α_{1D}-AR redistribution from internal vesicles to the plasma membrane (McCune et al., 2000). This suggests that constitutive activity could participate in the processes that define these receptors location, but other possibilities also exist.

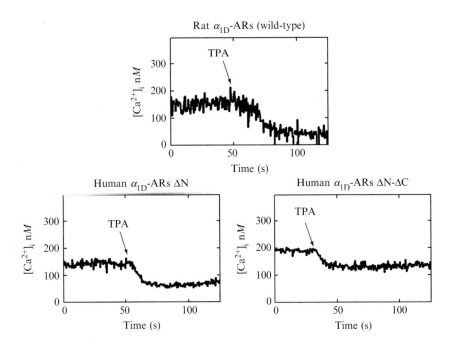

Figure 6.4 Representative tracings of the effect of tetradecanoyl phorbol acetate (TPA) on basal intracellular calcium in rat-1 fibroblasts expressing rat wild type, human amino-truncated (ΔN) and human amino- and carboxyl-truncated (ΔN-ΔC) α_{1D}-ARs are presented. Reprinted (modified) with permission (García-Sáinz and Torres-Padilla, 1999; Rodríguez-Pérez et al., 2009).

For example, it is known that misfolded receptors accumulate in vesicles and that some compounds act as pharmacological chaperones, allowing them to reach the plasma membrane (Conn and Janovick, 2009; Conn and Ulloa-Aguirre, 2010; Morello et al., 2000; Smith et al., 1998). Inverse agonist could have induced a change in α_{1D}-AR conformation that favored plasma membrane insertion.

A major finding to explain the internal location of α_{1D}-ARs was obtained by the group of Minneman (Hague et al., 2004a,b, 2006; Pupo et al., 2003). These authors observed that deletion of the initial 79 amino acids of these receptors enhances processing of a binding competent form in HEK293 cells; and show a clear dissociation between abundance of receptor protein and density of receptor binding sites. We have confirmed these findings using rat-1 fibroblasts. When we transfected these cells with the cDNA of wild-type human α_{1D}-ARs we were only able to isolate cell lines that express a very low density of receptors in their plasma membranes as reflected both by a very small Bmax (35 fmol/mg membrane protein) in radioligand binding saturation isotherms and by photoaffinity labeling studies with [aryl-^{125}I]azido-prazosin that required a large amount of membrane protein in the assays (ten-times the usual amount, i.e., 250 μg of membrane protein) (García-Sáinz et al., 2004). In contrast, using similar transfection conditions we were able to isolate cell lines expressing amino terminus-truncated human α_{1D}-ARs with densities as high as ≈2 pmol/mg membrane protein that were easily photoaffinity labeled using 25 μg of membrane protein (Rodríguez-Pérez et al., 2009). It is interesting to mention that amino terminus-truncation of human α_{1D}-ARs increases expression of binding sites but not protein, that is, a similar amount of receptor protein was observed in cells expressing the wild type and truncated forms, but a much higher density of sites by radioligand binding saturation isotherms (Pupo et al., 2003). The data clearly showed that such deletion enhances the processing of the receptor to a binding competent form rather than only its synthesis (Pupo et al., 2003). In addition, the α_{1D}-AR amino terminus act as a "transplantable" signal that, when expressed as chimeric constructions with other adrenergic receptors, decreased their membrane expression (Hague et al., 2004a). It is also worth noticing that when α_{1B}- and α_{1D}-ARs are co-expressed these receptors form heterodimers that reach the plasma membrane (Hague et al., 2006); plasma membrane α_{1D}-ARs were detected by various methods (including Western blotting and confocal microscopy) but not pharmacologically (i.e., they do not display the expected binding affinities for agonists and antagonists) which suggests that such heterodimers might form a single functional entity, with a novel pharmacological profile (Hague et al., 2006); this might explain the difficulties in detecting this subtype in native tissues (Yang et al., 1997).

5. Possible Physiological Implications

Most of the functional studies on α_{1D}-ARs have been performed in contractile models, particularly using isolated rat arteries that regulate vascular tone. These ARs participate in the function of arteries such as aorta, iliac, carotid, mesenteric, renal, and femoral (Arévalo-León et al., 2003; Gómez-Zamudio et al., 2002; Piascik et al., 1994, 1995; Tanoue et al., 2002; Villalobos-Molina et al., 1997a). Not surprisingly, they are also major players in the control of blood pressure. Previous to our work with transfected cells, Pilar D'Ocon and her group showed the presence of a population of constitutively active α_1-ARs in a native tissue: rat aorta (Noguera et al., 1996). Pharmacological studies evidenced that these receptors belong to the α_{1D} subtype (Gisbert et al., 2000) and are coupled to phosphoinositide turnover/calcium signaling (Gisbert et al., 2000, 2003). These authors (Gisbert et al., 2003) observed that following intracellular calcium depletion by incubating aorta with NA in medium without extracellular calcium, agonist removal, and restoration of extracellular calcium induced three responses: a biphasic increase in intracellular calcium, inositol phosphate accumulation, and contraction. None of these actions were observed when antagonists such as prazosin or BMY 7378 were present during the incubation indicating that all these responses were mediated through α_{1D}-ARs. Interestingly, when the experiments were performed in the presence of the calcium channel blocker, nimodipine, the sustained increase in calcium and the inositol phosphate accumulation were observed, but contraction was not. These data indicate that α_{1D}-AR-mediated intracellular calcium mobilization was insufficient to trigger contraction and that extracellular calcium entry through L-channels was required. The group extended their studies to iliac and mesenteric arteries, observing similar responses (Ziani et al., 2002). The data suggest that α_{1D}-ARs participate not only in adjusting increases in blood pressure but also play a very important role avoiding abrupt drops in blood pressure (Ziani et al., 2002).

The α_{1D}-AR knock-out mouse added considerable support to the role of these receptors in the control of blood pressure, that is, α_{1D}-AR knock-out mice showed hypotension and reduced pressor responses to phenylephrine and NA (Tanoue et al., 2002). In addition, recent data with $\alpha_{1A/B}$-AR knock-out mice, a "purer" system to study α_{1D}-AR pharmacology, clearly evidenced the roles of these receptors in different vascular territories and added the interesting fact that the different subtypes participate in the overall control of resistance arteries but that the lack of one receptor subtype is not compensated by the others (Methven et al., 2009).

Hypertension is a pathological state characterized by an increase in peripheral vascular resistance, a very important risk factor for cardiovascular and renal diseases, as well as stroke. It is certainly one of the major health problems worldwide and currently *in crescendo* (Whitworth, 2003). A large amount of evidence suggests a role for α_{1D}-ARs in age-dependent increases in blood pressure and progression to the hypertensive state (Gisbert et al., 2002; Ibarra et al., 1997; Rubio et al., 2002; Villalobos-Molina and Ibarra, 1999; Villalobos-Molina et al., 1999).

It has been suggested that a strong cross-talk exits in the actions of α_{1D}-ARs and angiotensin II AT_1 receptors (Abdulla et al., 2009; Villalobos-Molina and Ibarra, 2005; Villalobos-Molina et al., 2008). In addition, it has been shown that in preeclamptic hypertensive women, a significant increase in heterodimerization occurs between angiotensin II AT_1-receptor and bradykinin B_2 receptors (AbdAlla et al., 2001). Such heterodimerization enhances responsiveness to angiotensin II (AbdAlla et al., 2001). Interestingly, a very recent study, using pregnant hypertensive rats showed that angiotensin II AT_1 receptors and α_{1D}-ARs form heterodimers in aortic tissues and suggested that such heterodimerization may play a role in preeclampsia (González-Hernández et al., 2010).

Although data are very provocative, caution needs to be exercised. Current knowledge does not allow us to know to what extent constitutive activity of these receptors might play a role in the pathogenesis of these diseases and if there is any real therapeutic difference between classic antagonists and inverse agonists. This is an area of much debate in which much more solid data are urgently required.

Further words of caution seem necessary. Constitutive activity is a molecular property of some receptors. However, functional expression of such activity depends on many factors including receptor density (sufficient to induce a response in the absence of agonist), adequate coupling to G proteins (appropriate subunit amounts and subtypes), and on many other downstream effectors involved in the intracellular propagation of the signal. Therefore, its functional expression does relay on the cellular context and conditions. Absence of detectable constitutive activity does not mean that all the receptors remain in their inactive conformation in the absence of agonist (Kenakin, 2009). Inverse agonism is a phenotypic behavior, not a molecular property itself since the experimental conditions must be appropriate for the effect to be seen (Kenakin, 2004).

ACKNOWLEDGMENTS

The authors express their gratitude to Dr. Susanna Cotecchia for critical reading of the manuscript. This work was supported by grants from Consejo Nacional de Ciencia y Tecnología (79908) and Dirección General de Asuntos del Personal Académico (IN212609). Permission to reproduce and/or present modified versions of figures is gratefully acknowledged to FEBS Letters (Elsevier) and Naunyn Schmiedeberg's Archives of Pharmacology (Springer).

REFERENCES

AbdAlla, S., Lother, H., el Massiery, A., and Quitterer, U. (2001). Increased AT(1) receptor heterodimers in preeclampsia mediate enhanced angiotensin II responsiveness. *Nat. Med.* **7**, 1003–1009.

Abdulla, M. H., Sattar, M. A., Khan, M. A., Abdullah, N. A., and Johns, E. J. (2009). Influence of sympathetic and AT-receptor blockade on angiotensin II and adrenergic agonist-induced renal vasoconstrictions in spontaneously hypertensive rats. *Acta Physiol. (Oxf).* **195**, 397–404.

Allen, L. F., Lefkowitz, R. J., Caron, M. G., and Cotecchia, S. (1991). G-protein-coupled receptor genes as protooncogenes: Constitutively activating mutation of the alpha 1B-adrenergic receptor enhances mitogenesis and tumorigenicity. *Proc. Natl. Acad. Sci. USA* **88**, 11354–11358.

Arévalo-León, L. E., Gallardo-Ortiz, I. A., Urquiza-Marín, H., and Villalobos-Molina, R. (2003). Evidence for the role of alpha1D- and alpha1A-adrenoceptors in contraction of the rat mesenteric artery. *Vascul. Pharmacol.* **40**, 91–96.

Chalothorn, D., McCune, D. F., Edelmann, S. E., García-Cazarin, M. L., Tsujimoto, G., and Piascik, M. T. (2002). Differences in the cellular localization and agonist-mediated internalization properties of the alpha(1)-adrenoceptor subtypes. *Mol. Pharmacol.* **61**, 1008–1016.

Conn, P. M., and Janovick, J. A. (2009). Drug development and the cellular quality control system. *Trends Pharmacol. Sci.* **30,** 228–233.
Conn, P. M., and Ulloa-Aguirre, A. (2010). Trafficking of G-protein-coupled receptors to the plasma membrane: Insights for pharmacoperone drugs. *Trends Endocrinol. Metab.* **21,** 190–197.
Cotecchia, S. (2007). Constitutive activity and inverse agonism at the alpha1adrenoceptors. *Biochem. Pharmacol.* **73,** 1076–1083.
Forray, C., Bard, J. A., Wetzel, J. M., Chiu, G., Shapiro, E., Tang, R., Lepor, H., Hartig, P. R., Weinshank, R. L., Branchek, T. A., et al. (1994). The alpha 1-adrenergic receptor that mediates smooth muscle contraction in human prostate has the pharmacological properties of the cloned human alpha 1c subtype. *Mol. Pharmacol.* **45,** 703–708.
García-Sáinz, J. A., and Torres-Padilla, M. E. (1999). Modulation of basal intracellular calcium by inverse agonists and phorbol myristate acetate in rat-1 fibroblasts stably expressing alpha1d-adrenoceptors. *FEBS Lett.* **443,** 277–281.
García-Sáinz, J. A., and Villalobos-Molina, R. (2004). The elusive alpha(1D)-adrenoceptor: Molecular and cellular characteristics and integrative roles. *Eur. J. Pharmacol.* **500,** 113–120.
García-Sáinz, J. A., Vázquez-Prado, J., and Medina, L. C. (2000). Alpha 1-adrenoceptors: Function and phosphorylation. *Eur. J. Pharmacol.* **389,** 1–12.
García-Sáinz, J. A., Vázquez-Cuevas, F. G., and Romero-Ávila, M. T. (2001). Phosphorylation and desensitization of alpha1d-adrenergic receptors. *Biochem. J.* **353,** 603–610.
García-Sáinz, J. A., Rodríguez-Pérez, C. E., and Romero-Ávila, M. T. (2004). Human alpha-1D adrenoceptor phosphorylation and desensitization. *Biochem. Pharmacol.* **67,** 1853–1858.
Gisbert, R., Noguera, M. A., Ivorra, M. D., and D'Ocon, P. (2000). Functional evidence of a constitutively active population of alpha(1D)-adrenoceptors in rat aorta. *J. Pharmacol. Exp. Ther.* **295,** 810–817.
Gisbert, R., Ziani, K., Miquel, R., Noguera, M. A., Ivorra, M. D., Anselmi, E., and D'Ocon, P. (2002). Pathological role of a constitutively active population of alpha (1D)-adrenoceptors in arteries of spontaneously hypertensive rats. *Br. J. Pharmacol.* **135,** 206–216.
Gisbert, R., Pérez-Vizcaino, F., Cogolludo, A. L., Noguera, M. A., Ivorra, M. D., Tamargo, J., and D'Ocon, P. (2003). Cytosolic Ca2+ and phosphoinositide hydrolysis linked to constitutively active alpha 1D-adrenoceptors in vascular smooth muscle. *J. Pharmacol. Exp. Ther.* **305,** 1006–1014.
Gómez-Zamudio, J., Lázaro-Suárez, M. L., Villalobos-Molina, R., and Urquiza-Marín, H. (2002). Evidence for the use of agonists to characterize alpha 1-adrenoceptors in isolated arteries of the rat. *Proc. West. Pharmacol. Soc.* **45,** 159–160.
González-Hernández, M. L., Godinez-Hernández, D., Bobadilla-Lugo, R. A., and López-Sánchez, P. (2010). Angiotensin-II type 1 receptor (ATR) and alpha-1D adrenoceptor form a heterodimer during pregnancy-induced hypertension. *Auton. Autacoid Pharmacol.* 10.1111/j.1474-8673.2009.00446.x (in press).
Grynkiewicz, G., Poenie, M., and Tsien, R. Y. (1985). A new generation of Ca2+ indicators with greatly improved fluorescence properties. *J. Biol. Chem.* **260,** 3440–3450.
Hague, C., Chen, Z., Pupo, A. S., Schulte, N. A., Toews, M. L., and Minneman, K. P. (2004a). The N terminus of the human {alpha}1D-adrenergic receptor prevents cell surface expression. *J. Pharmacol. Exp. Ther.* **309,** 388–397.
Hague, C., Uberti, M. A., Chen, Z., Hall, R. A., and Minneman, K. P. (2004b). Cell surface expression of {alpha}1d-adrenergic receptors is controlled by heterodimerization with {alpha}1B-adrenergic receptors. *J. Biol. Chem.* **279,** 15541–15549.
Hague, C., Lee, S. E., Chen, Z., Prinster, S. C., Hall, R. A., and Minneman, K. P. (2006). Heterodimers of alpha1B- and alpha1D-adrenergic receptors form a single functional entity. *Mol. Pharmacol.* **69,** 45–55.

Han, C., Abel, P. W., and Minneman, K. P. (1987). Heterogeneity of alpha 1-adrenergic receptors revealed by chlorethylclonidine. *Mol. Pharmacol.* **32,** 505–510.

Hieble, J. P., Bylund, D. B., Clarke, D. E., Eikenburg, D. C., Langer, S. Z., Lefkowitz, R. J., Minneman, K. P., and Ruffolo, R. R., Jr. (1995). International Union of Pharmacology. X. Recommendation for nomenclature of alpha 1-adrenoceptors: Consensus update. *Pharmacol. Rev.* **47,** 267–270.

Ibarra, M., Terrón, J. A., López-Guerrero, J. J., and Villalobos-Molina, R. (1997). Evidence for an age-dependent functional expression of alpha 1D-adrenoceptors in the rat vasculature. *Eur. J. Pharmacol.* **322,** 221–224.

Kawanabe, Y., Hashimoto, N., Masaki, T., and Miwa, S. (2001). Ca(2+) influx through nonselective cation channels plays an essential role in noradrenaline-induced arachidonic acid release in Chinese hamster ovary cells expressing alpha(1A)-, alpha(1B)-, or alpha (1D)-adrenergic receptors. *J. Pharmacol. Exp. Ther.* **299,** 901–907.

Kenakin, T. (2004). Efficacy as a vector: The relative prevalence and paucity of inverse agonism. *Mol. Pharmacol.* **65,** 2–11.

Kenakin, T. P. (2009). Cellular assays as portals to seven-transmembrane receptor-based drug discovery. *Nat. Rev. Drug Discov.* **8,** 617–626.

Kjelsberg, M. A., Cotecchia, S., Ostrowski, J., Caron, M. G., and Lefkowitz, R. J. (1992). Constitutive activation of the alpha 1B-adrenergic receptor by all amino acid substitutions at a single site. Evidence for a region which constrains receptor activation. *J. Biol. Chem.* **267,** 1430–1433.

Leclerc, G., Rouot, B., Schwartz, J., Velly, J., and Wermuth, C. G. (1980). Studies on some para-substituted clonidine derivatives that exhibit an alpha-adrenoceptor stimulant activity. *Br. J. Pharmacol.* **71,** 5–9.

Lomasney, J. W., Cotecchia, S., Lorenz, W., Leung, W. Y., Schwinn, D. A., Yang-Feng, T. L., Brownstein, M., Lefkowitz, R. J., and Caron, M. G. (1991). Molecular cloning and expression of the cDNA for the alpha 1A-adrenergic receptor. The gene for which is located on human chromosome 5. *J. Biol. Chem.* **266,** 6365–6369.

McCune, D. F., Edelmann, S. E., Olges, J. R., Post, G. R., Waldrop, B. A., Waugh, D. J., Perez, D. M., and Piascik, M. T. (2000). Regulation of the cellular localization and signaling properties of the alpha(1B)- and alpha(1D)-adrenoceptors by agonists and inverse agonists. *Mol. Pharmacol.* **57,** 659–666.

Methven, L., Simpson, P. C., and McGrath, J. C. (2009). Alpha1A/B-knockout mice explain the native alpha1D-adrenoceptor's role in vasoconstriction and show that its location is independent of the other alpha1-subtypes. *Br. J. Pharmacol.* **158,** 1663–1675.

Morello, J. P., Petaja-Repo, U. E., Bichet, D. G., and Bouvier, M. (2000). Pharmacological chaperones: A new twist on receptor folding. *Trends Pharmacol. Sci.* **21,** 466–469.

Noguera, M. A., Ivorra, M. D., and D'Ocon, P. (1996). Functional evidence of inverse agonism in vascular smooth muscle. *Br. J. Pharmacol.* **119,** 158–164.

Perez, D. M., Piascik, M. T., and Graham, R. M. (1991). Solution-phase library screening for the identification of rare clones: Isolation of an alpha 1D-adrenergic receptor cDNA. *Mol. Pharmacol.* **40,** 876–883.

Perez, D. M., DeYoung, M. B., and Graham, R. M. (1993). Coupling of expressed alpha 1B- and alpha 1D-adrenergic receptor to multiple signaling pathways is both G protein and cell type specific. *Mol. Pharmacol.* **44,** 784–795.

Piascik, M. T., Smith, M. S., Soltis, E. E., and Perez, D. M. (1994). Identification of the mRNA for the novel alpha 1D-adrenoceptor and two other alpha 1-adrenoceptors in vascular smooth muscle. *Mol. Pharmacol.* **46,** 30–40.

Piascik, M. T., Guarino, R. D., Smith, M. S., Soltis, E. E., Saussy, D. L., Jr., and Perez, D. M. (1995). The specific contribution of the novel alpha-1D adrenoceptor to the contraction of vascular smooth muscle. *J. Pharmacol. Exp. Ther.* **275,** 1583–1589.

Pupo, A. S., Uberti, M. A., and Minneman, K. P. (2003). N-terminal truncation of human alpha1D-adrenoceptors increases expression of binding sites but not protein. *Eur. J. Pharmacol.* **462**, 1–8.

Rodríguez-Pérez, C. E., Romero-Ávila, M. T., Reyes-Cruz, G., and García-Sáinz, J. A. (2009). Signaling properties of human alpha(1D)-adrenoceptors lacking the carboxyl terminus: Intrinsic activity, agonist-mediated activation, and desensitization. *Naunyn Schmiedeberg's Arch. Pharmacol.* **380**, 99–107.

Rossier, O., Abuin, L., Fanelli, F., Leonardi, A., and Cotecchia, S. (1999). Inverse agonism and neutral antagonism at alpha(1a)- and alpha(1b)-adrenergic receptor subtypes. *Mol. Pharmacol.* **56**, 858–866.

Rubio, C., Moreno, A., Briones, A., Ivorra, M. D., D'Ocon, P., and Vila, E. (2002). Alterations by age of calcium handling in rat resistance arteries. *J. Cardiovasc. Pharmacol.* **40**, 832–840.

Schwinn, D. A., and Kwatra, M. M. (1998). Expression and regulation of alpha-1-adrenergic receptors in human tissues. *In* "Catecholamines. Bridging Basic Science with Clinical Medicine," (D. S. Goldstein, G. Eisenhofer, and R. McCarty, eds.), pp. 390–394. Academic Press, San Diego.

Schwinn, D. A., Johnston, G. I., Page, S. O., Mosley, M. J., Wilson, K. H., Worman, N. P., Campbell, S., Fidock, M. D., Furness, L. M., and Parry-Smith, D. J. (1995). Cloning and pharmacological characterization of human alpha-1 adrenergic receptors: Sequence corrections and direct comparison with other species homologues. *J. Pharmacol. Exp. Ther.* **272**, 134–142.

Shinoura, H., Shibata, K., Hirasawa, A., Tanoue, A., Hashimoto, K., and Tsujimoto, G. (2002). Key amino acids for differential coupling of alpha1-adrenergic receptor subtypes to Gs. *Biochem. Biophys. Res. Commun.* **299**, 142–147.

Smith, D. F., Whitesell, L., and Katsanis, E. (1998). Molecular chaperones: Biology and prospects for pharmacological intervention. *Pharmacol. Rev.* **50**, 493–514.

Tanoue, A., Nasa, Y., Koshimizu, T., Shinoura, H., Oshikawa, S., Kawai, T., Sunada, S., Takeo, S., and Tsujimoto, G. (2002). The alpha(1D)-adrenergic receptor directly regulates arterial blood pressure via vasoconstriction. *J. Clin. Invest.* **109**, 765–775.

Theroux, T. L., Esbenshade, T. A., Peavy, R. D., and Minneman, K. P. (1996). Coupling efficiencies of human alpha 1-adrenergic receptor subtypes: Titration of receptor density and responsiveness with inducible and repressible expression vectors. *Mol. Pharmacol.* **50**, 1376–1387.

Vázquez-Prado, J., and García-Sáinz, J. A. (1996). Effect of phorbol myristate acetate on alpha 1-adrenergic action in cells expressing recombinant alpha 1-adrenoceptor subtypes. *Mol. Pharmacol.* **50**, 17–22.

Vázquez-Prado, J., Medina, L. C., Romero-Ávila, M. T., González-Espinosa, C., and García-Sáinz, J. A. (2000). Norepinephrine- and phorbol ester-induced phosphorylation of alpha(1a)-adrenergic receptors. Functional aspects. *J. Biol. Chem.* **275**, 6553–6559.

Villalobos-Molina, R., and Ibarra, M. (1999). Vascular alpha 1D-adrenoceptors: Are they related to hypertension? *Arch. Med. Res.* **30**, 347–352.

Villalobos-Molina, R., and Ibarra, M. (2005). Increased expression and function of vascular alpha1D-adrenoceptors may mediate the prohypertensive effects of angiotensin II. *Mol. Interv.* **5**, 340–342.

Villalobos-Molina, R., Lopez-Guerrero, J. J., and Ibarra, M. (1997a). Alpha 1D- and alpha 1A-adrenoceptors mediate contraction in rat renal artery. *Eur. J. Pharmacol.* **322**, 225–227.

Villalobos-Molina, R., Vazquez-Prado, J., and Garcia-Sainz, J. A. (1997b). Chloroethylclonidine is a partial alpha1A-adrenoceptor agonist in cells expressing recombinant alpha1-adrenoceptor subtypes. *Life Sci.* **61**, PL 391–PL 395.

Villalobos-Molina, R., López-Guerrero, J. J., and Ibarra, M. (1999). Functional evidence of alpha1D-adrenoceptors in the vasculature of young and adult spontaneously hypertensive rats. *Br. J. Pharmacol.* **126,** 1534–1536.

Villalobos-Molina, R., Vazquez-Cuevas, F. G., Lopez-Guerrero, J. J., Figueroa-Garcia, M. C., Gallardo-Ortiz, I. A., Ibarra, M., Rodriguez-Sosa, M., Gonzalez, F. J., and Elizondo, G. (2008). Vascular alpha-1D-adrenoceptors are overexpressed in aorta of the aryl hydrocarbon receptor null mouse: Role of increased angiotensin II. *Auton. Autacoid Pharmacol.* **28,** 61–67.

Weinberg, D. H., Trivedi, P., Tan, C. P., Mitra, S., Perkins-Barrow, A., Borkowski, D., Strader, C. D., and Bayne, M. (1994). Cloning, expression and characterization of human alpha adrenergic receptors alpha 1a, alpha 1b and alpha 1c. *Biochem. Biophys. Res. Commun.* **201,** 1296–1304.

Whitworth, J. A. (2003). 2003 World Health Organization (WHO)/International Society of Hypertension (ISH) statement on management of hypertension. *J. Hypertens.* **21,** 1983–1992.

Yang, M., Verfurth, F., Buscher, R., and Michel, M. C. (1997). Is alpha1D-adrenoceptor protein detectable in rat tissues? *Naunyn Schmiedeberg's Arch. Pharmacol.* **355,** 438–446.

Ziani, K., Gisbert, R., Noguera, M. A., Ivorra, M. D., and D'Ocon, P. (2002). Modulatory role of a constitutively active population of alpha(1D)-adrenoceptors in conductance arteries. *Am. J. Physiol. Heart Circ. Physiol.* **282,** H475–H481.

CHAPTER SEVEN

Constitutive Activity of the Histamine H_1 Receptor

Saskia Nijmeijer, Rob Leurs, *and* Henry F. Vischer

Contents

1. Introduction	128
1.1. The inflammatory histamine H_1 receptor	128
1.2. Constitutive GPCR signaling	129
1.3. Receptor expression level determines constitutive signaling	129
1.4. G proteins determine constitutive receptor activity	133
2. Methods to Study Constitutive H_1R Signaling	133
2.1. Inositol phosphate accumulation assay	134
2.2. Reporter gene assays	136
2.3. Receptor selection and amplification technology (R-SAT)	138
2.4. Real-time cell electronic sensing (RT-CES) system	140
2.5. Determination of receptor up-regulation upon prolonged inverse agonist treatment	140
3. Constitutive Activity as Tool to Elucidate Receptor Activation and Crosstalk	142
3.1. Constitutive active mutants of the H_1R to investigate receptor activation	142
3.2. Constitutive activity of H_1R potentiates $G\alpha_i$-coupled signaling	142
4. Conclusion	144
Acknowledgments	144
References	145

Abstract

The histamine H_1 receptor (H_1R) is a key player in acute inflammatory responses. Antihistamines are widely used to relief the symptoms of allergic rhinitis by antagonizing histamine binding to the H_1R, without possessing intrinsic activity in classical assays such as guinea pig ileum contraction assays or intracellular Ca^{2+} mobilization. Overexpression of H_1R in heterologous cell lines unmasked the capacity of this receptor to signal in a histamine-independent manner. Moreover, a recent screen of therapeutic and reference antagonists on these H_1R-overexpressing cells revealed that the

Leiden/Amsterdam Center for Drug Research, VU University Amsterdam, Amsterdam, The Netherlands majority of these drugs are in fact inverse agonists, as they inhibit basal H_1R activity. In this chapter, we describe several approaches to study H_1R constitutive signaling that can be used to identify inverse agonists acting at this blockbuster target.

1. INTRODUCTION

1.1. The inflammatory histamine H_1 receptor

A plethora of physiological and pathological processes are associated with the local release of histamine, including inflammation, neurotransmission, and gastric acid secretion. Histamine exerts its bioactivity by acting on four G protein-coupled histamine receptor subtypes (Parsons and Ganellin, 2006). Allergen or pathogen recognition by mast cells or basophils triggers the rapid release of preformed histamine from granules in these cells. Subsequent activation of nearby H_1Rs that are present on sensory nerves, smooth muscle cells, airway epithelium, and vascular endothelium, results in several symptoms of (allergic) inflammatory responses, such as sneezing, itch, bronchoconstriction, vasodilation, increased vascular permeability, up-regulation of cell adhesion molecules, release of pro-inflammatory cytokines, prostaglandins, and nitric oxide (Hill et al., 1997). The search for compounds that counteract histamine-induced allergic disorders started already in the 1930s by Daniel Bovet, who received the Nobel Prize in Physiology or Medicine in 1957. The first generation antihistamines (e.g., mepyramine, diphenhydramine, and chlorpheniramine) were introduced in the 1940s and became widely used for the treatment of, for example, hay fever and allergic rhinitis. In the 1970s, nonsedative (second generation) antihistamines were identified, such as, for example, the "blockbuster" drug loratidine (Claritin®). These antihistamines were identified by measuring their capacity to antagonize the histamine-induced contractile response of isolated smooth muscles (e.g., guinea pig ileum and trachea) in ex vivo bioassays.

In 1993–1994, the human H_1R was cloned, allowing heterologous expression of this receptor protein in convenient cell lines for in vitro high throughput drug screening by measuring intracellular signaling (De Backer et al., 1993; Fukui et al., 1994; Moguilevsky et al., 1994). In addition, these heterologous expression systems allowed more detailed analysis of H_1R-induced signaling pathways by coexpressing signaling constituents and/or reporter genes under control of a specific transcription factor. Agonist binding to the H_1R induces coupling and activation of $G_{q/11}$ proteins (Leopoldt et al., 1997), leading to the stimulation of phospholipase C (PLC) enzyme and the subsequent formation of inositol-1,4,5-triphosphate ($InsP_3$) and diacylglycerol (DAG) from cell membrane phosphatidylinositol 4,5-bisphosphate (PIP_2), followed by the release of Ca^{2+} from endoplasmic reticulum stores, and the activation of protein kinase C (Hill et al., 1997). In addition, H_1R can activate several other pathways including adenylyl cyclase, phospholipase A2, and phospholipase D. H_1R signaling activates the transcription factors nuclear factor kappa B (NF-κB), cAMP response element-binding (CREB), Nuclear Factor of Activated T-Cells (NFAT), and serum response element (SRE) (Bakker et al., 2001; Boss et al., 1998; Hao et al., 2008; Notcovich et al., 2010). Interestingly, these heterologous expression systems also uncovered a hidden property of the H_1R, namely agonist-independent signaling (Bakker et al., 2000, 2001). Constitutive activity has also been shown for the other three histamine receptor subtypes (Morisset et al., 2000; Morse et al., 2001; Smit et al., 1996; Wieland et al., 2001).

1.2. Constitutive GPCR signaling

Traditionally, G protein-coupled receptors (GPCRs) were always seen as on/off switches that are inactive in the absence of agonists and trigger intracellular signaling upon agonist binding. However, in the last two decades more and more evidence was published that GPCRs can signal in an agonist-independent, constitutive manner. GPCRs coexist in a dynamic equilibrium between inactive (R) and active (R★) conformations. Agonists have higher affinity for the R★ states of the receptor and will shift the equilibrium toward R★. In contrast, the so-called inverse agonists display higher affinity for the R states of the receptor, thereby shifting the equilibrium toward an inactive conformation consequently resulting in a decrease of constitutive receptor signaling (Fig. 7.1). Finally, true antagonists do not change the equilibrium between R and R★ conformations and consequently display no efficacy with respect to receptor signaling. However, the probability that a ligand has similar affinities for the R and R★ conformations is rather unlikely and it has been shown that 86% of all previously identified antagonists are in fact inverse agonists (Kenakin, 2004). This also accounts for the majority of antihistamines that were originally classified as competitive antagonists in systems that display no apparent constitutive activity. These drugs, were recently reclassified as H_1R inverse agonist in recombinant H_1R overexpressing cells by measuring cell proliferation using receptor selection and amplification technology (R-SAT) or by investigating NF-κB activation (see Table 7.1; Bakker et al., 2007). So far, only histabudifen and histapendifen (Fig. 7.2) appeared to be true neutral H_1R antagonists (Govoni et al., 2003).

1.3. Receptor expression level determines constitutive signaling

Generally only a small proportion of receptors spontaneously isomerizes to a R★ conformation. Consequently constitutive activity of most receptors is low and in native cells often well below the detection sensitivity of most

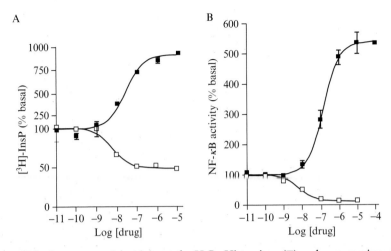

Figure 7.1 Inverse agonist activity at the H_1R. Histamine- (■) and mepyramine- (□) induced modulation of H_1R activation in COS7 cells. (A) inositol phosphate accumulation assay or (B) NF-κB luciferase reporter gene assay. Basal constitutive H_1R activity is set at 100%. This research was originally published by Bakker et al. (2004a), © the American Society for Biochemistry and Molecular Biology.

Table 7.1 Antagonists reclassified as inverse agonists on H_1R

Compound	Intrinsic activity (α) in NF-κB[a]	Intrinsic activity (α) in R-SAT[a]
Mepyramine	−0.9	−1.0
Acrivastine	−0.7	−1.0
Loratadine	−1.0	−0.8
Cyproheptadine	−0.9	−0.8
Ketotifen	−0.9	−1.0
Doxepin	−0.9	−1.0
Mianserin	−0.9	−1.0
Tripelennamine	−0.8	−1.0
D-Chlorpheniramine	−0.9	−1.0
Mirtazapine	−0.9	−0.9
Triprolidine	−0.7	−1.0
Levocabastine	−1.0	−1.0
Astemizole	−1.0	−1.0
Diphenhydramine	−0.7	−0.9

[a] Bakker et al. (2007, 2008).

signal transduction assays. However, heterologous overexpression of GPCRs increases the absolute numbers in the R★ conformation to such extent that agonist-independent signaling can be detected. Stable expression

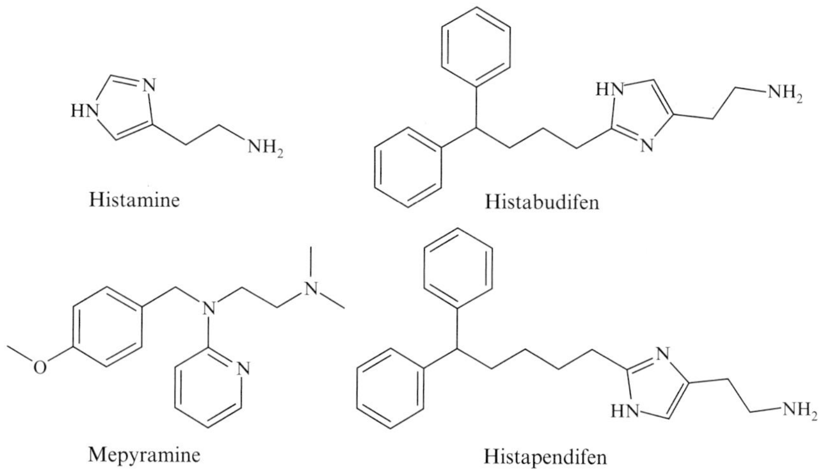

Figure 7.2 Structures of H_1R agonist, inverse agonist, and neutral antagonists. Chemical structures of the often used ligands histamine (endogenous agonist) and mepyramine (pyrilamine, inverse agonist). At the right hand side the two structures of the true neutral antagonists histabudifen and histapendifen are shown (Govoni et al., 2003).

of rat histamine H_2 receptor (H_2R) at a density of 96 ± 26 fmol receptor per mg protein did not significantly change basal cAMP levels in CHO cells. However, 3- and 10-fold higher H_2R numbers resulted in a significant 1.5- and 4.7-fold increase in agonist-independent signaling (Smit et al., 1996). Likewise, transient expression of H_1R at a density of 1 ± 0.1 to 4.2 ± 0.2 pmol receptors per mg protein unmasked constitutive H_1R activity (Fig. 7.3A), which could be fully inhibited by the inverse agonist mepyramine (Bakker et al., 2001). These expression levels are considerably higher than the 65 ± 7 and 227 ± 52 fmol H_1R per mg protein in guinea pig ileum and brain homogenates, respectively (Hill et al., 1977, 1978). Importantly, however, H_1R expression levels have been found upregulated under various pathophysiological conditions, such as in patients with allergic rhinitis (Dinh et al., 2005; Iriyoshi et al., 1996), which concomitantly might increase basal H_1R-mediated signaling. In fact, histamine-induced H_1R activation upregulates H_1R expression in HeLa cells (Das et al., 2007). Interestingly, short-term administration of the inverse H_1R agonist D-chlorpheniramine inhibited basal H_1R promoter activity and transcript levels, revealing that the H_1R constitutively stimulates its own gene expression. On the other hand, long-term treatment (i.e., 24 h) of HeLa cells with the inverse H_1R agonist tripelennamine results in a significant twofold increase of H_1R protein levels (Fig. 7.4A), which is associated with increased histamine-induced maximum signaling (Fig. 7.4B) (Govoni et al., 2003). Similar increase in receptor or G protein levels

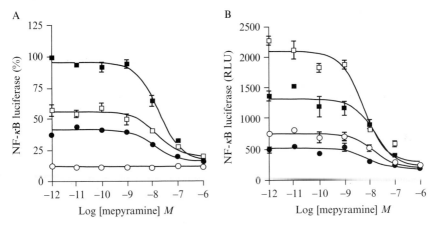

Figure 7.3 Receptor and G protein expression levels are important determinants of constitutive activity. (A) Effect of mepyramine on basal NF-κB reporter gene activity in COS7 cells transiently expressing 0 (mock) (○); 1 ± 0.1 (●); 2.8 ± 0.1 (□); or 4.2 ± 0.2 (■) pmol/mg H_1R. (B). Effects of mepyramine on the basal H_1R-induced NF-κB reporter gene activity in COS7 cells transiently transfected with H_1R in absence (●) or presence of either 1 (○), 5 (■), or 25 μg (□) pCMVGα$_{11}$ per 10^7 cells. RLU = relative light units. Graphs are adapted from Leurs et al. (2000).

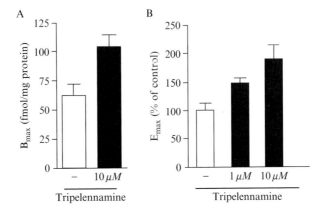

Figure 7.4 H_1R up-regulation after prolonged treatment with inverse agonist tripelennamine. HeLa cells endogenously expressing H_1R are treated for 24 h with 1 or 10 μM inverse agonist tripelennamine. (A) A significant up-regulation (B_{max}) of H_1R expression was determined by [^{3}H]-mepyramine binding. (B) Significant increase in histamine-induced inositol phosphate accumulation upon pretreatment with tripelennamine. Adapted from Govoni et al. (2003) with permission from ACS.

upon prolonged exposure to inverse agonists has been described for various other GPCRs (Berg et al., 1999; Smit et al., 1996).

1.4. G proteins determine constitutive receptor activity

Besides receptor numbers, also the expression levels of G proteins influence the level of constitutive receptor activity in different cell types. G proteins preferentially bind active (R★) receptor states, thereby shifting the equilibrium between R and R★ conformations toward the latter (Burstein et al., 1995a, 1997). Indeed, coexpressing extra $G\alpha_{11}$, $G\alpha_q$, or $G\beta\gamma$ proteins increased constitutive H_1R signaling, which could be repressed by the full inverse agonist mepyramine (Fig. 7.3B; Bakker et al., 2007; Fitzsimons et al., 2004; Leurs et al., 2000). Importantly, $G\alpha_q$ protein levels are upregulated in the nasal mucosa of a guinea pig model of nasal hyper-responsiveness (Chiba et al., 2002). This stabilization of the R★ conformation of H_1R by $G\alpha_q$ resulted in an increase of affinity for the agonist histamine. Surprisingly, overexpression of $G\alpha_{11}$ proteins has been reported to increase the binding affinity of the inverse agonist mepyramine to guinea pig H_1R (Fitzsimons et al., 2004), which suggests that G proteins can also couple to an inactive receptor conformation as predicted by the cubic ternary complex model (Weiss et al., 1996).

2. Methods to Study Constitutive H₁R Signaling

Heterologous overexpression of the H_1R or its cognate G proteins (i.e., $G_{q/11}$) in convenient cell lines, such as COS7, HEK293T, or CHO allows the detection of constitutive H_1R-mediated signaling. Constitutive activity can typically be quantified using accumulation assays such as the formation of second messengers in the presence of inhibitors that prevent degradation of these molecules, transcription of reporter genes, and cell proliferation. On the other hand, label-free impedance biosensors allow the measurement of transient agonist and inverse agonist responses. Commonly used readouts for constitutive activity along the H_1R signal transduction cascade are indicated in Fig. 7.5, and will be discussed below. Generation of H_1R mutants that display no or increased constitutive activity by site-directed mutagenesis allows the identification of amino acid residues that constrain the H_1R in inactive or active conformations, respectively. Importantly, detailed biochemical analysis of these constitutive inactive mutants (CIM) and constitutive active mutants (CAM) in combination with homology modeling and computational approaches provides valuable insight in the molecular activation mechanism of the H_1R.

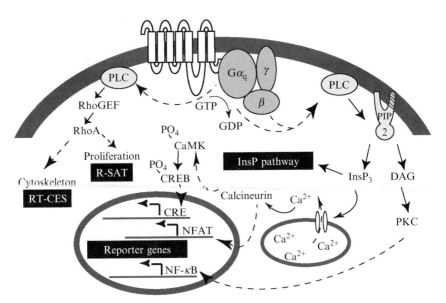

Figure 7.5 Signaling pathways modulated by the H_1R. Activated (constitutive or agonist-induced) H_1R initiates signaling via the $G\alpha_{q/11}$ family of heterotrimeric G proteins. Black boxes with white text indicate assays that are discussed in this review.

2.1. Inositol phosphate accumulation assay

PLC is the main intracellular effector enzyme of $G_{q/11}$-coupled GPCRs. PLC hydrolyzes PIP_2 into $InsP_3$ and DAG. The intracellular second messenger $InsP_3$ is rapidly degraded by phosphatases and recycled back via inositol into cell membrane PIP_2. The inositol phosphate accumulation assay utilizes the ability of lithium to inhibit the breakdown of inositol monophosphates ($InsP_1$), resulting in an accumulation of $InsP_3$, $InsP_2$, and $InsP_1$. In order to detect these accumulated inositol phosphates (InsP), cells are cultured in the presence of myo-$[2\text{-}^3H]$-inositol, which is metabolically incorporated into the cell membrane as $[^3H]$-PIP_2. PLC-mediated hydrolysis results in formation of $[^3H]$-$InsP_3$, $[^3H]$-$InsP_2$, and $[^3H]$-$InsP_1$ that can be quantified after separation from myo-$[2\text{-}^3H]$-inositol and $[^3H]$-PIP_2 via anion exchange chromatography.

2.1.1. Special materials and reagents required

- Cells expressing H_1R (either stable or transient).
  - HeLa, COS7, HEK293(T), CHO
- Cell culture medium
- Inositol-free medium (MEM with Earle's salts)
- myo-$[2\text{-}^3H]$-Inositol (Perkin Elmer/specific activity 17 Ci/mmol)
- Stimulation buffer; culture medium (DMEM), HEPES, LiCl

- Anion exchange columns (Dowex 1*8–400, 200–400 mesh, 1-chloride form)
- For regeneration, loading, washing, and elution of the dowex columns; sodium formate, formic acid, myo-inositol, borax, and ammonium formate
- Scintillation liquid
- 20 ml counting vials
- Packard 1900TM liquid scintillation counter

2.1.2. Description of the assay

H_1R-expressing cells are seeded in a 24- or 48-well plate in cell culture medium. In a transient expression system usually 625–1250 ng cDNA $hH_1R/10^6$ cells (e.g., COS7/HEK293(T)/CHO) is transfected. We generally pre-coat the well plates with poly-L-lysine to reduce cell loss during the assay. Cells are labeled with 1 μCi/ml *myo*-[2-^{3}H]inositol for 16 h in inositol-free culture medium supplemented with appropriate amounts of serum and antibiotics (depending on the cell line). To start the stimulation, medium is aspirated from the wells and appropriate compounds are added to the cells. Stimulation takes place in culture medium without serum supplemented with 25 mM HEPES pH 7.4 and 20 mM LiCl. Cells are incubated for 1 h (maximal 2 h) in a 37 °C and 5% CO_2 incubator.

Meanwhile, regenerate the Dowex columns for the anion exchange chromatography by washing the columns once with H_2O followed by solution I (3 M ammonium formate, 100 mM formic acid) again wash once with H_2O and finally with solution II (10 mM myo-inositol, 10 mM formic acid). The columns are now ready for loading of the samples.

Ligand stimulation is terminated by replacing stimulation medium with 10 mM formic acid (1.5 h at 4 °C or 0.5 h at room temperature) to extract [^{3}H]-inositol and accumulated [^{3}H]-InsP from the cytosol. Next, [^{3}H]-inositol is separated from [^{3}H]-InsP by loading the extractions on the columns followed by a wash with solution III (60 mM sodium formate, 5 mM borax). Elution conditions depend on the fraction of the $InsP_x$ mix (i.e., $InsP_3$, $InsP_2$, and/or $InsP_1$) you prefer to elute. The total [^{3}H]-InsP pool can be eluted by adding solution IV (1 M ammonium formate, 100 mM formic acid) to the columns. Eluents are collected in vials containing liquid scintillation fluid and quantified in a beta counter (Packard 1900 TR liquid scintillation counter).

2.1.3. Points of attention

Proper control conditions should always be taken along; that is, mock cells (cells expressing no H_1R) and nonstimulated H_1R expressing cells only incubated with stimulation buffer, to determine background and maximal basal constitutive activity, respectively.

To rule out any endogenously present agonist (histamine) in the assay that interferes with the determination of negative intrinsic activity, conditions can be taken along in which labeling is performed in *serum-free* culture medium. This rules out a possible histamine contamination in the incubation medium (Smit et al., 1996). In addition, to prevent endogenous synthesis of histamine in the Golgi apparatus by histidine decarboxylase, 10 or 100 μM S-(+)-α-fluoromethyl-histidine (irreversible inhibitor) can be added to the cells (Bakker et al., 2000; Hill et al., 1997).

Nowadays there are also high(er) throughput variants available for this assay. The anion exchange columns are available in miniaturized 96-wells format that allows the assay to be scaled down to a 96-well format.

Accumulated inositol phosphates can also be quantified via other methods than the time-consuming column chromatography. The scintillation proximity assay (SPA) uses positively charged yttrium silicate beads to separate $[^3H]$-InsP from $[^3H]$-inositol in cell extracts (Brandish et al., 2003). We have used this SPA-based method successfully to identify inverse agonists acting on the constitutive active herpesvirus-encoded GPCR US28 (Vischer et al., 2010).

A non-radioactive $InsP_1$ assay from CisBio based on their developed HTRF technology (HTRF®, CISbio International, Bagnols-sur-Cèze, France) is another option. Cells are lysed after the stimulation and subsequently an antibody-based ratiometric HTRF detection of accumulated $InsP_1$ is done using cryptate-labeled anti-$InsP_1$ monoclonal antibodies and d2-labeled $InsP_1$ (Bergsdorf et al., 2008).

2.2. Reporter gene assays

Receptor-mediated activation of transcription factors can be measured via a bioluminescent or enzymatic reporter gene (e.g., luciferase, β-galactosidase, or β-lactamase). The reporter gene is expressed upon binding of transcription factors to their response elements in a promoter region (Fig. 7.6A). Reporter gene assays have proven to be a useful readout for constitutive activity measurements because of their signal amplification. The H_1R is known to signal to NFAT, SRE, NF-κB, and CREB (Bakker et al., 2001; Smit et al., 2002).

2.2.1. Special materials and reagents required

- Reporter gene construct; TLN3xSRE-LUC, p5xNF-κB-LUC (Stratagene), pTLNC21xCRE-LUC
- Cells expressing both the H_1R and the reporter gene
- 96-wells tissue culture plates (black or white bottom)

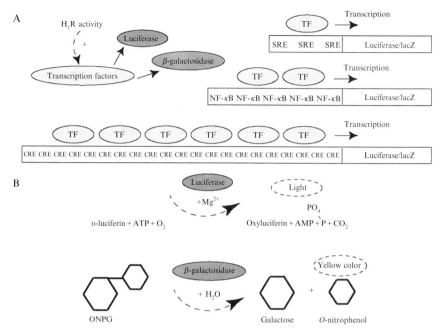

Figure 7.6 Principle of luciferase and β-galactosidase reporter gene assays. (A) Transciption factor (TF) responsive elements (SRE, NF-κB, CRE) are fused to a reporter gene encoding luciferase or β-galactosidase. The optimal amount of response elements are determined previously, 3× SRE, 5× NF-κB, and 21× CRE. H_1R activity leads to the activation of transcription factors and subsequent gene regulation. The transcription factors bind to their respective promoter regions and initiate gene transcription. The more transcription factors present, the more binding, activation, and hence translation, the higher the final signal. (B) Schematic representation of the components of the enzyme/substrate reactions. Luciferase reporter gene assays (B, top panel) measure light emission that is produced once D-luciferine is converted to oxyluciferin by the enzyme luciferase. *ortho*-Nitrophenyl-β-galactoside (ONPG) is converted to galactose and *o*-nitrophenol (B, bottom panel). The absorbance (yellow color of *o*-nitrophenol) can be measured on a plate reader.

- Cell culture medium
- Serum free cell culture medium
- For luciferase assay; D-luciferine (promega), $MgCl_2$, ATP, glycerol, Tris, sodium-pyrophoshate, Triton-X100, and dithiothreitol.
- For β-galactosidase assay; sodium phosphate, 2-nitrophenol β-D-pyranoside (ONPG), Triton X-100, $MgSO_4$, $MnCl_2$, and β-mercaptoethanol
- Victor^3 multilabel reader (or other plate reader capable of measuring luminescence).

2.2.2. Description of the assay

Cells are cotransfected in suspension with 625 ng H_1R and 3 μg reporter gene DNA per 10^6 cells. Transfected cells are seeded in 96-wells plates (white or black bottom) in serum-free culture medium.[1] Directly after transfection, cells are stimulated with compounds. This rapid addition of compounds is necessary to inhibit constitutive signaling right from appearance of H_1R protein at the cell membrane. Stimulation at a later time point will cause a significant increase in reporter protein in absence of inverse agonist, resulting in a smaller inverse agonist window.

Luciferase readout: After 48 h, cells are lysed and incubated for 30 min at 37 °C with the substrate D-luciferine, present in luciferase assay reagent (LAR; 80.83 mM ATP, 0.83 mM D-luciferine, 18.7 mM $MgCl_2$, 0.78 μM $Na_2H_2P_2O_7$, 38.9 mM Tris, pH 7.8, 0.39% (v/v) glycerol, 0.03% (v/v) Triton-X100, and 2.6 μM dithiothreitol). Luminescence can be measured in an appropriate plate reader for 1–3 s per well (Fig. 7.6B).

β-*Galactosidase readout:* After 48 h, cells are lysed in 100 μl assay buffer/well (100 mM sodium phosphate buffer, pH 8.0, 4 mM ONPG, 0.5% Triton X-100, 2 mM $MgSO_4$, 0.1 mM $MnCl_2$, and 40 mM β-mercaptoethanol). Cells are incubated at room temperature, and β-galactosidase activity is determined by measuring the absorbance at 420 nm in an appropriate plate reader 1 s per well (Fig. 7.6B).

2.2.3. Points of attention

In our experience, the transfection amount of reporter gene should be higher than the amount of receptor DNA. This will avoid a situation in which there is not enough reporter protein to detect further increase in activity.

Both white- and black-bottom plates can be used. White plates are generally more preferred in luminescence detection, despite their higher background autofluorescence and crosstalk.

A promoter region often consists of several (identical) response elements that increase the sensitivity compared to only a single element. Reporter genes are therefore under the transcriptional control of 3 SRE, 5 NF-κB, or 21 CRE enhancer elements.

2.3. Receptor selection and amplification technology (R-SAT)

R-SAT (ACADIA Pharmaceuticals, San Diego, CA) makes use of the property of NIH-3T3 cells to proliferate after GPCR-induced G_q activation (Croston, 2002; Weiner *et al.*, 2001). NIH-3T3 cells normally stop

---

[1] In case of HEK293(T) cells, DMEM without serum cannot be used since HEK293(T) cells will not survive without serum. However, during stimulation it is recommended to use less serum than in normal culture medium to prevent a-specific activation via serum components.

growing after forming a monolayer, due to contact inhibition. However, upon activation of pathways that promote cell growth, they can surmount their contact inhibition resulting in continuous proliferation. GPCR activity and stimulation can indirectly be assessed by quantifying the proliferative response of NIH-3T3 cells using a β-galactosidase marker gene (Brauner-Osborne and Brann, 1996; Burstein et al., 1995b), which allows graded responses to be measured, allowing accurate determinations of ligand potency and efficacy (Bakker et al., 2004b, 2007; Brauner-Osborne and Brann, 1996; Croston, 2002). In contrast to reporter gene assays, the β-galactosidase gene is not under direct transcriptional control of GPCR-activated transcription factors. R-SAT is applicable for both ligand-induced and constitutive H_1R signaling by discriminating between ligand efficacies. Results from R-SAT assays are comparable with other high throughput assays used in drug discovery (Burstein et al., 1997; Smit et al., 2002).

2.3.1. Special materials and reagents required

- NIH-3T3 cells
- Cell culture medium (Dulbecco's modified Eagles medium (A.B.I.)) supplemented with 4500 mg/l glucose, 4 nM L-glutamine, 50 U/ml penicillin G, 50 U/ml streptomycin, and 10% calf serum (Sigma)
- Cyto SF3 synthetic supplement (Kemp Laboratories)
- o-Nitrophenyl-β-D-galactopyranoside
- Nonidet P-40 (Sigma)
- Hanks balanced salt solution (HBSS) ($+MgCl_2$)
- $MgSO_4$
- $CaCl_2$
- Plate reader

2.3.2. Description of the assay

On the first day, NIH-3T3 cells are plated into 96-well plates at a density of 7500 cells/well and maintained in DMEM supplemented with 10% calf serum, 50 U/ml penicillin, 50 U/ml streptomycin, 4 nM L-glutamine, and 4.5 g/l glucose at 37 °C, 5% CO_2.

Next day, cells are transfected (Lipofectamine (GIBCO) or Superfect (Qiagen) according to manuals) with 10–25 ng H_1R DNA/well and 20 ng β-galactosidase/pSI (promega) DNA/well.

Twenty four hours post transfection, compounds (concentration curve) are added to the cells in DMEM containing 2% cyto SF3 synthetic supplement instead of calf serum.

Cells are cultured for 5 days, after which β-galactosidase levels are determined as follows: Cell culture media is removed and the cell layer is incubated with PBS containing 3.5 mM o-nitrophenyl-β-D-galactopyranoside and 0.5% nonidet P-40. The 96-wells plates are incubated at room temperature for 8 h and the colorimetric reaction is measured at 420 nm in an appropriate plate reader (Bakker et al., 2004b; Smit et al., 2002; Weiner et al., 2001).

Data are fitted to the following three parameter equation:

$$R = D + \frac{(A - D)}{1 + (X/C)}$$

In which R is the response, A is the minimum response, D is the maximum response, X is the concentration of ligand, and C is the EC_{50} value (Brauner-Osborne and Brann, 1996; Smit et al., 2002).

2.3.3. Points of attention

Other contact-inhibited cells lines, such as RAT1 or C3HIOTU2, can also be used for R-SAT, however in our hands, NIH-3T3 are transformed easily by transfected genes and are the preferred cell line of choice.

Since the compound incubation time is quite long compared to other assays, compounds should have a reasonable stability.

2.4. Real-time cell electronic sensing (RT-CES) system

Besides signaling to the heterotrimeric G proteins, activity of GPCRs also modulate the cytoskeleton (via actin) and subsequently the morphology of cells via small G proteins. The family of Rho GTPases plays an important role in the regulation and maintenance of the cytoskeleton framework (Etienne-Manneville and Hall, 2002). The actin cytoskeleton is a dynamic and plastic system that directly reflexes intracellular signaling (Dina et al., 2003). GPCRs signaling to Rho GTPases induce a specific morphological change that can be measured via a real-time cell electronic sensing (RT-CES) system. Electronic cell sensors are located in the bottom of the wells in an assay plate (96-wells) that measure very small changes in cell morphology (Solly et al., 2004). This is based on electric impedance and has the advantage that it is a label-free homogenous assay format. Interestingly, this technique is also able to identify inverse agonism for the H_1R (Yu et al., 2006).

2.5. Determination of receptor up-regulation upon prolonged inverse agonist treatment

A general feature of GPCRs is their desensitization and down-regulation upon prolonged agonist exposure. The reverse holds true for the treatment of constitutive active GPCRs with inverse agonists. After long-term exposure to inverse agonists, GPCRs may be sensitized to agonists (Berg et al., 1999) and even up-regulated, resulting in a higher protein amount (Milligan and Bond, 1997). Endogenous H_1R expression levels (B_{max}, 55 fmol/mg protein) and signaling properties in HeLa cells were studied after prolonged inverse agonist exposure. Importantly, the actual binding or signaling to determine the expression levels or E_{max}, respectively, should be done in absence of inverse agonists. A very quick dissociation of the initial inverse agonist is required. Tripelennamine was used as inverse agonist and after 24 h of treatment of HeLa cells a significant 2-fold increase in expression level was observed (Fig. 7.4A). Also a dose-dependent increase in E_{max} was observed (Fig. 7.4B), but not in EC_{50} value as determined in inositol phosphate accumulation assay upon histamine stimulation (Govoni et al., 2003).

2.5.1. Special material and reagents required

- [^{3}H] Mepyramine
- Tripelennamine
- Mianserin
- 50 mM Na$_2$/K phosphate buffer pH 7.4
- GF/C plates (Perkin Elmer)
- Polyethylenimine
- Beta counter

2.5.2. Description of the assay

HeLa cells endogenously expressing the H_1R are preincubated for 24 h with the inverse agonist tripelennamine, which displays one of the fastest dissociation kinetics of all inverse H_1R agonists. Full dissociation of tripelennamine from the pretreated cells is essential to avoid direct interference with radioligand binding resulting in an underestimation of H_1R expression levels. HeLa cells are harvested and homogenized in ice-cold binding buffer (50 mM Na$_2$/K-phosphate buffer, pH 7.4). Cell homogenates are centrifuged for 5 min at 1400× g and subsequently the supernatant for 20 min at 15,000×g. HeLa/H_1R cell membranes are incubated for 30 min at 30 °C in binding buffer containing 1 nM [^{3}H]-mepyramine. A specific binding is determined in the presence of 1 μM mianserin. Incubations are stopped by rapid dilution and subsequent filtration over 0.3% polyethyleneimine-coated Whatman GF/C filters using ice-cold binding buffer. Bound radioligand retained on the filters is measured by liquid scintillation counting.

3. Constitutive Activity as Tool to Elucidate Receptor Activation and Crosstalk

3.1. Constitutive active mutants of the H_1R to investigate receptor activation

Agonist-induced GPCR activation is associated with rigid-body rearrangements of transmembrane helices (TM) to allow the intracellular coupling of G proteins (Nygaard et al., 2009). Comparison of the inactive "dark-state" X-ray structure of bovine rhodopsin with the active "light-state" structure of bovine opsin in complex with the C-terminus of the $G\alpha$ protein transducin revealed considerable movements of TM5, TM6, and TM7 (Scheerer et al., 2008). GPCRs are constrained in an inactive conformation through intramolecular interactions that impede these TM motions. Agonist binding to the receptor releases these constrains (Jongejan et al., 2005). Constitutive activity-inducing mutations disrupt these constrains between microdomains as well, and may consequently provide valuable information on the role of these microdomains in receptor activation (Smit et al., 2007). Using site-directed (random saturation) mutagenesis of motifs that are potentially involved in receptor activation, our group identified several H_1R mutants that display a higher constitutive activity than the WT receptor or display no basal signaling (Fig. 7.7; Bakker et al., 2008; Jongejan et al., 2005). Especially, the orientation of TM6 was proven to be important in the maintenance of active or inactive conformations. $Ile^{6.40}$ was identified as constitutive active "hotspot", since mutating this residue into several other amino acids resulted in all cases in an highly active CAM. Basal signaling was even that strong that almost no additional histamine-induced signaling could be observed (Bakker et al., 2008). Based on these findings a computational working model for the H_1R activation was developed, showing the importance of CAM as tool to study receptor activity states (Bakker et al., 2008; Jongejan et al., 2005).

3.2. Constitutive activity of H_1R potentiates $G\alpha_i$-coupled signaling

An interesting feature of constitutive active H_1Rs is their ability to potentiate signaling of GPCRs that couple to a different heterotrimeric G protein (i.e., $G\alpha_i$). This is thought to be a general phenomenon since H_1Rs are capable of unmasking $G\alpha_i$ signaling of co-expressed serotonin $5HT_{1B}$, muscarinic M2, adenosine A1, and adrenergic α_{2A} receptors (Bakker et al., 2004a).

COS7 cells, co-expressing the H_1R and the $5HT_{1B}$ receptor proteins, show an increased level of inositol phosphate accumulation upon

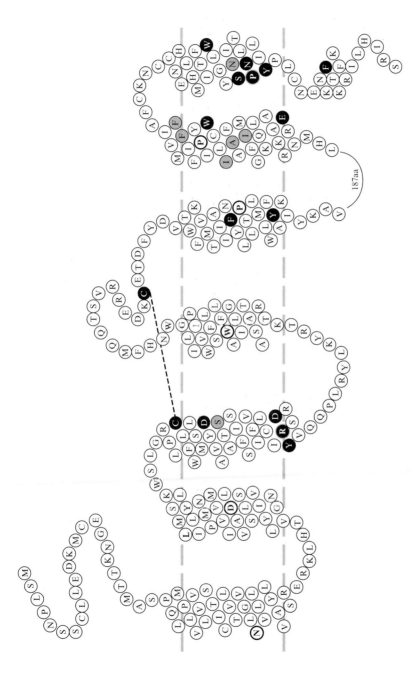

Figure 7.7 Snake plot of the human H$_1$R. Schematic representation showing the TM domains and connecting loops of the H$_1$R. Bold residues indicate the most conserved residue in each TM domain, numbered X.50 according to the Ballesteros and Weinstein scheme (Ballesteros and Weinstein, 1995), whereas other conserved residues that are involved in retaining the receptor in a specific conformation are indicated in black. Residues that give a CIM or CAM H$_1$R phenotype upon mutation(s) are depicted in gray.

CGS-12066A (5HT$_{1B}$ agonist) stimulation. Interestingly, in these cells the 5HT$_{1B}$ agonist CGS-12066A also shows a PTX-sensitive (not Gα_i-mediated) activation of NF-κB, which is not observed in cells that do not express 5HT$_{1B}$. In addition, inverse H$_1$R agonists not only inhibit the constitutive activity of H$_1$R but also inhibit CGS-12066A-induced signaling of 5HT$_{1B}$ receptors to PLC and NF-κB pathways. Binding studies confirming ligand specificity, proved that the interactions are not caused by a-specific interactions on either of the receptors.

A selection of Gα_i-coupled receptors was tested to check if the initial observations are a general feature of H$_1$R. Muscarinic M2 receptor coexpression and subsequent treatment with carbachol resulted in a PTX-sensitive activation of NF-κB. The same holds true for coexpression with adenosine A1 and adrenergic α_{2A} receptors (Bakker et al., 2004a).

This study shows a functional consequence of H$_1$R constitutive activity, by enabling Gα_i-coupled activation of PLC and NF-κB pathways. Since *in vivo* situations are likely to involve expression of multiple GPCRs on the cell membrane and the presence of several ligands at the same time, this new feature of H$_1$R constitutive activity illustrates the importance of the cellular context when monitoring GPCR pharmacology.

4. Conclusion

Overexpression of the histamine H$_1$ receptor in a variety of cell lines allows the detection of its constitutive activity. Reassessment of the intrinsic activity of previously identified "classical" H$_1$R antagonists on recombinant cells revealed that these widely used therapeutics act in fact as inverse agonists on the H$_1$R. Accurate characterization of the intrinsic activity of these drugs is crucial as we showed *in vitro* that prolonged treatment with inverse agonists, but not weak partial inverse agonists and neutral antagonists, actually upregulates histamine receptor expression resulting in an increased histamine-induced response (Govoni et al., 2003; Smit et al., 1996). Future evaluation of inverse agonists and neutral antagonists treatment in rhinitis disease models might reveal constitutive H$_1$R activity *in vivo*. In addition, detailed analysis of H$_1$R mutants displaying decreased or increased constitutive activity provides valuable insight in the activation mechanism of this receptor.

ACKNOWLEDGMENTS

The authors are supported by COST action BM0806.

REFERENCES

Bakker, R. A., et al. (2000). Constitutive activity of the histamine H(1) receptor reveals inverse agonism of histamine H(1) receptor antagonists. *Eur. J. Pharmacol.* **387,** R5–R7.

Bakker, R. A., et al. (2001). Histamine H_1-receptor activation of nuclear factor-kappa B: Roles for G beta gamma- and G alpha$_{q/11}$-subunits in constitutive and agonist- mediated signaling. *Mol. Pharmacol.* **60,** 1133–1142.

Bakker, R. A., et al. (2004a). Constitutively active Gq/11-coupled receptors enable signaling by co-expressed G(i/o)-coupled receptors. *J. Biol. Chem.* **279,** 5152–5161.

Bakker, R. A., et al. (2004b). 8R-lisuride is a potent stereospecific histamine H1-receptor partial agonist. *Mol. Pharmacol.* **65,** 538–549.

Bakker, R. A., et al. (2007). In vitro pharmacology of clinically used central nervous system-active drugs as inverse H(1) receptor agonists. *J. Pharmacol. Exp. Ther.* **322,** 172–179.

Bakker, R. A., et al. (2008). Constitutively active mutants of the histamine H1 receptor suggest a conserved hydrophobic asparagine-cage that constrains the activation of class A G protein-coupled receptors. *Mol. Pharmacol.* **73,** 94–103.

Ballesteros, J. A., and Weinstein, H. (1995). Integrated methods for the construction of three dimensional models and computational probing of structure-function relations in G-protein coupled receptors. *Methods Neurosci.* **25,** 366–428.

Berg, K. A., et al. (1999). Novel actions of inverse agonists on 5-HT_{2C} receptor systems. *Mol. Pharmacol.* **55,** 863–872.

Bergsdorf, C., et al. (2008). A one-day, dispense-only IP-One HTRF assay for high-throughput screening of Galphaq protein-coupled receptors: Towards cells as reagents. *Assay Drug Dev. Technol.* **6,** 39–53.

Boss, V., et al. (1998). Histamine induces nuclear factor of activated T cell-mediated transcription and cyclosporin A-sensitive interleukin-8 mRNA expression in human umbilical vein endothelial cells. *Mol. Pharmacol.* **54,** 264–272.

Brandish, P. E., et al. (2003). Scintillation proximity assay of inositol phosphates in cell extracts: High-throughput measurement of G-protein-coupled receptor activation. *Anal. Biochem.* **313,** 311–318.

Brauner-Osborne, H., and Brann, M. R. (1996). Pharmacology of muscarinic acetylcholine receptor subtypes (m1-m5): High throughput assays in mammalian cells. *Eur. J. Pharmacol.* **295,** 93–102.

Burstein, E. S., et al. (1995a). Constitutive activation of muscarinic receptor by the G-protein G_q. *FEBS Lett.* **363,** 261–263.

Burstein, E. S., et al. (1995b). Structure-function of muscarinic receptor coupling to G proteins. Random saturation mutagenesis identifies a critical determinant of receptor affinity for G proteins. *J. Biol. Chem.* **270,** 3141–3146.

Burstein, E. S., et al. (1997). Pharmacology of muscarinic receptor subtypes constitutively activated by G proteins. *Mol. Pharmacol.* **51,** 312–319.

Chiba, Y., et al. (2002). Elevated nasal mucosal G protein levels and histamine receptor affinity in a guinea pig model of nasal hyperresponsiveness. *Int. Arch. Allergy Immunol.* **127,** 285–293.

Croston, G. E. (2002). Functional cell-based uHTS in chemical genomic drug discovery. *Trends Biotechnol.* **20,** 110–115.

Das, A. K., et al. (2007). Stimulation of histamine H1 receptor up-regulates histamine H1 receptor itself through activation of receptor gene transcription. *J. Pharmacol. Sci.* **103,** 374–382.

De Backer, M. D., et al. (1993). Genomic cloning, heterologous expression and pharmacological characterization of a human histamine H_1 receptor. *Biochem. Biophys. Res. Commun.* **197,** 1601–1608.

Dina, O. A., et al. (2003). Role of the sensory neuron cytoskeleton in second messenger signaling for inflammatory pain. *Neuron* **39**, 613–624.

Dinh, Q. T., et al. (2005). Transcriptional up-regulation of histamine receptor-1 in epithelial, mucus and inflammatory cells in perennial allergic rhinitis. *Clin. Exp. Allergy* **35**, 1443–1448.

Etienne-Manneville, S., and Hall, A. (2002). Rho GTPases in cell biology. *Nature* **420**, 629–635.

Fitzsimons, C. P., et al. (2004). Mepyramine, a histamine H1 receptor inverse agonist, binds preferentially to a G protein-coupled form of the receptor and sequesters G protein. *J. Biol. Chem.* **279**, 34431–34439.

Fukui, H., et al. (1994). Molecular cloning of the human histamine H_1 receptor gene. *Biochem. Biophys. Res. Commun.* **201**, 894–901.

Govoni, M., et al. (2003). Synthesis and pharmacological identification of neutral histamine H1-receptor antagonists. *J. Med. Chem.* **46**, 5812–5824.

Hao, F., et al. (2008). Histamine induces Egr-1 expression in human aortic endothelial cells via the H1 receptor-mediated protein kinase Cdelta-dependent ERK activation pathway. *J. Biol. Chem.* **283**, 26928–26936.

Hill, S. J., et al. (1977). Specific binding of 3H-mepyramine to histamine H1 receptors in intestinal smooth muscle. *Nature* **270**, 361–363.

Hill, S. J., et al. (1978). The binding of [3H]mepyramine to histamine H1 receptors in guinea-pig brain. *J. Neurochem.* **31**, 997–1004.

Hill, S. J., et al. (1997). International Union of Pharmacology XIII. Classification of histamine receptors. *Pharmacol. Rev.* **49**, 253–278.

Iriyoshi, N., et al. (1996). Increased expression of histamine H-1 receptor mRNA in allergic rhinitis. *Clin. Exp. Allergy* **26**, 379–385.

Jongejan, A., et al. (2005). Linking agonist binding to histamine H1 receptor activation. *Nat. Chem. Biol.* **1**, 98–103.

Kenakin, T. (2004). Efficacy as a vector: The relative prevalence and paucity of inverse agonism. *Mol. Pharmacol.* **65**, 2–11.

Leopoldt, D., et al. (1997). G proteins endogenously expressed in Sf 9 cells: Interactions with mammalian histamine receptors. *Naunyn Schmiedeberg's Arch. Pharmacol.* **356**, 216–224.

Leurs, R., et al. (2000). Constitutive activity of G protein coupled receptors and drug action. *Pharm. Acta Helv.* **74**, 327–331.

Milligan, G., and Bond, R. A. (1997). Inverse agonism and the regulation of receptor number. *Trends Pharmacol. Sci.* **18**, 468–474.

Moguilevsky, N., et al. (1994). Stable expression of human H_1-histamine-receptor cDNA in Chinese hamster ovary cells–Pharmacological characterisation of the protein, tissue distribution of messenger RNA and chromosomal localisation of the gene. *Eur. J. Biochem.* **224**, 489–495.

Morisset, S., et al. (2000). High constitutive activity of native H3 receptors regulates histamine neurons in brain. *Nature* **408**, 860–864.

Morse, K. L., et al. (2001). Cloning and characterization of a novel human histamine receptor. *J. Pharmacol. Exp. Ther.* **296**, 1058–1066.

Notcovich, C., et al. (2010). Histamine acting on H1 receptor promotes inhibition of proliferation via PLC, RAC, and JNK-dependent pathways. *Exp. Cell. Res.* **316**, 401–411.

Nygaard, R., et al. (2009). Ligand binding and micro-switches in 7TM receptor structures. *Trends Pharmacol. Sci.* **30**, 249–259.

Parsons, M. E., and Ganellin, C. R. (2006). Histamine and its receptors. *Br. J. Pharmacol.* **147** (Suppl 1), S127–S135.

Scheerer, P., et al. (2008). Crystal structure of opsin in its G-protein-interacting conformation. *Nature* **455**, 497–502.

Smit, M. J., et al. (1996). Inverse agonism of histamine H_2 antagonists accounts for upregulation of spontaneously active histamine H_2 receptors. *Proc. Natl. Acad. Sci. USA* **93,** 6802–6807.
Smit, M. J., et al. (2002). G protein-coupled receptors and proliferative signaling. *Methods Enzymol.* **343,** 430–447.
Smit, M. J., et al. (2007). Pharmacogenomic and structural analysis of constitutive g protein-coupled receptor activity. *Annu. Rev. Pharmacol. Toxicol.* **47,** 53–87.
Solly, K., et al. (2004). Application of real-time cell electronic sensing (RT-CES) technology to cell-based assays. *Assay Drug Dev. Technol.* **2,** 363–372.
Vischer, H. F., et al. (2010). Identification of novel allosteric nonpeptidergic inhibitors of the human cytomegalovirus-encoded chemokine receptor US28. *Bioorg. Med. Chem.* **18,** 675–688.
Weiner, D. M., et al. (2001). 5-Hydroxytryptamine2A receptor inverse agonists as antipsychotics. *J. Pharmacol. Exp. Ther.* **299,** 268–276.
Weiss, J. M., et al. (1996). The cubic ternary complex receptor-occupancy model III. Resurrecting efficacy. *J. Theor. Biol.* **181,** 381–397.
Wieland, K., et al. (2001). Constitutive activity of histamine h(3) receptors stably expressed in SK-N-MC cells: Display of agonism and inverse agonism by H(3) antagonists. *J. Pharmacol. Exp. Ther.* **299,** 908–914.
Yu, N., et al. (2006). Real-time monitoring of morphological changes in living cells by electronic cell sensor arrays: An approach to study G protein-coupled receptors. *Anal. Chem.* **78,** 35–43.

CHAPTER EIGHT

CONSTITUTIVE ACTIVITY OF SOMATOSTATIN RECEPTOR SUBTYPES

Anat Ben-Shlomo, Kolja Wawrowsky, *and* Shlomo Melmed

Contents

1. Introduction	150
2. Choosing Cells to be Studied	151
3. Modifying SSTR Density	152
3.1. SSTR overexpression	153
3.2. SSTR knockdown	160
4. Summary	163
References	163

Abstract

The five somatostatin receptors (SSTR1–5) are G-protein-coupled receptors, coupling to $G_{\alpha i/o}$ subunits to regulate pathways including inhibiting adenylate cyclase activity and reduce intracellular cAMP levels and decrease intracellular calcium levels. In the pituitary gland, somatostatin actions, mediated through SSTR1, 2, 3, and 5, are inhibition of growth hormone, thyrotropin hormone, and adrenocorticotropin hormone release and to a lesser extent, inhibition of cell growth. Establishment of constitutive SSTRs action suggests that abundant pituitary SSTR expression contributes to pituitary function in maintaining homeostasis, aside from the SSTR response to episodic hypothalamic somatostatin release. In this chapter, we describe an experimental approach to directly and indirectly demonstrate constitutive SSTR activity by altering receptor density in AtT20 mouse pituitary corticotroph tumor cells, utilizing small interference RNA to knock receptor expression down or stable SSTRs transfection to overexpress selective receptor levels. We describe methodical validation for each of the approaches and the use of a sensitive cAMP assay to analyze consequences of changing membrane receptor number in the absence of an added ligand.

Pituitary Center, Department of Medicine, Cedars Sinai Medical Center, David Geffen School of Medicine at UCLA, Los Angeles, California, USA

1. Introduction

Somatostatin peptides are a phylogenetically ancient multigene family of small regulatory proteins produced by neurons and endocrine cells in the brain, gastrointestinal system, immune and neuroendocrine cells. Somatostatin-28 and 14 are enzymatically cleaved from the preprosomatostatin precursor molecule in somatostatin-producing cells to act as neurotransmitters or target cell regulators in an endocrine, paracrine, or autocrine manner. Cortistatins are related peptides produced mainly in brain and immune cells that bind somatostatin receptor subtypes with affinity equal to somatostatins. Somatostatins regulate diverse physiological processes including cell secretion and growth, neuromodulation, nutrient absorption, and smooth muscle contractility (Patel, 1999). In the pituitary, somatostatin-14 released from the hypothalamus inhibits growth hormone (GH), thyrotropin stimulating hormone (TSH), and adrenocorticotropin hormone (ACTH) release (Ben-Shlomo and Melmed, 2010). In pancreatic islet cells somatostatin inhibits insulin and glucagon release, and in the gastrointestinal tract somatostatin inhibits cholecystokinine, gastrin, secretin, vasoactive intestinal peptide (VIP) secretion as well as release of gastric acid, intestinal fluids, and pancreatic enzymes (Patel, 1999). Clinically available somatostatin agonists, octreotide, and lanreotide are mostly used for treatment of GH secreting pituitary adenomas, neuroendocrine tumors such as carcinoids, gastrinomas, insulinomas, glucagonomas and vipomas, and gastrointestinal disorders, including bleeding gastric varices, pancreatitis, and pancreatic surgical complications.

Somatostatin actions are mediated by five 7-transmembrane domain G-protein-coupled receptor subtypes (termed SSTR1–5) encoded by separate genes. SSTRs bind endogenous somatostatin with low nanomolar affinity to initiate a broad spectrum of biological effects. SSTRs, especially SSTR2, are ubiquitously expressed and different cell types often express more than one receptor subtype. Somatostatin binding to SSTRs regulates multiple downstream pathways through $G_{\alpha i}$ subunits of the G-protein tetramer including adenylate cyclase inhibition, opening of K^+ and closure of Ca^{2+} channels, regulation of Na^+/H^+ antiporter, guanylate cyclase, phospholipase C (PLC), phospholipase A2 (PLA2), MAP kinase (MAPK), and serine, threonine, and phosphotyrosyl protein phosphatase (PTP) (Patel, 1999). Upon stimulation, SSTR2 is phosphorylated, desensitized, and internalized. SSTR3 is also internalized, though to a lesser extent, SSTR1 does not internalize and whether or not SSTR5 is internalized is unclear (Ben-Shlomo et al., 2005).

Several hints deduced from somatostatin and its cognate receptor functions suggest that these GPCRs exhibit constitutive activity. First, as SSTR2 is ubiquitously expressed, octreotide, which exhibits high SSTR2 affinity, would be expected to cause significant side effects; yet adverse events caused by octreotide treatment are minimal and largely transient. Moreover, many endocrine and neuroendocrine tumors express SSTR2 but do not respond to somatostatin agonist treatment. It is unclear why cells express the receptor, if they do not respond to the ligand? Pituitary SSTR2 is similar to the receptor found in other tissues, and no mutation has been described. In addition, somatostatin is the major inhibitor of GH secretion, yet somatostatin null mice expected to develop gigantism or acromegaly due to unopposed GH secretion, have only slightly increased GH levels without showing excess growth. If somatostatin is the major inhibitor of GH release, it is unclear why somatotroph GH release is restrained in somatostatin null mice. Interestingly, several pituitary cell lines, that is, TtT/GF mouse pituitary folliculostellate cells, GH_3, and GH_4C_1 rat somatotroph tumor cells, are resistant to stable overexpression of SSTR2, but readily overexpress SSTR1, 3, and 5 (unpublished data). Lastly, transgenic mouse overexpressing SSTR2 or SSTR5 have not been reported, even though the SSTRs were cloned ~20 years ago and SSTR2 and SSTR5 null mice are readily available. These findings suggest that somatostatin receptors may function independently of somatostatin ligand. This constitutive activity may be important in maintaining cell secretion and growth homeostasis.

To our knowledge no natural constitutively active SSTR mutant has yet been discovered, and no established selective SSTR inverse agonist has been generated, making study of constitutive SSTR activity challenging.

This chapter focuses on approaches to unravel and analyze constitutive SSTR activity *in vitro* using intracellular cAMP measurements by altering receptor expression in the cell membrane (Ben-Shlomo *et al.*, 2007, 2009a).

2. Choosing Cells to be Studied

Many studies utilize "universal" cells like Chinese hamster ovary (CHO), human embryonic kidney (HEK293), or African green monkey kidney (COS7) cells to study receptor function. Advantages of using these cells include the experience that they are easy to grow and maintain, are highly receptive to transfected cDNA and functionally resilient. In addition, it has been assumed that these cells do not express endogenous receptors being studied, as they do not respond to treatment with a selective agonist, and are therefore "neutral" in their responses to the receptor and its ligands. However, HEK293 do in fact express endogenous SSTR2 (Law *et al.*, 1993).

We have chosen an approach utilizing AtT20 mouse pituitary corticotroph tumor cells. These cells secrete ACTH and were generated from anterior pituitary tumors in LAF_1 female mice that survived ionizing radiation (Furth *et al.*, 1953). These cells are relevant to the target organ, the pituitary, already express high levels of SSTRs enabling studying effects of both increasing and decreasing receptor expression, and are exquisitely responsive to somatostatin and its analogs. The AtT20 SSTR profile is similar to that of tumoral corticotroph cells in Cushing's disease, that is, SSTR2, 5, and 3, and may accurately reflect somatostatin system dynamics (Ben-Shlomo *et al.*, 2005).

AtT-20/D16v-F2 cells are purchased from the American Type Culture Collection (ATCC) and are suspended in serum-supplemented medium containing low glucose DMEM supplemented with 10% fetal bovine serum (Omega Scientifics, Tarzana, CA), with addition of 1% glutamine and 1% penicillin/streptomycin (Invitrogen, Carlsbad, CA), in 6% CO_2, 37 °C humidified incubator. Most cells are round but some are polygonal and somewhat elongated, they measure between 5 and 20 μm in diameter, up 10 μm in height, have a large nucleus, are adherent, and have a duplication time of $\sim$20 h. Cells should be passaged twice weekly and $\sim$1 million cells are plated in a 75 mL flask. AtT20 cell should not be too crowded or too sparsely plated. Attention should be given to proper cell separation after trypsinizing and washing, as these cells often adhere in clumps. We repeatedly pipette the washed cell pellet up to 15 times to ensure that cells are well separated before plating. Fresh AtT20 cells grow in ball-shaped colonies and do not sprout significantly. Colonies should not be allowed to grow extensively as these eventually detach from the flask and die. We monitor for mycoplasma contamination every few months using a quantitative real-time PCR assay. Fresh cells for plating should be thawed every 6 months.

As the absence of ligand is critical for determining constitutive receptor activity, we tested cells suspended with and without serum for somatostatin and cortistatin expression using TaqMan qRT-PCR. AtT20 cells were shown not to express endogenous somatostatin or cortistatin, therefore all cellular changes attributed to SSTRs are ligand independent.

3. Modifying SSTR Density

Changing cellular receptor expression levels by recombinant overexpression or silencing of endogenous receptor expression may unmask receptor constitutive activity. Studying live cells, as long as they do not express the ligand itself, is a superior approach to studying separated cell membranes that lack intracellular proteins interacting with the receptor (Wieland and Seifert, 2005). We describe effects of human SSTR2 and SSTR5 overexpression and SSTR2, SSTR5 and SSTR3 knockdown on intracellular cAMP levels in AtT20 mouse corticotroph tumor cells.

3.1. SSTR overexpression

3.1.1. Plasmids

To allow ready detection and accurate selection of exogenous SSTR-expressing cells we chose to express a plasmid that contained the ZsGreen fluorophor tag. The pIRES2-ZsGreen1 (Clontech, Mountain View, CA) contains the internal ribosome entry site (IRES) sequence between the SSTR and ZsGreen tag, allowing transcribed bicistronic mRNA to be translated to two separate proteins, SSTR and ZsGreen, enabling SSTR to traffic freely and bind its ligands while ZsGreen tags the cell for selection. Human SSTRs were kindly provided by Dr. Graeme Bell (University of Chicago, Chicago, IL) and exhibit high homology to mouse SSTRs, all previously shown to bind $G_{\alpha i}$ and inhibit adenylate cyclase activity, reducing generation of cAMP (Patel, 1999). Human and rodents have 94–99% sequence identity for SSTR1, 93–96% for SSTR2, and 82–83% for SSTR3 and SSTR5. Human SSTR2 (hSST2) and hSST5 cDNA were amplified from original plasmids and inserted into pIRES2-ZsGreen1 vector. The following primers were used for PCR of hSST2 and hSST5—hSST2 forward: 5′-CGG AAT TCA CCA TGG ACA TGG CGG ATG AGC CA-3′; hSST2 reverse: 5′-CGG GAT CCT CAG ATA CTG GTT TGG AGG TCT-3′; hSST5 forward: 5′-GAA GAT CTA TGG AGC CCC TGT TCC CAG C-3′; hSST5 reverse: 5′-CGG AAT TCT CAC AGC TTG CTG GTC TGC A-3′. All plasmids were verified by sequencing (Sequetech, Mountain View, CA). Even though empty vector containing IRES and ZsGreen should have been the ultimate control, we used a ZsGreen vector without IRES for our control cells. Unlike for cells with ZsGreen, IRES-ZsGreen containing cells appeared unhealthy and detached easily. Instability of 5′-unconjugated (with SSTR for example) IRES-ZsGreen mRNA was reported by the manufacturer.

3.1.2. Generating stable transfectants

One million ATt20 cells are plated in a 10 cm dish the night before transfection. The next morning cells are transfected using effectene transfection agent (Qiagen, Valencia, CA) as follows: 2 μg plasmid DNA, 300 μL EC buffer, 16 μL enhancer, 60 μL Effectene in 10 mL growing medium. Four hundred μg/mL selection antibiotics (geniticin) are added 2 days after transfection. Three monoclonal stable AtT20 transfectant subtypes were generated: AZsG (expressing ZsGreen only), AhSSTR2IZ (expressing hSSTR2-IRES-ZsGreen), and AhSSTR5IZ (expressing hSSTR5-IRES-ZsGreen). ZsGreen expression is followed using a fluorescent microscope, and serum-supplemented medium containing 400 μg/mL geniticin is replaced twice weekly. Monoclones expressing medium ZsGreen intensity are selected and used for all studies. Flow cytometry is used to quantify the percentage of cells expressing each of the human somatostatin receptor subtypes and ZsGreen intensity. AZsG, AhSSTR2IZ, and AhSSTR5IZ monoclones expressing similar ZsGreen intensity per cell are chosen. Once ZsGreen positive cell number drops below 90%, cells are enriched by reisolating ZsGreen positive colonies.

3.1.3. System validation

Validating receptor expression and function is important for accuracy of results. We recommend three levels of validation: (1) selective hSSTR mRNA expression, (2) selective hSSTR protein expression and localization, and (3) selective hSSTR function and coupling to the $G_{\alpha i}$ subunit.

For hSSTR mRNA expression: Approximately 200,000 cells per well are plated in six-well plates and grown for 48 h. RNA is collected, treated with DNase to eliminated genomic DNA and 3 µg purified RNA are checked on 2% agarose gels for quality control. Purified total RNA is reverse transcribed into first-strand cDNA using a kit which preferably includes genomic DNA elimination step. For measurements of SSTRs, we recommend using TaqMan assays with high specificity and sensitivity that do not require further validation on agarose gels. TaqMan Gene Expression Assays for receptor subtype can be purchased from Applied Biosystems (Foster City, CA) and real-time PCR is performed according to the manufacturer instruction. Standard curves for TaqMan assays are performed with human tissue including liver, breast, and pancreas, and pituitary. Tested sample signals are normalized to values obtained for GAPDH. As we found that 18S, actin, and RplP1 mRNA expression change with somatostatin treatment, in AtT20 cells these control housekeeping genes are not recommended for experiments with somatostatin. Importantly, as mouse and human SSTR2 are almost similar in sequence, mouse and human TaqMan gene expression assays may detect each other's SSTR2. Although we cannot overcome this cross-reaction with different SSTR2 TaqMan assays, a comparison of SSTR2 mRNA to AZsG control cells showed a 40-fold increase in AhSSTR2IZ confirming stable hSSTR2 overexpression.

Immunocytochemistry for hSSTRs allows both confirmation of protein presence and localization. Importantly, we cannot detect endogenous SSTR expression in pituitary cell lines with available SSTR antibodies, probably due to low receptor expression levels. However, SSTRs can be detected if overexpressed. Cells are plated in 12-well plates (150,000 cells per well) on sterile poly-D-lysine-coated 18 mm glass coverslips and left to attach. The next day cells are treated for 1 h with an agonist, washed with phosphate-buffered saline (PBS), fixed in 4% paraformaldehyde for 1 h at room temperature, treated with Triton X-100 0.1% for 15 min followed by 1 h incubation in PBS with 1% BSA. Cells are stained with specific receptor primary antibody. The following antibodies have proven successful for immunocytochemistry of hSSTR2 overexpressing cells: rabbit sera directed against human/mouse SSTR2 (obtained from

Dr. Stefan Schulz, Bayerische Julius-Maximilians-Universität, Würzburg, Germany), mouse anti-human SSTR2 monoclonal antibody from Life Span Biosciences and mouse anti-human SSTR2 monoclonal antibody from R&D Systems, all at a dilution of 1:1000. Rabbit sera against human/mouse SSTR5 (obtained from Dr. Stefan Schulz) is effective for hSSTR5 immunostaining in AhSSTR5IZ cells at a dilution 1:1000. Cells are suspended with primary antibody overnight in 4 °C, washed the next morning, and stained with Alexa-fluor-568 secondary antibody (Invitrogen). Cells are than washed, dried for 15 min in the dark at room temperature, and mounted. We recommend ProLong mounting medium (Invitrogen) and if one has a laser that detects Dapi we recommend to use ProLong mounting medium that also contains Dapi (Invitrogen) for DNA staining and nuclear visualization. It is critical to avoid bubble formation that alters protein visualization. A small drop of mounting medium should be placed on both slide and cell-containing coverslip after which coverslip is gently dropped on the slide to allow even bubbleless distribution of the mounting medium. As ZsGreen and Alexa-fluor-488 share similar excitation and emission wavelengths they cannot be used simultaneously.

A dynamic test can be performed at this stage for validating the ability of the receptor to internalize, thus supporting adequate receptor function. Utilizing two approaches, live cell and fixed cell analysis, cells are treated with either somatostatin-14 or SSTR selective receptor agonist. For fixed cell analysis, cells are plated as described 1 day prior to treatment. The next day cells are treated for 5, 15, 30, or 60 min, then washed, fixed, and immunostained as above. The presence of serum does not affect internalization. For live cell analysis, ~200,000 cells are plated on sterile poly-D-lysine-coated 25 mm glass coverslips in 6-well plates and left to attach. The next morning cells are washed and a coverslip containing attached cells is placed in a special chamber that fits the confocal microscope incubator. Cells are allowed to adjust to the new conditions for 5–10 min before the desired agonist concentration is carefully added. Ten microliters of the desired agonist (in concentration 100-fold higher that that desired) is added to 1 mL of medium in the chamber. Somatostatin that binds all SSTRs, and octreotide that binds SSTR2 > SSTR5 > SSTR3, are commercially available, however, other somatostatin receptor subtype selective agonists including BIM-23120 an SSTR2 agonist and BIM-23206 an SSTR5 agonist were obtained from the Ipsen Group (Milford, MA; Table 8.1).

The confocal microscope is useful to image and analyze receptor endocytosis by simultaneously imaging multiple channels (wavelength) to track fluorescent proteins in live cells. By repeatedly imaging a stack of optical slices over time we arrive at a 4D multi-channel dataset. Projections over space and time provide movies and still images for visualization of biological processes.

A careful selection and fine tuning of imaging parameter is essential for optimal results.

Table 8.1 Somatostatin agonist binding affinities to somatostatin receptor subtypes

Ligands		hSSTR1	hSSTR2	hSSTR3	hSSTR4	hSSTR5
		\multicolumn{5}{c}{Binding affinity (nM)[a]}				
Endogenous	SRIF14	0.1–2.26	0.2–1.3	0.3–1.6	0.3–1.8	0.2–0.9
	SRIF28	0.1–2.2	0.2–4.1	0.3–6.1	0.3–7.9	0.05–0.4
Synthetic (in clinical use or trial)	Pasireotide	9.3	1	1.5	>100	0.16
	Octreotide	280	0.4	7	>1000	6
	Lanreotide	180	0.5	14	230	17
Synthetic (experimental)	BIM-23120	>1000	0.3	412	>1000	190
	BIM-23206	>1000	128	>1000	>1000	2

[a] All tested in mono-receptor stable transfected cells including CHO-K1, COS7, and HEK293 cells. Adapted from Ben-Shlomo and Melmed (2010).

Temporal and spatial resolution and image intensity should be carefully balanced to prevent photobleaching. Scan speed should be high enough to resolve the endocytic process and the subsequent transport of cytoplasmic vesicles. Spatial resolution should be sufficient to resolve endocytic vesicles against the fluorescent background. The sample should be sufficiently bright to obtain a clear image and collect a sufficient number of photons emitted by the sample. Illumination intensity should be sufficient to prevent imaging noise but remain under the damage threshold. The excitation intensity should not induce photobleaching. Dynamic range of the photodetectors should be carefully adjusted because endocytic vesicles contain much higher concentrations of receptor molecules compared to the membrane. It is challenging to find the correct exposure latitude as vesicles are not present at the outset of the experiment. Complete integration of the receptor into the membrane with minimal retention in the endoplasmic reticulum (ER) or Golgi complex is important for optimal experimental results. The ideal cell appears as an even distribution of ZsGreen fluorescence throughout the cell body with a fine bright line around the edge (Fig. 8.1).

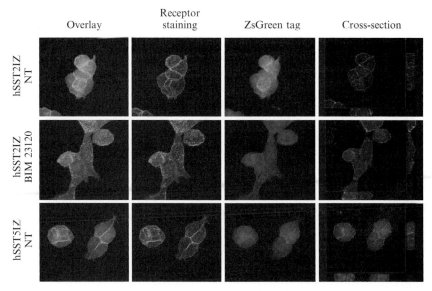

Figure 8.1 hSSTR protein expression and localization in AtT20 stable transfectants. Human SSTR2 and hSSTR5 localize to the cell membrane in AhSSTR2IZ and AhSSTR5IZ cells, respectively. One hour treatment with selective SSTR2 agonist, BIM-23120 (100 nM), causes hSSTR2 internalization. Cells were plated on coverslips, treated or not for 1 h at 37 °C, fixed in 4% paraformaldehyde, stained with the receptor-directed antibody and visualized with confocal microscopy. DNA, blue; ZsGreen, green; and receptors, red. Adapted from Ben-Shlomo *et al.* (2009a). (See Color Insert.)

Live cell receptor internalization (Ben-shlomo et al. 2009b) requires a heated CO_2 perfused incubation chamber, because endocytosis is sensitive to pH and temperature. However, an open chamber system can be used if receptor internalization takes only a few minutes. The microscope objective should also be warmed to prevent temperature gradients. Water immersion lens types are required for optimal results.

3.1.4. Receptor function: Intracellular cAMP measurements

The cAMP assay is useful and sensitive to verify adequate SSTR function. Determining cell number and comparable cell confluence are important for correct interpretation of absolute cAMP measurements. Cell growth rates should be determined as accurately as possible using growth curves spanning the time of experiment. We have found that over 2 days in culture there are no differences between AZsG, AhSSTR2IZ, and AhSSTR5IZ cell growth rates therefore comparison of absolute cAMP measurement in these cell groups is likely valid. Cells should be plated to a confluence suitable for normal growth over the time span of the experiment. AtT20 cells roughly ~50% confluence is acceptable for experiments of 2–3 days duration.

Thirty thousand cells per well are plated in 48-well plates with 4–8 wells used per sample. The next day cells are washed and treated with low glucose DMEM containing 0.3% BSA, 1% penicillin/streptomycin (serum-free medium), and 3-isobutyl-1-methylxanthine (IBMX, 1 mM, Sigma-Aldrich, St. Louis, MO) with or without an agonist for 30 min in a humidified incubator at 37 °C and 6% CO_2. At this cell number, cAMP assays available are not sufficiently sensitive to detect changes in baseline cAMP levels, therefore stimulation with forskolin (adenylate cyclase activator) or corticotropin releasing hormone (CRH) that stimulates cAMP through $G_{\alpha s}$ G-protein subunit is necessary. One to 10 μM forskolin or 1–100 nM CRH are used for this purpose. Changing pipette tips between dilutions is critical as some of somatostatin analogs are retained on the tip polymers, thereby producing artificial dilution values. Supernatants are aspirated after 30 min, cells washed gently to avoid detachment, and 0.5 mL of undiluted ethanol added quickly to each well. Ethanol opens up cell membranes and allows intracellular cAMP measurements; moreover this approach ensures cell treatment is performed in the incubator rather than on the bench, therefore is not influenced by changes in temperature or CO_2 content. In addition, reactions in all wells are halted simultaneously. Plates with ethanol are immediately placed in −20 °C for 24 h, after which content collected into eppendorf tubes and dried in a SpeedVac machine to obtain a pellet. Pellets can be stored at −20 °C for several years. Intracellular cAMP is assayed using the LANCE cAMP kit (Perkin Elmer, Waltham, MA) with a modification of manufacturer's instructions. cAMP pellets are resuspended in PBS at a dilution 1:100–200 and tubes carefully stirred to completely dissolve the pellet. At this stage, it is recommended to proceed with the analysis, otherwise samples can be stored for up to 24 h at 4 °C. The modified LANCE assay is performed in Corning White 96-well plates (Sigma-Aldrich). The sequence of incubations is as the manufacturer indicates; however, we change the working volumes as follows: 12 μL of sample (dissolved pellet) with 12 μL anti-cAMP antibody-Alexa 647 incubated for 1 h at room temperature, followed by addition of 24 μL of detection mix according to manufacturer's recommended dilutions for two more hours. Samples are measured in triplicate and read by a Victor 3 1420_015 spectrophotometer (Perkin Elmer). Results are compared to a standard curve with a range of 10^{-12} to 10^{-6} M. To assess potency, AZsG, AhSSTR2IZ, and AhSSTR5IZ are treated with somatostatin 14, BIM-23120, BIM-23206, octreotide to generate nine concentration dose response curves (1 fM to 100 nM). SSTR2 agonists octreotide and BIM-23120 are more potent in AhSSTR2IZ compared to AZsG and AhSSTR5IZ while SSTR5 agonist BIM-23206 is more potent in AhSSTR5IZ cells. Pretreatment with pertussis toxin (200 ng/mL for 3 h in serum replenished medium) blocks cAMP inhibition in treated AtT20 stable cell lines, indicating that both endogenous and exogenous receptors function through $G_{\alpha i}$ as expected.

3.1.5. Constitutive SSTR activity—an indirect approach

After careful validation of the above system, we tested constitutive SSTR activity *in vitro* utilizing mainly cAMP assays. The most important technical obstacle for cAMP measurements of constitutive SSTR activity is cell number. As changes in cAMP levels due to constitutive activity are expected to be modest, even small difference in plated cell numbers can appreciably alter measurements. After washing and calibration our cell counter is 1% intrasample variation which is sufficient to change measured cAMP values, and to measure artificial constitutive activity. Therefore, cAMP measurements should be repeated several times. In addition, extra caution should be placed into rigorously separating cells. Even though the cells generated are derived from a single clone and all clones studied have comparable ZsGreen intensity per cell, under the confocal microscope, we observed differences in ZsGreen expression inside colonies within the same recombinant cell line, and between three different recombinant cell lines. Therefore it should not be assumed that all cells in one line express similar amounts of SSTR, or that all cell lines express absolute comparable SSTR levels. In conclusion, we cannot compare different stable cell lines for absolute or relative changes in cAMP, and only compare changes within the same cell line.

cAMP levels do not differ between the three stable cell lines at baseline, or with forskolin or cholera toxin stimulation (1–10 ng/mL, 3 h in growing medium). Therefore we chose to demonstrate constitutive SSTRs activity using an indirect approach, that is, whether overexpression of SSTRs in a ligand-free cell alter cell response to factors mediating $G_{\alpha s}$ subunit that mediates adenylate cyclase activation and increase in cAMP levels. CRH, isoproterenol, and SKF38393 mediate CRH receptor subtype 1 (CRHR1), β2-adrenoreceptor, and dopamine receptor subtype 1 (D1R), respectively, to increase cAMP in corticotroph cells, and somatostatin inhibits this stimulation (Fig. 8.2). Cells are plated for dose response curves assessment and processed for cAMP assay as mentioned above. In AtT20 WT and AZsG cells, CRH stimulates cAMP up to 50-fold with EC50 $\sim$8 nM, isoproterenol up to 12-fold, and SKF38393 up to threefold with EC50 $\sim$500 nM. Overexpression of either hSSTR2 or hSSTR5 blunts cell response to CRH and SKF38393 and enhance response to isoproterenol. These responses correlate with changes in respective receptor mRNA levels, for example, decreased CRHR1 and D1R and increased β2-adrenoreceptor. CRHR1, D1R, and β2-adrenoreceptor are all $G_{\alpha s}$ coupled receptors and were shown to participate in the stress response to regulate corticotroph ACTH secretion emphasizing the relevance of ligand independent constitutively active SSTRs. In contrast, no effect is observed on vasopressin receptors subtypes 1 and 2, both of which are expressed in AtT20 cells, therefore serving as a negative control.

3.2. SSTR knockdown

A direct approach to unravel constitutive SSTR activity is achieved using small interference siRNA. AtT20 cells are the most suitable cell line for this approach as they express endogenous SSTR2, 3, and 5 and respond effectively to somatostatin and its analogs with reduction of cAMP. Other pituitary cell lines available including somatolactotrophs (rat GH3, GH4C1, GC), gonadotrophs (mouse αT3, LβT2), and lactotrophs (rat MMQ cells) respond modestly to somatostatin at concentrations $>$100 nM, and are therefore less suitable for analysis of physiologically relevant dose response curves; therapeutic levels of plasma octreotide, and lanreotide are $<$ 100 nM.

We use siRNA from Qiagen for transient transfections (Valencia, CA; www.qiagen.com/geneglobe/default.aspx). Four predesigned siRNAs are chosen randomly for each receptor subtype, two scrambled negative controls, and one somatostatin siRNA is also added as a negative control. Cells are transiently transfected with siRNA according to manufacturer recommendations using HiPerFect transfection reagent (Qiagen).

To determine the most effective conditions for inhibition of SSTR mRNA, we plate 300,000 cells per well suspended in growing medium in a 6-well plate the evening before transfection. The following morning cells are transfected with increasing concentrations of siRNA (between 5 and 20 nM in 2.5 nM increments) for 24, 48, and 72 h. All siRNAs are evaluated including scrambled RNA. Control scrambled RNA should not alter baseline or forskolin (10 μM) induced cAMP levels stimulation as compared to nontransfected cells treated with the transfection agent alone. The most

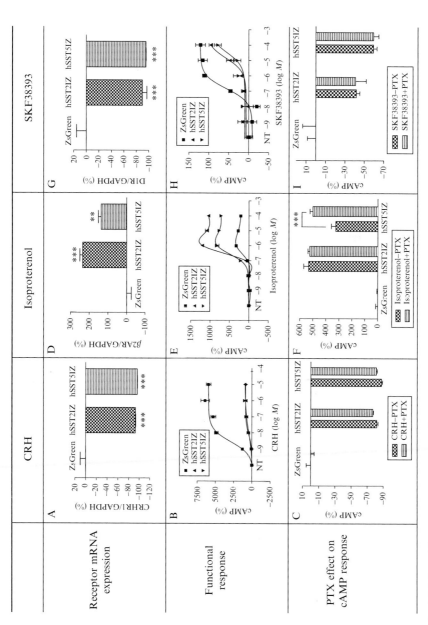

Figure 8.2 hSST2 and hSST5 overexpression determine corticotroph CRH, isoproterenol, and SKF38393 responses by regulating CRH, β2-AR, and D1R. (A, D, and G) CRHR1, β2AR, and D1R mRNA levels are measured using TaqMan expression assays. Y axis represents percent change in receptor levels from control ZsGreen cells ($p < 0.001$ and $p < 0.01$). (B, E, and H) Cells are treated with increasing

significant SSTR mRNA inhibition compared to control is observed at an siRNA concentration of 12.5 nM, and 24 h after transfection. For further analysis, siRNAs demonstrating the most pronounced reduction in mRNA levels are chosen. As cell number is critical for cAMP level comparison, the effect of each siRNA and scramble RNA on cell growth should be determined and can be accomplished using the cell proliferation reagent WST-1. AtT20 cell are plated (5000 cells per well) in a 96-well plate 2 days prior to transfection with 12.5 nM siRNA or scrambled RNA. There was no difference in cell proliferation between scrambled and any other siRNA tested over 2 days growth.

To further validate the system, the effect of SSTR mRNA level reduction on selective agonist potency can be studied. For cAMP assays cells are plated in 48-well plates (20,000 cell per well) and transfected with siRNA or control RNA the day after. Twenty four hours after transfection cells are treated for 30 min with a wide range of selective agonist concentrations (1 fM to 100 nM) in serum starved medium, IBMX, and forskolin as described above, expecting a left-shift in agonist potency. We observed that a reduction of 80% in mSSTR2 mRNA is not accompanied by reduced SSTR2 agonist potency, however, a reduction of 70% in mSSTR5 reduced SSTR5 selective agonist potency 23-fold. In evaluating these results it is important to understand that SSTR2 exhibits abundant spare receptors, suggesting that only small amounts of membranal SSTR2 are required to induce a maximal effect. Therefore, even large reductions in SSTR2 mRNA cannot shift SSTR2 agonist potency to the left. In contrast, SSTR5 does not exhibit spare receptors, and SSTR5 agonist potency was shifted significantly to the left after siRNA reduction of receptor mRNA expression.

3.2.1. Constitutive SSTR activity—a direct approach

To study constitutive SSTR activity, experiments were designed to demonstrate simultaneous reductions of mRNA levels, intracellular cAMP levels, and ACTH levels in the supernatant. Wells are transfected with 12.5 nM siRNA or control RNA for 24 h. mRNA measurements are performed as above. For cAMP and ACTH assay, cells are plated in 48-well plate (20,000 cell per well) and transfected the following day. Twenty four hours after transfection the supernatant is collected for ACTH assay, cells washed, and ethanol added for intracellular cAMP measurements.

---

concentrations of CRH or isoproterenol or SKF38393 and cAMP stimulation is measured. Y axis represents percent change in cAMP levels as compared to respective untreated (NT) cells. (C, F, and I) Cells were pretreated with PTX (200 ng/mL for 3 h) followed by serum deprived medium with or without CRH (10 nM), isoproterenol (1 μM), or SKF38393 (1 μM), respectively, for 30 min. Y axis represents percent change in cAMP levels from control ZsGreen cells. Results are depicted as mean ± SEM. Adapted from Ben-Shlomo et al. (2009a).

Secreted ACTH concentrations in the medium are measured by RIA with a measurable range of 10–1000 pg/mL (ICN Pharmaceuticals, Inc., Costa Mesa, CA). As transient transfection efficacy is variable, and cAMP changes due to constitutive SSTR activity are relatively small, multiple sampling and repeated experiments are required. As reduction of mRNA levels observed for SSTR2, 3, and 5 were different we cannot compare cAMP levels between receptors but can demonstrate that reducing endogenous SSTRs mRNA expression is followed by increased cAMP and increased ACTH release. As mentioned above, it is not possible to demonstrate SSTR protein level reduction using antibodies due to the low level of endogenous receptor expression.

4. Summary

We present here an experimental approach to study constitutive SSTR activity in live pituitary cells, by measuring intracellular cAMP levels. We recommend using cells that express the studied receptors endogenously, and readily respond to somatostatin so that the receptor will not be "unfamiliar" to the cell, and either receptor overexpression or receptor knockdown approaches can be employed. We chose AtT20 mouse pituitary corticotroph tumor cells that, like human tumoral pituitary corticotrophs express SSTR2, 3, and 5 and respond to somatostatin and its agonists. System generation (AtT20 cells stably overexpressing SSTRs or transiently transfected with siRNA to reduce SSTR expression), validation, and application for measuring constitutive SSTR activity are discussed.

REFERENCES

Ben-Shlomo, A., and Melmed, S. (2010). Pituitary somatostatin receptor signaling. *Trends Endocrinol. Metab.* **21,** 123–133.

Ben-Shlomo, A., Wawrowsky, K. A., Proekt, I., Wolkenfeld, N. M., Ren, S. G., Taylor, J., Culler, M. D., and Melmed, S. (2005). Somatostatin receptor type 5 modulates somatostatin receptor type 2 regulation of adrenocorticotropin secretion. *J. Biol. Chem.* **280,** 24011–24021.

Ben-Shlomo, A., Pichurin, O., Barshop, N. J., Wawrowsky, K. A., Taylor, J., Culler, M. D., Chesnokova, V., Liu, N. A., and Melmed, S. (2007). Selective regulation of somatostatin receptor subtype signaling: evidence for constitutive receptor activation. *Mol. Endocrinol.* **21,** 2565–2578.

Ben-Shlomo, A., Zhou, C., Pichurin, O., Chesnokova, V., Liu, N. A., Culler, M. D., and Melmed, S. (2009a). Constitutive somatostatin receptor activity determines tonic pituitary cell response. *Mol. Endocrinol.* **23,** 337–348.

Ben-Shlomo, A., Schmid, K., Wawrowsky, K., Pichurin, O., Hubina, E., Chesnokova, V., Liu, N. A., Culler, M. D., and Melmed, S. (2009b). Differential ligand-mediated pituitary Somatostatin receptor subtype signaling: inplications for corticotorph tumor therapy. *J. clin. Endocrinol. Metab.* **94,** 4342–4350.

Furth, J., Gadsen, E. L., and Upton, A. C. (1953). ACTH secreting transplantable pituitary tumors. *Proc. Soc. Exp. Biol. Med.* **84,** 253–254.

Law, S. F., Yasuda, K., Bell, G. I., and Reisine, T. (1993). Gi alpha 3 and G(o) alpha selectively associate with the cloned somatostatin receptor subtype SSTR2. *J. Biol. Chem.* **268,** 10721–10727.

Patel, Y. C. (1999). Somatostatin and its receptor family. *Front. Neuroendocrinol.* **20,** 157–198.

Wieland, T., and Seifert, R. (2005). Methodological approaches. *In* "G protein-coupled receptors as drug targets: Analysis of activation and constitutive activity" (R. Seifert and T. Wieland, eds.), Vol. 24, pp. 81–120. Wiley-VCH, Weinheim.

CHAPTER NINE

Assessment of Homologous Internalization of Constitutively Active N111G Mutant of AT$_1$ Receptor

Mohiuddin Ahmed Bhuiyan*,† *and* Takafumi Nagatomo†

Contents

1. Introduction	166
2. Preparation of Receptor Plasmid and Protein	167
2.1. Site-directed mutagenesis and plasmid preparation	167
2.2. Cell culture, transfection, and membrane preparation	168
3. Radioligand Binding Assay	168
4. Inositol Phosphate Accumulation Assay	170
5. Internalization Assay	171
6. Western Blot Analysis	173
7. Data Analysis	174
8. Concluding Remark	175
References	175

Abstract

Constitutively active mutants (CAMs) of G-protein-coupled receptors mimic the active conformation of the receptor in their ability to activate second messenger systems in the absence of agonist. They have revealed novel properties of drugs that reverse the basal levels of constitutive activity, indicating that the drugs have the inverse agonist activity. Internalization plays an important role in receptor endocytosis and signal transduction. The present chapter provides the investigation of the internalization behavior of CAM N111G of Angiotensin II type 1 (AT$_1$) receptor and correlates the result with the mechanism of constitutive activity of the mutant. Both wild-type (WT) and N111G mutant receptors were transiently expressed in COS-7 cells and total inositol phosphate production was measured in presence and absence of the angiotensin II receptor blockers (ARBs). The binding affinities toward agonist and ARBs were also determined. We found that the ARBs have the inverse agonist activity in CAM N111G of AT$_1$

---

* Department of Pharmacy, The University of Asia Pacific, Dhaka, Bangladesh
† Department of Pharmacology, Faculty of Pharmaceutical Sciences, Niigata University of Pharmacy and Applied Life Sciences, Niigatashi, Japan

Methods in Enzymology, Volume 484   © 2010 Elsevier Inc.
ISSN 0076-6879, DOI: 10.1016/S0076-6879(10)84009-1   All rights reserved.

receptor. The internalization of the mutant, which was much lower than WT receptor, was significantly increased in presence of the ARBs. The results indicate that internalization of CAM N111G of AT_1 receptor is induced by the ARBs, which may be an important characteristic of inverse agonist activities of the ARBs in N111G.

1. INTRODUCTION

G-protein-coupled receptors (GPCRs) form one of the largest protein families, with several hundred members in humans (Venter et al., 2001). Despite the wide variety of ligands and physiological roles, all these receptors are structurally characterized by seven transmembrane domains and most of them are thought to share common activation and desensitization mechanisms. Angiotensin II receptors are the members of the GPCR superfamily. Two subtypes of angiotensin II receptors have been identified (Chiu et al., 1989; Whitebread et al., 1989) and pharmacologically characterized, designated as angiotensin II type 1 and type 2 (AT_1 and AT_2) receptors (Bumpus et al., 1999). Interaction of angiotensin II (Ang II) with the AT_1 receptor induces vasoconstriction, sodium reabsorption, and stimulation of aldosterone release (Matsusaka and Ichikawa, 1997).

Mutagenesis studies of the AT_1 receptor have recently identified amino acid residues important in the binding of the natural ligand, Ang II and nonpeptide antagonists. A conserved residue, Lys^{199} in the fifth transmembrane domain (TMD V) of the AT_1 receptor, has recently been reported to be crucial for the binding of both peptide (Noda et al., 1995) and nonpeptide (Bhuiyan et al., 2009) ligands. Ang II contains two residues, Tyr^4 and Phe^8, which are essential for agonism (Noda et al., 1996). The activation of AT_1 receptor from the basal state requires an interaction between Asn^{111} in TMD III of the AT_1 receptor and the Tyr^4 residue of Ang II (Feng et al., 1998). This shows the importance of Asn^{111} residue in TMD III of the AT_1 receptor. No naturally occurring, constitutively active mutant (CAM) of AT_1 receptors have been reported, but engineered mutation of the Asn^{111} residue to glycine (N111G) results in constitutive activation of the AT_1 receptor (Noda et al., 1996). In the wild-type (WT) AT_1 receptor, interaction of Tyr^4 of Ang II with Asn^{111} in the receptor appears to act as the trigger to convert inactive (R) to active state (R^*) and allow receptor activation. Small side chain (glycine) substitution of Asn^{111} in the AT_1 receptor presumably releases this conformational switch, allowing constitutive activity, and removing the requirement of Tyr^4 in Ang II for maximal receptor activation.

Agonist binding to a GPCR induces conformational changes in the receptor, leading to activation of $G\alpha\beta\gamma$ heterotrimers (Feng et al., 2005).

One function of the activated G-proteins is to activate GPCR kinases (GRKs) that in turn phosphorylate the specific receptor for desensitization. Subsequently, β-arrestins bind to the GRK-phosphorylated motifs of the receptor and induce the receptor internalization. This homologous GPCR internalization is agonist specific and GRK dependent. This type of feedback regulation is conventional because it requires activation of classic G-proteins (Ferguson, 2001; Kohout and Lefkowitz, 2003; Lodowski et al., 2003). Homologous internalization of GPCRs can also take place through β-arrestin-independent pathway. Initially, internalization of the GPCRs was viewed as a mean to uncouple the receptor from its signaling components, thereby dampening the overall response (Gagnon et al., 1998; Hertel et al., 1985; Tsao et al., 2001; Waldo et al., 1983). The results of many studies indicate that the itinerary of the internalized GPCR is receptor- and cell-specific (Zhang et al., 1997). At least four pathways of agonist-induced internalization of GPCRs exist (Claing et al., 2000, 2002), and they may be cell type specific. The classical GPCR internalization pathway involves GRKs, β-arrestin, clathrin-coated pits, and the GTPase dynamin and is exemplified by the β_2-adrenergic receptor (Claing et al., 2000, 2002; Ferguson, 2001; Ferguson et al., 1996; Krupnick and Benovic, 1998; Lefkowitz, 1998; Pitcher et al., 1998). Thus in this chapter, we described the investigation about the binding profiles of both WT and N111G mutant of AT_1 receptors with AT_1 receptor agonist and ARBs, such as valsartan, losartan, candesartan, and telmisartan. We also determined total inositol phosphate (IP) accumulation by the cells expressing specified receptors and showed the inverse agonist activity of the ARBs in N111G mutant. Finally, we examined the internalization of the specified receptors and correlated the result with the mechanism of constitutive activity of N111G mutant.

2. Preparation of Receptor Plasmid and Protein

2.1. Site-directed mutagenesis and plasmid preparation

The synthetic rat AT_1 receptor gene, cloned in the shuttle expression vector pMT-2, was used for expression. Site-directed mutagenesis was performed on the WT AT_1 receptor by polymerase chain reaction (PCR) method with the QuichChange Site-Directed Mutagenesis Kit (Stratagene, CA, USA) according to the protocol of the manufacturer as described in earlier studies (Jongejan et al., 2005). Briefly, forward and reverse oligonucleotides were constructed to introduce desired mutation. PCR products were purified and finally DNA sequence analysis was done to confirm the site-directed mutation. Receptor plasmid was prepared using Midiprep kit (BIO-RAD, CA, USA) after transformation in XL1-Blue supercompetent cells through heat pulse according to manufacturer's protocol.

2.2. Cell culture, transfection, and membrane preparation

COS-7 cells were cultured in Dulbecco's Modified Eagle's medium (DMEM) supplemented with 10% fetal bovine serum, 100 U/ml penicillin, and 100 μg/ml streptomycin in 5% CO_2 at 37 °C. The WT and mutant AT_1 receptors were transfected transiently into COS-7 cells using Lipofectamine™ 2000 according to the manufacturer's protocol (Invitrogen Life Technologies, Rockville, MD, USA). To express the AT_1 receptor protein, 12 μg of purified plasmid DNA/10^7 cells was used in the transfection. Transfected COS-7 cells that had been cultured for 48 h were harvested with ice-cold phosphate buffer saline (PBS), pH 7.4; following two times wash by PBS. The cells were centrifuged at 3000 rpm for 10 min at 4 °C, and the cell pellets were washed by Hank's buffered salt solution (HBSS) with 1.5% 0.5 M EDTA, 0.15% 50 mg/ml PMSF, and 0.15% 2 mg/ml aprotinin, and finally suspended in 0.25 M sucrose solution containing 1.5% 0.5 M EDTA, 0.15% 50 mg/ml PMSF, and 0.15% 2 mg/ml aprotinin. The cells were then disrupted by Polytron Homogenizer (Kinematica, Switzerland) for 10 s. The mass was centrifuged at 4 °C for 5 min at 1260 × g and the supernatant containing the membrane fraction was ultra centrifuged at 4 °C for 20 min at 30,000 × g. The resulting pellets were suspended in binding assay buffer containing 20 mM phosphate buffer, 100 mM sodium chloride, 20 mM magnesium chloride, 1 mM EGTA, and 0.2% BSA, pH 7.4 and used for binding experiments. The protein contents of the membranes were measured by the method of Lowry et al. (1951) using bovine serum albumin as the standard.

3. Radioligand Binding Assay

Binding assays for WT and N111G mutant of AT_1 receptors were carried out in incubation tube that contained 10 μg of membrane protein, [^{125}I]-Sar^1-Ile^8-Ang II (Perkin Elmer, Inc., Boston, USA), unlabeled drug as required, and binding buffer in a final volume of 125 μl. Both saturation and competition binding assays were carried out as described previously (John et al., 2001; Miura et al., 2006). Briefly, for saturation binding studies, six to seven concentrations (5–800 pM) of [^{125}I]-Sar^1-Ile^8-Ang II were tested in duplicate. Nonspecific binding was defined as the amount of radioligand binding remaining in presence of 10 μM Ang II (Peptide Institute Inc. Japan). For competition binding studies, membranes were incubated with 250 pM of [^{125}I]-Sar^1-Ile^8-Ang II and different concentrations (10^{-4} to 10^{-11} M) of unlabelled drugs, such as candesartan (Takeda Chemical Industries Ltd., Japan), losartan (Merck Research Laboratories, NJ, USA), telmisartan (Nippon Boehringer Ingelheim, Japan), and valsartan (Novartis Institutes for BioMedical Research, Inc., Cambridge, MA, USA) for 1 h at 25 °C. The incubation was terminated by rapid filtration under vacuum through Whatman GF/C filters that had been presoaked in 0.5% polyethyleneimine followed by three times washing with ice-cold 50 mM Tris–HCl (pH 8.0). The bound ligand fraction was determined from the counts/min remaining on the membrane in γ-counter.

[^{125}I]-Sar^1-Ile^8-Ang II radioligand binding assay showed that the WT receptor and N111G mutant were bound as expected, with a dissociation constant (K_d) of 0.55 ± 0.02 and 0.78 ± 0.21 nM, respectively (Table 9.1). B_{max} values for the WT receptor and N111G mutant were calculated from the maximal specific binding of [^{125}I]-Sar^1-Ile^8-Ang II as 1.52 ± 0.07 and 0.68 ± 0.12 pmol/mg of protein, respectively (Table 9.1). The receptor expression for the mutant was decreased significantly compared to WT AT$_1$ receptors ($P < 0.001$). Table 9.2 shows the binding affinities (pK_i) for agonist and ARBs, valsartan, losartan, candesartan, and telmisartan, toward WT and N111G mutant of AT$_1$ receptors. The binding affinity of Ang II to the mutant N111G was markedly increased compared to WT receptor ($P < 0.05$) (Table 9.2, Fig. 9.1). Binding affinities of ARBs were two- to threefold decreased to the mutant N111G compared to AT$_1$ WT receptor

Table 9.1 Dissociation constant (K_d) and maximum binding sites (B_{max}) of [^{125}I]-Sar^1-Ilu^8-angiotensin II for wild-type and mutant N111G of AT$_1$ receptors

	AT$_1$ wild-type receptor	N111G mutant receptor
K_d value (nM)	0.55 ± 0.02	0.78 ± 0.21
B_{max} value (pmol/mg protein)	1.52 ± 0.07	0.68 ± 0.12**

[^{125}I]-Sar^1-Ile-Angiotensin II was used to label AT$_1$ wild-type and N111G mutant receptors transiently expressed in COS-7 cells. Data represent the mean ± SEM of four independent experiments, each performed in duplicate. **$P < 0.001$ versus wild-type.

Table 9.2 Binding affinities (K_i) in nM of agonists and antagonists to wild-type and N111G mutant of AT$_1$ receptors

	Wild type	N111G Mutant
Agonist		
Angiotensin II	30.58 ± 11.47	0.72 ± 0.09*
Antagonist		
Valsartan	8.21 ± 3.19	20.31 ± 3.75*
Losartan	59.41 ± 2.65	119.50 ± 5.01**
Candesartan	3.37 ± 0.27	3.23 ± 0.12
Telmisartan	2.96 ± 1.11	7.38 ± 0.82*

[^{125}I]-Sar^1-Ile^8-Angiotensin II (250 pM) was used to label AT$_1$ wild-type and N111G mutant receptors transiently expressed in COS-7 cells. Data represent the mean ± SEM of four independent experiments, each performed duplicate. *$P < 0.05$; **$P < 0.001$ versus wild type.

($P < 0.05$; Table 9.2) although candesartan showed almost no change in binding affinity to the mutant.

4. INOSITOL PHOSPHATE ACCUMULATION ASSAY

COS-7 cells at about 90% confluent in 10-cm dishes were seeded into 24-well plates taking about 10^5 cells per well. After 24 h the cells were transfected transiently using Lipofectamine™ 2000 with plasmid DNA of

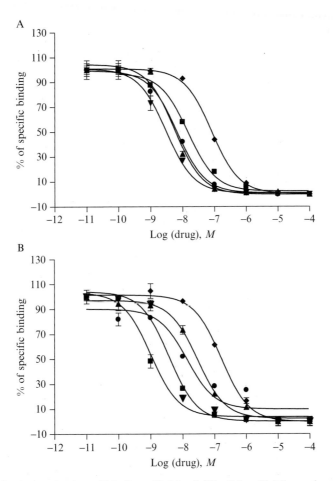

Figure 9.1 Determination of binding affinities (pK_i) of Ang II (■), candesartan (▼), losartan (♦), valsartan (▲), and telmisartan (●) to the (A) wild-type and (B) N111G mutant of AT_1 receptors as assessed by [^{125}I]-Sar^1-Ile^8-Angiotensin II radioligand.

both WT and N111G mutant of AT_1 receptors. Twenty-four hours after transfection the cells were labeled with 1 $\mu Ci/ml$ [^{3}H]myo-inositol (Amersham Biosciences, NJ, USA) in DMEM and incubated for 20 h at 37 °C in 5% CO_2. The cells were washed with HBSS and exposed with HBSS containing 20 mM phosphate buffer and 20 mM LiCl, pH 7.4 for 30 min at 37 °C. Agonist and four ARBs were added to each well and the incubation was continued for an additional 1 h at 37 °C. At the end of the incubation, the medium was removed, and the reaction was stopped by adding 1 ml of 10 mM formic acid (previously stored at 4 °C) to each well. The cells were then neutralized by 1 ml 500 mM KOH and 9 mM sodium tetraborate per well. The contents of each well were extracted and centrifuged for 5 min at 1400× g and the upper layer was transferred to a 1 ml AG1-X8 resin (100–200 mesh; BIO-RAD Laboratories, Inc., CA, USA) loaded column. The columns were washed two times with 5 ml 60 mM sodium formate and 5 mM borax. Total soluble IP was eluted with 5 ml 1 M ammonium formate and 0.1 M formic acid. Radioactivity was measured by liquid scintillation counter.

The cells expressing N111G mutant exhibited higher levels of agonist-independent (basal) IP production (Fig. 9.2A) showing constitutive activity. A decrease in the size of Asn^{111} side chain induces an intermediate activated receptor conformation, which may be responsible for the constitutive activity of N111G mutant (Noda et al., 1996). IP production by the cells expressing N111G mutant of AT_1 receptor was markedly decreased when incubated in presence of the ARBs used in the study (Fig. 9.2B) confirming the inverse agonist activity of the ARBs.

5. Internalization Assay

Internalization assay was performed as described previously (Modrall et al., 2001). Briefly, COS-7 cells in 12-well plates that had been transiently transfected were incubated separately at 37 °C in serum-free DMEM with 1 μM Ang II or 1 μM ARBs for 0, 2, 5, 10, 15, 30, and 45 min. The cells were washed twice by ice-cold PBS and incubated for 3 h at 4 °C with 0.1 nM [^{125}I]-Sar^1-Ile^8-Ang II in binding buffer. The cells were then washed with ice-cold PBS and surface bound [^{125}I]-Sar^1-Ile^8-Ang II was removed using the acid wash technique of Crozat et al. (1986) in which cells were exposed to 150 mM NaCl, 50 mM glycine, pH 3, for 10 min at 4 °C. For cell lysis, 300 μl of 1 M NaOH solution was added and the content was neutralized by 300 μl of 1 M HCl. The cell-associated radioactivity was measured by δ-counting. Percent of internalization was calculated from the difference in cell surface binding values at different time interval considering the cell surface binding value at zero as 100%.

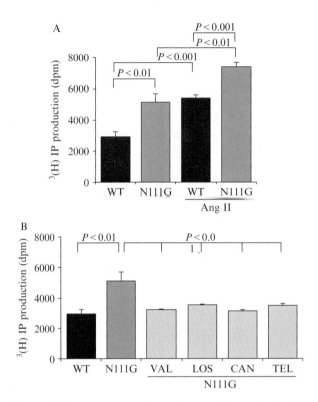

Figure 9.2 Effect of AT_1 receptor agonist and antagonists on inositol phosphate (IP) production by wild-type and N111G mutant of AT_1 receptors. (A) IP production in the absence (solid bar) and presence of 1 μM Ang II (gray bar) by COS-7 cells expressing wild-type and N111G mutant of AT_1 receptors. (B) Inhibition of IP production by 1 μM valsartan, losartan, candesartan, and telmisartan, respectively (light gray bars). Data are the mean ± SEM of 4–6 independent experiments, each performed in duplicate. Student's t-test was performed for statistical analyses.

Treatment of the WT and N111G mutant of AT_1 receptors with 1 μM Ang II for 45 min induced 74.63 ± 1.00% and 19.56 ± 2.87% receptor internalization, respectively (Table 9.3, Fig. 9.3). The mutant N111G showed very low internalization (only 26%) compared to WT AT_1 receptor after 45 min (Table 9.3, Fig. 9.3) and is unable to bind with adaptor protein that may cause the mutant receptor not to be internalized and increase in the receptor protein in the cell surface. This in turn promotes higher IP production without agonist stimulation. However, it is very interesting that the internalization of N111G mutant significantly increased in presence of the ARBs of this study. The ARBs may cause the change in the conformation of CAM N111G of AT_1 receptor from active to inactive state to promote internalization, which is consistent with the mechanism of inverse agonists of constitutive active GPCRs. On the other hand, there was

Table 9.3 Internalization of AT_1 wild-type and N111G mutant of AT_1 receptor without and with different AT_1 receptor antagonists

Receptor	Treatment	% of internalization (after 45 min)	K_e (h^{-1})
AT_1 WT	Ang II	74.63 ± 1.00	0.87 ± 0.37
	Valsartan	70.97 ± 1.35	0.65 ± 0.23
N111G	Ang II	19.56 ± 2.87*	0.41 ± 0.26
	Valsartan	63.22 ± 0.38*	0.24 ± 0.13
	Candesartan	62.43 ± 0.68*	0.29 ± 0.14
	Losartan	54.78 ± 0.61*	0.43 ± 0.21
	Telmisartan	60.85 ± 0.66*	0.25 ± 0.10

K_e is the rate constant of internalization. The data are the mean ± SEM of four independent experiments, each performed in duplicate. Student's t-test was performed for statistical analyses. *A P value of less than 0.01 compared to AT_1 wild-type receptor was taken as significant.

no significant change in the internalization of the WT receptor after the use of valsartan with the receptor indicating that the internalization of WT receptor is not induced by the ARBs. Some GPCRs, such as vasopressin V2, AT_1, and bradykinin B_2 receptor subtypes, were reported to internalize upon antagonist binding (Houle et al., 2000; Hunyady 1999; Pfeiffer et al., 1998). Pheng et al. (2003) also reported that the binding of Y_1 receptor antagonist, GR231118, induced time-dependent internalization of Y_1 receptors in HEK293 cells and this process was mediated in part by clathrin-dependent and G-protein independent mechanisms.

6. Western Blot Analysis

Equal amounts (20 μg) of whole cell lysates of the WT and mutant receptors were resolved by SDS–polyacrylamide gel electrophoresis and transferred onto Hybond ECL nitrocellulose membranes (Amersham Biosciences, DE, USA) using a semidry system (Trans-Blot SD Semi-dry Transfer Cell, BIO-RAD, USA) in immunotransfer buffer. The membranes were blocked with 10% skim milk in Tris-buffered saline containing 0.1% Tween 20 (TBS-T) by slow shaking for 1 h at room temperature. After blocking, membranes were exposed to AT_1 rabbit polyclonal IgG and actin rabbit polyclonal IgG (Santa Cruz Biotechnology, CA, USA) at 1:1000 dilutions in TBS-T with 1% milk for 1 h at room temperature, followed by incubation with HRP-conjugated anti-rabbit IgG (Promega, WI, USA) at 1:2000 dilutions in TBS-T with 1% milk overnight at 4 °C. The membranes were washed three times with TBS-T in each step and finally the blots on the membranes were visualized by adding Amersham ECL Western blotting detection reagent (GE Healthcare, UK). Western blot probed with anti-AT_1 antibody detected

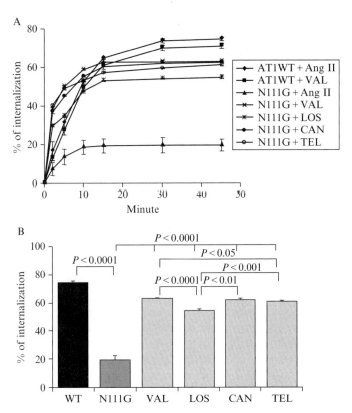

Figure 9.3 Internalization assay of wild-type and N111G mutant of AT_1 receptors. (A) % of internalization of wild-type AT_1 receptors treated with 1 μM Ang II and valsartan, and N111G mutant of AT_1 receptors treated with 1 μM Ang II and 1 μM ARBs. (B) % of internalization of wild-type (solid bar) and N111G mutant of AT_1 receptors in absence (gray bar) and presence of 1 μM ARBs (light gray bar) after 45 min of incubation. Internalization of ^{125}I-labeled Sar^1-Ile^8-angiotensin II at 37 °C by wild-type and N111G mutant of AT_1 receptors was determined as described in Materials and Methods. Data are the mean ± SEM of four independent experiments, each performed in duplicate. Student's t-test was performed for statistical analyses.

the specified protein of the receptors and immunoreactive bands were observed at 43 kDa both in WT and N111G mutant of AT_1 receptors (Fig. 9.4).

7. Data Analysis

Nonlinear regression analyses of saturation and competition binding assay were performed using GraphPad Prism software (San Diego, CA, USA). The results of the experiments were expressed as the mean ± SEM.

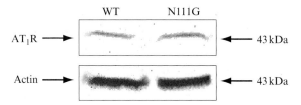

Figure 9.4 Western blot analysis showing the band of AT_1 receptor and actin for both wild-type and N111G mutant of AT_1 receptors.

In competition binding experiments, the values of inhibition constants (K_i) were calculated by the following equation (Cheng and Prusoff, 1973):

$$K_i = IC_{50}/\{1 + ([L]/K_d)\}$$

where, the inhibition concentrations (IC_{50}) were determined as the concentrations of ligands that inhibited $[^{125}I]$-Sar^1-Ile^8-Ang II binding by 50%; [L] = the concentration of $[^{125}I]$-Sar^1-Ile^8-Ang II used and K_d = the dissociation constant of $[^{125}I]$-Sar^1-Ile^8-Ang II for the receptor. Statistical analyses were performed by the Student's unpaired t-test (two tailed).

8. Concluding Remark

Homologous internalization of GPCRs is an active process that requires specific ligand binding, conformational changes of the receptor, and signal transduction initiated by the activated receptor. Internalization plays an important role in receptor endocytosis and signal transduction. The present study demonstrated the correlation between endocytosis and signal transduction of AT_1 receptors due to its site-directed mutagenesis. The results demonstrate that asparagine at position 111 of TMD III of AT_1 receptor is very important site for both agonist and antagonist binding. N111G mutant of AT_1 receptor can undergo ligand-induced internalization following stimulation of ARBs. If these results are applicable *in vivo*, the study can suggest that chronic treatment with the ARBs may induce cell surface receptor losses, leading to apparent conditional knock out of receptor activity, this possibly being of clinical significance of the present study.

REFERENCES

Bhuiyan, M. A., Ishiguro, M., Hossain, M., Nakamura, T., Ozaki, M., Miura, S., *et al.* (2009). Binding sites of valsartan, candesartan and losartan with angiotensin II receptor 1 subtype by molecular modeling. *Life Sci.* **85**(3–4), 136–140.

Bumpus, F. M., Catt, K. J., Chiu, A. T., DeGasparo, M., Goodfriend, T., Husain, A., et al. (1999). Nomenclature for angiotensin receptors. A report of Nomenclature Committee of the Council for High Blood Pressure Research. *Hypertension* **17,** 720–721.

Cheng, Y.-C., and Prusoff, W. H. (1973). Relationship between the inhibition constant (K_i) and the concentration of inhibition, which causes 50% inhibition (IC_{50}) of an enzymatic reaction. *Biochem. Pharmacol.* **22,** 3099–3108.

Chiu, A. T., Herblin, W. F., McDall, D. E., Ardecky, R. J., Carini, D. J., Duncia, J. V., et al. (1989). Identification of angiotensin receptor subtypes. *Biochem. Biophys. Res. Commun.* **165,** 196–203.

Claing, A., Perry, S. J., Achiriloaie, M., Walker, J. K., Albanesi, J. P., Lefkowitz, R. J., et al. (2000). Multiple endocytic pathways of G protein-coupled receptors delineated by GIT1 sensitivity. *Proc. Natl. Acad. Sci. USA* **97,** 1119–1124.

Claing, A., Laporte, S. A., Caron, M. G., and Lefkowitz, R. J. (2002). Endocytosis of G protein-coupled receptors: Roles of G protein-coupled receptor kinases and β arrestin proteins. *Prog. Neurobiol.* **66,** 61–79.

Crozat, A., Penhoat, A., and Saez, J. M. (1986). Processing of angiotensin II (A-II) and (Sar^1, Ala^8) A-II by cultured bovine adrenocortical cells. *Endocrinology* **118,** 2312–2318.

Feng, Y. H., Miura, S., Hussain, A., and Karnik, S. S. (1998). Mechanism of constitutive activation of the AT_1 receptor: Influence of the size of the agonist switch binding residue Asn(111). *Biochemistry* **37**(45), 15791–15798.

Feng, Y. H., Ding, Y., Ren, S., Zhou, L., Xu, C., and Karnik, S. S. (2005). Unconventional homologous internalization of the angiotensin II type-1 receptor induced by G-protein-independent signals. *Hypertension* **46**(2), 419–425.

Ferguson, S. S. (2001). Evolving concepts in G protein-coupled receptor endocytosis: The role in receptor desensitization and signaling. *Pharmacol. Rev.* **53,** 1–24.

Ferguson, S. S., Barak, L. S., Zhang, J., and Caron, M. G. (1996). G-protein-coupled receptor regulation: Role of G-protein-coupled receptor kinases and arrestins. *Can. J. Physiol. Pharmacol.* **74,** 1095–1110.

Gagnon, A. E., Kallal, L., and Benovic, J. L. (1998). Role of clathrin-mediated endocytosis in agonist-induced down-regulation of the beta2-adrenergic receptor. *J. Biol. Chem.* **273,** 6976–6981.

Hertel, C., Coulter, S., and Perkins, J. P. (1985). A comparison of catecholamine-induced internalization of beta-adrenergic receptors and receptor-mediated endocytosis of epidermal growth factor in human astrocytoma cells. Inhibition by phenylarsine oxide. *J. Biol. Chem.* **260,** 12547–12553.

Houle, S., Larrivee, J. F., Bachvarova, M., Bouthillier, J., Bachvarov, D. R., and Marceau, F. (2000). Antagonist-induced intracellular sequestration of rabbit bradykinin B(2) receptor. *Hypertension* **35,** 1319–1325.

Hunyady, L. (1999). Molecular mechanisms of angiotensin II receptor internalization. *J. Am. Soc. Nephrol.* **10**(11), S47–S56.

John, H., Jennifer, N. H., Steven, J. F., and Daniel, K. Y. (2001). Identification of angiotensin II type 2 receptor domains mediating high-affinity CGP 42112A binding and receptor activation. *J. Pharmacol. Exp. Ther.* **298**(2), 665–673.

Jongejan, A., Bruysters, M., Ballesteros, J. A., Haaksma, E., Bakker, R. A., Pardo, L., et al. (2005). Linking agonist-binding to histamine H_1 receptor activation. *Nat. Chem. Biol.* **1,** 98–103.

Kohout, T. A., and Lefkowitz, R. J. (2003). Regulation of G-protein couple receptor kinases and arrestins during receptor desensitization. *J. Mol. Pharmacol.* **63,** 9–18.

Krupnick, J. G., and Benovic, J. L. (1998). The role of receptor kinases and arrestins in G protein-coupled receptor regulation. *Annu. Rev. Pharmacol. Toxicol.* **38,** 289–319.

Lefkowitz, R. J. (1998). G protein-coupled receptors. III New roles for receptor kinases and β-arrestins in receptor signaling and desensitization. *J. Biol. Chem.* **273,** 18677–18680.

Lodowski, D. T., Pitcher, J. A., Capel, W. D., Lefkowitz, R. J., and Tesmer, J. J. (2003). Keeping G-proteins at bay: A complex between G-protein couple receptor kinase 2 and Gbetagamma. *Science* **300,** 1256–1262.

Lowry, O. H., Rosebrough, N. J., Farr, A. L., and Randall, R. J. (1951). Protein measurement with the Folin phenol reagent. *J. Biol. Chem.* **193,** 265–275.

Matsusaka, T., and Ichikawa, I. (1997). Biological functions of angiotensin and its receptors. *Annu. Rev. Physiol.* **59,** 395–412.

Miura, S., Fujino, M., Hanzawa, H., Kiya, Y., Imaizumi, S., Matsuo, Y., et al. (2006). Molecular mechanism underlying inverse agonist of angiotensin II type I receptor. *J. Biol. Chem.* **281**(28), 19288–19295.

Modrall, J. G., Nanamori, M., Sadoshima, J., Barnhart, C. B., Stanley, J. C., and Neubig, R. R. (2001). Ang II type 1 receptor downregulation does not require receptor endocytosis or G protein coupling. *Am. J. Physiol. Cell Physiol.* **281,** C801–C809.

Noda, K., Saad, Y., Kinoshita, A., Boyle, T. P., Graham, R. M., Husain, A., et al. (1995). Tetrazole and carboxylate groups of angiotensin receptor antagonists bind to the same subsite by different mechanisms. *J. Biol. Chem.* **270**(5), 2284–2289.

Noda, K., Feng, Y. H., Liu, X. P., Saad, Y., Husain, A., and Karnik, S. S. (1996). The active state of the AT_1 angiotensin receptor is generated by angiotensin II induction. *Biochemistry* **35**(51), 16435–16442.

Pfeiffer, R., Kirsch, J., and Fahrenholz, F. (1998). Agonist and antagonist-dependent internalization of the human vasopressin V2 receptor. *Exp. Cell Res.* **244,** 327–339.

Pheng, L. H., Dumont, Y., Fournier, A., Chabot, J. G., Beaudet, A., and Quirion, R. (2003). Agonist- and antagonist-induced sequestration/internalization of neuropeptide Y Y1 receptors in HEK293 cells. *Br. J. Pharmacol.* **139**(4), 695–704.

Pitcher, J. A., Freedman, N. J., and Lefkowitz, R. J. (1998). G protein-coupled receptor kinases. *Annu. Rev. Biochem.* **67,** 653–669.

Tsao, P., Cao, T., and Von Zastrow, M. (2001). Role of endocytosis in mediating downregulation of G-protein-coupled receptors. *Trends Pharmacol. Sci.* **22,** 91–96.

Venter, J. C., Adams, M. D., Myers, E. W., Li, P. W., Mural, R. J., Sutton, G. G., et al. (2001). The sequence of the human genome. *Science* **291,** 1304–1351.

Waldo, G. L., Northup, J. K., Perkins, J. P., and Harden, T. K. (1983). Characterization of an altered membrane form of the beta-adrenergic receptor produced during agonist-induced desensitization. *J. Biol. Chem.* **258,** 13900–13908.

Whitebread, S., Mele, M., Kamber, B., and DeGasparo, M. (1989). Preliminary biochemical characterization of two angiotensin II receptor subtypes. *Biochem. Biophys. Res. Commun.* **163,** 284–291.

Zhang, J., Barak, L. S., Winkler, K. E., Caron, M. G., and Ferguson, S. S (1997). A central role for beta-arrestins and clathrin-coated vesicle-mediated endocytosis in beta2-adrenergic receptor resensitization. Differential regulation of receptor resensitization in two distinct cell types. *J. Biol. Chem.* **272,** 27005–27014.

CHAPTER TEN

Methods to Detect Cell Surface Expression and Constitutive Activity of GPR6

Balakrishna M. Prasad,* Bettye Hollins,[†] *and* Nevin A. Lambert[†]

Contents

1. Introduction	180
2. GPR6 is Expressed in Intracellular Compartments	180
2.1. Methods of detecting cell surface expression of GPCRs	181
3. Comparison of the Cell Surface Protein Detection Methods	187
3.1. Methods of detecting constitutive activity of GPR6	189
4. Comparison of the Constitutive Gs-Activity Detection Methods	192
5. Conclusions	194
References	194

Abstract

GPR6 is a constitutively active Gs-coupled receptor that can signal from intracellular compartments. We present different methods used to study cell surface expression of receptors and other membrane proteins. A comparison of these methods shows that methods based on susceptibility to proteolytic enzymes are more efficient at providing estimates of cell surface expression than the commonly used cell surface biotinylation method. We also present different methods that can be used to detect constitutive activity of Gs-coupled receptors. Imaging-based assays to detect intracellular cyclic AMP accumulation are well suited to study signaling at a single cell level. These assays are particularly useful when the cells of interest form a small fraction of the culture such as primary cultures with low transfection efficiency.

---

* Department of Clinical Investigation, Eisenhower Army Medical Center, Georgia, USA
[†] Department of Pharmacology and Toxicology, Medical College of Georgia, Augusta, Georgia, USA

1. Introduction

GPR6 belongs to a family of constitutively active Gs-coupled G protein coupled receptors (GPCRs) that also includes GPR3 and GPR12 (Song et al., 1994). Depending on the anatomical location of their expression, these receptors are involved in a wide array of functions such as neurite growth, amyloid processing, appetitive learning behavior, and meiotic arrest of oocytes (Lobo et al., 2007; Mehlmann et al., 2004; Tanaka et al., 2007; Thathiah et al., 2009). The diverse functions of this family of receptors appear to be mediated by their constitutive activity and their ability to maintain elevated intracellular cyclic AMP concentration. The relative magnitude of their constitutive activity was reported to be the highest among all G-protein coupled receptors (Eggerickx et al., 1995; Seifert and Wenzel-Seifert, 2002). Although their constitutive activity was recognized more than a decade ago, very little information is available on the ligands and signaling mechanisms of these receptors. Sphingosine-1-phosphate was reported to be the endogenous agonist of GPR6, based on the finding that intracellular calcium concentrations were elevated by S1P in GPR6 expressing cells (Ignatov et al., 2003; Uhlenbrock et al., 2002). However, several recent observations do not support the notion that sphingosine-1-phosphate is an agonist of GPR6 (B. M. Prasad, unpublished data; Yin et al., 2009) and it can still be considered an orphan receptor. Given the importance of GPR6 family members in different functions in health and disease, identification of their ligands and understanding of their downstream signaling mechanisms are of keen interest.

As an initial approach to develop tools and to study signaling by GPR6 family members, we investigated the cellular localization of GPR6. Despite its constitutive activity, GPR6 appears to be predominantly located in intracellular compartments. Gs-coupled receptors that can signal from intracellular compartments are an emerging concept and may eventually expand our understanding of compartmentalization of GPCR-mediated signaling (Calebiro et al., 2009; Jalink and Moolenaar, 2010). Different methods used to measure cell surface expression and cAMP concentrations vary in their sensitivity and utility. We describe some of the available methods to study cell surface expression and Gs-mediated constitutive activity of heterologously expressed GPCRs.

2. GPR6 is Expressed in Intracellular Compartments

The first indication that GPR6 may be expressed in intracellular compartments was provided by imaging experiments in which green fluorescent protein (GFP) tagged GPR6 was found predominantly between plasma membrane markers and nuclei. Alexa Fluor-594 conjugated wheat germ agglutinin (WGA) is cell impermeant and binds to N-acetyl glucosamine and N-acetyl neuraminic acid moieties of cell surface proteins. Hoechst 33342 is a cell-permeant compound that binds to nucleic acids. These two compounds are used to label plasma membranes and nuclei of live cells, respectively. HEK 293 cells expressing GPR6 or dopamine (D1) receptors were labeled with WGA and Hoechst 33342 (obtained from Invitrogen, Carlsbad, CA, USA). As shown in Fig. 10.1, D1 receptor, a Gs-coupled receptor known to be expressed on the plasma membrane, is colocalized with WGA. On the other hand, GPR6 appears to be located in between WGA and Hoechst 33342 stains. Any attempts to measure the relative expression of receptors on the cell surface by fluorescence intensities and degree of colocalization would be hampered by the limited spatial resolution of light microscopy. Indeed, quantitative measures of colocalization were not useful in determining cell surface expression because of diffuse nature of WGA labeling. In order to obtain more accurate measurements of cell surface expression of receptors, we used different methods described below.

2.1. Methods of detecting cell surface expression of GPCRs

2.1.1. Cell surface biotinylation

Biotinylation of cell surface proteins with amine-reactive compounds is a commonly used technique to label cell surface proteins which can then be separated and quantified. Sulfo-NHS-SS biotin, which is frequently used for

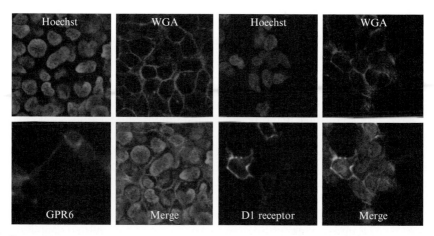

Figure 10.1 GPR6 appears to be predominantly located in the intracellular compartments. HEK293 cells expressing GFP-tagged GPR6 or D1 receptors were labeled with WGA (plasma membrane marker) and Hoechst 33342 (nuclear stain). These images show that D1 receptor is primarily located along the plasma membrane, while GPR6 fluorescence is mostly located between plasma membrane and nuclei. (See Color Insert.)

this purpose, is a membrane impermeant compound that reacts with amino groups of extracellular lysine side chains in proteins. We used sulfo-NHS-SS biotin to label cell surface proteins and measure the relative abundance of pHluorin-tagged GPR6 (pHGPR6) and D1R (pHD1) on the cell surface. As will be evident later in this chapter, the pHluorin tag at the N-terminus will be used to measure cell surface expression of GPR6 in different assays. Super ecliptic pHluorin is a pH-sensitive variant of GFP ($pK_a \sim 7.1$) which can easily be detected by commercially available GFP antibodies (Sankaranarayanan et al., 2000).

2.1.1.1. Methods 24-well tissue culture plates were coated with 100 ug/ml poly-D-lysine. HEK293 cells obtained from ATCC were seeded at a density of 200,000 cells per well. A day after plating, cells were transfected with pHGPR6 or pHD1 (0.3 μg/well using lipofectamine 2000 reagent, Invitrogen). Two days after transfection, cells were washed once with PBS and labeled with 1.5 mg/ml freshly prepared sulfo-NHS-SS biotin (Thermo Scientific, Rockford, IL, USA). Biotinylation was performed at 4 °C for 30 min in alkaline PBS (pH 8.5). Biotinylation reagent was quenched by two washes of 100 mM glycine buffer (5 min per wash). Cultures were extracted with 500 μl radioimmunoprecipitation assay (RIPA) buffer and incubated with 100 μl neutravidin agarose beads overnight at 4 °C with end over end mixing. Supernatant was separated from neutravidin beads by centrifugation and saved to assay for unbiotinylated proteins that represent the intracellular protein fraction. Beads were washed four times with RIPA buffer (1.0 ml per wash) and biotinylated proteins were eluted with 50 μl sample buffer with 50 mM dithiothreitol. Western blots with 25 μl of the biotinylated sample, and 25 μl of supernatant were probed with horse radish peroxidase conjugated GFP antibody (Invitrogen). Densitometric values of bands corresponding to GPR6 and D1 receptors were analyzed using NIH imageJ.

2.1.1.2. Results Data in Fig. 10.2 show that cell surface proteins can be selectively labeled, separated, and quantitated. Measurement of GPR6 and D1-receptor abundance in biotin labeled and supernatant fractions shows that approximately 12% of D1-receptor is expressed on the cell surface while GPR6 is undetectable at the surface. It is worth noting that the band intensities corresponding to biotinylated and supernatant fractions of D1 receptor in Fig. 10.2 and most biotinylated studies, reported in the literature, are somewhat similar. However, when the fraction of these samples loaded in Western blot analysis (25/50 μl for biotinylated and 25/500 μl for supernatant) is taken into account, the densitometric values of biotinylated sample are small compared to those of supernatant (Fig. 10.2C). Based on the staining pattern observed in Fig. 10.1, 12% appears to be an underestimation of true fraction of surface expression of D1 receptor. Thus, we used alternative assays described below to determine cell surface expression.

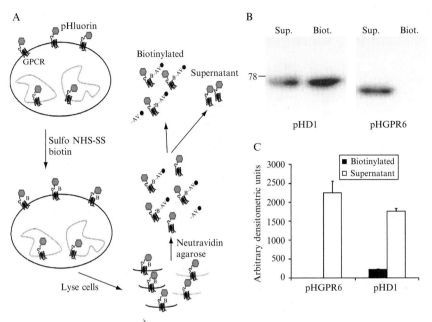

Figure 10.2 GPR6 is not detected at the cell surface by biotinylation assay. (A) Schematic representation of cell surface biotinylation assay. "B" represents biotin and "AV" represents neutravidin bound to agarose bead. (B) Representative blots pHD1 and GPR6 in biotinylation assay. (C) Densitometric analysis of bands corresponding to biotinylated (cell surface) and supernatant (intracellular) samples are presented. The densitometric values have been adjusted for differential loading of biotinylated (50%) and supernatant (5%) samples on the gel, $n = 4$.

2.1.2. Susceptibility to chymotrypsin treatment

Chymotrypsin, an endoprotease secreted by pancreas, cleaves proteins at aromatic amino acid residues (tyrosine, tryptophan, or phenylalanine). Brief exposure to this enzyme can be used to cleave cell surface proteins without dislodging cells from the culture dish. This enzyme is relatively large (~ 25 kDa) and does not have access to intracellular proteins of intact cells. We found that a pHlourin tag at the N-terminus of cell surface GPCRs is cleaved within a few seconds of exposure to chymotrypsin (see Fig. 10.4). The cleavage of pHluorin tag could thus be used as a readout for detection and quantification of receptors on the cell surface.

2.1.2.1. Methods HEK 293 cells were plated and transfected with pHGPR6 or pHD1 as described above. After one wash in PBS, cells were treated with 0.3 mg/ml of chymotrypsin (40 units/mg) for 5 min. Following two gentle washes in PBS, cells were extracted in RIPA buffer. In each experiment, cells without chymotrypsin treatment served as control to determine total amount of receptor. Western blots of lysates were probed with GFP antibody. Densitometric values of bands corresponding to total and chymotrypsin resistant samples were analyzed.

2.1.2.2. Results Data in Fig. 10.3 show that 71% of D1 receptors are present on the cell surface, while GPR6 is not susceptible to chymotrypsin treatment. This latter fact also demonstrates that residual enzyme activity in cell lysates is not responsible for increased sensitivity of chymotrypsin susceptibility method over the cell surface biotinylation assay.

2.1.3. Imaging based assays for cell surface expression

Although the data presented above show that GPR6 is located in the intracellular compartments of HEK293 cells, it is important to determine the expression pattern of GPR6 in neurons where it is normally expressed. It is conceivable that a missing subunit or chaperone of GPR6 is required for its surface expression in heterologous expression systems. We next assessed expression of GPR6 in neurons that are likely to have these putative GPR6 associated proteins. However, low transfection efficiency of GPR6 in neurons precluded us from using either of the above two methods.

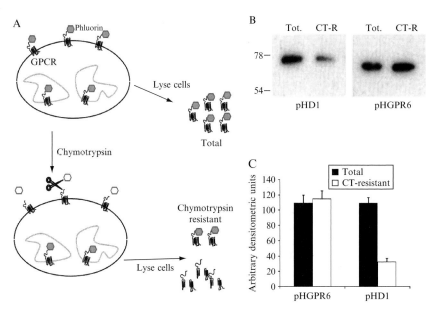

Figure 10.3 GPR6 is not detected at the cell surface chymotrypsin sensitivity assay. (A) Schematic representation of chymotrypsin sensitivity assay. (B) Representative blots pHD1 and GPR6 treated with (CT-resistant) or without (Total) 0.3 mg/ml chymotrypsin. (C) Densitometric analysis of bands corresponding to total and chymotrypsin resistant fraction are shown, $n = 4$.

We resorted to imaging based analysis of surface expression in individual cells that are transfected with pHGPR6 or pHD1. This assay takes advantage of the pH- and chymotrypsin-sensitivities of extracellular pHluorin.

2.1.3.1. Methods Striatal neurons were obtained from postnatal day 2–4 rat pups. After enzymatic dissociation in 20 units/ml papain solution, cells were plated on glass coverslips coated with 100 μg/ml poly-D-lysine and 5 μg/ml laminin. Neuronal medium conditioned over glia was used to maintain the cultures as previously described (Padmanabhan *et al.*, 2008). Neurons were transfected with 0.4 μg DNA constructs using the calcium phosphate method. Live cell images of transfected cells were obtained using laser scanning confocal microscope (Carl-Zeiss, Thornwood, NY, USA) using 488 nm excitation wavelength and 500–550 nm emission filter. Images were acquired at the rate of $1\ s^{-1}$ under continuous perfusion of control buffer (pH 7.4). Change in fluorescence caused by brief application of pH 6.0 buffer or 0.3 mg/ml chymotrypsin was measured. Analysis of fluorescence changes caused by these two treatments will be referred to as pH-imaging and chymotrypsin-imaging, respectively.

2.1.3.2. Results Brief exposure to pH 6.0 buffer caused a reversible decrease in pHD1 fluorescence of greater than 90% (Fig. 10.4). Chymotrypsin caused an irreversible decrease in fluorescence of comparable magnitude due to cleavage of the fluorophore on pHD1. Neither treatment caused a decrease in fluorescence of pHGPR6. Bright field images in Fig. 10.4 show that chymotrypsin treatment did not dislodge cells from glass coverslip. These data show that about 90% of pHD1 receptor is present on the cell surface, while pHGPR6 is predominantly located in the intracellular compartment of neurons.

2.1.4. Antibody labeling assay for cell surface expression

Internalization of GPCRs after being activated is one of the important mechanisms of receptor desensitization and subsequent down regulation (Ferguson, 2001). It is conceivable that, because of its high constitutive activity, GPR6 is continuously internalized and recycled to the surface. A small portion of GPR6 that is present on the cell surface may not be detected under the experimental conditions of the above three assays. Thus, we used an antibody labeling assay to determine if a small fraction of GPR6 is expressed at the cell surface and is being recycled. A high affinity GFP antibody is likely to bind all GPR6 receptors that are at the cell surface at any point during the period of antibody incubation. Furthermore the antibody can be labeled with fluorophores that have significantly greater quantum yield (brightness) than pHluorin, increasing the sensitivity of detection.

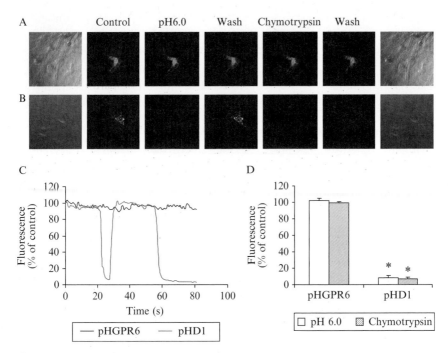

Figure 10.4 Surface expression of GPR6 could not be detected in striatal neurons by proton or chymotrypsin accessibility. (A) Representative fluorescence images of neurons expressing pHGPR6 in control, pH 6.0, 0.3 mg/ml chymotrypsin buffers. (B) Representative fluorescence images of pHD1 receptor in different buffers. (C) Normalized fluorescence intensities of pHGPR6 or pHD1 receptors are presented. (D) Average effects of pH 6.0 and chymotrypsin buffers on fluorescence of pHGPR6 and pHD1 receptors. *Significantly different from control buffer values ($P < 0.01$, $n = 4$). Reproduced with permission from Padmanabhan et al. (2009).

2.1.4.1. Methods Transfected HEK 293 cells were incubated with Alexa Fluor-594 conjugated polyclonal GFP antibody (10 μg/ml, Invitrogen) for 15 min at 37 °C. Z-stack images of cells were collected using 1.5 μm optical sections with 0.73 μm intervals, after washing out the labeling antibody. pHluorin was imaged with settings described above, while GFP antibody was imaged with 543 nm excitation wavelength and 590–720 nm emission filter. Fluorescence intensities were measured after the Z-stack images were compressed to obtain a projection image.

2.1.4.2. Results A comparison of fluorescence intensities of GFP antibody and pHluorin of cells expressing pHD1 receptors showed that antibody labeling is substantially more sensitive in detecting pHD1 receptors. Antibody labeling was clearly evident in cells where pHluorin fluorescence was below the limit of detection. Direct comparison of these two fluorescence intensities and minimum detectable pHluorin fluorescence of pHD1 (Fig. 10.5B) indicates that antibody labeling should have detected GPR6 even if less than 3% of total GPR6 is expressed on the surface. However, none of the pHGPR6 cells showed any antibody labeling, further demonstrating that GPR6 is located in intracellular compartments. Similar results were also obtained in antibody labeling of neurons expressing pHGPR6 (Padmanabhan *et al.*, 2009).

3. Comparison of the Cell Surface Protein Detection Methods

As is evident from the data presented above, different techniques have different efficiency, sensitivity, and utility to detect cell surface expression of membrane proteins. Of the different techniques used, the imaging based chymotrypsin and proton susceptibility methods showed best efficiency in detecting cell surface expression of D1 receptors. Thus, chymotrypsin sensitivity in imaging experiments was used as a bench mark (100% efficiency) to express relative efficiency to detect cell surface expression of pHD1 by different methods. In order to provide a reasonable comparison of different techniques, only data from HEK293 cells transfected with pHD1 receptors were used (Figs. 10.2, 10.3, 10.5 and imaging data in HEK293 cells from Padmanabhan *et al.*, 2009). Data in Table 10.1 show that cell surface biotinylation assay is only 15% as efficient as chymotrypsin-imaging assay. Similar low biotinylation efficiencies for other surface proteins have been reported, and several factors, such as buffer pH, cell type, available extracellular lysines, and membrane polarity, influence biotinylation rates (Gottardi and Caplan, 1992). Chymotrypsin sensitivity analysis by Western blots fared far better with 89% efficiency. Inherent differences in imaging and Western analysis of proteins might be responsible for the small differences in chymotrypsin sensitivities of these two methods (89% vs. 100%). The efficiency of the antibody labeling experiment could not be determined as intracellular labeling with antibody was not measured.

The sensitivity of detecting cell surface proteins by these methods is dependent on the epitope (pHluorin) detection methods. Both cell surface biotinylation and chymotrypsin susceptibility methods depend on antibody labeling of proteins on membranes. However, presumably due to poor labeling of the cell surface proteins, the biotinylation technique has relatively low sensitivity. Among the imaging methods, the antibody labeling has better sensitivity to detect cell surface proteins. This is primarily due to high quantum yield (brightness) of the fluorophore attached to the GFP antibody. As evident from Fig. 10.5B, GFP antibody fluorescence under our imaging conditions is several-fold higher than that of pHluorin

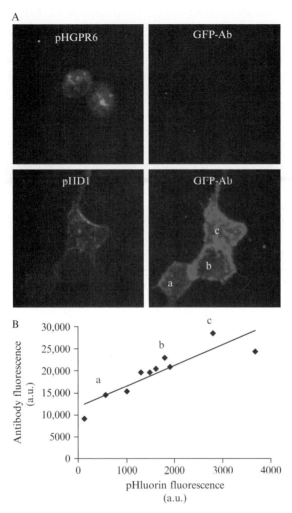

Figure 10.5 Surface expression of GPR6 in HEK293 cells was not detected by antibody labeling. (A) Compressed images constructed from Z-stacks of pHluorin fluorescence pictures of pHGPR6 or pHD1 are shown in green. Staining of the same cells with GFP antibody is shown in red. (B) pHluorin fluorescence intensities of individual cells expressing pHD1 are plotted against corresponding GFP antibody fluorescence values. Fluorescence data points corresponding to the three cells labeled in (A) are indicated by a, b, and c. The slope (4.7) and intercept (11,906) of linear regression line provide a measure of the sensitivity of this assay ($R^2 = 0.78$). None of the pHGPR6 expressing cells have any detectable antibody staining and are not shown in this graph. Reproduced with permission from Padmanabhan et al. (2009). (See Color Insert.)

Table 10.1 Comparison of cell surface expression assays

Method	Efficiency (%)	Sensitivity	Primary cul.	Endogenous rec.
Cell surface biotinylation	12	Low	No	Yes
Chymotrypsin	89	Moderate	No	Yes
Imaging	100	Moderate	Yes	No
Antibody labeling	ND	High	Yes	Yes

Efficiency is defined as ability to detect cell surface expression of D1 receptors relative to chymotrypsin-imaging technique.
Sensitivity is based on relative efficiencies of assays and detection limits of Western blots and imaging analysis.
Primary cul. refers to the adaptability of the technique for studying expressed receptors in primary cultures.
Endogenous rec. refers to the utility of the technique for studying endogenous receptors.

fluorescence. Several cells that had barely detectable pHluorin fluorescence readily showed antibody fluorescence as indicated by the Y axis intercept of the correlation line in Fig. 10.5B.

GPR6 is predominantly expressed in neurons in the striatum of brain (Lobo et al., 2007; Song et al., 1994). Thus, it is important to study the expression and signaling mechanisms of this receptor in its native environment. Cell surface biotinylation and chymotrypsin susceptibility methods are not suitable for studying expressed GPR6 because of poor transfection efficiency in neuronal cultures. However, imaging and antibody labeling can be easily used to determine surface expression in individual neurons. Future research on signaling mechanisms of GPR6 is required to validate the conclusions drawn from cloned GPR6 expressed in HEK293 cells and neurons. All the methods described here, with the exception of imaging analysis are likely to be useful in measuring surface expression of endogenous GPR6 in neuronal cultures. pH and chymotrypsin sensitivities in imaging assays rely on the pHluorin tag and thus is not likely to be useful in studying endogenous receptors without special labeling tools. The other three methods could easily be adapted to studying endogenous receptors if selective GPCR antibodies are available. Obviously, the use of antibody labeling method for endogenous GPCR labeling would require an antibody that selectively binds extracellular epitopes of receptors.

3.1. Methods of detecting constitutive activity of GPR6

As outlined in Section 1, GPR6 is a constitutively active Gs-coupled receptor. Heterologous expression of this receptor in different cell types significantly increases intracellular cyclic AMP. We used three different assays to assess the Gs-mediated constitutive activity of GPR6.

3.1.1. cAMP enzyme immunoassay

Enzyme immunoassays (EIAs) are simple and widely available assays that provide a direct measurement of cAMP in lysed cells. Constitutive activity of Gs-coupled receptors should be reflected in an elevated intracellular cAMP concentration, especially in the presence of a phosphodiesterase inhibitor.

3.1.1.1. Methods HEK293 cells plated in 24-well plates were transfected with pHGPR6 or pHD1 receptors. D1 receptors were used as a control for a Gs-coupled receptor that does not have appreciable constitutive activity. Two days after transfection, cells were washed once with PBS and then incubated in presence of 100 μM RO 20-1724 (a phosphodiesterase inhibitor) for 30 min. Dopamine (1 μM) was added to cells expressing pHD1 to test the effect of D1 receptor activation on cAMP content. Cells were then extracted into 300 μl of 0.1 N HCl. Samples were neutralized and assayed for cAMP content using an EIA kit from Biomedical Technologies, Inc. (Stoughton, MA, USA).

3.1.1.2. Results cAMP content of cells expressing pHGPR6 is significantly greater than that of pHD1 expressing cells (Fig. 10.6). As expected, D1 application stimulated cAMP production in pHD1 cells. pHGPR6 cells but not pHD1 cells had significantly greater cAMP content than untransfected cells, demonstrating the constitutive activity of GPR6 in absence of exogenous ligands (Eggerickx *et al.*, 1995; Padmanabhan *et al.*, 2009).

3.1.2. Measurement of cAMP changes in live cells using bioluminescence resonance energy transfer

In recent years, several genetically encoded indicators based on fluorescent and luminescent proteins have been developed for measurement of cAMP in living cells (Nikolaev and Lohse, 2006). Probes based on intramolecular changes in fluorescence- and bioluminescence resonance energy transfer (FRET and BRET) make use of cAMP-binding domains, which are placed between FRET or BRET donors and acceptors. Binding of cAMP changes the distance between and orientation of donors and acceptors, thus producing changes in FRET or BRET. For studies in easily-transfected heterologous cells we used the BRET sensor CAMYEL (Jiang *et al.*, 2007).

3.1.2.1. Methods HEK293 cells plated in 6-well plates were transfected with GPR6 and CAMYEL (ATCC, Manassas, VA, USA). After 24 h, cells were washed once with PBS–EDTA, detached, and resuspended in 0.9 ml of PBS. Suspended cells were distributed (200 μl per well) into black 96-well plates (Thermo Fisher Scientific Nunc, Rochester, NY, USA). Coelenterazine h (Nanolight, Pinetop, AZ, USA) was added to each well

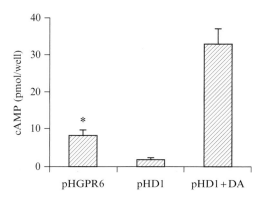

Figure 10.6 Constitutive activity of GPR6 can be detected by cAMP enzyme immunoassay. cAMP content of cultures expressing pHGPR6 or pHD1 are shown as mean ± S.E.M. pHD1 + DA represents cultures treated with 1 μM dopamine. *Significantly different from the other two groups ($P < 0.01$, $n = 4$). Reproduced with permission from Padmanabhan et al. (2009).

to a final concentration of 5 μM, and luminescence was measured using a multimode plate reader (Mithras LB940, Berthold, Bad Wildbad, Germany). The number of photons emitted at 535 nm divided by the number emitted at 480 nm was taken as the raw BRET ratio. This ratio decreases when cAMP binds to the CAMYEL indicator (Jiang et al., 2007).

3.1.2.2. Results As shown in Fig. 10.7, direct activation of adenylate cyclase with forskolin (10 μM) decreases CAMYEL BRET in untransfected cells, and this change is amplified by inhibiting phosphodiesterase activity with isobutylmethylxanthine (IBMX). Increasing the amount of GPR6 DNA transfected produces a graded increase in cAMP, as indicated by a graded decrease in BRET. Increases in cAMP induced by GPR6 expression were additive with those produced by forskolin. At the highest levels of GPR6 expression IBMX no longer decreased BRET signals in the presence of forskolin, suggesting possible saturation of the indicator ($K_d = 8.8$ μM cAMP; Jiang et al., 2007).

3.1.3. Phospho-CREB activity as a readout of constitutive activity of GPR6

As with the cell surface expression assays, it is important to determine the constitutive activity of GPR6 in neurons where it is normally expressed. However, low transfection efficiency in neurons limited us from using cAMP EIA or BRET assays in primary cells. We thus used phospho-cyclic AMP response element-binding (pCREB) protein as a downstream surrogate indicator of elevated cAMP concentrations.

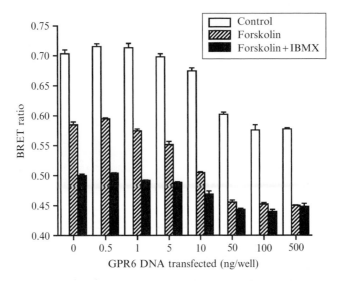

Figure 10.7 Constitutive activity of GPR6 can be detected using BRET. Raw BRET ratio is plotted against the amount of plasmid DNA encoding GPR6 used for transfection of each well of a 6-well plate. During detection, cells were also incubated with either 10 μM forskolin or forskolin + isobutylmethylxanthine (IBMX). Decreases in the BRET ratio indicate increases in [cAMP]. All experiments were performed in quadruplicate.

3.1.3.1. Methods Neuronal cultures transfected with pHGPR6 or pHD1 were fixed by incubating in 4% paraformaldehyde for 15 min. After two washes in PBS, cells were permeabilized and blocked using a buffer containing 1% triton X-100, 4% normal goat serum, and 1% bobine serum albumin. A rabbit pCREB antibody (1:1000, Millipore, Temecula, CA, USA) and Rhodamine Red-X conjugated anti-rabbit secondary antibody were used to immunostain the cells. pCREB fluorescence intensities of neurons overexpressing receptor constructs were compared to neighboring untransfected neurons.

3.1.3.2. Results pHGPR6 cells but not pHD1 cells has significantly greater pCREB staining intensity than untransfected neurons (Fig. 10.8). These data illustrate that pCREB immunostaining could be used as a surrogate down-stream indicator for constitutive activity of a Gs-coupled receptor.

4. COMPARISON OF THE CONSTITUTIVE Gs-ACTIVITY DETECTION METHODS

Of the three methods described here only the EIA and BRET assays are direct assays for cAMP, and thus capable of being calibrated to [cAMP]. Their sensitivity and dynamic range to detect cAMP depend on the cAMP antibody and reporter construct being used, respectively. In contrast, the pCREB assay is an indirect assay, and sensitivity cannot be measured for comparison with the other two methods. On the other hand, because of poor transfection efficiency, the EIA and BRET assays could not be used in primary cultures to measure the constitutive activity of expressed or endogenous GPR6. FRET indicators similar to CAMYEL (Nikolaev and Lohse, 2006) could be used for this purpose. Nevertheless, constitutive activity of expressed receptors could be easily detected using all three of these techniques. In the absence of suitable ligands, it is more difficult to document constitutive activity of endogenous GPR6 in native tissue. However, these

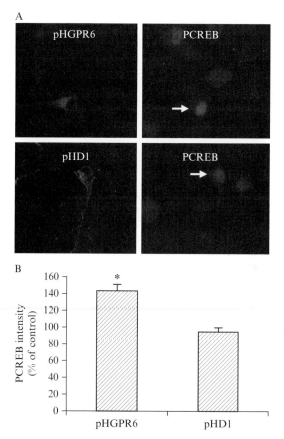

Figure 10.8 Constitutive activity of GPR6 in striatal neurons can be demonstrated by increase in CREB phosphorylation. (A) Representative confocal images of phospho-CREB immunostaining. Arrows indicate neurons transfected with pHGPR6 or pHD1. (B) Phospho-CREB fluorescence intensities of receptor transfected neurons are expressed as percent of untransfected controls. *Significantly different from controls ($P < 0.01$, $n = 6$). Reproduced with permission from Padmanabhan et al. (2009).

methods could be used to measure cAMP in GPR6-expressing and GPR6-null animals (Lobo et al., 2007), which are available from the Jackson Laboratory (Bar harbor, ME, USA).

5. CONCLUSIONS

Different assays have varying efficiencies and sensitivities to detect cell surface expression of membrane proteins. Of the different methods we used, those based on the chymotrypsin sensitivity appear to be most reliable in providing accurate estimates of relative cell surface expression. For measuring constitutive activity of Gs-coupled receptors, EIA or BRET assays are direct assays with good sensitivity. Development of GPR6 selective antibodies and different cAMP reported constructs targeted to subcellular compartments are likely to aid in studying constitutive signaling mechanisms of GPR6.

REFERENCES

Calebiro, D., Nikolaev, V. O., Gagliani, M. C., de Filippis, T., Dees, C., Tacchetti, C., Persani, L., and Lohse, M. J. (2009). Persistent cAMP-signals triggered by internalized G-protein-coupled receptors. *PLoS Biol.* **7**, e1000172.

Eggerickx, D., Denef, J. F., Labbe, O., Hayashi, Y., Refetoff, S., Vassart, G., Parmentier, M., and Libert, F. (1995). Molecular cloning of an orphan G-protein coupled receptor that constitutively activates adenylate cyclase. *Biochem. J.* **309**, 837–843.

Ferguson, S. S. (2001). Evolving concepts in G protein-coupled receptor endocytosis: The role in receptor desensitization and signaling. *Pharmacol. Rev.* **53**, 1–24.

Gottardi, C. J., and Caplan, M. J. (1992). Cell surface biotinylation in the determination of epithelial membrane polarity. *J. Tissue Cult. Methods* **14**, 173–180.

Ignatov, A., Lintzel, J., Kreienkamp, H. J., and Schaller, H. C. (2003). Sphingosine 1-phosphate is a high-affinity ligand for the G-protein coupled receptor GPR6 from mouse and induces intracellular Ca^{2+} release by activating the sphingosine-kinase pathway. *Biochem. Biophys. Res. Commun.* **311**, 329–336.

Jalink, K., and Moolenaar, W. H. (2010). G protein-coupled receptors: The inside story. *Bioessays* **32**, 13–16.

Jiang, L. I., Collins, J., Davis, R., Lin, K.-M., DeCamp, D., Roach, T., Hsueh, R., Rebres, R. A., Ross, E. M., Taussig, R., Fraser, I., and Sternweis, P. C. (2007). Use of a cAMP BRET sensor to characterize a novel regulation of cAMP by the sphingosine 1-phosphate/G13 pathway. *J. Biol. Chem.* **282**, 10576–10584.

Lobo, M. K., Cui, Y., Ostlund, S. B., Balleine, B. W., and Yang, X. W. (2007). Genetic control of instrumental conditioning by striatopallidal neuron-specific S1P receptor GPR6. *Nat. Neurosci.* **10**, 1395–1397.

Mehlmann, L. M., Saeki, Y., Tanaka, S., Brennan, T. J., Evsikov, A. V., Pendola, F. L., Knowles, B. B., Eppig, J. J., and Jaffe, L. A. (2004). The Gs-linked receptor GPR3 maintains meiotic arrest in mammalian oocytes. *Science* **306**, 1947–1950.

Nikolaev, V. O., and Lohse, M. J. (2006). Monitoring of cAMP synthesis and degradation in living cells. *Physiology (Bethesda)* **21**, 86–92.

Padmanabhan, S., Lambert, N. A., and Prasad, B. M. (2008). Activity-dependent regulation of the dopamine transporter is mediated by Ca(2+)/calmodulin-dependent protein kinase signaling. *Eur. J. Neurosci.* **28,** 2017–2027.

Padmanabhan, S., Myers, A. G., and Prasad, B. M. (2009). Constitutively active GPR6 is located in the intracellular compartments. *FEBS Lett.* **583,** 107–112.

Sankaranarayanan, S., De Angelis, D., Rothman, J. E., and Ryan, T. A. (2000). The use of pHluorins for optical measurements of presynaptic activity. *Biophys. J.* **79,** 2199–2208.

Seifert, R., and Wenzel-Seifert, K. (2002). Constitutive activity of G-protein-coupled receptors: Cause of disease and common property of wild-type receptors. *Naunyn Schmiedeberg's Arch. Pharmacol.* **366,** 381–416.

Song, Z. H., Young, W. S., Brownstein, M. J., and Bonner, T. I. (1994). Molecular cloning of a novel candidate G protein-coupled receptor from rat brain. *FEBS Lett.* **12,** 375–379.

Tanaka, S., Ishii, K., Kasai, K., Yoon, S. O., and Saeki, Y. J. (2007). Neural expression of G-protein coupled receptors GPR3, GPR6, and GPR12 up-regulates cyclic AMP levels and promotes neurite outgrowth. *J. Biol. Chem.* **282,** 10506–10515.

Thathiah, A., Spittaels, K., Hoffmann, M., Staes, M., Cohen, A., Horré, K., Vanbrabant, M., Coun, F., Baekelandt, V., Delacourte, A., Fischer, D. F., Pollet, D., et al. (2009). The orphan G protein-coupled receptor 3 modulates amyloid-beta peptide generation in neurons. *Science* **323,** 946–951.

Uhlenbrock, K., Gassenhuber, H., and Kostenis, E. (2002). Sphingosine 1-phosphate is a ligand of the human gpr3, gpr6 and gpr12 family of constitutively active G-protein coupled receptors. *Cell. Signal.* **14,** 941–953.

Yin, H., Chu, A., Li, W., Wang, B., Shelton, F., Otero, F., Nguyen, D. G., Caldwell, J. S., and Chen, Y. A. (2009). Lipid G protein-coupled receptor ligand identification using beta-arrestin PathHunter assay. *J. Biol. Chem.* **284,** 12328–12338.

CHAPTER ELEVEN

β_3-ADRENOCEPTOR AGONISTS AND (ANTAGONISTS AS) INVERSE AGONISTS: HISTORY, PERSPECTIVE, CONSTITUTIVE ACTIVITY, AND STEREOSPECIFIC BINDING

Maria Grazia Perrone *and* Antonio Scilimati

Contents

1. Introduction	198
2. β_3-Adrenoceptor	201
2.1. Therapeutic target	201
2.2. Structure and 7TD aminoacid–ligand specific interactions	205
2.3. Ligand selectivity	207
2.4. Stereospecific interactions and biological activity relationship	209
2.5. Inverse agonism and blockage of β_3-adrenoceptor activity not endogenous agonist-induced	210
2.6. Inverse agonists	212
3. Methodologies	214
3.1. Chemistry	214
3.2. Biological methods	223
Acknowledgments	224
References	225

Abstract

β_3-Adrenergic receptor (β_3-AR) is expressed in several tissues and is considered a drug target for the treatment of several pathologies such as obesity, type 2 diabetes, cachexia, metabolic syndrome, heart failure, anxiety and depressive disorders, preterm labor, overactive bladder, control colon motility, and of coadjuvants in colon cancer therapy. It is a seven-transmembrane domain (7TD) G-protein coupled receptor and is usually coupled to a Gs-protein (Gi-protein in very few cases), and its stimulation increases the production of cAMP. A lot of β_3-AR agonists have been uncovered and extensively characterized. Conversely, very little is known about β_3-AR inverse agonists that would suppress the agonist-independent activity (constitutive activity) of the receptor by

Department of Medicinal Chemistry, University of Bari "A. Moro", Bari, Italy stabilizing it in its inactive state. This chapter attempts to outline (a) the importance of the β_3-AR as a therapeutic target through the disquisition of its role in human health (physiology) and disease (pathology); (b) the description of β_3-AR structure [amino acid sequence and 7TD organization]; (c) the medicinal chemistry of β_3-AR: 7TD amino acid–ligand specific interactions, β-adrenoreceptor subtype selectivity, stereospecific interactions and biological activity relationships, inverse agonism and blockage of β_3-adrenoceptor constitutive activity; and (d) β_3-AR inverse agonists. The detailed procedure to prepare and assess the biological activity/selectivity of the more potent and selective β_3-AR inverse agonists (**SP-1e** and **SP-1g**) up to now known is also described.

1. INTRODUCTION

β-Adrenergic receptors (β-ARs) have been classified as β_1- and β_2-AR since 1967 (Lands et al., 1967). This subdivision led to the discovery of selective drugs for the treatment of hypertension and asthma. At the beginning of the 1980s, a new β-AR subtype (Arch, 1989), later called β_3-AR, was found in several species, including man, bovine, rat, and mouse (Strosberg and Pietri-Rouxel, 1996). The three β-AR subtypes (Table 11.1) share a certain functional homology (e.g., their stimulation leads to the activation of adenylyl cyclase), whereas their structural homology is restricted (e.g., introns, phosphorylation sites).

Table 11.1 Properties of the three human β-AR subtypes

	β_1-AR	β_2-AR	β_3-AR
Aminoacids, n	477	413	408
Introns, n	—	—	2
Phosphorylation by PKA and β-ARK	Yes	Yes	No
Endogenous agonist	Noradrenaline	Adrenaline	Noradrenaline
Selective antagonist	CGP 20712A, atenolol, metoprolol	ICI 118551	SR 59230A[a], DPJ 904, L-748,328, L-748,337
G protein	Gs	Gs	Gs, Gi
Effector	Adenylyl cyclase	Adenylyl cyclase	Adenylyl cyclase, NO synthase
Tissue distribution	Heart (mainly), adipose tissue, kidney	Lungs (mainly), heart, adipose tissue, bladder, uterus	Adipose tissue, heart, bladder, gut, uterus, pancreas, brain

[a] SR 59230A behaves as an antagonist in tissue (e.g., colon tissue) and as an agonist in isolated cells (e.g., CHO cells expressing human cloned β_3-AR) (Perrone et al. 2008b). β-ARK = β-AR kinase.

β_3-AR is expressed in several tissues (e.g., adipose tissue, heart, bladder, gut, uterus, colon, pancreas, and central nervous system), where it modulates different physiological functions. Hence, due to its wide tissue distribution, it can be considered a drug target for the treatment of several pathologies such as obesity, type 2 diabetes (Pang *et al.*, 2010), cachexia, metabolic syndrome, heart failure, anxiety and depressive disorders, preterm labor, overactive bladder (OAB), control colon motility, and of coadjuvants in colon cancer therapy (Table 11.2). The β_3-AR, like the other β_3-ARs, is a seven-transmembrane domain (7TD) G-protein coupled receptor (GPCR). It is usually coupled to a Gs protein and its stimulation increases the production of cAMP (Strosberg, 2000b). Recent evidences show that in the human heart, the signal of β_3-AR is transduced by the Gi-eNOS-NO-cGMP pathway and produce negative inotropic effect (Gan *et al.*, 2007; Pott *et al.*, 2006). A lot of β_3-AR agonists have been uncovered and extensively characterized (Hieble, 2007; Fig. 11.1). Conversely, very little is known about β_3-AR inverse agonists (Baker, 2005; Hoffmann *et al.*, 2004). Inverse agonism is not a new concept and it describes ligand behavior displaying negative efficacy (Costa and Cotecchia, 2005). In particular, for GPCRs, it has been widely assumed that inverse agonists suppress the agonist-independent activity (constitutive activity) of the receptor by stabilizing it in its inactive state (Strange, 2002). Novel findings suggest that some classical

Table 11.2 Possible clinical use of β_3-AR agents

Localization	Potential therapeutic indication	Type of drug
Adipose tissue	Cachexia	Antagonist, Inverse agonist
	Metabolic syndrome	Antagonist, Inverse agonist
	Obesity	Agonist
	Type 2 diabetes	Agonist
Heart	Cardiac diseases	Agonist, Antagonist, Inverse agonist[a]
Hippocampus, hypothalamus, amygdala and cerebral cortex	Anxiety and depressive disorders	Agonist
Myometrium	Preterm labour	Agonist
Bladder	Overactive bladder (OAB)	Agonist
Gut	Modulation of colonic motility Irritable Bowel Syndrome (IBS)	Agonist
	Colon cancer	Antagonist, Inverse agonist

[a] Agonists should serve in the first stage of the heart failure, whereas antagonists and inverse agonists in the late phase of the disease.

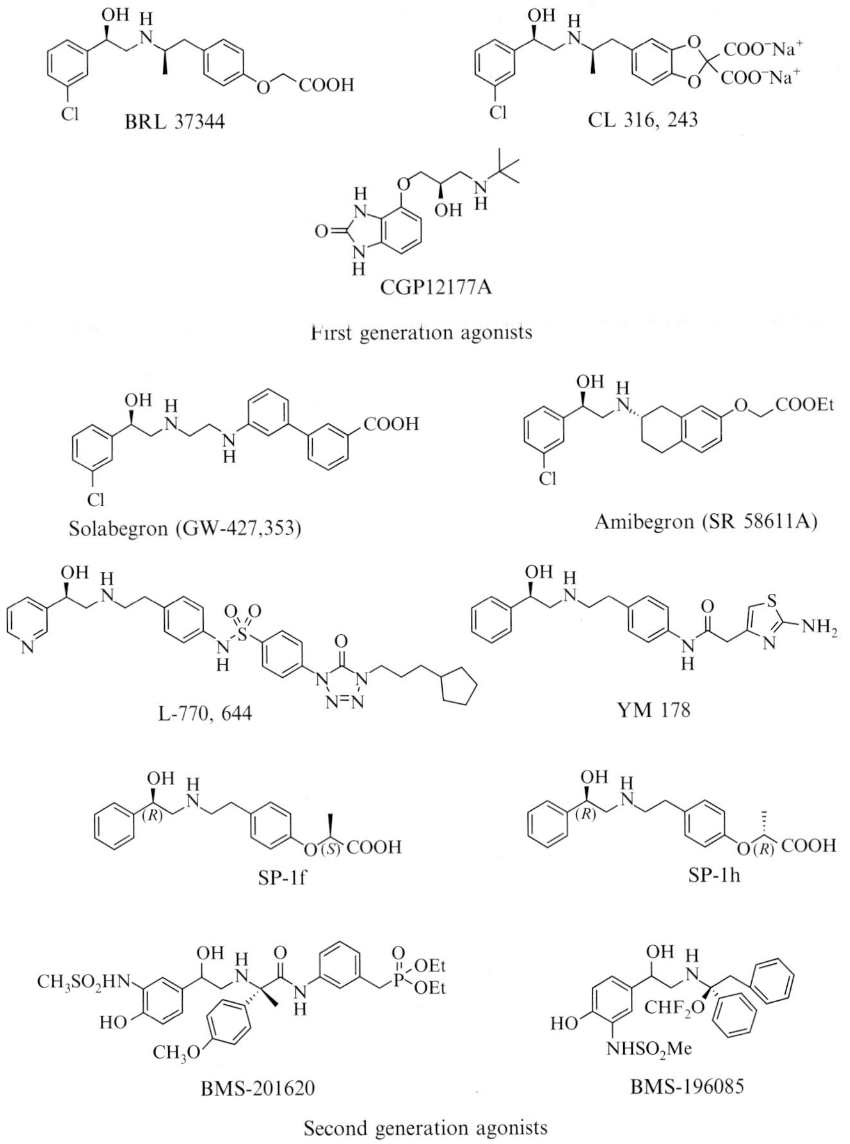

Figure 11.1 β_3-AR agonists.

β-AR antagonists behave either as partial agonists, neutral antagonists, or inverse agonists in cell systems expressing the wild type or a constitutively activated mutant of the human β-AR. For example, selective β_1-AR antagonists with significant inverse agonistic activity, such as metoprolol, have been proved to be safe in the treatment of heart failure patients.

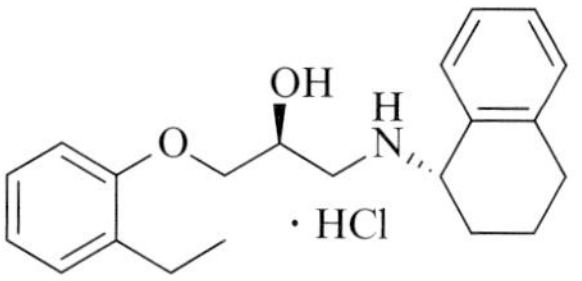

SR 59230 A

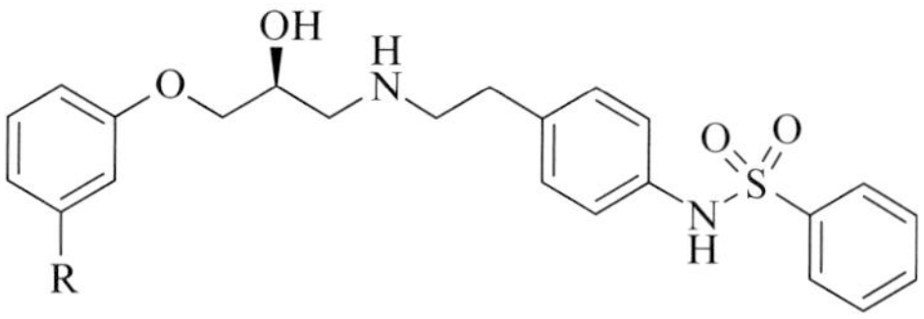

L-748,328 $R = SO_2NH_2$
L-748,337 $R = CH_2NHCOCH_3$

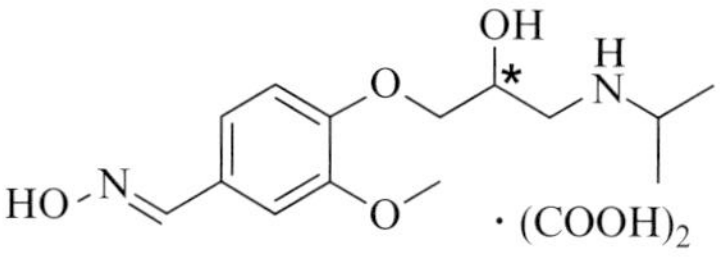

DPJ 904

Figure 11.2 β_3-AR antagonists.

Clinical studies with inverse agonists of the β_2- and β_3-AR have not been available yet (Fig. 11.2).

2. β_3-Adrenoceptor

2.1. Therapeutic target

β_3-AR is expressed in several tissues, including adipose tissue, heart, bladder, gut, uterus, pancreas, and central nervous system (CNS), where it is involved in a variety of pathophysiological processes. As a consequence, β_3-AR ligands constitute potential drugs useful to treat several diseases (Table 11.2).

Since their discovery, β_3-AR attracted the pharmaceutical company interest because it seemed to play a role in the regulation of energy balance and glucose homeostasis being expressed in human white as well as brown adipose tissue. In rodents adipocytes, the β_3-AR physiological role is well established; chronic treatments of β_3-AR agonists in obese diabetic animals were reported to reduce adiposity, and improve type 2 diabetes (Kato *et al.*, 2001; Liu *et al.*, 1998). The effects of β_3-AR agonists observed in rodents raised interest in the development of compounds for the treatment of obesity and type 2 diabetes in humans, but clinical studies were disappointing because of poor activity and selectivity of such compounds for the human β_3-AR (BRL 37344 in humans was found to be 60% less potent than in rats) (Blin *et al.*, 1994; Dolan *et al.*, 1994) and different contribution of white and brown adipocytes in rodents and humans. In healthy young adults, β_3-ARs play a minor role in the control of lipolysis and nutritive blood flow in human subcutaneous abdominal adipose tissue (Barbe et al., 1996), whereas in adult white adipocyte, a low β_3-AR expression was detected (Deng et al., 1996). The results of these studies confirm that the β-AR subtype distribution pattern in human brown and white adipose tissues and in rodents is different. In addition, very recently, it has been reported that the anti-diabetes effect of β_3-AR agonists CL 316,243 and SR 58611A seems due to the mechanistic link between the Free Fatty Acid receptor (GPR40) and adrenergic signalling in adipose tissues and pancreatic β-cell function (Pang et al., 2010).

Another possible therapeutic relevance of β_3-AR resides both in the increased lipolytic action of β_3-AR in visceral adipose tissue of obese subjects, and in the association between increased β_3-AR function and metabolic syndrome. In fact, in upper-body of obese subjects with signs of the metabolic syndrome, β_3-AR blockade might preferentially inhibit fatty acid release from visceral adipose tissue and improve some of the metabolic abnormalities associated with the high "portal" fatty acid flux (Strosberg, 2000d).

As reported in Table 11.2, β_3-AR antagonists and inverse agonists would also be useful to prevent or control cachexia, a *para*-neoplastic condition consisting of remarkable body weight loss. In cachectic oncologic patient urines, very high levels of a Zn-α2-glycoprotein known as lipid-mobilizing factor (LMF) and several cytokines (i.e., TNF-α) were found, and seems that at least in part the adipose tissue loss ($\approx 30\%$ of preillness stable weight at diagnosis) is induced from β_3-AR, activated by LMF (Tisdale, 2003).

In the human heart, β_3-AR coupling to a Gi-protein produces negative inotropic and chronotropic effects, via the activation of eNOS and the subsequent NO release that determines a consistent cGMP increase. In the failing human heart, there is a sustained release of catecholamines that induces a progressive downregulation of β_1- and β_2-AR, due to desensitization by PKA-mediated phosphorylation, whereas β_3-AR population grows or remains unchanged because it lacks the phosphorylation site (Table 11.1) responsible for desensitization and internalization of the other two β-AR subtypes. This may lead to an imbalance of the cardiac β-AR population with an excessive promotion of β_3-AR negative inotropic effect, which results in a potentially fatal reduction of cardiac contractility. As a consequence, β_3-AR becomes a new target for treatment and prevention of cardiac failure (Fedorov and Lozinsky, 2008; Rozec and Gauthier, 2006). Stimulation of β_3-AR subtype inhibits cardiac contractility, thus opposing to the response of β_1- and β_2-AR. In failing heart, β_3-AR is upregulated. It probably serves as a buffer, exerting a "rescue" function from the effects of high plasma levels of catecholamines, as those observed in hyperadrenergic states including heart failure. Upon disease progression, β_3-AR upregulation may produce a depression in contractility, which exacerbates heart failure (Gan et al., 2007; Pott et al., 2006; Rozec and

Gauthier, 2006). Hence, selective β_3-AR agonists should serve in the early stage of heart failure, whereas highly selective antagonists/inverse agonists might be useful in the advanced stage of the disease.

β_3-AR existence in the brain is still a matter of debate, since up to now binding studies aimed at detecting the presence of central β_3-AR have been inconclusive (Simiand et al., 1992). However, experiments using reverse transcription-PCR argue for the presence of β_3-AR mRNA in discrete regions of rat and human brain including the hippocampus, hypothalamus, amygdala, and cerebral cortex (Claustre et al., 2008; Rodriguez et al., 1995).

These brain regions participate in the control of emotions and are considered to be the main target for current treatments of depression that enhances serotonergic and/or noradrenergic transmissions (Nestler et al., 2002). Amibegron (SR58611A), a selective orally active and brain-penetrant β_3-AR agonist, has been shown to display broad anxiolytic and antidepressant-like effects in a variety of models in rodents comparable in terms of efficacy to the effects of those of a classical antidepressant fluoxetine (Stemmelin et al., 2010). Amibegron seems to lack important side effects such as the tachycardia or alteration of locomotor activity (Stemmelin et al., 2008). The effects of amibegron are mediated by the β_3-ARs, suggesting that this receptor subtype may represent an alternative target for novel antidepressant drugs. Phase III clinical trials on SR58611A for the treatment of anxiety and depression have been terminated (www.clinicaltrials.gov, 2008) and information about Phase IV is awaited.

β_3-Adrenoceptor is the predominant β-AR subtype in human pregnant and nonpregnant myometrium, where it reduces contractions by coupling to Gi protein and consequently inhibits the adenylyl cyclase/cAMP pathway. Hence, β_3-AR agonists may have a considerable future pharmacological use in the preterm labor clinical management, because BRL 37344, a β_3-AR agonist, induces relaxation of human myometrial contractions with similar potency to that of ritodrine (β_2-AR agonists), the most commonly used tocolytic agent (Dennedy et al., 2001), and reduced cardiovascular side effects (Rouget et al., 2005).

AR is widely distributed into the human detrusor smooth muscle. In general, the bladder base and urethra contain a high density of α-ARs, which stimulation results in contraction; bladder body contains a high density of β-ARs, which stimulation determines relaxation and urine storage; β_3-AR mRNA was also detected in human detrusor muscle (Fujimura et al., 1999) and in human urinary bladder urothelium (Otsuka et al., 2008), where it was found to be substantially more expressed than β_1- and β_2-AR mRNA.

The normal physiological contraction of the urinary bladder is mostly mediated by muscarinic receptors, primarily M_3 subtype. Bladder relaxation, required for urine storage, is predominantly, if not exclusively, mediated by β_3-AR subtype. An excessive stimulation of contraction or a reduced relaxation of the detrusor smooth muscle during the urine storage phase may determine OAB, a syndrome characterized by urinary frequency, nocturia, and urgency incontinence. The overexpression of β_3-AR subtype in human bladder suggested that agonists of this receptor could be effective in the treatment of OAB. Different studies were carried out aimed at demonstrating β_3-AR involvement in urinary frequency and at identifying potent relaxant agents of the human detrusor muscle as an alternative to the classical treatment with muscarinic receptor antagonists endowed with poor therapeutic index. Indeed, the β_3-AR agonists Solabegron and YM 178 are currently in Phase I and Phase III clinical trials, respectively, for the treatment of OAB (Takasu et al., 2007).

β_3-ARs are widely distributed in the gastrointestinal tract of several species, including humans (De Ponti et al., 1996; Perrone et al., 2008a) and rats (Bianchetti and Manara, 1990). In particular, they are expressed on gut vascular and nonvascular smooth muscle, where they mediate relaxation and are probably involved in the control of blood flow.

β_3-AR modulates colonic motility. In particular, isolated human colon elevated tone and spontaneous contractions are reduced and inhibited, respectively, by β_3-AR agonists such as SR 58611A or CGP 12177A (De Ponti et al., 1995). The relaxing effect of β_3-AR agonists on gut smooth muscle explains the increase in compliance (i.e., the ability of the gut to relax upon application of a distending stimulus) may turn out to be a useful approach in some functional gut disorders (e.g., functional dyspepsia), where decreased gastric accommodation is a pathophysiological feature (Bharucha et al., 1997).

In addition, activation of β_3-AR by their agonists leads to an inhibition of cholinergic contractions and evokes somatostatine release, resulting in a decrease of intestinal motility and secretion, and inducing analgesia (Cellek et al., 2007). Moreover, β_3-AR agonist administration confers gastroprotection in several models of gastric ulcer (Kuratani et al., 1994; Sevak et al., 2002), an effect that may be due to increased blood flow by vasodilatation and/or relaxation of the muscularis externa mediated by β_3-ARs. Recently, it has been shown that SR 58611A ameliorates dinitrobenzenesulfonic acid (DNBS)-induced colitis in rats and downregulates the biosynthesis of inflammatory cytokines. In this context, the finding of β_3-ARs on myenteric neurons and in nerve fibers clarifies the role of these receptors as a potential therapeutic target for gut inflammatory disease (Vasina et al., 2008).

A modified β-adrenergic function associated to proliferative alterations of numerous cancer cell lines seems to be involved in tumor cell proliferation and migration, and metastasis formation, which therefore seem to be involved in the most important aspects of malignant phenotype. Pharmacological modulation of β-ARs affects tumor cell growth in several experimental systems, and inhibition of metastasis formation by β-ARs antagonists in in vivo models has been ascertained. Initial epidemiological studies provided evidences that β-blockers can reduce cancer incidence, thus suggesting their possible role also in cancer prevention. A preliminary study, performed on human colon cancer and normal surrounding mucosa aimed at investigating the β_1-, β_2-, and β_3-ARs gene expression, showed significant difference of β_3-AR mRNA levels between normal mucosa and cancer tissue. A twofold higher expression of β_3-AR mRNA in cancer tissue than normal one was found, thus suggesting a β_3-AR possible involvement in the human colon tumor processing (Perrone *et al.*, 2008a).

2.2. Structure and 7TD aminoacid–ligand specific interactions

The human β_3-AR is composed of a single 408 amino acid residue peptide chain (Fig. 11.3, Table 11.1), belonging to the family of GPCRs. It has hydrophobic stretches of about 22–28 residues forming seven transmembrane (TM) segments. The TM regions are linked with three intracellular and three extracellular loops. The amino acid (N)-terminal of these receptors is located extracellularly and is glycosylated. The carboxy (C)-terminal is intracellular and in the case of the β_3-ARs does not have any phosphorylation site. The palmitoylated Cys^{360} residue is in the N-terminus of the i4 loop. A disulphide bond essential for the receptor activity (Cys^{110} and Cys^{189}) is also present. Among different species, the comparison of β_3-ARs reveals a high degree of sequence homology: approximately 80–90% between human, bovine, rodent,

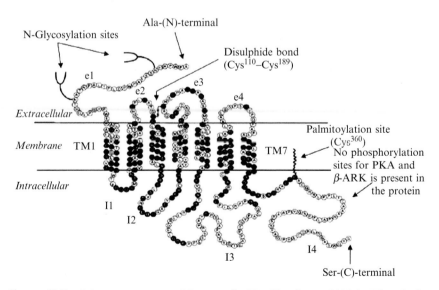

Figure 11.3 Primary structure of human β_3-AR (Strosberg, 2000a). The single polypeptide chain is arranged according to the model of rhodopsin. Residues in black circles are common to the three β-AR subtypes.

and canine. The human, monkey, and bovine β_3-ARs are closer to one another than any of the rodent sequences, particularly in the first TM segment.

Computer modeling has defined an image of the β_3-AR ligand-binding site (Furse and Lybrand, 2003; Kumar and Bharatam, 2005 and 2009). At least four of the 7TDs are essential for ligand binding. The crucial amino acids that are involved were identified by site-directed mutagenesis and photoaffinity labeling. Asp^{117} in TM3 (Fig. 11.4) is the residue found to be essential for binding all biogenic amines. The acidic side chain most likely forms a salt bridge with the basic group of the ligand. Indeed, substitution of this residue in the human β_3-AR completely suppresses agonist binding (Gros et al., 1998). Ser^{169} in TM4 is thought to form a hydrogen bond with the hydroxyl of the ethanolamine side-chain (Strader et al., 1989). Ser^{209} and Ser^{212} in TM5, also located in many biogenic amine receptors, are thought to form hydrogen bonds with the hydroxyl groups of the catechol moiety. Also, Phe^{309} in TM6 is involved in a hydrophobic interaction with the aromatic ring of catecholamines. Asp^{83} (TM2) and Tyr^{336} (TM7) are likely to be more important for signal transmission to Gs. The essential disulphide bond (S–S) linking Cys^{110} (e2) and Cys^{189} (e3) is also represented in Fig. 11.4.

The Gs interaction site on β_3-AR is situated in the intracellular region, mainly the membrane proximal regions of the second and third (i2, i3) intracellular loops and the carboxy-terminal domains. Deletion of small segments of the amino terminal and carboxy terminal regions of the i3 of the β_3-AR uncouples the human receptor from adenylyl cyclase upon agonist stimulation.

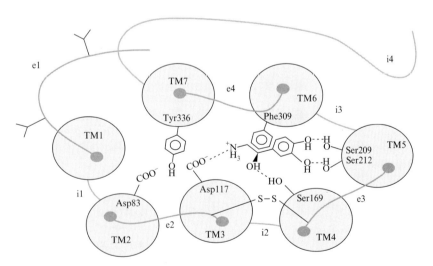

Figure 11.4 β_3-AR viewed from the outside of the cell: proposed interactions in the ligand binding region. Noradrenaline is shown surrounded by several of the amino acids involved in agonist binding.

2.3. Ligand selectivity

The basis of β-receptor subtype selectivity has been investigated in detail: a β_1- and β_3-AR model was generated by using the β_2-receptor model as a template. The high sequence similarity for these receptors, particularly in the TM helical bundle domain (69% sequence identity and 87% identity plus conservative substitution for β_1- vs. β_2-AR; 65% identity and 87% identity plus conservative substitution for β_1-AR vs. β_3-AR; 60% identity and 83% identity plus conservative substitution for β_2- vs. β_3-AR) suggests that they must possess quite similar three-dimensional structures. The residues lining the binding site region of the three receptors are nearly identical (75–85% sequence identity and 95–100% identity plus conservative substitution), so the binding site pockets in each receptor subtype must also have extremely similar structures and hydration characteristics. Despite the extremely high sequence similarity for all three receptor subtypes, there is emerging evidence that the β_3-AR may bind ligands somewhat differently than β_1- and β_2-AR (Fig. 11.5). A number of ligands that behaves as antagonists for β_1- and β_2-AR are partial to full agonists for the β_3-AR. Additionally, a number of ligands which exhibit impressive β_3-AR selectivity has been developed, confirming that exploitable differences do exist between these receptors (de Souza and Burkey, 2001).

Most β_3-AR selective agonists do not share adrenaline's catechol ring, but instead have a pyridine or m-chlorophenyl ring, or in some cases, a more extensive heteroaromatic ring system reminiscent of β_1-adrenergic receptor antagonists (β-blockers).

Support for the above binding site interactions (Fig. 11.4) is provided by massive studies of structure–activity relationships (SAR) on catecholamines (Fig. 11.5). These emphasize the importance of having both the alcohol group and the ionized amine in the side-chain, and also of the absolute configuration $(R)/(S)$ of the ethanolamine/propanolamine stereogenic center (Figs. 11.4 and 11.5). Some of the evidences supporting these conclusions are as follows:

- *The alcohol group*: the (R)-enantiomer of noradrenaline is more active than its (S)-enantiomer, indicating that the secondary alcohol is involved in a stereospecific hydrogen bonding interaction. Compounds lacking the hydroxyl group (e.g., dopamine) have a greatly reduced interaction. Some activity is retained, indicating that the alcohol group is important but not essential.
- *The amine* is normally protonated and ionized at physiological pH. This is important since replacing nitrogen with carbon results in a large drop in activity. Activity is also affected by the number of substituents on the nitrogen. Primary and secondary have good adrenergic activity, whereas tertiary amines and quaternary ammonium salts do not. Only a few examples of very good β_3-AR activity has been showed for tertiary amines in which the nitrogen atom is part of a piperazine ring and

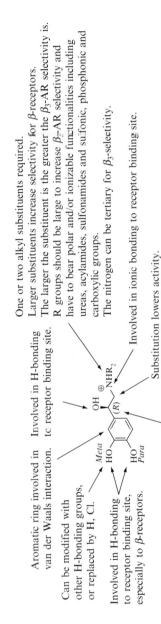

Figure 11.5 Structure–activity relationships: important β_3-AR binding groups.

Aromatic ring involved in van der Waals interaction.

Involved in H-bonding to receptor binding site.

Can be modified with other H-bonding groups, or replaced by H, Cl.

Involved in H-bonding to receptor binding site, especially to β-receptors.

One or two alkyl substituents required.
Larger substituents increase selectivity for β-receptors.
The larger the substituent is the greater the β_3-AR selectivity is. R groups should be large to increase β_2-AR selectivity and have to bear polar and/or ionizable functionalities including ureas, acylamides, sulfonamides and sulfonic, phosphonic and carboxylic groups.
The nitrogen can be tertiary for β_3-selectivity.

Involved in ionic bonding to receptor binding site.

Substitution lowers activity.

(R)-Enantiomer more active than,
(S)-Enantiomer in the arylethanolamine series,
Vice versa in the aryloxypropanolamine series.

the arylethanolamine replaced by an aryloxypropanolamine (Perrone et al., 2009).
- *The phenyl substituents* are important. Hydroxyl groups can be replaced by other groups capable of interacting with the binding site by hydrogen bonding. However, the *meta*-OH of the phenol group can be replaced by other groups such as Cl or can be the simple hydrogen.
- *Alkyl substitution*: it was discovered that adrenaline has the same potency for both types of adrenoceptor, whereas noradrenaline has a greater potency for α-adrenoceptors than for β-ARs. This indicates that N-alkyl substituent has a role to play in receptor selectivity. Increasing the size of the N-alkyl substituent determines the loss of potency at α-receptor and positively enhanced the potency towards β-receptors. The presence of a bulky N-alkyl group beneficial $β_3$-AR activity. These results indicate that $β_3$-AR should have a larger hydrophobic pocket than $β_1$- and $β_2$-ARs, into which a bulky alkyl/aryl/alkyaryl group can fit. Besides, $β_3$-AR activity is still exalted if the R (or Ar) group bears polar and/or ionizable functionalities including ureas, acylamides, sulfonamides and sulfonic, phosphonic, and carboxylic groups (Brockunier *et al.*, 2001; Dow *et al.*, 2004; Kashaw *et al.*, 2003; Mizuno *et al.*, 2004, 2005; Nakajima *et al.*, 2005; Tanaka *et al.*, 2003; Steffan *et al.*, 2002).

2.4. Stereospecific interactions and biological activity relationship

$β_3$-AR displays a different degree of stereoselectivity for several known traditional β-AR ligands. It shows both a lower degree of stereoselectivity for agonists such as isoprenaline and noradrenaline and a higher degree of stereoselectivity for antagonists (i.e., propranolol) than $β_1$- and $β_2$-AR. Thus, lower stereoselectivity should not be considered a defining characteristic of the $β_3$-AR, since it has a relatively high degree of stereoselectivity for high affinity β-AR antagonists. In addition, it is clear from the comparison of human, rat, and mouse $β_3$-AR that the endpoint extents vary between species. At the human $β_3$-AR, in particular, traditional β-AR ligands tend to show higher affinities and higher enantioselective indices than at mouse or rat receptors (Popp *et al.*, 2004).

There appears to be no simple, consistent explanation for ligand stereoselectivity for β-adrenergic agonists and antagonists. An attempt to rationalize the findings, strongly suggests that a single, static three-dimensional model is not adequate to explain stereoselective ligand binding (or other ligand-binding characteristics) for all β-adrenergic ligands. This is quite consistent with the idea that GPCRs undergo facile transitions between multiple conformational states under normal conditions (Kenakin, 1997), and that agonists, neutral antagonists, and inverse agonists preferentially bind and stabilize different receptor conformations (Dennedy *et al.*, 2001; Peleg *et al.*, 2001; Salamon *et al.*, 2002).

2.5. Inverse agonism and blockage of β_3-adrenoceptor activity not endogenous agonist-induced

The increased knowledge of GPCRs behavior in cellular systems led to the discovery of inverse agonism. GPCRs are allosteric proteins designed by nature to respond to small "drug-like" molecules (i.e., neurotransmitters) to affect changes in large protein–protein interaction (receptors and G-proteins). GPCRs can be spontaneously active in the absence of any ligand (Bond and IJzerman, 2006). Inverse agonist is an established drug class and possesses "negative efficacy."

Inverse agonists are ligands that, by binding to a receptor, have the capacity to modulate the basal activity of a cell signaling cascade in the opposite direction to that produced by an agonist.

Conceptually, agonists are envisaged to stimulate signaling cascade by binding to receptor higher affinity state, and thus preferentially stabilizing an active conformation of a GPCR, resulting in a more effective activation of the relevant G-protein. As a corollary, inverse agonists should interact preferentially with an inactive conformation of the GPCR and thus limiting G-protein activation. Such modulation implies that active and inactive conformations of a GPCR must have the thermodynamics capacity to interconvert spontaneously, a situation which has required extension and reanalysis of the ternary complex (TC) model. As such, basal activity of a GPCR-regulated signaling cascade is defined by the position of the equilibrium between these conformations.

Receptors can adopt numerous conformations, some of which are capable of producing a pharmacological effect ("active states") (Kenakin, 2002a,b). Proteins fold and unfold in different regions at different times (the same micro-regions undergoing a given conformational change will not coexist in a concerted way) (Kenakin, 2004). Therefore, the receptor conformation that exposes residues that are crucial for G-protein activation might be one of several conformations with different micro-conformations in other regions of the receptor, including those responsible for ligand recognition. Several receptor micro-conformations are capable of exposing crucial regions to activate G proteins. Within this context, ligand-selective receptor active states would be a natural consequence of such a network.

For GPCRs, it has been widely assumed that inverse agonists suppressed the agonist-independent activity of these receptors by stabilizing the receptor in an inactive state (R).

This depends on receptor activation (receptor-G protein coupling) occurring in the absence of the agonist, and the inverse agonist suppressing this activity in some way. In terms of TC model (Fig. 11.6), the inverse agonist stabilizes the R form of the receptor at the expense of the R⋆G form of the receptor, thus suppressing agonist-independent activity. This condition can be achieved with various action models: first, inverse agonist could act by binding to the R state of the receptor in preference to the R⋆ state

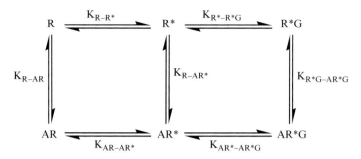

Figure 11.6 Ternary complex model: A, agonist; G, G-protein; K_{R-AR}, association constant for the binding of A to R; K_{R-R^*}, equilibrium constant governing the R:R* equilibrium; $K_{R^*-R^*G}$, equilibrium constant governing the R*:R*G equilibrium; R, receptor; R*, partially activated receptor.

(Model 1); alternatively, inverse agonist could bind to the uncoupled (R and R*) states of the receptor in preference to the coupled (R*G) state (Model 2); a third possibility is that the inverse agonist switches the receptor to an inactive conformation, different from those shown above, that can exist in G-protein-coupled and uncoupled forms but is inactive. This model is partly a restatement of the cubic TC model (Weiss et al., 1996a,b,c), which contains an inactive receptor conformation that can nevertheless couple G proteins. The possible effects of inverse agonists on the oligomerization state of the receptors should also be taken into account (Strange, 2002).

The extent to which inverse agonism could be a therapeutic advantage depends on the role of constitutive GPCR activity in pathology. One potential therapeutic area where this might have relevance is cancer. It has been shown that chronic elevation of second messengers in cells produced by constitutive G-protein activity can lead to cell transformation. Constitutive GPCR activity leading to chronic elevation of cell metabolism may also have a role in promoting the growth of tumors. There are examples of high levels of expression of specific GPCRs in tumor cells; it has been shown that endogenous ligands for these receptors are present at high levels in the tumor cells (self-regulation) and that they have proliferative properties. There also is evidence that inhibition of the cellular effects of these ligands can inhibit tumor growth (Kenakin, 2001). Constitutively active GPCRs may also be important in autoimmune diseases (de Ligt et al., 2000). Viral infection also can lead to constitutively active GPCR-mediated pathology.

In general, it still is not definitively clear to what extent GPCR constitutive activity plays a role in pathology. However, it is known that receptor expression and enzyme levels change in conditions of trauma (hypoxia, ischemia, physical damage), disease onset (inflammation, viral or bacterial infection), or disease development (Clifford et al., 1997; Donaldson et al., 1997; Kenakin, 2001).

2.6. Inverse agonists

Unlike β_1- (Lohse et al., 2003; Maack et al., 2001) and β_2-AR (Baker et al., 2003), few β_3-AR inverse agonists have been reported (Hoffmann et al., 2004; Perrone et al., 2008b; Fig. 11.7).

Novel data suggest that some classical β-AR antagonists behave either as partial agonists, neutral antagonists, or inverse agonists in cell systems expressing the wild type or a constitutively activated mutant of the human β-AR. For example, metoprolol, a β_1-selective AR antagonist with significant inverse agonistic activity, has been proven to be safe in the treatment of patients with heart failure. Unfortunately, up-to-date clinical studies with inverse agonists of the β_2- and β_3-AR are not yet available.

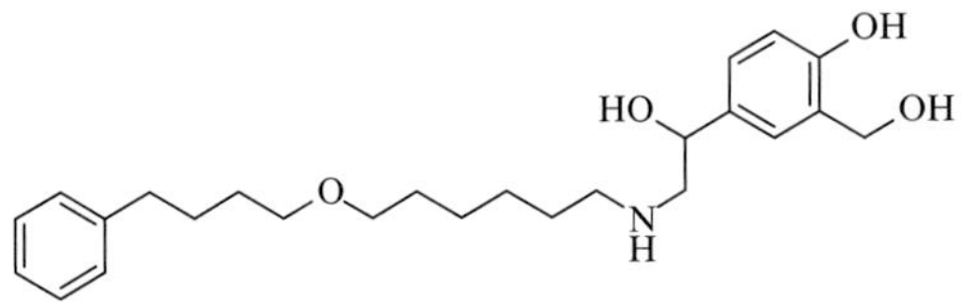

Salmeterol[a]
IA = −13% ± 16

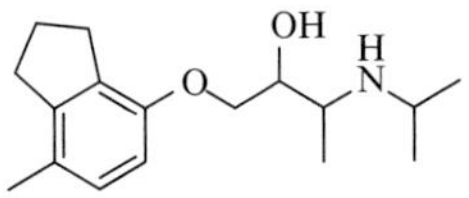

ICI 11855[a]
IA = −30% ± 3

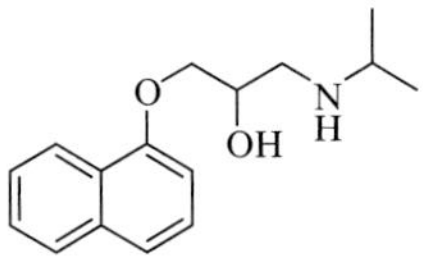

Propranol[a]
IA = −4% ± 3

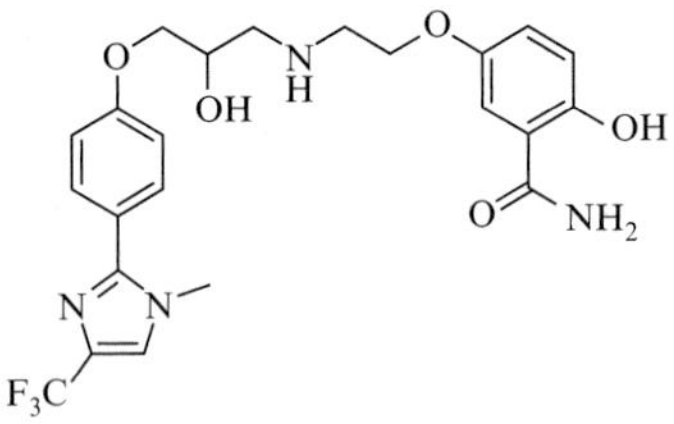

CGP 20712[a]
IA = −20% ± 6

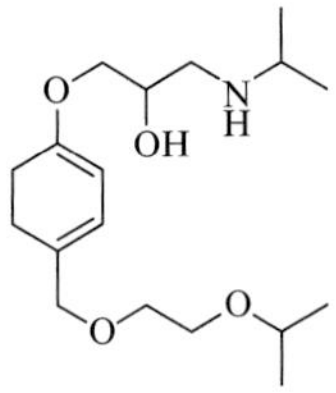

Bisoprolol[a]
IA = −1% ± 7

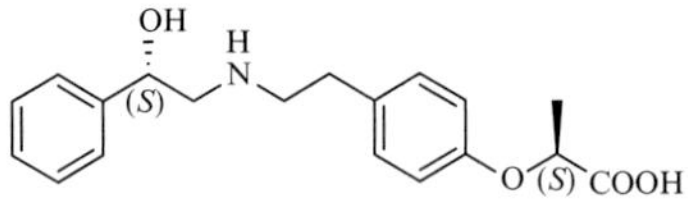

SP-1e[b]
IA = −64%

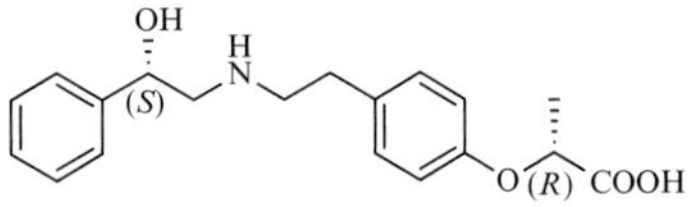

SP-1g[b]
IA = −73%

Figure 11.7 β_3-Inverse agonists. [a]Adenylyl cyclase stimulation represents percentage maximal stimulation achieved by isoproterenol (130 μM). Inverse agonist activities were calculated as % reduction of the basal adenylyl cyclase activity (Hoffmann et al., 2004). [b]IA is the fitted maximal value of the concentration–response curve, expressed as percentage of the maximal response to (R)-(−)-isoproterenol (10^{-4} M) (Perrone et al., 2008b).

Salmeterol is a long-acting β_2-AR agonist drug that is currently prescribed for the treatment of asthma and chronic obstructive pulmonary. It is also endowed with inverse agonist activity for both β_1- and β_3-AR (IA = −33% and −13%, respectively) (Hoffmann et al., 2004). ICI, 118551 was found to be an inverse agonist for all the three subtypes (IA = −22% for β_1-AR, −32% for β_2-AR, and −30% for β_3-AR). Propranolol is a nonselective β-blocker mainly used in the treatment of hypertension. It was the first β-blocker developed. In Hoffmann et al. (2004) experimental conditions, propranolol was found to be an inverse agonist for all the three β-AR subtypes (IA = −35% for β_1-AR, −35% for β_2-AR, and −4% for β_3-AR) as well as CGP20712, a highly selective and potent β_1-adrenoceptor antagonist (IA = −25% for β_1-AR, −30% for β_2-AR, and −20% for β_3-AR), and bisoprolol (IA = −33% for β_1-AR, −30% for β_2-AR, and −1% for β_3-AR).

Most of β_3-AR ligands reported so far, share a similar overall structure, in which three molecular portions can be identified: a left-hand side (LHS), a linker (LK), and a right-hand side (RHS). LHS is typically an arylethanolamine or aryloxypropanolamine (Perrone et al., 2006, 2008b), LK has various structures including both aromatic and aliphatic moieties (Perrone et al., 2009), RHS typically contains polar and/or ionizable functionalities including ureas, acylamides, sulfonamides and sulfonic, phosphonic, and carboxylic groups (Dow et al., 2004; Nakajima et al., 2005).

β_3-ARs exhibit a different extent degree of stereoselectivity for several β-adrenoceptor ligands. They display both a lower degree of stereoselectivity for agonists such as isoprenaline and noradrenaline and show a higher degree of stereoselectivity for β-blockers than β_1- and β_2-AR (Popp et al., 2004). Thus, low stereoselectivity should not be considered a defining characteristic of the β_3-AR since there is a relatively high degree of stereoselectivity for high affinity β-AR antagonists. In addition, it is clear from the comparison of human, rat, and mouse β_3-AR that the extent of the effect varies between species. At the human β_3-AR, in particular, β-AR ligands tend to show higher affinities and higher enantioselectivity indices than at mouse or rat receptors (Popp et al., 2004).

Up to now, several studies also dealing with the relationship between stereochemical demand and β_3-AR agonistic activity have been carried out using optically active compounds, proving that the activity toward the receptor is higher when the stereogenic center bearing the hydroxyl group at the LHS has (R)-absolute configuration in the series of arylethanolamines, (S) in that of aryloxypropanolamines. Besides, several research groups synthesized and evaluated new optically active β_3-AR agonists, bearing additional stereogenic centers. They were almost exclusively located on the carbon atoms adjacent to the aminic nitrogen, in both LHS and LK (Dallanoce et al., 2007; Fisher, et al., 1996; Gavai et al., 2001; Harada et al., 2003; Shearer et al., 2007; Sher et al., 1997; Tanaka et al., 2003;

Uehling et al., 2002; Weber et al., 1998). Such compounds resulted full or partial agonists and antagonists of the β_3-AR (Fig. 11.1 and 11.2).

To gain further insight into the influence of stereochemistry on β_3-AR activity, new β_3-AR agonists bearing stereogenic centers in both the left- and the right-hand sides have been prepared (Table 11.3, Schemes 11.1–11.4; Perrone et al., 2008b).

Compounds bearing two methyl groups at the β-position (**SP-1a–c**) showed a similar β_3-AR agonistic activity. Racemic **SP-1a** had an $EC_{50} = 4.9$ nM, with 68% intrinsic activity with respect to maximal effect (100%) by isoproterenol. Surprisingly, its enantiomers **SP-1b** and **SP-1c** proved equally active as the racemic form, irrespective of the absolute configuration at Cα, which is generally (R) in β_3-AR agonists ($EC_{50} = 3.9$ and 3.4 nM; IA = 72% and 76%, respectively). A methyl group was then removed from Cβ leading to the racemic **SP-1d**. In order to evaluate the effect of this structural modification and the corresponding stereochemical implication on β_3-AR agonistic activity, all the four possible stereoisomers of **SP-1d** were separately tested. The racemic form **SP-1d** behaved as a strong agonist at the β_3-AR, with a potency ($EC_{50} = 3.8$ nM, IA = 65%) comparable to that observed for compounds **SP-1a–c**.

Among **SP-1d** stereoisomers, the most potent one has absolute configuration (R) at Cα (**SP-1f**, $EC_{50} = 2.7$ nM, IA = 50%). This result is in agreement with data published for many other β_3-AR agonists. Opposite configuration at Cβ, as for **SP-1h**, dramatically decreased β_3-AR agonistic potency ($EC_{50} = 235$ nM, IA = 34%). In fact, **SP-1h** was found 87-fold less active than **SP-1f**. Interesting results were found for **SP-1e** and **SP-1g**. These compounds differ from the previous two (**SP-1f** and **SP-1h**) in the stereochemistry at Cα (Fig. 11.8).

In this case, the compounds behaved as inverse agonists on β_3-AR. Such a finding seems crucial, since few β_3-AR inverse agonists have been identified to date (Hoffmann et al., 2004; Soudijn et al., 2005). The most potent inverse agonist was (αS, βR)-**SP-1g** with an $EC_{50} = 136$ nM and IA = -73%. Its epimer (αS, βS)-**SP-1e** had a comparable potency and intrinsic activity ($EC_{50} = 181$ nM, IA = -64%).

3. METHODOLOGIES

3.1. Chemistry

Compounds **SP-1a–h** were prepared by *N*-alkylation of the corresponding racemic-, (R)- or (S)-2-amino-1-phenylethanol with phenethyl bromide derivatives **3a–d**, followed by hydrolysis of the ester intermediates **2a–h** (not shown), as depicted in Scheme 11.1.

Table 11.3 Evaluation of cAMP accumulation in CHO cells expressing human β_3-AR by **SP-1a–h**

Compound	Structure	EC_{50}^{a} (nM ± S.E.M.[b])	IA (%)[c]
SP-1a	α,β unspecified	4.9 ± 0.25	68
SP-1b	(R) at α	3.9 ± 2.1	72
SP-1c	(S) at α	3.4 ± 0.8	76
SP-1d	α,β unspecified	3.8 ± 0.7	65
SP-1e	(S) at α, (S) at β	181 ± 19	−64[d]
SP-1f	(R) at α, (S) at β	2.7 ± 0.7	50
SP-1g	(S) at α, (R) at β	136 ± 20	−73[d]
SP-1h	(R) at α, (R) at β	235 ± 37	34
ISO[e]	(R)-isoproterenol	5.8 ± 1.2	100

[a] EC_{50} = substance concentration which produces a cAMP response equal to 50% of its maximal response.
[b] S.E.M., standard error mean from at least three experiments ($n \geq 3$).
[c] IA is the fitted maximal value of the concentration–response curve, expressed as percentage of the maximal response to (R)-(−)-isoproterenol (10^{-4} M).
[d] Some further experiments were conducted in the presence of different concentrations (ranging from 10^{-9} to 10^{-7} M) of SR 59230A, a well-known neutral antagonist of β_3-AR expressed in rat and human colon (Manara et al., 1996). Unfortunately, the curve did not shift to the right, as expected, because SR 59230A behaves as a partial agonist in CHO expressing human cloned β_3-AR (Perrone et al., 2008b; Strosberg, 2000c). On the other hand, very preliminary experiments conducted on rat proximal colon showed that **SP-1e** and **SP-1g** are endowed with antagonist activity (in functional test in which colon motility is evaluated, it is not possible to discriminate antagonism from inverse agonism).
[e] EC_{50} = 3.9 nM for isoproterenol obtained by using [^{3}H]-cyclic AMP assay system (Emorine et al., 1989) instead of DELFIA (Perrone et al., 2008b). In both assay systems, the same intact cells expressing human β_3-AR were used.

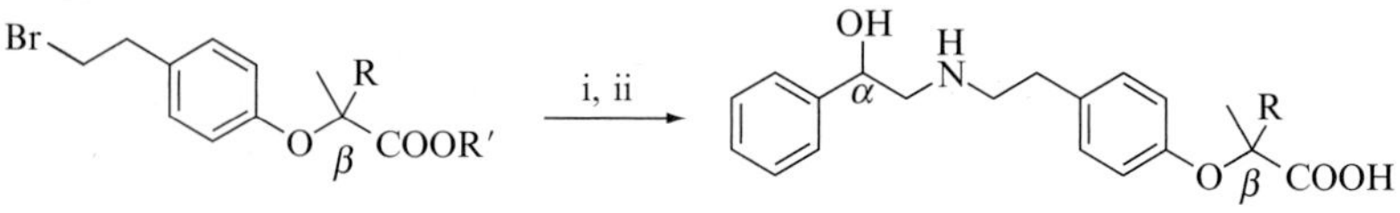

3a: R = CH$_3$, R' = C$_2$H$_5$
3b: β-rac, R = H, R' = C$_2$H$_5$
3c: (R), R = H, R' = C$_2$H$_5$
3d: (S), R = H, R' = CH$_3$

SP-1a: α-rac, R = CH$_3$
SP-1b: (αR), R = CH$_3$
SP-1c: (αS), R = CH$_3$
SP-1d: α-rac, β-rac, R = H
SP-1e: (αS, βS), R = H
SP-1f: (αR, βS), R = H
SP-1g: (αS, βR), R = H
SP-1h: (αR, βR), R = H

Scheme 11.1 General procedure for the preparation of **SP-1a–h**. Reagents and conditions: (i) racemic-, (R)- or (S)-2-amino-1-phenylethanol, DMF, 70 °C; (ii) NaOH, THF/H$_2$O, rt.

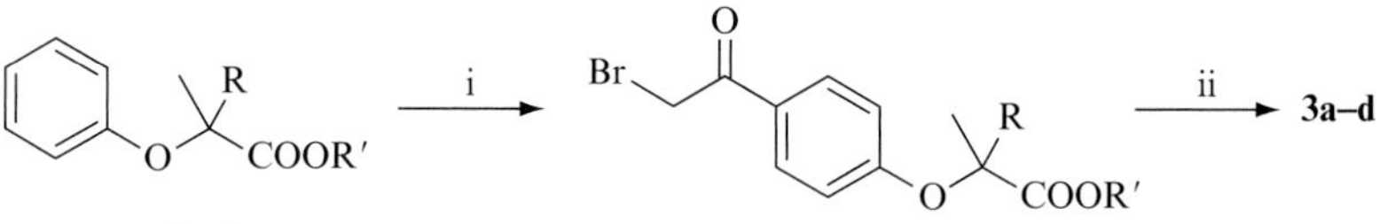

5a–d   **4a–d**   **3a–d**

Scheme 11.2 Preparation of **3a–d**. Reagents and conditions: (i) bromoacetyl bromide, AlCl$_3$, CH$_2$Cl$_2$, reflux; (ii) Et$_3$SiH, TFA, 70 °C.

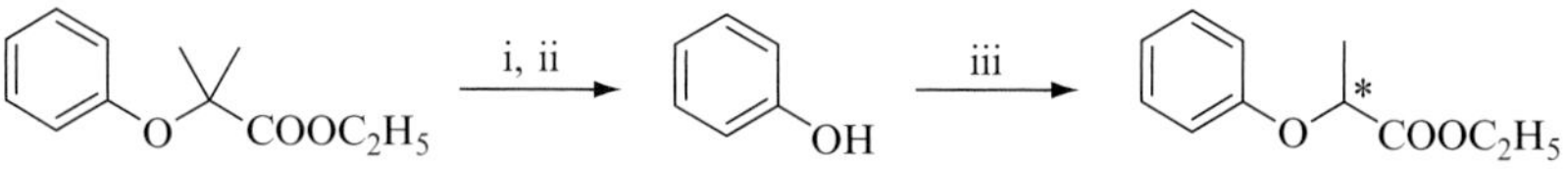

5a   **5b**

Scheme 11.3 Preparation of **5a–b**. Reagents and conditions: (i) Chloretone®, NaOH, acetone, reflux; (ii) SOCl$_2$, EtOH, reflux; (iii) ethyl 2-bromopropanoate, K$_2$CO$_3$, acetone, reflux.

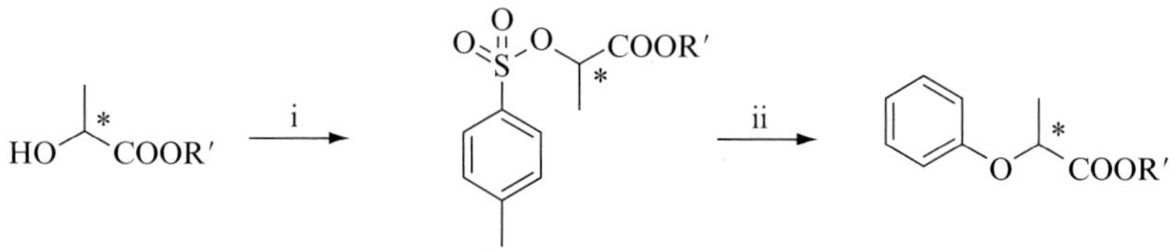

Ethyl (S)-lactate, R' = C$_2$H$_5$
Methyl (R)-lactate, R' = CH$_3$

6c: (S), R' = C$_2$H$_5$
6d: (R), R' = CH$_3$

5c: (S), R' = C$_2$H$_5$
5d: (R), R' = CH$_3$

Scheme 11.4 Preparation of **5c–d**. Reagents and conditions: (i) p-toluenesulfonyl chloride, CH$_2$Cl$_2$, pyridine, rt; (ii) phenol, CsF, DMF, rt.

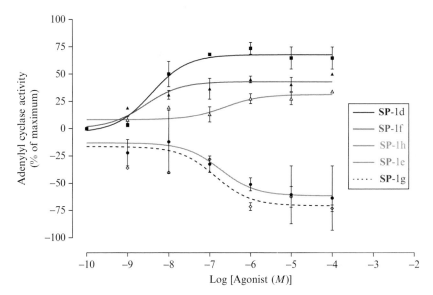

Figure 11.8 Representative curves of adenylyl cyclase activity evaluation in CHO cells stably expressing human β_3-AR subtype. Adenylyl cyclase stimulation was calculated as percentage (%) of the maximum effect by isoproterenol (ISO, Table 11.3). (See Color Insert.)

Compounds **3a–d** were obtained by Et$_3$SiH reduction of their precursors **4a–d**, that in turn were prepared by Friedel–Crafts acylation of 2-methyl-2-phenoxypropanoates (**5a**, R = CH$_3$) and racemic-, (R)- or (S)-2-phenoxypropanoates (**5b–d**, R = H) with bromoacetyl bromide (Scheme 11.2).

Different routes were followed for **5a–d** preparation. **5a** was obtained in a two-step process, in which phenol was first reacted with 1,1,1-trichloro-2-methyl-2-propanol (Chloretone®) in alkaline conditions and then treated with thionyl chloride in ethanol under reflux (Scheme 11.3).

Compound **5b** was obtained from the reaction of phenol with racemic ethyl 2-bromopropanoate. Its optically active analogues (R)-**5c** and (S)-**5d** were prepared by reacting the phenol with the p-toluenesulfonyl derivatives **6c–d** of commercially available ethyl (S)-lactate and methyl (R)-lactate, respectively, in the presence of CsF (Scheme 11.4).

3.1.1. Procedure for the preparation of SP-1e and SP-1g

3.1.1.1. 1-(Alkyloxycarbonyl)ethyl 4-methylbenzenesulfonates (6c–d): General procedure

Alkyl lactate [(S)-(−)-ethyl lactate or (R)-(+)-methyl lactate] (25 mmol) was dissolved in CH$_2$Cl$_2$ (10 mL), and p-toluenesulfonyl chloride (50 mmol) and pyridine (5 mL) were added. The reaction mixture was stirred at room temperature (rt) for 6 h. The solvent was removed under reduced pressure, the residue was dissolved in ethyl acetate and washed three times with 1 N HCl and three times with water. The product was isolated by chromatography (silica gel; mobile phase: petroleum ether/ethyl acetate = 8:2).

(S)-(−)-1-(Ethoxycarbonyl)ethyl 4-methylbenzenesulfonate (6c): Mp: 33.5–34.6 °C, white solid (85% yield). $[\alpha]_D$ −36.6 (ca. 1.45, $CHCl_3$). FT-IR (KBr): 3075, 2988, 2935, 2877, 1756, 1598, 1495, 1449, 1370, 1307, 1191, 1179, 1122, 1083, 1029, 944, 888, 818, 785, 665 cm^{-1}. ^1H NMR (300 MHz, $CDCl_3$, δ): 1.18 (t, $J = 7.1$ Hz, 3H, OCH_2CH_3), 1.48 (d, $J = 6.9$ Hz, 3H, $CHCH_3$), 2.42 (s, 3H, $CH_3C_6H_4SO_3$), 4.09 (q, $J = 7.1$ Hz, 2H, OCH_2CH_3), 4.90 (q, $J = 6.9$ Hz, 1H, $CHCH_3$), 7.31–7.34 (m, 2H, aromatic protons), 7.78–7.82 (m, 2H, aromatic protons). ^{13}C NMR (75 MHz, $CDCl_3$, δ): 14.15, 18.62, 21.88, 62.04, 74.39, 128.23, 130.02, 133.56, 145.31, 169.29. GC–MS (70 eV) m/z (rel. int.): 274 [$(^{34}S)M^+$, 1], 272 [$(^{32}S)M^+$, 11], 201 (3), 199 (48), 157 (9), 156 (18), 155 (100), 139 (5), 92 (22), 91 (71), 65 (21). Anal. Calcd. for $C_{12}H_{16}O_5S$: C, 52.91; H, 5.93. Found: C, 52.88; H, 5.89.

(R)-(+)-1-(Methoxycarbonyl)ethyl 4-methylbenzenesulfonate (6d): Colorless oil (quantitative yield). $[\alpha]_D$ +35.1 (ca. 1.05, $CHCl_3$). FT-IR (neat): 3101, 2993, 2957, 2928, 1762, 1598, 1458, 1433, 1369, 1311, 1221, 1193, 1179, 1083, 1027, 975, 945, 824 cm^{-1}. ^1H NMR (300 MHz, $CDCl_3$, δ): 1.46 (d, $J = 6.9$ Hz, 3H, $CHCH_3$), 2.41 (s, 3H, $CH_3C_6H_4$), 3.62 (s, 3H, OCH_3), 4.90 (q, $J = 6.9$ Hz, 1H, $CHCH_3$), 7.30–7.33 (m, 2H, aromatic protons), 7.75–7.78 (m, 2H, aromatic protons). ^{13}C NMR (75 MHz, $CDCl_3$, δ): 18.56, 21.84, 52.77, 74.26, 128.18, 130.04, 133.51, 145.36, 169.70. GC–MS (70 eV) m/z (rel. int.): 260 [$(^{34}S)M^+$, 1], 258 [$(^{32}S)M^+$, 13], 199 (45), 157 (7), 156 (12), 155 (100), 139 (10), 91 (79), 65 (23). Anal. Calcd. for $C_{11}H_{14}O_5S$: C, 51.13; H, 5.47. Found: C, 51.11; H, 5.46.

3.1.1.2. Alkyl 2-phenoxypropanoates (5c–d): General procedure

CsF (11 mmol) was dried at 200 °C in vacuum for 20 min, and it was then cooled at room temperature under a nitrogen stream. Anhydrous DMF (34 mL) was then added to CsF and the resulting suspension was stirred at room temperature under nitrogen atmosphere. Phenol (11 mmol) was added and the resulting reaction mixture was stirred at room temperature for 20 min. 1-(Alkoxycarbonyl)ethyl 4-methylbenzenesulfonate (6c–d, 3.67 mmol) in anhydrous DMF (8.5 mL) was added dropwise. The reaction mixture was stirred at room temperature for 32 h. The mixture was then diluted with ethyl acetate, washed three times with a saturated solution of $NaHCO_3$ and with a saturated solution of NaCl. The organic layer was dried over anhydrous Na_2SO_4. The solvent was removed under reduced pressure. The product was isolated as a colorless oil by chromatography (petroleum ether/ethyl acetate = 10:1).

(R)-(+)-Ethyl 2-phenoxypropanoate (5c): Yield 77%. [α]$_D$ +27.9 (c, 1.4, CHCl$_3$). FT-IR (neat): 3065, 3042, 2987, 2939, 2902, 2876, 1763, 1588, 1496, 1457, 1375, 1343, 1272 1050, 1020, 946, 885, 860, 802, 758, 691 cm^{-1}. ^{1}H NMR (300 MHz, CDCl$_3$, δ): 1.24 (t, J = 7.1 Hz, 3H, OCH$_2$CH$_3$), 1.62 (d, J = 6.7 Hz, 3H, CHCH$_3$), 4.21 (q, J = 7.1 Hz, 2H, OCH$_2$CH$_3$), 4.75 (q, J = 6.7 Hz, 1H, CHCH3), 6.87–6.90 (m, 2H, aromatic protons), 6.94–6.99 (m, 1H, aromatic proton), 7.23–7.30 (m, 2H, aromatic protons). ^{13}C NMR (75 MHz, CDCl$_3$, δ): 14.33, 18.78, 61.44, 72.82, 115.33, 121.76, 129.74, 157.85, 172.44. GC–MS (70 eV) m/z (rel. int.): 194 (M$^+$, 49), 122 (13), 121 (100), 103 (4), 94 (27), 93 (18), 77 (39), 65 (7), 51 (11). Anal. Calcd. for C$_{11}$H$_{14}$O$_3$: C, 68.00; H, 7.27. Found: C, 67.98; H, 7.24.

(S)-(−)-Methyl 2-phenoxypropanoate (5d): Yield 77%. [α]$_D$ −22.5 (ca. 1.15, CHCl$_3$). FT-IR (neat): 3065, 3042, 2993, 2955, 2848, 1760, 1601, 1589, 1495, 1456, 1376, 1287, 1244, 1205, 1136, 1100, 1053, 979, 754, 692 cm^{-1}. ^{1}H NMR (300 MHz, CDCl$_3$, d): 1.62 (d, J = 6.9 Hz, 3H, CHCH$_3$), 3.74 (s, 3H, OCH$_3$), 4.77 (q, J = 6.9 Hz, 1H, CHCH$_3$), 6.87–6.90 (m, 2H, aromatic protons), 6.95–7.00 (m, 1H, aromatic proton), 7.25–7.30 (m, 2H, aromatic protons). ^{13}C NMR (75 MHz, CDCl$_3$, δ): 18.81, 52.49, 77.73, 115.29, 121.84, 129.80, 157.79, 172.93. GC–MS (70 eV) m/z (rel. int.): 180 (M$^+$, 59), 122 (13), 121 (100), 94 (37), 93 (17), 91 (9), 77 (41), 65 (11), 59 (7), 51 (11). Anal. Calcd. for C$_{10}$H$_{12}$O$_3$: C, 66.63; H, 6.72. Found: C, 66.61; H, 6.69.

3.1.1.3. Alkyl 2-[4-(bromoacetyl)phenoxy]propanoates (4c–d): General procedure

Anhydrous AlCl$_3$ (11.54 g, 86.5 mmol) was suspended in anhydrous CH$_2$Cl$_2$ (65 mL) under nitrogen atmosphere. Bromoacetyl bromide (7.5 mL, 86.5 mmol) was carefully added and the mixture was stirred at 0 °C for 1 h. Then, alkyl 2-phenoxypropanoate (5c–d, 28.8 mmol) in anhydrous CH$_2$Cl$_2$ (25 mL) was added. The resulting solution was stirred overnight at rt under nitrogen atmosphere, and then refluxed for 6 h. The mixture was then poured into crushed ice and extracted three times with ethyl acetate. The combined extracts were washed three times with a saturated solution of NaHCO$_3$, three times with water, and dried over anhydrous Na$_2$SO$_4$. The solvent was removed under reduced pressure. The product was isolated as a yellow oil by chromatography (silica gel; mobile phase: petroleum ether/ethyl acetate = 1:1).

(R)-(+)-Ethyl 2-[4-(bromoacetyl)phenoxy]propanoate (4c): Yield 79%. [α]$_D$ +29.8 (ca. 1.1, CHCl$_3$). FT-IR (neat): 3073, 3048, 2987, 2941, 2875, 1755, 1682, 1594, 1575, 1509, 1446, 1431, 1393, 1378, 1285, 1190, 1136, 1097, 1050, 1017, 949, 843, 763, 606 cm^{-1}. ^{1}H NMR (300 MHz, CDCl$_3$, δ): 1.22 (t, J = 7.1 Hz, 3H, OCH$_2$CH$_3$), 1.62 (d, J = 6.7 Hz, 3H, CH$_3$CH), 4.19 (q, J = 7.1 Hz, 2H, OCH$_2$CH$_3$), 4.36 (s, 2H, CH$_2$Br), 4.80

(q, J = 6.7 Hz, 1H, CH$_3$CH), 6.86–6.91 (m, 2H, aromatic protons), 7.89–7.93 (m, 2H, aromatic protons). ^{13}C NMR (75 MHz, CDCl$_3$, δ): 14.32, 18.59, 31.01, 61.78, 72.77, 115.13, 127.77, 131.54, 162.26, 171.48, 190.04. GC–MS (70 eV) m/z (rel. int.): 316 [(^{81}Br)M$^+$, 10], 314 [(^{79}Br) M$^+$, 10], 243 (15), 241 (15), 222 (19), 221 (100), 163 (8), 121 (46), 107 (17), 104 (8), 90 (7), 77 (4), 76 (7). Anal. Calcd. for C$_{13}$H$_{15}$BrO$_4$: C, 49.51; H, 4.80. Found: C, 49.53; H, 4.84.

(S)-(−)-Methyl 2-[4-(bromoacetyl)phenoxy]propanoate (4d): Yield 84%. [α]$_D$ −32.7 (c 1.65, CHCl$_3$). FT-IR (neat): 3101, 3043, 2995, 2953, 2848, 1754, 1676, 1600, 1509, 1434, 1375, 1285, 1256, 1202, 1176, 1135, 1100, 1051, 978, 843, 606 cm^{-1}. ^{1}H NMR (300 MHz, CDCl$_3$, δ): 1.63 (d, J = 6.9 Hz, 3H, CH$_3$CH), 3.74 (s, 3H, OCH$_3$), 4.37 (s, 2H, CH$_2$Br), 4.84 (q, J = 6.9 Hz, 1H, CH$_3$CH), 6.87–6.90 (m, 2H, aromatic protons), 7.91–7.94 (m, 2H, aromatic protons). ^{13}C NMR (75 MHz, CDCl$_3$, δ): 18.63, 30.96, 52.75, 72.69, 115.10, 127.84, 131.59, 162.19, 171.97, 190.04. GC–MS (70 eV) m/z (rel. int.): 302 [(^{81}Br)M$^+$, 6], 300 [(^{79}Br) M$^+$, 6], 243 (7), 241 (8), 222 (6), 208 (14), 207 (100), 163 (8), 121 (46), 107 (15), 93 (6), 76 (6). Anal. Calcd. for C$_{12}$H$_{13}$BrO$_4$: C, 47.84; H, 4.35. Found: C, 47.86; H, 4.38.

3.1.1.4. Alkyl 2-[4-(2-bromoethyl)phenoxy]propanoates (3c–d): General procedure

(R)-(+)- or (S)-(−)-Alkyl 2-[4-(bromoacetyl)phenoxy]propanoate (4c–d, 1 g, 3.18 mmol) was dissolved in trifluoroacetic acid (1 or 15 mL). Triethylsilane (1.12 mL, 7 mmol) was added and the reaction mixture was stirred at 70 °C for 4 h. It was, then, cooled at rt and ethyl acetate was added. The organic phase was washed several times with a saturated solution of NaHCO$_3$, three times with water, and then dried over anhydrous Na$_2$SO$_4$. The solvent was removed under reduced pressure and the product was isolated by chromatography (petroleum ether/ethyl acetate = 9:1).

(R)-(+)-Ethyl 2-[4-(2-bromoethyl)phenoxy]propanoate (3c): Yellow oil (81% yield). [α]$_D$ +22.0 (ca. 1.1, CHCl$_3$). FT-IR (neat): 3033, 2986, 2938, 2874, 1752, 1612, 1585, 1512, 1448, 1376, 1297, 1265, 1243, 1199, 1135, 1098, 1052, 1016, 971, 823, 734 cm^{-1}. ^{1}H NMR (300 MHz, CDCl$_3$, δ): 1.24 (t, J = 7.1 Hz, 3H, OCH$_2$CH$_3$), 1.60 (d, J = 6.9 Hz, 3H, CHCH$_3$), 3.08 (t, J = 7.7 Hz, 2H, CH$_2$CH$_2$Br), 3.50 (t, J = 7.7 Hz, 2H, CH$_2$Br), 4.21 (q, J = 7.1 Hz, 2H, OCH$_2$CH$_3$), 4.71 (d, J = 6.9 Hz, 1H, CHCH$_3$), 6.80–6.84 (m, 2H, aromatic protons), 7.08–7.11 (m, 2H, aromatic protons). ^{13}C NMR (75 MHz, CDCl$_3$, δ): 14.36, 18.78, 33.41, 38.80, 61.48, 72.91, 115.45, 129.95, 132.18, 156.82, 172.40. GC–MS (70 eV) m/z (rel. int.): 302 [(^{81}Br)M$^+$, 61], 300 [(^{79}Br)M$^+$, 60], 230 (12), 229 (100), 228 (13), 227 (99), 221 (24), 207 (38), 185 (4), 183 (5), 147 (20), 121 (23), 120 (14), 119 (26), 107 (70), 104 (21), 103 (21), 91 (22),

78 (11), 77 (22), 65 (7). Anal. Calcd. for $C_{13}H_{17}BrO_3$: C, 51.82; H, 5.69. Found: C, 51.84; H, 5.70.

(S)-(−)-Methyl 2-[4-(2-bromoethyl)phenoxy]propanoate (3d): Yellow oil (75% yield). $[\alpha]_D$ −22.0 (ca. 1.1, $CHCl_3$). FT-IR (neat): 3033, 2993, 2953, 1761, 1613, 1586, 1507, 1448, 1376, 1231, 1134, 1100, 1053, 979, 823, 638 cm^{-1}. ^{1}H NMR (300 MHz, $CDCl_3$, δ): 1.60 (d, J = 6.9 Hz, 3H, $CHCH_3$), 3.08 (t, J = 7.7 Hz, 2H, CH_2CH_2Br), 3.51 (t, J = 7.7 Hz, 2H, CH_2Br), 3.75 (s, 3H, OCH_3), 4.73 (q, J = 6.9 Hz, 1H, $CHCH_3$), 6.80–6.84 (m, 2H, aromatic protons), 7.09–7.13 (m, 2H, aromatic protons). ^{13}C NMR (75 MHz, $CDCl_3$, δ): 18.81, 33.36, 38.80, 52.54, 72.84, 115.41, 130.01, 132.26, 156.75, 172.92. GC–MS (70 eV) m/z (rel. int.): 288 [(^{81}Br) M$^+$, 70], 286 [(^{79}Br)M$^+$, 70], 229 (94), 227 (93), 207 (33), 193 (74), 147 (21), 121 (31), 120 (18), 119 (28), 107 (100), 104 (21), 103 (23), 91 (32), 90 (9), 89 (8), 78 (14), 77 (26), 65 (11), 59 (13). Anal. Calcd. for $C_{12}H_{15}BrO_3$: C, 50.15; H, 5.27. Found: C, 50.14; H, 5.31.

3.1.1.5. Alkyl 2-{4-[2-(2-hydroxy-2-phenylethylamino)ethyl]phenoxy}propanoates (2e and 2g): General procedure

A mixture of alkyl 2-[4-(2-bromoethyl)phenoxy]propanoate (1 mol/L for **3c** and 0.85 mol/L for **3d**), (S)-2-amino-1-phenylethanol (1.3 mol/L for **3c** and 0.93 mol/L for **3d**) in anhydrous DMF was stirred under nitrogen atmosphere at 70 °C for 70 h. The reaction mixture was then diluted with ethyl acetate, washed with a saturated solution of NaCl and dried over anhydrous Na_2SO_4. The solvent was removed under reduced pressure and the product was isolated as yellow oil by chromatography (silica gel; mobile phase: dichloromethane/ethanol = 40:1).

(2S)-Methyl 2-{4-[2-((2S)-2-hydroxy-2-phenylethylamino)ethyl]phenoxy}propanoate (2e): Yield 31%. $[\alpha]_D$ +11.8 (ca. 0.95, $CHCl_3$). FT-IR (neat): 3600–3200, 3029, 2987, 2937, 2885, 1752, 1612, 1585, 1512, 1451, 1426, 1377, 1348, 1299, 1250, 1207, 1181, 1137, 1100, 1071, 1047, 825, 758, 700 cm^{-1}. ^{1}H NMR (400 MHz, $CDCl_3$, δ): 1.57 (d, J = 6.7 Hz, 3H, $CHCH_3$), 2.69–2.93 (m, 6H, CH_2CHOH, CH_2NH, CH_2CH_2NH), 3.45–3.64 (b s, 2H, OH and NH: exchange with D_2O), 3.71 (s, 3H, OCH_3), 4.70 (q, J = 6.7 Hz, 2H, $CHCH_3$), 4.74 (dd, J = 9.3 and 3.2 Hz, 1H, CHOH), 6.75–6.77 (m, 2H, aromatic protons), 7.03–7.05 (m, 2H, aromatic protons), 7.21–7.24 (m, 1H, aromatic proton), 7.27–7.32 (m, 4H, aromatic protons). ^{13}C NMR (100 MHz, $CDCl_3$, δ): 18.83, 35.05, 50.79, 52.56, 56.91, 71.53, 72.81, 115.36, 126.03, 127.80, 128.63, 129.99, 132.50, 142.51, 154.31, 173.01. GC–MS (70 eV) m/z (rel. int.): 325 [(M − 18)$^+$, 9], 325 (41), 324 (25), 238 (31), 206 (24), 193 (40), 147 (30), 134 (30), 132 (100), 130 (21), 120 (20), 107 (44), 106 (21), 105 (50), 104 (45), 103 (30), 91 (69), 90 (20), 78 (20), 77 (38), 70 (25), 51 (19). MS-ESI m/z (%): 344 [M + H]$^+$ (100%). Anal. Calcd. for $C_{20}H_{25}NO_4$: C, 69.93; H, 7.34; N, 4.08. Found: C, 69.95; H, 7.31; N, 4.06.

(2R)-(+)-Ethyl 2-{4-[2-((2S)-2-hydroxy-2-phenylethylamino)ethyl]phenoxy} propanoate (2g): Yield 66%. [α]$_D$ +55.6 (ca. 1.01, CHCl$_3$). FTIR (neat): 3600–3100, 3058, 3022, 2986, 2935, 2855, 1750, 1669, 1612, 1581, 1511, 1449, 1375, 1292, 1239, 1135, 1050, 1014, 826, 733, 702 cm^{-1}. ^{1}H NMR (300 MHz, CDCl$_3$, δ): 1.23 (t, J = 7.1 Hz, 3H, OCH$_2$CH$_3$), 1.58 (d, J = 6.8 Hz, 3H, CHCH$_3$), 2.66–3.00 (m, 6H, CH$_2$CHOH, CH$_2$NH, CH$_2$CH$_2$NH), 4.20 (q, J = 7.1 Hz, 2H, OCH$_2$CH$_3$), 4.70 (q, J = 6.8 Hz, 2H, CHCH$_3$), 4.86 (dd, J = 9.3 and 3.3 Hz, 1H, CHOH), 5.10–5.30 (b s, 2H, OH and NH: exchange with D$_2$O), 6.74–6.82 (m, 2H, aromatic protons), 6.97–7.12 (m, 2H, aromatic protons), 7.24–7.35 (m, 5H, aromatic protons). ^{13}C NMR (75 MHz, CDCl$_3$, δ): 14.35, 34.29, 50.50, 56.52, 61.47, 66.05, 72.90, 115.51, 126.03, 127.86, 128.66, 129.94, 131.84, 142.28, 156.52, 172.47. GC–MS (70 eV) m/z (rel. int.): 339 [(M − 18)$^+$, 1], 284 (4), 250 (100), 221 (24), 207 (7), 176 (10), 150 (14), 147 (14), 132 (69), 121 (19), 107 (18), 105 (14), 104 (10), 103 (11), 91 (10), 77 (14), 43 (9). Anal. Calcd. for C$_{21}$H$_{27}$NO$_4$: C, 70.55; H, 7.72; N, 3.92. Found: C, 70.58; H, 7.74; N, 3.90.

3.1.1.6. 2-{4-[2-(2-Hydroxy-2-phenylethylamino)ethyl]phenoxy} propanoic acid (*SP-1e, SP-1g*): General procedure

1 N NaOH (8.4 mL, 8.4 mmol) was added to a solution of alkyl 2-{4-[2-(2-hydroxy-2-phenylethylamino)ethyl]phenoxy}-propanoate (**2e** or **2g**, 4.2 mmol) in THF (10 mL). The reaction mixture was stirred at rt for 1 h. Then, THF was removed under reduced pressure and 2 N HCl was then added to pH 6. A precipitate formed that was filtered and washed with water. The residue was treated with hot acetone to afford a crystalline product.

*(2S)-(+)-2-{4-[2-((2S)-2-Hydroxy-2-phenylethylamino)ethyl]phenoxy}propanoic acid (**SP-1e**)*: Mp: 202 °C (dec.), white solid (45% yield), which was judged by HPLC peak integrations to be a 90:10 mixture of (S,S)/minor diastereoisomers. [α]$_D$ +0.68 (c 1.19, CH$_3$COOH). FT-IR (KBr): 3600–3200, 2998, 2807, 1613, 1585, 1511, 1453, 1423, 1227, 1138, 1099, 1037, 932, 816, 747, 698 cm^{-1}. ^{1}H NMR (500 MHz, acetic acid-d_4, δ): 1.61 (d, J = 6.8 Hz, 3H, CHCH$_3$), 3.06–3.08 (m, 2H, CH$_2$CH$_2$NH), 3.27 (dd, J = 12.5 and 10.7 Hz, 1H, CH$_2$CHOH), 3.31–3.37 (m, 2H, CH$_2$NH), 3.41 (dd, J = 12.5 and 2.5 Hz, 1H, CH$_2$CHOH), 4.84 (q, J = 6.8 Hz, 1H, CHCH$_3$), 5.26 (dd, J = 10.7 and 2.5 Hz, 1H, CHOH), 6.86–6.88 (m, 2H, aromatic protons), 7.18–7.19 (m, 2H, aromatic protons), 7.28–7.31 (m, 1H, aromatic proton), 7.34–7.37 (m, 2H, aromatic protons), 7.39–7.41 (m, 2H, aromatic protons). ^{13}C NMR (125 MHz, CDCl$_3$, δ): 17.84, 31.24, 49.56, 54.15, 69.45, 72.18, 115.50, 126.04, 128.43, 128.78, 129.76, 130.12, 140.26, 156.91, 176.99. MS-ESI m/z (%): 330 [M + H]$^+$ (100%), 328 [M − H]$^−$ (100%). Anal. Calcd. for C$_{19}$H$_{23}$NO$_4$: C, 69.26; H, 7.04; N, 4.25. Found: C, 69.28; H, 7.06, N, 4.22.

(2R)-(+)-2-{4-[2-((2S)-2-Hydroxy-2-phenylethylamino)ethyl]phenoxy} propanoic acid (**SP-1g**): Mp: 193–194.5 °C, white solid (65% yield), which was judged by HPLC peak integrations to be a 87:13 mixture of (R,S)/minor diastereoisomers. [α]$_D$ +44.9 (ca. 1.00, CH$_3$COOH). Spectroscopic data are identical to those ones reported for **SP-1e**.

3.2. Biological methods

3.2.1. General

Chinese hamster ovary (CHO) cell lines transfected with human cloned $β_3$- and $β_2$-ARs were kindly provided by the Institute Cochin de Génétique Moléculaire, Paris, France. The CHO cell line transfected with human cloned $β_1$-ARs was kindly provided by Institut für Pharmakologie und Toxikologie, Universität Würzburg, Germany. For receptor binding and activity studies, compounds were dissolved in absolute ethanol. Forskoline was purchased from Tocris Cookson Ltd., UK; [^{3}H]-dihydroalprenolol (3,59TBq) was obtained from Amersham Biosciences (Milan, Italy). Protein concentration was determined by commercial protein determination kit based on the Lowry method (Lowry et al., 1951), using a Perkin–Elmer UV/vis LAMBDA BIO 20 spectrophotometer. All binding and activity data obtained were analyzed by Graph-Pad Prism program.

3.2.2. Cell culture and membrane preparation

CHO cells expressing each subtype of human cloned $β_1$-, $β_2$-, or $β_3$-ARs were grown in an atmosphere of 5% CO$_2$ in air at 37 °C in Dulbecco's modified Eagle's medium with nutrient mixture F12 (DMEM/F12) supplemented with 10% Foetal calf serum, 2 mM L-glutamine, 100 U/mL of penicillin G, and 100 U/mL of streptomycin. Preconfluent cells were washed with ice-cold PBS, scraped from the plate surface, collected in ice-cold lysis buffer (10 × 10^6/mL; 5 mM Tris/HCl, 2 mM EDTA, pH 7.4 at 4 °C) and homogenized with a Brinkman politron (5 for 3 × 10 s). The cell membrane suspension was centrifuged for 10 min at 4 °C at 1000×g. Supernatant was centrifuged at 10,000×g for 30 min at 4 °C. The resultant membrane pellet was resuspended in ice cold incubation buffer (50 mM Tris/HCl, 10 mM MgCl$_2$, pH 7.4, for $β_1$-AR binding experiments; 50 mM Tris/HCl, pH 7.4, for $β_2$-AR binding experiments) and protein content was measured. The membrane suspension was used immediately or stored frozen at −80 °C, for radioligand binding experiments.

3.2.3. $β_1$- and $β_2$-AR binding experiments

Saturation binding experiments were performed by incubating cell membranes (50 μg of protein) in a total volume of 500 μL incubation buffer, containing increasing concentrations of [^{3}H]-dihydroalprenolol (0.1, 0.5, 1, 3, 5, 10 nM). Incubations were carried out at 30 °C for 30 min for $β_1$- or

90 min for β_2-AR binding assay. Nonspecific binding was determined in the presence of 10 μM alprenolol. Reactions were terminated by rapid filtration through Whatman GF/C glass fibre filters that had been soaked for 60 min in 0.5% polyethyleneamine for β_1- or 0.3% polyethyleneamine for β_2-AR binding assay. The filters were washed with 3 × 1 mL of ice-cold incubation buffer. The radioactivity bound to the filters was measured using LS6500 Multi-Purpose scintillation Counter, Beckman. Competition experiments were performed by incubating 50 μg of protein with increasing amounts of test compound (from 10^{-9} to 10^{-5} M) and 4 nM [^{3}H]-dihydroalprenolol for β_1- or 0.4 nM for β_2-AR binding assay, in a final incubation volume of 500 μL incubation buffer. Nonspecific binding was determined in the presence of 10 μM alprenolol. Reactions were terminated and radioactivity quantified as previously described. In CHO-β_1 cells, K_d value for alprenolol was 12.49 nM and B_{max} 2970 fmol/mg of protein while in CHO-β_2-AR cells K_d value for alprenolol was 0.50 nM and B_{max} 540 fmol/mg of protein. K_d and B_{max} values have been obtained as a mean of two experiments with samples in duplicate.

3.2.4. β_3-Adrenoceptor activity by DELFIA cAMP-Eu assay

The DELFIA (dissociation enhanced lanthanide fluoroimmuno assay) cAMP-binding assay was performed according to technical data sheet by Perkin–Elmer Life Science. The optimization of experimental conditions (amount of cell, incubation times, and other parameters) is reported below.

One confluent plate of cells was trypsinised and resuspended in the above medium and cultured overnight at a concentration of 50,000 cells/200 μL per well into 96-well flat-bottomed plates. The medium was aspirated from each well and replaced with 100 μL of preheated (37 °C) medium without serum. The plate was then placed back in the CO_2-incubator for 30 min at 37 °C. About 50 μL of 1 mM IBMX (3-isobutyl-1-methylxanthine), phosphodiesterase inhibitor was added to each well. Fifty microliters of test compounds at different concentrations (100, 500 nM, 1, 10, 100 μM) were then added to the wells and the plate was incubated for 30 min at 37 °C. Cells were then lysed and incubated at rt for 5 min. The plate was immediately used for measuring cAMP levels or stored at 4 °C until the assay was performed. Measure of the samples was carried out in time resolved fluorometer using 1420 Multilabel Counter Victor3, Perkin–Elmer. The wavelengths were 340 and 615 nm in excitation and emission, respectively.

ACKNOWLEDGMENTS

The research relevant to this chapter was supported by the University of Bari "A. Moro"-Bari (Italy); InterUniversity Research Center "Per la Ricerca e la Sperimentazione di Biotecnologie nel Trattamento dello Scompenso Cardiaco Avanzato"-Bari (Italy).

REFERENCES

Arch, J. R. S. (1989). The brown adipocyte beta-adrenoceptor. *Proc. Nutr. Soc.* **48**, 215–223.

Baker, J. G. (2005). Evidence for a secondary state of the human β3-adrenoceptor. *Mol. Pharmacol.* **68**, 1645–1655.

Baker, J. G., Hall, I. P., and Hill, S. (2003). Agonist and inverse agonist actions of β-blockers at the human β_2-adrenoceptor provide evidence for agonist-directed signaling. *J. Mol. Pharmacol.* **64**, 1357–1369.

Barbe, P., Millet, L., Galitzky, J., Lafontan, M., and Berlan, M. (1996). In situ assessment of the role of the beta 1-, beta 2- and beta 3-adrenoceptors in the control of lipolysis and nutritive blood flow in human subcutaneous adipose tissue. *Br. J. Pharmacol.* **117**, 907–913.

Bharucha, A. E., Camilleri, M., Zinsmeister, A. R., and Hanson, R. B. (1997). Adrenergic modulation of human colonic motor and sensory function. *Am. J. Physiol.* **273**, 997–1006.

Brockunier, L. L., Candelore, M. R., Cascieri, M. A., Liu, Y., Tota, L., Wyvratt, M. J., Fisher, M. H., Weber, A. E., and Parmee, E. R. (2001). Human beta3 adrenergic receptor agonists containing cyanoguanidine and nitroethylenediamine moieties. *Bioorg. Med. Chem. Lett.* **11**, 379–382.

Bianchetti, A., and Manara, L. (1990). In vitro inhibition of intestinal motility by phenylethanolaminotetralines: Evidence of atypical beta-adrenoceptors in rat colon. *Br. J. Pharmacol.* **100**, 831–839.

Blin, N., Nahmias, C., Drumare, M. F., and Strosberg, A. D. (1994). Mediation of most atypical effects by species homologues of the beta 3-adrenoceptor. *Br. J. Pharmacol.* **112**, 911–919.

Bond, R. A., and IJzerman, Ad. P. (2006). Recent developments in constitutive receptor activity and inverse agonism, and their potential for GPCR drug discovery. *Trends Parmacol. Sci.* **27**, 92–96.

Cellek, S., Thangiah, R., Bassil, A. K., Campbell, C. A., Gray, K. M., Stretton, J. L., Lalude, O., Vivekanandan, S., Wheeldon, A., Winchester, W. J., Sanger, G. J., Schemann, M., et al. (2007). Demonstration of functional neuronal beta(3)-adrenoceptors within the enteric nervous system. *Gastroenterology* **133**, 175–183.

Claustre, Y., Leonetti, M., Santucci, V., Bougault, I., Desvignes, C., Rouquier, L., Aubin, N., Keane, P., Busch, S., Chen, Y., Palejwala, V., Tocci, M., et al. (2008). Effects of the beta(3)-adrenoceptor agonist SR58611A (amibegron) on serotonergic and noradrenergic transmission in the rodent: Relevance to its antidepressant/anxiolytic-like profile. *Neuroscience* **156**, 353–364.

Clifford, E. E., Martin, K. A., Dalal, P., Thomas, R., and Dubyak, G. R. (1997). Stage-specific expression of P2Y receptors, ecto-apyrase and ecto-S'-nucleotidase in myeloid leukocytes. *Am. J. Physiol.* **273**, C973–C987.

Costa, T., and Cotecchia, S. (2005). Historical review: Negative efficacy and the constitutive activity of G-protein-coupled receptors. *Trends Pharmacol. Sci.* **26**, 618–624.

Dallanoce, C., Frigerio, F., De Amici, M., Dorsch, S., Klotz, K.-N., and De Micheli, C. (2007). Novel chiral isoxazole derivatives: Synthesis and pharmacological characterization at human beta-adrenergic receptor subtypes. *Bioorg. Med. Chem.* **15**, 2533–2543.

de Ligt, R. A. F., Kourounakis, A. P., and IJzerman, A. P. (2000). Inverse agonism at G protein-coupled receptors: (Patho)physiological relevance and implications for drug discovery. *Br. J. Pharmacol.* **130**, 1–12.

De Ponti, F., Cosentino, M., Costa, A., Girani, M., Gibelli, G., D'Angelo, L., Frigo, G., and Crema, A. (1995). Inhibitory effects of SR 58611A on canine colonic motility: Evidence for a role of beta 3-adrenoceptors. *Br. J. Pharmacol.* **114**, 1447–1453.

De Ponti, F., Gibelli, G., Croci, T., Arcidiaco, M., Crema, F., and Manara, L. (1996). Functional evidence of atypical beta 3-adrenoceptors in the human colon using the beta 3-selective adrenoceptor antagonist, SR 59230A. *Br. J. Pharmacol.* **117,** 1374–1376.
de Souza, C. J., and Burkey, B. F. (2001). Beta(3)-adrenoceptor agonists as anti-diabetic and anti-obesity drugs in humans. *Curr. Pharm. Des.* **7,** 1433–1449.
Deng, C., Paoloni-Giacobino, A., Kuehne, F., Boss, O., Revelli, J.-P., Moinat, M., Cawthorne, M. A., Muzzin, P., and Giacobino, J.-P. (1996). Respective degree of expression of beta 1-, beta 2- and beta 3-adrenoceptors in human brown and white adipose tissues. *Br. J. Pharmacol.* **118,** 929–934.
Dennedy, M. C., Friel, A. M., Gardeil, F., and Morrison, J. J. (2001). Beta-3 versus beta-2 adrenergic agonists and preterm labour: In vitro uterine relaxation effects. *Br. J. Obstet. Gynaecol.* **108,** 605–609.
Dolan, J. A., Muenkel, H. A., Burns, M. G., Pellegrino, S. M., Fraser, C. M., Pietri, F., Strosberg, A. D., Largis, E. E., Dutia, M. D., Bloom, J. D., Bass, A. S., Tanikella, T. K., et al. (1994). Beta-3 adrenoceptor selectivity of the dioxolane dicarboxylate phenethanolamines. *J. Pharmacol. Exp. Ther.* **269,** 1000–1006.
Donaldson, L. F., Hanley, M. R., and Villablanca, A. C. (1997). Inducible receptors. *Trends Pharmacol. Sci.* **18,** 171–181.
Dow, R. L., Paight, E. S., Schneider, S. R., Hadcock, J. R., Hargrove, D. M., Martin, K. A., Maurer, T. S., Nardone, N. A., Tess, D. A., and DaSilva-Jardine, P. (2004). Potent and selective, sulfamide-based human beta 3-adrenergic receptor agonists. *Bioorg. Med. Chem. Lett.* **14,** 3235–3240.
Emorine, L. J., Marullo, S., Briend-Sutren, M. M., Patey, G., Tate, K., Delavier-Klutchko, C., and Strosberg, A. D. (1989). Molecular characterization of the human beta 3-adrenergic receptor. *Science* **245,** 1118–1121.
Fedorov, V. V., and Lozinsky, I. T. (2008). Is the beta 3-adrenergic receptor a new target for treatment of post-infarct ventricular tachyarrhythmias and prevention of sudden cardiac death? *Heart Rhythm* **5,** 298–299.
Fisher, L. G., Sher, P. M., Skwish, S., Michel, I. M., Seiler, S. M., and Dickinson, K. E. J. (1996). BMS-187257, a potent, selective, and novel heterocyclic β3 adrenergic receptor agonist. *Bioorg. Med. Chem. Lett.* **6,** 2253–2258.
Fujimura, T., Tamura, T., Tsutsumi, T., Yamamoto, T., Nakamura, K., Koibuchi, Y., Kobayashi, M., and Yamaguchi, O. (1999). Expression and possible functional role of the β3-adrenoceptor in human and rat detrusor muscle. *J. Urol.* **161,** 680–685.
Furse, K. E., and Lybrand, T. P. (2003). Three-dimensional models for β-adrenergic receptor complexes with agonists and antagonists. *J. Med. Chem.* **46,** 4450–4462.
Gan, R. T., Li, W. M., Xiu, C. H., Shen, J. X., Wang, X., Wu, S., and Kong, Y. H. (2007). Chronic blocking of ß3-adrenoceptor ameliorates cardiac function in rat model of heart failure. *Chin. Med. J. (Eng)* **120**(24), 2250–2255.
Gavai, A. V., Sher, P. M., Mikkilineni, A. B., Poss, K. M., McCann, P. J., Girotra, R. N., Fisher, L. G., Wu, G., Bednarz, M. S., Mathur, A., Wang, T. C., Sun, C. Q., et al. (2001). BMS-196085: A potent and selective full agonist of the human beta(3) adrenergic receptor. *Bioorg. Med. Chem. Lett.* **11,** 3041–3044.
Gros, J., Manning, B., Pietrì-Rouxel, F., Guillaume, J. L., Drumare, M. F., and Strosberg, A. D. (1998). Site directed mutagenesis of the human beta-3 adrenoreceptor: Transmembrane residues involved in ligand binding and signal transductyion. *Eur. J. Biochem.* **251,** 590–596.
Harada, H., Hirokawa, Y., Suzuki, K., Hiyama, Y., Oue, M., Kawashima, H., Yoshida, N., Furutani, Y., and Kato, S. (2003). Novel and potent human and rat beta3-adrenergic receptor agonists containing substituted 3-indolylalkylamines. *Bioorg. Med. Chem. Lett.* **13,** 1301–1305.

Hieble, J. P. (2007). Recent advances in identification and characterization of beta-adrenoceptor agonists and antagonists. *Curr. Top. Med. Chem.* **7**, 207–216.

Hoffmann, C., Leitz, M. R., Obendorf-Maass, S., Lohse, M. J., and Klotz, K.-N. (2004). Comparative pharmacology of human β-adrenergic receptor subtypes-characterization of stably transfected receptors in CHO cells. *Naunyn Schmiedebergs Arch. Pharmacol.* **369**, 151–159.

Kashaw, S. K., Rathi, L., Mishra, P., and Saxena, A. K. (2003). Development of 3D-QSAR models in cyclic ureidobenzenesulfonamides: human beta3-adrenergic receptor agonist. *Bioorg. Med. Chem. Lett.* **13**, 2481–2484.

Kato, H., Ohue, M., Kato, K., Nomura, A., Toyosawa, K., and Furutani, Y. (2001). Mechanism of amelioration of insulin resistance by β3-adrenoceptor agonist AJ-9677 in the KK-Ay/Ta diabetic obese mouse model. *Diabetes* **50**, 113–122.

Kenakin, T. (1997). Agonist-specific receptor conformations. *Trends Pharmacol. Sci.* **18**, 416–417.

Kenakin, T. (2001). Inverse, protean, and ligand-selective agonism: Matters of receptor conformation. *FASEB* **15**, 598–611.

Kenakin, T. (2002a). Drug efficacy at G protein-coupled receptors. *Annu. Rev. Pharmacol. Toxicol.* **42**, 349–379.

Kenakin, T. (2002b). Efficacy at G-protein-coupled receptors. *Nat. Rev. Drug Discov.* **1**, 103–110.

Kenakin, T. (2004). Principles: Receptor theory in pharmacology. *Trends Pharmacol. Sci.* **25**, 186–192.

Kumar, P. S., Bharatam P. V. (2005). CoMFA study on selective human β3-adrenoceptor agonists. *Arkivoc* xiii:67–79.

Kumar, P. S., and Bharatam, P. V. (2009). Comparative 3D QSAR study on β1-, β2-, and β3-adrenoceptor agonists. *Med. Chem. Res.* DOI: 10.1007/s00044-009-9257-x.

Kuratani, K., Kodama, H., and Yamaguchi, I. (1994). Enhancement of gastric mucosal blood flow by beta-3 adrenergic agonists prevents indomethacin-induced antral ulcer in the rat. *J. Pharmacol. Exp. Ther.* **270**, 559–565.

Lands, A. M., Arnold, A., McAuliff, J. P., Ludaena, F. P., and Brown, T. G., Jr. (1967). Differentiation of receptor systems activated by sympathomimetic amines. *Nature* **214**, 597–598.

Liu, X., Pérusse, F., and Bukowiecki, L. J. (1998). Mechanisms of the antidiabetic effects of the β3-adrenergic agonist CL-316243 in obese Zucker-ZDF rats. *Am. J. Physiol.* **274**, R1212–R1219.

Lohse, M. J., Hoffmann, C., and Engelhardt, S. (2003). Inverse agonism at β1-adrenergic receptors. *Int. Congr. Ser.* **1249**, 55–61.

Lowry, O. H., Rosebrough, N. J., Farr, A. L., and Randall, R. J. (1951). Protein measurement with the Folin phenol reagent. *J. Biol. Chem.* **193**, 265–273.

Maack, C., Tyroller, S., Schnabel, P., Cremers, B., Dabew, E., Südkamp, M., and Böhm, M. (2001). Characterization of β1-selectivity, adrenoceptor-Gs-protein interaction and inverse agonism of nebivolol in human myocardium. *Br. J. Pharmacol.* **132**, 1817–1826.

Manara, L., Badone, D., Baroni, M., Boccardi, G., Cecchi, R., Croci, T., Giudice, A., Guzzi, U., Landi, M., and Le Fur, G. (1996). Functional identification of rat atypical ß-adrenoceptors by the first β3-selective antagonists, aryloxypropanolaminotetralins. *Br. J. Pharmacol.* **117**, 435–442.

Mizuno, K., Sawa, M., Harada, H., Taoka, I., Yamashita, H., Oue, M., Tsujiuchi, H., Arai, Y., Suzuki, S., Furutani, Y., and Kato, S. (2005). Discovery of 1, 7-cyclized indoles as a new class of potent and highly selective human beta3-adrenergic receptor agonists with high cell permeability. *Bioorg. Med. Chem.* **13**, 855–868.

Mizuno, K., Sawa, M., Harada, H., Tateishi, H., Oue, M., Tsujiuchi, H., Furutani, Y., and Kato, S. (2004). Tryptamine-based human beta3-adrenergic receptor agonists. Part 1: SAR studies of the 7-position of the indole ring. *Bioorg. Med. Chem. Lett.* **14,** 5959–5962.

Nakajima, Y., Hamashima, H., Washizuka, K., Tomishima, Y., Ohtake, H., Imamura, E., Miura, T., Kayakiri, H., and Kato, M. (2005). Discovery of a novel, potent and selective human beta3-adrenergic receptor agonist. *Bioorg. Med. Chem. Lett.* **15,** 251–254.

Nestler, E. J., Barrot, M., DiLeone, R. J., Eisch, A. J., Gold, S. J., and Monteggia, L. M. (2002). Neurobiology of depression. *Neuron* **34,** 13–25.

Otsuka, A., Shinbo, H., Matsumoto, R., Kurita, Y., and Ozono, S. (2008). Expression and functional role of β-adrenoceptors in the human urinary bladder. *Naunyn Schmiedebergs Arch. Pharmacol.* **377,** 473–481.

Pang, Z., Wu, N., Zhang, X., Avallone, R., Croci, T., Dressler, H., Palejwala, V., Ferrara, P., Tocci, M. J., and Polites, H. G. (2010). GPR40 is partially required for insulin secretion following activation of β3 adrenergic receptors. *Mol. Cell Endocrinol.* **325**(1–2), 18–25.

Peleg, G., Ghanouni, P., Kobilka, B. K., and Zare, R. N. (2001). Singlemolecule spectroscopy of the beta(2) adrenergic receptor: Observation of conformational substates in a membrane protein. *Proc. Natl. Acad. Sci. USA* **98,** 8469–8474.

Perrone, M. G., Santandrea, E., Giorgio, E., Bleve, L., Scilimati, A., and Tortorella, P. (2006). A chemoenzymatic scalable route to optically active (R)-1-(pyridin-3-yl)-2-aminoethanol, valuable moiety of β_3-adrenergic receptor agonists. *Bioorg. Med. Chem.* **14,** 1207–1214.

Perrone, M. G., Notarnicola, M., Caruso, M. G., Tutino, V., and Scilimati, A. (2008a). Upregulation of β_3-adrenergic receptor mRNA in human colon cancer: A preliminary study. *Oncology* **75,** 224–229.

Perrone, M. G., Santandrea, E., Bleve, L., Vitale, P., Colabufo, N. A., Jockers, R., Milazzo, F. M., Sciarroni, A. F., and Scilimati, A. (2008b). Stereospecific synthesis and bio-activity of novel β_3-adrenoceptor agonists and inverse agonists. *Bioorg. Med. Chem.* **16,** 2473–2488.

Perrone, M. G., Bleve, L., Santandrea, E., Vitale, P., Niso, M., and Scilimati, A. (2009). The tertiary amine nitrogen atom of piperazine sulfonamides as a novel determinant of potent and selective β_3-adrenoceptor agonists. *Chem. Med. Chem.* **4,** 2080–2097.

Popp, B. D., Hutchinson, D. S., Evans, B. A., and Summers, R. J. (2004). Stereoselectivity for interactions of agonists and antagonists at mouse, rat and human β3-adrenoceptors. *Eur. J. Pharm.* **484,** 323–331.

Pott, C., Brixius, K., Bloch, W., Ziskoven, C., Napp, A., and Schwinger, R. H. G. (2006). Beta3-adrenergic stimulation in the human heart: Signal transduction, functional implications and therapeutic perspectives. *Pharmazie* **61,** 255–260.

Rodriguez, M., Carillon, C., Coquerel, A., Le Fur, G., Ferrara, P., Caput, D., and Shire, D. (1995). Evidence for the presence of beta 3-adrenergic receptor mRNA in the human brain. *Brain. Res. Mol. Brain Res.* **29,** 369–375.

Rouget, C., Bardou, M., Breuiller-Fouché, M., Loustalot, C., Qi, H., Naline, E., Croci, T., Cabrol, D., Advenier, C., and Leroy, M. J. (2005). β3-Adrenoceptor is the predominant β-adrenoceptor subtype in human myometrium and its expression is up-regulated in pregnancy. *J. Clin. End. Met.* **90,** 1644–1650.

Rozec, B., and Gauthier, C. (2006). Beta 3-adrenoceptors in the cardiovascular system: Putative roles in human pathologies. *Pharmacol. Ther.* **111,** 652–673.

Salamon, Z., Hruby, V. J., Tollin, G., and Cowell, S. (2002). Binding of agonists, antagonists, and inverse agonists to the human deltaopioid receptor produces distinctly different conformationalstates distinguishable by plasmon-waveguide resonance spectroscopy. *J. Pept. Res.* **60,** 322–328.

Sevak, R., Paul, A., Goswami, S., and Santani, D. (2002). Gastroprotective effect of beta3 adrenoreceptor agonists ZD 7114 and CGP 12177A in rats. *Pharmacol. Res.* **46,** 351–356.

Shearer, B. G., Chao, E. Y., Uehling, D. E., Deaton, D. N., Cowan, C., Sherman, B. W., Milliken, T., Faison, W., Brown, K., Adkisond, K. K., and Leed, F. (2007). Synthesis and evaluation of potent and selective β_3 adrenergic receptor agonists containing heterobiaryl carboxylic acids. *Bioorg. Med. Chem. Lett.* **17,** 4670–4677.

Sher, P. M., Mathur, A., Fisher, L. G., Wu, G., Skwish, G. S., Michel, I. M., Seiler, S. M., and Dickinson, K. E. J. (1997). Carboxyl-promoted enhancement of selectivity for the β_3-Adrenergic receptor. Negative charge of the sulfonic acid BMS-187413 introduces β_3 binding selectivity. *Bioorg. Med. Chem. Lett.* **7,** 1583–1588.

Simiand, J., Keane, P. E., Guitard, J., Langlois, X., Gonalons, N., Martin, P., Bianchetti, A., Le Fur, G., and Soubrié, P. (1992). Antidepressant profile in rodents of SR 58611A, a new selective agonist for atypical beta-adrenoceptors. *Eur. J. Pharmacol.* **219,** 193–201.

Soudijn, W., van Wijngaarden, I., and Ijzerman, A. P. (2005). Structure-activity relationships of inverse agonists for G-protein-coupled receptors. *Med. Res. Rev.* **25,** 398–426.

Steffan, R., Ashwell, M. A., Solvibile, W. R., Matelan, E., Largis, E., Han, S., Tillet, J., and Mulvey, R. (2002). Novel substituted 4-aminomethylpiperidines as potent and selective human beta3-agonists. Part 1: aryloxypropanolaminomethylpiperidines. *Bioorg. Med. Chem. Lett.* **12,** 2957–2961.

Stemmelin, J., Cohen, C., Terranova, J.-P., Lopez-Grancha, M., Pichat, P., Bergis, O., Decobert, M., Santucci, V., Françon, D., Alonso, R., Stahl, S. M., Keane, P., et al. (2008). Stimulation of the beta3-Adrenoceptor as a novel treatment strategy for anxiety. *Neuropsychopharmacology* **33,** 574–587.

Stemmelin, J., Cohen, C., Yalcin, I., Keane, P., and Griebel, G. (2010). Implication of β_3-adrenoceptors in the antidepressant-like effects of amibegron using Adrb3 knockout mice in the chronic mild stress. *Behav. Brain Res.* **206,** 310–312.

Strader, C. D., Candelore, M. R., Hill, W. S., Sigal, I. S., and Dixon, R. A. (1989). Identification of two serine residues involved in agonist activation of the beta-adrenergic receptor. *J. Biol. Chem.* **264,** 13527–13528.

Strange, P. G. (2002). Mechanisms of inverse agonism at G-protein-coupled receptors. *Trends Pharmacol. Sci.* **23,** 89–95.

Strosberg, A. D. (2000a). The β3-Adrenoreceptor. Taylor & Francis, New York, p. 5.

Strosberg, A. D. (2000b). The β3-Adrenoreceptor. Taylor & Francis, New York, p. 11.

Strosberg, A. D. (2000c). The β3-Adrenoreceptor. Taylor & Francis, New York, p. 54.

Strosberg, A. D. (2000d). The β3-Adrenoreceptor. Taylor & Francis, New York, p. 84–85.

Strosberg, A. D., and Pietri-Rouxel, F. (1996). Function and regulation of β3-adrenoceptor. *Trends Pharmacol. Sci* **17,** 373–381.

Takasu, T., Ukai, M., Sato, S., Matsui, T., Nagase, I., Maruyama, T., Sasamata, M., Miyata, K., Uchida, H., and Yamaguchi, O. (2007). Effect of (R) 2-(2-aminothiazol-4-yl)-4'-{2-[(2-hydroxy-2-phenylethyl)amino]ethyl} acetanilide (YM178), a novel selective beta3-adrenoceptor agonist, on bladder function. *J. Pharmacol. Exp. Ther.* **321,** 642–647.

Tanaka, N., Tamai, T., Mukaiyama, H., Hirabayashi, A., Muranaka, H., Ishikawa, T., Kobayashi, J., Akahane, S., and Akahane, M. (2003). Relationship between stereochemistry and the β_3-adrenoceptor agonistic activity of 4'-hydroxynorephedrine derivative as an agent for treatment of frequent urination and urinary incontinence. *J. Med. Chem.* **46,** 105–112.

Tisdale, M. J. (2003). Pathogenesis of cancer cachexia. *J. Support. Oncol.* **1,** 159–168.

Uehling, D. E., Donaldson, K. H., Deaton, D. N., Hyman, C. E., Sugg, E. E., Barrett, D. G., Hughes, R. G., Reitter, B., Adkison, K. K., Lancaster, M. E., Lee, F., Hart, R., et al. (2002). Synthesis and evaluation of potent and selective beta(3) adrenergic receptor agonists containing acylsulfonamide, sulfonylsulfonamide, and sulfonylurea carboxylic acid isosteres. *J. Med. Chem.* **45,** 567–583.

Vasina, V., Abu-gharbieh, E., Barbara, G., Degiorgio, R., Colucci, R., Blandizzi, C., Bernardini, N., Croci, T., Del Tacca, M., and Deponti, F. (2008). The β_3-adrenoceptor agonist SR58611A ameliorates experimental colitis in rats. *Neurogastroenterol. Motil.* **20,** 1030–1041.

Weber, A. E., Mathvink, R. J., Perkins, L., Hutchins, J. E., Candelore, M. R., Tota, L., Strader, C. D., Wyvratt, M. J., and Fisher, M. H. (1998). Potent, selective benzenesulfonamide agonists of the human β_3 adrenergic receptor. *Bioorg. Med. Chem. Lett.* **8,** 1101–1106.

Weiss, J. M., Morgan, P. H., Lutz, M. W., and Kenakin, T. (1996a). The cubic ternary complex receptor-occupancy model I. Model description. *J. Theor. Biol.* **178,** 151–167.

Weiss, J. M., Morgan, P. H., Lutz, M. W., and Kenakin, T. (1996b). The cubic ternary complex receptor-occupancy model. II. Understanding apparent affinity. *J. Theor. Biol.* **178,** 169–182.

Weiss, J. M., Morgan, P. H., Lutz, M. W., and Kenakin, T. (1996c). The cubic ternary complex receptor-occupancy model III. Resurrecting efficacy. *J. Theor. Biol.* **181,** 381–397.

CHAPTER TWELVE

Constitutive Activity of the Lutropin Receptor and Its Allosteric Modulation by Receptor Heterodimerization

Deborah L. Segaloff

Contents

1. Introduction — 232
2. General Principles for Quantifying Receptor Activation — 235
3. Modifying Cell Surface Expression Levels of Recombinant hLHR and Mutants Thereof — 239
   3.1. Growth of HEK293 cells — 240
   3.2. cDNA encoding LHR — 240
   3.3. Transient transfections of HEK293 cells — 241
   3.4. Manipulating transient transfection conditions to yield cells with comparable cell surface densities of wt and mutant LHR — 241
   3.5. Manipulating transient transfection conditions to yield cells with a similar range of cell surface levels of wt and mutant LHR — 242
4. Quantifying Cell Surface hLHR Expression — 243
   4.1. Measurement of maximal ^{125}I-hCG binding to intact cells — 243
   4.2. Measurement of cell surface epitope-tagged hLHR by flow cytometry — 244
5. Quantifying cAMP Production in Cells Expressing the hLHR — 245
   5.1. Cell incubations for the measurement of basal and/or hormone-stimulated cAMP — 245
6. Experimental Strategies for Characterizing the Attenuating Effects of a Signaling Inactive hLHR on a Coexpressed wt or CAM hLHR — 246

Acknowledgment — 250
References — 250

Department of Physiology and Biophysics, The University of Iowa Roy J. and Lucille A. Carver College of Medicine, Iowa City, Iowa, USA

Methods in Enzymology, Volume 484 © 2010 Elsevier Inc.
ISSN 0076-6879, DOI: 10.1016/S0076-6879(10)84012-1 All rights reserved.

Abstract

The lutropin receptor (LHR) is a G protein-coupled receptor (GPCR) that mediates the actions of pituitary LH in males and females and that of placental hCG in pregnant women and, therefore, plays an essential role in reproductive physiology. Mutations of the *lhcgr* gene that result in constitutive activation of the LHR have been shown to be causative of gonadotropin-independent precocious puberty in young boys. Studies on constitutively active mutants (CAMs) of the LHR have been extremely informative in elucidating the roles of the LHR in reproductive physiology as well as in understanding the molecular basis underlying activation of this GPCR. The constitutive activities of hLHR CAMs can be attenuated by introducing mutations into the CAMs that stabilize the resting state of the hLHR or by coexpressing the hLHR CAMs with an hLHR mutant that is stabilized in the resting state, allowing the two forms of the hLHR to heterodimerize. This chapter describes the experimental methods and strategies underlying studies of hLHR CAMs.

1. Introduction

The luteinizing hormone or lutropin receptor (LHR) is a G protein-coupled receptor (GPCR) that plays a pivotol role in reproductive physiology in males and females. In females, the LHR is expressed primarily in the ovaries, where it is found in theca cells and in mature granulosa cells of follicles. In response to pituitary LH, the LHR in theca cells stimulates the synthesis of androgens, which are used as precursors for the FSH-stimulated synthesis of estrogen by neighboring granulosa cells. Importantly, the LHR in preovulatory follicles mediates follicular rupture and ovulation in response to the midcycle surge of LH. Interestingly, the LHR also recognizes a highly homologous hormone, human chorionic gonadotropin (hCG), which is secreted during pregnancy by the blastocyst and then the chorion of the placenta. Therefore, if the released oocyte is fertilized, the subsequent secretion of hCG rescues the corpus luteum derived from the ruptured follicle, permitting the continuation of pregnancy. In males, the LHR is expressed primarily in the testes, where it is localized to the interstitial Leydig cells. These cells, in response to pituitary LH, synthesize and secrete testosterone, which is essential for normal spermatogenesis and for the development of secondary sexual characteristics.

In both males and females, the release of the gonadotropins LH and FSH are normally suppressed until the time of puberty, at which time the regular pulsatile secretion of these hormones mediates sexual development and gametogenesis. In 1993, Shenker *et al.* (1993) reported the first constitutively activate LHR mutant in a young boy with familial male-limited gonadotropin-independent precocious puberty and Leydig cell hyperplasia (also referred to as testotoxicosis). Since then, numerous other constitutively activating mutants (CAMs) of the LHR have been identified in boys with gonadotropin-independent precocious puberty (Latronico and Segaloff, 1999; Segaloff, 2009; Themmen and Huhtaniemi, 2000). These mutations arise either as sporadic mutations or are inherited in an autosomal dominant manner. Interestingly, mothers and female siblings of affected males who are heterozygous for an activating mutation of the *lhcgr* gene do not display precocious puberty and appear to have a normal reproductive phenotype, underscoring the necessity of activation of both the LHR as well as FSHR pathways in females for pubertal development.

The LHR is a member of Family A (rhodopsin-like) GPCRs. Structurally, the LHR, like the related FSHR and thyrotropin receptor (TSHR), is composed of the prototypical serpentine domain containing seven transmembrane helices attached to a large extracellular domain (Ascoli *et al.*, 2002). The binding of hormone to the extracellular domain stabilizes the receptor in an active conformation such that the serpentine domain then activates G proteins. Many of the actions of the LHR can be attributed to its activation of Gs. At high receptor densities and high hormone concentrations, Gq is also stimulated, suggesting that this pathway may be involved in the stimulation of ovulation by the LHR (Gudermann *et al.*, 1992). Recent studies further suggest that the LH-stimulated activation of the inositol phosphate pathway in mature granulosa cells (mediated by the high levels of LHR in these cells) attenuates the FSH-mediated induction of aromatase (mediated by relatively low levels of FSHR in these cells stimulating Gs), suggesting a role for the LHR stimulation of Gq during the ovarian periovulatory and luteal phases (Andric and Ascoli, 2008a; Donadeu and Ascoli, 2005; Hirakawa *et al.*, 2002). It has also been shown that $G_{i/o}$ is activated by the LHR; however, the physiological consequences of this are not known.

All of the naturally occurring activating mutations of the hLHR are localized to the transmembrane helices of the serpentine domain (Segaloff, 2009). A particularly large cluster is found in transmembrane helix 6, a region implicated as critical to the signal transduction properties of the receptor. Importantly, all naturally occurring LHR CAMs have been shown to cause an elevation in the basal production of cAMP, consistent with their constitutive activation of Gs. The elevation of cAMP mediated by LHR CAMs is not as great, however, as that observed by the wild-type (wt) LHR challenged with a maximally stimulatory concentration of LH or hCG. Nonetheless, the extent of activation by the CAMs is clearly sufficient to mediate a physiological response. Although many of the naturally occurring LHR CAMs respond further to hormonal stimulation, not all do (Hirakawa *et al.*, 2002; Latronico and Segaloff, 2007; Latronico *et al.*, 1998; Zhang *et al.*, 2007). Only one CAM, D578H, has thus far been found to be associated with Leydig cell tumor formation in addition to gonadotropin-independent precocious puberty and this mutation, as reported independently by three groups, was identified as a somatic mutation (Kiepe et al., 2008; Liu et al., 1999; Richter-Unruh et al., 2002). This is in contrast to all other LHR CAMs, which are germline in origin and are not associated with tumors. Although it was originally reported that hLHR(D578H) activated signaling pathways different than those activated by the wt receptor and other LHR CAMs (Liu et al., 1999), subsequent studies have shown this not to be the case (Hirakawa and Ascoli, 2003; Hirakawa et al., 2002). Therefore, it is most likely that the oncogenic properties of the D578H mutation are attributable to its somatic origin.

LHR CAMs, both naturally occurring and of laboratory design, have proven to be valuable tools toward elucidating the structural features of the receptor that are associated with increased activation of Gs. In collaboration with our laboratory and others, Dr. Francesca Fanelli has utilized molecular dynamic simulations to mark those features common to LHR CAMs that set them apart from the wt receptor. These studies suggest that the active state of the LHR (with respect to Gs activation) is associated with the disruption of a salt bridge that normally links the cytoplasmic ends of TM3 and TM6, concomitant with an increase in the solvent accessible surface area of residues at the cytoplasmic ends of these helices, suggesting an opening in the intracellular crevice between TM3 and TM6 (Ascoli et al., 2002; Hirakawa et al., 2002; Latronico and Segaloff, 2007; Latronico et al., 1998; Zhang et al., 2007). It is presumed that the LH or hCG-mediated activation of the LHR would lead to similar structural changes, although it should be pointed out that the state of activation of the LHR CAMs is intermediate in nature compared with a maximally stimulated wt receptor.

As has been reported for many other GPCRs, the LHR self-associates into dimers and oligomers (referred to herein as dimers for simplicity; Guan et al., 2009; Urizar et al., 2005). LHR dimers are present on the cell surface, where they have been shown to be functional and to display allosteric properties. LHR dimers can be detected in the ER as well, suggesting that they form early in the biosynthetic pathway and are transported to the cell surface already preassociated. Because the LHR dimerization initiates in the ER, the coexpression of a misfolded mutant that is retained in the ER with a wt receptor results in a dominant-negative effect on cell surface expression of the wt LHR (Guan et al., 2009; Tao et al., 2004). A dominant-negative effect on wt LHR expression would also be predicted if one or more splice variants of the LHR known from expression studies of cDNAs in heterologous cells to be retained in the ER were physiologically expressed (Apaja et al., 2006; Dickinson et al., 2009; Minegishi et al., 2007; Nakamura et al., 2004).

Once at the cell surface, several lines of evidence suggest that the propensity for LHR dimerization is not affected by the activation state of the LHR (Guan et al., 2009). One line of evidence utilized BRET saturation curves to compare LHR CAMs with wt receptor and with a signaling inactive mutant (D405N,Y546F) that maintains normal cell surface expression and hormone binding affinity. It was found that the $BRET_{50}$'s, reflective of the relative affinities of receptor self-association, were comparable. However, the ability of the LHR to dimerize may lead to alterations in activity resulting from the ability of receptor protomers with a dimeric complex to exert allosteric effects on each other (Zhang et al., 2009).

LHR CAMs have been an extremely valuable tool for elucidating the mechanisms of underlying LHR signaling. In the following sections, general guidelines and specific methods for analyzing the activities of CAMs are described.

2. General Principles for Quantifying Receptor Activation

The definition of a GPCR CAM is that it possesses higher basal activity than the cognate wt receptor. However, the basal, hormone-stimulated, and constitutive activities of the LHR (or any other GPCR) are highly dependent on cell surface receptor numbers (Feng et al., 2008; Zhang et al., 2007). Therefore, in order to analyze the potential constitutive activity of an LHR mutant, it is critical that the experiment be performed in a manner that controls for the cell surface expression densities of wt and mutant receptors. For example, if the cell surface expression of the putative CAM were greater than the wt receptor, it would not be possible to unambiguously conclude that an observed increased in basal activity was due to constitutive activity because the increased basal activity could be a result of the higher cell surface expression of the mutant. One method that has been used in some cases to compare basal activities of a GPCR mutant with the wt receptor when the receptor numbers do not match has been to compare the ratios of cAMP/cell surface receptor expression. However, this is not valid unless one can be certain that the cAMP production of the wt and mutant receptors are linear with respect to cell surface receptor density. As shown in Fig. 12.1, the basal cAMP levels measured levels in cells expressing different densities of the wt hLHR cannot be fit by a straight line. Thus, this approach would not work for the hLHR. More valid results are obtained if one sets up the experiments so that either (i) the transfections are performed to achieve the same cell surface densities of the wt and putative CAM within that experiment, or (ii) the transfections are performed so that the activities of each receptor are measured over a range of cell surface receptor densities. Of these, the latter approach yields the most robust results and, if one is only measuring basal activity, it is actually easier

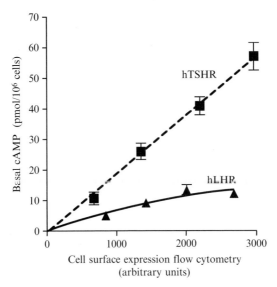

Figure 12.1 Basal cAMP as a function of cell surface density of the hLRH and hTSHR. HEK293 cells were transiently transfected with increasing amounts of the indicated plasmids, which were myc-tagged. Within a given experiment, both cell surface receptor expression, as determined by flow cytometry, and basal cAMP production were assayed. The cAMP levels of cells transfected with empty vector have been subtracted from the cAMP values shown. The data shown are the mean ± S.E.M. of triplicate determinations within a given experiment that is representative of at least three independent experiments. Adapted with permission from Feng *et al.* (2008). Copyright 2008, The Endocrine Society.

to perform. However, if one is going to perform dose–response curves on the wt and putative CAMs, then it becomes essential to utilize the first approach and match the cell surface receptor levels. Although matching cell surface receptor numbers sounds straightforward, it can be difficult to achieve and, unless the receptor numbers turn out to be tightly matched in the experiment, the experiment may not be usable.

Examples of each of the two strategies are shown below. In Table 12.1, three different experiments are shown in which cells expressing hLHR(wt) and hLHR(L457R) were transfected into HEK293 cells to yield similar levels of maximal cell surface ^{125}I-hCG binding. Because the binding affinities had been shown to be the same (Latronico *et al.*, 1998), the maximal cell surface binding capacities could be interpreted as proportional to the numbers of cell surface receptors. Under these conditions, basal and maximal hCG-stimulated cAMP levels were determined. Because the cell surface densities on L457R cells were, if anything, slightly lower than the hLHR(wt) cells, the increased basal cAMP levels in the L457R cells could readily be interpreted as increased basal activity of L457R as compared to

Table 12.1 Basal and hCG-stimulated cAMP in cells designed to express similar cell surface densities of hLHR(wt) and hLHR(L457R)

hLHR	Cell surface ^{125}I-hCG (ng bound/10^6 cells)	Basal cAMP (pmol/10^6 cells)	hCG-stimulated cAMP (pmol/10^6 cells)
wt	3.25	2.94	156
L457R	2.89	40.4	35.8
wt	3.18	5.19	124
L457R	2.74	47.4	32.7
wt	2.10	9.10	186
L457R	1.90	66.4	69.5

HEK293 cells were transiently transfected using plasmid concentrations that would yield similar cell surface densities of wt and L457R receptors within a given experiment. In each experiment, maximal ^{125}I-hCG binding to intact cells was determined as well as basal and hCG-stimulated cAMP. Adapted with permission from Latronico et al. (1998). Copyright 1998, The Endocrine Society.

wt hLHR. Note that even though L457R is a relatively strong CAM as compared to other hLHR CAMs, its activity is nonetheless not as great as the wt receptor challenged with a stimulatory concentration of hCG. The data in Table 12.1 also show that hLHR(L457R) does not respond further to hCG stimulation. In this respect, it is one of a few hLHR CAMs that are hormonally unresponsive (Hirakawa et al., 2002; Latronico and Segaloff, 2007; Latronico et al., 1998; Zhang et al., 2007). Using this paradigm of matching cell surface receptor numbers between cells expressing a mutant (in this case L457R) and wt hLHR, the experiment shown in Fig. 12.2 was performed in which full dose–response curves for cells expressing the two receptors were generated. These data (which are expressed as fold increases in cAMP relative to wt basal levels) further support the conclusions drawn from Table 12.1.

The experiments depicted in Fig. 12.3 were obtained using the other experimental paradigm in which the basal and hCG-stimulated cAMP levels were determined in cells expressing varying cell surface densities of wt hLHR or four different hLHR CAMs. The data in Fig. 12.3 readily depict how the basal levels of cAMP elicited by the four hLHR CAMs are elevated when compared to wt receptor over the range of cell surface densities tested. However, whereas the basal cAMP levels as a function of cell surface receptor densities yield straight lines for three of the CAMs, they yield a hyperbolic curve for the L457R CAM. Therefore, while it is possible to rank the relative activities of the M398T, I542L, and T557I CAMs (i.e., I542I > M398T > T577I), a comparison of L457R with three other

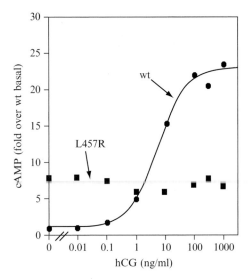

Figure 12.2 Dose–response curves for wt hLHR and the hLHR CAM L457 as determined in cells expressing matching levels of cell surface receptors. HEK293 cells were transiently transfected using plasmid concentrations that would yield similar cell surface densities of wt and L457R receptors. Basal and hCG-stimulated cAMP were determined and are expressed as the fold increase relative to the basal levels of cAMP elicited by hLHR(wt). Adapted with permission from Latronico *et al.* (1998). Copyright 1998, The Endocrine Society.

CAMs depends upon the cell surface density examined (i.e., L457R activity > T557I activity at lower receptor densities, but L45R activity is comparable to T557I activity at higher receptor densities).

Regardless of which strategy is employed to quantify constitutive activity, one will need to be able to quantify cell surface receptor expression and to quantify cAMP production. Although other methods have also been described, we typically use either of two approaches to quantify cell surface receptors. If the binding affinities are known to be the same as wt LHR, then one can measure the maximal binding of ^{125}I-hCG to intact cells expressing wt or mutant receptor. The other approach we use to quantify cell surface receptor is flow cytometry performed on nonpermeabilized cells, which entails the use of LHR constructs with small epitope tags such as myc, HA, or FLAG fused to the N-terminus of the mature protein (i.e., inserted between the signal peptide and the N-terminus of the mature protein). For the LHR, the placement of small epitope tags such as these on the N-terminus does not adversely affect cell surface expression, binding affinity, or signal transduction. (Oddly, the placement of these small tags on the C-terminus does, however, impede cell surface receptor expression whereas the placement of large additions, such as GFP or luciferase, on the C-terminus does not impair cell surface LHR expression.)

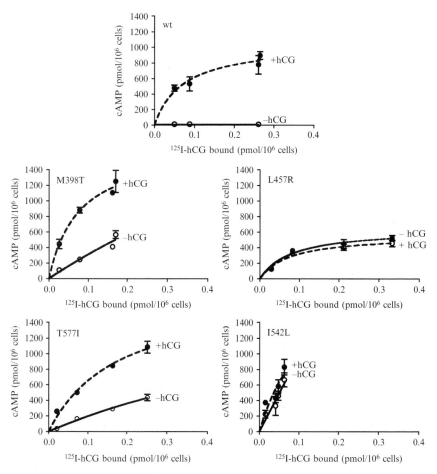

Figure 12.3 Basal activities of wt hLHR and hLHR CAMs as determined over a range of cell surface receptor densities. HEK293 cells were transiently transfected using plasmid concentrations designed to yield a range of cell surface receptor expression. Basal cAMP levels as a function of cell surface receptor densities are shown. This research was originally published by Zhang et al. (2007). Copyright 2007, The American Society for Biochemistry and Molecular Biology.

3. Modifying Cell Surface Expression Levels of Recombinant hLHR and Mutants Thereof

Recombinant hLHR can be introduced into cultures of gonadal cells by infection with adenovirus or they can be introduced into heterologous cells such as HEK293 cells, CHO cells, or COS-7 cells by transfections with expression plasmids. Unlike the more complex activation of the ERK1/2 pathway by LHR stimulation, which is cell-type specific (Andric and Ascoli, 2008b; Shiraishi and Ascoli, 2008), the direct activation of Gs by the LHR does not appear to be dependent on the cell type in which the LHR is expressed. Indeed, hLHR mutants thought to be CAMs due to their identification in boys with gonadotropin-independent precocious puberty (resulting from Gs activation by the endogenous hLHR in Leydig cells) exhibit constitutive activity when expressed in heterologous cells. Similarly, mutants defined as constitutively active in transfected heterologous cells exhibit constitutive activity when infected into gonadal cells (Hirakawa et al., 2002). Given the validation of the equivalency of quantifying recombinant hLHR CAM activity in gonadal versus heterologous cells and the greater ease and reduced cost of using heterologous cells, the latter is the system of choice for evaluating hLHR CAMs. Because the heterologous expression of the LHR in HEK293 cells has most typically been used for the evaluation of hLHR CAMs and it is routinely used in our laboratory, it is this system that will be described herein. Although other methods can be used to transiently transfect 293 cells, we typically use a calcium phosphate method because it is inexpensive and yields sufficiently high percentages of transfected cells to readily allow for receptor characterization. Therefore, it is this method that is described herein.

3.1. Growth of HEK293 cells

HEK293 cells are available from the American Type Culture Collection (ATCC catalog number CRL-1573). They are maintained in growth media consisting of high-glucose DMEM containing 50 µg/ml gentamicin, 10 mM HEPES, and 10% newborn calf serum. For experiments, cells are plated onto 6-well dishes at a density (worked out empirically) such that at the time of transfection (24–72 h after plating) the cells are 50–80% confluent. To ensure that cells remain attached throughout the experiment, the wells are precoated for 15 min in 1% gelatin dissolved in calcium and magnesium free Dulbecco's phosphate buffered saline, pH 7.1 (CMF-D-PBS) prior to plating the cells.

3.2. cDNA encoding LHR

Human LHR cDNA's were kindly given to us independently by Ares Advanced Technology (Ares-Serono Group, Randolph, MA) and by Dr Aaron Hsueh (Stanford University; Jia et al., 1991). For expression studies (such as those used to determined constitutive activity), the receptor cDNA's are cloned into the expression vector pcDNA3.1/neo (Invitrogen). Mutations are introduced using standard methodologies and colonies of transformed bacteria are screened using automated DNA sequencing and then recloned to ensure true clonality. Sequences of the full coding regions are verified by sequencing. Highly purified plasmids are prepared using Qiagen kits and the full coding sequences of each plasmid preparation are always determined.

3.3. Transient transfections of HEK293 cells

Cells are transfected using the calcium phosphate method of Chen and Okayama (1987), with the modification that the overnight transfection is performed with cells incubated at 5% CO_2 rather than 2.5% CO_2. Cells should be 50–80% confluent on the day of transfection. The following transfection cocktail volumes are written per one well of a 6-well plate. One would multiply these volumes by the numbers of wells to be transfected plus one or two extra. In a sterile plastic tube in the tissue culture hood, combine 86 μl sterile H_2O, 10 μl 2.5 M $CaCl_2$, and 4 μl plasmid DNA (1 μg/μl pcDNA3.1 containing the wt or mutant hLHR cDNA in sterile H_2O) and swirl gently. Placing a pipette at the bottom of the tube, add 100 ml of a 2× preparation of BSS (280 mM NaCl, 1.5 mM Na_2HPO_4, 50 mM BES, pH 6.95). Expel the liquid and gently bubble air as the pipette is slowly lifted. Incubate the transfection cocktail 15 min in the hood without disturbing. In the meantime, prepare another sterile tube containing 1.8 ml of growth medium per number of wells being prepared and keep it in the incubator loosely capped until needed. Add the transfection cocktail dropwise to the medium and mix by pipeting gently up and down or by swirling. Add 2 ml of the transfection mix to each well, mixing gently between additions. Place the cells in a 5% CO_2 humidified incubator and incubate overnight 16–20 h. The next morning a fine precipitate should be visible. Wash the cells twice and then add 2 ml growth media. Place the cells back in the incubator for another 24 h, at which time they can be used for experiments.

3.4. Manipulating transient transfection conditions to yield cells with comparable cell surface densities of wt and mutant LHR

As discussed above, one means of determining if a mutant LHR is constitutively active is to compare the basal cAMP levels in cells expressing the mutant to cells expressing the same cell surface density of wt LHR. Typically, we first perform pilot experiments in which the concentrations of plasmids encoding the mutant and wt receptors are varied and cell surface receptor expression is measured. In all cases, the total amount of plasmid transfected is held constant by the addition of empty vector. Importantly, we start these analyses with large preparations of purified plasmids because a change in the plasmid preparation can result in altered levels of receptor expression. Initially, we would test a wide range of plasmid concentrations for both the wt and mutant LHRs. Then, several concentrations narrowly flanking the concentrations that seem to give somewhat similar cell surface expression levels are tested until concentrations of plasmids encoding wt and mutant LHRs are identified that yield closely matched levels of cell surface receptor expression. It has been our experience that cell surface receptor densities need to be very tightly matched because just slight differences in cell surface expression can translate into notable differences in basal and/or hormone-stimulated receptor activities.

Once these conditions have been found, one is ready to perform the actual analyses of CAM activities. However, one cannot assume that the levels of receptor expression (and the matching of cell surface densities of wt and mutant receptor) will be the same from experiment to experiment. Unfortunately, it has been our experience that, even when using the same plasmid preparations and carefully controlled conditions, there is inevitably some variability in cell surface receptor expression between experiments. Therefore, each experiment should be set up such that cell surface receptor expression and signaling activities are determined within the same experiment. The reason that this approach can be problematic is precisely because the wt and mutant LHR cell surface expression may not necessarily match in each experiment even when great care is taken to replicate conditions tightly. If the receptor density of the mutant turns out to be somewhat less than the wt receptor and the basal cAMP of the mutant-expressing cells is greater, then it is still possible to conclude that the mutant is constitutively active. However, if the mutant cell surface expression is greater than the wt receptor, the data would be ambiguous and the experiment would not be usable.

3.5. Manipulating transient transfection conditions to yield cells with a similar range of cell surface levels of wt and mutant LHR

If one is not performing full dose–response curves, then the most rigorous (and most time efficient) method for analyzing constitutive activity is to perform the studies where the cell surface receptor expression and basal activities for the wt and mutant receptor are analyzed over a range of receptor densities (as in Fig. 12.3). Although a given experiment may be larger than if one were matching a given receptor density as described above, in the long run it is more time efficient because one does not incur the risk of having to discard experiments that ultimately do not yield a sufficiently close match.

Here too one initially would examine the cell surface receptor expression of the wt and mutant(s) receptors over a range of plasmid concentrations (always using empty vector to maintain a constant total amount of plasmid). If, for example, the mutant were expressed at much lower cell surface levels than the wt receptor, one would then select lower concentrations of wt receptor plasmid so that the basal activities of the receptors could be compared over the same range of receptor expression. Once one has identified the plasmid concentrations of each that would yield several data points in the receptor range selected, then the experiments can be set up where cell surface receptor expression and basal cAMP levels were analyzed in the same experiment. A comparison of the curves will readily and convincingly demonstrate whether or not the mutant exhibited a greater basal activity.

4. Quantifying Cell Surface hLHR Expression

4.1. Measurement of maximal ^{125}I-hCG binding to intact cells

Highly purified hCG (purchased from Dr Al Parlow and the National Hormone and Pituitary Program of NIDDK/NIH; http://www.humc.edu/hormones/) is iodinated by lactoperoxidase (a detailed protocol for the iodination is available upon request) and stored at $-80\ °C$. Cells are plated on 6-well plates that have been precoated with gelatin and transfected as described above. Typically, three wells are used per binding point where two are for total binding and one for nonspecific binding. For each set of cells (i.e., wt or mutant), two additional wells are plated with cells to be used to determine the number of cells per well. The optimal ^{125}I-hCG binding conditions differ for LHR of different species. The protocol described herein is for the hLHR. A stock solution of ^{125}I-hCG is prepared in buffered isotonic saline (0.15 N NaCl, 20 mM HEPES, pH 7.4) containing 1% BSA (BIS/BSA). For nonspecific binding, a crude preparation of hCG (Sigma catalog #CG-10) is prepared such that one vial of 10,000 I.U. is dissolved in 10 ml BIS/BSA. On the day of the experiment, the cells per well are determined for each group of cells. For the binding assay, cells are first washed at RT twice with Waymouth's media lacking sodium biocarbonate and containing 1% BSA and 1.9 ml of the same media is added to each well. To the wells determining nonspecific binding, 50 μl of crude hCG is added and to the well determining total binding 50 μl of BIS/BSA is added. All wells then receive 50 μl of stock ^{125}I-hCG (to yield a final concentration of 500 ng/ml) and the cells are incubated 1 h at RT. To terminate the binding reaction, the cells are placed on ice and washed three times with cold Hanks Balanced Salt Solution containing 1% BSA. After aspiration of the wash media, 100 ml 0.5 N NaOH is added to each well and the solubilized cells are picked up with a Q-tip (cut in half) and transferred to a plastic tube with the cotton swab at the bottom of the tube. Each well is wiped with another cut Q-tip, which is added to the same tube. The tubes are counted in a gamma counter and the cpm of the nonspecific binding well are subtracted from the average of the total cpm bound to yield specifically bound cpm. The specific cpm are divided by the specific activity of the ^{125}I-hCG on that date to yield the ng ^{125}I-hCG bound per well. These data are then corrected for the cells per well to yield ng ^{125}I-hCG bound/10^6 cells. By reducing the data in this manner, it is then possible to (at least roughly) compare data between experiments.

4.2. Measurement of cell surface epitope-tagged hLHR by flow cytometry

As noted earlier, measurement of cell surface hLHR by flow cytometry is performed using hLHR constructs modified so that the N-terminus of the mature protein contains an epitope tag such as myc, FLAG, or HA. Cells are plated in gelatin-coated 6-well plates, using two wells per sample, and transfected as described above. On the day of the experiment, the cells are washed once with filtered PBS for immunohistochemistry (PBS-IH, 137 mM NaCl, 2.7 mM KCl, 1.4 mM KH$_2$PO$_4$, 4.3 mM Na$_2$HPO$_4$, pH 7.4). After adding 1 ml PBS-IH to each well, detach the cells by gently pipetting up and down. Transfer the contents to 12 × 75 plastic tubes on ice, combining the contents of the two wells for a given sample into one tube. Centrifuge the cells for 5 min at 4 °C at 500 × g and decant the supernatant by turning the test tube upside down on a paper towel. Do not aspirate the supernatant because the pellet is soft and it can easily be aspirated. Resuspend the cell pellet 1 ml blocking solution (10% BSA in PBS-IH) and incubate 1 h at RT. Centrifuge and decant the supernatant as described. Resuspend the cells in PBS-IH containing 0.5% BSA (PBS-IH/BSA) containing the appropriate dilution of primary antibody and incubate 1 h at 4 °C in the dark. For the HA tag, we use HA.11 monoclonal antibody from Covance diluted 1:500 and for the myc tag we use 9E10 monoclonal antibody from Santa Cruz diluted 1:20. Cells are washed with PBS-IH/BSA and incubated an additional 1 h at 4 °C in the dark with secondary antibody in PBS-IH/BSA (typically phycoerythrin conujugated goat antimouse IgG from BD Biosciences diluted 1:500). Cells are washed with PBS-IH/BSA and resuspended in 1 ml PBS-IH/BSA. One tube at a time, gently vortex the sample for about 5 s. Using a P1000 Pipetman with the tip having been enlarged by cutting the end off, draw up the cell suspension and force the sample through a 70-μm BD Falcon Cell Strainer into a clean plastic tube on ice. Use a clean filter for each sample. The filters can be washed, dried, and reused for subsequent experiments. Keep the samples covered with foil to prevent bleaching. Prior to sorting, dead cells are stained by the addition of 50 μl propidium iodine (50 μg/ml) to 1 ml of cells. After vortexing, A Becton Dickinson DiVa fluorescence-activated cell sorter with a 488-nm laser is used to quantify cell surface expression in 10,000 cells from each transfection. Control gating is set using cells transfected with empty vector and stained with antibody as described. Arbitrary fluorescence units describing total cell surface fluorescence are determined as the product of the percent cells gated and the geometric mean fluorescence of the sample minus the respective product in the control (empty vector) group.

5. Quantifying cAMP Production in Cells Expressing the hLHR

There are a number of methods by which cAMP can be measured. We utilize an RIA to quantify acetylated samples because this is an extremely sensitive measure of cAMP levels that allows for the determination of cAMP over a very wide range of concentrations.

5.1. Cell incubations for the measurement of basal and/or hormone-stimulated cAMP

Cells are plated onto gelatin-coated 6-well plates and transfected as described earlier. Typically, we use two or three wells per variable for cAMP determinations, with each well assayed in duplicate in the RIA. In addition, two wells per variable are plated for determination of the numbers of cells per well. On the day of the experiment, wash the cells twice with 2 ml portions of Waymouth's media containing sodium bicarbonate and 1% BSA. Add 1 ml of the same media containing 0.5 mM MIX (isomethylbutylxanthine) to each well and set the cells in a 37 °C humidified CO_2 incubator for 15 min. Remove the plates and add 50 μl BIS/BSA to each well for the determination of basal cAMP and 50 μl hCG in BIS/BSA (to give a final concentration of 100 ng/ml) for the determination of hormone-stimulated cAMP. Incubate the cells a further 60 min in a 37 °C humidified CO_2 incubator.

If a cAMP RIA is to be used for the measurement of cAMP, the samples need to be extracted as follows. At the end of the incubation, set the cells on ice, quickly aspirate the media, and add 1.5-ml ice-cold 0.5 N perchloric acid containing 180 μl/ml theophylline to each well. Using a Pasteur pipette, transfer the contents of each well to a 12 × 75 glass tube in an ice bath. Using a Pipetman, transfer 1.0 ml of each sample to a fresh 12 × 75 glass tube on ice. To this, add 500 μl 0.72 M KOH/0.6 M KHCO$_3$ and vortex. Centrifuge for 10 min at 4 °C. Pipet 1.2 ml of each supernatant to a fresh 12 × 75 glass tube. Once extracted, the samples can be used for acetylation and setting up the RIA that day or stored −20 °C.

Although there are commercial kits available for cAMP RIA's, we hold costs down by preparing our own reagents rather than purchasing a kit. The cAMP RIA that we employ utilizes ^{125}I-cAMP that we iodinate and commercially purchased anti-cAMP polyclonal antibody (Strategic Biosolutions). For greater sensitivity, the samples and standards are acetylated. Detailed protocols describing the iodination of cAMP and the establishment of the RIA can be obtained upon request. If using a kit, the methodology included in the kit should be followed for performing the RIA.

6. Experimental Strategies for Characterizing the Attenuating Effects of a Signaling Inactive hLHR on a Coexpressed wt or CAM hLHR

Two mutants of the hLHR (D405N and Y546F) have previously been shown to have normal cell surface expression and hormone binding affinities, but partially attenuated responses to hCG, as indicated by increased EC_{50}'s and decreased R_{max}'s for hCG-stimulated cAMP production (Min and Ascoli, 2000). We showed that, over a range of cell surface receptor expression, the basal cAMP levels in cells expressing each of these mutants were also decreased as compared to the wt receptor (Zhang et al., 2009). Furthermore, when the D405N or Y546F mutations were introduced into hLHR CAMs, the constitutive activities of the CAMS were reduced. As shown in Fig. 12.4, the basal activities of the three hLHR CAMs are reduced when D405N is introduced into the CAM. Similar results were obtained for Y546F. These results suggest that D405N and Y546F each stabilize a resting state of the hLHR, mimicking the effects of the binding of an inverse agonist to the receptor.

Although D405N and Y546F each exhibited a partial attenuation of hCG-stimulated cAMP as compared to cells expressing the same cell surface density of wt hLHR (Zhang et al., 2009), a much more profoundly signaling inactive hLHR mutant was identified by merging the two mutations in one receptor construct. Interestingly, hLHR(D405N,Y546F) was expressed at the cell surface at levels similar to that of the wt receptor and it bound hCG with the same affinity. However, dose–response curves examining hCG-stimulated cAMP performed using cells transfected to yield closely matched cell surface levels of wt or mutant hLHR indicated little or no cAMP production in the cells expressing hLHR(D405N,Y546F). Thus, this mutant represents a unique hLHR in that it is profoundly signaling impaired due to stabilization of a resting conformation of the hLHR and not due to impairments in cell surface trafficking or hCG binding.

In order to determine what effects, if any, coexpression of the signaling inactive hLHR(D405N,Y546F) might have on wt hLHR or CAMs of the hLHR, we created 293 cells stably transfected with either hLHR(D405N, Y546F) or with empty pcDNA3.1 vector. In one set of experiments, each

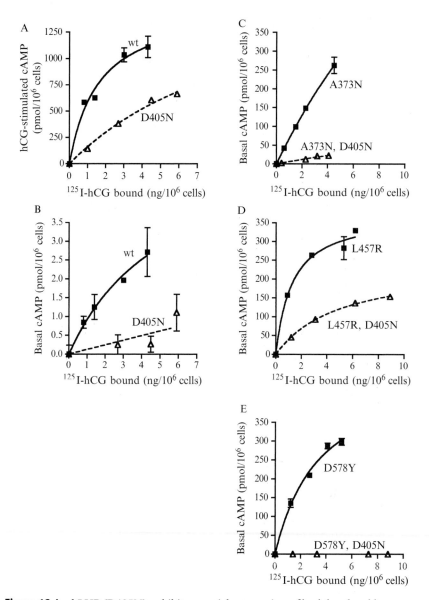

Figure 12.4 hLHR(D405N) exhibits a partial attenuation of both basal and hormone-stimulated cAMP over a range of receptor densities and, when introduced into hLHR CAMs, attenuates their constitutive activities. HEK293 cells were transiently transfected with varying plasmid concentrations of the indicated constructs. Intracellular cAMP concentrations as a function of increasing cell surface receptor density (determined in the same experiment by ^{125}I-hCG binding to intact cells) were measured in response to a saturating concentration of hCG (Panel A) and under basal conditions (Panels B, C, D, and E). Data shown are the mean ± S.E.M. of triplicate determinations within a single experiment that contained all panels. The experiment shown is representative of two independent experiments. This research was originally published by Zhang *et al.* (2009). Copyright 2009, Elsevier.

of these cell lines was then transiently transfected with HA-hLHR(wt) or with empty vector under conditions that would match the cell surface expression of HA-hLHR(wt) in the hLHR(405N,Y546F) cell line and the empty vector cell line. The results of these experiments are shown in Fig. 12.5, which show that hCG-stimulated cAMP by wt hLHR is attenuated by the coexpression of the signaling inactive hLHR. Because these experiments were performed under conditions where the cell surface expression of wt receptor was the same in the cells with or without the signaling inactive mutant, we can conclude that the attenuation of signaling was independent of any changes in trafficking of the wt receptor. It is possible that the signaling inactive mutant might sequester Gs without causing its activation, thus effectively decreasing the availability of Gs for activation by the wt hLHR. However, when the same experiment was performed under conditions where $G\alpha_s$, $G\beta$, and $G\gamma$ were cotransfected

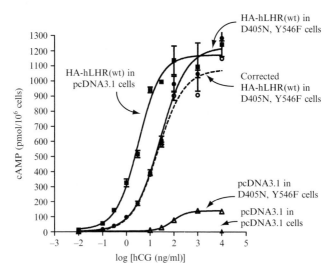

Figure 12.5 Attenuation of hCG-stimulated cAMP by hLHR(wt) when coexpressed with the signaling inactive hLHR(D405N,Y546F). In the same experiment, HEK293 cells stably expressing hLHR(D405N,Y546F) or empty vector pcDNA3.1 were each transfected with HA-hLHR(wt). Flow cytometry data for cell surface HA-hLHR(wt) expressed in the hLHR(D405N,Y546) cells versus empty vector cells were 227 and 240, respectively. Accumulation of intracellular cAMP in response to increasing concentrations of hCG was determined and is shown by the solid symbols and solid lines as indicated. The open circles and dashed line depict corrected hCG-stimulated cAMP values for D405N,Y546F cells transfected with HA-hLHR(wt) in which the values obtained for hCG-stimulated cAMP in D405N,Y546F cells transfected with empty vector were subtracted. Data shown are the mean ± S.E.M. of triplicate determinations within a single experiment, representative of three independent experiments. This research was originally published by Zhang *et al.* (2009). Copyright 2009, Elsevier.

with hLHR(wt), a similar attenuation of signaling through the wt receptor was observed (Zhang et al., 2009).

In another set of experiments, one of three different myc-tagged hLHR CAMs was transiently transfected into the hLHR(D405N,Y546F) cells or the empty vector cells under conditions that would yield varying degrees of cell surface expression of the CAM. As shown in Fig. 12.6, there was a decreased constitutive activity of each of the CAMs when coexpressed in the cells stably expressing the signaling impaired hLHR mutant.

It was determined by BRET saturation curves that the signaling inactive hLHR mutant forms heterodimers with the wt or constitutively active forms of the hLHR (Zhang et al., 2009). It was further determined that

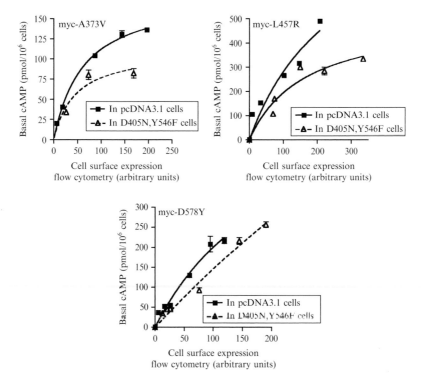

Figure 12.6 Coexpression of the signaling inactive hLHR(D405N,Y546F) with hLHR CAMs causes an attenuation of their constitutive activities. HEK293 cells stably expressing hLHR(D405N,Y546F) or empty vector were each transiently transfected with increasing concentrations of myc-hLHR(A373V; Panel A), myc-hLHR(L457R; Panel B), or with myc-hLHR(D578Y; Panel C). Basal levels of cAMP were measured as a function of cell surface receptor density of the CAM (determined by flow cytometry). Data shown are the mean ± S.E.M. of triplicate determinations within a single experiment, representative of three independent experiments. This research was originally published by Zhang et al. (2009). Copyright 2009, Elsevier.

the MC3R, another GPCR that signals through Gs, does not heterodimerize with the hLHR or hLHR(D405N,Y546F), and the signaling activity of the MC3R is not attenuated when transiently transfected into the hLHR (D405N,Y546F) cells. Again, this was determined by comparing the dose–response curves of hLHR(D405N,Y546F) cells or empty vector cells expressing carefully matched cell surface receptor levels of myc-MC3R.

Taken altogether, these data suggest that there is an allosteric communication between receptor protomers within an hLHR dimeric complex such that the signaling properties of hLHR(D405N,Y546F), which is stabilized in a resting state, attenuates the signaling properties of an associated hLHR in the wt or constitutively active state.

ACKNOWLEDGMENT

Studies discussed from the author's laboratory were supported by NIH grants DK068614 and HD22196.

REFERENCES

Andric, N., and Ascoli, M. (2008a). Mutations of the lutropin/choriogonadotropin receptor that do not activate the phosphoinositide cascade allow hCG to induce aromatase expression in immature rat granulosa cells. *Mol. Cell. Endocrinol.* **285**, 62–72.

Andric, N., and Ascoli, M. (2008b). The luteinizing hormone receptor-activated extracellularly regulated kinase-1/2 cascade stimulates epiregulin release from granulosa cells. *Endocrinology* **149**, 5549–5556.

Apaja, P. M., Tuusa, J. T., Pietila, E. M., Rajaniemi, H. J., and Petaja-Repo, U. E. (2006). Luteinizing hormone receptor ectodomain splice variant misroutes the full-length receptor into a subcompartment of the endoplasmic reticulum. *Mol. Biol. Cell* **17**, 2243–2255.

Ascoli, M., Fanelli, F., and Segaloff, D. L. (2002). The lutropin/choriogonadotropin receptor, a 2002 perspective. *Endocr. Rev.* **23**, 141–174.

Chen, C., and Okayama, H. (1987). High-efficiency transformation of mammalian cells by plasmid DNA. *Mol. Cell. Biol.* **7**, 2745–2752.

Dickinson, R. E., Stewart, A. J., Myers, M., Millar, R. P., and Duncan, W. C. (2009). Differential expression and functional characterization of luteinizing hormone receptor splice variants in human luteal cells: Implications for luteolysis. *Endocrinology* **150**, 2873–2881.

Donadeu, F. X., and Ascoli, M. (2005). The differential effects of the gonadotropin receptors on aromatase expression in primary cultures of immature rat granulosa cells are highly dependent on the density of receptors expressed and the activation of the inositol phosphate cascade. *Endocrinology* **146**, 3907–3916.

Feng, X., Muller, T., Mizrachi, D., Fanelli, F., and Segaloff, D. L. (2008). An intracellular loop (IL2) residue confers different basal constitutive activities to the human lutropin receptor and human thyrotropin receptor through structural communication between IL2 and helix 6, via helix 3. *Endocrinology* **149**, 1705–1717.

Guan, R., Feng, X., Wu, X., Zhang, M., Zhang, X., Hebert, T. E., and Segaloff, D. L. (2009). Bioluminescence resonance energy transfer studies reveal constitutive dimerization of the human lutropin receptor and a lack of correlation between receptor activation and the propensity for dimerization. *J. Biol. Chem.* **284**, 7483–7494.

Gudermann, T., Birnbaumer, M., and Birnbaumer, L. (1992). Evidence for dual coupling of the murine luteinizing hormone receptor to adenylyl cyclase and phosphoinositide breakdown and Ca^{+2} mobilization. *J. Biol. Chem.* **267,** 4479–4488.

Hirakawa, T., and Ascoli, M. (2003). A constitutively active somatic mutation of the human lutropin receptor found in Leydig cell tumors activates the same families of G proteins as germ line mutations associated with Leydig cell hyperplasia. *Endocrinology* **144,** 3872–3878.

Hirakawa, T., Galet, C., and Ascoli, M. (2002). MA-10 cells transfected with the human lutropin/choriogonadotropin receptor (hLHR): A novel experimental paradigm to study the functional properties of the hLHR. *Endocrinology* **143,** 1026–1035.

Jia, X. C., Oikawa, M., Bo, M., Tanaka, T., Ny, T., Boime, I., and Hsueh, A. J. (1991). Expression of human luteinizing hormone (LH) receptor: Interaction with LH and chorionic gonadotropin from human but not equine, rat, and ovine species. *Mol. Endocrinol.* **5,** 759–768.

Kiepe, D., Richter-Unruh, A., Autschbach, F., Kessler, M., Schenk, J. P., and Bettendorf, M. (2008). Sexual pseudo-precocity caused by a somatic activating mutation of the LH receptor preceding true sexual precocity. *Horm. Res.* **70,** 249–253.

Latronico, A., and Segaloff, D. (1999). Naturally occurring mutations of the luteinizing hormone receptor: Lessons learned about reproductive physiology and G protein-coupled receptors. *Am. J. Hum. Genet.* **65,** 949–958.

Latronico, A. C., and Segaloff, D. L. (2007). Insights learned from the L457(3.43)R, an activating mutant of the human lutropin receptor. *Mol. Cell. Endocrinol.* **260–262,** 287–293.

Latronico, A. C., Abell, A. N., Arnhold, I. J. P., Liu, X., Lins, T. S. S., Brito, V. N., Billerbeck, A. E., Segaloff, D. L., and Mendonca, B. B. (1998). A unique constitutively activating mutation in the third transmembrane helix of the luteinizing hormone receptor causes sporadic male gonadotropin independent precocious puberty. *J. Clin. Endocrinol. Metab.* **83,** 2435–2440.

Liu, G., Duranteau, L., Carel, J.-C., Monroe, J., Doyle, D. A., and Shenker, A. (1999). Leydig-cell tumors caused by an activating mutation of the gene encoding the luteinizing hormone receptor. *N Engl. J. Med.* **341,** 1731–1736.

Min, L., and Ascoli, M. (2000). Effect of activating and inactivating mutations on the phosphorylation and trafficking of the human lutropin/choriogonadotropin receptor. *Mol. Endocrinol.* **14,** 1797–1810.

Minegishi, T., Nakamura, K., Yamashita, S., and Omori, Y. (2007). The effect of splice variant of the human luteinizing hormone (LH) receptor on the expression of gonadotropin receptor. *Mol. Cell. Endocrinol.* **260–262,** 117–125.

Nakamura, K., Yamashita, S., Omori, Y., and Minegishi, T. (2004). A splice variant of the human luteinizing hormone (LH) receptor modulates the expression of wild-type human LH receptor. *Mol. Endocrinol.* **18,** 1461–1470.

Richter-Unruh, A., Wessels, H. T., Menken, U., Bergmann, M., Schmittmann-Ohters, K., Schaper, J., Tappeser, S., and Hauffa, B. P. (2002). Male LH-independent sexual precocity in a 3.5-year-old boy caused by a somatic activating mutation of the LH receptor in a Leydig cell tumor. *J. Clin. Endocrinol. Metab.* **87,** 1052–1056.

Segaloff, D. L. (2009). Diseases associated with mutations of the human lutropin receptor. *Prog. Mol. Biol. Transl. Sci.* **89C,** 97–114.

Shenker, A., Laue, L., Kosugi, S., Merendino, J. J., Jr., Minegishi, T., and Cutler, G. B., Jr. (1993). A constitutively activating mutation of the luteinizing hormone receptor in familial male precocious puberty. *Nature* **365,** 652–654.

Shiraishi, K., and Ascoli, M. (2008). A co-culture system reveals the involvement of intercellular pathways as mediators of the lutropin receptor (LHR)-stimulated ERK1/2 phosphorylation in Leydig cells. *Exp. Cell Res.* **314,** 25–37.

Tao, Y. X., Johnson, N. B., and Segaloff, D. L. (2004). Constitutive and agonist-dependent self-association of the cell surface human lutropin receptor. *J. Biol. Chem.* **279,** 5904–5914.

Themmen, A. P. N., and Huhtaniemi, I. T. (2000). Mutations of gonadotropins and gonadotropin receptors: Elucidating the physiology and pathophysiology of pituitary-gonadal function. *Endocr. Rev.* **21,** 551–583.

Urizar, E., Montanelli, L., Loy, T., Bonomi, M., Swillens, S., Gales, C., Bouvier, M., Smits, G., Vassart, G., and Costagliola, S. (2005). Glycoprotein hormone receptors: Link between receptor homodimerization and negative cooperativity. *EMBO J.* **24,** 1954–1964.

Zhang, M., Tao, Y. X., Ryan, G. L., Feng, X., Fanelli, F., and Segaloff, D. L. (2007). Intrinsic differences in the response of the human lutropin receptor versus the human follitropin receptor to activating mutations. *J. Biol. Chem.* **282,** 25527–25539.

Zhang, M., Feng, X., Guan, R., Hebert, T. E., and Segaloff, D. L. (2009). A cell surface inactive mutant of the human lutropin receptor (hLHR) attenuates signaling of wild-type or constitutively active receptors via heterodimerization. *Cell. Signal.* **21,** 1663–1671.

CHAPTER THIRTEEN

Assessing Constitutive Activity of Extracellular Calcium-Sensing Receptors *In Vitro* and in Bone

Wenhan Chang, Melita Dvorak, *and* Dolores Shoback

Contents

1. Introduction	254
2. Materials and Methods	258
2.1. Construction of activating CaSR mutant constructs	258
2.2. Cell transfection	258
2.3. Inositol phosphate (InsP) assays	258
2.4. Expression of Act-CaSR in osteoblasts	259
2.5. Assessment of Act-CaSR expression in transgenic osteoblasts: Immunoblotting	259
2.6. Assessment of skeletal phenotypes in transgenic mice	261
2.7. Statistics	261
3. Results	262
3.1. Assessment of activity of Act-CaSR *in vitro*	262
3.2. Transgenic expression of Act-CaSR in osteoblasts	263
3.3. Skeletal phenotype of transgenic Act-CaSR mice	263
4. Conclusions	263
Acknowledgments	265
References	265

Abstract

Constitutive activity of the extracellular calcium-sensing receptor (CaSR) has been studied in kindreds with the human disorder autosomal dominant hypocalcemia (ADH) and in an animal model called the *Nuf* mouse. These families generally showed reduced parathyroid hormone (PTH) secretion and excessive renal calcium (Ca^{2+}) excretion. Soft tissues calcifications in the kidney and basal ganglia are frequent (10–50% of ADH cases), and there is a single report of skeletal abnormalities in a family resulting in short stature and premature osteoarthritis. In the latter, a causative mechanism could not be determined.

Endocrine Research Unit, Department of Veterans Affairs Medical Center, Department of Medicine, University of California, San Francisco, California, USA

The phenotype of the *Nuf* mouse is one of ectopic calcifications and cataracts in addition to biochemical abnormalities (low serum Ca^{2+} and high serum phosphate concentrations). To better understand the role of CaSRs in the control of osteoblastic function, we generated a transgenic mouse model with constitutively active CaSRs in mature osteoblasts. An analysis of the skeletal phenotype of that mouse indicates that strong signaling by CaSRs in this cell lineage induces alterations in the bone homeostasis reflected in mild osteopenia in male and female mice during growth and in adulthood. These studies indicate that this approach can be readily adapted to assess CaSR actions in other cell systems.

1. INTRODUCTION

Studying the relationship between the serum $[Ca^{2+}]$ and parathyroid hormone (PTH) gave rise to the concept of a membrane mechanism for mediating sensing changes in the extracellular Ca^{2+} concentration ($[Ca^{2+}]_e$). Small increases in the $[Ca^{2+}]_e$ within the physiological range inhibit PTH secretion and cell proliferation. These events are transduced via multiple G protein-mediated signaling pathways, including the stimulation of phospholipase C, A2, and D (PLC, PLA2, and PLD) activities, intracellular Ca^{2+} release, and mitogen-activated protein (MAP) kinase activity. High $[Ca^{2+}]_e$ also inhibit adenylate cyclase activity in acutely dispersed bovine parathyroid cells (PTCs; Hofer and Brown, 2003). Cloning of the extracellular calcium-sensing receptor (CaSR) cDNA from bovine parathyroid glands (PTGs) established the molecular basis for extracellular Ca^{2+}-sensing. When expressed in exogenous cell systems after cDNA transfection, CaSRs couple minute changes in $[Ca^{2+}]_e$, in the sub-millimolar range, to signaling cascades similar to those in PTCs (Brown *et al.*, 1993).

Generalized or conditional knockout in PTCs of the *Casr* gene produces hypercalcemia and hyperparathyroidism (HPT) in mice due to the inability of the cells to respond to elevated serum $[Ca^{2+}]$ (Ho *et al.*, 1995). In humans, activating and inactivating CaSR mutations induce hypocalcemia and hypercalcemia, respectively. PTH levels are low (or inappropriately normal) in patients with activating mutations and high (or inappropriately normal) in patients with inactivating mutations (Egbuna and Brown, 2008). These observations clearly support the critical functional role of the CaSR in mediating Ca^{2+}-sensing and secretion in PTCs. The phenotype of the global CaSR knockout mouse and these genetic disorders of Ca^{2+}-sensing further establish the central role of the CaSR in renal Ca^{2+} handling. Inactivating mutations in the CaSR are associated with hypocalciuria, while activating mutations are accompanied by often profound hypercalciuria.

A mouse model of *Casr* activation has suggested potential roles for activated CaSRs *in vivo*. The Nuf model comprises *Nuf/+* (heterozygous) and *Nuf/Nuf* (homozygous) mice with one or both *Casr* alleles carrying a substitution of Leu at position 723 with Gln, respectively (Hough *et al.*, 2004). *Nuf/Nuf* mice demonstrate severe hypocalcemia, hyperphosphatemia, and low PTH levels. These animals also manifest sudden death, cataracts, and ectopic calcifications throughout its tissues. An elevated Ca × phosphate product may be responsible for these calcifications, or alternatively, excessive and uncontrolled CaSR signaling may produce these pathologic soft tissue calcifications. Families with heterozygous activating CaSR mutations also show basal ganglia and renal calcifications which may be due to the same factors. These observations shed light on the phenotype of global CaSR activation but do address the role of CaSRs in specific tissues.

CaSRs are broadly expressed in tissues outside the PTG, including bone, cartilage, brain, kidney, skin, breast, gastrointestinal tract, and smooth and cardiac muscle. Understanding the role of CaSRs in these tissues using generalized CaSR knockout mice (CaSR+/− or CaSR−/−) is challenging. At baseline, these mice have chronic mild to severe HPT (Ho *et al.*, 1995). Their HPT obscures the role of CaSRs in tissues like bone and cartilage because high PTH levels can affect those target organs. Assessment of CaSR functions selectively in these and other tissues is now feasible, owing to a newly developed floxed CaSR mouse model (Chang *et al.*, 2008). Using this mouse model, we demonstrated that CaSRs in osteoblasts and chondrocytes are critical regulators of skeletal development.

Based on its topology and sequence homology (Fig. 13.1), the CaSR is classified in family C of the G protein-coupled receptor (GPCR) superfamily. This receptor has a large extracellular domain (ECD, ≈600 amino acids), the classic seven-transmembrane domain (7-TMD, ≈250 amino acids), and a long carboxyl-terminal tail (C-tail, ≈200 amino acids) (Brown *et al.*, 1993; Chang and Shoback, 2004; Hofer and Brown, 2003). The CaSR likely functions as a dimer (homo- or heterodimer) like other family C GPCRs. Both dimerization and glycosylation are critical steps for efficient cell-surface targeting and ligand binding to the receptor.

As a typical GPCR, the 7-TMD of the CaSR has three extracellular (EC1–3) and three intracellular (IC1–3) loops that are responsible for transducing the extracellular stimulus (after ligand binding) into intracellular signals by interacting with different G-protein subunits (Gαq, Gαi, Gα11, G$\beta\gamma$, etc.). Mutational analysis of CaSR mutants revealed amino acid residues in IC2 and IC3 that are critical for PLC signaling and efficient cell-surface expression of the receptor (Chang *et al.*, 2000). Substitutions of specific residues in these receptor domains with alanine or nonconserved amino acids altered efficiency of the CaSR's coupling to downstream effectors. Studies of CaSRs in which the carboxyterminal tail is either truncated or mutated revealed the importance of an alpha-helical structure

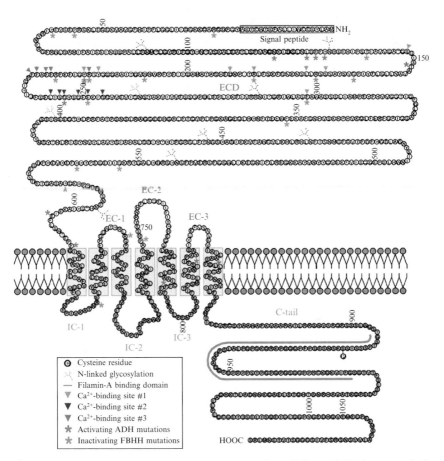

Figure 13.1 Topology of the bovine CaSR. The legend (lower left) shows symbols representing naturally occurring activating (gain-of-function) mutations causing ADH and inactivating (loss-of-function) mutations responsible for FBHH. Modified, with permission, from Brown *et al.* (1993). (See Color Insert.)

in the N-terminus of the tail, for both signaling and cell-surface expression of the receptor (Chang *et al.*, 2001). Others have shown that there is a filamin-A binding domain in the middle of the C-tail that is required for interaction with the cytoskeleton and for the formation of signaling complexes with other intracellular proteins (Hofer and Brown, 2003).

Family C GPCRs are thought to have evolved from the bacterial periplasmic proteins (PBPs) based on the homology of protein sequences in their ECDs. The PBPs contain a Venus Flytrap (VFT) domain that binds and transports nutrients such as Ca^{2+}, Mg^{2+}, and amino acids across the periplasm and the peptidoglycan mesh (Brown *et al.*, 1993; Chang and Shoback, 2004; Hofer and Brown, 2003; Huang *et al.*, 2009; Silve *et al.*, 2005; Wellendroph and Brauner-Osborne, 2009). Crystallography and computer modeling reveal a VFT-like structure in the ECD of the CaSR (residues 1–540). The CaSR-VFT domain contains three Ca^{2+}-binding sites that include several amino acids in the N-terminus of the ECD (between residues #100 and #300) (Fig. 13.1). The efficiency of Ca^{2+} binding to these sites depends on the overall tertiary structure of the ECD that is maintained by intra- and intermolecular disulfide bonding, interactions of hydrophobic and hydrophilic domains, and more importantly, local ionic strength. Any substitution of an amino acid residue within or adjacent to Ca^{2+}-binding sites that alters the conformation of the ECD will likely alter the Ca^{2+} binding of the receptor. Furthermore, amino acids, aminoglycosides, polyamines, pH, and salinity, which alter local ionic strength, can also potentiate or suppress the response of receptor to Ca^{2+}, potentially by altering the conformation of ECD and its affinity to Ca^{2+}.

The biological significance of mutations in the CaSR is clearly demonstrated by three genetic disorders—familial benign hypocalciuric hypercalcemia (FBHH), neonatal severe HPT (NSHPT), and autosomal dominant hypocalcemia (ADH). Heterozygous and homozygous inactivating (loss-of-function) mutations produce FBHH and NSHPT, respectively, while activating (gain-of-function) mutations cause ADH (Egbuna and Brown, 2008; Huang et al., 2009; Pidasheva et al., 2004; http://www.casrdb.mcgill.ca). The majority of these naturally occurring mutations are found in the ECD of the CaSR (Fig. 13.1). In contrast, the *Nuf* mouse has mutations in a CaSR residue in the fourth transmembrane domain of the receptor (Hough et al., 2004). These CaSR mutations, in general, cause a shift (to the right for the inactivating mutations and to the left for the activating mutations) in their Ca dose–response curves when expressed in the exogenous cell systems, further underscoring the role of ECD in Ca^{2+}-binding.

The above disorders predominantly affect parathyroid and renal function. The CaSR is, however, expressed in many other tissues, but it has been difficult to determine what role the CaSR plays in these tissues on the basis of these disease phenotypes. Growing interest has focused on exploring the functional impact of the CaSR in different systems using cultured cells. Many of these studies support a role for CaSRs in modulating distinct cell functions depending on the system. Translating the findings from *in vitro* studies to the biological actions of the CaSR *in vivo* has been challenging. To address this, we and others have begun to study gain-of-function and loss-of-function in mouse models. The tissue-specific, conditional knock-out approach using Cre-lox recombination is being used to delete CaSRs from specific cell populations *in vivo*. Using another strategy, we have developed a transgenic mouse model with tissue-specific overexpression of the CaSR in mature osteoblasts to study the gain-of-function of CaSR signaling in this cell type. This report describes methods to generate mice that overexpress a constitutively active CaSR in mature osteoblasts to investigate the role of CaSR in bone remodeling.

2. Materials and Methods

2.1. Construction of activating CaSR mutant constructs

A pcDNA3.1 vector (Wt-CaSR/pcDNA3.1) containing a bovine CaSR cDNA was subjected to site-directed mutagenesis using Chameleon Mutagenesis Kit (Stratagene, La Jolla, CA) and a set of selection (5'-CCGCTGGA-GAAGACGACAAT-3') and mutagenic (5'-CCACCCTG AGTTTTGTG GCCCCGATCAAGATTGACCCTTTGAAGCTTGGTGAGTTCTGC A-3') primers to make the Act-CaSR/pcDNA3.1 vector which encodes five point mutations [Q118P; N119I; S123P; N125K; and D127G] in the ligand-binding region of the receptor. This mutant was previously identified by Jensen et al. (2000) using a random saturation mutagenesis approach and shown to exert constitutive activity in transfected NIH-3T3 and tsA (a transformed HEK-293 cell) cell lines. Wt-CaSR/pcDNA3.1 and Act-CaSR/pcDNA3.1 vectors were amplified in TOP-10 *Escherichia coli*, purified using a Qiagen Maxi-Prep Kit (Qiagen, Valencia, CA), and stored in a sterile Tris–HCl buffer (10 mM, pH 8.5) at a concentration of 1 μg/μl.

2.2. Cell transfection

HEK-293 cells were cultured in T75 flasks containing growth medium (Dulbecco's Minimal Essential Medium, containing 10% fetal calf serum and penicillin (100 U/ml) and streptomycin (100 μg/ml) at 37 °C in 5% CO_2 until confluence). Confluent cells were split (1:6) into fresh T75 flasks containing 10 ml of growth medium, cultured for 4–6 h, and then transfected with CaSR cDNA using Ca–phosphate precipitation method (Chang et al., 2000, 2001). Ten micrograms of vector DNA was first mixed with 0.5 ml of 0.25 M $CaCl_2$. The DNA/$CaCl_2$ mixture was then added dropwise over 60 s to 0.5 ml of twofold concentrated HES buffer (in mM, 280 NaCl, 10 KCl, 1.5 $NaH_2PO_4 \cdot 2H_2O$, 12 dextrose, 50 HEPES (pH 7.5)) in a 15 ml test tube with gentle agitation by swirling the tube continually. The resulting mixture was incubated for 15 min at room temperature and then added to the culture flasks. After 18–24 h incubation at 37 °C in 5% CO_2, transfected cells were trypsinized and plated into wells of a 6-well Falcon culture plate (Nunc, Rochester, NY). Cells from one T75 flask were dispensed into a total of 24 wells and cultured for 24–48 h before assays.

2.3. Inositol phosphate (InsP) assays

Transfected HEK-293 cells were prelabeled with [^{3}H]myoinositol (2 mCi/ml; Amersham, Piscataway, NJ) for 18–24 h in the growth media (Chang et al., 2000, 2001). After three washes with fresh growth media, cells were pretreated with LiCl (to block Ins polyphosphate 1-phosphatase activity) in a MEM-LiCl solution [Eagle's modified essential medium with Earle's salts custom-prepared without $CaCl_2$, $MgSO_4$, and $NaHCO_3$ and supplemented with 0.5 mM $CaCl_2$, 0.5 mM $MgSO_4$, and 10 mM LiCl] for 15 min at 37 °C. For experimental treatments, media were replaced with fresh MEM-LiCl media supplemented with different concentrations of $CaCl_2$ (0.1–10 mM) and incubated for 60 min at 37 °C. Total ^{3}H-InsPs were extracted from cells and quantitated by anion-exchange chromatography (Chang et al., 1998).

2.4. Expression of Act-CaSR in osteoblasts

In vivo expression of the Act-CaSR transgene was driven by the 3.5 kb human osteocalcin (OC) promoter which is active in mature osteoblasts (Dvorak et al., 2007). Act-CaSR cDNA was subcloned from the Act-CaSR/pcDNA3.1 vector into ClaI/XbaI sites between the OC promoter and a bovine growth hormone poly A (bGH-poly A) sequence in the pBluescript II vector (Fig. 13.2). The transgene containing the OC promoter, Act-CaSR, and bGH-poly A was released by digestion with Kpn I restriction enzyme, purified using Qiagen Gel Extraction Kit (Qiagen, Valencia, CA), and stored in 10 mM Tris–HCl buffer (pH 8.5). The DNA was microinjected into the pronuclei of fertilized ova harvested from the oviducts of newly plugged FVB/N mice by standard techniques at the University of California San Francisco Transgenic Core Facility. Surviving microinjected embryos were reimplanted into the oviduct of pseudopregnant female recipients. After birth, transgenic FVB/N mice were maintained under standardized environmental conditions, with access to food and water *ad libitum*. Genotyping of tail DNA was performed using two sets of transgene-specific primers (set 1: 5'-GTGCTGCCTCCGCCACTGAT-3' and 5'-GCCCA-CCTGCTGCTTTGAGT-3', spans end of OC promoter and 5' end of transgene amplifying a 1379 bp fragment; set 2: 5'-GGGTATGGTGCG-GAGGAAGG-3' and 5'-TTGTGCTGCCCGAC CCTTTC-3', spans end of OC promoter and 3' end of transgene amplifying a 1539 bp fragment). All protocols were approved by the Animal Care Committee of the San Francisco Veterans Affairs Medical Center.

2.5. Assessment of Act-CaSR expression in transgenic osteoblasts: Immunoblotting

Crude membrane proteins were prepared from cultured bone marrow stromal cells from Wt and transgenic Act-CaSR mice as described previously (Dvorak et al., 2007). Immunoblotting with a polyclonal anti-CaSR antiserum and peroxidase-conjugated anti-rabbit IgG secondary antibodies (GE Healthcare, Little Chalfont, UK) was performed as described previously (Chang et al., 1998). Specificity of the antiserum was confirmed by the

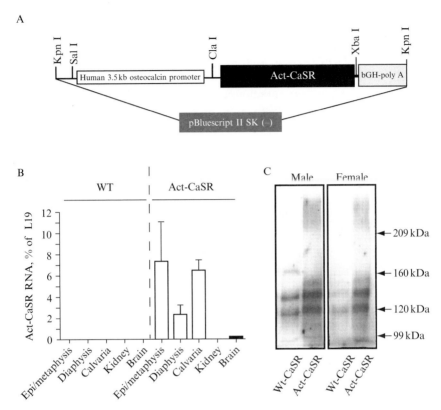

Figure 13.2 Expression of Act-CaSR transgene in osteoblasts. (A) Act-CaSR transgenic construct containing a 3.5 kb fragment of the human OC promoter, Act-CaSR coding sequence, and bovine growth hormone poly A (bGH-poly A) signal sequence in a pBlusescript II SK vector. (B) qPCR confirmed the expression of Act-CaSR RNA in bone (distal femur, epiphysis/metaphysics, femoral midshaft, and calvaria) but not in kidney and brain in transgenic Act-CaSR mice or in any tissue tested in Wt mice. RNA expression was normalized to the expression of the housekeeping gene L19. (C) Immunoblotting for the CaSR was performed on crude membrane proteins of bone marrow stromal cells (75 μg) cultured from Wt and Act-CaSR mice, using polyclonal antibodies against the ECD of the bovine parathyroid CaR (Brown et al., 1993). Reproduced with permission from Dvorak et al. (2007).

absence of signal after preincubating anti-CaSR antiserum with the peptide against which it was raised.

2.5.1. Quantitative PCR (qPCR)

Femora of 6-week-old Wt and Act-CaR littermates were isolated and separated into epiphyseal/metaphyseal and diaphyseal compartments. Bone marrow was flushed out with PBS before bones were frozen in liquid nitrogen. For RNA isolation, the bones were powdered using a multi-sample biopulverizer (RPI, Mt. Prospect, IL) and homogenized with a rotor-stator homogenizer (Polytron PT 3000; Brinkmann Instruments, Westbury, NY) in RNA-Stat reagent (Tel-Test, Friendswood, TX). cDNAs were synthesized with Moloney murine leukemia virus reverse transcriptase (Invitrogen, Carlsbad, CA), and expression of Act-CaSR transgene was quantified using TaqMan qPCR kits and ABI PRISM 7900HT Sequence Detection System with SDS software (Applied Biosystems Foster City, CA). A threshold cycle (number of PCR cycles required to generate a fluorescent signal exceeding a preset threshold) was determined for the Act-CaSR transgene and normalized to the threshold cycle for a housekeeping gene (L19) in the same sample. Primers and probes for Act-CaR (forward primer: 5″-AGGCCAGCTGCTCGAGAGT-3″, reverse primer: 5″-CTTGAGTCTTCAGAAGTCACATCATG-3″, and probe: 5′-FAM-ACTCAGCTCAGCACGACTGGGAAGC-BHQ-3′) were custom made by Integrated DNA Technologies (Skokie, IL) according to published nucleotide sequences.

2.6. Assessment of skeletal phenotypes in transgenic mice

The approach used to monitor temporal changes in bone mass and structure was *in vivo* microcomputed tomography (μCT) because it allows quantitation of mineralized tissue and assessment of three-dimensional structural parameters in live animals over time. μCT scans were performed on transgenic Act-CaSR versus Wt mice using a SCANCO VivaCT 40 scanner (SCANCO Medical, Bassersdorf, Switzerland). Animals were anesthetized with 2% isoflurane (Baxter Healthcare Corporation, Deerfield, IL), mixed in oxygen, and scanned every 6 weeks from 6 to 30 weeks of age. To examine trabecular bone, 100 serial cross-sectional scans (1.05 mm) of the secondary spongiosa of the left distal femoral metaphysis were obtained, from the end of the growth plate extending proximally to the shaft. The isotropic voxel (volumic pixel) size was 10.5 μm, and X-ray energy was 55 kV. A global threshold (27.5% of the grayscale) was applied to segment mineralized from soft tissue in the μCT images. Linear attenuation was calibrated using preset hydroxyapatite blocks. 3-D image reconstruction and analysis were performed using software provided by the manufacturer.

2.7. Statistics

Data from two groups were compared using unpaired Student's *t* test. Data from multiple groups were compared using ANOVA with Tukey's *post hoc* analysis. Significance was assigned for $p < 0.05$.

3. RESULTS

3.1. Assessment of activity of Act-CaSR *in vitro*

In principle, the activity of mutant CaSRs can be assessed by their ability to stimulate PLC, raise the intracellular $[Ca^{2+}]$, enhance MAPK activity, and inhibit cAMP production. However, measurement of PLC activity and intracellular Ca^{2+} mobilization are more standard methods for assessing receptor activation because they are straightforward, economical, and adaptable to high-throughput assays. More importantly, these methods can quantify changes in the efficacy of CaSR's response to its ligand, in this case, high $[Ca^{2+}]_e$.

By measuring the accumulation of total InsPs as an index of PLC activation, we observed different signaling responses in HEK-293 cells expressing Wt-CaSR versus Act-CaSR cDNAs (Fig. 13.3). We observed a clear left-shift in the EC_{50} of the Ca dose–response curves from ≈ 3.5 mM in Wt-CaSR-expressing cells to <0.3 mM in Act-CaSR-expressing cells. We also observed an elevation of basal activity at 0.1 mM Ca^{2+} in Act-CaSR-expressing versus Wt-CaSR-expressing cells, compatible with the notion that this was a constitutively active receptor mutant.

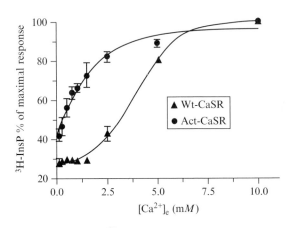

Figure 13.3 Effects of raising $[Ca^{2+}]_e$ from 0.1 mM to the concentrations shown on total ^{3}H-InsP accumulation in HEK-293 cells transiently expressing cDNAs encoding Wt-CaSR and Act-CaSR. Results are expressed as the percentage of the maximal response at 10 mM Ca^{2+} in cells expressing either Wt-CaSR or Act-CaSR as noted ($N=4$ independent transfections). Data were reanalyzed from experiments previously reported (Dvorak *et al.*, 2007).

3.2. Transgenic expression of Act-CaSR in osteoblasts

Extensive biochemical and mutational analyses have identified CaSR mutants that demonstrate constitutive activity (Egbuna and Brown, 2008; Pidasheva et al., 2004). Little is known about whether and how these mutants impact cellular functions outside the PTGs in vivo, especially in bone and cartilage, because the resulting alterations in parathyroid function (hypoparathyroidism) can affect those tissues directly. Our transgenic approach, using a highly cell-specific OC promoter to express the Act-CaSR in mature osteoblasts, allowed us to target the mutant receptor to bone and avoid the metabolic changes of a more generalized promoter. Indeed in two transgenic Act-CaSR mouse lines that we established, we observed no significant differences in levels of serum Ca^{2+} or PTH between Wt and Act-CaSR transgenic mice (Dvorak et al., 2007).

We further confirmed targeting of the transgene to bone by qPCR analysis of RNA extracted from femurs and calvariae but not in the other tissues tested (Fig. 13.2). Both the epiphyseal/metaphyseal and diaphyseal femoral compartments, which primarily contain cancellous and cortical bone, respectively, exhibited significant transgene expression (Fig. 13.2). Immunoblotting of crude membrane fractions of cultured bone marrow stromal cells from Act-CaSR mice revealed immunoreactive bands corresponding to the bovine CaR (120, 140, and 150 kDa) (Fig. 13.2).

3.3. Skeletal phenotype of transgenic Act-CaSR mice

By in vivo µCT scanning, we observed profound changes in trabecular bone of the transgenic Act-CaSR mice, from 6 to 30 weeks of age (Fig. 13.4). Act-CaSR mice consistently showed reduced trabecular bone mass (Fig. 13.4) as indicated by the decreased ratio of total bone volume (BV) over total tissue volume (TV) (BV/TV) and an osteopenic phenotype supported by the reduced bone mineral density (BMD; Fig. 13.4). Reduced bone mass in the Act-CaSR mice was mainly due to decreases in the number and connectivity of trabecular bones and increases in trabecular spacing. Similar bone phenotypes were observed in both male and female mice. These data suggest that constitutive activity of CaSRs in osteoblasts is detrimental to skeletal function and causes osteopenia likely due to increased bone breakdown. Bone histomorphometric parameters (increased osteoclast surface per eroded surface) further support that conclusion (Dvorak et al., 2007).

4. Conclusions

While human activating mutations uncovered during the genetic evaluation of individuals with ADH (hypocalcemia, low PTH levels, and hypercalciuria) demonstrate the prominence of the CaSR in regulating

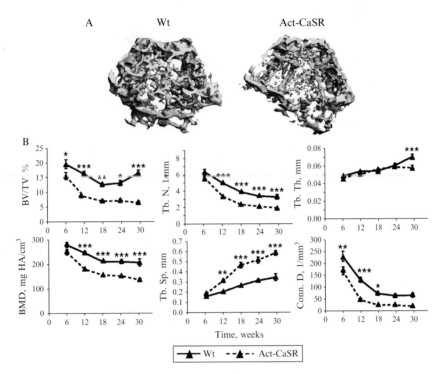

Figure 13.4 *In vivo* μCT scans to investigate changes in trabecular bone mass and structure in Wt versus Act-CaSR mice. μCT analysis was performed on the distal femur of 6–30-week-old female mice. (A) Three-dimensional μCT image reconstruction of trabecular bones from 12-week-old Wt and transgenic Act-CaSR mice. (B) Structural parameters derived from the μCT analyses, including trabecular bone volume fraction (BV/TV), bone mineral density (BMD), trabecular number (Tb.N), trabecular spacing (Tb.Sp), trabecular thickness (Tb.Th), and connectivity density (Conn. D). Data represent mean ± SE ($n = 8$ mice per group; ANOVA; *, $p < 0.05$; **, $p < 0.01$; ***, $p < 0.001$). Reproduced with permission from Dvorak *et al.* (2007).

parathyroid and renal function, it has been difficult to discern the role of CaSRs in other tissues from studying these families. Mouse models of generalized knockout of the CaSR are dominated by the mild to severe disturbances in serum [Ca^{2+}] and PTH levels that have secondary effects on skeletal tissues. This report describes the analysis in transgenic mice of a constitutively activated CaSR mutant targeted to mature mineralizing osteoblasts through the use of a well-defined OC promoter. In the absence of abnormalities in serum [Ca^{2+}] and PTH, this strategy demonstrated that the CaSR is likely playing a role in controlling the balance of bone remodeling—modulating levels of bone resorption most likely. Use of this and other well-defined mutants in transgenic models should give additional information relevant to understanding CaSR function outside the PTG.

ACKNOWLEDGMENTS

The authors acknowledge the technical skills of Ms. Tsui-Hua Chen and administrative support of Ms. Janelle Mendiola of the San Francisco VA Medical Center Endocrine Research Unit Laboratory and support from the Department of Veterans Affairs Research service and the NIH and the Department of Defense.

REFERENCES

Brown, E. M., Gamba, G., Riccardi, D., Lombardi, M., Butters, R., Kifor, O., Sun, A., Hediger, M. A., Lytton, J., and Hebert, S. C. (1993). Cloning and characterization of an extracellular Ca^{2+}-sensing receptor from bovine parathyroid. *Nature* **366**, 575–580.

Chang, W., and Shoback, D. (2004). Extracellular Ca^{2+}-sensing receptor—An overview. *Cell Calc.* **35**, 183–186.

Chang, W., Pratt, S., Chen, T. H., Nemeth, E., Huang, Z., and Shoback, D. (1998). Coupling of calcium receptors to inositol phosphate and cyclic AMP generation in mammalian cells and *Xenopus laevis* oocytes and immunodetection of receptor protein by region-specific antipeptide antisera. *J. Bone Min. Res.* **13**, 570–580.

Chang, W., Chen, T. H., Pratt, S., and Shoback, D. (2000). Amino acids in the second and third intracellular loops of the parathyroid Ca2+-sensing receptor mediate efficient coupling to phospholipase C. *J. Biol. Chem.* **275**, 19955–19963.

Chang, W., Pratt, S., Chen, T. H., and Shoback, D. (2001). Amino acids in the cytoplasmic C terminus of the parathyroid Ca2+-sensing receptor mediate efficient cell-surface expression and phospholipase C activation. *J. Biol. Chem.* **276**, 44129–44136.

Chang, W., Tu, C., Chen, T. H., Bikle, D., and Shoback, D. (2008). The extracellular calcium-sensing receptor (CaSR) is a critical modulator of skeletal development. *Sci. Signal.* **1**(35), ra1.

Dvorak, M. M., Chen, T. H., Orwoll, B., Garvey, C., Chang, W., Bikle, D. D., and Shoback, D. M. (2007). Constitutive activity of the osteoblast Ca2+-sensing receptor promotes loss of cancellous bone. *Endocrinology* **148**, 3156–3163.

Egbuna, O. I., and Brown, E. M. (2008). Hypercalcaemic and hypocalcemic conditions due to calcium-sensing receptor mutations. *Best Pract. Res. Clin. Rheumatol.* **22**, 129–148.

Ho, C., Conner, D. A., Pollak, M. R., Ladd, D. J., Kifor, O., Warren, H. B., Brown, E. M., Seigman, J. G., and Seidman, C. E. (1995). A mouse model of human familial hypocalciuric hypercalcemia and neonatal severe hyperparathyroidism. *Nat. Genet.* **11**, 389–394.

Hofer, A. M., and Brown, E. M. (2003). Extracellular calcium and signaling. *Nat. Rev. Mol. Cell Biol.* **4**, 530–538.

Hough, T. A., Bogani, D., Cheeseman, M. T., Favor, J., Nesbit, M. A., Thakker, R. V., and Lyon, M. F. (2004). Activating calcium-sensing receptor mutation in the mouse is associated with cataracts and ectopic calcification. *Proc. Natl. Acad. Sci. USA* **101**, 13566–13571.

Huang, Y., Zhou, Y., Castiblanco, A., Yang, W., Brown, E. M., and Yang, J. J. (2009). Multiple Ca(2+)-binding sites in the extracellular domain of the Ca(2+)-sensing receptor corresponding to the cooperative Ca(2+) response. *Biochemistry* **48**, 388–398.

Jensen, A. A., Spalding, T. A., Burstein, E. S., Sheppard, P. O., O'Hara, P. J., Brann, M. R., Krogsgaard-Larsen, P., and Brauner-Osborne, H. (2000). Functional importance of the Ala(116)-Pro(136) region in the calcium-sensing receptor. Constitutive activity and inverse agonism in a family C G-protein-coupled receptor. *J. Biol. Chem.* **275**, 29547–29555.

Pidasheva, V., D'Souza-Li, L., Canaff, L., Cole, D. E. C., and Hendy, G. N. (2004). CASRdb: Calcium-sensing receptor locus-specific database for mutations causing familial (benign hypocalciuric hypercalcemia, neonatal severe hyperparathyroidism and autosomal dominant hypocalcemia). *Hum. Mutat.* **24,** 107–111.

Silve, C., Petrel, C., Leroy, C., Bruel, H., Mallet, E., Rognan, D., and Ruat, M. (2005). Delineating a Ca2+ binding pocket with the venus flytrap module of the human calcium-sensing receptor. *J. Biol. Chem.* **280,** 37917–37923.

Wellendroph, P., and Brauner-Osborne, H. (2009). Molecular basis for amino acid sensing by family C G-protein-coupled receptors. *Br. J. Pharm.* **156,** 869–884.

CHAPTER FOURTEEN

CONSTITUTIVE ACTIVITY OF NEURAL MELANOCORTIN RECEPTORS

Ya-Xiong Tao,[*] Hui Huang,[*] Zhi-Qiang Wang,[*,†] Fan Yang,[*] Jessica N. Williams,[*] *and* Gregory V. Nikiforovich[‡]

Contents

1. Introduction	268
2. Signaling Assay for the Neural Melanocortin Receptors	268
2.1. Culture and transfection of HEK293T cells	268
2.2. Signaling assay for measuring receptor activation	269
3. Naturally Occurring Constitutively Active MC4R Mutants	270
3.1. Inverse agonism of AgRP at the MC4R	271
3.2. Small molecule MC4R inverse agonists	273
4. Inverse Agonism of AgRP at the MC3R	273
5. Computational Modeling of the Constitutively Active MC4R Mutants	273
Acknowledgments	277
References	277

Abstract

The two neural melanocortin receptors (MCRs), melanocortin-3 and -4 receptors (MC3R and MC4R), are G protein-coupled receptors expressed primarily in the brain that regulate different aspects of energy homeostasis. The MCRs are unique in having endogenous antagonists, agouti and agouti-related protein (AgRP). These antagonists were later shown to be inverse agonists. The MC3R has little or no constitutive activity, whereas the MC4R has significant constitutive activity that can easily be detected. We describe herein methods for detecting constitutive activities in these receptors and small molecule ligands as inverse agonists. AgRP is an inverse agonist for both MC3R and MC4R. We also provide models for the constitutively active MC4R mutants.

[*] Department of Anatomy, Physiology and Pharmacology, College of Veterinary Medicine, Auburn University, Auburn, Alabama, USA
[†] College of Veterinary Medicine, Yangzhou University, Jiangsu, People's Republic of China
[‡] MolLife Design LLC, St. Louis, Missouri, USA

1. Introduction

Obesity is an epidemic in the United States and an increasing health problem worldwide. Current therapeutic approaches, including pharmaceutical, surgical, and lifestyle interventions, are not effective or associated with significant side effects. Identifying novel targets for obesity treatment is a very active area of research.

The two neural melanocortin receptors (MCRs), the melanocortin-3 and -4 receptors (MC3R and MC4R), were recently found to regulate different aspects of energy homeostasis (Cone, 2005; Tao, 2005). The primary role of the MC3R is regulation of feed efficiency, the amount of energy ingested that is stored as fat in the body (Butler *et al.*, 2000; Chen *et al.*, 2000). The MC4R regulates both food intake and energy expenditure (Huszar *et al.*, 1997), with the effect of food intake accounting for 60% of the effect on body weight (Balthasar *et al.*, 2005; reviewed in Tao, 2010). Human genetic studies provided further evidence that these two receptors are important in maintaining energy homeostasis in humans. The role of *MC3R* in human obesity pathogenesis is controversial, with some studies supporting (Feng *et al.*, 2005; Lee *et al.*, 2002, 2007; Mencarelli *et al.*, 2008; Tao, 2007; Tao and Segaloff, 2004) and another study refuting (Calton *et al.*, 2009) a causal relationship (earlier studies reviewed in (Tao, 2005)). The role of *MC4R* in human obesity pathogenesis is undisputed. Since the original reports of *MC4R* mutations associated with childhood obesity (Vaisse *et al.*, 1998; Yeo *et al.*, 1998), more than 150 distinct mutations in the *MC4R* have been identified from populations of different ethnic backgrounds (reviewed in Tao, 2009). Mutations in the *MC4R* are the most common monogenic form of obesity (Farooqi *et al.*, 2003).

The MC3R and MC4R are G protein-coupled receptors that are positively coupled to the adenylyl cyclase system. Therefore, receptor activation will lead to increased cyclic AMP (cAMP) production. Receptor activation can be either due to ligand binding or mutation. A mutation that causes the receptor being activated in the absence of ligand is called constitutively active mutation. Indeed, the wild-type (WT) MC4R has some basal activity, whereas the MC3R has little or no basal activity (Tao, 2007). We describe here the methods used for measuring the constitutive activities of WT and mutant receptors.

2. Signaling Assay for the Neural Melanocortin Receptors

2.1. Culture and transfection of HEK293T cells

Human embryonic kidney (HEK) 293T cells, highly transfectable derivative of the HEK293 cell line, constitutively express the simian virus 40 (SV40) large T antigen. These cells, obtained from American Type Culture

Collection, were maintained in Dulbecco's modified Eagle's medium supplemented with 10% newborn calf serum, 100 units/ml penicillin, and 100 μg/ml streptomycin, at 37 °C in humidified air containing 5% CO_2. Although fetal bovine serum was suggested by American Type Culture Collection, we found that newborn calf serum can fully support these cells, resulting in significant savings in cell culture cost. For maintenance of culture, cells were cultured in 75 cm^2 flasks. For transfection, cells were plated into 6-well clusters (Corning, Corning, NY, cat # 3506) coated with 0.1% gelatin.

Transient transfection was routinely performed using the calcium phosphate precipitation method (Chen and Okayama, 1987). Other investigators have used various commercial transfection reagent kits. We have found that in HEK293T cells, the calcium phosphate precipitation method works well with minimal cost compared with commercial kits. For each well in the 6-well cluster, add the following sequentially:

86 μl sterile double-distilled water
10 μl of 2.5 M calcium chloride
4 μl of plasmid DNA (0.25 mg/ml in sterile water)

After mixing these components by gentle swirling, 100 μl of 2× BSS is added to the bottom of the tube (2× BSS consists of 280 mM sodium chloride, 1.5 mM Na_2HPO_4, and 50 mM of BES (N,N-bis[2-hydroxyl]-2-aminoethane sulfonic acid, Sigma cat # B9879), pH adjusted to 6.95 with sodium hydroxide). After these components are mixed, the tube is left in the hood for 15 min. Then, the content is combined with 1.8 ml growth media and added into each well.

Cells can be used for signaling assays 24–96 h after transfection. We have routinely used cells at 48 h after transfection when maximal expression of the receptors is frequently observed.

2.2. Signaling assay for measuring receptor activation

Because the primary signaling pathway of the MC3R and the MC4R is activation of the stimulatory G protein G_s, resulting in activation of the adenylyl cyclase and increased intracellular cAMP levels, direct or indirect measurement of cAMP levels is the most commonly used method for measuring signaling activities of these receptors. Indirect measurement of cAMP measures the reporter gene activity driven by increased intracellular cAMP levels. A commonly used reporter gene is luciferase. Before the assay, cells are serum starved for several hours (for example, 8 h), and then different concentrations of ligands are added to the cells and incubated for 16 additional hours to allow the gene transcription and translation to occur. Finally, the enzyme activity is measured, usually using a commercially available kit. If no ligand is added, the activity measured is the basal activity of the receptor. This assay is very sensitive due to additional amplification, from cAMP to increased transcription and translation of luciferase.

Our lab has been using direct measurement of cAMP to monitor signaling. Cells stimulated with ligands are lysed with perchloric acid. For measurement of intracellular cAMP, 0.5 N perchloric acid with 180 µg/ml theophylline (an inhibitor of phosphodiesterase to block the breakdown of cAMP) is used (Tao and Segaloff, 2003). Lysate is neutralized with 0.72 M KOH/0.6 M KHCO$_3$. After a 10-min centrifugation at 4 °C, the supernatant is saved for radioimmunoassay.

Before assay, cAMP samples are acetylated to increase sensitivity and specificity and to reduce interference. This is done as follows: add 20 µl of 5 N KOH to a 12 × 75 mm glass tube with 500 µl standards or samples, add 5 µl of acetic anhydride, immediately vortex for 3–5 s, and let it sit at room temperature for 30 min. Then the samples are placed on ice and used for assay within an hour.

The cAMP standard used is from Sigma (cat # A9501). Succinyl cAMP tyrosine methyl ester (Sigma cat # M2257) is iodinated with chloramines T method and used as the tracer. Bulk cAMP–antibody can be purchased from various commercial sources. We obtained our antibody from Strategic Biosolutions (Newark, DE). The radioimmunoassay is performed as originally described in Steiner et al. (1969) except that polyethylene glycol 8000 is used for precipitation of the cAMP–antibody complex (Fan et al., 2008b). This method involves the use of radioactive material. Appropriate local regulations regarding safety training and proper disposal should be strictly followed.

3. NATURALLY OCCURRING CONSTITUTIVELY ACTIVE MC4R MUTANTS

MC4R activation results in decreased food intake and increased energy expenditure. Loss-of-function mutations in the *MC4R* result in obesity (Tao, 2009). Loss of constitutive activity is suggested to be one mechanism that mutations in *MC4R* cause obesity (Srinivasan et al., 2004). We showed that indeed some MC4R mutants have decreased basal activities whereas some mutants retain normal basal activities (Fan and Tao, 2009; Rong et al., 2006; Roth et al., 2009; Tao and Segaloff, 2005). Constitutively active mutations in the *MC4R* are expected to be associated with a constitutional lean phenotype, perhaps even anorexia nervosa. Indeed, the higher constitutive activity of I251L was proposed to be responsible for the negative association of this variant with body mass index (Xiang et al., 2006). Paradoxically, six constitutively active mutations were identified from obese patients, including H76R, S127L, D146N,

H158R, P230L, and L250Q (reviewed in Tao, 2010). The reason for these mutations in causing obesity is not well established (Tao, 2008). The constitutive activities of S127L and P230L are modest, less than threefold (Fan and Tao, 2009). However, the constitutive activities of the other four mutants are substantial, 6- to 15-fold higher than the basal activity of the WT MC4R (Hinney et al., 2006; Tao, 2010; Vaisse et al., 2000). We transfected HEK293T cells with increasing concentrations of plasmids of H76R, D146N, and L250Q. As shown in Fig. 14.1, the basal cAMP levels increase with increasing concentrations of plasmids transfected, reaching a plateau at higher concentrations for L250Q.

3.1. Inverse agonism of AgRP at the MC4R

The MCRs are unique in having two endogenous antagonists, agouti and AgRP (agouti-related protein). Agouti is an antagonist for melanocortin-1 receptor, whereas AgRP is an antagonist for MC3R and MC4R (reviewed in Cone, 2006). Previous studies have shown that a fragment of AgRP, AgRP (83–132), is an inverse agonist for WT human MC4R (Nijenhuis et al., 2001) and a constitutively active mouse MC4R mutant (Haskell-Luevano and Monck, 2001). Later studies showed that full-length AgRP, a smaller fragment of AgRP, AgRP (87–120), as well as two short peptides derived from AgRP, are all inverse agonists at the MC4R (Chai et al., 2003).

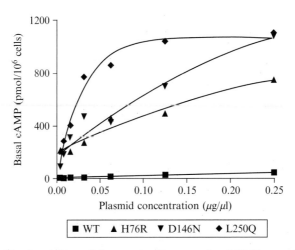

Figure 14.1 Basal activities of three naturally occurring *MC4R* mutations, H76R, D146N, and L250Q. Different concentrations of plasmids were transfected into HEK293T cells. Empty vector pcDNA3 was used to normalize the amount of plasmid DNA added to each well. Basal cAMP levels were measured 48 h after transfection in the presence of phosphodiesterase inhibitor for 75 min. Cyclic AMP levels were measured with radioimmunoassay (Fan et al., 2008a).

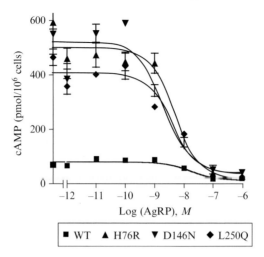

Figure 14.2 Inverse agonism of AgRP at the MC4R. Cells expressing WT or constitutively active MC4R mutants, H76R, D146N, and L250Q, were treated with different concentrations of AgRP. Cyclic AMP levels were measured with radioimmunoassay (Fan et al., 2008a).

Table 14.1 Inverse agonism of AgRP and Ipsen 5i at the MC4R

MC4R	AgRP		Ipsen 5i	
	EC_{50}	Basal activity remaining at 10^{-6} M AgRP	EC_{50}	Basal activity remaining at 10^{-6} M Ipsen 5i
WT	23.46 ± 4.56	21 ± 3	17.46 ± 3.23	33 ± 5
H76R	5.56 ± 0.51	4 ± 0	9.28 ± 1.00	44 ± 4
D146N	3.50 ± 1.00	6 ± 1	5.43 ± 1.19	36 ± 3
L250Q	2.22 ± 0.93	6 ± 1	16.32 ± 6.32	35 ± 1

Data shown are from three or more experiments, with the mean ± standard error of the mean listed in the table.

In our experiments with WT and three constitutively active mutant (CAM) MC4Rs, we showed that AgRP is a potent inverse agonist (Fig. 14.2 and Table 14.1). In WT MC4R, AgRP is a full inverse agonist, decreasing basal cAMP level to the detection limit of our radioimmunoassay. In the CAM MC4Rs, AgRP also decreased the basal cAMP levels to just above the detection limit of our assays (Fig. 14.2 and Table 14.1). Therefore, AgRP can also be considered as a full inverse agonist in the CAM MC4Rs.

3.2. Small molecule MC4R inverse agonists

We recently reported that ML00253764, 2-[2-[2-(5-bromo-2-methoxyphenyl)-ethyl]-3-fluorophenyl]-4,5-dihydro-1H-imidazole, is a MC4R inverse agonist, decreasing the basal activities of two constitutively active MC4R mutants, H76R and D146N (Tao, 2010). However, it seems that ML00253764 is only a partial inverse agonist. At 10 μM concentration, the WT MC4R does not have any residual constitutive activity. However, the two mutants still retain high constitutive activities. The low affinity of ML00253764 for the MC4R might be responsible for the partial inverse agonism. Originally described by Vos et al. (2004) at Millennium Pharmaceuticals, the Ki of ML00253764 for human MC4R is 0.16 μM.

We recently tested the inverse agonist activity of another small molecule antagonist, Ipsen 5i. Described by Roubert and colleagues at Ipsen, the Ki of this compound for the MC4R is 2 nM (Poitout et al., 2007). Enzo Life Sciences International, Inc. (Plymouth Meeting, PA) synthesized the compound for us. We transfected HEK293T cells with WT or mutant MC4Rs. Forty-eight hours after transfection, cells were incubated with different concentrations of Ipsen 5i for 1 h. As shown in Fig. 14.3, intracellular cAMP levels decrease with increasing concentrations of Ipsen 5i for both WT and mutant MC4Rs, H76R, D146N, and L250Q. The maximal inhibition ranged from 56% to 63% (Table 14.1). Therefore, Ipsen 5i is also a partial inverse agonist for the MC4R.

4. INVERSE AGONISM OF AGRP AT THE MC3R

Although it is well known that AgRP is an inverse agonist for the MC4R (see above), it was not known whether this endogenous antagonist is also an inverse agonist for the MC3R. The WT MC3R has little or no constitutive activity (Tao, 2007) and no constitutively active MC3R mutant had been reported. Therefore, there was no tool to study this question. We recently showed that F347A has increased basal activity, with a basal activity about 6.9-fold that of the WT MC3R (Wang and Tao, 2010). Using this newly found tool, we showed herein that AgRP is also an inverse agonist for the MC3R, with EC_{50} of 0.17 nM (Fig. 14.4). When the cells were treated with 1 μM AgRP, the basal cAMP of F347A was reduced to 33% that of the untreated cells.

5. COMPUTATIONAL MODELING OF THE CONSTITUTIVELY ACTIVE MC4R MUTANTS

The recent X-ray studies revealed the structure of opsin in complex with the C-terminal peptide of transducin stabilizing the photoactivated structure of rhodopsin (Scheerer et al., 2008). This structure may be

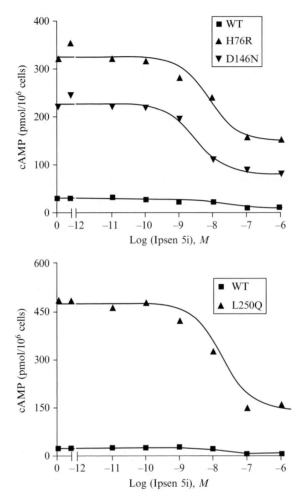

Figure 14.3 Partial inverse agonism of Ipsen 5i at the MC4R. Cells expressing WT or constitutively active MC4R mutants, H76R, D146N, and L250Q, were treated with different concentrations of Ipsen 5i. Cyclic AMP levels were measured with radioimmunoassay (Fan et al., 2008a).

regarded as a prototype for the general activated structure of the rhodopsin-like GPCRs. We assumed that the X-ray structure of β_2-adrenoreceptor (Cherezov et al., 2007; the PDB entry 2RH1) may be a template for the ground state of MC4R and the X-ray structure of opsin (the PDB entry 3DQB) may be a template for the activated state of MC4R.

Details of the modeling protocol are described elsewhere (Nikiforovich and Baranski, 2010). Briefly, modeling of the transmembrane (TM) regions of MC4R and the mutants involved several main steps: sequence alignment

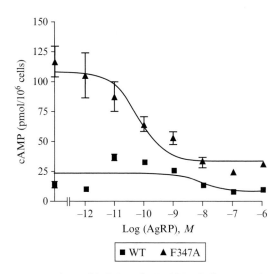

Figure 14.4 Inverse agonism of AgRP at the MC3R. Cells expressing WT or constitutively active MC3R mutant, F347A, were treated with different concentrations of AgRP. AgRP decreases the basal activities of F347A to that of the WT MC3R, which is devoid of basal activity. Cyclic AMP levels were measured with radioimmunoassay (Fan et al., 2008a).

to the selected template to define boundaries of TM helices; conformational calculations for individual TM helices; structural alignment to the selected templates; and final energy minimization with optimization of the side chain packing.

Our modeling results revealed specific changes in system of contacts between the cytoplasmic parts of TM6 and TM5 upon transition from the presumed ground state to the presumed activated state of MC4R. Namely, the cytoplasmic part of TM6 moved away from TM7 and toward TM5. One more structural difference between the ground and activated states of MC4R was the orientation of the flexible side chain of R147 located in TM3. While in the ground state this orientation is determined almost exclusively by the salt bridge between the side chains of D146 and R147, in the activated state, the side chain of D146 is involved in the strong salt bridge with the side chain of R165 in TM4. As a result, interaction D146–R147 is significantly weakened, and the R147 side chain may change orientation, possibly contacting L250 in TM6 (compare Fig. 14.5A and B). This specific orientation of the side chain of R147 may be an important structural feature characteristic for activation of MC4R.

If these results are valid, conformational transitions from the ground state of MC4R to the activated state may be facilitated either by weakening the salt bridge D146–R147 (destabilization of the ground state) or by enhancing interaction between the side chain of R147 and that in position 250

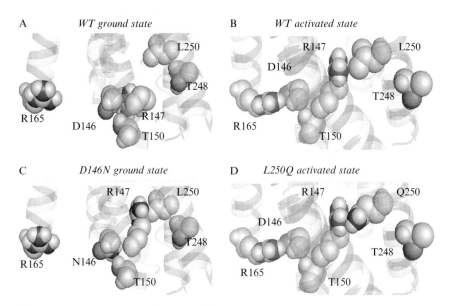

Figure 14.5 Sketches of the cytoplasmic portion of the TM region of MC4R. Presented are MC4R ground state (A), MC4R activated state (B), mutant D146N ground state (C), and mutant L250Q activated state (D). Side chains of D/N146, R147, T150, R165, T248, and L/Q250 are shown as space-filled models. TM helices are shown as semitransparent cartoons.

(stabilization of the activated state). Indeed, further modeling showed that in mutant D146N, where the salt bridge D146–R147 is replaced by a weaker hydrogen bonding, orientation of the R147 side chain characteristic for the activated state may be already adopted in the ground state (Fig. 14.5C). On the other hand, according to our modeling, the salt bridge D146–R147 remained unchanged in the ground state of mutant L250Q, but the side chain of R147 is involved in hydrogen bonding with the side chain of Q250 in the activated state (Fig. 14.5D). The same hydrogen bonding was suggested for the activated state of mutant L250Q in the earlier modeling study (Proneth et al., 2006). Our further modeling showed that the discussed specific orientation of R147 was stabilized in L250N by interactions with N250, but was destabilized by steric clashes with the side chain of F250 in L250F. Since L250N and L250Q are known as strong CAMs, and L250F is not (Proneth et al., 2006), that structural feature found by modeling may be considered significant for displaying constitutive activity in the L250x mutant series.

In summary, our molecular modeling employed an assumption that the ground state of MC4R may be based on the X-ray structure of β_2-adrenoreceptor and the activated state may be based on the X-ray structure of opsin cocrystallized with the transducin peptide. Modeling successfully rationalized the mutagenic data on several CAMs of MC4R by suggesting that transition to the activated state may be facilitated by mutations that either destabilize the ground state (D146N) or stabilize the activated state (L250Q), two ways to generate CAMs (Tao et al., 2000). These results, in turn, confirm our general assumption.

ACKNOWLEDGMENTS

These studies were supported by National Institutes of Health (R15DK077213) and intramural support from Auburn University, including Animal Health and Disease Research Program and Interdisciplinary Research Program. J.N.W. was partially supported by Auburn University Undergraduate Research Fellowship.

REFERENCES

Balthasar, N., Dalgaard, L. T., Lee, C. E., Yu, J., Funahashi, H., Williams, T., Ferreira, M., Tang, V., McGovern, R. A., Kenny, C. D., Christiansen, L. M., Edelstein, E., et al. (2005). Divergence of melanocortin pathways in the control of food intake and energy expenditure. Cell **123**, 493–505.

Butler, A. A., Kesterson, R. A., Khong, K., Cullen, M. J., Pelleymounter, M. A., Dekoning, J., Baetscher, M., and Cone, R. D. (2000). A unique metabolic syndrome causes obesity in the melanocortin-3 receptor-deficient mouse. Endocrinology **141**, 3518–3521.

Calton, M. A., Ersoy, B. A., Zhang, S., Kane, J. P., Malloy, M. J., Pullinger, C. R., Bromberg, Y., Pennacchio, L. A., Dent, R., McPherson, R., Ahituv, N., and Vaisse, C. (2009). Association of functionally significant melanocortin-4 but not melanocortin-3 receptor mutations with severe adult obesity in a large North American case-control study. Hum. Mol. Genet. **18**, 1140–1147.

Chai, B. X., Neubig, R. R., Millhauser, G. L., Thompson, D. A., Jackson, P. J., Barsh, G. S., Dickinson, C. J., Li, J. Y., Lai, Y. M., and Gantz, I. (2003). Inverse agonist activity of agouti and agouti-related protein. Peptides **24**, 603–609.

Chen, A. S., Marsh, D. J., Trumbauer, M. E., Frazier, E. G., Guan, X. M., Yu, H., Rosenblum, C. I., Vongs, A., Feng, Y., Cao, L., Metzger, J. M., Strack, A. M., et al. (2000). Inactivation of the mouse melanocortin-3 receptor results in increased fat mass and reduced lean body mass. Nat. Genet. **26**, 97–102.

Chen, C., and Okayama, H. (1987). High-efficiency transformation of mammalian cells by plasmid DNA. Mol. Cell. Biol. **7**, 2745–2752.

Cherezov, V., Rosenbaum, D. M., Hanson, M. A., Rasmussen, S. G., Thian, F. S., Kobilka, T. S., Choi, H. J., Kuhn, P., Weis, W. I., Kobilka, B. K., and Stevens, R. C. (2007). High-resolution crystal structure of an engineered human β_2-adrenergic G protein-coupled receptor. Science **318**, 1258–1265.

Cone, R. D. (2005). Anatomy and regulation of the central melanocortin system. Nat. Neurosci. **8**, 571–578.

Cone, R. D. (2006). Studies on the physiological functions of the melanocortin system. Endocr. Rev. **27**, 736–749.

Fan, Z. C., Sartin, J. L., and Tao, Y. X. (2008a). Molecular cloning and pharmacological characterization of porcine melanocortin-3 receptor. J. Endocrinol. **196**, 139–148.

Fan, Z. C., Sartin, J. L., and Tao, Y. X. (2008b). Pharmacological analyses of two naturally occurring porcine melanocortin-4 receptor mutations in domestic pigs. *Domest. Anim. Endocrinol.* **34,** 383–390.

Fan, Z. C., and Tao, Y. X. (2009). Functional characterization and pharmacological rescue of melanocortin-4 receptor mutations identified from obese patients. *J. Cell. Mol. Med.* **13,** 3268–3282.

Farooqi, I. S., Keogh, J. M., Yeo, G. S., Lank, E. J., Cheetham, T., and O'Rahilly, S. (2003). Clinical spectrum of obesity and mutations in the melanocortin 4 receptor gene. *N. Engl. J. Med.* **348,** 1085–1095.

Feng, N., Young, S. F., Aguilera, G., Puricelli, E., Adler-Wailes, D. C., Sebring, N. G., and Yanovski, J. A. (2005). Co-occurrence of two partially inactivating polymorphisms of MC3R is associated with pediatric-onset obesity. *Diabetes* **54,** 2663–2667.

Haskell-Luevano, C., and Monck, E. K. (2001). Agouti-related protein functions as an inverse agonist at a constitutively active brain melanocortin-4 receptor. *Regul. Pept.* **99,** 1–7.

Hinney, A., Bettecken, T., Tarnow, P., Brumm, H., Reichwald, K., Lichtner, P., Scherag, A., Nguyen, T. T., Schlumberger, P., Rief, W., Vollmert, C., Illig, T., *et al.* (2006). Prevalence, spectrum, and functional characterization of melanocortin-4 receptor gene mutations in a representative population-based sample and obese adults from Germany. *J. Clin. Endocrinol. Metab.* **91,** 1761–1769.

Huszar, D., Lynch, C. A., Fairchild-Huntress, V., Dunmore, J. H., Fang, Q., Berkemeier, L. R., Gu, W., Kesterson, R. A., Boston, B. A., Cone, R. D., Smith, F. J., Campfield, L. A., *et al.* (1997). Targeted disruption of the melanocortin-4 receptor results in obesity in mice. *Cell* **88,** 131–141.

Lee, Y. S., Poh, L. K., Kek, B. L., and Loke, K. Y. (2007). The role of melanocortin 3 receptor gene in childhood obesity. *Diabetes* **56,** 2622–2630.

Lee, Y. S., Poh, L. K., and Loke, K. Y. (2002). A novel melanocortin 3 receptor gene (MC3R) mutation associated with severe obesity. *J. Clin. Endocrinol. Metab.* **87,** 1423–1426.

Mencarelli, M., Walker, G. E., Maestrini, S., Alberti, L., Verti, B., Brunani, A., Petroni, M. L., Tagliaferri, M., Liuzzi, A., and Di Blasio, A. M. (2008). Sporadic mutations in melanocortin receptor 3 in morbid obese individuals. *Eur. J. Hum. Genet.* **16,** 581–586.

Nijenhuis, W. A., Oosterom, J., and Adan, R. A. (2001). AgRP(83-132) acts as an inverse agonist on the human melanocortin-4 receptor. *Mol. Endocrinol.* **15,** 164–171.

Nikiforovich, G. V., and Baranski, T. J. (2010). Structural modeling of CA mutants of GPCRs: C5a receptor. *Methods Enzymol.* **484,** 267–279.

Poitout, L., Brault, V., Sackur, C., Bernetiere, S., Camara, J., Plas, P., and Roubert, P. (2007). Identification of a novel series of benzimidazoles as potent and selective antagonists of the human melanocortin-4 receptor. *Bioorg. Med. Chem. Lett.* **17,** 4464–4470.

Proneth, B., Xiang, Z., Pogozheva, I. D., Litherland, S. A., Gorbatyuk, O. S., Shaw, A. M., Millard, W. J., Mosberg, H. I., and Haskell-Luevano, C. (2006). Molecular mechanism of the constitutive activation of the L250Q human melanocortin-4 receptor polymorphism. *Chem. Biol. Drug Des.* **67,** 215–229.

Rong, R., Tao, Y. X., Cheung, B. M., Xu, A., Cheung, G. C., and Lam, K. S. (2006). Identification and functional characterization of three novel human melanocortin-4 receptor gene variants in an obese Chinese population. *Clin. Endocrinol.* **65,** 198–205.

Roth, C. L., Ludwig, M., Woelfle, J., Fan, Z. C., Brumm, H., Biebermann, H., and Tao, Y. X. (2009). A novel melanocortin-4 receptor gene mutation in a female patient with severe childhood obesity. *Endocrine* **36,** 52–59.

Scheerer, P., Park, J. H., Hildebrand, P. W., Kim, Y. J., Krauss, N., Choe, H. W., Hofmann, K. P., and Ernst, O. P. (2008). Crystal structure of opsin in its G-protein-interacting conformation. *Nature* **455,** 497–502.

Srinivasan, S., Lubrano-Berthelier, C., Govaerts, C., Picard, F., Santiago, P., Conklin, B. R., and Vaisse, C. (2004). Constitutive activity of the melanocortin-4 receptor is maintained by its N-terminal domain and plays a role in energy homeostasis in humans. *J. Clin. Invest.* **114,** 1158–1164.

Steiner, A. L., Kipnis, D. M., Utiger, R., and Parker, C. (1969). Radioimmunoassay for the measurement of adenosine 3′,5′-cyclic phosphate. *Proc. Natl. Acad. Sci. USA* **64,** 367–373.

Tao, Y. X. (2005). Molecular mechanisms of the neural melanocortin receptor dysfunction in severe early onset obesity. *Mol. Cell. Endocrinol.* **239,** 1–14.

Tao, Y. X. (2007). Functional characterization of novel melanocortin-3 receptor mutations identified from obese subjects. *Biochim. Biophys. Acta* **1772,** 1167–1174.

Tao, Y. X. (2008). Constitutive activation of G protein-coupled receptors and diseases: Insights into mechanism of activation and therapeutics. *Pharmacol. Ther.* **120,** 129–148.

Tao, Y. X. (2009). Mutations in melanocortin-4 receptor and human obesity. *Prog. Mol. Biol. Transl. Sci.* **88,** 173–204.

Tao, Y. X. (2010). The melanocortin-4 receptor: Physiology, pharmacology, and pathophysiology. *Endocr. Rev.* **31,** 506–543.

Tao, Y. X., Abell, A. N., Liu, X., Nakamura, K., and Segaloff, D. L. (2000). Constitutive activation of G protein-coupled receptors as a result of selective substitution of a conserved leucine residue in transmembrane helix III. *Mol. Endocrinol.* **14,** 1272–1282.

Tao, Y. X., and Segaloff, D. L. (2003). Functional characterization of melanocortin-4 receptor mutations associated with childhood obesity. *Endocrinology* **144,** 4544–4551.

Tao, Y. X., and Segaloff, D. L. (2004). Functional characterization of melanocortin-3 receptor variants identify a loss-of-function mutation involving an amino acid critical for G protein-coupled receptor activation. *J. Clin. Endocrinol. Metab.* **89,** 3936–3942.

Tao, Y. X., and Segaloff, D. L. (2005). Functional analyses of melanocortin-4 receptor mutations identified from patients with binge eating disorder and nonobese or obese subjects. *J. Clin. Endocrinol. Metab.* **90,** 5632–5638.

Vaisse, C., Clement, K., Guy-Grand, B., and Froguel, P. (1998). A frameshift mutation in human MC4R is associated with a dominant form of obesity. *Nat. Genet.* **20,** 113–114.

Vaisse, C., Clement, K., Durand, E., Hercberg, S., Guy-Grand, B., and Froguel, P. (2000). Melanocortin-4 receptor mutations are a frequent and heterogeneous cause of morbid obesity. *J. Clin. Invest.* **106,** 253–262.

Vos, T. J., Caracoti, A., Che, J. L., Dai, M., Farrer, C. A., Forsyth, N. E., Drabic, S. V., Horlick, R. A., Lamppu, D., Yowe, D. L., Balani, S., Li, P., et al. (2004). Identification of 2-[2-[2-(5-bromo-2-methoxyphenyl)-ethyl]-3-fluorophenyl]-4,5-dihydro-1H-imidazole (ML00253764), a small molecule melanocortin 4 receptor antagonist that effectively reduces tumor-induced weight loss in a mouse model. *J. Med. Chem.* **47,** 1602–1604.

Wang, S. X., and Tao, Y. X. (2010). The functions of DPLIY motif and helix 8 in human melanocortin-3 receptor. *Diabetes* **59**(Suppl), A531–A532 (Abstract 2031–P).

Xiang, Z., Litherland, S. A., Sorensen, N. B., Proneth, B., Wood, M. S., Shaw, A. M., Millard, W. J., and Haskell-Luevano, C. (2006). Pharmacological characterization of 40 human melanocortin-4 receptor polymorphisms with the endogenous proopiomelanocortin-derived agonists and the agouti-related protein (AGRP) antagonist. *Biochemistry* **45,** 7277–7288.

Yeo, G. S., Farooqi, I. S., Aminian, S., Halsall, D. J., Stanhope, R. G., and O'Rahilly, S. (1998). A frameshift mutation in MC4R associated with dominantly inherited human obesity. *Nat. Genet.* **20,** 111–112.

CHAPTER FIFTEEN

MEASUREMENT OF CONSTITUTIVE ACTIVITY OF BMP TYPE I RECEPTORS

David J. J. de Gorter, Maarten van Dinther, *and* Peter ten Dijke

Contents

1. Introduction	281
2. Determining ALK2 Constitutive Activity	284
2.1. Western blot analysis of Smad1 phosphorylation	284
2.2. Determining BMP-Smad-dependent transcriptional activity	287
3. Determining the Effects of ALK2 Constitutive Activity on Osteoblast Differentiation	289
3.1. Determining the effect on alkaline phosphatase induction in ALK2 mutant-transduced C2C12 cells	289
4. Concluding Remarks	291
Acknowledgments	292
References	292

Abstract

Bone morphogenetic proteins (BMPs) are pleiotropic cytokines controlling a multitude of processes, among which bone formation. BMPs function by binding and activating BMP type I and type II receptors, resulting in activation of Smad transcription factors and expression of BMP target genes. Mutations in the BMP type I receptor ALK2 were identified in patients suffering from Fibrodysplasia Ossificans Progressiva (FOP). The mutation found in all patients displaying classical FOP symptoms, the ALK2 R206H mutation, renders ALK2 constitutively active. Here, we provide a detailed description of how to determine whether other ALK2 mutations will confer constitutive activity to the receptor.

1. INTRODUCTION

Bone morphogenetic proteins (BMPs) are dimeric-secreted cytokines that were originally identified by their ability to induce ectopic bone and cartilage formation (Urist, 1965; Wozney *et al.*, 1988). However, currently it is established that BMPs are pleiotropic cytokines with pivotal roles in embryonic pattern formation, skeletogenesis, pulmonary, cardiovascular, reproductive and urogenital organs, and the nervous system (Sieber et al., 2009). Furthermore, BMP signaling was shown to affect stem cell maintenance and cancer progression (Piccirillo et al., 2006; Xu et al., 2008).

BMPs exert their effects by binding distinct combinations of two different types of serine/threonine kinase receptors, that is, type I and type II receptors (Sieber et al., 2009). Three BMP type II receptors, BMPR-II, ActRII, and ActRIIB, and four BMP type I receptors, ALK1, ALK2, ALK3, and ALK6, have been identified. In the ligand-induced hetero-oligomeric complex of type I and type II receptors, the type I receptors act as a substrate for the constitutively active type II receptors. The type I receptors are the predominant determinants of the signaling response induced by the BMP receptor complex (Fujii et al., 1999). Upon activation, the type I receptors propagate the signal through C-terminal phosphorylation of BMP receptor-regulated Smads (BMP R-Smads), Smad1, -5, and -8, which then associate with the Co-Smad, Smad4, and translocate into the nucleus, where they together with other transcription factors bind promoters of target genes and control their expression (Sieber et al., 2009; Fig. 15.1A). BMP-induced osteoblast differentiation requires Smad signaling, and overexpression of Smad proteins in the premyoblast cell line C2C12 is sufficient to drive differentiation toward the osteoblast lineage (Fujii et al., 1999).

Type II receptor-induced serine and threonine phosphorylation occurs in the glycine/serine residue-rich domain (GS domain) of the type I receptors (Wrana et al., 1994). Specific artificial mutations in this domain, such as Q207D in ALK2, render the type I receptors constitutively active, that is, they can signal in the absence of ligand stimulation and type II receptors (Wieser et al., 1995; Fig. 15.1B). The negative regulator FK506 binding protein 12 kDa (FKBP12) can bind to the type I receptors via the GS domain and inhibits signaling by shielding the serine and threonine residues from phosphorylation by the type II receptor (Chen et al., 1997; Huse et al., 1999; Okadome et al., 1996; Wang et al., 1996).

Interestingly, a single point-mutation in the ALK-2 GS domain, ALK2 R206H, was found to underlie the rare autosomal dominant disorder Fibrodysplasia Ossificans Progressiva (FOP), which is characterized by skeletal malformations and progressive ossification of muscles, tendons, ligaments, and other connective tissues (Shore et al., 2006). We demonstrated that this mutation causes ALK2 to be resistant to FKBP12-mediated inhibition, rendering the type I receptor constitutively active and able to increase differentiation toward the osteoblast lineage when expressed in mesenchymal stem cells (van Dinther et al., 2010). In addition to the ALK2 R206H mutation found in all FOP patients with classical symptoms, some in-frame missense mutations in ALK2 different from R206H were identified in

Measurement of Constitutive Activity of BMP Type I Receptors

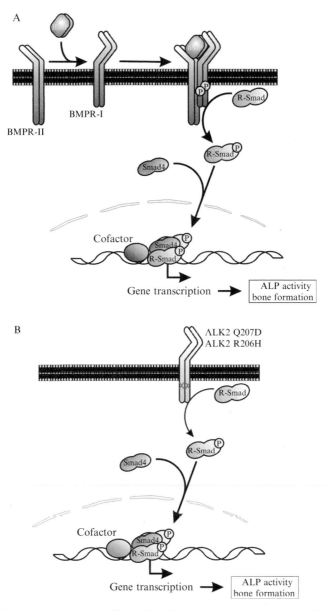

Figure 15.1 Schematic model illustrating the Smad signaling pathway induced by BMP ligands or constitutive active ALK2 in osteoblasts. (A) BMP signals via heteromeric complexes of BMP type I and type II transmembrane serine/threonine kinase receptors and specific intracellular Smad effector proteins. Activated heteromeric complexes between BMP R-Smads (Smad1, -5, and -8) and Smad4 can act as transcription factors and regulate gene transcriptional responses, ultimately stimulating alkaline phosphatase activity (early marker for osteoblast differentiation) and bone formation. (B) Constitutive active ALK2 activates the BMP-Smad pathway in a ligand-independent manner.

patients suffering from atypical FOP with heterotopic ossification combined with unusual additional clinical symptoms (Kaplan *et al.*, 2009). Here, we describe techniques to examine whether mutations found in ALK2 confer constitutive activity to the BMP type I receptor and can sensitize mesenchymal cells to undergo BMP-induced osteoblast differentiation.

2. Determining ALK2 Constitutive Activity

In this section, we describe methods to examine putative constitutive activity of ALK2 by performing Western blotting analysis and luciferase reporter assays. To examine the activity of BMP type I receptor mutants, and compare their activity to the wildtype form, they have to be introduced in cells. This can be achieved by transient transfection procedures or by viral transduction, as described in the following section.

2.1. Western blot analysis of Smad1 phosphorylation

The most specific manner to examine activation of BMP type I receptors is to determine the phosphorylation of their direct substrates, the BMP R-Smads. Activated ALK2, and also activated ALK1, ALK3, and ALK6, phosphorylates the Smad1/5/8 proteins at their extreme C-terminal SXS motif, which can easily be detected by immunoblotting total cell lysates using phospho-specific antibodies raised against this motif (Fig. 15.2; Persson *et al.*, 1998). An additional option to study increased kinase activity of ALK2, which is not described in detail here, is to determine the level of autophosphorylation of the mutated type I receptor compared to that of the wildtype variant by performing an *in vitro* kinase assay (Wieser *et al.*, 1995).

A very valuable cell line to study BMP signal transduction is the murine premyoblast cell line C2C12. C2C12 cells are easy-to-culture highly proliferative cells, display low basal BMP signaling activity and good BMP responsiveness, and can be transfected with high efficiency using most transfection reagents. In response to BMP signal, the default myoblast differentiation will be blocked and the C2C12 cells will differentiate toward the osteoblast lineage, a property which as discussed below can be used to determine BMP receptor activity. Also, cells other than C2C12 can be used, as long as they show a clear BMP response. Furthermore, instead of introducing the transgenes by means of transient transfection, one could express ALK2 mutants by adenoviral transduction of the cells, as described in Section 3.

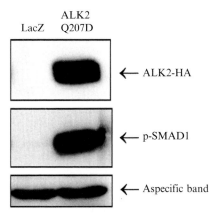

Figure 15.2 Western blot analysis of Smad1 phosphorylation. C2C12 cells were transduced with adenoviruses expressing hemagglutinin (HA)-tagged ALK2 Q207D. The LacZ adenovirus expressing β-galactosidase was used as a negative control. Cell extracts were fractionated by SDS-PAGE and Western blotted. The membranes were incubated with P-SMAD1 antibody, which specifically recognizes phosphorylated Smad1/5/8. Cell lysates were also subjected to immunoblotting to check expression of the transduced ALK2 Q207D using HA-tag antibody. An aspecific band served as a loading control.

Required materials

- C2C12 cells can be obtained from the American Type Culture Collection (ATCC, ATCC Number CRL-1772)
- Transfection reagents
- Expression plasmids to express ALK2, ALK2 containing the mutation(s) of interest, and an empty vector control
- Growth medium: The growth medium for the cells contains Dulbecco's Modified Eagle's Medium (DMEM) supplemented with 10% FCS. Antibiotics (e.g., penicillin, streptomycin) could be used during cell growth, but should in most cases be avoided during the transfection procedure (see manufacturer's protocol)
- Serum-free medium: DMEM without serum, antibiotics can be present
- SDS-polyacrylamide gel electrophoresis (SDS-PAGE) and Western blotting equipment
- SDS-PAGE sample buffer (75 mM Tris–HCl pH 6.8, 5% glycerol, 2.5% β-mercapto-ethanol, 2% SDS, 0.005% bromophenol blue)
- Phosphate-buffered Saline (PBS)
- Cell scrapers
- Ponceau S (200 mg Ponceau S, 15 ml trichloric acetic acid (20%), add water up to 100 ml)
- Smad1/5 phospho-specific antibodies

- Secondary antibodies
- Low-fat dry milk powder
- TBS-T buffer (50 mM Tris–HCl pH 7.4, 150 mM NaCl, 0.1% Tween 20)
- Detection reagents and equipment

Cells should be seeded in 6-well plates and transiently transfected with the different ALK2 expression plasmids according to manufacturer's instructions. Notably, in order to being able to draw any conclusions it is important to compare transfection efficiencies of the various conditions. Therefore, it is recommended to transfect tagged ALK2 mutants (e.g., HA-tagged ALK2 mutants) and determine the expression of these receptors. Alternatively, one can consider cotransfection of GFP, which allows to estimate transfection efficiency by using a fluorescent microscope before cell lysis and given the availability of high affinity GFP antibodies also by immunoblotting.

One day after transfection the cells have to be washed with PBS and serum-starved for at least 2–3 h (overnight serum starvation is recommended since serum contains BMPs). Then, the cells should be washed with (preferably ice-cold) PBS and lysed by adding $\sim$300–400 µl SDS sample buffer, scraping the cells, and boiling the samples for 5–10 min. If desirable, cells can be stimulated with BMPs for 60 min before lysis. After boiling, the samples can either be stored at $-20\,^\circ$C or 20–30 µl can be subjected to SDS-PAGE and Western blotting directly. Since ALK2 has a molecular weight of 65 kDa and Smad1/5 of 60 kDa, it is recommended to make use of 10% SDS-polyacrylamide gels. To our experience good results will be obtained when the proteins are transferred to nitrocellulose membranes, however, PVDF membranes can also be used without any problems. For how to perform SDS-PAGE and Western blotting, the reader is referred to the manufacturer's instructions and "Molecular Cloning: a Laboratory Manual" (Sambrook and Russel, 2001).

In order to confirm whether protein transfer during Western blotting occurred successfully and whether equal amounts of protein is present in the different samples, it is recommended to perform Ponceau staining on the blots. Ponceau S is a negative stain which reversibly binds to the positively charged amino groups of the protein, which become visible as red bands. The blots should be incubated with Ponceau S for approximately 1 min and then destained with water until the proteins are clearly visible (the Ponceau S can be used multiple times). Next, the blots should be blocked by incubating them for at least 20 min in TBS-T containing 5% milk powder, followed by incubation with the Smad1/5 phospho-specific antibodies (1:1000) in TBS-T containing 0.5% milk powder for at least 30 min (or with antibodies directed against the tag of the ALK2 variant under its optimal conditions). Notably, when blocking or primary antibody incubation is performed overnight it is recommended to incubate the blots at 4 $^\circ$C.

Next the blots should be washed three times in TBS-T for 5–10 min, prior to incubation for 30–60 min with the secondary antibodies. Subsequently, the blots should be washed four times in TBS-T for 5–10 min, and detection of antibody binding should be carried out according to the manufacturer's instructions.

2.2. Determining BMP-Smad-dependent transcriptional activity

As mentioned before, active BMP type I receptors phosphorylate Smad1/5/8 proteins at their C-termini, which results in their translocation to the nucleus where they bind promoters of BMP target genes and control their transcription (Fig. 15.1). Work from our laboratory identified the BMP responsive element (BRE) in the promoter of the BMP-target gene *Id1*, and by cloning multimerized BREs upstream of a minimal promoter driving expression of a *Luciferase* reporter gene, a highly sensitive and specific BMP-reporter was obtained (Fig. 15.3; Korchynskyi and Ten Dijke, 2002). Luciferase reporter assays are very sensitive assays enabling subtle differences in transcriptional activity to be measured. Notably, it is important to cotransfect an internal control plasmid, such as *LacZ* or *Renilla luciferase*, enabling to normalize the results and eliminate eventual variations due to differences in transfection efficiency or cell viability. It is essential that the promoter driving expression of the internal control reporter gene is not affected by the stimulus of interest, in this case, ALK2 activity. CMV or thymidine kinase promoter-based vectors can be used as an internal control without any problems.

Required materials

- C2C12 cells can be obtained from the American Type Culture Collection (ATCC, ATCC Number CRL-1772)
- Transfection reagents
- Expression plasmids to express ALK2, ALK2 containing the mutation(s) of interest, and an empty vector control.
- BRE-luciferase reporter construct (Korchynskyi and Ten Dijke, 2002)
- Internal control plasmid (e.g., *LacZ* or *Renilla luciferase* plasmid)
- Growth medium: The growth medium for the cells contains DMEM supplemented with 10% FCS. Antibiotics (e.g., penicillin, streptomycin) could be used during cell growth, but should in most cases be avoided during the transfection procedure (see manufacturer's protocol).
- Serum-free medium: DMEM without serum, antibiotics can be present
- PBS
- Luciferase lysis buffer
- Luciferase reporter assay system (Promega)

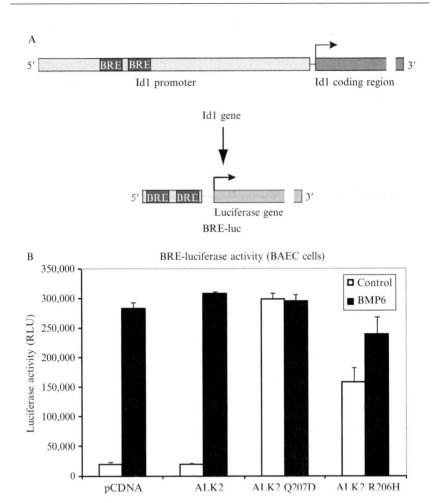

Figure 15.3 BMP-Smad-dependent transcriptional activity determined by luciferase assay. (A) Schematic representation of the generation of the highly sensitive and specific BMP-reporter BMP responsive element (BRE) luciferase construct. Multimerized BREs derived from the promoter of the BMP-target gene *Id1* were cloned upstream of a minimal promoter driving expression of a *Luciferase* reporter gene. (B) BAEC cells were transfected with pcDNA3 or the indicated ALK2-pcDNA3-based expression plasmids (i.e., ALK2, ALK2 Q207D, and ALK2 R206H), together with the BRE-luciferase transcriptional reporter. The cells were not stimulated or stimulated with 100 ng/ml BMP6 for 16 h before the cells were lysed and luciferase activity was measured. RLU, Relative light units.

Cells should be seeded in 24- or 12-well plates and according to the manufacturer's protocol transiently cotransfected in triplicate with the different ALK2 expressing plasmids, the BRE-luciferase reporter (Korchynskyi and Ten Dijke, 2002), and a plasmid that can serve as an internal control, for example, β-*galactosidase* or *Renilla* plasmids. After 1–2 days, the cells should be incubated in serum-free medium for at least 8 h (overnight serum starvation is also an option). It is recommended to transfect 6-wells with each ALK2 mutant and incubate 1 triplo in serum-free medium supplemented with BMP6 or -7 (50 ng/ml). Then, the cells have to be washed with ice-cold PBS and lysed by adding ice-cold luciferase lysis buffer (150 μl in case of 24-well plates, 300 μl in case of 12-well plates), followed by scraping the cells. Next, the activity of luciferase and β-galactosidase/Renilla can be determined directly or the lysates can be stored at $-80\,°C$. Measuring activity of the reporter genes should be carried out according to the reporter assay manufacturer's instructions.

If desired, it is possible to determine the expression levels of the ALK2 receptors in the lysates of the samples by Western blotting, provided that tagged versions of the ALK2 variants were cotransfected. For this purpose, a fraction of the lysates should be mixed with 1/3 volume 4× SDS-PAGE sample buffer (300 mM Tris–HCl pH 6.8, 20% glycerol, 10% β-mercapto-ethanol, 8% SDS, 0.02% bromophenol blue), boiled for 5–10 min, and immunoblotted using primary antibodies directed against the tagged proteins.

3. Determining the Effects of ALK2 Constitutive Activity on Osteoblast Differentiation

In order to determine whether the putative constitutive activity of ALK2 will sensitize mesenchymal cells to undergo osteoblastic differentiation, one can measure alkaline phosphatase activity (Fig. 15.4). Alkaline phosphatase is a marker of early osteoblast differentiation, and measuring ALP activity is a frequently used approach to measure BMP activity in the past. To measure the ability of ALK2 mutants to induce ALP activity in C2C12 cells, we recommend to introduce the transgenes into cells by adenoviral transduction. C2C12 can be transduced efficiently and, unlike what is frequently the case when the cells are transfected with plasmid DNA, adenoviral transduction does hardly affect the differentiation potential of the cells. It is noteworthy that when culturing C2C12 cells it must be avoided that the cells grow confluent, since this will negatively affect their differentiation potential.

3.1. Determining the effect on alkaline phosphatase induction in ALK2 mutant-transduced C2C12 cells

Required materials

- Cells of interest, for example, C2C12
- Adenoviruses; LacZ, ALK2, and ALK2 mutants

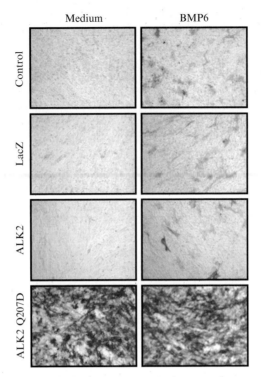

Figure 15.4 The effect on alkaline phosphatase induction in ALK2 mutant-transduced C2C12 cells. C2C12 cells were transduced with adenoviruses encoding hemagglutin (HA)-epitope tagged ALK2 or ALK2 Q207D, or LacZ (expressing β-galactosidase used as a negative control), and incubated for 3 days in the absence or presence of 100 ng/ml BMP6. ALP activity in and/or associated with cells was measured histochemically.

- Growing media: DMEM with 10% FBS and antibiotics (penicillin and streptomycin)
- Recombinant BMP6 (R&D systems)
- PBS
- Naphtol AS-MX phosphate disodium salt (N500, Sigma-Aldrich)
- Fast Blue RR salt (F0500, Sigma-Aldrich)
- $MgSO_4$
- 1 M TRIS pH 8.9
- 1 M NaOH
- Ethanol
- Adeno-X™ Rapid Titer Kit (Clontech)

Equipment

- Spectrophotometer

3.1.1. Production of adenoviruses and transduction of C2C12 cells

Adenoviruses have to be produced in 293T cells according to the Invitrogen ViraPower™ Adenoviral Expression System protocol. The crude viral lysates can then be used to determine the titer using the Adeno-X™ Rapid Titer Kit (Clontech).

C2C12 cells should be seeded in 24-well plates and grown until a confluence of 40–50% is reached at the day of transduction. Adenoviruses have to be added to the cells at a multiplicity of infection (MOI) of 250 in normal culture media. The cells should be incubated with the adenoviruses for 16 h. After the 16 h incubation, the media containing the adenoviruses has to be replaced with normal culture media. Approximately 24 h later, the cells can be stimulated or not simulated with BMP6 in normal media (media change). The cells have to be incubated for 3 or 4 days, after which alkaline phosphatase activity can be determined.

3.1.2. Alkaline phosphatase assay

The transduced and stimulated cells have to be washed once with PBS and then fixed in 3.7% formalin (in PBS) for 5 min at room temperature. Next, the cells should be washed two times with PBS, followed by incubation with ALP staining solution (see recipe below) for 5–30 min, until you can see the purple cells appearing. When incubating too long, you will have a yellow/brownish background staining which will result in high basal values when quantifying the staining. After the incubation with the ALP staining solution, the cells should be washed once with PBS.

For quantifying the staining, remove the PBS and incubate the cells with NaOH/EtOH solution (50 mM NaOH in 100% ethanol). When all the color from the cells has dissolved into the NaOH/EtOH solution, the solutions can be measured using a spectrophotometer at a wavelength of 550 nm. Do not wait too long before measuring the samples because a precipitate may form that can interfere with the measurement.

ALP staining solution

- 0.5 mM Naphtol AS-MX phosphate disodium salt (N500, Sigma-Aldrich)
- 1.5 mM Fast Blue RR salt (F0500, Sigma-Aldrich)
- 0.01% MgSO$_4$
- 100 mM TRIS pH 8.9

4. CONCLUDING REMARKS

Here, we described techniques enabling to demonstrate increased activity of BMP type I receptors. However, to formally show that ALK2 displays constitutive activity, that is, signals in a ligand-independent manner, we suggest the inclusion of some extra conditions in the experimental setup. Extracellular BMP antagonists, such as Noggin and DAN family members, block ligand binding to the receptors and thereby inhibit signal initiation (Gazzerro and Canalis, 2006). Since constitutively active ALK2 by definition should be able to signal independent of BMPs, incubation with Noggin should not (completely) block receptor activity. On the other hand, intracellular antagonists like the inhibitory Smads, Smad6 and -7, or pharmacological BMP type I receptor kinase inhibitors such as dorsomorphin (also known as Compound C) and LDN-193189 are able to inhibit constitutively active ALK2-induced responses (Fujii *et al.*, 1999; van Dinther *et al.*, 2010; Yu *et al.*, 2008a,b). Thus, by also examining the effect of extracellular antagonists and inhibitors of downstream signal transduction on the activity of the ALK2 mutants, it can be determined whether the receptors meet all the criteria for being considered constitutively active.

ACKNOWLEDGMENTS

The research relevant to this chapter was supported by the Centre for Biomedical Genetics, Dutch Organization for Scientific Research (NWO 918.66.606), and IOP Genomics grant IGE07001.

REFERENCES

Chen, Y. G., Liu, F., and Massague, J. (1997). Mechanism of TGFβ receptor inhibition by FKBP12. *EMBO J.* **16**, 3866–3876.

Fujii, M., Takeda, K., Imamura, T., Aoki, H., Sampath, T. K., Enomoto, S., Kawabata, M., Kato, M., Ichijo, H., and Miyazono, K. (1999). Roles of bone morphogenetic protein type I receptors and Smad proteins in osteoblast and chondroblast differentiation. *Mol. Biol. Cell* **10**, 3801–3813.

Gazzerro, E., and Canalis, E. (2006). Bone morphogenetic proteins and their antagonists. *Rev. Endocr. Metab. Disord.* **7**, 51–65.

Huse, M., Chen, Y. G., Massague, J., and Kuriyan, J. (1999). Crystal structure of the cytoplasmic domain of the type I TGFβ receptor in complex with FKBP12. *Cell* **96**, 425–436.

Kaplan, F. S., Xu, M., Seemann, P., Connor, J. M., Glaser, D. L., Carroll, L., Delai, P., Fastnacht-Urban, E., Forman, S. J., Gillessen-Kaesbach, G., Hoover-Fong, J., Koster, B., *et al.* (2009). Classic and atypical fibrodysplasia ossificans progressiva (FOP) phenotypes are caused by mutations in the bone morphogenetic protein (BMP) type I receptor ACVR1. *Hum. Mutat.* **30**, 379–390.

Korchynskyi, O., and Ten Dijke, P. (2002). Identification and functional characterization of distinct critically important bone morphogenetic protein-specific response elements in the Id1 promoter. *J. Biol. Chem.* **277**, 4883–4891.

Okadome, T., Oeda, E., Saitoh, M., Ichijo, H., Moses, H. L., Miyazono, K., and Kawabata, M. (1996). Characterization of the interaction of FKBP12 with the transforming growth factor-β type I receptor in vivo. *J. Biol. Chem.* **271**, 21687–21690.

Persson, U., Izumi, H., Souchelnytskyi, S., Itoh, S., Grimsby, S., Engstrom, U., Heldin, C. H., Funa, K., and Ten Dijke, P. (1998). The L45 loop in type I receptors for TGF-β family members is a critical determinant in specifying Smad isoform activation. *FEBS Lett.* **434,** 83–87.

Piccirillo, S. G., Reynolds, B. A., Zanetti, N., Lamorte, G., Binda, E., Broggi, G., Brem, H., Olivi, A., Dimeco, F., and Vescovi, A. L. (2006). Bone morphogenetic proteins inhibit the tumorigenic potential of human brain tumour-initiating cells. *Nature* **444,** 761–765.

Sambrook, J., and Russel, D. W. (2001). Commonly Used Techniques in Molecular Cloning. Molecular Cloning: A Laboratory Manual. Cold Spring Harbor Press, Cold Spring Harbor, New York, pp. A8.40–A8.55.

Shore, E. M., Xu, M., Feldman, G. J., Fenstermacher, D. A., Cho, T. J., Choi, I. H., Connor, J. M., Delai, P., Glaser, D. L., LeMerrer, M., Morhart, R., Rogers, J. G., et al. (2006). A recurrent mutation in the BMP type I receptor ACVR1 causes inherited and sporadic fibrodysplasia ossificans progressiva. *Nat. Genet.* **38,** 525–527.

Sieber, C., Kopf, J., Hiepen, C., and Knaus, P. (2009). Recent advances in BMP receptor signaling. *Cytokine Growth Factor Rev.* **20,** 343–355.

Urist, M. R. (1965). Bone: Formation by autoinduction. *Science* **150,** 893–899.

van Dinther, M., Visser, N., de Gorter, D. J., Doorn, J., Goumans, M. J., de Boer, J., and Ten Dijke, P. (2010). ALK2 R206H mutation linked to fibrodysplasia ossificans progressiva confers constitutive activity to the bmp type I receptor and sensitizes mesenchymal cells to BMP-induced osteoblast differentiation and bone formation. *J. Bone Miner. Res.* **25,** 1208–1215.

Wang, T., Li, B. Y., Danielson, P. D., Shah, P. C., Rockwell, S., Lechleider, R. J., Martin, J., Manganaro, T., and Donahoe, P. K. (1996). The immunophilin FKBP12 functions as a common inhibitor of the TGFβ family type I receptors. *Cell* **86,** 435–444.

Wieser, R., Wrana, J. L., and Massague, J. (1995). GS domain mutations that constitutively activate TβR-I, the downstream signaling component in the TGF-β receptor complex. *EMBO J.* **14,** 2199–2208.

Wozney, J. M., Rosen, V., Celeste, A. J., Mitsock, L. M., Whitters, M. J., Kriz, R. W., Hewick, R. M., and Wang, E. A. (1988). Novel regulators of bone formation: Molecular clones and activities. *Science* **242,** 1528–1534.

Wrana, J. L., Tran, H., Attisano, L., Arora, K., Childs, S. R., Massague, J., and O'Connor, M. B. (1994). Two distinct transmembrane serine/threonine kinases from *Drosophila melanogaster* form an activin receptor complex. *Mol. Cell. Biol.* **14,** 944–950.

Xu, R. H., Sampsell-Barron, T. L., Gu, F., Root, S., Peck, R. M., Pan, G., Yu, J., Antosiewicz-Bourget, J., Tian, S., Stewart, R., and Thomson, J. A. (2008). NANOG is a direct target of TGFβ/activin-mediated SMAD signaling in human ESCs. *Cell Stem Cell* **3,** 196–206.

Yu, P. B., Deng, D. Y., Lai, C. S., Hong, C. C., Cuny, G. D., Bouxsein, M. L., Hong, D. W., McManus, P. M., Katagiri, T., Sachidanandan, C., Kamiya, N., Fukuda, T., et al. (2008a). BMP type I receptor inhibition reduces heterotopic ossification. *Nat. Med.* **14,** 1363–1369.

Yu, P. B., Hong, C. C., Sachidanandan, C., Babitt, J. L., Deng, D. Y., Hoyng, S. A., Lin, H. Y., Bloch, K. D., and Peterson, R. T. (2008b). Dorsomorphin inhibits BMP signals required for embryogenesis and iron metabolism. *Nat. Chem. Biol.* **4,** 33–41.

CHAPTER SIXTEEN

PROBING THE CONSTITUTIVE ACTIVITY AMONG DOPAMINE D1 AND D5 RECEPTORS AND THEIR MUTANTS

Bianca Plouffe, Jean-Philippe D'Aoust, Vincent Laquerre, Binhui Liang, *and* Mario Tiberi

Contents

1. Introduction	296
2. Design of Genetically Modified D1-Like Receptor Constructs and Cloning Strategy	299
3. Transfection of D1R and D5R Expression Constructs in HEK293 Cells	301
3.1. Making and propagating HEK293 cell stocks	301
3.2. Transfection of HEK293 cells	302
4. Radioligand-Binding Assays	303
4.1. Crude membrane preparation	304
4.2. Saturation and competition curves	304
5. Whole Cell cAMP Assays	305
5.1. Preparation of 6- and 12-well plates for whole cell cAMP assays	306
5.2. Metabolic labeling with [^{3}H]-adenine	307
5.3. Assessment of [^{3}H]-ATP conversion to [^{3}H]-cAMP in intact cells	307
6. Results Validating Experimental Approaches	313
6.1. Serum has no impact on constitutive activity of D1R and D5R	313
6.2. Quantification of constitutive activity of mutant forms of D1R and D5R expressed at low levels	313
6.3. Regulation of forskolin stimulation by D5R constitutive activity and inverse agonists	317
6.4. HA and Flag-tagged D5R exhibit similar CAM phenotype relative to untagged D5R	319
6.5. Phenotypic expression of constitutive activity of human D5R is potentially dependent on cellular factors sensitive to phorbol esters	321

Ottawa Hospital Research Institute (Neurosciences), Departments of Medicine/Cellular and Molecular Medicine/Psychiatry, University of Ottawa, Ontario, Canada

Methods in Enzymology, Volume 484 © 2010 Elsevier Inc.
ISSN 0076-6879, DOI: 10.1016/S0076-6879(10)84016-9 All rights reserved.

7. Concluding Remarks   324
Acknowledgments   326
References   327

Abstract

Dopamine D1 and D5 receptors are prototypical cell-surface seven-transmembrane (TM) G protein-coupled receptors (GPCRs) mediating elevation of intracellular cAMP levels. The high level of constitutive activity of D5 receptor mediating intracellular cAMP production is one of the functional hallmarks distinguishing the closely related D1-like dopaminergic subtypes (D1 and D5). D1-like subtypes share over 80% identity within their TM regions. Thus, D1 and D5 receptors can serve as unparalleled and useful molecular tools to gain structural and mechanistic insights into subtype-specific determinants regulating GPCR constitutive activation and inverse agonism. A method has been developed that relies on the use of transfected human embryonic kidney 293 cells with wild-type (WT), epitope-tagged, chimeric, truncated, and mutant forms of mammalian D1 and D5 receptors using a modified DNA and calcium phosphate precipitation procedure. Receptor expression levels are quantified by a radioligand binding using [^{3}H]-SCH23390, a D1-like selective drug. Regulation of ligand-independent and dependent activity of WT and mutated D1 and D5 receptors is determined by whole cell cAMP assays using metabolic [^{3}H]-adenine labeling and sequential purification radiolabeled nucleotides over Dowex and alumina resin columns. Results on the regulation of D1 and D5 constitutive activity are presented here. Our studies indicate that dopamine-mediated D5 receptor stimulation in a dose-dependent manner is not always detectable, suggesting that D5 receptors can exist in a "locked" constitutively activated state. This "locked" constitutively active state of D5 receptor is not linked to aberrant high receptor expression levels or cell behavior, as D1 receptor function remains essentially unchanged in these cells. In fact, we show that phorbol ester treatment of cells harboring "locked" constitutively active D5 receptors abrogates constitutive activation of D5R to allow its stimulation by dopamine in a dose-dependent manner.

1. Introduction

G protein-coupled receptors (GPCRs) are cell-surface seven-transmembrane (TM) proteins that serve as specialized biosensors for discriminating among a wide array of extracellular cues. Typically, GPCR signal transduction across the plasma membrane is initiated through binding to and activation of receptors by extracellular ligands and sensory stimuli (agonists) that disrupt stabilizing receptor intramolecular interactions (Rosenbaum et al., 2009). This process promotes receptor TM helix movements and conformational changes of TM helix intracellular ends culminating in

GPCR interaction with and activation of heterotrimeric GTP binding proteins (G proteins) localized at the cytoplasmic face of the plasma membrane (Rosenbaum et al., 2009). This mechanistic scheme leads ultimately to the regulation of distinct subsets of downstream effector systems. Ligand binding is then a critical functional property of GPCRs for triggering intracellular signaling through G proteins. However, seminal GPCR studies using reconstituted β2-adrenergic receptors (β2-ARs) and NG108 cell membrane preparations expressing naturally δ-opioid receptors (DORs) have demonstrated that GPCRs can spontaneously adopt active states independently of agonists (Cerione et al., 1984; Costa and Herz, 1989). The active "agonist-free" β2-ARs and DORs were capable of stimulating G proteins and increased basal activity in their respective experimental systems. The constitutive activity measured in these experimental paradigms was selectively blocked by ligands, thereafter named inverse agonists (ligands displaying negative efficacy). The molecular basis for the constitutive activity of GPCRs was initially explored using mutagenesis studies on recombinant adrenergic receptors (reviewed in Cotecchia, 2007). Mutations in the cytoplasmic end of TM6 of α1B-adrenergic receptors (α1B-AR) led to its constitutive activation, a finding that was also recapitulated in β2-ARs and α2A-adrenergic receptors similarly mutated (reviewed in Costa and Cotecchia, 2005).

The discovery of constitutively activating mutations in adrenergic receptors was instrumental in formulating a revised version of the GPCR activation model known as the allosteric ternary complex model and in stimulating research endeavors on inverse agonists (Chidiac et al., 1994; Costa and Cotecchia, 2005; Samama et al., 1993, 1994). The allosteric ternary complex model postulates that GPCRs exist in equilibrium between "*inactive*" (R) and "*active*" (R*) states (Cotecchia, 2007; Samama et al., 1993). In the absence of agonists, GPCRs are predominantly maintained in R state by molecular constraints prohibiting the interaction with G proteins. These molecular constraints are released following agonist binding or by activating mutations, which promote conformational changes in GPCRs. In the two-state model R $\Leftrightarrow$ R*, an isomerization step regulates the most favorable GPCR conformation (R*) capable of interacting with G proteins. Hence, the extent of GPCR constitutive activity is controlled by a shift of the conformational equilibrium toward and stabilization of either R (displaying no or low constitutive activity) or R* (displaying higher constitutive activity). Agonists and inverse agonists stabilize R* and R, respectively. This is achieved through high affinity binding to their preferential state. Meanwhile, classical antagonists exhibit equal affinity at R and R* and are not capable of altering the thermodynamic equilibrium underlying constitutive activity of GPCR systems. In fact, these antagonists do not inhibit basal activity of GPCRs and are thus referred to as neutral ligands.

Difference in the extent of constitutive activity between subtypes of the same GPCR family was first demonstrated with human and rat Gs-coupled

D1-like dopaminergic receptor subtypes (D1R and D5R) (Tiberi and Caron, 1994). Interestingly, D5R naturally displayed the pharmacological properties of constitutively active mutant (CAM) GPCRs (higher constitutive activity, increased agonist affinity, and decreased inverse agonist affinity) when compared with those of D1R expressed in human embryonic kidney 293 (HEK293) cells (Tiberi and Caron, 1994). Overall, these findings suggest that variations in the extent of constitutive activity among highly homologous GPCR subtypes may explain the existence of multiple receptors and the molecular complexity involved in GPCR coupling to the same G protein-effector systems.

The potential physiological relevance of GPCR spontaneous activity was first highlighted with CAM α1B-ARs exhibiting agonist-independent oncogenic property in transfected fibroblasts and nude mice (Allen et al., 1991). Notably, the physiological importance of constitutive activity of GPCRs in human health was soon after recognized with studies reporting that activating mutations in GPCRs was associated with various pathological conditions (Tao, 2008). Likewise, several studies suggest that constitutive activity of GPCRs plays an important role in normal physiological functions such as regulation of food intake (Costa and Cotecchia, 2005). Meanwhile, studies have proposed that constitutive activity of D5R may regulate burst-firing pattern of subthalamic neurons and atrial natriuretic factor release from hypothalamic cell cultures via the Gs-adenylyl cyclase pathway (Baufreton et al., 2005; Lee et al., 1999). Yet, further studies are needed to establish unequivocally that constitutive activity of D5 receptors plays a critical role in normal and pathophysiological conditions. Meanwhile, by virtue of their subtype-specific pharmacological phenotype, D1R and D5R may be considered as prototypical GPCRs existing predominantly in R and R* states, respectively. Consequently, because D1-like dopaminergic receptors share over 80% identity within their TM regions, D1R and D5R have been instrumental in probing molecular and structural determinants underlying constitutive activity among subtypes of the same GPCR family as well as formation of R and R* states. Swapping specific regions of the third intracellular loop (IL3) and cytoplasmic tail (CT) of D1-like subtypes have disclosed crucial determinants that dictate pharmacological properties associated with D1R and CAM phenotype of D5R (Charpentier et al., 1996; Iwasiow et al., 1999; Jackson et al., 2000; Tumova et al., 2003, 2004). Most importantly, these studies have also allowed revisiting notions put forward from studies of CAM adrenergic receptors (Cotecchia, 2007). For instance, mutations in the critical C-terminal region of IL3 (also referred to as the cytosolic extension of TM6) of adrenergic receptors have systematically promoted constitutive activation of adrenergic receptors (Cotecchia, 2007). However, an exchange of divergent residues found in the C-terminal region of IL3 of human D1-like subtypes promoted higher constitutive activation of D1R while reducing

D5R spontaneous activity (Charpentier et al., 1996). Interestingly, we have recently reported that D1R and D5R chimeras retaining constitutive activity of their respective wild-type (WT) counterpart can bind to inverse agonists with similar affinity, which is somewhat opposed to the notion of inverse agonists displaying lower affinity at CAM (D'Aoust and Tiberi, 2010). Overall, studies with D1R and D5R have been valuable in establishing that pharmacological properties reported for CAM cannot merely be explained by an aberrant phenotype of mutated GPCRs. Importantly, studies using D1R and D5R have demonstrated that ligand affinity and agonist-dependent G protein-coupling properties of GPCRs can be dissociated from the constitutive activation status.

To help researchers with structure–activity relationships and regulation studies on constitutive activity among subtypes of the same GPCR family, such as D1R and D5R, we describe an experimental framework and detailed procedures for using mutagenesis, cell culture, and functional assays to explore in a mechanistic fashion how constitutive activity regulates GPCR function. We also report results validating the successful application of different approaches as well as preliminary findings about the potential existence of cellular factors controlling the extent of constitutive activity among GPCRs.

2. Design of Genetically Modified D1-Like Receptor Constructs and Cloning Strategy

In this section, we briefly discuss conditions and procedures for making epitope-tagged receptors using recombinant DNAs encoding D1R and D5R. A more detailed account of experimental procedures for generating genetically modified D1R and D5R using single-point mutations, chimeras and truncations can be found elsewhere (Chaar et al., 2001; Charpentier et al., 1996; Iwasiow et al., 1999; Jackson et al., 2000; Tumova et al., 2003, 2004). Our main cloning strategy is based on subcloning of DNA fragments into mammalian expression vector pCMV5 (Andersson et al., 1989) containing WT form of human or rat D1-like receptors. For the sake of simplicity, we will not present cloning strategy of other mammalian expression vectors such as pcDNA3 family (Invitrogen), which we have also utilized in our laboratory. As for pCMV5, note that this expression vector includes the promoter-enhancer region of the major immediate early gene of the human cytomegalovirus, a synthetic polylinker region containing 11 unique restriction sites (*Eco*RI, *Bgl*II, *Kpn*I, *Mlu*I, *Cla*I, *Hin*dIII, *Pst*I, *Sal* I, *Xba*I, *Bam*HI, and *Sma*I), the transcription termination and polyadenylation region of the bovine growth hormone gene (for greater stability of mRNA transcript), and the SV40 virus DNA replication origin and early region enhancer from plasmid pcD-X.

For the epitope tagging of the amino terminus of human D1R and D5R, we have relied on the widely used HA (Tyr-Pro-Tyr-Asp-Val-Pro-Asp-Tyr-Ala) and Flag (Asp-Tyr-Lys-Asp-Asp-Asp-Asp-Lys) peptide sequences, for which antibodies are commercially available. A site-directed polymerase chain reaction (PCR)-based overlap extension procedure was employed to generate HA- and Flag-tagged D1R and D5R using universal HA and Flag oligonucleotide primers (Sigma Genosys) containing EcoRI restriction site, ribosome binding region (CGCCGCCACC), ATG, and partial epitope sequences along with specific DNA oligonucleotides for each of D1-like subtypes, as listed in Table 16.1. In brief, WT human D1R and D5R DNA sequences subcloned in pCMV5 were used as templates in a two-step PCR procedure.

In the first step, DNA sequences are amplified in separate PCRs using forward primers, HD1HA1, HD5HA1, HD1FLAG1, and HD5FLAG1, with the reverse pCMV5-B2 primer (Table 16.1). PCR products are separated on 1% agarose gel, excised, and bands purified on Qiaex II resin (Qiagen) according to the manufacturer's protocol. Purified fragments are

Table 16.1 Oligonucleotide sequences of PCR primers for the construction of HA- and Flag-tag at the amino terminus of human wild-type D1R and D5R

Primer name	Primer sequence 5′ → 3′	1st Step PCR	2nd Step PCR
UHA1 (forward)	ggAATTcgccgccAccATgTAcccATAcgAcgTc (34-mer oligonucleotide)		√
HD1HA1 (forward)	TAcgAcgTcccAgAcTAcgcTAggAcTccTgAAc (34-mer oligonucleotide)	√	
HD5HA1 (forward)	TAcgAcgTcccAgAcTAcgcTcTgccgccAggc (33-mer oligonucleotide)	√	
UFLAG1 (forward)	ggAATTcgccgccAccATggAcTAccAggAcgAT (34-mer oligonucleotide)		√
HD1FLAG1 (forward)	AAggAcgATgAcgAcAAgAggAcTcTgAAc (30-mer oligonucleotide)	√	
HD5FLAG (forward)	AAggAcgATgAcgAcAAgcTgccgccAggc (30-mer oligonucleotide)	√	
pCMV5-B2 (reverse)	TTAggAcAAggcTggTgg (18-mer oligonucleotide)	√	
pCMV5-3′ (reverse)	ggccAggAgAggcA (14-mer oligonucleotide)		√

Standard PCRs were done using the following conditions: 1 cycle (94 °C for 3 min, 50 °C for 1 min, 72 °C for 3 min) and 25 cycles (94 °C for 45 s, 50 °C for 1 min, 72 °C for 1 min) completed by an anneal extension step at 72 °C for 8 min. Primers used in the 1st and 2nd steps of PCR are indicated.

then used as DNA templates in the second PCR step using universal HA and Flag primers with the pCMV5-3′ primer (Table 16.1). The final PCR products are purified, digested with appropriate DNA restriction enzymes (human D1R: *Eco*RI and *Bgl*II; human D5R: *Eco*RI and *Apa*I), purified again on Qiaex II resin, and ligated in-frame with respective linearized human D1R and D5R-pCMV5 constructs. The integrity of epitope and coding sequences is confirmed using automated fluorescent DNA sequencing. The HA- and Flag-tagged D1R and D5R are available upon request.

3. Transfection of D1R and D5R Expression Constructs in HEK293 Cells

In this section, the procedures used in our laboratory for the propagation and transfection of human adenovirus type 5-transformed HEK293 cells (CRL-1573; ATCC, Manassas, VA, USA) are described, which we found highly suitable in terms of transfection efficiency, cell viability, and costs.

3.1. Making and propagating HEK293 cell stocks

Nalgene cryovials containing 1 ml of cells ($\sim 5 \times 10^6$ cells in frozen medium: 10% cell culture grade and sterile Sigma Hybri-Max dimethylsulfoxide (DMSO; Cat. No. D2650); 20% heat-inactivated fetal bovine serum (FBS); 70% minimal essential medium with Earle's salts (MEM, Invitrogen, Cat. No. 11095080) containing 40 µg/ml of gentamicin sulfate (10 mg/ml stock solution; Invitrogen, Cat. No. 15710-064)) are stored in a liquid nitrogen Dewar. Note ATCC recommends growing HEK293 cells in horse serum (15%), which we have replaced with heat-inactivated FBS without any noticeable effect on cell proliferation. Our studies are routinely done with cells from one of three different frozen stocks of HEK293 cell isolates, all of which are at the starting cell passage 38 (P38). To make lab stocks of cells, one frozen vial of cells (1 ml) is briefly "warmed up" on ice and rapidly thawed in a 37 °C water bath, cells added to one polystyrene 75 cm^2 flasks (with 0.2 µm vented blue plug seal cap; BD Falcon, Cat. No. 137787) containing 20 ml of complete MEM (10% FBS and 40 µg/ml gentamicin), and flasks are put at 37 °C in humidified 5% CO_2 incubators. The next day cells are fed with fresh complete MEM. Cells are grown for 3–4 days when reaching 80–100% confluency and reseeded in new flasks. To reseed and expand the flask stocks of HEK293 cells, media is aspirated and cells washed in 5 ml of Ca^{++}- and Mg^{++}-free phosphate buffered saline 1× (PBS) using gentle hand rocking. PBS should not be added directly on cells as they may detach prior to trypsinization. PBS is then aspirated and 1 ml of trypsin–EDTA

1× solution (trypsin 0.25%, EDTA 0.05%; Invitrogen, Cat. No. 25200114) is added per flask. Cells are incubated for <1 min at room temperature, harvested with 20 ml of complete MEM, and gently mixed by trituration using 10 ml pipette (10 times) to dissociate cells and reduce clump formation. During trypsin incubation, flasks can be gently rocked to speed up cell dissociation. Following trituration, 20 ml of complete MEM is added to each flask (40 ml total) and cells counted using hemacytometer. New stocks are prepared using $\sim 2 \times 10^6$ cells per flask and cells cultivated for a week during which media is replaced with fresh complete MEM once either at culture day 3 or 4 postseeding.

Preparation of HEK293 cells for transfection is performed as follows. Polystyrene tissue culture dishes (100 × 20 mm) are seeded with $\sim 2.5 \times 10^6$ cells in a final volume of 10 ml of complete MEM and grown until the next day (transfection day). Cells can also be seeded at a lower density (2×10^6 cells/dish) for transfection to be done 2 days after seeding. However, in this case, it is recommended that media be replaced the day before the transfection.

3.2. Transfection of HEK293 cells

Procedure for transfection of HEK293 cells with high quality plasmid DNA (Qiagen kit procedure) using a DNA-calcium phosphate precipitation approach has been optimized for cells seeded in 100 × 20 mm tissue culture dishes. For studies requiring transfection of the receptor alone, 5 μg of DNA per dish is the amount in our hands that gives maximal receptor expression. In our studies, higher DNA amounts have not yielded to any significant augmentation in WT D1-like receptor numbers (WT human and rat D1-like typically expressed at a ~ 15 and ~ 30 pmol/mg membrane proteins, respectively). Moreover, we have not observed any increase in the low-expressing mutant forms of D1R and D5R following the use of higher amount of plasmid DNA (10–20 μg/dish). However, if cotransfection of different expression plasmids is planned (e.g., regulation by specific kinases or accessory proteins), it may be necessary to employ a higher total DNA amount (>5 μg).

Plasmid DNA transfection mixture is prepared in sterile 13-ml plastic tube (100 × 16 mm). A total of 10 μg of plasmid DNA (in volume less than 50 μl) is first added to tubes. Volume in each tube is adjusted to 900 μl using sterile milli-Q-water, followed by the addition of 100 μl of 2.5 M calcium chloride. For this step, it may be necessary to remove mixture drips left on side of tubes by gentle tapping on a hard surface. Next, 1 ml of 2× HEPES-buffered saline solution (0.28 M NaCl, 0.05 M HEPES, pH 7.0, 1.5 mM Na$_3$PO$_4$, pH 7.1) is added dropwise to DNA-calcium phosphate solution to a final volume of 2 ml and mixed by gentle flicking of tubes. The transfection mixture (2 ml) is used to transfect two 100 × 20 mm dishes using 1 ml per dish added dropwise to the whole surface of media. HEK293 cells are then incubated with DNA-calcium phosphate precipitates overnight at 37 °C in humidified 5% CO_2 incubators. If studies require lower receptor expression (e.g., dose–responses curves for cAMP production or normalization of receptor levels between WT and mutant forms), plasmid DNA quantities for receptor constructs can be titrated accordingly to obtain similar expression levels. Meanwhile, researchers will ensure that the total amount of plasmid DNA is normalized at a constant concentration between conditions using empty plasmid (e.g., pCMV5). This will reduce variations in transfection efficiency between experimental conditions.

Most importantly, we have consistently measured similar transfection efficiencies. This was done using cells transfected with β-arrestin 2 or different WT and mutant receptor constructs (tagged with green fluorescent protein or small epitope such as HA and Flag). Cells processed for immunofluorescence microscopy were scored for expression of fluorescent-tagged constructs and nuclear (Hoechst) staining relative to cells displaying only nuclear staining. Experiments performed using fluorescence-activated cell sorter gave essentially similar results. We established that in our hands, the transfection efficiency of HEK293 cells is ∼75% using the experimental approach described above. In the following sections, we describe experimental procedures to assess ligand-binding properties, agonist-independent and dependent activities of WT, and mutant forms of D1R and D5R.

Note: It is worth mentioning that in our experience, proliferation rate of HEK293 cells gradually increased up to 52 passages. Beyond 52 passages, cells grow significantly faster, display morphological changes (form foci), and are less adherent, which lead to lower expression of transfected D1-like receptor constructs and unreliable whole cell cAMP assays. We have thus set a range of 40–52 passages for experimental use.

4. Radioligand-Binding Assays

In this section, we describe experimental approaches utilizing membrane preparations from transfected cells and radioligand-binding assays to measure affinity of dopaminergic ligands and receptor expression of WT and mutant forms of D1R and D5R. These studies represent the first step for probing and comparing ligand-binding properties of WT and mutant forms of D1R and D5R with respect to the acquisition or loss of CAM phenotype by mutated D1-like receptors.

4.1. Crude membrane preparation

Following an overnight incubation with DNA–calcium phosphate precipitates, culture medium in 100 × 20 mm dishes is aspirated and cells washed with room temperature PBS (~5 ml). Note that PBS should not be added directly onto cells but on the side of dishes to avoid detaching cells. Thereafter, cells are trypsinized, triturated, and pooled into 150 × 25 mm polystyrene tissue culture dishes. Typically, for WT D1R and D5R, we pooled four 100 × 20 mm dishes of HEK293 cells transfected with 5 μg or more of plasmid DNA. In our hands, one 150 × 25 mm dish yields enough membrane receptor preparation to perform several radioligand-binding assays (one saturation binding experiments and displacement curves to test 4 drugs). When larger amount of membrane preparations are required, the number of 100 × 20 mm dishes transfected will be increased accordingly always using 3–4 small dishes (100 × 20 mm) pooled in one large dish (150 × 25 mm) to prevent over confluent cell cultures. Cells pooled in 150 × 25 mm dishes are grown for ~48 h at 37 °C in a humidified 5% CO_2 environment to allow optimal receptor expression. On the day of experiment, dishes are put on an ice tray and the culture medium is removed. Then, 10 ml of cold PBS is added to the side of dishes to wash cells, PBS is aspirated, and cells are harvested in 15 ml of ice-cold lysis buffer (10 mM Tris–HCl, pH 7.4; 5 mM EDTA, pH 8.0) using cell lifter and transferred to 50 ml polycarbonate centrifuge tubes (29 × 104 mm). Harvested dishes are washed once with 15 ml of lysis buffer and the volume is added to cell lysates in centrifuge tubes. Samples are centrifuged at 40,000×g for 20 min at 4 °C and supernatants are discarded. Membrane pellets are detached, homogenized in a centrifuge tube containing 3 ml of lysis buffer using Brinkmann Polytron (17,000 rpm for 15 s), and the final volume is adjusted to 30 ml of lysis buffer prior to final centrifugation (40,000×g for 20 min at 4 °C). The supernatants are discarded and pellets are homogenized in 3 ml of cold lysis buffer. A fraction of membrane homogenates (0.6 ml) is diluted (1:6) with 3 ml resuspension buffer (62.5 mM Tris–HCl, pH 7.4; 1.25 mM EDTA, pH 8.0) and used immediately for saturation studies. The remaining lysates are divided into two aliquots (~1.2 ml) added to microfuge tubes, frozen in liquid nitrogen, and stored at −80 °C until used for competition studies.

4.2. Saturation and competition curves

Binding reactions are carried out in polystyrene culture tubes (12 × 75 mm, VWR International, Cat. No. CA60830-021) using 100 μl of membranes and 50 μl of [^{3}H]-SCH23390 (60–90 Ci/mmol; Perkin–Elmer) in the absence and presence of cold competing drugs in a final volume of 500 μl of assay buffer (final concentration in assays: 50 mM

Tris–HCl, pH 7.4; 120 mM NaCl; 5 mM KCl; 4 mM MgCl$_2$; 1.5 mM CaCl$_2$; 1 mM EDTA, pH 8.0) at room temperature ($\sim$20 °C) for 90–120 min. Radioligand and drugs are made in milli-Q-water.

For saturation curves, fresh membranes are incubated with increasing concentrations of [^{3}H]-SCH23390 (final in assays: $\sim$0.01–10 nM) in the absence and presence of a final concentration of 10 µM of cis-flupenthixol (Sigma-Aldrich, Cat. No. F-114) to measure the total and nonspecific binding, respectively. For competition curves, frozen membranes (1.2 ml per aliquot) are thawed on ice, mixed in 5.5 ml (to test two cold drugs) or 11 ml (to test 4 cold drugs) of resuspension buffer using Brinkmann Polytron (17,000 rpm for 15 s) and incubated with increasing concentrations of cold competing drugs and a constant concentration of [^{3}H]-SCH23390. Determination of the constant concentration of radioligand is based on K_d of [^{3}H]-SCH23390 for the receptor under study (final concentration usually ranged from $\sim$0.5 to 1.5 nM). Competition studies using dopamine are performed in the presence of a final concentration of 0.1 mM ascorbic acid (to reduce dopamine oxidation).

At the end of incubation period, binding reactions are terminated using rapid filtration through Whatman glass fiber filters (GF/C, VWR International, Cat. No. 28497-619). Filters are washed three times with 5 ml of cold washing buffer (50 mM Tris–HCl, pH 7.4; 100 mM NaCl), put in plastic scintillation vials, and tritium-bound radioactivity is assessed by liquid scintillation counting with 30–40% efficiency (Beckman Counter LS6500). Protein concentration of membranes used in saturation studies is measured using Bio-Rad assay kit with bovine serum albumin (BSA) as standard. Binding isotherms are analyzed using a nonlinear curve-fitting program from GraphPad Prism (GraphPad Software, San Diego, CA, USA, www.graphpad.com) to determine equilibrium dissociation constant (K_d, nM) and maximal binding capacity (B_{max}, pmol/mg of membrane proteins; an index of receptor expression) of radioligand (saturation studies) and equilibrium dissociation constant of unlabeled drugs (K_i, nM; competition studies).

5. Whole Cell cAMP Assays

In this section, we describe an experimental approach based on whole cell cAMP assays to probe constitutive activity and agonist-dependent activity of WT, and mutant D1R and D5R. To begin with the assessment of constitutive activity, WT and mutant forms are transfected in HEK293 cells with 5 µg of plasmid DNA per dish (as described above) to reach highest receptor expression. Importantly, separate dishes are transfected with empty vector for mock condition. The mock condition is critical in determining the amount of endogenous basal adenylyl cyclase activity and intracellular cAMP levels in cells not heterologously transfected with Gs-coupled GPCRs (e.g., D1-like receptors). A high baseline value observed in mock condition will likely interfere with detection of robust constitutive activity of transfected WT and mutant receptors even at high receptor expression. Generally, we observe that transfected cells expressing D1R (displaying lower constitutive activity than D5R) at receptor densities of 15–20 pmol/mg membrane proteins have $\sim$10–15-fold higher basal cAMP levels when compared with mock condition. As constitutive activity of GPCRs is linearly correlated with receptor densities (Samama et al., 1993; Tiberi and Caron, 1994), the issue of cells with high baseline value for intracellular cAMP will be relevant for mutant GPCRs displaying reduced expression. Additionally, the choice of cell culture dishes for whole cAMP assays is also critical. The 12-well plates are routinely used for generating dose–response curves to agonists as well as testing inverse agonists with whole cell cAMP assays (D'Aoust and Tiberi, 2010). However, we strongly recommend that 6-well plates be employed initially to characterize constitutive activity of WT and mutant D1R and D5R versus mock condition. Seeding higher amounts of cells per well will be possible using these plates and hence help achieving greater assay sensitivity for assessment of constitutive activity levels.

5.1. Preparation of 6- and 12-well plates for whole cell cAMP assays

Following an overnight incubation with DNA-calcium phosphate precipitates, HEK293 cells are reseeded in 6- or 12-well culture plates. Typically, either using 6- or 12-well plates, four transfected 100 × 20 mm dishes per experimental conditions are utilized to measure metrics for constitutive activity and agonist-dependent G protein-coupling properties. Cells in dishes are washed with PBS, incubated with trypsin (0.5 ml per dish), harvested in complete MEM, pooled, and mixed by gentle trituration, as described above (Sections 3 and 4). For studies probing constitutive activity and dopamine-mediated maximal stimulation of adenylyl cyclase (E_{max}) in cells with highest receptor expression (achieved using 5 μg of plasmid DNA), one 6-well plate per experimental condition is seeded using 4 ml per well ($\sim$200,000 cells). For experiments assessing potency and E_{max} of dopaminergic agonists with dose–responses curves, cells expressing moderate levels of receptors ($\sim$1–3 pmol/mg membrane proteins) are seeded in two 12-well plates using 1 ml per well ($\sim$50,000 cells). Additionally, for each experimental condition, one 100 × 20 mm dish is seeded with 10–15 ml of cells, which is referred to as a binding dish. The binding dish is used to prepare crude membranes to assess B_{max} value of receptors with [^{3}H]-SCH23390. Cells seeded in culture plates are grown in complete MEM at 37 °C in humidified 5% CO_2 environment for $\sim$24 h prior to metabolic labeling with [^{3}H]-adenine, while cells in binding dishes are kept in complete MEM until the day of experiments (~48 h) without changing media.

5.2. Metabolic labeling with [^{3}H]-adenine

The metabolic labeling of transfected HEK293 cells with [^{3}H]-adenine (24–27 Ci/mmol, Perkin–Elmer, Cat. No. NET063) is adapted from a cell monolayer method for measurement of [^{3}H]-ATP conversion to [^{3}H]-cAMP described before in this series (Salomon, 1991). The day following cell reseeding in culture plates, medium is replaced with fresh MEM containing 5% FBS, 40 µg/ml gentamicin, and [^{3}H]-adenine. For 6- and 12-well plates, cells in each well are labeled overnight with 2 ml (2 µCi/ml) and 1 ml (1 µCi/ml) of [^{3}H]-adenine medium, respectively. We noted that higher [^{3}H]-adenine concentration does not significantly improve the assay sensitivity with respect to measuring constitutive activity in studies using 6-well plates. Moreover, metrics (potency and E_{max}) assessed from dose–response curves in 12-well plates remains essentially unchanged whether cells are labeled with 1 or 2 µCi/ml of [^{3}H]-adenine. Consequently, the use of 1 µCi/ml of [^{3}H]-adenine will significantly reduce the amount of radioactivity and cost of experiments. Alternatively, if cells are not labeled on the day following reseeding (day 2 posttransfection), it is possible to label cells on the day of experiment (day 3 posttransfection) using a metabolic labeling period of 4 h prior to performing whole cell cAMP assays. However, detectable [^{3}H]-cAMP levels will be reduced accordingly, and we thus suggest using higher [^{3}H]-adenine concentration to circumvent low signal detection. Additionally, the potential role of FBS contaminating catecholamines as a confounding factor for constitutive activity of D1-like receptors has been previously addressed by replacing serum in labeling medium with a supplement containing insulin, transferrin and selenite (Tiberi and Caron, 1994). Herein, we provide additional data supporting the lack of serum contribution to the basal activity of D1R and D5R (see Section 6).

5.3. Assessment of [^{3}H]-ATP conversion to [^{3}H]-cAMP in intact cells

5.3.1. Whole cell cAMP assay medium

On the day of the experiment, 20 mM HEPES-buffered MEM containing 1 mM isobutyl-1-methylxanthine (IBMX, Sigma-Aldrich, Cat. No. I5879; a nonselective phosphodiesterase inhibitor), referred to as cAMP assay medium, is freshly prepared as follows: a stock solution of 200 mM IBMX prepared in DMSO and stored at 4 °C, is thawed in a 37 °C water bath and diluted in 20 mM HEPES-buffered MEM to a final concentration of 1 mM. The HEPES-buffered MEM (without IBMX) is usually made in advance and kept refrigerated until used for several weeks. The stock solution of IBMX (generally in 5 ml of DMSO) is prepared in a 50 ml polypropylene conical tube to facilitate rapid thawing of IBMX at 37 °C. We recommend making a small volume of solution to avoid frequent "freeze–thaw" cycles. If an "unfrozen" IBMX solution kept at 4 °C is noted, it should be disposed of. The cAMP assay medium is kept at 37 °C until completion of experiments and leftover discarded.

5.3.2. Setup of cAMP assays and sample preparation using cell monolayers

Labeling medium in culture plates is aspirated and HEK293 cells incubated in 2 ml (6-well plates) or 1 ml (12-well plates) of cAMP assay medium in the absence or presence of drugs for varying time periods (0–30 min) at 37 °C. Ascorbic acid 0.1 mM final in well or milli-Q-water is used as a control vehicle in the absence of drugs. The volume of drugs added per well is 20 and 10 µl for 6- and 12-well plates, respectively. At the end of the incubation period, the culture plates are put on ice, the cAMP assay medium is removed, and each well is filled with 1 ml of lysis solution containing 2.5% perchloric acid (PCA), 0.1 mM cAMP (Sigma-Aldrich, Cat. No. A6885), and [^{14}C]-cAMP (250–275 mCi/mmol, Moravek, Cat. No. MC157) using bottletop dispenser. The amount of [^{14}C]-cAMP in lysis solution is $\sim$3.3 nCi per ml ($\sim$10,000 dpm). This tracer is used for measuring [^{3}H]-cAMP recovery by chromatography column procedure. Plates are incubated with lysis solution at 4 °C for 30 min. Lysates in each well are transferred to culture tubes (12 × 75 mm) containing 0.1 ml of neutralizing solution (4.2 M KOH) and briefly vortexed. Samples can either be subjected immediately to sequential chromatography on Dowex and alumina columns or be stored at 4 °C up to 5 days. We have not detected any differences between the results obtained from the samples processed immediately or after being stored at 4 °C. If longer time of storage is needed, samples should be frozen at -20 °C. Importantly, samples are centrifuged at 1500 rpm (400–500×g, 4 °C; Beckman Coulter Allegra^{TM} 6R centrifuge) for 10 min to precipitate potassium perchlorate salts before proceeding with the sequential chromatography procedure described in the following section.

In the meantime, cells in binding dish are harvested with 10 ml lysis buffer in a polycarbonate 15 ml centrifuge tube (18 × 100 mm, Beckman, Cat. No. 34080) and membranes prepared essentially, as described in Section 4.1. Final pellets are homogenized in 0.6–1 ml of resuspension buffer and membranes used immediately. Receptor expression is determined by incubating membrane preparations with a saturating concentration of [^{3}H]-SCH23390 in the absence or presence of 10 µM cis-flupenthixol as detailed above (Section 4.2).

5.3.3. Preparation of chromatography columns for sequential purification of [^{3}H]-cAMP

Procedures described herein have been adapted from those previously reported in this series (Johnson et al., 1994). Different sets of 0.8 × 4 cm polypropylene chromatography columns (Poly-Prep Columns with 10 ml reservoir and graduated volume markings, Bio-Rad Laboratories, Inc., Cat. No. 731–1550) are packed with Dowex AG 50W-4X resin (Hydrogen form, 200–400 dry mesh, 63–150 μm wet bead; Bio-Rad laboratories, Inc., Cat. No. 142-1351) and alumina (Alumina N Super I, MP Biomedicals, Cat. No. 04583). Sets of Dowex (100) and alumina (100) columns are separately mounted on custom-made Plexiglas racks with equal spacing. These racks must be designed such that Dowex columns can be directly eluted into alumina columns and elution from alumina columns can be carried out through direct dripping in racks of scintillation vials.

Prior to mounting columns on Plexiglas racks, the column tip closure is snapped off and 1 ml volume marking highlighted by drawing a line using a black ink marker. The polyethylene bed support inside the column is also removed and a small piece of glass wool (Pyrex Brand Glass Wool, Fisher Scientific, Cat. No. 11-388) is put at the bottom of the columns for resin packing. Separate sets of columns mounted on racks are filled with Dowex and alumina as follows: Dowex resin is mixed with milli-Q-water in a beaker and Dowex slurry added to individual columns up to black line (1 ml) using P-1000 or P-5000 pipette with a cutoff tip. Let water drain completely and add more Dowex suspension if the amount of resin is lower than 1 ml marking. Likewise, another set of columns is filled with alumina. Alumina is mixed with 0.1 M imidazole pH 7.5 (Sigma-Aldrich, Cat. No. I0125) in a beaker and the resin suspension poured into columns up to 1 ml volume marking (black line) using a pipette as described for Dowex procedure. Alternatively, dry alumina powder is put into a beaker and columns filled up to 1 ml volume marking (black line). However, if dry powder is added to columns, more washes will be needed prior to testing elution profile. Once columns are filled up with wet resin to 1 ml volume marking, Dowex and alumina are then subjected to a series of washes prior to determining elution profiles. Liquid wastes from column racks are collected in polypropylene tray (30.5 × 40.5 × 20.25 cm; Thermo Fisher Scientific Inc., Cat. No. 15-236-2D). Note that the Dowex procedure described before (Johnson et al., 1994) recommended that resin be first washed sequentially with 0.1 N NaOH and water prior to acid wash. In our experience, NaOH makes no difference in cAMP elution profile of Dowex columns and we now omit this step. We routinely calibrate Dowex and alumina columns as follows: Dowex columns are washed twice with 10 ml of 0.1 N HCl. Let column drain completely before proceeding with second HCl wash. Dowex is then rinsed twice with 10 ml of distilled water. For alumina columns, wet resin is washed twice with 10 ml of 0.1 M imidazole pH 7.5. Columns are then ready for measurement of elution profile using [^{14}C]-cAMP.

The elution profile of [^{14}C]-cAMP is first performed on Dowex using 3 or 4 columns. Tubes containing 1 ml of KOH-neutralized lysis solution are subjected to low-speed centrifugation as stated in previous section (Section 5.3.2). Test columns are placed above plastic scintillation vials and 0.85 ml of [^{14}C]-cAMP supernatants ($\sim$8000–10,000 dpm) are applied to Dowex. Elution flow through is saved in vials and Dowex columns are washed sequentially with nine volumes of 1 ml distilled water and eluates (1 ml) collected in different vials. Vials are filled with 10 ml of scintillation counting fluid (Bio-Safe II^{TM}, Biodegradable Counting Cocktail, Research Products International, Inc., Cat. No. 111195) and [^{14}C]-cAMP radioactivity per vial is quantified to establish Dowex elution profile. Figure 16.1A depicts a representative example of elution profiles obtained with Dowex columns. We routinely obtained the following Dowex profile: 1 ml for wash and 4 ml for elution onto alumina columns. Following the determination of wash and elution volumes for Dowex columns, elution profile can be assessed with different alumina columns (3 or 4). To do so, [^{14}C]-cAMP supernatants (0.85 ml) are first loaded on different test Dowex columns and washed with an appropriate volume of distilled water. Test columns are then fitted on top of alumina columns, [^{14}C]-cAMP bound to Dowex resin eluted into alumina columns using correct water volume (e.g., 4 ml) and the alumina column flow through collected in scintillation vials. In a similar fashion to Dowex, elution profiles of alumina columns are done with repeated 1 ml washes (nine times) using 0.1 M imidazole pH 7.5, eluted samples mixed with 10 ml of scintillation cocktail, radioactivity counted in a Beckman LS6500 scintillation counter, and typical alumina elution profiles are determined, as shown in Fig. 16.1B (e.g., 1 and 3 ml imidazole for alumina wash and elution volumes, respectively). It is worth mentioning that elution volume for alumina columns should not exceed 4 ml of imidazole as larger aqueous volumes have led to reduced miscibility with biodegradable scintillation cocktails tested. In fact, most biodegradable scintillation fluids we used have a maximal sample holding capacity of up to 25% aqueous. However, we have noted that 4 ml of imidazole is not always miscible with 16–17 ml of scintillation cocktail. Miscibility of aqueous samples may be dependent on scintillation fluid and/or alumina resin batches. Notably, the size of precipitates may impact detection sensitivity of the whole cell cAMP assays. Unless critical for improving sample recovery of sequential chromatography, experimenters should use 3 ml imidazole to elute most of radioactive cAMP (Fig. 16.1B). Indeed, eluting with 4 ml imidazole had only a small incremental effect on the overall sample recovery when compared with 3 ml (Fig. 16.1C). Miscibility of 3 ml aqueous samples using scintillation cocktail (16–17 ml) is not an issue.

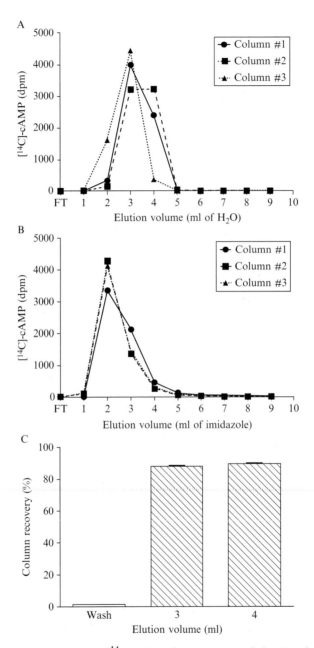

Figure 16.1 Elution profiles of [^{14}C]-cAMP from Dowex and alumina chromatography columns. Representative examples of [^{14}C]-cAMP flow through (FT) and elution profiles obtained with Dowex (A) and alumina (B) columns using sequential addition of 1 ml of distilled water and imidazole, respectively. Comparison of the overall column recovery of [^{14}C]-cAMP obtained from alumina columns using 3 and 4 ml imidazole is shown in (C).

5.3.4. Quantification of intracellular [^{3}H]-cAMP levels produced by agonist-independent and dependent activation of WT and mutant forms of D1R and D5R

Clarified supernatants (0.85 ml) from PCA lysates of 6- or 12-well culture plates are subjected to sequential purification using Dowex and alumina columns, as previously described in Section 5.3.3. (*Note*: the choice of a volume of 0.85 ml applied on Dowex columns is to avoid disturbing and pipetting salt sediments at the bottom of the tube.) Imidazole eluates from alumina columns are mixed with scintillation cocktail to determine [^{3}H]-cAMP amounts (CA) in each well. Importantly, a 50 μl aliquot from an unused volume of each PCA lysates is added to separate scintillation vials to assess the total amount of [^{3}H]-adenine uptake (TU) in each well. Additionally, triplicate samples of 0.85 ml neutralized lysis solution ([^{14}C]-cAMP) are prepared by mixing with imidazole (∼2 ml) and scintillation cocktail. The average [^{14}C]-cAMP counts in lysis solution (C14$_{total}$) will be used to calculate column recovery and determine the appropriate correction factor (CF) for each column. Radioactivity is quantified with dual ^{3}H and ^{14}C liquid scintillation counting using Beckman Counter LS6500. Counts in dpm for ^{3}H and ^{14}C are entered in Excel spreadsheet, and intracellular [^{3}H]-cAMP levels generated by constitutive activity and ligand-mediated regulation of WT and mutant forms of D1R and D5R are calculated using the following formulas:

1. CF = C14$_{total}$/^{14}C dpm sample recovered from each column.
2. [^{3}H]-cAMP counts corrected for column efficiency (^{3}H$_{corrected}$): ^{3}H dpm sample recovered from each column × CF.
3. [^{3}H]-cAMP produced in total volume of neutralized lysates (1 ml lysate + 0.1 ml KOH = 1.1 ml) or CA: (1.1/0.85) × ^{3}H$_{corrected}$ (measured from 0.85 ml lysates).
4. Total amount of [^{3}H]-adenine uptake or TU from 50 μl lysate aliquots: ^{3}H$_{50μl}$ dpm sample × 22 (1.1 ml/0.05 ml).
5. Intracellular cAMP levels or adenylyl cyclase activity in arbitrary units: CA/TU × 1000.

Alternatively, [^{3}H]-cAMP levels can be expressed as a percent conversion (PC) of the total uptake of [^{3}H]-adenine using the formula CA/TU × 100.

Data can then be normalized to a particular condition if required. Moreover, results obtained with dose–response curves can be analyzed by a four-logistic parameter equation using a nonlinear curve-fitting software from GraphPad Prism. In the following section, we report unpublished data validating experimental procedures described herein.

6. Results Validating Experimental Approaches

6.1. Serum has no impact on constitutive activity of D1R and D5R

To assess the impact of serum in labeling medium, HEK293 cells transfected with empty expression vector (pCMV5) or human D1-like receptors grown in 6-well culture plates were labeled overnight with [^{3}H]-adenine (2 µCi/ml) in our standard labeling MEM medium (5% FBS) or UltraCULTURE^{TM} general purpose serum-free medium (Lonza, Cat. No. 12-769F) supplemented with L-glutamine (2 mM). The next day, cells were incubated in cAMP assay medium containing IBMX with or without dopamine for 30 min at 37 °C. Our results show that constitutive activity of human D1R and D5R in cells labeled with 5% FBS was not higher than that of cells labeled in UltraCULTURE^{TM} serum-free medium (Fig. 16.2A, left panel). Likewise, the extent of dopamine-mediated maximal activation of adenylyl cyclase was essentially unchanged in cells labeled in the absence or presence of serum (Fig. 16.2A, right panel). In addition, we tested the extent of constitutive activity of rat D5R expressed at low and high levels in cells labeled overnight with [^{3}H]-adenine (2 µCi/ml) in serum-free MEM (5% FBS replaced with 0.1% BSA). Whole cell cAMP assays performed in 6-well culture plates show that the extent of constitutive activity of rat D5R expressed at low and high levels is not modulated by serum (Fig. 16.2B).

6.2. Quantification of constitutive activity of mutant forms of D1R and D5R expressed at low levels

The comparison of constitutive activity between WT and mutant forms of D1R and D5R can be hindered by intrinsic differences between receptor expression levels in HEK293 cells. Some CAMs are inherently unstable and expressed at significantly lower levels than their WT counterparts (Gether et al., 1997; Milligan and Bond, 1997). Typically, a meaningful way to circumvent this problem is to perform experiments relying on the linear relationship that exists between the degree of constitutive activation of G protein-regulated effectors and cellular GPCR levels using cells transfected with increasing amounts of plasmid DNA (Jackson et al., 2000; Samama et al., 1993; Tiberi and Caron, 1994). Slope of curves can be used to assess differences in constitutive activation of mutated or chimeric GPCRs vis-à-vis WT counterparts. Alternatively, CAM GPCRs can be further subjected to mutagenesis for achieving higher expression in cells. Importantly,

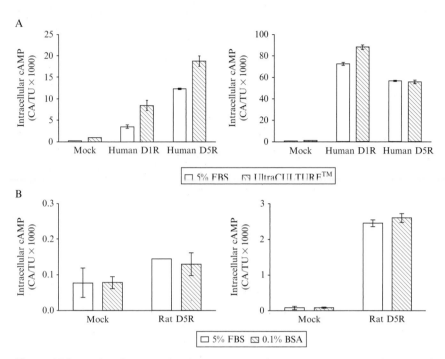

Figure 16.2 Role of FBS in the determination of constitutive activity of D1R and D5R using whole cell cAMP assays. Constitutive activity of D1R and D5R was assessed following an overnight metabolic labeling with [^{3}H]-adenine using MEM containing 5% FBS and serum-free UltraCULTURE^{TM} (A) or 0.1% BSA (B). (A) Shown is a representative example of constitutive activity (left panel) and stimulation of adenylyl cyclase (right panel) in cells transfected with empty pCMV5 (mock), human D1R and D5R incubated in 20 mM HEPES-buffered MEM containing 1 mM IBMX in the absence and presence of 10 μM dopamine for a 30 min. (B) Constitutive activity of rat D5R expressed at low (left panel) and high (right panel) receptor levels was assessed in 20 mM HEPES-buffered MEM containing 1 mM IBMX for 10 min. Constitutive activity is expressed as [^{3}H]-cAMP produced (CA) over the total amount of [^{3}H]-adenine uptake (TU) × 1000 (CA/TU × 1000) done in triplicate determinations. B_{max} values (in pmol/mg membrane proteins) were 18.2 (human D1R, 5% FBS), 19.5 (human D1R, UltraCULTURE^{TM}), 20.6 (human D5R, 5% FBS), 23.6 (human D1R, UltraCULTURE^{TM}), 0.29 (rat D5R low, 5% FBS), 0.53 (rat D5R high, 0.1% BSA), 38.8 (rat D5R high, 5% FBS), and 30.3 (rat D5R high, 0.1% BSA).

one needs to verify that the genetic modification by itself does not significantly alter constitutive activation of the WT receptor. In support of this approach, we have observed that the drastic reduction of B_{max} value of a constitutively activated chimeric D1R harboring CT of D5R (D1-CT$_{D5}$) was restored to WT level by an exchange of the third extracellular loop

(EL3) of D1R with that of D5R (Jackson et al., 2000; Tumova et al., 2003). Notably, EL3 has a limited role in promoting subtype-specific constitutive activity of D1R and D5R (Iwasiow et al., 1999). Importantly, this complementation mutational approach highlighted the important role of CT in mediating constitutive activity phenotype of D1-like subtypes using CAM with low and unaltered B_{max} values. Notwithstanding the difference in B_{max} values between these constitutively active chimeric D1-like receptors, D1-CT_{D5} and D1-EL3CT$_{D5}$ also exhibited distinct agonist-dependent G protein-coupling properties, suggesting that structurally related CAMs adopt different R* conformations (Iwasiow et al., 1999; Jackson et al., 2000; Tumova et al., 2003). Here, we take into consideration the linear relationship existing between B_{max} values and the extent of GPCR constitutive activity to establish a simple procedure to compare the constitutive activity of new human D5R mutants (manuscript in preparation) displaying lower expression levels with that of WT receptors in HEK293 cells transfected with 5 μg of plasmid DNA per 100 × 20 mm dish, as described in Section 3.2. At this DNA concentration, we achieve the maximal receptor expression in this cellular system, as discussed in Section 3. Constitutive activity of WT and mutant forms of D5R were measured in 6-well culture plates using whole cell cAMP assays (Fig. 16.3A). In parallel experiments, we set up transfection dishes with plasmid DNA quantities that lead to similar B_{max} values between WT and mutant D5R (Fig. 16.3B). The extent of constitutive activity is reported using averaged raw data and normalized values relative to WT (Fig. 16.3A–C). Our data demonstrate that normalization of constitutive activity relative to B_{max} provides a convenient and easy method (constitutive activity/B_{max}) to compare the degree of constitutive activation between GPCRs displaying differences in their expression levels (Fig. 16.3D). It could be argued that this calculation method may lead to an overestimation of the constitutive activity of GPCRs expressing at lower B_{max}. This may be true if the relationship between GPCR constitutive activity and B_{max} is nonlinear because of the limiting amount of G proteins and effectors. Consequently, it is important to be knowledgeable about how GPCR constitutive activity behaves in a given cellular system prior to using this calculation method. So far, in our experience with HEK293 cells transfected with D1R and D5R or other Gs-linked GPCRs, a nonlinear relationship between constitutive activity and receptor levels has not been observed. In fact, our data show that when expressed at similar B_{max} values, similar conclusions can be seemingly drawn using normalized constitutive activity values. Importantly, this methodological approach does not unduly confer a constitutively activated status to GPCRs. Indeed, mutant D5R displays a similar degree of constitutive activity relative to WT when comparing normalized values using different B_{max} and data obtained at similar receptor levels (Fig. 16.3).

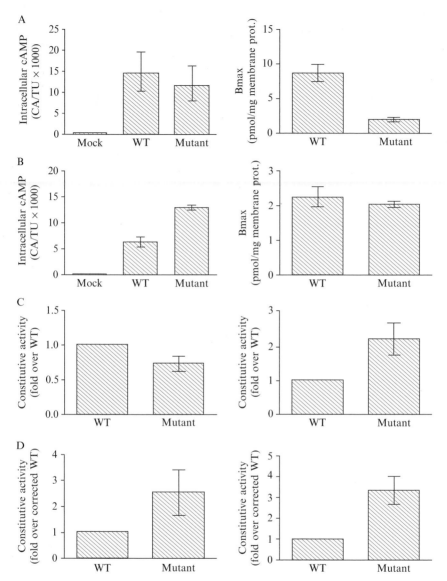

Figure 16.3 Comparison of constitutive activity of human wild-type and mutant D5R in HEK293 cells expressing different receptor levels. HEK293 cells were transfected with wild-type (WT) and mutant D5R using 5 μg of receptor construct DNA per dish for studies shown in A panels or a lower amount of WT construct (0.02 μg per dish supplemented with empty pCMV5) for experiments reported in B panels. Basal intracellular cAMP levels in single wells of 6-well culture plates incubated in the presence of 1 mM IBMX for 30 min are expressed as the arithmetic mean ± S.E. of [^{3}H]-cAMP produced (CA) over the total amount of [^{3}H]-adenine uptake (TU) × 1000 (CA/TU × 1000) from 3 to 7 experiments done in triplicate determinations (A, B).

6.3. Regulation of forskolin stimulation by D5R constitutive activity and inverse agonists

Studies assessing constitutive activity described in above sections have been performed with 6-well culture plates. As stated previously, the 6-well culture plates will provide the best signal-to-noise ratio (D1R/D5R-transfected cells vs. mock-transfected cells). However, it would be convenient to explore the regulation of constitutive activity by inverse agonists using dose–response curves in 12-well culture plates. Indeed, whole cell cAMP assays using 12-well culture plates may prove useful in measuring relative effective concentration inhibiting 50% of the constitutive activity or IC_{50} (index of potency) of different inverse agonists at WT and mutant forms of D1R and D5R. In fact, while feasible, relying on 6-well culture plates for testing multiple inverse agonists with dose–response curves in a single experiment would require a larger number of transfected cells and culture plates (typically three culture plates per drug). Before considering the use of 12-well culture plates for inverse agonist dose–response curves, we tested whether inhibition of constitutive activity of D1R and D5R by inverse agonists is dependent on receptor levels using 6-well culture plates. This issue is important in determining the appropriate transfection condition for receptor expression and hence obtaining optimal signal-to-noise ratio. Here we show that constitutive activity of rat D5R in cells labeled overnight with [^{3}H]-adenine (2 µCi/ml) is completely blocked following a 15 min exposure to fluphenazine (Sigma-Aldrich, Cat. No. F-101) and *cis*-flupenthixol (Fig. 16.4A). Constitutive activity of D5R in cells treated with these two inverse agonists was indistinguishable from intracellular cAMP levels measured in mock condition. Importantly, these two dopaminergic drugs evoked inverse agonism both at low and high D5R expression levels (Fig. 16.4A). Meanwhile, a low signal-to-noise ratio in cells expressing low receptor levels remains a concern for measuring robust inhibition of constitutive activity by inverse agonists and detecting drugs that are partial inverse agonists. One way to address this potential experimental caveat is to perform whole cell cAMP assays in the presence of forskolin (FSK), an activator of adenylyl cyclase. Indeed, activated Gαs subunits augment FSK-stimulated adenylyl cyclase (Harry *et al.*, 1997; Sunahara *et al.*, 1997). Therefore, consistent with the idea that constitutive

---

The B_{max} values in pmol/mg of membrane proteins are shown in the right A and B panels. (C) The constitutive activity of mutant (expressed as arithmetic mean ± S.E.) was calculated relative to WT from cells expressing different (left panel) or similar (right panel) receptor levels, respectively. (D) Each data expressed as CA/TU × 1000 (shown in A and B) were divided by their respective B_{max} and corrected constitutive activity normalized relative to WT value measured in cells expressing different (left panel) or similar (right panel) receptor levels, respectively.

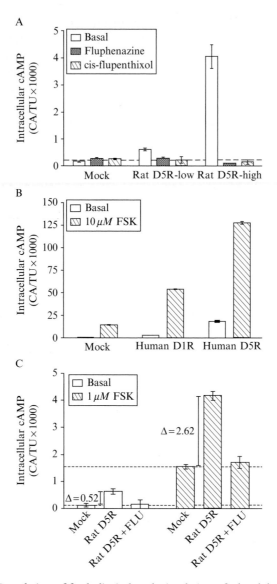

Figure 16.4 Regulation of forskolin-induced stimulation of adenylyl cyclase by constitutive activity and inverse agonism at D1R and D5R. (A) Intracellular cAMP levels in single wells of 12-well culture plates incubated in the absence and presence of fluphenazine (2 µM) or cis-flupenthixol (FLU, 1 µM) for 10 min are expressed as the arithmetic mean ± S.E. of [^{3}H]-cAMP produced (CA) over the total amount of [^{3}H]-adenine uptake (TU) × 1000 (CA/TU × 1000) done in triplicate determinations. Arithmetic mean ± S.E. of B_{max} values (in pmol/mg of membrane proteins) were 2.19 ± 0.84 (rat D5R-low) and 12.7 ± 3.3 (rat D5R-high). (B) Intracellular cAMP levels were measured in single wells of 6-well culture plates incubated in the absence and presence

activity is associated with ligand-independent activation of G proteins by GPCRs, cells transfected with Gs-linked receptors will exhibit an increased FSK stimulation of adenylyl cyclase relative to mock-transfected in the absence of agonists and in a receptor level-dependent manner, as shown previously with H1 histaminergic and 5-HT7 serotonergic receptors (Alewijnse et al., 1997; Sheng et al., 2005). As a proof of principle, we show that the extent of FSK stimulation in cells expressing human D1R (lower constitutive activity) is increased in comparison with mock condition. In agreement with the higher constitutive activity of D5R, the extent of FSK stimulation is significantly larger in cells transfected with D5R relative to that of cells expressing D1R at similar receptor levels (Fig. 16.5B). HEK293 cells grown in 12-well culture plates expressing low levels of rat D5R (~ 0.4 pmol/mg membrane proteins) and labeled overnight with [^{3}H]-adenine (2 μCi/ml) were used to test inverse agonism in the absence and presence of 1 μM FSK. Our data show that cis-flupentixol (inverse agonist) robustly inhibits FSK stimulation linked to D5R expression as compared with mock and does so through reduction of its constitutive activity (Fig. 16.5C). Therefore, regulation of the extent of FSK stimulation mediated by constitutive activity of D1R and D5R represents an alternative and useful experimental approach to explore the underlying molecular mechanisms of inverse agonism at WT and mutant forms of D1-like receptors.

6.4. HA and Flag-tagged D5R exhibit similar CAM phenotype relative to untagged D5R

The epitope-tagging approach has been widely used to explore posttranslational modifications and intracellular sorting properties of GPCRs. An important issue with this approach is to document that the epitope-tagged receptor retains the functional characteristics of its untagged WT counterpart. We have introduced HA or Flag epitope at the amino terminus of human D1R and D5R using a PCR-based approach. Recently, we have

---

of 10 μM forskolin (FSK) for 30 min using triplicate determinations of CA/TU × 1000. The B_{max} values in pmol/mg of membrane proteins were 1.7 (human D1R, no FSK), 1.4 (human D1R, 10 μM FSK), 2.4 (human D5R, no FSK), and 2.6 (human D5R, 10 μM FSK). (C) FSK-independent and dependent intracellular cAMP formation was measured in single wells of 12-well culture plates incubated in the absence and presence of FLU (10 μM) for 10 min using triplicate determinations. Broken lines set the limits of intracellular cAMP levels in mock-transfected cells incubated in the absence and presence of FSK (1 μM) for 10 min. The net changes (Δ) in intracellular cAMP levels mediated by rat D5R in the absence and presence of FSK relative to mock-transfected cells are also depicted. The B_{max} value of rat D5R was 0.44 pmol/mg of membrane proteins.

Table 16.2 Ligand-binding properties of untagged and epitope-tagged wild-type human dopaminergic D1R and D5R

Receptor	Saturation studies [³H]-SCH23390		Competition studies K_i (nM)			
	K_d (nM)	B_{max} (pmol/mg prot.)	SCH23390	Dopamine	cis-Flupenthixol	(+)-Butaclamol
WT-D1R	0.62 (0.49–0.78)	12.1 (7.86–16.3)	0.59 (0.49–0.71)	7260 (5530–9530)	6.28 (4.87–8.10)	3.31 (2.05–5.35)
Flag-D1R	0.62 (0.53–0.72)	13.9 (9.20–18.6)	0.56 (0.39–0.80)	8270 (6440–10,610)	6.61 (5.05–8.66)	3.68 (2.39–5.66)
HA-D1R	0.72 (0.53–0.96)	15.2 (10.0–20.4)	0.58 (0.51–0.66)	8060 (5290–12,290)	7.04 (5.51–8.99)	3.46 (2.06–5.35)
WT-D5R	1.18 (0.95–1.47)	13.2 (9.87–16.5)	1.01 (0.65–1.56)	760 (550–1070)	12.7 (8.78–18.5)	25.8 (19.6–33.9)
Flag-D5R	1.02 (0.75–1.39)	15.7 (11.7–19.7)	1.17 (1.03–1.32)	910 (660–1260)	13.5 (9.44–19.3)	29.9 (21.8–38.4)
HA-D5R	1.38 (0.94–2.02)	15.5 (11.6–19.4)	1.18 (1.03–1.36)	790 (540–1150)	15.3 (10.4–22.5)	32.7 (23.5–42.5)

Data are expressed as geometric (K_d, K_i) and arithmetic (B_{max}) means with 95% lower and upper confidence intervals from 5 to 6 experiments done in duplicate determinations. Best-fitted parameters were obtained from binding isotherms analyzed using nonlinear curve regression programs from GraphPad Prism. WT, wild-type K_d, equilibrium dissociation constant; B_{max}, maximal binding capacity; and K_i, equilibrium dissociation constant of unlabeled drug.

demonstrated using a chimerical approach that the amino terminus and TM1 regions of D1-like receptors may play a role in regulating ligand binding and G protein-coupling properties (D'Aoust and Tiberi, 2010). Studies were performed with untagged and tagged versions of human D1R and D5R in HEK293 cells to verify whether the addition of HA or Flag epitope at the amino terminus alters ligand binding, constitutive activity, and dopamine-dependent G protein-coupling properties using radioligand binding and whole cell cAMP assays. Saturation and competition studies were performed, as described in Section 4. Results indicate that affinity of dopaminergic ligands for HA- and Flag-tagged D1R and D5R are not significantly changed in comparison to their respective WT counterparts (Table 16.2). Likewise, receptor expression (B_{max}) of D1R and D5R was not modulated by HA or Flag (Table 16.2). Constitutive activity and dopamine dose–response curves were determined using 6-well and 12-well culture plates, respectively. Whole cell cAMP assays suggest that epitope tagging of the amino terminus of D1R and D5R has no significant effect on the extent of constitutive activity (Fig. 16.5), EC_{50} (index of potency), and E_{max} values of dopamine (Fig. 16.6).

6.5. Phenotypic expression of constitutive activity of human D5R is potentially dependent on cellular factors sensitive to phorbol esters

Studies suggest that intrinsic allosteric mechanisms in cells potentially regulate GPCR constitutive activity and inverse agonism (Chidiac et al., 1994; Cotecchia, 2007; Kenakin, 2005; Milligan, 2003). Meanwhile, these cellular and allosteric determinants remain to be fully appreciated. While dopamine dose–response curves have established distinguishing agonist-dependent G protein-coupling properties of D1R and D5R (e.g., lower E_{max} for dopamine in D5R relative to D1R-expressing cells), we have also observed that HEK293 cells transfected with D5R is from time to time (∼10% of dose–response studies) insensitive to increasing concentrations of dopamine (Fig. 16.7). These observations are seemingly not linked to the "health status" or passage of HEK293 cells, as in parallel experiments using the same cells, D1R always display a graded response to increasing concentrations of dopamine (Fig. 16.7). Most importantly, D5R and D1R were expressed at similar B_{max} values (∼2 pmol/mg membrane proteins). Data expressed as percentage of maximal activation show that the extent of constitutive activity of hD1R (∼1% vs. ∼30%) and hD5R (∼7% vs. ∼73%) is strikingly different in these two cell populations (compare Fig. 16.7A and B). Meanwhile, a graded response to dopamine is observed in both groups of cells expressing hD1R while being seemingly converted into a switch-like response in cells transfected with hD5R (Fig. 16.7B). In fact, the absence of robust dopamine effect may apparently be explained by an almost

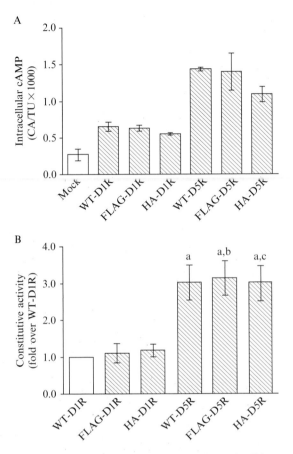

Figure 16.5 Comparison of constitutive activity of human wild-type and epitope-tagged D1R and D5R in HEK293 cells. Constitutive activity of human wild-type (WT) and epitope-tagged D1R and D5R was assessed in single wells of 6-well culture plates following an overnight metabolic labeling with [^{3}H]-adenine using MEM containing 5% FBS. (A) Shown is a representative example of basal intracellular cAMP levels measured in mock and receptor-transfected HEK293 cells using triplicate determinations and expressed as [^{3}H]-cAMP produced (CA) over the total amount of [^{3}H]-adenine uptake (TU) × 1000 (CA/TU × 1000). (B) Intracellular cAMP levels (CA/TU × 1000) were divided by B_{max} and values calculated relative to WT-D1R. Data are expressed as arithmetic means ± S.E. of 5–6 experiments. Arithmetic means ± S.E. of B_{max} values (in pmol/mg membrane proteins) were 14.6 ± 2.3 (WT-D1R), 14.2 ± 1.8 (Flag-D1R), 13.4 ± 1.8 (HA-D1R), 11.9 ± 1.2 (WT-D5R), 10.7 ± 2.2 (Flag-D5R), and 10.8 ± 2.4 (HA-D5R). [a]$p < 0.05$ when compared with a value of 1 (WT-D1R) using one-sample t test; [b]$p < 0.05$ when compared with Flag-D1R and [c]$p < 0.05$ when compared with HA-D1R using one-way ANOVA followed by Newman–Keuls posttest.

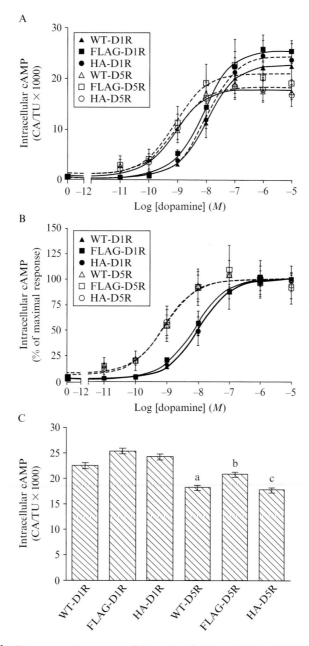

Figure 16.6 Dose–response curves of dopamine for intracellular cAMP formation by wild-type and epitope-tagged human D1R and D5R expressed in HEK293 cells. Intracellular cAMP levels were measured in single wells of 12-well culture plates incubated in the absence or presence of increasing concentrations of dopamine for 30 min and plotted as a function of log of dopamine concentrations. Each point is the

full constitutive activation of D5R in this cell population (Fig. 16.7B). Therefore, the thermodynamic equilibrium of D5R may be predominantly "locked" in R* state (R ⇒ R*). Previously, we have shown that treatment of HEK293 cells with phorbol-12-myristate-13-acetate (PMA, an activator of protein kinase C) leads to sensitization and desensitization of dopamine-induced D1R and D5R responsiveness, respectively (Jackson et al., 2005). Additionally, we have shown that PMA treatment completely inhibits D5R constitutive activity (Jackson et al., 2005). Interestingly, PMA treatment abrogates drastically D5R constitutive activity, which in turn restores the graded response to dopamine in these hD5R-expressing cells (Fig. 16.7B). These findings imply that PMA treatment may lead to a shift in D5R thermodynamic equilibrium through the stabilization of R state, as do inverse agonists. Overall, these results support the presence of potential inducible allosteric factors in HEK293 cells regulating the constitutive activity levels of D5R and to a lesser extent those of D1R. Moreover, our studies indicate that constitutive activity of D5R is controlled through the PMA-induced regulation of specific receptor determinants and cellular allosteric factors, which remain to be elucidated.

7. CONCLUDING REMARKS

D1R and D5R were the first reported native closely related GPCRs to be differentiated on the basis of displaying distinct constitutive activity levels. Furthermore, studies with D1R and D5R have been instrumental in validating many of the hallmark pharmacological properties of R and R* states established with constitutively active mutant forms of GPCRs. Different experimental approaches based on radioligand binding on membrane

---

arithmetic mean ± S.E. of three experiments done in triplicate determinations and expressed as [^{3}H]-cAMP produced (CA) over the total amount of [^{3}H]-adenine uptake (TU) × 1000 (shown in A) or the percentage of maximal activation obtained with respective wild-type or epitope-tagged D1R and D5R (shown in B). Curves were analyzed by simultaneous nonlinear curve fitting using GraphPad Prism and statistical significance determined using unconstrained and constrained curve fits. The best-fitted values (± approximate S.E.) for dopamine-mediated maximal intracellular cAMP formation (CA/TU × 1000) or E_{max} obtained from curves depicted in (A) are reported in (C). The EC_{50} values (in nM with 95% lower and upper confidence intervals) are as follows: WT-D1R, 10.2 (4.3–24.5); Flag-D1R, 7.6 (3.5–16.6); HA-D1R, 10.7 (4.7–24.2); WT-D5R, 0.85^a (0.27–2.7); Flag-D5R, 0.88^b (0.32–2.4); and HA-D5R, 0.77^c (0.24–2.5). B_{max} values (in pmol/mg membrane proteins) were 2.7 ± 0.7 (WT-D1R), 3.1 ± 1.0 (Flag-D1R), 2.3 ± 0.8 (HA-D1R), 2.8 ± 0.5 (WT-D5R), 3.0 ± 0.5 (Flag-D5R), and 2.1 ± 0.3 (HA-D5R). $^a p < 0.05$ when compared with WT-D1R, $^b p < 0.05$ when compared with Flag-D1R, and $^c p < 0.05$ when compared with HA-D1R.

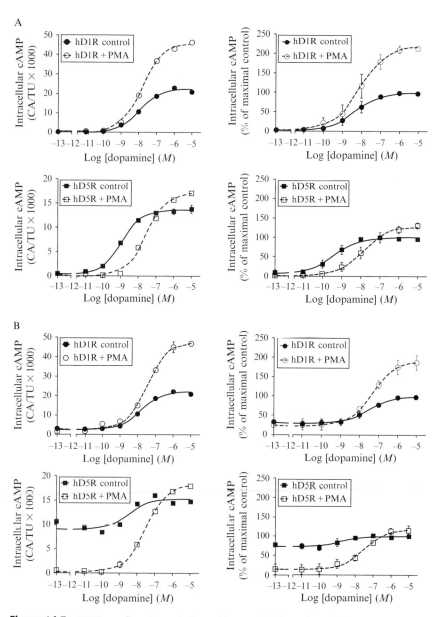

Figure 16.7 PMA-mediated regulation of human D1R and D5R in HEK293 cells. Intracellular cAMP levels were measured in single wells of 12-well culture plates incubated in the absence or presence of increasing concentrations of dopamine for 30 min with or without 1 μM phorbol-12-myristate-13-acetate (PMA) and plotted as a function of log of dopamine concentrations. Each point is the arithmetic mean ± S.E. of three experiments done in triplicate determinations and expressed as [^{3}H]-cAMP produced (CA) over the total amount of [^{3}H]-adenine uptake (TU) × 1000 (shown

preparations and whole cell cAMP assays have been used to probe structure–activity relationships of D1-like subtype-specific regions and determinants responsible for the constitutive activity among D1R and D5R. These diverse methods have also allowed gaining insight into the molecular complexity and receptor region interplays involved in the regulation of distinct pharmacological properties of D1R and D5R. Therefore, our studies performed with these experimental procedures underscore the importance of D1R and D5R as critical tools to explore molecular processes implicated in the agonist-independent and dependent activation of GPCRs as well as their silencing through inverse agonist binding and phorbol ester-dependent posttranslational modification. Refinements in the use of whole cell approaches will prove useful in identifying potential intrinsic cellular factors and allosteric mechanism controlling D5R constitutive activity.

ACKNOWLEDGMENTS

Mario Tiberi expresses his sincere gratitude to former laboratory members, particularly Rafal Iwasiow, Adele Jackson, and Katerina Tumova, for their hard work, dedication, and for providing invaluable insights into this research topic. The authors also thank Dr. Kursad Turksen and Andrew Charrette for reading the manuscript and helpful comments. Bianca Plouffe and Jean-Philippe D'Aoust received graduate scholarships from Fonds de la recherche en santé du Québec and Canadian Health Research Institutes (CIHR). Operating grants from CIHR (MOP-81341) and Natural Sciences and Engineering Research Council of Canada (DG#203694) supported the MT research relevant to this chapter.

---

in left panels of A and B) or the percentage of maximal activation obtained with wild-type human D1R (hD1R) and D5R (hD5R) control conditions (without PMA), respectively (shown in the right panels of A and B). Curves were analyzed by a simultaneous nonlinear curve fitting using GraphPad Prism and the statistical significance was determined using unconstrained and constrained curve fits. B_{max} values (in pmol/mg membrane proteins) were 2.20 ± 0.19 (hD1R, A), 2.06 ± 0.42 (hD1R, B), 2.19 ± 0.09 (hD5R, A), and 2.61 ± 0.64 (hD5R, B). HEK293 cells used for the assessment of hD1R and hD5R responsiveness were grouped into two distinct classes based on the low (A) and high (B) constitutive activity of hD5R measured in these cells. Similar passage of cells was used for studies shown in A (45, 49, and 51) and B (43, 50, and 52).

REFERENCES

Alewijnse, A. E., Smit, M. J., Rodriguez Pena, M. S., Verzijl, D., Timmerman, H., and Leurs, R. (1997). Modulation of forskolin-mediated adenylyl cyclase activation by constitutively active G(s)-coupled receptors. *FEBS Lett.* **419**, 171–174.

Allen, L. F., Lefkowitz, R. J., Caron, M. G., and Cotecchia, S. (1991). G-protein-coupled receptor genes as protooncogenes: Constitutively activating mutation of the alpha 1B-adrenergic receptor enhances mitogenesis and tumorigenicity. *Proc. Natl. Acad. Sci. USA* **88**, 11354–11358.

Andersson, S., Davis, D. L., Dahlback, H., Jornvall, H., and Russell, D. W. (1989). Cloning, structure, and expression of the mitochondrial cytochrome P-450 sterol 26-hydroxylase, a bile acid biosynthetic enzyme. *J. Biol. Chem.* **264**, 8222–8229.

Baufreton, J., Zhu, Z. T., Garret, M., Bioulac, B., Johnson, S. W., and Taupignon, A. I. (2005). Dopamine receptors set the pattern of activity generated in subthalamic neurons. *FASEB J.* **19**, 1771–1777.

Cerione, R. A., Codina, J., Benovic, J. L., Lefkowitz, R. J., Birnbaumer, L., and Caron, M. G. (1984). The mammalian beta 2-adrenergic receptor: Reconstitution of functional interactions between pure receptor and pure stimulatory nucleotide binding protein of the adenylate cyclase system. *Biochemistry* **23**, 4519–4525.

Chaar, Z. Y., Jackson, A., and Tiberi, M. (2001). The cytoplasmic tail of the D1A receptor subtype: Identification of specific domains controlling dopamine cellular responsiveness. *J. Neurochem.* **79**, 1047–1058.

Charpentier, S., Jarvie, K. R., Severynse, D. M., Caron, M. G., and Tiberi, M. (1996). Silencing of the constitutive activity of the dopamine D1B receptor. Reciprocal mutations between D1 receptor subtypes delineate residues underlying activation properties. *J. Biol. Chem.* **271**, 28071–28076.

Chidiac, P., Hebert, T. E., Valiquette, M., Dennis, M., and Bouvier, M. (1994). Inverse agonist activity of beta-adrenergic antagonists. *Mol. Pharmacol.* **45**, 490–499.

Costa, T., and Cotecchia, S. (2005). Historical review: Negative efficacy and the constitutive activity of G-protein-coupled receptors. *Trends Pharmacol. Sci.* **26**, 618–624.

Costa, T., and Herz, A. (1989). Antagonists with negative intrinsic activity at delta opioid receptors coupled to GTP-binding proteins. *Proc. Natl. Acad. Sci. USA* **86**, 7321–7325.

Cotecchia, S. (2007). Constitutive activity and inverse agonism at the α_1 adrenoceptors. *Biochem. Pharmacol.* **73**, 1076–1083.

D'Aoust, J. P., and Tiberi, M. (2010). Role of the extracellular amino terminus and first membrane-spanning helix of dopamine D1 and D5 receptors in shaping ligand selectivity and efficacy. *Cell. Signal.* **22**, 106–116.

Gether, U., Ballesteros, J. A., Seifert, R., Sanders-Bush, E., Weinstein, H., and Kobilka, B. K. (1997). Structural instability of a constitutively active G protein-coupled receptor. Agonist-independent activation due to conformational flexibility. *J. Biol. Chem.* **272**, 2587–2590.

Harry, A., Chen, Y., Magnusson, R., Iyengar, R., and Weng, G. (1997). Differential regulation of adenylyl cyclases by Galphas. *J. Biol. Chem.* **272**, 19017–19021.

Iwasiow, R. M., Nantel, M. F., and Tiberi, M. (1999). Delineation of the structural basis for the activation properties of the dopamine D1 receptor subtypes. *J. Biol. Chem.* **274**, 31882–31890.

Jackson, A., Iwasiow, R. M., and Tiberi, M. (2000). Distinct function of the cytoplasmic tail in human D1-like receptor ligand binding and coupling. *FEBS Lett.* **470**, 183–188.

Jackson, A., Sedaghat, K., Minerds, K., James, C., and Tiberi, M. (2005). Opposing effects of phorbol-12-myristate-13-acetate, an activator of protein kinase C, on the signaling of structurally related human dopamine D1 and D5 receptors. *J. Neurochem.* **95**, 1387–1400.

Johnson, R. A., Alvarez, R., and Salomon, Y. (1994). Determination of adenylyl cyclase catalytic activity using single and double column procedures. *Methods Enzymol.* **238,** 31–56.

Kenakin, T. (2005). The physiological significance of constitutive receptor activity. *Trends Pharmacol. Sci.* **26,** 603–605.

Lee, D., Dong, P., Copolov, D., and Lim, A. T. (1999). D5 dopamine receptors mediate estrogen-induced stimulation of hypothalamic atrial natriuretic factor neurons. *Mol. Endocrinol.* **13,** 344–352.

Milligan, G. (2003). Constitutive activity and inverse agonists of G protein-coupled receptors: A current perspective. *Mol. Pharmacol.* **64,** 1271–1276.

Milligan, G., and Bond, R. A. (1997). Inverse agonism and the regulation of receptor number. *Trends Pharmacol. Sci.* **18,** 468–474.

Rosenbaum, D. M., Rasmussen, S. G., and Kobilka, B. K. (2009). The structure and function of G-protein-coupled receptors. *Nature* **459,** 356–363.

Salomon, Y. (1991). Cellular responsiveness to hormones and neurotransmitters: Conversion of [3H]adenine to [3H]cAMP in cell monolayers, cell suspensions, and tissue slices. *Methods Enzymol.* **195,** 22–28.

Samama, P., Cotecchia, S., Costa, T., and Lefkowitz, R. J. (1993). A mutation-induced activated state of the β2-adrenergic receptor. Extending the ternary complex model. *J. Biol. Chem.* **268,** 4625–4636.

Samama, P., Pei, G., Costa, T., Cotecchia, S., and Lefkowitz, R. J. (1994). Negative antagonists promote an inactive conformation of the β2-adrenergic receptor. *Mol. Pharmacol.* **45,** 390–394.

Sheng, Y., Wang, L., Liu, X. S., Montplaisir, V., Tiberi, M., Baltz, J. M., and Liu, X. J. (2005). A serotonin receptor antagonist induces oocyte maturation in both frogs and mice: Evidence that the same G protein-coupled receptor is responsible for maintaining meiosis arrest in both species. *J. Cell. Physiol.* **202,** 777–786.

Sunahara, R. K., Dessauer, C. W., Whisnant, R. E., Kleuss, C., and Gilman, A. G. (1997). Interaction of Gsα with the cytosolic domains of mammalian adenylyl cyclase. *J. Biol. Chem.* **272,** 22265–22271.

Tao, Y. X. (2008). Constitutive activation of G protein-coupled receptors and diseases: Insights into mechanisms of activation and therapeutics. *Pharmacol. Ther.* **120,** 129–148.

Tiberi, M., and Caron, M. G. (1994). High agonist-independent activity is a distinguishing feature of the dopamine D1B receptor subtype. *J. Biol. Chem.* **269,** 27925–27931.

Tumova, K., Iwasiow, R. M., and Tiberi, M. (2003). Insight into the mechanism of dopamine D1-like receptor activation. Evidence for a molecular interplay between the third extracellular loop and the cytoplasmic tail. *J. Biol. Chem.* **278,** 8146–8153.

Tumova, K., Zhang, D., and Tiberi, M. (2004). Role of the fourth intracellular loop of D1-like dopaminergic receptors in conferring subtype-specific signaling properties. *FEBS Lett.* **576,** 461–467.

CHAPTER SEVENTEEN

IDENTIFICATION OF GAIN-OF-FUNCTION VARIANTS OF THE HUMAN PROLACTIN RECEPTOR

Vincent Goffin,* Roman L. Bogorad,*,[1] and Philippe Touraine*,†

Contents

1. Introduction 330
   1.1. Context 330
   1.2. Missense polymorphisms of the human PRL receptor in breast tumors 331
   1.3. Experimental models: Various options 333
2. Experimental Procedures 333
   2.1. Generation of cell models 333
   2.2. Bioassays and readouts 341
3. Identification of Constitutive Activity: Results and Discussion 345
4. Conclusions 353
Acknowledgments 353
References 353

Abstract

There is currently no known genetic disease linked to prolactin (PRL) or its receptor (PRLR) in humans. Recently, we identified three missense variants of the PRLR in patients presenting with breast tumors. Two of them (named $PRLR_{I146L}$ and $PRLR_{I76V}$) had been reported earlier, but failed to draw much attention because the eventual impact of these substitutions on receptor properties remained unknown. In this chapter, we describe the various bioassays (cell types and readouts) that led to the discovery that both variants exhibit gain-of-function properties. Reconstituted cell models involving Ba/F3, HEK293, and MCF-7 cell lines all highlighted the constitutive, PRL-independent potency of $PRLR_{I146L}$ to trigger downstream signaling, leading to antiapoptotic

* Inserm, Unit 845, Research Center Growth and Signaling, Team "PRL/GH Pathophysiology," University Paris Descartes, Faculty of Medicine, Paris, France
† Assistance Publique—Hôpitaux de Paris, Department of Endocrinology and Reproductive Medicine, GH Pitié Salpêtrière, Paris, France
[1] Current address: The David H. Koch Institute for Integrative Cancer Research, Massachusetts Institute of Technology, Cambridge, Massachusetts, USA

Methods in Enzymology, Volume 484 © 2010 Elsevier Inc.
ISSN 0076-6879, DOI: 10.1016/S0076-6879(10)84017-0 All rights reserved.

and proliferation properties. The lower level of basal activity of $PRLR_{I76V}$ could be demonstrated only in the very sensitive Ba/F3 cell assay. While comparative analysis of ligands is a routine procedure in many labs, comparison of receptor variants *de facto* imposes the use of different cell clones (or population) in which each receptor variant is expressed individually. This is more delicate, as one must ensure that differences in biological responses really reflect differences in the intrinsic properties of receptor variants, and not any feature of cell clones/populations that are used, which could bias the interpretation.

1. INTRODUCTION

1.1. Context

Prolactin (PRL) was discovered 80 years ago, based on the ability of pituitary extracts to stimulate lactation in rabbit (Riddle *et al.*, 1933; Stricker and Grueter, 1928). Nowadays, endocrinology textbooks define PRL as a pituitary-secreted polypeptide hormone which was named for its lactogenic properties. Various models of permanent PRLR signaling failure (Horseman *et al.*, 1997; Ormandy *et al.*, 1997) widely contributed to understanding the functional pleiotropy of PRL as this hormone appears to be a modulator of metabolic, endocrine, immune, neuronal, and osmoregulatory systems (Ben Jonathan *et al.*, 2008; Bole-Feysot *et al.*, 1998). Despite the fact that almost 300 functions or targets could be identified for PRL in various species, the question remains open which of them are really relevant in humans. Clearly, the mammary gland is and remains its main target in mammals.

PRL pathophysiology is currently restricted to hyperprolactinemia, the unique pathology that is unanimously recognized to be linked to excess PRL due to a pituitary adenoma or could be drug-induced. This disease is defined as circulating PRL levels above normal range that occurs in conditions other than pregnancy and lactation. It should be noticed that hyperprolactinemia is not *per se* a pathology of PRL (the hormone secreted is perfectly normal), but it rather reflects dysfunctions in the mechanisms regulating PRL synthesis. The main outcome of pathological hyperprolactinemia is hypogonadism. This condition is efficiently cured using synthetic analogs of dopamine, the physiological negative regulator of PRL production by lactotroph cells in the pituitary (Molitch, 2003). Beyond hyperprolactinemia, one of the difficulties preventing a clear understanding of the involvement of PRL in human pathophysiology is that no mutation of the genes encoding PRL or its PRLR has been identified yet. Therefore, we lack a definitive clinical model of isolated PRL deficiency/resistance that could be used to identify the functions that depend on, or are modulated by PRL. As a good example, the involvement of PRL in mammary cancer proposed several decades ago based on observations involving rodent models (Clevenger *et al.*, 2003; Welsch and Nagasawa, 1977) anxiously awaits the identification of PRLR mutation in breast cancer to definitely support its participation in the disease. It is in this context that we recently discovered two gain-of-function PRLR variants.

1.2. Missense polymorphisms of the human PRL receptor in breast tumors

As all members of class I cytokine receptor superfamily (Bazan, 1989), the PRLR is single-pass transmembrane receptor devoid of intrinsic enzymatic activity. It signals via various kinase-associated pathways, the main of which include Jak/Stat, MAPK, Src, and PI3K/Akt pathways (Clevenger *et al.*, 2003; Goffin *et al.*, 2005; Fig. 17.1A). The active PRLR is a homodimer, which is presumably triggered through subtle conformational changes.

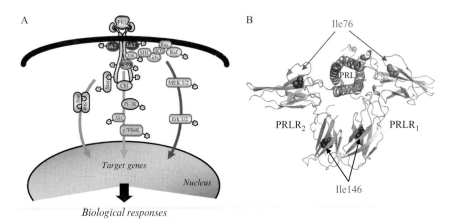

Figure 17.1 Molecular features of the PRLR and its constitutively active variants. (A) This figure represents a simplified view of PRLR signaling. The main cascades involve the tyrosine kinase Jak2, which in turn activates three members of the STAT (signal transducers and activators of transcription) family, Stat1, Stat3 and, mainly, Stat5. The MAP Kinase pathway is another important cascade activated by the PRLR. It involves the Shc/Grb2/Sos/Ras/Raf intermediaries upstream to MEK and Erk-1/-2 kinases. Activation of other signaling pathways have been reported, among which Src and Akt kinases are known to play important roles in some responses such as cell proliferation. Yellow boxes symbolize phosphorylated tyrosine. Reproduced with permission from Goffin *et al.* (2005) (Copyright 2010, The Endocrine Society). (B) The crystal structure of human PRL (green) bound to two moieties of the rat PRLR extracellular domain (orange) has been solved recently (Broutin *et al.*, 2010). Isoleucine 76 (red) and Isoleucine 146 (blue) are represented on the receptor. (See Color Insert.)

Studies performed using the rat PRLR identified phosphorylation of the C-terminal tyrosine as one of the earliest event following PRL stimulation (Lebrun et al., 1995). Tyrosine phosphorylation is induced by Jak2 kinase and is essential for triggering Stat5 cascade. Alternative splicing of primary PRLR transcripts leads to the translation of many receptor isoforms, which differ by the length of their cytoplasmic domains and are thus referred to as short, intermediate, or long receptors (Clevenger et al., 2003). Short and intermediate isoforms have a truncated C-terminal tail and are not tyrosine-phosphorylated, which correlates with their inability to activate Stat5.

Considering that the human PRLR was cloned more than 20 years ago (Boutin et al., 1989), the paucity of known missense polymorphisms affecting this receptor in diseases is somehow surprising. Actually, we are aware of only four studies that investigated this issue, which all involved breast cancer patients (Canbay et al., 2004; Glasow et al., 2001; Lee et al., 2007; Vaclavicek et al., 2006). While several noncoding single nucleotide polymorphisms (SNPs) were reported in promoters and introns, only two missense SNPs could be identified in the PRLR gene, which both involved codons encoding residues of the extracellular domain (Fig. 17.1B). The first one substitutes a Valine for Isoleucine 76 ($PRLR_{I76V}$), and the second Leucine for Isoleucine 146 ($PRLR_{I146L}$). At the epidemiological level, none of the three studies in which one or both SNPs were identified involved a sufficient number of patients to propose any association with the disease on a statistical evidence (Canbay et al., 2004; Lee et al., 2007; Vaclavicek et al., 2006). At the functional level, the eventual impact of amino acid substitutions on PRLR properties was not investigated. Last but not least, $PRLR_{I76V}$ had been reported as a polymorphism in the population (NCBI database), suggesting it had no relevance in human disease.

We recently reported a study focused on the analysis of PRLR SNPs in a rare benign disease of unknown etiology named multiple breast fibroadenoma. The main features of this pathology and the clinical profile of patients are detailed elsewhere (Courtillot et al., 2010). We sequenced the coding regions of PRLR gene (11 exons) in 106 patients, which identified 3 germline, heterozygous missense SNPs. Interestingly, the two SNPs mentioned above were found again in our study. While I76V SNP was found in both patient and control populations with no statistic difference, I146L SNP was identified only in patients (4/106) and not in control subjects ($n = 194$; $p < 0.01$; Courtillot et al., 2010). We also identified E554Q substitution in the intracellular domain of the receptor as a new SNP ($n = 1$). We decided to characterize the functional properties of these natural PRLR variants in order to support (or not) their potential involvement in breast tumorogenesis. This chapter focuses on the two extracellular variants that were shown to exhibit constitutive activity. Sections below address the pro and cons of the various biological approaches that were used, and possible bias of interpretation.

1.3. Experimental models: Various options

First, it is worth emphasizing that comparative functional analyses of receptor variants are much more difficult to ascertain than that of ligands. Comparative analysis of ligands (classically mutants of a given protein produced by recombinant technology) can be performed using a single cell bioassay expressing the PRLR, providing these cells respond to the ligand by any measurable readout (proliferation, reporter gene, etc.). Once the bioassay is established (and/or the appropriate cell line identified among all available cell lines), it can be used for all further studies involving any putative ligand. As an illustration, we characterized >100 PRL analogs using three cell bioassays involving the rat or the human PRLR (Bernichtein et al., 2003; Goffin et al., 1996; Kinet et al., 2001). In contrast, comparison of receptor variants de facto imposes the use of different cell clones (or population) that each express one of the receptors to be compared. This makes all the difference. First, the cell models that are needed for such comparative studies never exist naturally, hence they have to be generated. Second, once cell models have been obtained, the optimal readout and the reliability of its monitoring have to be determined, which can be time-consuming and face unexpected problems. Third, when these issues have been solved, one must ensure that differences in biological responses monitored between cell models that are compared really reflect differences in the intrinsic properties of receptor variants, and not any feature of cell clones/populations that are used, which could bias the interpretation.

At the beginning of our study, we did not anticipate $PRLR_{I146L}$ and $PRLR_{I76V}$ to have major functional impact on receptor function, given the conservative character of residue substitutions and their location outside PRL binding sites (Fig. 17.1B). We thus selected three cell types and three readouts that had been largely validated in the past for comparing ligands of the human PRLR (Bernichtein et al., 2003; Llovera et al., 2000). The cell models involved HEK-293, Ba/F3, and MCF-7 cell lines. Readouts involved the major PRLR signaling cascades analyzed by immunoblot, a luciferase gene reporter assay, and a cell survival/proliferation assay. Each of them is detailed in the following sections.

2. Experimental Procedures

2.1. Generation of cell models

2.1.1. HEK-293 cells
Cell line: The human embryonic kidney (HEK) fibroblast 293 cell line is a classical model for analyzing cytokine receptors. This adherent cell line expresses very little amounts of endogenous PRLR, therefore transfection

of PRLR expression vectors ensures that the receptors to be analyzed will be expressed at a tremendously higher level compared to endogenous PRLR. In other words, the events that are measured can be considered as mediated by the ectopic receptor, which can be assessed by comparing transfected versus nontransfected cells. HEK-293 cell line can be used for monitoring different readouts (e.g., signaling, reporter genes, receptor internalization, etc.; there is no proliferation response to PRL), using various transfected cell formats (transient transfections, stable clones, or stable populations). Classical culture conditions are reported in Table 17.1.

Table 17.1 Summary of culture conditions

	HEK-293	MCF-7	Ba/F3
Cell type	Human embryonic kidney (adherent)	Human breast cancer (adherent)	Mouse pro-B lymphoid (suspension)
Routine culture conditions (growth medium)	DMEM 10% FCS 50 U/ml penicillin, 50 µg/ml streptomycin 2 mM glutamine	DMEM or RPMI 1640 10% FCS 50 U/ml penicillin, 50 µg/ml streptomycin 2 mM glutamine	RPMI 1640 10% FCS (heat-inactivated) 50 U/ml penicillin, 50 µg/ml streptomycin 2 mM glutamine 10% conditioned medium of WEHI-3B cell as a source of IL-3
Passages	At confluency, trypsinize, centrifuge, and dilute 1/10 (passage every 3–4 days)	At confluency, trypsinize, centrifuge, and dilute 1/4 every (passage every 3–4 days)	Centrifuge and dilute 1/5 every other day (cells should be maintained at density between 400,000 and 2,000,000 ml^{-1})
Starvation	Idem growth medium *except* only 0–0.5% FCS Cells maintained alive >24 h	Idem growth medium *except* only 0–1% FCS Cells maintained alive >24 h	Idem growth medium *except* only 1–2% FCS Apoptosis detected after 6 h in 1% FCS
Specific device	Classical incubators, 37 °C, 5% CO_2		

Generation and selection of stable clones: Transfection can be performed in various plate formats, from 96-well to P-150 Petri dishes, depending on the purpose of the experiment (see Section 2.2). Transfection procedures are summarized in Table 17.2. When no stable clone is necessary, transiently transfected cells can be used (stimulated) 24–48 h after transfection. When stable clones need to be obtained, transfected cells are diluted 24 h after transfection in 5–10 dishes containing growth medium, and the selection antibiotic is added 24 h later. Resistance to G-418 (Geneticin®) is conferred by aminoglycoside-3′phosphotransferase encoded by *neo* gene in the pcDNA3-neo series of expression vectors (Invitrogen). The antibiotic was added into growth medium at a concentration of 500 μg/ml active G-418 per ml (note that Geneticin® is provided as a powder that usually contains <80% active compound). When starting with a new cell line, a range of G-418 concentration (e.g., 200–1000 μg/ml) should be first tested using the parental cell line, and the lowest concentration that kills cells within 7–10 days should be used for clonal selection after transfection. From this step, G-418 is systematically added to all culture media to maintain selection pressure. After ~2 weeks, single G-418-resistant colonies can be localized under microscope, and when visible by eye, they can be picked out individually by local trypsinization (e.g., using sterilized Cloning Cylinder, Bellco Glass) and amplified (starting with 12- or 24-well plates) before screening. It is worth noting that G-418 resistance does not automatically mean PRLR expression, as only part of the plasmid containing G-418 resistance gene, but not the entire PRLR expression cassette, can be integrated into genome. Clone screening can be performed by immunoblot to assess PRLR expression (see Table 17.3 for antibodies). If the purpose is to obtain stable clones to be used for luciferase assays (see below), one must ensure that both receptor and luciferase reporter plasmids are present in selected clones. To that end, ligand-induced luciferase responsiveness can be used for initial screening, providing the receptor of interest is known to be active. This procedure is especially recommended when the luciferase reporter plasmid to be cointegrated into the genome does not contain any antibiotic resistance (which was our case for the LHRE-luc). Practically, each clone is distributed in a 96-well plate and responsiveness is evaluated by comparing cells treated or not with 1 μg/ml PRL (use duplicate or triplicate wells).

2.1.2. MCF-7 cells

Cell line: The human mammary tumor MCF-7 cell line is a classical model for analyzing PRLR signaling in breast cancer cell. MCF-7 is an adherent, epithelial luminal cell line positive for estrogen and progesterone receptors. Although they are referred to as hormone-dependent, the minimal amount of steroid hormones present in serum is sufficient to ensure cell growth. This cell line was used to evaluate the functional impact of PRLR variants in a mammary cell context. In addition, as these cells express endogenous

Table 17.2 Summary of transfection conditions

Type of transfection	HEK-293 Transient	HEK-293 Stable clones	MCF-7 Stable clones	Ba/F3 Stable populations
Procedure(s) Specific device	Fugene or lipofectamine (strictly follow manufacturer recommendations) —	—	—	Electroporation Electroporator
Cell density, plate, and medium	96-well plates 50,000 cells/well 0.5% FCS medium for transfection	Petri dishes 5.10^6 cells/P100 0.5% FCS medium for transfection (adapt to smaller/larger dishes)	Petri dishes 5.10^6 cells/P100 0.5% FCS medium for transfection (adapt to smaller/larger dishes)	Gene Pulser (BIO-RAD): 400 µl at $1–2.5 \times 10^6$ cells/ml 200 V and 950 µF Amaxa® Nucleofector® Technology (Lonza): 400 µl at 2.5×10^6 cells/ml use Cell Line Nucleofector® Kit V for Ba/F cells (Lonza)
Transfected plasmids	20 ng/well pcDNA3(+) hPRLR 4 ng/well LHRE-firefly Luc 4 ng/well CMV-Renilla Luc	2 µg/dish pcDNA3(+) hPRLR 0.4 µg/dish LHRE-firefly Luc	2 µg/dish pcDNA3(+) hPRLR	2 µg (Amaxa) to 10–30 µg (Gene Pulser) of pcDNA3(+)hPRLR (vector can be linearized within AmpR cassette to increase chances of proper insertion)
Selection for stable clones/population		– 24 h after transfection: trypsinize cells, split in 5–10 Petri dishes with fresh growth medium. Let cells adhere		– Immediately after transfection, dilute transfected cells in

Table 17.2 (continued)

Type of transfection	HEK-293 Transient	HEK-293 Stable clones	MCF-7 Stable clones	Ba/F3 Stable populations
		– 24 h later, add 500 μg/ml active G-418, and maintain always G-418 in growth medium for later passages	– 24 h later, add 500 μg/ml active G-418, and maintain always G-418 in growth medium for later passages	– 5–10 ml (Gene Pulser) or 2 ml (Amaxa) of growth medium with IL-3 – 24 h later, add 500 μg/ml active G-418, and maintain always G-418 in growth medium for later passages
Second round selection				Substitute PRL (100–1000 ng/ml) for IL-3 (=do not add WEHI conditioned medium in growth medium)
Screening		Test >5 clones per receptor variant Use 1–2 clones/receptor, keep all positive clones as frozen backups		No screening, use the population of resistant cells
Perform Bioassay	24–48 h after transfection	When clones are selected and characterized for PRLR expression, membrane export, etc.		When populations are characterized for PRLR expression, membrane export, etc.
Readouts	Luciferase assay (96-well) Signaling studies (Petri dishes)		Proliferation (96-well)	Survival/proliferation using WST-1 (96-well) or FACS analysis Signaling studies (6-well)

Table 17.3 Summary of immunoblotting conditions

Antigen	Clone	Provider	Stock concentration	Epitope	Type	Immunoprecipitation	Immunoblotting
Primary antibodies							
PRLR	H-300 (sc-20992)	Santa Cruz Biotechnology	200 µg/ml	Intracellular domain (residues 323–622 of hPRLR).	Rabbit polyclonal	Ab: 0.4–1 µg/ml Cell lysate: >300 µg protein	1:1000 (preferably use 1A2B1)
	1A2B1 (Cat #35-9200)	Invitrogen (Zymed)	500 µg/ml	PRLR extracellular domain	Mouse monoclonal	Ab: 1–5 µg/ml Cell lysate: >500 µg protein (preferably use H-300)	1:1000
pho-Tyr	4G-10	Upstate cell signaling	1000 µg/ml	Phosphorylated tyrosines	Mouse monoclonal	Not recommended	1:10,000
Stat5 a/b	C-17 (sc-835)	Santa Cruz biotechnology	200 µg/ml	Peptide mapping the C-terminus of mouse Stat5	Rabbit polyclonal	Ab: 0.4–2 µg/ml Cell lysate: >300 µg protein	1:1000
pho-STAT5	AX-1	Advantex BioReagents	Lyophilised, resuspend at 1000 µg/ml in H_2O/glycerol (50/50)	Phosphorylated Y694/99 of Stat5a/b	Mouse monoclonal	Not recommended	1:5000–1:10,000 (more potent than phosphotyrosine mAb)

Table 17.3 (continued)

Antigen	Clone	Provider	Stock concentration	Epitope	Type	Immunoprecipitation	Immunoblotting
Erk1/2	CT (Cat #06-182)	Millipore	~350 µg/ml	C-terminus of rat 44 kDa Erk2	Rabbit polyclonal	Not recommended	1:1000
pho-Erk 1/2	12D4 (Cat #91065)	Upstate cell signaling	Not provided	Phosphorylated Thr202/Tyr204 of Erk1 and Erk2	Mouse monoclonal	Not recommended	1:1000
Secondary antibodies							
Anti-rabbit-HRP	Anti-rabbit IgG, HRP-linked antibody	Upstate cell signaling	Not provided	Rabbit IgG (heavy and light chains)	Goat affinity purified polyclonal	—	1:2000
Anti-mouse-HRP	Anti-mouse IgG, HRP-linked antibody	Amersham (GE Healthcare)	Not provided	Mouse IgG	Sheep affinity purified polyclonal	—	1:2000–1:4000

PRLR (~7000 receptors per cell), this places the ectopic receptor in a "heterozygous" context found in patients (Courtillot et al., 2010).

Generation and selection of stable clones: Transfection was performed using experimental procedures very similar to those described above for HEK-293 cells (see Table 17.2). Clone screening involved monitoring the level of PRLR expression, which was performed using both immunoblot analysis and radio-receptor assay (see below).

2.1.3. Ba/F3 cells

Cell line: Ba/F3 is a pro-B murine cell line dependent on interleukin-3 (IL-3) for growth. It is a classical cell line for investigations of cytokine receptors. It does not endogenously express the PRLR, and is absolutely dependent on cytokine receptor signaling for survival/growth. Their high propensity to undergo cell death in the absence of appropriate stimuli was a key feature for identifying the constitutive properties of PRLR variants of interest (see below). These cells grow in suspension.

Generation and selection of stable populations: Electroporation is the method of choice for transfecting cells growing in suspension. We have used two types of eletroporator (Gene Pulser® from Bio-Rad, and more recently Amaxa® Nucleofector® Technology from Lonza). Transfection conditions are reported in Table 17.2 (for additional details, follow recommendations of the manufacturers). Cells stably expressing the PRLR of interest were selected by several passages in G418-containing medium. As Ba/F3 cells are intrinsically dependent on cytokine signaling for survival/growth, a second selection round can be performed by substituting PRL for IL-3 (i.e., use growth medium without adding WEHI conditioned medium; see Table 17.2). This step eliminates cells that may have integrated only the antibiotic resistance cassette (but not the receptor cDNA). Also, as we experienced that the level of PRLR expression was higher after PRL selection compared to Geneticin® selection, this suggests that this second selection step may also eliminate cells that do not express enough PRLR to ensure PRL-dependent survival/proliferation, although this is speculative. Of course, this second selection is only successful when the ectopic PRLR is functional, therefore it should be viewed as mean not only to select a highly responsive cell population, but also to identify low/nonfunctional receptors (evidenced when cells die/hardly survive and do not generate fast growing populations). In our hands, stable populations were sufficient to achieve reliable conclusions regarding PRLR properties without the need to generate stable clones by limited dilution of G-418-resistant populations.

2.1.4. Pros and cons of transient versus stable transfections

Transient transfection presents the disadvantage that experiments are most of the time performed in blind regarding the amount of receptor expressed in the different wells to be compared. Although transfecting equivalent amounts of expression plasmids is usually considered to lead to expression of similar protein receptor amounts, this is rarely checked experimentally. Transfection efficiency and expression of receptor variants can be affected by various parameters (plasmid quality, stability of mutated mRNA, etc.). We advise using transient transfections for preliminary assays (e.g., qualitative screening of receptor variant properties, determine whether they are active, etc.), and to try minimize artifacts by any possible means to avoid misinterpretations. The gold standard is to cotransfect cells using Renilla luciferase plasmid (p-RL-TK, Promega), as the latter is assumed to not be regulated by classical stimuli (i.e., it acts as an invariable internal control). Caution is required, however, as some hormonal stimuli (e.g., dihydrotestosterone and dexamethasone) were shown to regulate p-RL-TK luciferase expression (Ibrahim et al., 2000). Our preference goes to stable clones, as these tools are (assumed to be) stable in time regarding their main features, which allows initial characterization of the various clones that will be compared in later studies. When comparing various clones expressing different receptors, it is mandatory to use clones that express similar levels of the latter. This can be performed by ligand binding assay or immunoblot (see Section 3). It must be stressed that, as observed for all types of *in vitro* cell cultures, the expression of some specific markers (including the "transfected" protein) can slightly decrease along cell passages even when using stable clones. Of course, adenoviral/lentiviral constructs can also be advised as the efficacy of infections is clearly higher than classical plasmid transfection; at the time we performed our studies, these tools were not available in our laboratory.

2.2. Bioassays and readouts

2.2.1. Signaling

Principle: The PRLR activates several cascades, the main of which are represented in Fig. 17.1A. Classically, Jak2/Stat5 and MAPK have been used to reflect the ability of the PRLR to transmit a signal into target cells. Stat5 is probably the most widely distributed PRL-responsive cascade (LeBaron et al., 2007). It contains an SH2 domain that binds to the C-terminal tyrosine (Y587) of the PRLR once the latter is phosphorylated by Jak2 (Lebrun et al., 1995). In contrast, MAPK cascade can be triggered irrespective of PRLR phosphorylation, including by the short PRLR (Bole-Feysot et al., 1998). It must be stressed that signaling cascade display cell-specificity, meaning that the cascades schematized in Fig. 17.1A are not necessarily activated by the PRLR in all cell types. Therefore, it is mandatory to first identify in the cell line of interest (the most) PRL-responsive cascades before using it/them as (a) readout(s).

Assay: Transiently or stably transfected cells are plated in culture plates/wells sized according to the number of immunoblot experiments to be performed (immunoprecipitation (IP) is more protein-consuming than analyses of total lysates). Usually, P100 Petri dishes ($\sim 5 \times 10^6$ cells) are appropriate, but other plate format can be used (adapt cell number to plate/well surface). Cells are starved overnight in 0–0.5% FCS medium (HEK293, MCF-7) or 4–6 h in medium containing 1–2% FCS medium (Ba/F3), before being stimulated or not (time-course and dose–responses are recommended to identify the most responsive conditions). Cell lysates are prepared using chilled lysis buffer (10 mM Tris–HCl pH 7.5, 5 mM EDTA, 150 mM NaCl, 10% glycerol, and 0.5% Triton X-100) containing protease inhibitors (Roche, Complete mini, 1 tablet/10 ml lysis buffer) and 1 mM sodium orthovanadate as a tyrosine phosphatase inhibitor. Efficient lysis and appropriate final protein concentration will be obtained for 5×10^6 (HEK, MCF-7) or $10\text{–}20 \times 10^6$ (Ba/F) cells/ml of lysis buffer. Efficient detection of constitutive or ligand-induced activation of downstream PRLR cascades depends on several parameters, including the amount of proteins (cell lysate) analyzed, the amount of antigen expressed by the cells used, the level of activation, the quality of the antibodies, etc. IP will be used to concentrate the antigen, and/or to minimize background on immunoblots. Classically, Stat5 phosphorylation is monitored after IP to reduce nonspecific background using AX1 antibody; MAPK phosphorylation can be monitored using total lysates as p-Erk antibody gives lower background. PRLR phosphorylation can also be analyzed, although the lack of phospho-specific PRLR antibody imposes to immunoprecipitate the receptor before immunoblotting using a phosphotyrosine monoclonal antibody (mAb). While 4G-10 mAb has been successfully used for years in our lab, we recently faced disappointing results using this mAb, potentially reflecting clone derivation. The most appropriate conditions that we recommend for IP/immunoblotting are summarized in Table 17.3. Proteins on membrane are detected by autoradiography using enhanced chemiluminescence (ECL, Amersham/GE Healthcare, or LumiGLO™ Reagent, cell signaling).

Data analysis: Blots are not quantitative, therefore overinterpretation should be avoided. When data are obvious from the autoradiography, and reproducible, it can be considered as reliable. When variations between different conditions are less obvious, densitometric quantification can be performed using the free software Quantity One (Bio-Rad). However, one should avoid interpreting borderline differences only revealed by densitometric quantification. Classically, phospho-specific signals are normalized against total amount of the cognate antigen, which gives an idea of the relative phosphorylation induced by PRLR activation. Importantly, conditions to be directly compared should be run on the same gel to ensure that membrane transfer, primary/secondary immunoblotting, and ECL reaction do not bias the result.

2.2.2. LHRE-Luciferase reporter assay

Principle: The lactogenic hormone response elements (LHRE)-luciferase reporter gene is an expression vector encoding the firefly luciferase under the control of an artificial promoter containing a six-repeat LHRE from the β-casein promoter, a classical PRL target gene. It is mainly mediated by Stat5 cascade, although participation of other signaling molecules cannot be formally excluded. The principle of this bioassay is thus to monitor the ability of any PRLR (WT or variant) to induce transcriptional activation of LHRE artificial promoter by measuring the level of luciferase produced by the cell. It should be used only as a "reporter" or receptor triggering, that is, that any pathophysiological interpretations should be avoided.

Bioassay: After trypsinization, cells are counted and aliquoted in 96-well plates at a density of 50,000 cells/100 μl/well. Although this assay has been classically run in low serum condition (0.5% FCS) to minimize luciferase background induced by any serum component (Bernichtein et al., 2003), growth medium (10% FCS) can be recommended as (i) it ensures better cell adhesion, which will minimize variability resulting from cell aspiration when medium is removed prior the lysis step, and (ii) bovine PRL (potentially present in FCS) was recently shown to not activate the human PRLR (Utama et al., 2009). Eighteen hours after plating (overnight), 100 μl of twofold concentrated hormones diluted in same medium are added to each well (add only medium in negative control). After another 18–24-h period, culture medium is aspirated and cells are lysed under gentle agitation for > 10 min in 50 μl of lysis buffer (Promega). Ten to 20 μl of lysate is then transferred into white 96 well plates and luciferase activity is monitored in a bioluminescence multiplate reader (e.g., LB 940 Mithras, Berthold technologies). When using transient transfection, firefly luciferase (PRLR-induced) and Renilla luciferase (PRLR-nonresponsive, used as transfection control) will be measured using Dual-Luciferase® Reporter Assay System (Promega). If stable clones (receptor + LHRE-luciferase) are used, only the former is measured.

Data analysis: Raw data (relative light units, RLU) can vary depending on transfection efficiency, cell number, lysis efficiency, sensitivity of plate reader, etc. When using transient transfections, normalization of firefly/Renilla luciferase RLU is performed to normalize all values with respect to these variable parameters. Transcriptional activity can be expressed either as renilla-normalized raw data, or as fold induction, which is the ratio between RLU obtained for stimulated versus nonstimulated conditions. Fold induction presents the advantage to facilitate comparison (as nonstimulated conditions always has value = 1), but it masks part of the information as the efficacy of the system reflected in absolute values is not displayed (see discussion section below).

2.2.3. Survival/proliferation

2.2.3.1. WST-1 approach *Principle*: While HEK-293 cells do not exhibit proliferation response under PRL stimulation, breast cancer cells (MCF-7) can proliferate when PRL-stimulated, although one must emphasized that the amplitude of the proliferation response to PRL is mild, which often requires repeating the experiment several times to achieve significance. In contrast, PRLR-expressing Ba/F3 cells are an excellent system to monitor the proliferation/antiapoptotic effect of PRLR signaling, since these cells die rapidly in the absence of cytokine stimulus. As indicated below, we used WST-1 tetrazolium salt method which is metabolized by living cells (Bernichtein et al., 2003). It is noteworthy that this assay reflects both mitogenic and antiapoptotic (survival) activity. The use of stable populations, initially selected or not by several passages in medium containing PRL instead of IL-3, is particularly appropriate for this routine survival/proliferation assay.

Bioassay: Cells are starved (no PRL or IL-3) for 4–6 h in 1–2% FCS RPMI medium with glutamine and antibiotics (see Table 17.1), then distributed in 96-well plates at a density of 50,000 cells/well in a final volume of 100 μl. One hundred microliters of [2×] hormones diluted in the same medium are added after starvation period. Cell proliferation is estimated after 2 or 3 days by adding 10 μl of WST-1 tetrazolium salt. Optical density at 450 nm (OD_{450}) is measured after 1–4 h of colorimetric reaction using a multiplate reader. The experiments were routinely performed at least three times in triplicate or quadruplicate (variability usually does not exceed 10% provided peripheral wells are avoided to circumvent edge effects). Preliminary experiments of cell dilutions showed that the ratio cell density/optical density is almost linear in the range of ∼ 50,000–200,000 cells per well, which corresponds to cell densities achieved in this bioassay (Bernichtein et al., 2003). It must be stressed that nonstimulated cells that have remained 2–3 days without PRL (negative control) are for most dead or engaged in the apoptotic process; the WST-1 approach is thus convenient to distinguish PRLR-responsive versus nonresponsive cells, but should not be used to analyze in detail apoptotic versus proliferating responses.

Data analysis: Optical density values measured at any time point (1–4 h after WST-1 addition) are monitored and analyzed by any appropriate software allowing to calculate EC_{50} (EXCEL, GraphPad, etc.). Time course experiments can be performed by plating the same quantity of cells on day 0, then monitoring living cell number every day and comparing values obtained after the same time of colorimetric WST-1 reaction.

2.2.3.2. FACS analysis

Fluorescence-activated cell sorting (FACS) analysis complements WST-1 assay as deciphers the different phases of cell cycle. It is less suitable than WST-1 procedures for analyzing more than a few samples (e.g., dose–responses, etc.). Briefly, cell cycle and apoptosis were assessed by DNA content analysis after staining with the DNA intercalator propidium iodide (PI). One million cells were harvested by centrifugation and permeabilized using 30 μl of COULTER® DNA-Prep LPR reagent, followed by addition of 0.5 ml of DNA-Prep stain PI solution (DNA-Prep reagents, Beckman Coulter). After vortexing, samples were incubated at 37 °C for

30 min, and then analyzed by flow cytometry on a FACScan (Becton Dickinson) using low flow rate. Cell cycle distribution was determined using CellQuest software (Becton Dickinson) and manual gating. Apoptosis was estimated as the percentage of DNA localized in the hypo-diploid peak (sub-G_0/G_1) of the cell cycle, as earlier described (Jeay et al., 2000).

3. Identification of Constitutive Activity: Results and Discussion

As already mentioned, we first aimed at generating stable clones (HEK-293, MCF-7) or populations (Ba/F3) in order to minimize bias when comparing PRLR variants to the wild-type receptor using transient transfection. We first screened different clones for PRLR expression based on semiquantitative PRLR immunoblotting. As shown in Fig. 17.2A, this type of experiments assessed that clones that were used to compare the properties of $PRLR_{WT}$ and $PRLR_{I146L}$ expressed similar amounts of receptor, which in the case of MCF-7, corresponds approximately to twice the level of $PRLR_{WT}$ expressed in the parental cell line (endogenous PRLR is not detected in HEK-293 cells). As the majority of the PRLR is known to reside inside the cell (reticulum, *trans*-Golgi network, endocytotic vesicles), performing immunoblot analysis of total lysates is not sufficient to certify similar membrane expression, and one cannot exclude that the mutations affect export of some PRLR variants at cell surface, for example, as observed for cytokine receptors in which the WS motif has been mutated. Therefore, we completed this preliminary analysis by radioligand receptor assay using ^{125}I-labeled PRL. For determining the number of cell surface receptors, binding assays were performed using living cells, as described (Berlanga et al., 1997). Displacement curves were analyzed using nonlinear curves fitting (model: homologous competitive binding curves, one class of binding sites, and PRISM software). Both HEK-$PRLR_{WT}$ and HEK-$PRLR_{I146L}$ clones were found to express ~5,000 surface receptors/cell, while Ba/F–$PRLR_{WT}$ and Ba/F–$PRLR_{I146L}$ populations expressed many fewer surface receptors (~500 receptors/cell, which were not detectable by immunoblot), which correlates earlier observation using this cell line (Bernichtein et al., 2003). The same verifications were made for cells expressing $PRLR_{I76V}$. These experiments indicated that substitutions I76V and I146L did not alter receptor export at cell membrane. Binding affinities of the two receptor variants determined using cell homogenates prepared from stable HEK-293 clones and Ba/F3 populations were also found to be identical (Courtillot et al., 2010; Fig. 17.2B). This is in agreement with the fact that both residues 76 and 146 are outside ligand binding site (Fig. 17.1B).

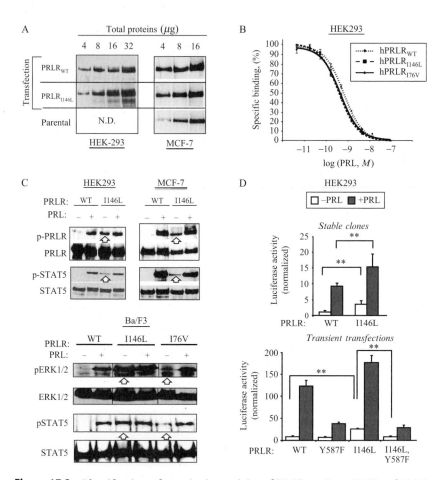

Figure 17.2 Identification of constitutive activity of PRLR variants I76V and I146L. (A) To compare the expression levels of PRLR$_{WT}$ and PRLR$_{I146L}$ in stable clones (HEK-293 and MCF-7), serial dilutions of cell lysates (protein amounts indicated) were prepared from each clone and parental cell line, then analyzed by PRLR immunoblotting (1A2B1, see Table 17.3). For both cell types, selected clones expressed similar amount of PRLR. (B) Competition of ^{125}I-PRL by unlabeled hPRL using stable HEK-293 clones expressing either receptor (as indicated). Data are shown as means ± SD ($n = 3$, in duplicates). There was no difference of binding affinity between PRLR$_{WT}$ and receptor variants. (C) HEK-293, MCF-7, and Ba/F3 cells stably expressing PRLR$_{WT}$, PRLR$_{I76V}$, or PRLR$_{I146L}$ were serum-starved, then stimulated or not using 1 μg/ml recombinant human PRL (15 min). PRLR and Stat5 were immunoprecipitated, while MAPK (Erk1/2) was analyzed in total lysates. Phosphorylated and total PRLR, Stat5, and MAPK were analyzed by immunoblotting (see Table 17.3 for experimental conditions). White arrows identify basal phosphorylations observed in the absence of PRL in clones expressing PRLR$_{I146L}$ and, to a lesser extent, PRLR$_{I76V}$. (D) Top panel: two of the available HEK-293-PRLR$_{WT}$ and HEK-293-PRLR$_{I146L}$ clones also stably incorporated the PRL-responsive LHRE-luciferase gene. Luciferase activity was measured after 24-h treatment with (gray bars) or without (open bars)

The first evidence for constitutive activity of $PRLR_{I146L}$ came from the analysis of its phosphorylation status. As shown in Fig. 17.2C, the PRLR immunoprecipitated from lysates of stable clones expressing $PRLR_{I146L}$ appeared to be phosphorylated in the absence of PRL stimulation. As the sole ectopically expressed receptor is detectable using HEK-293, this indicated that $PRLR_{I146L}$ was phosphorylated. In MCF-7, some level of phosphorylation was also observed, albeit to lower levels compared to HEK-293 clones. In addition, as we do not hold a specific antibody for $PRLR_{I146L}$, this experiment did not permit to determine whether PRLR phosphorylation in MCF-7 implied the mutant only, or also the endogenous WT receptor. Whatever, this finding highlighted that receptor phosphorylation persisted in the "heterozygous" context. No phosphorylation was observed for cells expressing only $PRLR_{WT}$, confirming that spontaneous phosphorylation of the PRLR was a specific feature of cells expressing $PRLR_{I146L}$ variant, and not an artifact due to (over-)expression of the $PRLR_{WT}$. Importantly, $PRLR_{I146L}$ phosphorylation was confirmed using another stable HEK clone (not shown). Interestingly, no constitutive phosphorylation of $PRLR_{I76V}$ variant could be observed in this type of assay (not shown).

We next investigated the main PRLR signaling cascade, that is, Stat5. In agreement with the constitutive phosphorylation of $PRLR_{I146L}$, Stat5 was phosphorylated (activated) in the absence of PRL only in cells expressing this receptor variant. However, we noticed that there was no direct correlation between phosphorylation levels of Stat5 and PRLR; in other words, apparent similar level of PRLR phosphorylation led to very different levels of Stat5 phosphorylation (compare stimulated versus nonstimulated conditions in HEK clones in Fig. 17.2C). It is currently unknown whether this observation reflects a technical limit of the experimental approach (mAb sensitivity, etc.), or whether it really reflects specific molecular features of signaling molecules involved. Constitutive signaling induced by $PRLR_{I146L}$ variant was confirmed using stable Ba/F populations. In this cell line, MAPK pathway was also constitutively activated, indicating that nonstimulated $PRLR_{I146L}$ triggers the two main PRLR signaling cascades in this cell lines.

---

10 µg/ml PRL. Data normalized to basal luciferase activity of HEK-293-$PRLR_{WT}$ (see Fig. 17.3 for explanations) show significantly higher background in HEK-293-$PRLR_{I146L}$ cells (means ± S.D., $n = 6$, **$p < 0.01$). *Bottom panel*: HEK-293 cells were transiently cotransfected using expression plasmids encoding PRLR (WT or I146L), LHRE-luciferase (firefly) reporter gene, and Renilla luciferase gene. Luciferase activities were measured after 24-h treatment with (gray bars) or without (open bars) PRL. Data normalized to basal firefly/Renilla luciferase activity ratio of cells expressing $PRLR_{WT}$ show that Y587F substitution abolishes the higher background of $PRLR_{I146L}$ (means ± SD, $n = 3$, **$p < 0.01$). Panels A, C (top), and D are reproduced with minor modifications from Bogorad *et al.* (2008) (Copyright 2008, National Academy of Sciences, USA), and panels B and C (bottom) from Courtillot *et al.* (2010) (Copyright 2010, The Endocrine Society).

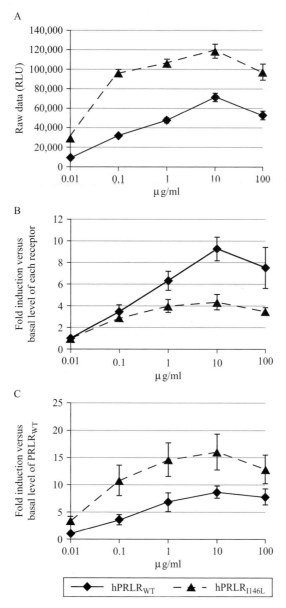

Figure 17.3 Three ways of expressing luciferase data. This figure involves HEK-293-PRLR$_{WT}$ and HEK-293-PRLR$_{I146L}$ clones that stably incorporated the PRL-responsive LHRE-luciferase gene (see Fig. 17.2D, top panel). Raw data (panel A) clearly show that stimulated values are much higher than nonspecific background, indicating that experiment is reliable. In panel B, stimulated luciferase values were normalized to the basal (not stimulated) level obtained for each clone; using such a representation, the fold induction induced by PRL is approximately twice lower for PRLR$_{I146L}$ compared to

In addition, Ba/F–PRLR$_{I76V}$ population also revealed that this receptor variant exhibited some level of constitutive activation, as both Stat5 and MAPK displayed basal phosphorylation, albeit lower compared to BA/F–PRLR$_{I146L}$ cells. As Stat5 phosphorylation was not detected using HEK clones expressing PRLR$_{I76V}$, this suggests that intrinsic features of cell lines/stable clones can hamper the detection of mild constitutive activity. In this matter, Ba/F cells appeared to be the most sensitive model we used, which may relate to the selection process used to enrich PRL-responsive cells by substituting PRL for IL-3 in culture medium. Interestingly, among the three cell lines we used, Ba/F cells expressed the lowest number of receptors, arguing against the dogma that "the more receptor, the stronger response."

We next turned to functional bioassays involving a cell response, in order to complement the analysis of signaling cascades. We first used the LHRE-luciferase reporter assay. We used two clones that had stably integrated this reporter plasmid (or at least part thereof). Despite similar level of PRLR expression in the stable clones to be compared, HEK-PRLR$_{I146L}$ cells exhibited significantly higher basal luciferase activity compared to HEK-PRLR$_{WT}$ cells (Fig. 17.2D, top). To discard the possibility that this effect could have resulted from the integration of a higher number of LHRE-Luciferase copies in the genome of HEK-PRLR$_{I146L}$ cells, a PCR (25–29 cycles) was performed using 50 ng genomic DNA for both clones (extracted from 1×10^6 cells) and specific primers for firefly luciferase gene. β-Globin-specific primers were used as gene copy number control. This experiment revealed that the stable clones to be compared contained similar copy numbers of the luciferase sequence in their genome. In addition, the fact that higher basal activity was also observed using cell populations (not shown) as well as transient transfections (Fig. 17.2D, bottom) strongly argued for the reliability of this finding. The higher basal activity of HEK-PRLR$_{I146L}$ was reversed back to the level observed for PRLR$_{WT}$ when the C-terminal tyrosine of PRLR$_{I146L}$ was mutated into Phe (Y587F). As this tyrosine residue is critical for receptor phosphorylation and downstream activation of STAT5 and target genes (Lebrun et al., 1995), this experiment not only validated the increased basal level of activity of PRLR$_{I146L}$, but also revealed that this constitutive activity involved phosphorylation of this key residue. Besides constitutive activity, we also noticed

---

PRLR$_{WT}$, which may lead to the interpretation that the former is less potent than the latter. However, when all data are normalized to the same experimental condition, that is, basal level of HEK-293-PRLR$_{WT}$ clone (panel C), the higher background and stimulated values of PRLR$_{I146L}$ are now displayed. Thus the lower fold induction observed for PRLR$_{I146L}$ (panel B) is only the consequence of its higher background, and not of poor PRL-responsiveness. Panel A is reproduced with minor modifications from Courtillot et al. (2010) (Copyright 2010, The Endocrine Society), and panel B from Bogorad et al. (2008) (Copyright 2008, National Academy of Sciences, USA).

that the activity of $PRLR_{I146L}$ in the presence of any concentration of PRL was systematically higher than that displayed by $PRLR_{WT}$ (Bogorad et al., 2008). In this luciferase assay, the variant $RRLR_{I76V}$ behaved exactly as $PRLR_{WT}$, in agreement with signaling studies performed in HEK cells.

We must emphasize that care should be taken when analyzing luciferase data, as interpretations can be very misleading. First, one should ensure that experimental values are not in the same range of nonspecific background values, which would mean that any luciferase activity induction/repression are not reliable; obviously, this was not the case for PRLR studies (Fig. 17.3A). One classical way to present the results is by "fold induction", which normalizes stimulated RLU values obtained for each clone/receptor versus the nonspecific background (no PRL added). When this calculation was applied to HEK-$PRLR_{I146L}$ cells, the fold induction appeared to be significantly reduced compared to HEK-$PRLR_{WT}$ cells, which may be erroneously interpreted as lower activity of the variant (Fig. 17.3B). When one looks at raw data (Fig. 17.3A), however, it clearly appears that absolute values including basal activity are always higher for HEK-$PRLR_{I146L}$, indicating that this receptor is at least as potent as the WT receptor, and that reduced fold induction actually results from higher basal level. Hence, when dose–response curves of both receptors are normalized to background of HEK-$PRLR_{WT}$ cells, the higher basal and PRL-induced level of activity of $PRLR_{I146L}$ cells is displayed (Fig. 17.3C). As receptor levels and PRL binding activities were identical between HEK-293-$PRLR_{WT}$ and HEK-293-$PRLR_{I146L}$ clones (see above), this indicates that the higher basal in the latter reflects an intrinsic biological property of $PRLR_{I146L}$. Caution is thus required when analyzing such a bioassay.

The second assay involved proliferation/survival of Ba/F3 cells, which was performed using WST-1 and FACS approaches. According to the fact that the IL-3-dependence of Ba/F3 cells for proliferation/survival can be shifted to any other cytokine providing they express the cognate receptor, Ba/F–$PRLR_{WT}$ cells grew in the presence of PRL (no IL-3), but underwent massive apoptosis in the absence thereof within 24 h (Fig. 17.4A and B). In sharp contrast, Ba/F–$PRLR_{I146L}$ survived without any cytokine, and even proliferated autonomously. Consequently, these cells could never be synchronized in G_0/G_1 phase by PRL deprivation (Fig. 17.4B), which uses to be a classical step before performing proliferation assays. Nevertheless, the ability of $PRLR_{I146L}$ to shift cells to the S/M phase was further increased by PRL stimulation, and as observed in the luciferase assay (Fig. 17.2D), it attained higher level than for stimulated Ba/F–$PRLR_{WT}$ cells. Interestingly, the stimulatory effect of PRL observed by FACS (Fig. 17.4B) was not detected using the WST-1 assay (Fig. 17.4A), which may reveal differences in assay sensitivity; clearly, WST-1 cannot detect small variations of viable cell numbers at low and high cell densities. The Ba/F assay also confirmed the mild basal activity of $PRLR_{I76V}$ identified in signaling studies using the same cells.

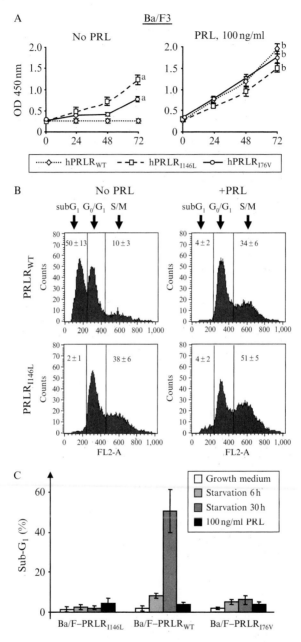

Figure 17.4 Complementarities of WST-1 and FACS analyses for characterizing Ba/F3–PRLR cells (A) Basal (no PRL) and PRL-induced growth/survival of Ba/F–PRLR$_{WT}$ (diamonds), Ba/F–PRLR$_{I146L}$ (squares), and Ba/F–PRLR$_{I76V}$ (circles) populations were monitored using WST-1 reagent (OD$_{450}$). Error bars indicate SD from one representative experiment (out of three) performed in triplicate. *Symbols*: a, $p < 0.05$

Although the number of Ba/F–PRLR$_{I76V}$ cells spontaneously entering in S/M phase was less spectacular compared to Ba/F–PRLR$_{I146L}$ cells (not shown), the former were clearly still protected from apoptosis after 30 h PRL deprivation, when 50% of Ba/F–PRLR$_{WT}$ already died (Fig. 17.4C). This highlights that the monitoring of Ba/F cells using FACS can be very informative and complementary to WST-1 assay, which will be preferentially recommended to distinguish cells presenting with different mitogenic potency (Fig. 17.4A).

As functional investigation of PRLR variants was linked to their potential involvement in breast tumorogenesis, they were also analyzed in a mammary cell context. Although conventional wisdom implies using cells not expressing the receptor that is to be ectopically expressed after transfection (as are HEK-293 and Ba/F3 cells), the use of MCF-7 cells was relevant in our case as endogenous expression of PRLR$_{WT}$ led to investigate the behavior of PRLR$_{I146L}$ in the "heterozygous" context met in patients (Courtillot et al., 2010). In contrast to Ba/F–PRLR$_{WT}$ cells, breast cancer cells are not dependent on PRL for growth. In addition, the amplitude of PRL-induced cell proliferation is mild, therefore we experienced that proliferation assays had to be repeated several times to achieve statistical difference, especially regarding spontaneous, PRL-independent proliferation. In the absence of PRL, MCF-7 expressing only PRLR$_{WT}$ survived but failed to proliferate as assessed by WST-1 assay. In the same conditions, MCF7 stably expressing PRLR$_{I146L}$ cells proliferated to a submaximal level, which could be further enhanced by adding PRL (not shown). This assay demonstrated that the growth-promoting effect of PRLR$_{I146L}$, as observed for PRLR/Stat5 phosphorylation (Fig. 17.2C), occurs irrespectively of coexpression of the WT receptor. Unfortunately, as MCF-7 karyotype is pluriploid, we could not complement these data using FACS analysis. Therefore, one should be cautious when

---

versus Ba/F–PRLR$_{WT}$; b, $p < 0.05$ for each population versus its nonstimulated counterpart. This assay displays the autonomous survival/proliferation of Ba/F–PRLR$_{I146L}$ and, to a lesser extent, of Ba/F–PRLR$_{I76V}$ cells. In the same time, Ba/F–PRLR$_{WT}$ undergone massive apoptosis over the 3-day assay; due to low sensitivity of WST-1 colorimetric reaction at low-cell density, this assay does not permit detection of cell death in control wells. (B) Cell cycle distribution of Ba/F–PRLR$_{WT}$ and Ba/F–PRLR$_{I146L}$ cells was monitored by FACS analysis. Cells were starved by PRL depletion and stimulated with 200 ng/ml recombinant human PRL (right panels) or not (left panels) for 24-h. DNA content analysis of PI-stained cells is represented. Numbers indicate the average proportion of cells $\pm$ SD (>7 independent series) exhibiting $<2n$ (subG$_1$), $2n$ (G$_0$/G$_1$), and $>2n$ (S/M). In the absence of PRL, Ba/F–PRLR$_{WT}$ cells undergo massive apoptosis, while expression of PRLR$_{I146L}$ protects Ba/F–PRLR$_{I146L}$ cells from death and stimulates autonomous proliferation (bottom left panel). (C) Using the same assay, we were able to demonstrate the protective effect of PRLR$_{I76V}$ variant against apoptosis (monitored as Sub-G$_1$ cells in FACS profile as shown on panel B), despite the fact that its mild ability to stimulate spontaneous growth. This underscores the complementarities of these two types of assay.

interpreting the actions of $PRLR_{I46L}$ on MCF-7 based on the sole WST-1 assay, as it may reflect a balance of antiapoptotic/proliferation effects, although breast cancer cells are known to not be prone to apoptosis.

4. CONCLUSIONS

This chapter focuses on the three cell models and readouts that were used to identify the first natural PRLR variant presenting constitutive activity. None of the bioassays used was novel or particularly sophisticated, as we used widely validated cell lines and molecular/biological responses (Bernichtein et al., 2003). The main challenge was to distinguish differences actually reflecting specific properties of mutated versus WT receptors from artifactual differences intrinsic to the biological tools that were generated for this study. We all know that ectopic receptor overexpression can lead to "nonspecific" elevated background, which in our case had to be demonstrated as a "specific" feature of $PRLR_{I146L}$ variant. This was possible thanks to the use of different cell lines that all pointed to the constitutive activity of this variant. On this basis, we speculated on the involvement of this receptor in breast tumorogenesis (Bogorad et al., 2008; Courtillot et al., 2010). This remains more tricky for the I76V variant, as its level of basal activity was clearly lower than that of $PRLR_{I146L}$. Nevertheless, this study confirmed the very high sensitivity of Ba/F3 cells. This cell line, used in the past to reveal the residual activity of PRL variants otherwise considered as PRLR antagonists (Jomain et al., 2007), was demonstrated here to be the only one able to display the low constitutive activity of $PRLR_{I76V}$.

ACKNOWLEDGMENTS

We thank all current and past members of the laboratory who contributed to the establishment of the protocols detailed in this chapter. We specifically acknowledge Paul Kelly, Sophie Bernichtein, Isabelle Fernandez, Estelle Tallet, and Chi Zhang for helpful discussions during the writing of this chapter. V. G. was awarded an Institut National de la Santé et de la Recherche Médicale—Assistance Publique—Hôpitaux de Paris Interface contract, and R. L. B. received an unrestricted grant from Pfizer Company.

REFERENCES

Bazan, F. (1989). A novel family of growth factor receptors: A common binding domain in the growth hormone, prolactin, erythropoietin and IL-6 receptors, and p75 IL-2 receptor β-chain. *Biochem. Biophys. Res. Commun.* **164**, 788–795.
Ben Jonathan, N., Lapensee, C. R., and Lapensee, E. W. (2008). What can we learn from rodents about prolactin in humans? *Endocr. Rev.* **29**, 1–41.

Berlanga, J. J., Gualillo, O., Buteau, H., Applanat, M., Kelly, P. A., and Edery, M. (1997). Prolactin activates tyrosyl phosphorylation of insulin receptor substrate-1 and phosphatidylinositol-3-OH kinase. *J. Biol. Chem.* **272,** 2050–2052.

Bernichtein, S., Jeay, S., Vaudry, R., Kelly, P. A., and Goffin, V. (2003). New homologous bioassays for human lactogens show that agonism or antagonism of various analogs is a function of assay sensitivity. *Endocrine* **20,** 177–190.

Bogorad, R. L., Courtillot, C., Mestayer, C., Bernichtein, S., Harutyunyan, L., Jomain, J. B., Bachelot, A., Kuttenn, F., Kelly, P. A., Goffin, V., and Touraine, P. (2008). Identification of a gain-of-function mutation of the prolactin receptor in women with benign breast tumors. *Proc. Natl. Acad. Sci. USA* **105,** 14533–14538.

Bole-Feysot, C., Goffin, V., Edery, M., Binart, N., and Kelly, P. A. (1998). Prolactin and its receptor: actions, signal transduction pathways and phenotypes observed in prolactin receptor knockout mice. *Endocr. Rev.* **19,** 225–268.

Boutin, J. M., Edery, M., Shirota, M., Jolicoeur, C., Lesueur, L., Ali, S., Gould, D., Djiane, J., and Kelly, P. A. (1989). Identification of a cDNA encoding a long form of prolactin receptor in human hepatoma and breast cancer cells. *Mol. Endocrinol.* **3,** 1455–1461.

Broutin, I., Jomain, J. B., Tallet, E., Van Agthoven, J., Raynal, B., Hoos, S., Kragelund, B. B., Kelly, P. A., Ducruix, A., England, P., and Goffin, V. (2010). Crystal structure of an affinity-matured prolactin complexed to its dimerized receptor reveals the topology of hormone binding site 2. *J. Biol. Chem.* **285,** 8422–8433.

Canbay, E., Degerli, N., Gulluoglu, B. M., Kaya, H., Sen, M., and Bardakci, F. (2004). Could prolactin receptor gene polymorphism play a role in pathogenesis of breast carcinoma? *Curr. Med. Res. Opin.* **20,** 533–540.

Clevenger, C. V., Furth, P. A., Hankinson, S. E., and Schuler, L. A. (2003). The role of prolactin in mammary carcinoma. *Endocr. Rev.* **24,** 1–27.

Courtillot, C., Chakhtoura, Z., Bogorad, R., Genestie, C., Bernichtein, S., Badachi, Y., Janaud, G., Akakpo, J. P., Bachelot, A., Kuttenn, F., Goffin, V., and Touraine, P. (2010). Characterization of two constitutively active prolactin receptor variants in a cohort of 95 women with multiple breast fibroadenomas. *J. Clin. Endocrinol. Metab.* **95,** 271–279.

Glasow, A., Horn, L. C., Taymans, S. E., Stratakis, C. A., Kelly, P. A., Kohler, U., Gillespie, J., Vonderhaar, B. K., and Bornstein, S. R. (2001). Mutational analysis of the PRL receptor gene in human breast tumors with differential PRL receptor protein expression. *J. Clin. Endocrinol. Metab.* **86,** 3826–3832.

Goffin, V., Bernichtein, S., Touraine, P., and Kelly, P. A. (2005). Development and potential clinical uses of human prolactin receptor antagonists. *Endocr. Rev.* **26,** 400–422.

Goffin, V., Shiverick, K. T., Kelly, P. A., and Martial, J. A. (1996). Sequence-function relationships within the expanding family of prolactin, growth hormone, placental lactogen and related proteins in mammals. *Endocr. Rev.* **17,** 385–410.

Horseman, N. D., Zhao, W., Montecino-Rodriguez, E., Tanaka, M., Nakashima, K., Engle, S. J., Smith, F., Markoff, E., and Dorshkind, K. (1997). Defective mammopoiesis, but normal hematopoiesis, in mice with a targeted disruption of the prolactin gene. *EMBO J.* **16,** 6926–6935.

Ibrahim, N. M., Marinovic, A. C., Price, S. R., Young, L. G., and Frohlich, O. (2000). Pitfall of an internal control plasmid: Response of Renilla luciferase (pRL-TK) plasmid to dihydrotestosterone and dexamethasone. *Biotechniques* **29,** 782–784.

Jeay, S., Sonenshein, G., Postel-Vinay, M. C., and Baixeras, E. (2000). Growth hormone prevents apoptosis through activation of nuclear factor-κB in interleukin-3-dependent Ba/F3 cell line. *Mol. Endocrinol.* **14,** 650–661.

Jomain, J. B., Tallet, E., Broutin, I., Hoos, S., Van Agthoven, J., Ducruix, A., Kelly, P. A., Kragelund, B. B., England, P., and Goffin, V. (2007). Structural and thermodynamical bases for the design of pure prolactin receptor antagonists. X-ray structure of Del1-9-G129R-hPRL. *J. Biol. Chem.* **282,** 33118–33131.

Kinet, S., Bernichtein, S., Llovera, M., Kelly, P. A., Martial, J. A., and Goffin, V. (2001). Molecular basis of the interaction between human prolactin and its membrane receptor: A ten year study. *Recent Res. Devel. Endocrinol.* **2,** 1–24.

LeBaron, M. J., Ahonen, T. J., Nevalainen, M. T., and Rui, H. (2007). In vivo response-based identification of direct hormone target cell populations using high-density tissue arrays. *Endocrinology* **148,** 989–1008.

Lebrun, J. J., Ali, S., Goffin, V., Ullrich, A., and Kelly, P. A. (1995). A single phosphotyrosine residue of the prolactin receptor is responsible for activation of gene transcription. *Proc. Natl. Acad. Sci. USA* **92,** 4031–4035.

Lee, S. A., Haiman, C. A., Burtt, N. P., Pooler, L. C., Cheng, I., Kolonel, L. N., Pike, M. C., Altshuler, D., Hirschhorn, J. N., Henderson, B. E., and Stram, D. O. (2007). A comprehensive analysis of common genetic variation in prolactin (PRL) and PRL receptor (PRLR) genes in relation to plasma prolactin levels and breast cancer risk: The multiethnic cohort. *BMC Med Genet.* **8,** 72.

Llovera, M., Pichard, C., Bernichtein, S., Jeay, S., Touraine, P., Kelly, P. A., and Goffin, V. (2000). Human prolactin (hPRL) antagonists inhibit hPRL-activated signaling pathways involved in breast cancer cell proliferation. *Oncogene* **19,** 4695–4705.

Molitch, M. E. (2003). Dopamine resistance of prolactinomas. *Pituitary* **6,** 19–27.

Ormandy, C. J., Camus, A., Barra, J., Damotte, D., Lucas, B. K., Buteau, H., Edery, M., Brousse, N., Babinet, C., Binart, N., and Kelly, P. A. (1997). Null mutation of the prolactin receptor gene produces multiple reproductive defects in the mouse. *Genes Dev.* **11,** 167–178.

Riddle, O., Bates, R. W., and Dykshorn, S. W. (1933). The preparation, identification and assay of prolactin—A hormone of the anterior pituitary. *Am. J. Physiol.* **105,** 191–216.

Stricker, P., and Grueter, R. (1928). Action du lobe antérieur de l'hypophyse sur la montée laiteuse. *C. R. Soc. Biol.* **99,** 1978–1980.

Utama, F. E., Tran, T. H., Ryder, A., LeBaron, M. J., Parlow, A. F., and Rui, H. (2009). Insensitivity of human prolactin receptors to nonhuman prolactins: Relevance for experimental modeling of prolactin receptor-expressing human cells. *Endocrinology* **150,** 1782–1790.

Vaclavicek, A., Hemminki, K., Bartram, C. R., Wagner, K., Wappenschmidt, B., Meindl, A., Schmutzler, R. K., Klaes, R., Untch, M., Burwinkel, B., and Forsti, A. (2006). Association of prolactin and its receptor gene regions with familial breast cancer. *J. Clin. Endocrinol. Metab.* **91,** 1513–1519.

Welsch, C. W., and Nagasawa, H. (1977). Prolactin and murine mammary tumorigenesis: A review. *Cancer Res.* **37,** 951–963.

CHAPTER EIGHTEEN

INVESTIGATIONS OF ACTIVATED ACVR1/ALK2, A BONE MORPHOGENETIC PROTEIN TYPE I RECEPTOR, THAT CAUSES FIBRODYSPLASIA OSSIFICANS PROGRESSIVA

Frederick S. Kaplan,[*,†,‡] Petra Seemann,[§] Julia Haupt,[§] Meiqi Xu,[*,‡] Vitali Y. Lounev,[*,‡] Mary Mullins,[∥] and Eileen M. Shore[*,¶,‡]

Contents

1. Introduction	358
2. Patient Methodologies	359
2.1. Precise definition of FOP phenotype	359
2.2. FOP mutational analysis	360
2.3. ACVR1/ALK2 mutational analysis	360
3. Cellular Methodologies	362
3.1. BMP signaling in FOP cells	362
3.2. Expression plasmid constructs	363
3.3. Cell culture methodologies	363
3.4. Cell transfections and luciferase assays	364
3.5. Immunoblot analysis methodologies	364
3.6. Immunoprecipitation methodologies	364
3.7. Connective tissue progenitor cells	365
3.8. SHED cell isolation and culture	365
3.9. SHED cell differentiation assays	366
3.10. SHED cell mineralization assays	366

[*] Department of Orthopaedic Surgery, The University of Pennsylvania School of Medicine, Philadelphia, USA
[†] Department of Medicine, The University of Pennsylvania School of Medicine, Philadelphia, USA
[‡] The Center for Research in FOP and Related Disorders, The University of Pennsylvania School of Medicine, Philadelphia, USA
[§] Berlin Brandenburg Center for Regenerative Therapies (BCRT), Charité-Universitätsmedizin Berlin, Berlin, Germany
[∥] Department of Cell and Developmental Biology, The University of Pennsylvania School of Medicine, Philadelphia, USA
[¶] Department of Genetics, The University of Pennsylvania School of Medicine, Philadelphia, USA

Methods in Enzymology, Volume 484 © 2010 Elsevier Inc.
ISSN 0076-6879, DOI: 10.1016/S0076-6879(10)84018-2 All rights reserved.

4. Tissue Methodologies 366
   4.1. Studying FOP in the chicken 366
   4.2. Micromass culture system 367
   4.3. Constructs and virus production 367
   4.4. Isolation and cultivation of micromass cultures 367
   4.5. Alcian blue staining 368
   4.6. ALP staining after PFA-fixation 368
   4.7. qPCR for chondrogenic marker genes 368
5. *In Vivo* Methodologies 369
   5.1. *In vivo* analysis of zebrafish embryos 369
   5.2. Zebrafish embryo methodologies 369
   5.3. Murine BMP overactivity and overexpression models 370
   5.4. A conditional/constitutively active ACVR1/ALK2 model 370
   5.5. A knock-in mouse chimera for FOP 371
Acknowledgments 371
References 371

Abstract

Bone morphogenetic protein (BMP) type I receptors are serine-threonine kinase transmembrane signal transduction proteins that regulate a vast array of ligand-dependent cell-fate decisions with temporal and spatial fidelity during development and postnatal life. A recent discovery identified a recurrent activating heterozygous missense mutation in a BMP type I receptor [Activin receptor IA/activin-like kinase 2 (ACVR1; also known as ALK2)] in patients with the disabling genetic disorder fibrodysplasia ossificans progressiva (FOP). Individuals with FOP experience episodes of tissue metamorphosis that convert soft connective tissue such as skeletal muscle into a highly ramified and disabling second skeleton of heterotopic bone. The single nucleotide ACVR1/ALK2 mutation that causes FOP is one of the most specific disease-causing mutations in the human genome and to date the only known inherited activating mutation of a BMP receptor that causes a human disease. Thus, the study of FOP provides the basis for understanding the clinically relevant effects of activating mutations in the BMP signaling pathway. Here we briefly review methodologies that we have applied to studying activated BMP signaling in FOP.

1. INTRODUCTION

Fibrodysplasia ossificans progressiva (FOP), a rare and illustrative genetic disorder of skeletal malformations and progressive heterotopic ossification, is the most disabling condition of ectopic skeletogenesis in humans (Kaplan *et al.*, 2005). Typically, during the first decade of life, episodes of inflammatory soft tissue swellings (flare-ups) seize the body's skeletal muscles and connective tissue, and transform them through an endochondral process into an immobilizing second skeleton of heterotopic bone (Kaplan *et al.*, 2005). Most cases are caused by spontaneous new mutations. Inheritance, when observed, is autosomal dominant with variable expression (Shore *et al.*, 2005). At present, there is no definitive treatment (Kaplan *et al.*, 2008a).

A large body of work has supported dysregulated bone morphogenetic protein (BMP) signaling in the pathogenesis of FOP (Ahn *et al.*, 2003; Billings *et al.*, 2008; Fiori *et al.*, 2006; Glaser *et al.*, 2003; Kaplan *et al.*, 2007a; Lounev *et al.*, 2009; Serrano de la Peña *et al.*, 2005; Shafritz *et al.*, 1996; Shen *et al.*, 2009). Heterozygous missense mutations were identified in the glycine-serine activation domain of Activin receptor IA/Activin-like kinase 2 (ACVR1; also known as ALK2), a BMP type I receptor, in all affected individuals of several multigenerational families and in all sporadically affected individuals with features of classic FOP (Shore *et al.*, 2006). Remarkably the same single nucleotide mutation and substituted amino acid residue (c.617G>A; R206H) are altered in all classically affected FOP patients, providing the basis for elucidating the dysregulated BMP signaling that underlies the catastrophic phenotype of FOP (Groppe *et al.*, 2007; Shen *et al.*, 2009; Shore *et al.*, 2006). Conversely, FOP provides an important clinical and scientific window through which to understand the complexities of dysregulated BMP signaling relevant to human pathology and regenerative medicine (Kaplan *et al.*, 2007b, 2009a; Shore *et al.*, 2006). To date, FOP is the only human condition caused by inherited activating mutations in a BMP receptor. Recent detailed articles on FOP (Kaplan *et al.*, 2008a) and the BMP signaling pathway (Sieber *et al.*, 2009) provide an in-depth review of these critical background topics for readers.

Here, we briefly review methodologies that have been useful in understanding activated BMP signaling in FOP. While it may be possible in some instances to indicate optimal techniques, there are other instances, such as in the development of relevant animal models, where methodologies are rapidly evolving.

2. Patient Methodologies

2.1. Precise definition of FOP phenotype

In all laboratory studies in which FOP patient samples are used, it is critical that the clinical FOP phenotype be clearly and unequivocally established before proceeding with molecular and biochemical analyses (Kaplan *et al.*, 2005, 2009b; Shore *et al.*, 2006). The diagnosis of FOP is made on a clinical basis, however, mutational analysis can be confirmative, aid in early diagnosis, and be extremely valuable in cases of suspected variants (Kaplan *et al.*, 2008b, 2009b).

Patients with classic FOP have two defining clinical features: characteristic congenital malformations of the great toes and progressive heterotopic ossification in characteristic anatomic patterns (Kaplan et al., 2009b; Shore et al., 2006). In addition, common but variable features are seen in most individuals with FOP including proximal medial tibial osteochondromas, cervical spine malformations, short, broad femoral necks, conductive hearing impairment, and malformations of the thumbs (Kaplan et al., 2009b).

Some patients with FOP-type heterotopic ossification have more or less severe clinical presentations. Individuals with FOP-plus have classic features of FOP plus one or more atypical features (Kaplan et al., 2009b). Individuals with FOP variants have major variations in one or both of the classic defining features of FOP (Kaplan et al., 2009b).

2.2. FOP mutational analysis

Mutational analysis of ACVR1/ALK2 in affected members of the few identified multigenerational families and in all sporadically affected individuals worldwide identified a recurrent missense activating mutation in codon 206 of the cytoplasmic glycine-serine domain of the receptor (Shore et al., 2006). Follow-up studies identified alternate mutations in ACVR1/ALK2 in patients with highly variable and extremely rare clinical variants of the condition (Kaplan et al., 2009b). While the recurrent ACVR1/ALK2 (R206H) mutation was found in all cases of classic FOP and most cases of FOP-plus, mutations at different positions occur in ACVR1/ALK2 in all FOP variants and in rare cases of FOP-plus (Kaplan et al., 2009b).

2.3. ACVR1/ALK2 mutational analysis

In order to determine the genotype of individuals suspected of having classic FOP, genomic DNA is commonly screened from buccal swabs, blood, or lymphoblastoid cell lines for the canonical R206H mutation in ACVR1/ALK2. For individuals with classic FOP, genomic DNA is screened for mutations in ACVR1 by PCR amplification using primers that flank the recurrent c.617G>A mutation. For individuals with a suspected FOP variant phenotype, mutations in ACVR1/ALK2 can be screened by PCR amplification of the nine exons containing protein-coding sequences (ACVR1/ALk2 transcript report ensemble v35, accessions number ENST00000263640, GenBank RefSeq NM_001105.4 and NP_001096.1) using exon-flanking primers (Shore et al., 2006). When available, DNA samples from parents are also examined. To date, no mutations have been found in unaffected parents.

PCR amplification is performed using Amplitaq Gold enzyme on PE2700 automated thermocyclers (Perkin Elmer, ABI, Waltham, MA). The PCR volume is 25 μl, with 60 ng of genomic DNA, 10 pmol each of forward and reverse primer (Table 18.1), 200 μM of deoxyribonucleotide triphosphate, 1.5 mM of MgCl$_2$, and 10× PCR buffer. The general cycling

Table 18.1 ACVR1 primers for human genomic DNA PCR amplification

Protein-coding exon #	Forward primer	Reverse primer	PCR product size (bp)
Exon 1	5'-AAGTAAGGCAATATATCTGAGG-3'	5'-GAGTGTTTAAGTTTGATAGGC-3'	307
Exon 2	5'-ATATGAACACCACAGGGGG-3'	5'-CCTTCTGGTAGACGTGGAAG-3'	449
Exon 3	5'-TTTTTCCCCTTCCTTTCTCTC-3'	5'-CAGGGTGACCTTCCTTGTAG-3'	438
Exon 4	5'-AATTCCCCCTTTTCCCTCCAAC-3'	5'-TAAGAACGTGTCTCCAGACACC-3'	300
Exon 5	5'-CCAGTCCTTCTTCCTTCTTCC-3'	5'-AGCAGATTTTCCAAGTTCCATC-3'	350
Exon 6	5'-TCCCAAGCTGAGTTTCTCC-3'	5'-AGAGCAAAGGCAGACAATTG-3'	346
Exon 7	5'-GACATTTACTGTGTAGGTCGC-3'	5'-AGAGATGCAACTCACCTAACC-3'	438
Exon 8	5'-TGGGGTTGGTTAAAATCCTTC-3'	5'-AGGTAGCTGGATCAAGAGAAC-3'	337
Exon 9	5'-CACATTATAACCTGTGACACCC-3'	5'-ATACCAGTTGAAACTCAAAGGG-3'	299
	5'-GTATTGCTGCTTTTGGCAAC-3'	5'-CAGTCCCTACCTTTGCAAC-3'	700

Mutations in *ACVR1* are detected by DNA sequencing following PCR-amplification of genomic DNA corresponding to the nine exons containing protein-coding sequences (ACVR1 Transcript Report, Ensembl v35), using exon-flanking primers. Protein-coding exon 1 contains the ATG protein start codon. The R206H mutation is in protein-coding exon 4. Additional/alternate exons containing 5'UTRs are reported. (Gene ID ENSG00000115170; transcript ID ENST00000263640) reports 11 exons for ACVR1 (with the first two exons containing only 5' untranslated sequences), consistent with GenBank BC033867, full-length cDNA clone. Note: all primers are the same as reported in Shore *et al.* (2006) Nature Genetics Supplemental Methods except for the Exon 1 primer pair.

conditions are 94 °C for 5 min, followed by 30 cycles of 94 °C for 30 s, 60 °C for 30 s, and 72 °C for 60 s, then 1 cycle of 72 °C for 5 min. The PCR product amplification is verified by agarose gel electrophoresis. The amplicons are subsequently treated with Shrimp Alkaline Phosphatase and Exonuclease (USB, Cleveland, OH) to eliminate unincorporated nucleotides and primers before DNA sequence analysis. The products are purified on a 96-well purification plate (Edge Biosystems, Gaithersburg, MD), dissolved in water, and analyzed by dye terminator cycle sequencing with an ABI 3700XL sequencer (ABI) using either the reverse or forward PCR primer (Kaplan et al., 2009b; Shore et al., 2006). Sequence data are analyzed using 4Peaks software v.1.6 (available online: www.mekentosj.com/4peaks). Mutations are identified by nucleotide numbering that reflects the cDNA sequence, with +1 corresponding to the A of the ATG translation initiation codon in the reference sequence, according to gene nomenclature guidelines (www.hgvs.org/mutnomen). The protein initiation codon is codon 1 (Kaplan et al., 2009b; Shore et al., 2006).

As an alternative to DNA sequence analysis, the c.617G>A; R206H ACVR1 mutation can be readily identified through differences in restriction endonuclease recognition (Shore et al., 2006). Genomic DNA (0.1 μg) is amplified using primers for protein-coding exon 4 (Table 18.1). Following agarose gel electrophoresis, PCR products (350 bp) are recovered from agarose using QIAquick Gel Extraction reagents (Qiagen). Purified PCR products are digested with either HphI (5 U/μl) or Cac8I (4 U/μl) (New England Biolabs) at 37 °C for 2 h and fragments resolved on 3% NuSieve 3:1 agarose (FMC BioProducts) gels with 100 bp ladder (New England Biolabs) as size markers.

Similarly, some FOP variant ACVR1 mutations can be detected by differential restriction endonuclease digestion (Kaplan et al., 2009b). New cleavage sites are created by the c.619C>G (NruI) and c.1067G>A (DrdI) nucleotide substitutions. A StyI digestion site is eliminated by each of the single nucleotide substitutions identified in codon 328.

3. CELLULAR METHODOLOGIES

Studies performed prior to the discovery of the causative mutation for FOP revealed overactive BMP signaling in FOP cells through both the canonical SMAD pathway and the p38 MAPK pathway (Ahn et al., 2003; Fiori et al., 2006; Serrano de la Peña et al., 2005; Shafritz et al., 1996).

3.1. BMP signaling in FOP cells

Following the discovery of the FOP ACVR1 mutation, we performed numerous cell-based assays to evaluate the effects of the mutation on BMP signaling (Shen et al., 2009). In these in vitro assays, we examined

BMP pathway-specific SMAD phosphorylation and downstream transcriptional targets of BMP signaling. We determined that mutant R206H ACVR1/ALK2 activated BMP signaling in the absence of BMP and enhanced BMP signaling in the presence of BMP. We further investigated the interaction of mutant R206H ACVR1/ALK2 with FKBP1A (also known as FKBP12), a glycine-serine domain binding protein that prevents leaky type I BMP receptor activation in the absence of ligand (Kaplan et al., 2007b; Shore et al., 2006). We found reduced FKBP1A binding to the mutant protein suggesting that increased BMP pathway activity in cells with R206H ACVR1/ALK2 is due at least in part to decreased binding of this inhibitory factor (Shen et al., 2009). Relevant methodologies are outlined below.

3.2. Expression plasmid constructs

A human ACVR1/ALK2 expression vector was generated by insertion of the hACVR1/ALK2 protein-coding sequence (GenBank accession number NM_001105.4) into the pcDNA 3.1 D V5-His-TOPO vector (Invitrogen). An FOP mutant ACVR1/ALK2 expression vector was generated by site directed mutagenesis of the wild-type ACVR1/ALK2 sequence at cDNA position 617 (from G to A) using the GeneTailor Site-Directed Mutagenesis System (Invitrogen), and the oligonucleotides: forward 5'-GTACAAAGAACAGTGGCTCaCCAGATTACACTG-3'; reverse 5'-GTGAGCCACTGTTCTTTGTACCAGAAAAGGAAG-3'. The FKBP1A/FKBP12 expression vector is from Origene. The human *ID1* promoter ($-985/+94$) luciferase reporter construct was previously described and is a standard BMP early response gene used to monitor downstream BMP transcriptional activity (Katagiri et al., 2002). The *ID1* gene is a direct target of BMP signaling and encodes a dominant-negative inhibitor of basic helix-loop-helix transcription factors, including members of the MyoD family that are important in myoblast differentiation.

3.3. Cell culture methodologies

We transfected a variety of mammalian cell lines to investigate the effects of ACVR1/ALK2 (R206H) on BMP signaling (Shen et al., 2009). COS-7 African green monkey kidney cells, C2C12 mouse myoblastic cells, MC3T3-E1 human osteoblastic cells, and U-2 OS human osteosarcoma cells are obtained from ATCC. Cells are cultured in DMEM (COS-7 and C2C12), α-MEM (MC3T3-E1), or McCoy's 5A medium (U-2 OS) plus 10% FBS (all from Invitrogen). All cells are cultured in a humidified atmosphere of 5% CO_2 at 37 °C.

3.4. Cell transfections and luciferase assays

COS-7 cells are seeded into 24-well plates at 7×10^4 cells per well in culture medium without antibiotics. After 24 h, expression vectors are transfected into the cells using FuGene 6 (Roche) according to the manufacturer's protocol. Our efficiency of transfection, as assessed by cotransfection with GFP constructs and subsequent GFP detection, is estimated at 60–70%. At 48 h, cells are washed twice with PBS and lysed in 1× passive lysis buffer (Dual-Luciferase Reporter Assay, Promega). Luciferase activity is assayed following the recommended protocol and normalized to pRL-TK-*Renilla* (Promega) luciferase signals (Promega).

3.5. Immunoblot analysis methodologies

COS-7 cells, plated at 70% confluence in 100-mm tissue culture dishes, are transfected with vector alone or pcDNA3 constructs with wild-type or mutant ACVR1. Total proteins are harvested in lysis buffer (20 mM Tris–HCl; pH 8.0), 150 mM NaCl, phosphatase inhibitors (Pierce), protease inhibitors (C complete protease inhibitor cocktail, Roche), and 1% Triton X-100. For immunoblot analysis, 50 μg of protein from each total cell lysate is electrophoresed through 10% SDS-polyacrylamide gels, transferred to nitrocellulose membranes (iBlot membranes; Invitrogen). Membranes are incubated overnight at 4 °C with antibodies specific for phospho-Smad1/5/8 and Smad1 (Cell Signaling Technology), V5 (Invitrogen), or β-actin (Santa Cruz Biotechnology, Inc.) in PBS containing 5% nonfat milk and 0.5% BSA. Membranes are washed with PBS and incubated for 1 h with the corresponding secondary antibody conjugated with horseradish peroxidase. The enhanced chemiluminescent Immobilon Western blotting detection system (Millipore) is used to detect the antigen–antibody complex. Similar protocols are used for immunoblot analysis of cell protein extracts from MC3T3-E1 (mouse preosteoblasts), U-2 OS (human osteosarcoma), and C2C12 (mouse myoblasts with osteogenic potential). Western blotting of phospho-Smad1/5/8 is performed according to standard methodology (Shen *et al.*, 2009).

3.6. Immunoprecipitation methodologies

To examine the interaction between FKBP1A and ACVR1/ALK2 in the absence or presence of BMPs, COS-7 cells are cotransfected with normal (c.617G) or mutant (c.617A) ACVR1/ALK2 expression constructs and the FKBP1A expression construct. After 48 h of transfection, cells are starved for 2 h in serum-free medium and then treated for 1 h with 100 ng/ml BMP4 or BMP7 (R&D Systems). Total proteins are isolated, and protein concentration is detected by the Bradford assay.

Immunoprecipitation assays use 500 µg of protein from each experimental sample and 2 µg of FKBP1A or ACVR1/ALK2 antibody (both from Santa Cruz Biotechnology, Inc.) at 4 °C overnight, followed by treatment with 30 µl of Protein A/G agarose beads (Pierce) at 4 °C for 1 h and centrifugation at 800×g for 5 min. The immunoprecipitated complex is dissociated by 12% SDS-PAGE and detected with V5 monoclonal antibody (Invitrogen) or FKBP1A antibody (N19; Santa Cruz Biotechnology, Inc.) (Shen et al., 2009). Additional studies on FOP cells (Fukuda et al., 2009; van Dinther et al., 2009) also describe useful techniques related to the original approaches noted above.

3.7. Connective tissue progenitor cells

While transfected cells (Shen et al., 2009) and immortalized lymphoblastoid cells (Fiori et al., 2006; Serrano de la Peña et al., 2005; Shafritz et al., 1996) have been helpful in deciphering BMP pathway pathology in FOP, the study of FOP is hampered by the lack of readily available connective tissue progenitor cells that reflect the pathophysiology of the disease. In order to overcome this technical difficulty, we isolate such cells from discarded primary teeth of patients with FOP and controls (Billings et al., 2008). Using these primary cell strains, we discovered dysregulation of BMP signaling and rapid osteoblast differentiation in FOP cells compared with control cells. Tissue progenitor cell lines from exfoliated and discarded primary teeth of patients and controls are extremely valuable, and have several advantages for studying activated BMP signaling in FOP over transfected cells because they are obtained directly from FOP patients and thus preserve the normal stoichiometry of BMP receptor copy number and locus fidelity.

3.8. SHED cell isolation and culture

Naturally exfoliated teeth are obtained from children. SHED (stem cells from human exfoliated deciduous teeth) cell strains are established from patients with FOP and unaffected age- and sex-matched controls. Cells are isolated as previously reported (Miura et al., 2003) with minor protocol modifications. The dental pulp is digested with 2 mg/ml type II collagenase for 1 h (37 °C) in serum-free DMEM and filtered through a 100 µm cell strainer (BD Falcon, Franklin Lakes, NJ, USA). Cells in the filtrate are recovered by centrifugation (400×g, 10 min) and plated in DMEM with 10% FCS, GlutaMAX supplement, and antibiotics. The presence of ACVR1/ALK2 mutations in codon 206 (R206H) in FOP cells is confirmed by DNA sequence analysis (as described above).

For experimental treatments, cells are plated in 6-well plates (5×10^4 cells/well) in DMEM/10% FBS and grown for 4–6 days

(80–90% confluence). Cells are washed with PBS, incubated for 1 h in serum-free medium, and treated with 100 ng/ml BMP4 in serum-free medium for 1.5 h. For transfection experiments, SHED cells are seeded into 6-well plates and transfected with expression constructs (1 µg/well) for 48 h using TransIT-LT1 transfection reagent (Mirus, Madison, WI, USA) following the recommended protocol (Billings et al., 2008; Miura et al., 2003).

3.9. SHED cell differentiation assays

Alkaline phosphatase (ALP) activity is detected histochemically with BCIP (5-bromo-4-chloro-3-indolyl phosphate)/NBT (nitroblue tetrazolium)-plus substrate (Moss Substrates, Pasadena, MD, USA) at 37 °C. To quantify ALP enzyme activity, cells are lysed in 0.1 M Tris (pH 7.5) and 0.1% Triton X-100. Cell debris is removed by centrifugation (5000×g, 5 min, 4 °C) and 50 µl aliquots of supernatant are assayed in 0.5 ml of 1 mg/ml Sigma Phosphatase Substrate in 0.1 M Tris (pH 9.5), 1 mM MgCl$_2$, p-nitrophenol (p-NP), and detected by spectrophotometry at 405 nm. Protein is determined with a BCA (bicinchoninic acid) protein assay (Pierce, Rockford, IL, USA), using BSA as standard. Enzyme activity is expressed as micromoles p-NP per minute per microgram protein (Billings et al., 2008).

3.10. SHED cell mineralization assays

To induce mineralization, SHED cells are plated (5×10^4 cells/well, 24-well plates) and after 24 h are treated with osteogenic medium (OM; DMEM/10% FCS, supplemented with 50 µg/ml ascorbate, 10 mM β-glycerophosphate, and 10 nM dexamethasone) for 14–21 days. The medium is changed twice weekly. To detect calcium mineralization, cells are stained with 1% Alizarin red (LabChem, Pittsburgh, PA, USA) for 30 min. To quantify mineralization, Alizarin red-stained cells are solubilized in 0.5 N HCl and 5% SDS for 30 min and detected at 405 nm using a Bio-Tek Synergy HT microplate reader (Billings et al., 2008).

4. Tissue Methodologies

4.1. Studying FOP in the chicken

Studies using chick assays established ACVR1/ALK2 as a BMP type I receptor (Zhang et al., 2003). ACVR1/ALK2 expression was examined *in vitro* in isolated chick chondrocytes and osteoblasts and *in vivo* in the developing chick limb bud (Zhang et al., 2003).

4.2. Micromass culture system

Micromass cultures are used to investigate the effects of activated BMP signaling on tissue morphogenesis, particularly chondrogenesis. Micromass culture experiments were used to confirm that mutant R206H ACVR1/ALK2 activates BMP signaling in the absence of BMP ligand. It was further determined that expression of early markers of chondrogenesis was enhanced by BMP in cells expressing mutant ACVR1/ALK2 compared to wild type (Shen et al., 2009). We prepare micromass cultures from primary mesenchymal cells isolated from embryonic chicken limb buds. The advantage of using chicken cells is that the gene of interest can be efficiently overexpressed by using an avian-specific replication competent retrovirus as a shuttle.

4.3. Constructs and virus production

The avian retrovirus RCAS (for more information please visit: http://home.ncifcrf.gov/hivdrp/RCAS/index.html) is used to overexpress chicken ACVR1/ALK2 or activated variants of it in micromass cultures. Viral constructs were prepared by PCR amplification of the coding sequence of chicken ACVR1/ALK2 from chicken embryo cDNA using the following primer pair: chAcvr1-NcoI-fwd, 5′-accATGGCT CTCCCCGTGCTGCTG-3′, and chAcvr1-BamHI-rev, 5′-aggatcc-TCACCAGTCAGCCTTCAGTTT-3′. The PCR product was digested with *NcoI* and *BamHI* and subcloned into the pSLAX-13 shuttle vector. This construct is used to introduce the corresponding human FOP mutation (R206H) and the constitutively active (ca) variant of the receptor (Q207D) by Site-Directed Mutagenesis (QuikChange, Stratagene).

Inserts were subcloned by *ClaI* into the avian-specific viral vector RCASBP(A). To produce the virus, RCASBP plasmids are transfected into DF1 cells, culture medium is harvested, and viral particles are concentrated by ultracentrifugation. Titers of all receptor-expressing RCASBP(A) viruses are determined, and equal concentrations of all constructs are used to infect micromass cultures (Shen et al., 2009).

4.4. Isolation and cultivation of micromass cultures

Micromass cultures are prepared from dissected limb buds from chicken embryos at Hamburger-Hamilton stage HH22-24 (after 4.5 days incubation at 37.5 °C). Limb buds are incubated with Dispase II (Roche) in HBSS (3 mg/ml) in a 37 °C waterbath for 15 min to remove the ectoderm. Next, limb buds are further digested in digestion solution (0.1% collagenase type Ia (Sigma), 0.1% trypsin, 5% FCS or CS in PBS without Ca/Mg) for 30 min at 37 °C. Prewarmed growth media (DMEM:F12, 10% FCS, 0.2% CS, Pen/Strep) is added, and the cell suspension is passed through a Falcon cell strainer (40 μm) to obtain a single cell solution. The final concentration of the cell suspension is adjusted to 2×10^7 cells/ml. For each culture, 10 μl of cells (2×10^5) are mixed with the virus and plated in a 24-well plate and incubated for 1 h in a humidified chamber at 37 °C and 5% CO_2. After attachment of the cells, 1 ml of growth medium is added to each well and replaced three times weekly.

4.5. Alcian blue staining

Differentiation into cartilage can be evaluated by Alcian blue staining of proteoglycans in the extracellular matrix. Micromass cultures are fixed with Kahle's fixative (28.9% [v/v] EtOH; 0.37% formaldehyde; 3.9% [v/v] acetic acid) and stained with 1% Alcian blue in 0.1 N HCl overnight. Excess dye is removed by washing with water, and cultures are dried before photographs are taken. For quantification of incorporated Alcian blue into the proteoglycan-rich extracellular matrix, cultures are incubated with 6 M guanidine hydrochloride overnight, followed by photometric measurement at OD_{595}.

4.6. ALP staining after PFA-fixation

ALP activity will serve as a marker for prehypertrophic chondrocytes. Micromass cultures are incubated with NBT/BCIP. The reaction is stopped with TE buffer, and photographs are taken. For quantification, histomorphometric analysis is performed using Autmess AxioVision 4.6 software (Zeiss) (Shen et al., 2009). Quantitative determination of ALP activity was previously described (see Section 3.9).

4.7. qPCR for chondrogenic marker genes

Additionally, downstream genes can be analyzed by qPCR. RNA from micromass cultures is extracted using peqGold Trifast (peqLab Biotechnologie GmbH) following manufacturer's instructions. One microgram of RNA is transcribed into cDNA by using the TaqMan Reverse Transcription Kit (Applied Biosystems). Relative expression levels are determined by SYBR Green-based qPCR on a ABIPrism 7900HT cycler (Applied Biosystems) with primer pairs for chicken collagen type II (marker for early chondrogenesis), chicken aggrecan (marker for extracellular matrix of chondrocytes), chicken Ihh (marker for prehypertrophic chondrocytes), and chicken collagen type X (marker for hypertrophic chondrocytes) (Shen et al., 2009). To normalize gene expression, we strongly recommend chicken 18S rRNA or 28S rRNA as reference genes because other standard reference genes such as β-actin, gapdh, β-2-microtubulin, tbp, or hprt are regulated by the constitutively active variant of ACVR1/ALK2 (Q207D) and to a lesser degree by the FOP mutation ACVR1/ALK2 (R206H) (Li *et al.*, 2005). 18S and 28S rRNA are the most abundant RNAs in an mRNA preparation; therefore it is necessary to reduce the amount of cDNA amplified by the primer pairs for the 18S and 28S rRNA relative to other genes. Thus, we use a dilution of 1:10 from the cDNA for qPCR of a standard gene, compared to a dilution of 1:50,000 to detect 18S or 28S rRNA. Relative expression levels normalized to 28S rRNA expression can be calculated using the qBase software (Hellemans *et al.*, 2007).

5. *In Vivo* Methodologies

Various animal models in different species (including *Drosophila*, zebrafish, chicken, and mice) have been useful in understanding *in vivo* effects of BMPs in ectopic bone formation and have shed light on the pathophysiology of FOP as well as nonsyndromic forms of posttraumatic heterotopic ossification (Fukuda *et al.*, 2006; Glaser *et al.*, 2003; Kan *et al.*, 2004, 2009; Lounev *et al.*, 2009; Shen *et al.*, 2009; Twombly *et al.*, 2009; Urist, 1965; Wozney *et al.*, 1998; Yu *et al.*, 2008b). Here we will focus on zebrafish and mouse models.

5.1. *In vivo* analysis of zebrafish embryos

In zebrafish embryonic development, a gradient of BMP signaling patterns the dorsal–ventral axis: slight perturbations in this BMP signaling gradient cause easily recognized mutant phenotypes making this system a sensitive assay for increased or decreased BMP signaling (Little and Mullins, 2006). *In vivo* analysis of the ACVR1/ALK2 FOP mutation (R206H) in zebrafish embryos reveals BMP-independent hyperactivation of BMP signaling (Shen *et al.*, 2009). These studies support that the mutant R206H ACVR1/ALK2 receptor in FOP patients is encoded by an activating mutation that induces BMP signaling in both a BMP-independent and a BMP-responsive manner. Zebrafish studies have been helpful in establishing the *in vivo* effects of the FOP mutation during vertebrate embryogenesis (Shen *et al.*, 2009) and in screening small molecule libraries for BMP receptor antagonists (Yu *et al.*, 2008a).

5.2. Zebrafish embryo methodologies

PCR-amplified cDNA encoding control (c.617G) or mutant (c.617A) hACVRI is inserted into a derivative of pCS2+ vector containing an in-frame C-terminal FLAG epitope. mRNA is *in vitro* transcribed using the SP6 mMessage mMachine kit (Ambion) from plasmids linearized with

NotI. Bmp7 and Bmp2b (ortholog of BMP2) are the BMPs that pattern the embryonic dorsal–ventral axis in zebrafish. Morpholinos against genes of interest are injected at 1 ng per embryo. A morpholino mixture of 2 ng smad5MO1 (5′-ATGGAGGTCATAGTGCTGGGCTGC-3′) and 2.5 ng smad5MO3 (5′-GCAGTGTGCCAGGAAGATGATTATG-3′) per embryo was used to knock down *smad5* translation. All injections are performed at the one-cell stage. For treatment with dorsomorphin (DM), a BMP signaling inhibitor that dorsalizes zebrafish embryos (Yu *et al.*, 2008a), embryos were placed in E3 embryo medium containing DMSO either alone or with 40 μM DM prior to the first cell cleavage. Embryo imaging is performed on a MZ12.5 stereomicroscope (Leica) with a ColorSNAP-cf digital camera (Photometrics) and processed using Adobe Photoshop. *In situ* hybridization and subsequent imaging were carried out as described (Shen *et al.*, 2009).

5.3. Murine BMP overactivity and overexpression models

Models for stimulating BMP-induced heterotopic ossification have traditionally involved the implantation of recombinant BMP into soft connective tissue (Glaser *et al.*, 2003; Urist, 1965; Wozney *et al.*, 1998) or the transgenic overexpression of BMP using the tissue-specific promoters (Kan *et al.*, 2004). These animal models are useful in examining overactivity of the BMP signaling pathway through a ligand-based approach, and the reader is directed to these important *in vivo* studies.

Studies using these models showed that dysregulation of BMP signaling in connective tissue progenitor cells contribute to all cell lineages of FOP-like heterotopic ossification (Kan *et al.*, 2009; Lounev *et al.*, 2009). Further, translational studies in an FOP patient who received bone marrow transplantation for an intercurrent condition established that even wild-type immune cells could induce heterotopic ossification in a genetically susceptible host (Kan *et al.*, 2009; Kaplan *et al.*, 2007a; Lounev *et al.*, 2009).

5.4. A conditional/constitutively active ACVR1/ALK2 model

Generation of a mouse with conditional activation of the BMP type I receptor ACVR1/ALK2 provides a model for investigating activated ACVR1/ALK2 gene function in a tissue-specific manner in a mammalian system (Fukuda *et al.*, 2006). Although the ca mutation (ACVR1/ALK2; Q207D) does not cause FOP (Fukuda *et al.*, 2006), it reproduces some of the features of the disease (Yu *et al.*, 2008b). Use of this animal model to induce heterotopic ossification through adenovirus-mediated Cre induction of a floxed ca ACVR1/ALK2 transgene was recently reported (Yu *et al.*, 2008b). We have found that modifications of the published protocol including using mice younger than 4 weeks of age when inducing heterotopic ossification by adenovirus-mediated Cre induction of caALK2 improve the reproducibility of heterotopic ossification in this model (our unpublished data).

5.5. A knock-in mouse chimera for FOP

The more recent development of a knock-in mouse model of the ACVR1/ALK2 (R206H) mutation in FOP provides the most compelling validation of the role of ACVR1/ALK2 in FOP and its myriad phenotypic effects in development and postnatal life (Chakkalakal et al., 2008). This mouse model is still under development and analysis.

ACKNOWLEDGMENTS

This work was supported in part by the Center for Research in FOP and Related Disorders, the International FOP Association, the Ian Cali Endowment, the Weldon Family Endowment, the Penn Center for Musculoskeletal Disorders, and the Isaac & Rose Nassau Professorship of Orthopaedic Molecular Medicine and by grants from the Rita Allen Foundation and the NIH (R01-AR40196 to F. S. K. and E. M. S.; R01-GM056326 to M. C. M.). We thank our collaborators and members of our laboratory for valuable assistance and discussion of these methodologies.

REFERENCES

Ahn, J., Serrano de la Peña, L., Shore, E. M., and Kaplan, F. S. (2003). Paresis of a bone morphogenetic protein antagonist response in a genetic disorder of heterotopic skeletogenesis. *J. Bone Joint Surg.* **85-A**, 667–674.

Billings, P. C., Fiori, J. L., Bentwood, J. L., O'Connell, M. P., Jiao, X., Nussbaum, B., Caron, R. J., Shore, E. M., and Kaplan, F. S. (2008). Dysregulated BMP signaling and enhanced osteogenic differentiation of connective tissue progenitor cells from patients with fibrodysplasia ossificans progressiva. *J. Bone Miner. Res.* **23**, 305–313.

Chakkalakal, S. A., Zhang, D., Raabe, T., Richa, J., Hankenson, K., Kaplan, F. S., and Shore, E. M. (2008). ACVR1 knock-in mouse model for fibrodysplasia ossificans progressiva. *J. Bone Miner. Res.* **23**, s57.

Fiori, J. L., Billings, P. C., Serrano de la Peña, L., Kaplan, F. S., and Shore, E. M. (2006). Dysregulation of the BMP-p38 MAPK signaling pathway in cells from patients with fibrodysplasia ossificans progressiva (FOP). *J. Bone Miner. Res.* **21**, 902–909.

Fukuda, T., Scott, G., Komatsu, Y., Araya, R., Kawano, M., Ray, M. K., Yamada, M., and Mishina, Y. (2006). Generation of a mouse with conditionally activating signaling through the BMP receptor ALK2. *Genesis* **44**, 159–167.

Fukuda, T., Kohda, M., Kanomata, K., Nojima, J., Nakamura, A., Kamizono, J., Noguchi, Y., Iwakiri, K., Kondo, T., Kurose, J., Endo, K. I., Awakura, T., *et al.* (2009). Constitutively activated ALK2 and increased SMAD 1/5 cooperatively induce bone morphogenetic protein signaling in fibrodysplasia ossificans progressiva. *J. Biol. Chem.* **284**, 7149–7156.

Glaser, D. L., Economides, A. N., Wang, L., Liu, X., Kimble, R. D., Fandl, J. P., Wilson, J. M., Stahl, N., Kaplan, F. S., and Shore, E. M. (2003). In vivo somatic cell gene transfer of an engineered noggin mutein prevents BMP4-induced heterotopic ossification. *J. Bone Joint Surg.* **85-A,** 2332–2342.

Groppe, J. C., Shore, E. M., and Kaplan, F. S. (2007). Functional modeling of the ACVR1 (R206H) mutation in FOP. *Clin. Orthop. Rel. Res.* **462,** 87–92.

Hellemans, J., Mortier, G., DePaepe, A., Speleman, F., and Vandesompele, J. (2007). QBase relative quantification framework and software for management and automated analysis of real-time quantitative PCR data. *Genome Biol.* **8**(2), R19.

Kan, L., Hu, M., Gomes, W. A., and Kessler, J. A. (2004). Transgenic mice overexpressing BMP4 develop a fibrodysplasia ossificans progressiva (FOP)-like phenotype. *Am. J. Pathol.* **165,** 1107–1115.

Kan, L., McGuire, T. L., Berger, D. M., Awatramani, R. B., Dymeki, S. M., and Kessler, J. A. (2009). Dysregulation of local stem/progenitor cells as a common cellular mechanism for heterotopic ossification. *Stem Cells* **27,** 150–156.

Kaplan, F. S., Glaser, D. L., Shore, E. M., Deirmengian, G. K., Gupta, R., Delai, P., Morhart, P., Smith, R., Le Merrer, M., Rogers, J. G., Connor, J. M., and Kitterman, J. A. (2005). The phenotype of fibrodysplasia ossificans progressiva. *Clin. Rev. Bone Miner. Metab.* **3,** 183–188.

Kaplan, F. S., Glaser, D. L., Shore, E. M., Pignolo, R. J., Xu, M., Zhang, Y., Senitzer, D., Forman, S. J., and Emerson, S. G. (2007a). Hematopoietic stem-cell contribution to ectopic skeletogenesis. *J. Bone Joint Surg. Am.* **89,** 347–357.

Kaplan, F. S., Groppe, J., Pignolo, R. J., and Shore, E. M. (2007b). Morphogen receptor genes and metamorphogenes: Skeleton keys to metamorphosis. *Ann. NY Acad. Sci.* **1116,** 113–133.

Kaplan, F. S., LeMerrer, M., Glaser, D. L., Pignolo, R. J., Goldsby, R. E., Kitterman, J. A., Groppe, J., and Shore, E. M. (2008a). Fibrodysplasia ossificans progressiva. *Best Pract. Res. Clin. Rheumatol.* **22,** 191–205.

Kaplan, F. S., Xu, M., Glaser, D. L., Collins, F., Connor, M., Kitterman, J., Sillence, D., Zackai, E., Ravitsky, V., Zasloff, M., Ganguly, A., and Shore, E. M. (2008b). Early diagnosis of fibrodysplasia ossificans progressiva. *Pediatrics* **121,** e1295–e1300.

Kaplan, F. S., Pignolo, R. J., and Shore, E. M. (2009a). The FOP metamorphogene encodes a novel type I receptor that dysregulates BMP signaling. *Cytokine Growth Factor Rev.* **20,** 399–407.

Kaplan, F. S., Xu, M., Seemann, P., Connor, J. M., Glaser, D. L., Carroll, L., Delai, P., Xu, M., Seemann, P., Connor, J. M., Glaser, D. L., Carroll, L., *et al.* (2009b). Classic and atypical fibrodysplasia ossificans progressiva (FOP) phenotypes are caused by mutations in the bone morphogenetic protein (BMP) type I receptor ACVR1. *Hum. Mutat.* **30**(3), 379–390.

Katagiri, T., Imada, M., Yanai, T., Suda, T., Takahashi, N., and Kamijo, R. (2002). Identification of a BMP-responsive element in Id1, the gene for inhibition of myogenesis. *Genes Cells* **7,** 949–960.

Li, Y. P., Bang, D. D., Handberg, K. J., Jorgensen, P. H., and Zhang, M. F. (2005). Evaluation of the stability of six host genes as internal control in real-time RT-PCR assays in chicken embryo cell cultures infected with infectious bursal disease virus. *Vet. Micrbiol.* **110,** 155–165.

Little, S. C., and Mullins, M. C. (2006). Extracellular modulation of BMP activity in patterning the dorsoventral axis. *Birth Defects Res. C. Embryo Today* **78,** 224–242.

Lounev, V., Ramachandran, R., Wosczyna, M. N., Yamamoto, M., Maidment, A. D. A., Shore, E. M., Glaser, D. L., Goldhamer, D. J., and Kaplan, F. S. (2009). Identification of progenitor cells that contribute to heterotopic skeletogenesis. *J. Bone Joint Surg. Am.* **91,** 652–663.

Miura, M., Gronthos, S., Zhao, M., Lu, B., Fisher, L. W., Robey, P. G., and Shi, S. (2003). SHED: Stem cells from human exfoliated deciduous teeth. *Proc. Natl. Am. Sci. USA* **100**, 5807–5812.

Serrano de la Peña, L., Billings, P. C., Fiori, J. L., Ahn, J., Shore, E. M., and Kaplan, F. S. (2005). Fibrodysplasia ossificans progressiva (FOP), a disorder of ectopic osteogenesis, misregulates cell surface expression and trafficking of BMPRIA. *J. Bone Miner. Res.* **20**, 1168–1176.

Shafritz, A. B., Shore, E. M., Gannon, F. H., Zasloff, M. A., Taub, R., Muenke, M., and Kaplan, F. S. (1996). Over-expression of an osteogenic morphogen in fibrodysplasia ossificans progressiva. *N. Engl. J. Med.* **335**, 555–561.

Shen, Q., Little, S. C., Xu, M., Haupt, J., Ast, C., Katagiri, T., Mundlos, S., Seemann, P., Kaplan, F. S., Mullins, M. C., and Shore, E. M. (2009). The fibrodysplasia ossificans progressiva R206H ACVR1 mutation activates BMP-independent chondrogenesis and zebrafish embryo ventralization. *J. Clin. Invest.* **119**(11), 3462–3472.

Shore, E. M., Feldman, G. J., Xu, M., and Kaplan, F. S. (2005). The genetics of fibrodysplasia ossificans progressiva. *Clin. Rev. Bone Miner. Metab.* **3**, 201–204.

Shore, E. M., Xu, M., Feldman, G. J., Fenstermacher, D. A., Cho, T.-J., Choi, I. H., Connor, J. M., Delai, P., Glaser, D. L., Le Merrer, M., Morhart, R., Rogers, J. G., *et al.* (2006). A recurrent mutation in the BMP type I receptor ACVR1 causes inherited and sporadic fibrodysplasia ossificans progressiva. *Nat. Genet.* **38**, 525–527.

Sieber, C., Kopf, J., Hiepen, C., and Knaus, P. (2009). Recent advances in BMP receptor signaling. *Cytokine Growth Factor Rev.* **20**, 353–355.

Twombly, V., Bangi, E., Le, V., Malnic, B., Singer, M. A., and Wharton, K. A. (2009). Functional analysis of saxophone, the drosophila gene encoding the BMP type I receptor ortholog of human ALK1/ACVRL1A and ACVR1/ALK2. *Genetics* **83**, 563–579.

Urist, M. (1965). Bone formation by autoinduction. *Science* **150**(698), 893–899.

van Dinther, M., Visser, N., de Gorter, D. J. J., Doorn, J., Goumans, M.-J., de Boer, J., and ten Dijke, P. (2009). ALK2 R206H mutation linked to fibrodysplasia ossificans progressiva confers constitutive activity to the BMP type I receptor and sensitizes mesenchymal cells to BMP-induced osteoblast differentiation and bone formation. *J. Bone Miner. Res.* **25**, 1208–1215.

Wozney, J. M., Rosen, V., Celeste, A. J., Mitsock, L. M., Whitters, M. J., Kriz, R. W., Hewick, R. M., and Wang, E. A. (1998). Novel regulators of bone formation: Molecular clones and activities. *Science* **242**(4885), 1528–1534.

Yu, P. B., Hong, C. C., Sachidanandan, C., Babitt, J. L., Deng, D. Y., Hoyng, S. A., Lin, H. Y., Bloch, K. D., and Peterson, R. T. (2008a). Dorsomorphin inhibits BMP signals required for embryogenesis and iron metabolism. *Nat. Chem. Biol.* **4**, 33–41.

Yu, P. B., Deng, D. Y., Lai, C. S., Hong, C. C., Cuny, G. D., Bouxsein, M. L., Hong, D. W., McManus, P. M., Katagiri, T., Sachidanandan, C., Nobuhiro, K., Fukuda, T., *et al.* (2008b). BMP type I receptor inhibition reduces heterotopic ossification. *Nat. Med.* **14**, 1363–1369.

Zhang, D., Schwarz, E. M., Rosier, R. N., Zuscik, M. J., Puzas, J. E., and O'Keefe, R. J. (2003). ALK2 functions as a BMP type I receptor and induces Indian hedgehog in chondrocytes during skeletal development. *J. Bone Miner. Res.* **18**, 1593–1604.

CHAPTER NINETEEN

IDENTIFICATION AND EVALUATION OF CONSTITUTIVELY ACTIVE THYROID STIMULATING HORMONE RECEPTOR MUTATIONS

Joaquin Lado-Abeal,*,†,‡ Leah R. Quisenberry,† and Isabel Castro-Piedras*,‡

Contents

1. Introduction	376
2. *TSHR* Gene Mutational Screening	382
2.1. DNA extraction from paraffin-embedded tissues	382
2.2. *TSHR* gene amplification for sequencing and identification of mutations	383
3. Determination of TSHR Constitutive Activity *In Vitro*	386
3.1. Required material	386
3.2. Transfection and luciferase assay	387
4. Measurement of TSHR Expression at Cell Surface by Flow Cytometry Analysis	388
4.1. Required material	388
4.2. Flow cytometry analysis	389
5. TSH–TSHR Binding Assays	389
5.1. Required material	389
5.2. Binding assay	390
6. TSHR Phosphorylation Analysis	391
6.1. Required material	391
6.2. Cell phosphorylation	392
Acknowledgments	392
References	392

* Department of Internal Medicine, Tech University Health Sciences Center-SOM, Lubbock, Texas, USA
† Department of Cell Biology and Biochemistry, Texas Tech University Health Sciences Center-SOM, Lubbock, Texas, USA
‡ UETeM, Department of Medicine, University of Santiago de Compostela-SOM, Santiago de Compostela, Spain

Methods in Enzymology, Volume 484 © 2010 Elsevier Inc.
ISSN 0076-6879, DOI: 10.1016/S0076-6879(10)84019-4 All rights reserved.

Abstract

Thyroid stimulating hormone receptor (TSHR) is a guanine nucleotide-binding protein-coupled seven-transmembrane-domain receptor that controls the differentiation, growth, and function of the thyroid gland through stimulation of adenylyl cyclase and phospholipase C pathways. Thyroid stimulating hormone (TSH) is the main TSHR ligand, and unliganded receptor remains silent due to the interaction of its large extracellular domain with the extracellular loops of the serpentine. The TSHR gene is highly mutagenic and constitutively active mutations have been extensively described. Naturally occurring TSHR-activating mutations can affect any part of the receptor, but most activating mutations affect the serpentine region, and the majority of these are located in the third intracellular loop or transmembrane domain six. We describe several simple and relatively cheap methods used in our laboratory to study constitutive TSHR mutations that include (1) screening of TSHR gene mutations in paraffin-embedded thyroid tissue samples, (2) measurement of TSHR constitutive activity *in vitro*, (3) measurement of TSHR expression at cell surface by flow cytometry analysis, (4) TSH binding to TSHR, and (5) TSHR phosphorylation analysis.

1. Introduction

Differentiation, growth, and function of the thyroid gland are primarily controlled by thyroid stimulating hormone (TSH) through activation of the TSH receptor (TSHR), a guanine nucleotide-binding protein (G protein)-coupled seven-transmembrane-domain receptor (GPCR) anchored to the basolateral plasma membrane of thyrocytes. TSHR is a 764 amino acid glycoprotein codified by the *TSHR* gene which is located on the long arm of chromosome 14 (14q31) and contains 10 exons. Exons 1–9 codify a long extracellular amino-terminal domain responsible for TSH binding. Exon 10 codifies the seven-transmembrane and carboxy-terminal intracellular domains, the parts of the receptor that interact with G proteins (Vassart and Dumont, 1992).

TSHR signaling is regulated by TSH binding and by several posttranslational processes (Kursawe and Paschke, 2007). The unliganded receptor has a silent conformation due to the interaction of its large extracellular domain with the extracellular loops of the serpentine, and enzymatic amputation of the amino-terminal segment increases basal activity of TSHR (Vlaeminck-Guillem *et al.*, 2002). On binding of TSH, the structure of TSHR changes not only in its extracellular but also in its transmembrane and intracellular domains, leading to G protein coupling (Vlaeminck-Guillem *et al.*, 2002; Zhang *et al.*, 2000). TSHR resides at the plasma membrane as a single polypeptide chain and as two subunits (α or A and β or B) linked by a disulfide bond. The α and β subunits are a result of a proteolytic cleavage in the extracellular domain. The α-subunit is released from the membrane-bound receptor due to a disulfide bond reduction known as receptor shedding (Davies et al., 2002; Kursawe and Paschke, 2007). The β-subunit contains the transmembrane and carboxy-terminal intracellular domains of TSHR. TSHR dimerizes at the plasma membrane (Urizar et al., 2005), and upon binding of TSH, the receptor's multimers rapidly dissociate into active open monomers. The monomers are recruited more efficiently into lipid rafts, a mechanism that can explain why monomers couple preferentially with G proteins (Davies et al., 2002). Other posttranslational processes important to TSHR signaling are palmitoylation, sulfation, and glycosylation (Kursawe and Paschke, 2007).

In humans, TSHR regulates thyroid gland function, stimulating adenylyl cyclase (AC) and phospholipase C (PLC) activity; however, activation of PLC requires a concentration of TSH 10 times greater than that which activates AC (Vassart and Dumont, 1992). TSHR activation induces coupling of various Gα proteins, that is, Gs, Gq/11, Gi, and G12 (Laugwitz et al., 1996; Selzer et al., 1993), although TSHR signaling is mainly mediated by Gs and Gq/11 (Allgeier et al., 1994). Gs coupling to TSHR activates the AC/cAMP cascade that regulates thyroid hormone secretion. The increase in intracellular concentration of cAMP activates protein kinase A (PKA) which phosphorylates the cAMP-response-element binding protein (CREB) that, together with the coactivator CRE-binding protein (CBP), binds to cAMP-response elements (CRE) to stimulate expression of thyroglobulin (TG), thyroid peroxidase (TPO), sodium iodide symporter (NIS), and the thyroid transcription factors TTF1/NKx2.1, TTF2/FoxE1, and PAX8 (Postiglione et al., 2002; Vassart and Dumont, 1992). TSHR signaling, directly via cAMP stimulation of the exchange nucleotide protein which is activated by cAMP (Epac1), or indirectly via PKA, activates ERK1/2 and p38MAPK pathways as well as Ras, all of which are involved in thyroid cell proliferation. TSHR activation of the PLC pathway is mediated by Gq/11 coupling that leads to hydrolysis of phosphatidyl inositols and thus generates diacylglycerol (DAG) and inositoltriphosphate (IP3) as second messengers. The PLC-IP3-DAG pathway controls thyroid hormone production. DAG stimulates protein kinase C (PKC), while IP3 increases cytosolic Ca^{2+} levels, thereby activating production of H_2O_2 and subsequent iodination of thyroglobulin, a required step in thyroid hormone synthesis (Song et al., 2010). TSH also couples Gi to TSHR, which partially inhibits stimulation through Gs, a mechanism which desensitizes TSHR (Kimura et al., 2001).

Homologous desensitization of the TSHR is an early regulatory mechanism activated after TSHR stimulation by TSH binding (Wynford-Thomas et al., 1983). Although desensitization of TSHR occurs at several levels, it is probably most relevant at the receptor level. The main mechanism for GPCR desensitization is receptor phosphorylation by GRKs and subsequent binding of β-arrestins, along with PKA- and PKC-dependent phosphorylation events at other sites on the receptor (Ferguson, 2007). The human thyroid expresses GRKs 2–6 (Nagayama et al., 1996) which in vitro can desensitize TSHR (Iacovelli et al., 1996), an effect that is amplified by overexpression of β-arrestins 1 and 2 (Voigt et al., 2004). Although GRKs phosphorylate GPCRs at both serine and threonine residues localized within either the third intracellular loop or carboxyl-terminal tail domains, it is not known where TSHR is phosphorylated by GRKs. However, truncated receptors in which the serine/threonine-rich cytoplasmic C-terminal tail of the receptor is deleted undergo homologous desensitization, suggesting that GRKs target the third intracellular loop (Haraguchi et al., 1994; Nagayama et al., 1994). Arrestins are a family of proteins that block GRK-phosphorylated GPCRs, uncoupling receptors from heterotrimeric G protein and targeting GPCRs for clathrin-mediated endocytosis (Ferguson, 2007). In response to GPCR activation, cytosolic β-arrestin proteins translocate to the plasma membrane and then redistribute to clathrin-coated pits bound to receptors. The thyroid gland expresses β-arrestins 1 and 2, and both of these can internalize and desensitize TSHR in vitro, with β-arrestin 2 having a faster action, stronger internalization effect, and higher affinity than β-arrestin 1 (Frenzel et al., 2006). β-Arrestins and TSHR colocalize near the plasma membrane but not within endosomes (Frenzel et al., 2006). The role of β-arrestins in the thyroid gland as adaptor proteins that modulate the downstream activities of various signaling networks and gene transcription remains unknown (Rajagopal et al., 2010). TSHR is internalized by clathrin-coated vesicles and, although TSHR has a low constitutive endocytosis level in the absence of TSH (10% of cell surface receptor molecules) (Baratti-Elbaz et al., 1999), receptor endocytosis increases in the presence of TSH (Baratti-Elbaz et al., 1999). The TSHR–TSH complex accumulates in endosomal vesicles (Singh et al., 2004) from where the majority of internalized receptor molecules (90%) are recycled to the cell surface, an important step in restoring receptor responsiveness after desensitization, while the hormone itself is degraded in lysosomes (Baratti-Elbaz et al., 1999). Receptors that are not recycled are also targeted to lysosomes and degraded, contributing to receptor downregulation.

Phosphorylation is not absolutely required for GPCR desensitization and phosphorylation-independent desensitization of GPCR can be mediated by GRKs and other GPCR-interacting proteins. Among these are regulators of G protein signaling (RGS), a family of guanosine triphosphatase-activating proteins (GAP) that blunt G protein signaling and thus induce the reassociation of G proteins α and βγ. The human thyroid gland expresses several RGS, and in primary cultures of thyroid epithelial cells TSH increases expression of RGS-2 (Eszlinger et al., 2004), an inhibitor of TSHR-mediated IP3 signaling under specific conditions (Eszlinger et al., 2004). RGS-2 lacks Gsα GAP activity, but it can inhibit Gsα signaling by direct inhibition of several adenylyl cyclases, although in the thyroid gland such an effect has not yet been described. Phosphodiesterases (PDE) 4, 7, and 8 recognize and hydrolyze cAMP, and the thyroid gland expresses high levels of PDE8 (Hayashi et al., 1998). In thyroid cell cultures, TSH stimulation induces activation of PDE4 via PKA activation, a mechanism that decreases cAMP accumulation (Oki et al., 2000) and which might explain the finding that toxic adenomas bearing TSH-activating mutations have higher PDE4 activity (Persani et al., 2000). Finally, after TSH stimulation, levels of TSHR mRNA decrease within a few hours, leading to long-term TSHR desensitization. PKA-mediated phosphorylation in the TSH-stimulated cAMP cascade also activates the cAMP-responsive element modulator (CREM), stimulating expression of inducible cAMP early repressor (ICER), which in turn binds to a CRE-like sequence in the TSHR promoter, and blocks TSHR expression (Lalli and Sassone-Corsi, 1995). Downregulation of TSHR can also be mediated by calcium ionophores that do not alter cAMP levels (Saji et al., 1991).

The *TSHR* gene is highly mutagenic and both loss-of-function (Sunthornthepvarakui et al., 1995) and constitutively active mutations (Parma et al., 1993) have been extensively described. Germline mutations in the *THSR* gene that cause constitutive activity of the receptor are a recognized cause of neonatal nonautoimmune hyperthyroidism (Duprez et al., 1994). Thyroid toxic adenomas, a frequent cause of nonautoimmune hyperthyrodism in iodine-deficient areas, bear a TSHR-activating mutation in up to 80% of cases in some European studies (Gozu et al., 2006; Palos-Paz et al., 2008).

Naturally occurring TSHR-activating mutations can affect any part of the receptor, although the extracellular domain is more prone to inactivating than to constitutively activating mutations. To date, only four amino acids in the extracellular domain have been linked to naturally occurring constitutively activating mutations: lysine183 (K183R), serine281 (S281T, S281N, S281I), arginine301 (R301C), and cysteine390 (C390W). Most activating mutations affect the serpentine region codified by exon 10 of *TSHR*, and the majority of these are located in the third intracellular loop or transmembrane domain 6 (Palos-Paz et al., 2008). Interestingly, there are also several artificially induced activating mutations located in amino acid residues where naturally occurring mutations have not yet been found (Table 19.1; http://gris.ulb.ac.be). The reason why naturally occurring activating mutations are not found in certain TSHR amino acids that are capable of constitutively activating the receptor when artificially mutated remains unknown.

In a series of 85 toxic adenomas (TA) that we studied previously (Palos-Paz et al., 2008), the most frequently found mutations were C → T and T → C transitions and G → T transversions. CpG dinucleotides were found to undergo TG transversions with a high frequency, and since methylated CpG islands are prone to mutation due to deamination of 5-methylcytosine, we performed an *in silico* analysis of the *TSHR* gene coding sequence that showed a hypothetical 760 bp CpG island inside exon 10. However, only a T632I mutation was located in this region, suggesting that

Table 19.1 Reported *TSHR* gene naturally occurring constitutively active mutations

Gene mutation	TSHR domain	Gene mutation	TSHR domain
A428V	First transmembrane domain	M626I	Sixth transmembrane domain
S425I	First transmembrane domain	A627V	Sixth transmembrane domain
G431S	First transmembrane domain	L629F	Sixth transmembrane domain
M453T	Second transmembrane domain	I630M	Sixth transmembrane domain
M463V	Second transmembrane domain	I630L	Sixth transmembrane domain
I486F	First extracellular loop	F631C	Sixth transmembrane domain
I486M	First extracellular loop	F631L	Sixth transmembrane domain
I486N	First extracellular loop	F631I	Sixth transmembrane domain
S505R	Third transmembrane domain	F631S	Sixth transmembrane domain
S505N	Third transmembrane domain	F631V	Sixth transmembrane domain
V509A	Third transmembrane domain	T632A	Sixth transmembrane domain
L512R	Third transmembrane domain	T632I	Sixth transmembrane domain
L512Q	Third transmembrane domain	D633E	Sixth transmembrane domain
I568T	Second extracellular loop	D633A	Sixth transmembrane domain
I568V	Second extracellular loop	D633H	Sixth transmembrane domain
V597L	Fifth transmembrane domain	D633Y	Sixth transmembrane domain
V597F	Fifth transmembrane domain	I635V	Sixth transmembrane domain
Y601N	Fifth transmembrane domain	P639S	Sixth transmembrane domain
D617Y	Third extracellular loop	I640K	Sixth transmembrane domain
D619G	Third extracellular loop	N650Y	Third extracellular loop
T620I	Third extracellular loop	V656F	Third extracellular loop
A623F	Third extracellular loop	F666L	Seventh transmembrane domain
A623I	Third extracellular loop	N670S	Seventh transmembrane domain
A623V	Third extracellular loop	C672Y	Seventh transmembrane domain
A623S	Third extracellular loop	L677V	Seventh transmembrane domain

deamination of 5-methylcytosine is not a common mutagenic mechanism in TA. Replication slippage in quasipalindromic DNA sequences causes frame-shift and base-substitution mutations (Lado-Abeal et al., 2005), and a computational analysis of *TSHR* indicated that imperfect DNA complementarity was not a cause for the high frequency of M453T mutations found in our studies (Palos-Paz et al., 2008). Oxidative damage caused by an increase in H_2O_2 generation in thyroid glands under conditions of iodine or selenium deficiency could be a major cause of high mutational rates at identified codons (Krohn et al., 2007).

Constitutively active TSHR mutations generally activate the Gs-AC-cAMP cascade. Some mutations, however, also activate the PLC-dependent pathway (Parma et al., 1995), due not to a higher constitutive activity but rather to qualitative differences (Parma et al., 1995). To date, there have been no reports of phenotypic differences between toxic adenomas due to the intracellular cascade activated. Constitutively active mutations can either disrupt interaction within the inactive TSHR conformation or form a new interaction that stabilizes the active state (Parnot et al., 2002). Mutations in the extracellular domain of the receptor can change the interaction between the ectodomain and the extracellular loops of the serpentine responsible for receptor silencing (Farid and Szkudlinski, 2004; Ho et al., 2001) and/or can simulate TSH binding (Ho et al., 2001; Vlaeminck-Guillem et al., 2002). As mentioned above, the regions of TSHR that accumulate most constitutively active mutations are the third intracellular loop and the sixth transmembrane segment (Fig. 19.1 and Table 19.1; Palos-Paz et al., 2008), two regions shown to be critical for TSHR's interaction with Gs and Gq proteins (Claus et al., 2006; Kosugi et al., 1993). Naturally occurring activating mutations have not yet been found in the first or second intracellular loop. Several TSHR-activating mutations show decreased expression at the plasma membrane and are retained inside the cell where they continue to activate the cAMP cascade pathway (Castro et al., 2009). In agreement with those observations, it has recently been found that TSH–TSHR complexes continue to stimulate cAMP production after internalization and that signaling from the internalized receptor continues after removal of TSH, causing persistent signaling (Calebiro et al., 2010).

The mechanisms that allow constitutively active TSHR mutations to apparently escape desensitization and downregulation and to continue signaling remain unknown. Hyperfunctioning thyroid nodules show higher expression levels of β-arrestin 2, and GRKs 3 and 4 (Voigt et al., 2000, 2004) as well as higher PDE4 activity (Persani et al., 2000) than normal thyroid tissue, suggesting that some fundamental mechanisms involved in negative feedback regulation of activated cAMP pathways are induced by activating mutations. Expression of RGS2, an inhibitor of TSHR-mediated IP3 signaling, is increased by TSH, but hot nodules have decreased RSG2 mRNA expression, suggesting that hyperfunctioning thyroid nodules might have a defect in the RGS regulatory pathway (Eszlinger et al., 2004).

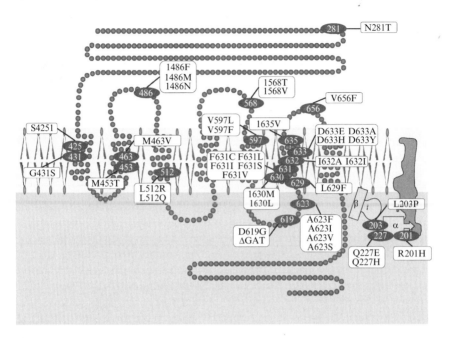

Figure 19.1 TSHR constitutively active mutations identified in a series of 85 toxic adenomas from an iodine-deficient region located in NW Spain, studied by us between 2004 and 2007. The mutated amino acids are represented in the circles and the mutations are described in the white boxes. Most of the mutations are located in the third intracellular loop and the sixth transmembrane segment, two critical regions for TSHR's interaction with Gs and Gq proteins.

In the following sections, we describe some of the methods used in our laboratory to study constitutively active TSHR mutation. We describe the methods used by us (1) to extract DNA samples from paraffin-embedded tissues, a common way to archive thyroid surgical specimens, followed by our method to detect mutations once DNA has been extracted (2) to investigate the mutant capacity to increase cellular cAMP as a probe of constitutive activity, (3) to evaluate TSH binding to TSHR, (4) to measure the expression of TSHR mutants at the plasma membrane, and (5) to evaluate the phosphorylation state of the mutant receptor.

2. *TSHR* Gene Mutational Screening

2.1. DNA extraction from paraffin-embedded tissues

TSHR gene mutational screening is usually performed on DNA samples extracted from thyroid tissue surgical specimens. Although DNA extraction from fresh samples is straightforward, extraction of good quality DNA from paraffin-embedded thyroid tissue samples can be a challenge. We have been routinely using the following protocol of deparaffinization in Xilol and a graded series of Ethanol prior to DNA extraction in order to get a good yield.

2.1.1. Required material

Devices

- Water bath shaker, centrifuge (Eppendorf Centrifuge 5415R, Hamburg, Germany), and a vortex mixer.

Reagents and solutions

- Xilol (Mallinckrodt Baker, Phillipsburg, NJ, USA), ethanol (Fisher Scientific, Atlanta, GA, USA).

Additional materials

- Paraffin-embedded thyroid samples.

Disposables

- 1.5 ml and 2 ml Eppendorf tubes.

2.1.2. Sample deparaffinization

- Add 1 ml Xilol per 100 mg of sample in a 1.5 ml Eppendorf tube. Rock at 250 rpm for 1 h at 56 °C. Centrifuge at 12,000 rpm and discard the supernatant. Repeat this step twice.
- Add 1 ml of 100% ethanol. Rock at 250 rpm for 1 h at 37 °C. Centrifuge at 12,000 rpm and discard the supernatant. Repeat step once.
- Add 1 ml of 90% ethanol. Rock at 250 rpm for 30 min at 37 °C. Centrifuge at 12,000 rpm, pouring the supernatant.
- Add 1 ml of 80% ethanol. Rock at 250 rpm for 30 min at 37 °C. Centrifuge at 12,000 rpm and discard the supernatant.
- Add 1 ml of 70% ethanol. Rock at 250 rpm for 30 min at 37 °C. Centrifuge at 12,000 rpm and discard the supernatant.
- Air dry the sample and proceed with DNA extraction (we use Real pure extraction DNA kit, Durviz, Barcelona, Spain).

2.2. *TSHR* gene amplification for sequencing and identification of mutations

After DNA extraction, amplification of *TSHR* gene is performed by PCR as follows.

2.2.1. Required material

Devices

- Thermocycler, microwave, gel electrophoresis system, ultraviolet light, ABI 3730× DNA sequencer (Applied Biosystems, Foster City, CA, USA).

Reagents and solutions

- PCR reaction: Ecotaq polymerase (Ecogen, Barcelona, Spain), 10× Ecotaq reaction buffer (Ecogen), 10 mM dNTP mix (Ecogen), oligonucleotide primers (Sigma, St. Louis, MO, USA), and DNase, RNase, protease free water (Acros Organics, Geel, Belgium).
- Agarose gel: Agarose D1 Low EEo (Conda, Madrid, Spain); TAE 50× buffer: 242 g Tris base (EMD Biosciences, San Diego, CA, USA), 57.1 ml glacial acetic acid (Mallinckrodt Baker) and 100 ml of 0.5 M EDTA (EMD Biosciences) pH 8.0 adjusted to a final volume of 1 l; ethidium bromide (Sigma); 6× DNA loading buffer: 50% glycerol (Sigma), 1 mM EDTA pH 8.0 and 0.4% bromophenol blue (Sigma) and 100 pb DNA ladder (New England Biolabs, MA).
- PCR product purification: ExoSAP-IT (USB Corporation, Cleveland, OH, USA).
- Sequencing reaction: BigDye® Terminator v3.1 cycle sequencing Kit (Applied Biosystems) containing BigDye® Terminator v3.1 ready reaction mix and 5× sequencing buffer; 1 μM oligonucleotide primer (Sigma) and water, DNase, RNase, protease free (Acros Organics).
- Ethanol/EDTA purification: 125 mM EDTA (EMD Biosciences) and ethanol (Fisher Scientific, Atlanta, GA, USA).
- Hi-Di™ Formamide (Applied Biosystems).

Additional materials

- DNA extracted from surgical thyroid tissue specimens.

Disposables

- 1.5 ml and 0.2 ml Eppendorf tubes.

2.2.2. PCR and gene sequencing

TSHR gene, including exon–intron boundaries, is amplified by PCR from thyroid tissue DNA samples. Oligonucleotide primers and lengths of amplified DNA fragments are shown in Table 19.2. PCR amplifications are performed in a 50 μl volume containing: 100 ng DNA, 10 mmol/l Tris–HCl (pH 8.3), 1.5 mmol/l MgCl2, 50 mmol/l KCl, 200 μmol/l deoxi-NTPs, 1 U of Ecotaq polymerase and 0.25 μmol/l of each oligonucleotide

Table 19.2 Sequences of the oligonucleotide primers used for PCR amplification of *TSHR* gene and length of the PCR products submitted for sequencing

Exon	PCR product length (pb)	Sense oligonucleotides	Antisense oligonucleotides
10 (1489 pb)	391	GTCATGAGCCACTGCGCC	GGTTGAACTCATCGGACTTGG
	380	CCCAGGAAGAGACTCTACAAGCT	GAAACCAGCCGTGTTGCAC
	380	CCATCGACTGGCAGACAGG	CGGATTTCGGACTGTGATGT
	380	ACATAGTTGCCTTCGTCATCGT	GCTGTTCTTTGGAGGAACCC
	380	GCCTTCAGAGGGATGTGTTC	CCATGAAACATTGAAACATCGC
9 (189 pb)	380	CATCTCCCAATTAACCTCAGG	CAAACCAGGAAGCATCTTCCC
8 (78 pb)	370	GGGACTTGCAGAAGCCTTTAC	GTGCTCAAGCCAGAAGAAGATA
7 (69 pb)	300	TGGGACCTGAAAAACCTTTATG	TCTGGCGTGGAAGATGCT
6 (78 pb)	300	GCGGTTAAGAAGGGTCAGTG	TGGATGGTCTGTAAACATATGAAAA
5 (75 pb)	300	TTAGGGAAGGTGTTGGGAG	TTATCTTCAACCTACCCTCATGA
4 (75 pb)	300	GGTACCCTGTGGCGTAAATG	CCTGGACCACATCATCTAGG
3 (75 pb)	300	TCCATGAGGGTTGTACATGTT	CAATCAAGAAACCAGCTCCC
2 (62 pb)	300	CCAACATATTGTGAAAACTGTCA	AATATAACTGCCATTGATTTATGC
1 (269 pb)	380	AGCCCCTTGGAGCCCTC	CTGCTTTTGTGCACTTCGGG

primer. Cycling conditions are an initial denaturation at 94 °C for 10 min, 35 cycles each at 94 °C for 30 s, 57 °C for 30 s (exons 1–9) or 58 °C for 30 s (five sets of primers for exon 10), 72 °C for 40 s and a final extension at 72 °C for 5 min.

PCR products are assessed by 1.5% agarose gel electrophoresis. Before sequencing, the contaminating dNTPs and oligonucleotide primers are removed by an enzymatic method, ExoSAP-IT, following the manufacturer instructions.

The sequencing reaction is prepared with 2.5 μl of purified PCR product, 1 μl of BigDye® Terminator v3.1 ready reaction mix, 2 μl of 5× sequencing buffer, and 1 μM of oligonucleotide primer in a total volume of 10 μl. The sequencing reaction is submitted to PCR as follows: 96 °C for 5 min plus 35 cycles of 96 °C 10 s, 50 °C 5 s and 60 °C 4 min.

Samples are purified with EDTA/ethanol by adding 2.5 μl EDTA and 30 μl of 100% Ethanol to sequencing PCR product. Then mix and incubate samples for 15 min at room temperature, and centrifuge for 20 min at 13,000×g. Discard the supernatant and add 70% Ethanol. Centrifuge again for 10 min at 13,000×g. Discard the supernatant and allow samples to air dry, and then resuspend in 10 μl of Hi-Di™ Formamide.

3. DETERMINATION OF TSHR CONSTITUTIVE ACTIVITY IN VITRO

To evaluate the constitutive activity of the TSHR mutant, we measure the activation of adenylyl cyclase/cAMP cascade with a relatively cheap and easy to perform method. COS-7 cells are transfected with both a plasmid containing the *TSHR* gene cDNA and a cAMP-response element (CRE)-luciferase reporter plasmid. The luciferase activity is measured 48 h after transfection as an indirect way to evaluate cellular cAMP levels.

3.1. Required material

Devices

- CO_2 Incubator, biological safety cabinet, and luminometer.

Reagents and solutions

- Cell growth medium: Dulbecco's modified eagle's medium (Sigma) supplemented with 10% of heat inactivated fetal bovine serum (FBS) (Sigma) and 50 μg/ml gentamicin (Sigma).
- Phosphate buffer saline (PBS).
- TrypLE™ Express as trypsin replacement (Gibco, Grand Island, NY, USA).
- Lipofectamine™ (Invitrogen, Carlsbad, CA, USA).

- Luciferase assay reagent (Promega, Madison, WI, USA).
- Lysis buffer: 25 mM Tris (EMD Biosciences) pH 8.0, 2 mM dithiothreitol (Sigma), 2 mM 1,2-diaminocyclohexane-N,N,N',N'-tetracetic acid (Sigma), 10% glycerol (Sigma), and 1% Triton X-100 (Sigma).

Additional materials

- COS-7 cells (African green monkey kidney fibroblast-like cell line obtained from the European Collection of Cell Cultures and purchased from Sigma–Aldrich, Dorset, England).
- Plasmids: pSVL-TSHR wild-type or pSVL-TSHR mutant constructs and a cAMP-response element (CRE)-luciferase reporter plasmid.

Disposables

- Fifty-milliliter centrifuge tubes (VWR, Suwane, GA, USA), 100 mm cell culture dish and 24-well cell culture plates (BD Falcon, Franklin Lakes, NJ, USA), 9″ Pasteur glass pipettes (VWR), and 75 × 12 mm polypropylene tubes (VWR).

3.2. Transfection and luciferase assay

COS-7 cells are grown in 100 mm culture dishes in DMEM containing 10% fetal bovine serum and 50 μg/l gentamicin, at 37 °C in 100% humidity and 10% CO_2. At 100% confluence cells are detached with TrypLE^{TM} Express and transferred to 24-well cell culture plates for transient transfection. To detach the cells, the media is removed by gentle aspiration with a Pasteur pipette, and the cell monolayer is washed twice with 2 ml of PBS for 30 s, removing the PBS each time by gentle aspiration. Then, 1 ml of TrypLE^{TM} Express warmed to room temperature is added, and the cells are incubated for 2–5 min at 37 °C, using inverted microscopy to check for cell detachment. Once cells are detached, 9 ml of DMEM are added to the TrypLE^{TM} Express solution and the plate is washed twice to recover the cells. The cell suspension is transferred to a sterile 50 ml conical tube and 20 ml of growth medium is added for a total volume of 30 ml. After mixing thoroughly, 0.5 ml of the cell suspension is added to each well of a 24-well culture plate.

Transfection is performed at 70–80% cell confluence as follows. On the day of transfection, the growth medium is removed by gentle aspiration with a Pasteur glass pipette. Cells are washed twice with 0.5 ml of PBS and 0.5 ml of fresh growth medium is added immediately to each well. One hour later, 100 μl of a solution mixture containing DMEM, lipofectamine (1.25 μl/well), CRE-luciferase reporter plasmid (250 ng/well), and either pSVL alone, pSVL-TSHR-WT, or mutant expression vectors (each at a final concentration of 250 ng/well) are added to each well.

After 48 h, the cells are harvested as follows. Medium is aspirated and cells are washed twice with PBS as described above. Lysis buffer is added (50 μl) and cells are kept at room temperature, covered with aluminum foil, for 15 min. Cell detachment is checked under inverted microscopy, and once detached, 20 μl of cells lysate is mixed with 50 μl of luciferase reagent in a polypropylene tube and luciferase activity is measured in a luminometer over a 10-s time interval. Cells transfected with pSVL and CRE-luciferase reporter plasmid are used as controls.

4. Measurement of TSHR Expression at Cell Surface by Flow Cytometry Analysis

Constitutively active TSHR mutants often show lower expression at the plasma membrane. The following is a description of a simple flow cytometry analysis (FACS) method that we use to evaluate TSHR expression at the cell surface.

4.1. Required material

Devices

- Flow cytometer (we use the BD FACSCalibur, Becton Dickinson, Heidelberg, Germany).

Reagents and solutions

- Cell growth medium, PBS, TrypLE™ Express, lipofectamine assay reagent as described above.
- Buffer A: 1 mM EDTA (EMD Biosciences) and 1 mM EGTA (Acros Organics) in PBS.
- Buffer B: 0.1% bovine serum albumin fraction V protease free (BSA) and 0.1% NaN_3 (Sigma) in PBS.
- Mouse anti-human TSHR antibody (TSHR 2C11: sc-32263-PE, Santa Cruz Biotechnology, Santa Cruz, CA, USA).
- Paraformaldehyde (Sigma).

Additional materials

- COS-7 cells and pSVL-TSHR plasmids as described above.

Disposables

- 100 mm Cell culture dish, 12-well cell culture plates, and 9″ Pasteur glass pipettes as described above, 15 ml centrifuge tubes (BD Falcon).

4.2. Flow cytometry analysis

COS-7 cells are transfected as described above with pSVL-TSHR-WT (500 ng/well) or pSVL-TSHR mutant expression vectors (500 ng/well). Forty-eight hours after transfection, the media is removed by aspiration with a Pasteur pipette and the cells are washed twice with 2 ml of PBS. To detach the cells, 2 ml of buffer A are added, the culture plate is incubated, and cell detachment is checked under inverted microscopy. When the cells are detached, 2 ml of Buffer B are added, and then the medium and cells are recovered and transferred to a 15 ml centrifuge tube and centrifugated at $1000 \times g$ at 4 °C for 5 min. The supernatant is discarded and the cell pellet is resuspended in 2 ml of buffer B and 10 μl of TSHR-phyco-eritrin (PE) antibody to a final dilution of 1:200. The suspension is incubated in the dark for 1 h at 4 °C, then centrifuged at $1000 \times g$ at 4 °C, the supernatant is discarded, and the cell pellet is washed with 4 °C PBS three times, each time removing the PBS by centrifuging ($1000 \times g$ at 4 °C) and discarding the supernatant. Finally, the cells are resuspended in 500 μl of 1% paraformaldehyde in PBS and submitted to FACS.

5. TSH–TSHR Binding Assays

Constitutively active mutants often show changes in TSH binding affinity. The method that we use to evaluate changes in binding affinity between TSH and TSHR is described below.

5.1. Required material

Devices

- Automatic γ-counter.

Reagents and solutions

- Bovine TSH (bTSH) (Sigma).
- 0.05 M Sodium borate–KCl buffer: 0.05 M H_3BO_3 (Sigma), 0.05 M KCl (EMD Biosciences), pH 8.6.
- Na–^{125}I (Perkin–Elmer, Madrid, Spain).
- 0.05 M $NaPO_4$ (Sigma), pH 7.6.
- Chloramine-T (Merck KGaA, Darmstadt, Germany).
- Sodium metabisulfite (Merck KGaA).
- 0.05 M Sodium barbital buffer (pH 8.6) (Merck KGaA).
- Bovine serum albumin fraction V protease free (BSA) (Sigma).

- Modified Hank's buffer: 5.36 mM KCl (EMD Biosciences), 0.44 mM KH$_2$PO$_4$ (Sigma), 0.41 mM MgSO$_4$ (Sigma), 0.33 mM Na$_2$HPO$_4$ (Sigma), 5.55 mM glucose (Sigma) supplemented with 220 mM sucrose (Sigma).
- NaOH (Sigma).
- Cell growth medium, PBS, TryPLE™ Express, Lipofectamine™ as described above.

Additional materials

- COS-7 cells and plasmid vectors as described above.
- Sephadex G-200, Sephadex G-50 (Pharmacia, Barcelona, Spain).

Disposables

- 100 mm Cell culture dish, 12-well cell culture plates, 9″ Pasteur glass pipettes as described above.

5.2. Binding assay

bTSH radio-iodination is carried out by the chloramine-T method (Greenwood *et al.*, 1963). For each iodination, 7.5 µg of bTSH are diluted in 125 µl of 0.5 M NaPO$_4$, pH 7.6. Under a hood and with radioactive protection, 20 µl of chloramine-T (at concentration 1 mg/ml in 0.05 M N$_a$PO$_4$, pH 7.6) and 10 µl of ^{125}I are added to the bTSH solution. The mixture is incubated for 20 s and the reaction is stopped by adding 50 ml of sodium metabisulfite (at concentration 1 mg/ml in 0.05 M NaPO$_4$, pH 7.6). The labeled hormone is purified from free ^{125}I by gel filtration chromatography on a Sephadex G-50 column (1 × 40 cm), at a flow rate of 20 ml/h. The column is prepared by equilibrating the resin in 0.05 M sodium barbital buffer (pH 8.6), followed by successive elutions with 10 ml of 0.05 M sodium barbital buffer (pH 8.6) and 2% BSA. The labeled bTSH is eluted with 0.8 ml of 0.05 M sodium barbital buffer into a tube containing 1.2 ml of PBS and 0.1% BSA (pH 7.4).

Binding studies are carried out in COS-7 cells transfected as described for FACS. Forty-eight hours after transfection, cells are washed twice with modified Hank's buffer supplemented with 220 mM sucrose, each time removing the buffer by aspiration with a Pasteur pipette. Labeling buffer is prepared by adding 2.5% low-fat milk and 100,000 counts/ml of ^{125}I-bTSH to modified Hank's buffer supplemented with 220 mM sucrose. Cells are incubated at room temperature with 500 µl of the labeling buffer and increasing concentrations of unlabeled bTSH (0, 0.01, 0.03, 0.1, 0.3, 1, 3, 10, 30, and 100 mU/ml). Four hours later, cells are washed twice with 1 ml of modified Hank's buffer and solubilized with 500 µl of 1 N NaOH. The radioactivity is measured in a γ-counter.

6. TSHR Phosphorylation Analysis

The main mechanism for TSHR desensitization is receptor phosphorylation. The method that we use for evaluation of receptor phosphorylation is described next.

6.1. Required material

Devices

- Eppendorf centrifuge 5415R
- Tube rotator
- Thermoblock
- Vertical gel electrophoresis system
- Vacuum and gel dryer
- Ultralow freezer ($-86\ °C$)
- Developer

Reagents and solutions

- Cell growth medium, PBS, TrypLE™ Express, Lipofectamine™ as described above.
- Phosphate-free medium: phosphate-free Dulbecco's Modified Medium (Gibco) with 20 mM HEPES (Sigma).
- Phosphorus-32 (32P) (Perkin–Elmer).
- Okadaic acid (Sigma).
- RIPA buffer: 150 mM NaCl, 50 mM Tris–HCl (EMD Biosciences), 5 mM EDTA (EMD Biosciences), 1% NP-40 (Sigma), 0.5% sodium-deoxycholate (Sigma), 10 mM NaF (Sigma), 10 mM disodium pyrophosphate (Sigma), and protease inhibitors (Thermo Scientific, IL).
- Mouse IgG antibody (Cell Signaling, Beverly, MA, USA), this secondary antibody corresponds to the host species of TSHR primary antibody.
- Protein A/G PLUS-Agarose (Santa Cruz Biotechnology).
- TSHR antibody (Santa Cruz Biotechnology).
- 1× Electrophoresis sample buffer: 2% SDS (Sigma), 10% Glycerol (Sigma), 0.1% bromphenol-blue (Sigma), 75 mM Tris–HCl (EMD Biosciences) pH 6.8, and 1% β-mercaptoethanol (Sigma).

Additional materials

- COS-7 cells and plasmid vectors as described above.

Disposables

- Whatman paper (VWR), 21 gauge needle (VWR), X-ray film (Fisher Scientific), 35× 10 mm cell culture dish (BD Falcon), and 1.5 ml Eppendorf tubes.

6.2. Cell phosphorylation

COS-7 cells are transfected in 35 × 10 mm dishes as described above. Twenty-four hours later, growth medium is removed by aspiration with a Pasteur pipette and cells are washed twice with phosphate-free medium. The labeling media is prepared adding 200 µCi of 32P/ml and 1 µM okadaic to phosphate-free medium. One milliliter of labeling media is added to each plate and cells are incubated for 3 h. Labeling is stopped by transferring the plates to ice, followed by aspiration of the medium with a Pasteur pipette and washing twice with cold PBS. After PBS removal, cells are lysed in 1 ml of RIPA buffer with protease inhibitors for 1 h at 4 °C followed by repeated aspiration (8–12 times) through a 21 gauge needle. The cellular debris is pelleted by centrifugation at 13,000×g for 15 min at 4 °C and the recovered supernatant is transferred to a 1.5 ml Eppendorf tube. The cellular lysate is precleared by adding 1 µg of mouse IgG antibody and 20 µl of Protein A/G PLUS Agarose for 30 min at 4 °C on a rotating device. The suspension is centrifuged at 1000×g for 5 min at 4 °C and the supernatant is recovered and transferred to 1.5 ml Eppendorf tube. TSHR antibody (2 µg) is added to the supernatant and incubated for 1 h at 4 °C on a rotator. After this incubation period, 20 µl of Protein A/G PLUS-Agarose are added and the solution is incubated overnight at 4 °C on a rotator. Immunoprecipitates are collected by centrifugation at 1000×g at 4 °C. The pellet is washed four times with RIPA buffer, and supernatant is discarded. The pellet is resuspended in 1× electrophoresis sample buffer then heated at 65 °C for 10 min. The sample is resolved by SDS-PAGE in a 10% gel. The gel is transferred to Whatman paper and dried at 80 °C in a vacuum drier for 2 h. Finally, the dried gel is exposed to X-ray film at −80 °C for 72 h and developed.

ACKNOWLEDGMENTS

This work was supported by the Fondo de Investigaciones Sanitarias FIS (grant PI030401 to J. L.-A.), Ministerio de Educación (grant SAF2006-02542 to J. L.-A.), and Xunta de Galicia (grant PGIDIT04PXIC20801PN to J. L.-A., PGIDIT06PXIB 208360PR to J. L.-A.).

REFERENCES

Allgeier, A., Offermanns, S., Van Sande, J., Spicher, K., Schultz, G., and Dumont, J. E. (1994). The human thyrotropin receptor activates G-proteins Gs and Gq/11. *J. Biol. Chem.* **269,** 13733–13735.

Baratti-Elbaz, C., Ghinea, N., Lahuna, O., Loosfelt, H., Pichon, C., and Milgrom, E. (1999). Internalization and recycling pathways of the thyrotropin receptor. *Mol. Endocrinol.* **13,** 1751–1765.

Calebiro, D., Nikolaev, V. O., Persani, L., and Lohse, M. J. (2010). Signaling by internalized G-protein-coupled receptors. *Trends Pharmacol. Sci.* **31,** 221–228.

Castro, I., Lima, L., Seoane, R., and Lado-Abeal, J. (2009). Identification and functional characterization of two novel activating thyrotropin receptor mutants in toxic thyroid follicular adenomas. *Thyroid* **19,** 645–649.

Claus, M., Neumann, S., Kleinau, G., Krause, G., and Paschke, R. (2006). Structural determinants for G-protein activation and specificity in the third intracellular loop of the thyroid-stimulating hormone receptor. *J. Mol. Med.* **84,** 943–954.

Davies, T., Marians, R., and Latif, R. (2002). The TSH receptor reveals itself. *J. Clin. Invest.* **110,** 161–164.

Duprez, L., Parma, J., Van Sande, J., Allgeier, A., Leclere, J., Schvartz, C., Delisle, M. J., Decoulx, M., Orgiazzi, J., Dumont, J., et al. (1994). Germline mutations in the thyrotropin receptor gene cause non-autoimmune autosomal dominant hyperthyroidism. *Nat. Genet.* **7,** 396–401.

Eszlinger, M., Holzapfel, H. P., Voigt, C., Arkenau, C., and Paschke, R. (2004). RGS 2 expression is regulated by TSH and inhibits TSH receptor signaling. *Eur. J. Endocrinol.* **151,** 383–390.

Farid, N. R., and Szkudlinski, M. W. (2004). Minireview: Structural and functional evolution of the thyrotropin receptor. *Endocrinology* **145,** 4048–4057.

Ferguson, S. S. (2007). Phosphorylation-independent attenuation of GPCR signalling. *Trends Pharmacol. Sci.* **28,** 173–179.

Frenzel, R., Voigt, C., and Paschke, R. (2006). The human thyrotropin receptor is predominantly internalized by beta-arrestin 2. *Endocrinology* **147,** 3114–3122.

Gozu, H. I., Bircan, R., Krohn, K., Muller, S., Vural, S., Gezen, C., Sargin, H., Yavuzer, D., Sargin, M., Cirakoglu, B., and Paschke, R. (2006). Similar prevalence of somatic TSH receptor and Gsalpha mutations in toxic thyroid nodules in geographical regions with different iodine supply in Turkey. *Eur. J. Endocrinol.* **155,** 535–545.

Greenwood, F. C., Hunter, W. M., and Glover, J. S. (1963). The preparation of I-131-labelled human growth hormone of high specific radioactivity. *Biochem. J.* **89,** 114–123.

Haraguchi, K., Saito, K., Kaneshige, M., Endo, T., and Onaya, T. (1994). Desensitization and internalization of a thyrotrophin receptor lacking the cytoplasmic carboxy-terminal region. *Mol. Endocrinol.* **13,** 283–288.

Hayashi, M., Matsushima, K., Ohashi, H., Tsunoda, H., Murase, S., Kawarada, Y., and Tanaka, T. (1998). Molecular cloning and characterization of human PDE8B, a novel thyroid-specific isozyme of $3',5'$-cyclic nucleotide phosphodiesterase. *Biochem. Biophys. Res. Commun.* **250,** 751–756.

Ho, S. C., Van Sande, J., Lefort, A., Vassart, G., and Costagliola, S. (2001). Effects of mutations involving the highly conserved S281HCC motif in the extracellular domain of the thyrotropin (TSH) receptor on TSH binding and constitutive activity. *Endocrinology* **142,** 2760–2767.

Iacovelli, L., Franchetti, R., Masini, M., and De Blasi, A. (1996). GRK2 and beta-arrestin 1 as negative regulators of thyrotropin receptor-stimulated response. *Mol. Endocrinol.* **10,** 1138–1146.

Kimura, T., Van Keymeulen, A., Golstein, J., Fusco, A., Dumont, J. E., and Roger, P. P. (2001). Regulation of thyroid cell proliferation by TSH and other factors: A critical evaluation of *in vitro* models. *Endocr. Rev.* **22,** 631–656.

Kosugi, S., Okajima, F., Ban, T., Hidaka, A., Shenker, A., and Kohn, L. D. (1993). Substitutions of different regions of the third cytoplasmic loop of the thyrotropin (TSH) receptor have selective effects on constitutive, TSH-, and TSH receptor autoantibody-stimulated phosphoinositide and $3',5'$-cyclic adenosine monophosphate signal generation. *Mol. Endocrinol.* **7,** 1009–1020.

Krohn, K., Maier, J., and Paschke, R. (2007). Mechanisms of disease: Hydrogen peroxide, DNA damage and mutagenesis in the development of thyroid tumors. *Nat. Clin. Pract. Endocrinol. Metab.* **3,** 713–720.

Kursawe, R., and Paschke, R. (2007). Modulation of TSHR signaling by posttranslational modifications. *Trends Endocrinol. Metab.* **18,** 199–207.

Lado-Abeal, J., Dumitrescu, A. M., Liao, X. H., Cohen, R. N., Pohlenz, J., Weiss, R. E., Lebrethon, M. C., Verloes, A., and Refetoff, S. (2005). A de novo mutation in an already mutant nucleotide of the thyroid hormone receptor beta gene perpetuates resistance to thyroid hormone. *J. Clin. Endocrinol. Metab.* **90,** 1760–1767.

Lalli, E., and Sassone-Corsi, P. (1995). Thyroid-stimulating hormone (TSH)-directed induction of the CREM gene in the thyroid gland participates in the long-term desensitization of the TSH receptor. *Proc. Natl. Acad. Sci. USA* **92,** 9633–9637.

Laugwitz, K. L., Allgeier, A., Offermanns, S., Spicher, K., Van Sande, J., Dumont, J. E., and Schultz, G. (1996). The human thyrotropin receptor: A heptahelical receptor capable of stimulating members of all four G protein families. *Proc. Natl. Acad. Sci. USA* **93,** 116–120.

Nagayama, Y., Chazenbalk, G., Takeshita, A., Kimura, H., Ashizawa, K., Yokoyama, N., Rapoport, B., and Nagataki, S. (1994). Studies on homologous densensitization of the thyrotropin receptor in 293 human embryonal kidney cells. *Endocrinology* **135,** 1060–1065.

Nagayama, Y., Tanaka, K., Hara, T., Namba, H., Yamashita, S., Taniyama, K., and Niwa, M. (1996). Involvement of G protein-coupled receptor kinase 5 in homologous desensitization of the thyrotropin receptor. *J. Biol. Chem.* **271,** 10143–10148.

Oki, N., Takahashi, S. I., Hidaka, H., and Conti, M. (2000). Short term feedback regulation of cAMP in FRTL-5 thyroid cells. Role of PDE4D3 phosphodiesterase activation. *J. Biol. Chem.* **275,** 10831–10837.

Palos-Paz, F., Perez-Guerra, O., Cameselle-Teijeiro, J., Rueda-Chimeno, C., Barreiro-Morandeira, F., and Lado-Abeal, J.The Galician Group for the Study of Toxic Multinodular Goitre (2008). Prevalence of mutations in TSHR, GNAS, PRKAR1A and RAS genes in a large series of toxic thyroid adenomas from Galicia, an iodine-deficient area in NW Spain. *Eur. J. Endocrinol.* **159,** 623–631.

Parma, J., Duprez, L., Van Sande, J., Cochaux, P., Gervy, C., Mockel, J., Dumont, J., and Vassart, G. (1993). Somatic mutations in the thyrotropin receptor gene cause hyperfunctioning thyroid adenomas. *Nature* **365,** 649–651.

Parma, J., Van Sande, J., Swillens, S., Tonacchera, M., Dumont, J., and Vassart, G. (1995). Somatic mutations causing constitutive activity of the thyrotropin receptor are the major cause of hyperfunctioning thyroid adenomas: Identification of additional mutations activating both the cyclic adenosine 3′,5′-monophosphate and inositol phosphate-Ca^{2+} cascades. *Mol. Endocrinol.* **9,** 725–733.

Parnot, C., Miserey-Lenkei, S., Bardin, S., Corvol, P., and Clauser, E. (2002). Lessons from constitutively active mutants of G protein-coupled receptors. *Trends Endocrinol. Metab.* **13,** 336–343.

Persani, L., Lania, A., Alberti, L., Romoli, R., Mantovani, G., Filetti, S., Spada, A., and Conti, M. (2000). Induction of specific phosphodiesterase isoforms by constitutive activation of the cAMP pathway in autonomous thyroid adenomas. *J. Clin. Endocrinol. Metab.* **85,** 2872–2878.

Postiglione, M. P., Parlato, R., Rodriguez-Mallon, A., Rosica, A., Mithbaokar, P., Maresca, M., Marians, R. C., Davies, T. F., Zannini, M. S., De Felice, M., and Di Lauro, R. (2002). Role of the thyroid-stimulating hormone receptor signaling in development and differentiation of the thyroid gland. *Proc. Natl. Acad. Sci. USA* **99,** 15462–15467.

Rajagopal, S., Rajagopal, K., and Lefkowitz, R. J. (2010). Teaching old receptors new tricks: Biasing seven-transmembrane receptors. *Nat. Rev. Drug Discov.* **9,** 373–386.

Saji, M., Ikuyama, S., Akamizu, T., and Kohn, L. D. (1991). Increases in cytosolic Ca++ down regulate thyrotropin receptor gene expression by a mechanism different from the cAMP signal. *Biochem. Biophys. Res. Commun.* **176,** 94–101.

Selzer, E., Wilfing, A., Schiferer, A., Hermann, M., Grubeck-Loebenstein, B., and Freissmuth, M. (1993). Stimulation of human thyroid growth via the inhibitory guanine nucleotide binding (G) protein Gi: Constitutive expression of the G-protein alpha subunit Gi alpha-1 in autonomous adenoma. *Proc. Natl. Acad. Sci. USA* **90,** 1609–1613.

Singh, S. P., McDonald, D., Hope, T. J., and Prabhakar, B. S. (2004). Upon thyrotropin binding the thyrotropin receptor is internalized and localized to endosome. *Endocrinology* **145,** 1003–1010.

Song, Y., Massart, C., Chico-Galdo, V., Jin, L., De Maertelaer, V., Decoster, C., Dumont, J. E., and Van Sande, J. (2010). Species specific thyroid signal transduction: Conserved physiology, divergent mechanisms. *Mol. Cell. Endocrinol.* **319,** 56–62.

Sunthornthepvarakui, T., Gottschalk, M. E., Hayashi, Y., and Refetoff, S. (1995). Brief report: Resistance to thyrotropin caused by mutations in the thyrotropin-receptor gene. *N. Engl. J. Med.* **332,** 155–160.

Urizar, E., Montanelli, L., Loy, T., Bonomi, M., Swillens, S., Gales, C., Bouvier, M., Smits, G., Vassart, G., and Costagliola, S. (2005). Glycoprotein hormone receptors: Link between receptor homodimerization and negative cooperativity. *EMBO J.* **24,** 1954–1964.

Vassart, G., and Dumont, J. E. (1992). The thyrotropin receptor and the regulation of thyrocyte function and growth. *Endocr. Rev.* **13,** 596–611.

Vlaeminck-Guillem, V., Ho, S. C., Rodien, P., Vassart, G., and Costagliola, S. (2002). Activation of the cAMP pathway by the TSH receptor involves switching of the ectodomain from a tethered inverse agonist to an agonist. *Mol. Endocrinol.* **16,** 736–746.

Voigt, C., Holzapfel, H., and Paschke, R. (2000). Expression of beta-arrestins in toxic and cold thyroid nodules. *FEBS Lett.* **486,** 208–212.

Voigt, C., Holzapfel, H. P., Meyer, S., and Paschke, R. (2004). Increased expression of G-protein-coupled receptor kinases 3 and 4 in hyperfunctioning thyroid nodules. *J. Endocrinol.* **182,** 173–182.

Wynford-Thomas, D., Stringer, B. M., Harach, H. R., and Williams, E. D. (1983). Control of growth in the rat thyroid—An example of specific desensitization to trophic hormone stimulation. *Experientia* **39,** 421–423.

Zhang, M., Tong, K. P., Fremont, V., Chen, J., Narayan, P., Puett, D., Weintraub, B. D., and Szkudlinski, M. W. (2000). The extracellular domain suppresses constitutive activity of the transmembrane domain of the human TSH receptor: Implications for hormone–receptor interaction and antagonist design. *Endocrinology* **141,** 3514–3517.

CHAPTER TWENTY

Assessment of Constitutive Activity of a G Protein-Coupled Receptor, Cpr2, in *Cryptococcus neoformans* by Heterologous and Homologous Methods

Chaoyang Xue,[*,†,1] Yina Wang,[*] *and* Yen-Ping Hsueh[‡]

Contents

1. Introduction of Receptors and Constitutive Receptors	398
2. Identification of Cpr2 as a Natural Occurring Constitutively Active Receptor	399
2.1. GPCR and G protein signaling in *C. neoformans*	399
2.2. Identification of Cpr2 protein	400
2.3. Expression of *CPR2* in a yeast heterologous expression system	401
3. Additional Constitutively Active Receptors Identified in Fungi	409
Acknowledgments	409
References	409

Abstract

G protein-coupled receptors (GPCRs) comprise the largest superfamily of cell surface receptors and are primary targets for drug development. A variety of detection systems have been reported to study ligand–GPCR interactions. Using *Saccharomyces cerevisiae* to express foreign proteins has long been appreciated for its low cost, simplicity, and conserved cellular pathways. The yeast pheromone-responsive pathway has been utilized to assess a range of different GPCRs. We have identified a pheromone-like receptor, Cpr2, that is located outside of the *MAT* locus in the human fungal pathogen *Cryptococcus neoformans*. To characterize its function and potential ligands, we expressed *CPR2* in a yeast heterologous expression system. To optimize for *CPR2* expression in this

[*] Public Health Research Institute, University of Medicine and Dentistry of New Jersey, Newark, New Jersey, USA
[†] Department of Microbiology and Molecular Genetics, University of Medicine and Dentistry of New Jersey, Newark, New Jersey, USA
[‡] Division of Biology, California Institute of Technology, Pasadena, California, USA
[1] Corresponding author.

system, pheromone receptor Ste3, regulator of G protein signaling (RGS) Sst2, and the cyclin-dependent kinase inhibitor Far1 were mutated. The *lacZ* gene was fused with the promoter of the *FUS1* gene that is activated by the yeast pheromone signal and then introduced into yeast cells. Expression of *CPR2* in this yeast heterologous expression system revealed that Cpr2 could activate the pheromone-responsive pathway without addition of potential ligands, suggesting it is a naturally occurring, constitutively active receptor. Mutation of a single amino acid, Leu^{222}, was sufficient to reverse the constitutive activity of Cpr2. In this chapter, we summarize methods used for assessing the constitutive activity of Cpr2 and its mutants, which could be beneficial for other GPCR studies.

1. Introduction of Receptors and Constitutive Receptors

All living organisms must sense environmental signals and respond appropriately to survive. Sensing the environment depends on cell surface receptors, which include the family of G protein-coupled receptors (GPCRs; Bahn et al., 2007). GPCRs comprise the largest superfamily of cell surface receptors, which respond to a variety of extracellular stimuli as diverse as hormones, neurochemicals, odorants, nutrients, and light (Bockaert and Pin, 1999). Despite exhibiting diversity in primary sequence and biological function, all GPCRs possess the same fundamental architecture, consisting of seven transmembrane domains (7TMs) and share common mechanisms of signal transduction. These receptors are also of great pharmacological importance (Drews, 1996). To date, almost half of all clinically relevant drugs act as agonist or antagonists of GPCRs (Drews, 1996).

Generally, GPCRs function as guanine nucleotide exchange factors (GEFs) to activate heterotrimeric G proteins that play a central role in transducing extracellular signals into intrinsic signals and effecting appropriate biochemical and physiological responses. Ligand activation of GPCRs triggers the coupled Gα subunit to release GDP and bind GTP, transforming them into the Gα–GTP active state. The GTP binding results in a conformational change of the Gα subunit, which promotes its dissociation from the Gβγ complex. The liberated Gα or Gβγ subunits then interact with downstream effectors to switch on or off signaling cascades. RGS (regulators of G protein signaling) proteins, however, negatively regulate the G protein signaling by converting Gα protein from GTP-bound state to GDP-bound state.

Much of GPCR research has focused on the effects of agonist stimulation of these GPCRs, but it has become clear that many GPCRs exist in a constitutively active state, where activation occurs in the absence of an agonist. These GPCRs are locked in an active state and active downstream signal in a ligand-independent fashion, and continuously stimulate their intracellular signaling pathways (Rosenkilde et al., 2006). The hypothesis that some GPCRs can signal in the absence of any external ligand was first supported by studies of the opioid and β2-adrenergic receptors (Cerione et al., 1984; Koski et al., 1982). Constitutively active GPCRs have been generated artificially by mutagenesis, and also have been reported to exist as naturally occurring point mutations that increase constitutive activity (Van Sande et al., 1995). Moreover, approximately 60 wild-type GPCRs from human, mouse, or rat have been shown to exhibit considerable constitutive activity and many of them found to be linked to human diseases (Seifert and Wenzel-Seifert, 2002). Therefore, constitutively active GPCRs may be of great physiological importance in both natural physiology and disease states.

A pheromone receptor-like gene, *CPR2*, was identified in *Cryptococcus neoformans*, a human fungal pathogen that commonly causes meningoencephalitis mostly in immunocompromised populations (Hsueh et al., 2009). Cpr2 appears to be a naturally occurring, constitutively active GPCR based on biochemical and genetic evidence: first, expression of *CPR2* in a yeast heterologous expression system revealed that Cpr2 activates downstream signals without addition of potential ligands; second, a single amino acid, Leu^{222}, was found to produce its constitutive activity. The constitutively active status of Cpr2 was further confirmed in phenotypic analysis and protein–protein interaction assays using a split–ubiquitin system (Hsueh et al., 2009). In this chapter, we will discuss several methods used to assess GPCR constitutive activity in fungi using Cpr2 as an example.

2. IDENTIFICATION OF CPR2 AS A NATURAL OCCURRING CONSTITUTIVELY ACTIVE RECEPTOR

2.1. GPCR and G protein signaling in *C. neoformans*

GPCRs have been identified in a variety of fungi and play important roles in cell development and fungal pathogenesis. The model ascomycete *Saccharomyces cerevisiae* has only three GPCRs: two pheromone receptors (Ste2 and Ste3) and one sugar sensor (Gpr1; Burkholder and Hartwell, 1985; Hagen et al., 1986; Kraakman et al., 1999). The pheromones and pheromone receptors are expressed in a cell type-specific manner in *S. cerevisiae*. **a** cells express the lipopeptide pheromone **a**-factor and the α-factor receptor Ste2, whereas α cells express the peptide pheromone α-factor and the **a**-factor receptor Ste3. Ste2 and Ste3 activate Gpa1 G protein to regulate the pheromone-responsive pathway, while Gpr1 is coupled to Gpa2 to control the nutrient sensing of yeast cells (Harashima and Heitman, 2004; Lengeler et al., 2000). The Gpa1-pheromone response pathway is among the best understood signaling pathways in eukaryotes and serves as a model for GPCR-mediated signaling. Studies of this model have contributed enormously to the understanding of the mechanisms of G protein signaling and regulation. The yeast pheromone pathway has also been developed as a heterologous expression system to identify the potential agonists and antagonists for a receptor (Mentesana et al., 2002; Xue et al., 2008b).

In the basidiomycete C. neoformans, three G protein α submits control two major G protein-signaling pathways that are important for cell development and fungal virulence. One Gα protein, Gpa1, activates the nutrient sensing pathway by regulating the activity of protein kinase A (PKA) and controls fungal virulence (Alspaugh et al., 1997; Xue et al., 2008a). The receptors for activating this nutrient signaling pathway remain to be understood even though one GPCR, Gpr4, has been found to be involved in this activation, probably by sensing amino acids (Xue et al., 2006). Two other Gα subunits, Gpa2 and Gpa3, are both involved in the pheromone sensing pathway in C. neoformans. C. neoformans has a defined sexual cycle and a simple bipolar mating system (**a** and α) in which a single mating-type (*MAT*) locus spans an approximately 120 kb recombinationally suppressed region (Hull et al., 2002). Only **a**-factor-like lipopeptide pheromones and Ste3-like pheromone receptors are known in C. neoformans; neither an α-like pheromone gene nor a Ste2-like ortholog is apparent in any of sequenced genomes. Pheromone receptors Ste3α and Ste3**a** are encoded in the *MAT* locus and sense pheromones from cells of the opposite mating type and activate a G protein complex that includes Gpa2 and Gpa3, the Gβ subunit Gbp1, and the Gγ subunits Gpg1 and Gpg2 (Hsueh et al., 2007; Li et al., 2007). Following activation of Gpa2 and Gpa3, the Gβ subunits Gbp1 is released to activate the downstream MAPK cascade to trigger mating responses.

2.2. Identification of Cpr2 protein

A Ste3-like protein (CNAG_03938) was identified in our search for potential GPCR proteins in C. neoformans H99 genome and was named as Cpr2 (*Cryptococcus* Pheromone Receptor 2) (Hsueh et al., 2009). Cpr2 contains 414 amino acids and seven predicted transmembrane domains (TMs), a structural hallmark of GPCRs. The *MAT* locus of C. neoformans has been extensively studied, and among 20 or so genes encoded by this locus, the homeodomain proteins and the pheromone/pheromone receptor genes establish the sexual identity of a cell (Chung et al., 2002; Hull et al., 2005). Interestingly, Cpr2 is not located in the *MAT* locus. A search for Ste3-like receptors in other basidiomyceteous fungi revealed that many species also have additional pheromone receptor paralogs that are not part of *MAT* (Aimi et al., 2005; James et al., 2006), suggesting basidiomyceteous fungi may have an additional mechanism for pheromone sensing and regulation.

Functional studies revealed that Cpr2 expression is regulated by pheromone and highly induced during mating. A strain carrying a deletion mutation of *CPR2* is still fertile but has a defect in cell fusion. However, overexpression of *CPR2* elicits unisexual mating (Hsueh et al., 2009). To understand whether this protein functions as a pheromone sensor, we expressed *CPR2* gene in a yeast heterologous expression system.

2.3. Expression of *CPR2* in a yeast heterologous expression system

In order to better understand the function of GPCRs, researchers attempt to express the proteins in less sophisticated organisms. Even though GPCRs in fungi have not yet been developed as drug targets for disease control, GPCR-related signaling pathways have been utilized to deorphanize GPCRs (Mentesana et al., 2002). A bioassay based on the *S. cerevisiae* pheromone response pathway has been developed to characterize heterologous GPCRs. However, several preconditions must be met to utilize this pathway. First, proper plasma membrane expression of the target GPCR is needed; second, the foreign receptor must properly couple to the $G\alpha$ protein Gpa1 to activate the downstream signal; third, a reporter gene in this pathway is needed to monitor GPCR expression and signal activation. Finally, the pathway needs to be optimized to enhance signal sensitivity.

Some foreign GPCR genes can be directly expressed in *S. cerevisiae* and coupled to the Gpa1 G protein to activate the yeast pheromone-responsive pathway, including some pheromone receptors from *Schizophyllum commune* (Fowler et al., 1999), the mammalian $SSTR_2$ receptor (Price et al., 1995), the adenosine A_2 receptor (Price et al., 1996), the melatonin Mel_{1a} receptor (Kokkola et al., 1998), and the UDP-glucose receptor (Chambers et al., 2000). We also expressed the *C. neoformans CPR2* gene in this *Saccharomyces* heterologous system to identify its potential ligands (Hsueh et al., 2009).

2.3.1. Establishment of the yeast heterologous expression system

To engineer the *S. cerevisiae* pheromone response pathway as a heterologous expression system, several modifications are necessary to optimize signal output. The cyclin-dependent kinase inhibitor Far1 promotes yeast cell arrest in G1 in response to pheromone (Chang and Herskowitz, 1990), and *FAR1* is deleted to allow continued cell division in cells responding to pheromone or a heterologous ligand. The RGS protein Sst2 functions as a negative regulator of Gpa1 activity and deletion of *SST2* significantly increases the sensitivity of pheromone response (Dietzel and Kurjan, 1987) and the yeast heterologous GPCR expression signal. To monitor the pheromone response signal for large-scale screens, the pheromone inducible gene *FUS1* (Hagen et al., 1991) is commonly fused with either the *lacZ* gene or

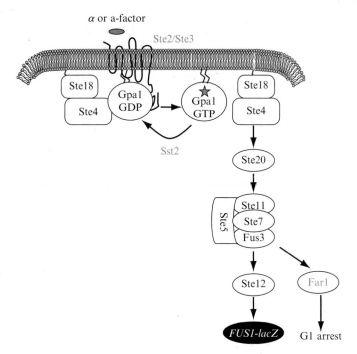

Figure 20.1 The yeast pheromone-responsive pathway and generation of the yeast heterologous expression system. The pheromone receptor, Ste2 or Ste3, activates the Gα protein Gpa1 by converting GDP-bound to GTP-bound and promoting the release of the Gβγ subunits (Ste4 and Ste18) from the heterotrimeric G protein complex. Ste4 in turn activates PAK kinase Ste20, and MAP kinase cascade including MAPKKK (Ste11), MAPKK (Ste7), and MAPK (Fus3). The activation of transcription factor Ste12 by Fus3 promotes mating. To generate a yeast heterologous system for GPCRs using this signaling pathway, the pheromone receptor Ste3 is deleted to eliminate signal interference by endogenous receptor. The RGS protein Sst2 and the cyclin-dependent kinase inhibitor Far1 are deleted to optimize the signal readout. The promoter of the pheromone inducible gene *FUS1* is fused with a reporter gene, such as *lacZ* or *HIS3* to quantitatively measure the pheromone response signal. In the Cpr2 study, the *lacZ* gene was used.

the *HIS3* gene. Finally, to avoid interference by the endogenous receptors, the yeast pheromone receptor gene *STE2* or *STE3* is deleted (Fig. 20.1).

2.3.1.1. Required materials Vectors and strains used are listed in Table 20.1.

Vectors: pTCFL1 is a *S. cerevisiae* expression vector that contains a *FUS1–lacZ* fusion (Chen and Kurjan, 1997). When the pheromone-responsive pathway is activated, the *FUS1–lacZ* fusion will express and the β-galactosidase activity can be measured. pPGK is a *S. cerevisiae* over-expression vector that contains the yeast phosphoglycerate kinase (PGK) promoter (Kang et al., 1990).

Table 20.1 Plasmids and strains used for Cpr2 assessment

Strains	Genotype	References
W303-1A	MATa ade2-1 his3-11,15 leu2-3,112 trp-1 ura3-1 can1-100	Fowler et al. (1999)
W303-1B	MATα ade2-1 his3-11,15 leu2-3,112 trp-1 ura3-1 can1-100	Fowler et al. (1999)
SDK45	MATα ade2-1 his3-11,15 leu2-3,112 trp-1 ura3-1 can1-100 ste3::ADE2	Fowler et al. (1999)
SDK47	MATα ade2-1 his3-11,15 leu2-3,112 trp-1 ura3-1 can1-100 ste3::ADE2 sst2::LEU2	Fowler et al. (1999)
YDX4	MATα ade2-1 his3-11,15 leu2-3,112 trp-1 ura3-1 can1-100 ste3::ADE2 sst2::LEU2 far1::HYG	Xue et al., unpublished
YDX10	MATα ade2-1 his3-11,15 leu2-3,112 trp-1 ura3-1 can1-100 ste3::ADE2 sst2::LEU2 far1::HYG FUS1-lacZ-TRP	Xue et al., unpublished
Plasmids	Description	References
pTCFL1	FUS1::lacZ TRP1 AmpR	Chen and Kurjan (1997)
pPGK	Yeast overexpression vector, PGK promoter URA3 AmpR	Kang et al. (1990)
pYH58	pPGK::CPR2 cDNA URA3 AmpR	Hsueh et al. (2009)
pYH59	pPGK::CPR2^{L222P} cDNA URA3 AmpR	Hsueh et al. (2009)

S. cerevisiae strains: W303-1A (*MATa*) and W303-1B (*MATα*) are laboratory strains used in this assay. SDK45 was generated based on W303-1B by deleting pheromone receptor *STE3*, and SDK47 was generated based on SDK45 by further deleting RGS gene *SST2* (Fowler et al., 1999). The *FAR1* gene was mutated in SDK47 to generate YDX4.

Media. YPD rich medium, SD-Trp, and SD-Trp Ura dropout media were prepared as described (Aimi et al., 2005). To prepare 200 ml of 5× Z-buffer, 16.1 g $Na_2HPO_4 \cdot 7H_2O$, 6.23 g $NaH_2PO_4 \cdot 2H_2O$, 0.75 g KCl, and 0.246 g $MgSO_4 \cdot 7H_2O$ are added and adjusted to pH 7.0; then add H_2O to a total volume of 200 ml. Z-buffer should be stored at 4 °C.

Chemicals: Lithium acetate (LiOAc, Fisher Scientific), Salmon sperm single-stranded DNA (Sigma), CPRG (chlorophenolred-β-D-galactopyranoside, CalBiochem), and β-mercaptoethanol (Sigma). Synthetic mating pheromones MF**a** and MFα (Heitman lab).

Equipment: Microcentrifuge (Qiagen), Water incubator (Fisher Scientific), Culture incubator (Sanyo), Shaker incubator (New Brunswick Scientific), Spectrometer (Bio-Rad).

Disposables: 1.5 ml microfuge tubes (USA Scientific), 15 ml falcon tissue culture tubes (BD Bioscience), 50 ml centrifuge tubes (Denville), Petri dishes (Fisher Scientific), 1.5 ml cuvette (VWR).

In order to utilize the yeast system, we modified the *S. cerevisiae* strains (original provided by Dr. Thomas Fowler, Southern Illinois University, Edwardsville) using the following yeast transformation protocol. We generated a strain YDX4, in which three genes in pheromone pathway (*STE3*, *SST2*, and *FAR1*) were deleted to optimize signal output. The vector pTCFL1 containing a *FUS1–lacZ* fusion cassette was used to transform YDX4 to generate YDX10, in which the *lacZ* gene will be activated by pheromone response signals.

2.3.1.1.1. Protocol for yeast transformation Day 1: Inoculate a single yeast colony into a 5-ml YPD culture and incubate overnight (24 h) with shaking at 30 °C.

Day 2: Take out 120 µl overnight yeast culture and reinoculate into 5 ml fresh YPD liquids and incubate for 5 h. After 5-h incubation, centrifuge the culture at 2000 rpm for 3 min, and resuspend the pellet in 1 ml sterile deionized water (dH$_2$O) and transfer cells to 1.5 ml Eppendorf tubes. Centrifuge again for 10 s fast spin, wash pellets twice with 1 ml dH$_2$O and wash once with 1 ml 0.1 M LiOAc. Resuspend washed pellets into 25 µl 0.1 M LiOAc. Add the following reagents to each tube: 200 µl 50% PEG (MW3640), 25 µl 1 M LiOAc, 25 µl dH$_2$O, 5 µl 10 mg/ml ssDNA, <10 µl plasmid DNA. Mix each reaction gently and incubate at 30 °C for 30 min. Transfer tubes to 42 °C for 15 min. Centrifuge and resuspend pellets in 100 µl dH$_2$O and spread on selective medium, SD-Trp-Ura, for pTCFL1 and pPGK based vectors. Incubate at 30 °C for 2–3 days.

Day 5: Streak out single colonies on a fresh SD-Trp-Ura plate. Typically at least three colonies should be streaked out for each transformation.

Yeast transformation is a routine technique and many modified protocols are available (Amberg, 2005). We found the above protocol provided a simple method with consistent high transformation efficiency. Our experience with yeast transformation showed that it is very important to not overgrow the culture after the second inoculation. Also always purify yeast colonies by restreaking on selective medium to make sure only transformants expressing selective markers are obtained.

Upon the generation of the yeast strain, we first validated the sensitivity of the system by measuring the β-galactosidase activity (see below for protocol) for all three yeast strains, SDK45, SDK47, and YDX4. The supernatant from an overnight culture of W303-1A, which contains **a**-factor secreted by this *MATa* strain, was used as a ligand to activate the system. Strong induction of β-galactosidase activity was observed in all three strain backgrounds when native Ste3 was expressed; almost no signal was detected when the empty vector pPGK was introduced. These results indicate the system is working.

2.3.1.1.2. Protocol for measuring the β-galactosidase activity in liquid cultures Day 1: Start a 2-ml overnight culture in yeast dropout medium and let it grow at 30 °C with shaking for about 24 h.

Day 2: Centrifuge 1 ml of culture in a microfuge tube for about 5 s. Cell pellets were resuspended in 0.5 ml 1× Z-buffer (add 67 μl β-mercaptoethanol, 19 μl 10% SDS, 5 ml 5× Z-buffer, and 19.914 ml ddH$_2$O to make 25 ml of 1× Z-buffer) and vortexed for 1 min at room temperature. Add 25 μl chloroform and vortex for 5 min at room temperature. Add 100 μl CPRG solution (final concentration of CPRG is 4 mM) and incubate at 28 °C until red color appears (8–20 min). Record the time of incubation. Put the sample on ice for 5 min to "stop" the reaction. Spin out debris for 5 min at full speed at 4 °C and transfer 500 μl to cuvette and read O.D.$_{595}$ in a spectrometer. Prepare appropriate dilution for measurement if color is too intense. Also read O.D.$_{600}$ for the remaining 1 ml of overnight culture (usually needs 1:5 dilution). Calculate the enzyme activity using the following formula: β-galactosidase activity (Miller units) = 1000 × (O.D.$_{595}$)/((volume assayed) × time × (O.D.$_{600}$)). At least three replicates are needed for enzyme activity quantification because of the potential variation between different replications.

This is a simple protocol to quickly determine the β-galactosidase enzyme activity. It is important to measure the absorbance quickly to obtain representative results, because the color will increase over time even when tubes are kept on ice to "stop" the reaction.

2.3.2. Assessment of constitutive activity of Cpr2 in the yeast system

The establishment of a yeast system that can be used to quantify the activity of pheromone-responsive pathway by measuring the β-galactosidase activity allowed us to test the expression and activity of Cpr2 in this system. The *CPR2* overexpression construct YPH58 was used to transform SD45, SDK47, and YDX4. The expression of *CPR2* was confirmed by Northern blot. In this experiment, pPGK empty vector was used as a negative control and a vector expressing native Ste3 from *S. cerevisiae* was used as the positive control. Yeast cells expressing *CPR2* and *FUS1–lacZ* were cultured on a yeast dropout medium (SD-Trp-Ura) and β-galactosidase activities were measured following induction with different extracellular signals. Surprisingly, high β-galactosidase activity readouts were generated without exposing cells to putative ligands such as *C. neoformans* MF**a** or MFα synthetic pheromones (Hsueh *et al.*, 2009; Fig. 20.2). Cpr2 induced high levels of β-galactosidase activity even in strain SDK45 background, which still has the wild-type *SST2* and *FAR1* alleles, and therefore, we have carried out

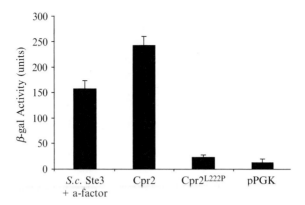

Figure 20.2 The constitutive activity of Cpr2 is dependent on the Leu^{222} residue based on assays in a yeast heterologous expression system. Cpr2 is heterologously expressed in SDK45 carrying the *FUS1–lacZ* reporter. Cells expressing *S. cerevisiae* Ste3 treated with yeast a-cell supernatant served as a positive control. Cells containing the empty vector pPGK served as a negative control. Cells expressing Cpr2 or Cpr2^{L222P} were measured for the β-galactosidase activity without treatment of potential ligands. All results are based on at least three repeats. Error bars indicate standard deviations.

the rest of the assays in strain SDK45 (Fig. 20.2). The *S. cerevisiae* reporter strains SDK45, SDK47, and YDX10 are MATα; therefore, they produce only α factor, which should not act on GPCRs related to Ste3 (respond only to lipid-modified peptides).

The highly induced β-galactosidase activity in *CPR2*-expressing cells suggested that Cpr2 was either activated by a ubiquitous ligand present in the culture supernatants or that Cpr2 is a constitutively active GPCR that functions in a ligand-independent manner. The further identification of a single amino acid that is responsible for the enzyme activity without additional ligands confirms that the Cpr2 wild-type form is a naturally occurring, constitutively active receptor (Hsueh *et al.*, 2009).

2.3.3. Identification of Leu^{222} as the cause of constitutive activity of Cpr2

It is known that alteration of amino acid sequences in certain conserved positions in some GPCRs results in constitutively active mutants (CAMs). Ninety percent of GPCRs contain a conserved proline in the sixth transmembrane domain (TM-VI), and substitution of this proline with leucine increases the ligand-independent activity of many GPCRs (Baldwin, 1993). CAMs have been generated using such point mutations in the pheromone receptors of *S. cerevisiae* (Ste2, Ste3) and *Schizosaccharomyces pombe* (Mam2; Konopka *et al.*, 1996; Ladds *et al.*, 2005; Stefan *et al.*, 1998). Surprisingly, this conserved proline is absent in the TM-VI of wild-type Cpr2, and instead, a leucine is substituted at this position Leu^{222}. To test the possibility that this leucine in TM-VI might be responsible for the constitutive activity of Cpr2, a construct (YPH59) expressing the $CPR2^{L222P}$ allele was introduced in *S. cerevisiae* SDK45. As shown in Fig. 20.2, ligand-independent induction of β-galactosidase activity was reduced to the basal level in cells expressing $CPR2^{L222P}$, compared to the empty vector control. This result provides evidence that Cpr2 activates the pheromone response pathway via an intrinsic constitutive activity instead of responding to a ubiquitous ligand present in the culture.

Because CPR2 was overexpressed in the *S. cerevisiae* heterologous system, one possibility is that the overexpression might cause the constitutive activity of Cpr2. It has been reported that even some wild-type ligand-dependent GPCRs, a small fraction of the total receptor pool may exist in the active form in the absence of agonist; thus, simply overexpressing some GPCRs could elicit a constitutively active response (Milano *et al.*, 1994). However, we do not think that overexpression alone is sufficient in the case of Cpr2-induced pheromone signaling because overexpressing either the canonical pheromone receptor Ste3α or the Cpr2^{L222P} allele did not activate the *FUS1–lacZ* reporter gene in *S. cerevisiae* (Fig. 20.2).

Several other lines of evidence, including phenotypic analysis of *cpr2* mutants in *C. neoformans*, confirmed our conclusion. Overexpression of the Cpr2^{L222P} allele in the *C. neoformans ste3α* mutant did not cause a hyperfilamentation phenotype as did of Cpr2 overexpression. Overexpression of Cpr2 in a *C. neoformans mfα 1,2,3* pheromoneless mutant still conferred a hyperfilamentous phenotype, eliminating the possible involvement of an autocrine-signaling loop. In protein–protein interaction assays using a split-ubiquitin yeast two-hybrid system, only the interactions of Cpr2^{L222P} with Gpa2 and Gpa3, but not of Cpr2 with these G proteins, were detected, implying the interaction between wild-type Cpr2 and the Gα subunits could be transient due to its constitutive activity (Hsueh *et al.*, 2009).

Although in many cases, including Cpr2, substitution of proline with leucine increases constitutive activity of GPCRs, not all GPCRs become constitutively active when this proline residue is missing or mutated. The Ste3**a** receptor in *C. neoformans* is one such example, as it lacks a conserved proline in TM-VI but does not display any constitutive activity.

We have also tried to identify the potential inverse agonist agents for Cpr2 by testing the synthetic mating pheromones MF**a** and MFα, as well as a range of farnesol derivatives (pheromones are farnesylated) in the yeast heterologous expression system. None of these chemicals significantly altered the readout of β-galactosidase activity. Therefore, inverse agonists for Cpr2, if they exist, remain to be identified. Their existence is an intriguing hypothesis for future investigation.

2.3.4. Limitation of the yeast heterologous expression system and alternatives

Not all GPCRs can be expressed correctly, couple with *S. cerevisiae* Gpa1 and activate the downstream pheromone pathway. One example is Gpr4 in *C. neoformans*. We cloned *GPR4* in pPGK and expressed it in YDX10, but did not observe any pheromone-induced lacZ activity (Xue *et al.*, unpublished). So modifying the system, such as making chimeric G proteins or chimeric GPCRs, may expand the system for the analysis of other receptors that do not directly express well in yeast.

Actually, in most cases, modifications of Gpa1 or the GPCR itself are necessary to promote proper GPCR-G protein interaction and pathway activation. For example, most GPCRs couple to the C terminus of the G protein Gpa1, and one way to improve coupling between foreign receptors and Gpa1 without interfering Gpa1 activity is to generate a chimeric Gα by replacing the Gpa1 C-terminal region with the corresponding protein of the foreign Gα. Some studies indicate that replacing only five amino acids is sufficient to promote specific GPCR–Gα coupling (Komatsuzaki *et al.*, 1997; Kostenis *et al.*, 1997; Mentesana *et al.*, 2002). We have constructed three chimeric G proteins where C-terminal 30 amino acids of *S. cerevisiae* Gpa1 were replaced by the corresponding C-terminal region of Gα proteins (Gpa1, Gpa2, and Gpa3) in *C. neoformans*, respectively. These chimeric G proteins have been successfully expressed in a yeast strain with *STE3*, *SST2*, and *FAR1* all mutated (Xue *et al.*, unpublished). Another approach that has been successfully applied is generating chimeric proteins by fusing adrenergic receptors and their cognate Gα subunits (Bertin *et al.*, 1994; Wise *et al.*, 1997). However, this approach is not suitable for all receptors, suggesting the mechanisms for the pathway activation may be more complex.

Modifying the receptor itself is another way to improve foreign receptor expression and signal activation in yeast. Foreign GPCRs fused with the cytoplasmic domain of the Ste2 or Ste3 pheromone receptor serve to enforce coupling between receptor and Gpa1 (King *et al.*, 1990). This approach is particularly useful for GPCRs with unknown cognate Gα subunits (Yin *et al.*, 2004).

As an alternative, there are other systems available to serve similar purposes. In the fission yeast *S. pombe*, the GPCR Stm1 has been identified as a nutrient sensor. Overexpression of *STM1* promotes uncontrolled cell division of yeast cells, which triggers a severe growth defect and also conversion of diploid to haploid cells (Chung *et al.*, 2003). This phenotype has been successfully utilized for high throughput screens of Stm1 inhibitors. Chemical compounds were applied to an *STM1* overexpression strain and inhibitors of Stm1 rescued the growth defect of the test strain (Chung *et al.*, 2007). Because Stm1 interacts with the G protein Gpa2 to exert its effect on cell growth (Chung *et al.*, 2003), this system can also be used to test heterologous GPCRs for their interactions with Gpa2 and to screen for potential modulators.

3. ADDITIONAL CONSTITUTIVELY ACTIVE RECEPTORS IDENTIFIED IN FUNGI

Although no naturally occurring constitutively active GPCR has been reported in ascomycetes, mutagenesis has been used to introduce mutant receptors with constitutive activity. The pheromone receptor Ste2 has been most thoroughly studied to identify amino acids important for function. Both site-direct mutagenesis and PCR-derived random mutagenesis approaches have been applied to select point mutations that cause the receptor to be constitutively active or dominantly negative (Konopka et al., 1996; Lee et al., 2006; Sommers et al., 2000). A number of sites in Ste2 have been identified as critical for ligand binding and protein function using these approaches. Also in the fission yeast S. pombe, a CAM of P-factor receptor Mam2 (P262L) has also been reported (Ladds et al., 2005).

Many basidiomyceteous fungi, such as mushrooms, contain multiple sexes controlled by many different pheromone receptors. For example, in the mushroom *Coprinus cinereus*, the multiallelic B mating-type genes encode a large family of 7TM receptors and CaaX-modified pheromones. According to results obtained using a yeast heterologous expression assay, the *C. cinereus* pheromone receptor acts as a GPCR and self-compatible mutations cause its constitutive activation (Olesnicky et al., 1999). In the laboratory, two TM-VI mutations of pheromone receptors in *S. commune* have been identified to produce constitutive receptor activity (Tom Fowler, personal communication). Similar to Cpr2, many of these receptors are not located in the *MAT* locus. It is possible that naturally occurring constitutively active receptors may exist to regulate the pheromone sensing response in other fungi. Thus, the methods used to assess constitutive activity of Cpr2 will be beneficial for the studies of other potential constitutively active receptors and their mutants.

ACKNOWLEDGMENTS

We thank Joe Heitman, Carol Newlon, and Tom Fowler for critical reading and comments on the chapter and Tom Fowler and James Konopka for *S. cerevisiae* strains and vectors. The original research on Cpr2 functional study was supported by National Institute of Health R21 grant (AI070230) to Joe Heitman. This work was supported by the new PI institutional start-up fund from UMDNJ to C. X.

REFERENCES

Aimi, T., et al. (2005). Identification and linkage mapping of the genes for the putative homeodomain protein (*hox1*) and the putative pheromone receptor protein homologue (*rcb1*) in a bipolar basidiomycete, *Pholiota nameko*. Curr. Genet. **48**, 184–194.

Alspaugh, J. A., et al. (1997). Cryptococcus neoformans mating and virulence are regulated by the G-protein alpha subunit GPA1 and cAMP. Genes Dev. **11**, 3206–3217.
Amberg, D. C. (2005). Methods in Yeast Genetics. Cold Spring Harbor Laboratory Press.
Bahn, Y. S., et al. (2007). Sensing the environment: Lessons from fungi. Nat. Rev. Microbiol. **5**, 57–69.
Baldwin, J. M. (1993). The probable arrangement of the helices in G protein-coupled receptors. EMBO J. **12**, 1693–1703.
Bertin, B., et al. (1994). Cellular signaling by an agonist-activated receptor/Gs alpha fusion protein. Proc. Natl. Acad. Sci. USA **91**, 8827–8831.
Bockaert, J., and Pin, J. P. (1999). Molecular tinkering of G protein-coupled receptors: An evolutionary success. EMBO J. **18**, 1723–1729.
Burkholder, A. C., and Hartwell, L. H. (1985). The yeast alpha-factor receptor: Structural properties deduced from the sequence of the STE2 gene. Nucleic Acids Res. **13**, 8463–8475.
Cerione, R. A., et al. (1984). The mammalian beta 2-adrenergic receptor: Reconstitution of functional interactions between pure receptor and pure stimulatory nucleotide binding protein of the adenylate cyclase system. Biochemistry **23**, 4519–4525.
Chambers, J. K., et al. (2000). A G protein-coupled receptor for UDP-glucose. J. Biol. Chem. **275**, 10767–10771.
Chang, F., and Herskowitz, I. (1990). Identification of a gene necessary for cell cycle arrest by a negative growth factor of yeast: FAR1 is an inhibitor of a G1 cyclin, CLN2. Cell **63**, 999–1011.
Chen, T., and Kurjan, J. (1997). Saccharomyces cerevisiae Mpt5p interacts with Sst2p and plays roles in pheromone sensitivity and recovery from pheromone arrest. Mol. Cell. Biol. **17**, 3429–3439.
Chung, S., et al. (2002). Molecular analysis of CPRalpha, a MATalpha-specific pheromone receptor gene of Cryptococcus neoformans. Eukaryot. Cell **1**, 432–439.
Chung, K. S., et al. (2003). Functional over-expression of the Stm1 protein, a G-protein-coupled receptor, in Schizosaccharomyces pombe. Biotechnol. Lett. **25**, 267–272.
Chung, K. S., et al. (2007). Yeast-based screening to identify modulators of G-protein signaling using uncontrolled cell division cycle by overexpression of Stm1. J. Biotechnol. **129**, 547–554.
Dietzel, C., and Kurjan, J. (1987). Pheromonal regulation and sequence of the Saccharomyces cerevisiae SST2 gene: A model for desensitization to pheromone. Mol. Cell. Biol. **7**, 4169–4177.
Drews, J. (1996). Genomic sciences and the medicine of tomorrow. Nat. Biotechnol. **14**, 1516–1518.
Fowler, T. J., et al. (1999). Multiple sex pheromones and receptors of a mushroom-producing fungus elicit mating in yeast. Mol. Biol. Cell. **10**, 2559–2572.
Hagen, D. C., et al. (1986). Evidence the yeast STE3 gene encodes a receptor for the peptide pheromone a factor: Gene sequence and implications for the structure of the presumed receptor. Proc. Natl. Acad. Sci. USA **83**, 1418–1422.
Hagen, D. C., et al. (1991). Pheromone response elements are necessary and sufficient for basal and pheromone-induced transcription of the FUS1 gene of Saccharomyces cerevisiae. Mol. Cell. Biol. **11**, 2952–2961.
Harashima, T., and Heitman, J. (2004). Nutrient control of dimorphic growth in Saccharomyces cerevisiae. In "Nutrient Induced Responses in Eukaryotic Cells," (J. Winderickx and P. M. Taylor, eds.), Vol. 7, pp. 131–169. Springer-Verlag, Berlin, Germany.
Hsueh, Y. P., et al. (2007). G protein signaling governing cell fate decisions involves opposing Galpha subunits in Cryptococcus neoformans. Mol. Biol. Cell. **18**, 3237–3249.
Hsueh, Y. P., et al. (2009). A constitutively active GPCR governs morphogenic transitions in Cryptococcus neoformans. EMBO J. **28**, 1220–1233.

Hull, C. M., et al. (2002). Cell identity and sexual development in *Cryptococcus neoformans* are controlled by the mating-type-specific homeodomain protein Sxi1alpha. *Genes Dev.* **16**, 3046–3060.

Hull, C. M., et al. (2005). Sex-specific homeodomain proteins Sxi1alpha and Sxi2a coordinately regulate sexual development in *Cryptococcus neoformans. Eukaryot. Cell* **4**, 526–535.

James, T. Y., et al. (2006). Evolution of the bipolar mating system of the mushroom *Coprinellus disseminatus* from its tetrapolar ancestors involves loss of mating-type-specific pheromone receptor function. *Genetics* **172**, 1877–1891.

Kang, Y. S., et al. (1990). Effects of expression of mammalian G alpha and hybrid mammalian–yeast G alpha proteins on the yeast pheromone response signal transduction pathway. *Mol. Cell. Biol.* **10**, 2582–2590.

King, K., et al. (1990). Control of yeast mating signal transduction by a mammalian beta 2-adrenergic receptor and Gs alpha subunit. *Science* **250**, 121–123.

Kokkola, T., et al. (1998). Mutagenesis of human Mel1a melatonin receptor expressed in yeast reveals domains important for receptor function. *Biochem. Biophys. Res. Commun.* **249**, 531–536.

Komatsuzaki, K., et al. (1997). A novel system that reports the G-proteins linked to a given receptor: A study of type 3 somatostatin receptor. *FEBS Lett.* **406**, 165–170.

Konopka, J. B., et al. (1996). Mutation of Pro-258 in transmembrane domain 6 constitutively activates the G protein-coupled alpha-factor receptor. *Proc. Natl. Acad. Sci. USA* **93**, 6764–6769.

Koski, G., et al. (1982). Modulation of sodium-sensitive GTPase by partial opiate agonists. An explanation for the dual requirement for Na^+ and GTP in inhibitory regulation of adenylate cyclase. *J. Biol. Chem.* **257**, 14035–14040.

Kostenis, E., et al. (1997). Genetic analysis of receptor-Galphaq coupling selectivity. *J. Biol. Chem.* **272**, 23675–23681.

Kraakman, L., et al. (1999). A *Saccharomyces cerevisiae* G-protein coupled receptor, Gpr1, is specifically required for glucose activation of the cAMP pathway during the transition to growth on glucose. *Mol. Microbiol.* **32**, 1002–1012.

Ladds, G., et al. (2005). A constitutively active GPCR retains its G protein specificity and the ability to form dimers. *Mol. Microbiol.* **55**, 482–497.

Lee, Y. H., et al. (2006). Interacting residues in an activated state of a G protein-coupled receptor. *J. Biol. Chem.* **281**, 2263–2272.

Lengeler, K. B., et al. (2000). Signal transduction cascades regulating fungal development and virulence. *Microbiol. Mol. Biol. Rev.* **64**, 746–785.

Li, L., et al. (2007). Canonical heterotrimeric G proteins regulating mating and virulence of *Cryptococcus neoformans. Mol. Biol. Cell.* **18**, 4201–4209.

Mentesana, P. E., et al. (2002). Functional assays for mammalian G-protein-coupled receptors in yeast. *Methods Enzymol.* **344**, 92–111.

Milano, C. A., et al. (1994). Enhanced myocardial function in transgenic mice overexpressing the beta 2-adrenergic receptor. *Science* **264**, 582–586.

Olesnicky, N. S., et al. (1999). A constitutively active G-protein-coupled receptor causes mating self-compatibility in the mushroom *Coprinus. EMBO J.* **18**, 2756–2763.

Price, L. A., et al. (1995). Functional coupling of a mammalian somatostatin receptor to the yeast pheromone response pathway. *Mol. Cell. Biol.* **15**, 6188–6195.

Price, L. A., et al. (1996). Pharmacological characterization of the rat A2a adenosine receptor functionally coupled to the yeast pheromone response pathway. *Mol. Pharmacol.* **50**, 829–837.

Rosenkilde, M. M., et al. (2006). Molecular pharmacological phenotyping of EBI2. An orphan seven-transmembrane receptor with constitutive activity. *J. Biol. Chem.* **281**, 13199–13208.

Seifert, R., and Wenzel-Seifert, K. (2002). Constitutive activity of G-protein-coupled receptors: Cause of disease and common property of wild-type receptors. *Naunyn Schmiedebergs Arch. Pharmacol.* **366,** 381–416.

Sommers, C. M., et al. (2000). A limited spectrum of mutations causes constitutive activation of the yeast alpha-factor receptor. *Biochemistry* **39,** 6898–6909.

Stefan, C. J., et al. (1998). Mechanisms governing the activation and trafficking of yeast G protein-coupled receptors. *Mol. Biol. Cell.* **9,** 885–899.

Van Sande, J., et al. (1995). Somatic and germline mutations of the TSH receptor gene in thyroid diseases. *J. Clin. Endocrinol. Metab.* **80,** 2577–2585.

Wise, A., et al. (1997). Measurement of agonist efficacy using an alpha2A-adrenoceptor-Gi1alpha fusion protein. *FEBS Lett.* **419,** 141–146.

Xue, C., et al. (2006). G protein-coupled receptor Gpr4 senses amino acids and activates the cAMP-PKA pathway in *Cryptococcus neoformans*. *Mol. Biol. Cell.* **17,** 667–679.

Xue, C., et al. (2008a). The RGS protein Crg2 regulates both pheromone and cAMP signalling in *Cryptococcus neoformans*. *Mol. Microbiol.* **70,** 379–395.

Xue, C., et al. (2008b). Magnificent seven: Roles of G protein-coupled receptors in extracellular sensing in fungi. *FEMS Microbiol. Rev.* **32,** 1010–1032.

Yin, D., et al. (2004). Probing receptor structure/function with chimeric G-protein-coupled receptors. *Mol. Pharmacol.* **65,** 1323–1332.

CHAPTER TWENTY-ONE

IN VITRO AND IN VIVO ASSESSMENT OF MU OPIOID RECEPTOR CONSTITUTIVE ACTIVITY

Edward J. Bilsky,* Denise Giuvelis,* Melissa D. Osborn,* Christina M. Dersch,[†] Heng Xu,[†] *and* Richard B. Rothman[†]

Contents

1. Introduction	414
2. Measuring Opioid Receptor Constitutive Activity *In Vitro*	415
2.1. Cell culture	415
2.2. [^{35}S]-GTP-γ-S binding assay	417
3. cAMP Quantification Assay in CHO Cells Expressing Cloned Opioid Receptors	425
3.1. Materials and equipment required	425
3.2. Stimulation	426
3.3. [^{3}H]cAMP competition binding assay	426
3.4. Data analysis using GraphPad prism	427
3.5. Examples of using the cAMP assay to measure constitutively active mu opioid receptors	427
4. *In Vivo* Assessment of Antagonist Potency in Opioid Naïve Subjects	430
4.1. Materials and supplies required	431
4.2. Determination of antagonist potency and time of peak effect	431
5. *In Vivo* Assessment of Antagonist Potency to Precipitate Withdrawal	435
5.1. Materials and supplies required	437
5.2. Models of acute and chronic opioid physical dependence	438
5.3. Acute physical dependence protocol	438
5.4. Chronic physical dependence protocol	439
5.5. Competition studies	440
6. Summary	440
Acknowledgments	441
References	441

---

* Department of Biomedical Sciences, College of Osteopathic Medicine, University of New England, Biddeford, Maine, USA
[†] Clinical Psychopharmacology Section, IRP/NIDA/NIH, Baltimore, Maryland, USA

Methods in Enzymology, Volume 484 © 2010 Elsevier Inc.
ISSN 0076-6879, DOI: 10.1016/S0076-6879(10)84021-2 All rights reserved.

Abstract

Constitutive (basal) signaling has been described and characterized for numerous G protein coupled receptors (GPCRs). The relevance of this activity to disease, drug discovery and development, and to clinical pharmacotherapy is just beginning to emerge. Opioid receptors were the first GPCR systems for which there was definitive evidence presented for constitutive activity, with numerous studies now published on the regulation of this activity (e.g., *structure/activity* of the receptor as it relates to basal activity, pharmacology of ligands that act as agonists, inverse agonists and "neutral antagonists," etc.). This chapter summarizes some of the methods used to characterize constitutive activity at the mu opioid receptor (MOR) in preclinical *in vitro* and *in vivo* model systems. This includes cell-based systems that are useful for higher throughput screening of novel ligands and for studying variables that can impact basal tone in a system. *In vivo* assays are also described in which constitutive activity is increased in response to acute or chronic opioid agonist exposure and where withdrawal is precipitated with antagonists that may function as inverse agonists or "neutral" antagonists. The methods described have inherent advantages and disadvantages that need to be considered in any drug discovery/development program. A brief discussion of progress toward understanding the clinical implications of MOR constitutive activity in the management of opioid addiction and chronic pain is also included in this chapter.

1. INTRODUCTION

Evidence for spontaneous (constitutive or basal) signaling of G protein coupled receptors (GPCRs) was first presented in 1989 (Costa and Herz, 1989). Using NG108-15 neuroblastoma–glioma cells that expressed delta opioid receptors, Costa and Herz were able to differentiate agonists versus inverse agonists by measuring GTPase activity. Critically, they identified a "neutral" antagonist (MR2266) that minimally affected GTPase activity by itself but was able to block both the agonist effects of DADLE and the inverse agonist effects of ICI174,864. As technologies developed for creating recombinant receptor systems, routinely manipulating receptor and signaling proteins to study structure/function, and rapidly screen thousands of ligands in high-throughput assays, the evidence accumulated for constitutive activity in many different GPCR systems including all three opioid receptors (Liu and Prather, 2001; Sadée *et al.*, 2005). The importance of considering constitutive activity in the drug discovery and development process is highlighted by several recent reviews (Aloyo *et al.*, 2009; Covel *et al.*, 2009; Lunn, 2010). In this chapter, we present some of the methodology that we have used *in vitro* and *in vivo* to study constitutive activity at MOR and screen for inverse agonist versus "neutral" antagonist profiles at MOR.

2. Measuring Opioid Receptor Constitutive Activity *In Vitro*

The opioid receptors, as members of the 7-transmembrane GPCR family, are coupled primarily to G proteins of the Gi/Go family and classically modulate the function of effector molecules, such as adenylate cyclase and protein kinases (Standifer and Pasternak, 1997; Williams et al., 2001). Agonist activation of opioid receptors *in vitro* is typically assessed with two functional assays: forskolin stimulation of cAMP accumulation and agonist stimulation of [^{35}S]-GTP-γ-S binding. Agonists inhibit forskolin-stimulated cAMP accumulation in intact cells that express opioid receptors. Constitutively active receptors should produce a decreased forskolin-stimulated cAMP accumulation under control conditions, and an increased naloxone-induced cAMP overshoot in dependent cells. Measurement of cAMP can be accomplished in a variety of ways, ranging from antibody-based methods to the original cAMP competition binding assay. Opioid agonists stimulate [^{35}S]-GTP-γ-S binding (see, e.g., Breivogel et al., 1997; Traynor and Nahorski, 1995), and constitutive receptor activity is detected both by increased basal [^{35}S]-GTP-γ-S binding and by the ability of inverse agonists to inhibit basal [^{35}S]-GTP-γ-S binding (for review, see Sadée et al., 2005). Opioid receptors, when expressed in cells, normally exhibit low to undetectable levels of constitutive activity. For these receptors, treating cells chronically with an agonist will induce constitutive activity (Wang et al., 2001, 2007). [^{35}S]-GTP-γ-S binding assays are typically conducted with membranes prepared from brain or cultured cells which stably express opioid receptors. In some cases, [^{35}S]-GTP-γ-S binding has been conducted with cells rendered permeable by detergents (Jordan and Devi, 1999). Given the widespread use of the [^{35}S]-GTP-γ-S binding assay, there are variations in the methods used by different laboratories. In the descriptions provided herein, we detail the methods employed by our lab, which are typical of those used by other labs.

2.1. Cell culture

The cAMP and [^{35}S]-GTP-γ-S assays use cells that express opioid receptors. This section describes the methods we use for preparing and maintaining these cells.

Tissue source: Chinese Hamster Ovary (CHO) cells that stably express the cloned human mu (hMOR-CHO), delta (hDOR-CHO), or kappa (hKOR-CHO) opioid receptors. These cells were obtained from Dr. Larry Toll, SRI International, and have been extensively used to characterize opioid ligands in both radioligand binding assays and [^{35}S]-GTP-γ-S binding assays (Toll et al., 1998). For many years, we used these cells for both radioligand binding and [^{35}S]-GTP-γ-S binding assays. Eventually, the hDOR-CHO cells suffered from decreased efficacy in the [^{35}S]-GTP-γ-S assay, so we now use the NG108-15 neuroblastoma × glioma cells, which express a native delta opioid receptor (Klee and Nirenberg, 1974), for the delta receptor [^{35}S]-GTP-γ-S assay. These can be obtained from the American Type Culture Collection (ATCC).

Growth medium supplies

- DMEM: 4500 mg/L D-glucose, L-glutamine, and 25 mM HEPES, no sodium phosphate or sodium pyruvate.
- DMEM/F-12: 1:1 Nutrient mixture (Ham) contains 15 mM HEPES buffer and L-glutamine.
- Trypsin–EDTA: 0.25% trypsin and 0.38 g/L ethylenediaminetetraacetic acid, tetrasodium salt in Hank's balanced salt solution without $CaCl_2$, $MgCl_2 \cdot 6H_2O$ and $MgSO_4 \cdot 7H_2O$; contains phenol red.
- Fetal bovine serum, heat inactivated.
- Hyclone FetalClone II (Thermo-Fisher).
- Geneticin selective antibiotic (G418 sulfate), powder; made in DMEM and sterile filtered; liquid stored in a sterile 15 ml polypropylene conical tube and kept refrigerated.
- HAT cell culture supplement.
- Phosphate-buffered saline (PBS), pH 7.4.

The cells are grown in 175 cm^2 cell culture flasks (USA Scientific) under 95% air/5% CO_2 at 37 °C. Culture medium and trypsin–EDTA should be warmed up in a 37 °C water bath before use. When the cells reach 80–90% confluency, 10 ml of trypsin–EDTA is added into a flask and allowed to incubate at room temperature until the cells come off the bottom. The cell suspension is then pipetted into a sterile 15 ml polypropylene conical tube along with 5 ml of DMEM. The cells are centrifuged in a Beckman GPR refrigerated (10 °C) centrifuge at 2000 rpm for 5 min. The supernatant is aspirated and the cell pellet is resuspended in 5 ml PBS and centrifuged at 2000 rpm for 5 min. For the [^{35}S]-GTP-γ-S assays, the supernatant is removed from the pellet and the tubes are placed in a −80 °C freezer until the day of the assay. For the cAMP assays, the cells are resuspended in growth medium and plated out in 24-well cell culture plates (0.5 ml/well), and incubated under 95% air/5% CO_2 at 37 °C. Cells should be confluent but not overgrown the day of the stimulation.

Disposables for cell culture

- 175 cm^2 cell culture flasks
- 1, 5, 10, and 25 ml sterile pipettes
- 15 ml sterile polypropylene conical test tubes

- 50 ml sterile polypropylene conical test tubes
- 24-well of cell culture plates from Costar (catalog no. 3524)

2.2. [^{35}S]-GTP-γ-S binding assay

2.2.1. Disposables

- Polystyrene 17 × 100 mm test tubes
- Polystyrene 12 × 75 mm test tubes
- 24-well NUNC plates (PerkinElmer LAS, Waltham, MA)
- Plateseal, permanent seal for microplates (PerkinElmer LAS)
- Whatman GF/B filters (Brandel, Gaithersburg, MD)

2.2.2. Reagents

- Trizma preset crystals, pH 7.0 and 7.4 (Sigma-Aldrich)
- NaCl
- $MgCl_2$ hexahydrate
- EDTA disodium salt
- Dithiothreitol (DTT: Cleland's reagent)
- Bovine serum albumin (BSA) minimum 98% purity
- Guanosine 5′-diphosphate, sodium salt, type III
- Guanosine 5′-[γ-thio]triphosphate, tetralithium salt
- [^{35}S]-Guanosine 5′(γ-thio)triphosphate (PerkinElmer LAS)
- [D-Ala^2,N-Me-Phe^4,Gly-ol^5]-enkephalin (DAMGO)
- (+)-4-[(αR)-α-((2S,5R)-4-allyl-2,5-dimethyl-1-piperazinyl)-3-methoxybenzyl]-N,N-diethylbenzamide (SNC80)
- (−)-trans-(1S,2S)-U-50488 hydrochloride ((−)U50488)
- Leupeptin: (acetyl-L-leucyl-L-leucyl-L-arginal hemisulfate monohydrate) (Peptides International)
- Chymostatin: A mixture of Types A, B, C. [(S)-1-carboxy-2-phenylethyl]carbamoyl-α-[2-iminohexahydro-4-(S)-pyrimidyl]-(S)-glycyl-X-phenylalaninal, X = L-Leu (Type A), L-Val (Type B), L-Ile (Type C) (Peptides International)
- Ubenimex (Bestatin): [(2S,3R)-3-amino-2-hydroxy-4-phenyl-butanoyl]-L-Leu (Peptides International)
- Bacitracin: 16.67G solid; 75,000 units/G solid
- CytoScint scintillation cocktail (MP Biomedicals)

2.2.3. [^{35}S]-GTP-γ-S binding assay: Detailed information

In this section, we explain in detail the methods and conditions that we use for our [^{35}S]-GTP-γ-S binding assay. This assay can be used to characterize agonist, inverse agonist, and antagonist compounds. The basic techniques and conditions are the same for all three assays; several minor variations on the agonist assay allow us to determine inverse agonist and antagonist activity.

Stimulation or inhibition curves are generated by displacing a single concentration of [^{35}S]-GTP-γ-S by 9 or 10 concentrations of a test compound or a standard. To determine agonist activity, stimulation curves with 9 concentrations of the compound of interest are generated. This curve is followed by a single concentration of a standard agonist which produces a maximum stimulation (1 μM DAMGO, 0.5 μM SNC80, or 0.5 μM (−)-U50488). Thus, the experimental compound's efficacy as an agonist can be determined by comparison with the standard, the standard's efficacy being 100%. As described in the Section 2.2.4 on data analysis, these curves are fit to a dose–response equation for the best-fit estimates of the EC_{50} and E_{MAX} values. Antagonist activity is determined using a "shift" paradigm similar in design to classic Schild analysis. In this design, 10-point dose–response curves of the appropriate standard (DAMGO, SNC80, or (−) U50,488) are generated in the absence and presence of a fixed concentration of the test compound. The shift in the EC_{50} value of the standard produced by the test agent is then used to calculate a functional Ki (Ke) value. Inverse agonist dose–response curves are generated in a manner similar to that used for an agonist, except that the response is inhibition of [^{35}S]-GTP-γ-S binding rather than stimulation of binding. As will be described below, inverse agonist activity is more readily detected when constitutively active receptors are generated by treating the cells chronically with an agonist (Sally et al., 2009).

As described in Section 2.1, cells that express the receptor of interest are grown and frozen as pellets to be used for the membrane protein in the binding assays. All cell membrane pellets remain frozen in a −80 °C freezer until the day of the assay.

Buffers are made fresh, in polypropylene beakers, on the day of the assay.

Buffer A: 50 mM Trizma 7.0 preset crystals, containing a protease inhibitor cocktail of 4 μg/ml leupeptin, 2 μg/ml chymostatin, 10 μg/ml bestatin, and 100 μg/ml bacitracin. Buffer A is placed on ice and is used for the cell membrane preparation. At this temperature, the pH is 7.4.

Buffer B: 50 mM Trizma 7.4 preset crystals containing 100 mM NaCl, 10 mM MgCl$_2$, and 1 mM EDTA. Since the EDTA is slow to go into solution, this buffer is allowed to spin for 15 min at room temperature. After the EDTA is completely dissolved, B is divided into two components.

Component 1 becomes binding buffer (BB) by the addition of 0.1% BSA. BB is used in the assay to prepare all standards, compounds to be tested, the nonspecific (NS) binding as well as for any other assay additions. BB is maintained at room temperature. The inclusion of BSA in this buffer is essential, since it limits the absorption of test compounds to the assay tubes. Although such "stickiness" may not be an issue with some compounds, it is a problem with other compounds. Since this is not generally known in advance, our experience has shown that including the BSA is a wise precaution.

Component 2 becomes the tissue buffer (TB) by the addition 1.67 mM DTT and 0.15% BSA. BB is maintained at room temperature. The reader should note that although some labs (see, e.g., Traynor et al., 2002), like ours, typically use DTT in our assays, other investigators do not (Wang et al., 2007). Recent unpublished experiments conducted in our lab show that, for the hMOR-CHO cells, a higher degree of agonist-stimulated [^{35}S]-GTP-γ-S is achieved in the absence of DTT.

Wash buffer (WB) (ice-cold (4 °C) 10 mM Trizma 7.0 preset crystals) is used in the last step of the assay to wash the Whatman GF/B filters.

The assay setup proceeds as follows. Compounds to be tested are serially diluted in BB from a 10 mM stock (made in DMSO) into 9 or 10 concentrations in 17 × 100 mm polystyrene test tubes. An appropriate volume of the 10 mM stock is added to the first tube of the dilution curve. The final concentration of test drug in these tubes is 10 times the final intended concentration. A new pipette tip is used for each serial dilution. GDP is likewise prepared from a 100 mM stock (made in BB) to yield a final concentration of 40 μM. NS binding is determined using a final concentration of 40 μM GTP-γ-S also made in BB. The reader should note that the concentration of GDP should be determined when first setting up the assay with a particular cell line. For this type of experiment, agonist-stimulated [^{35}S]-GTP-γ-S binding is determined in the absence and presence of increasing concentrations of GDP. The concentration of GDP that leads to the highest degree of agonist-induced stimulation is then selected for general use (see Traynor and Nahorski, 1995).

The [^{35}S]-GTP-γ-S is purchased as a fresh lot from PerkinElmer Life Sciences. We receive 250 μCi, which is thawed and diluted into 2 ml of buffer containing 10 mM Tris–HCl, pH 7.4, and 10 mM DTT. The Tris–DTT dilution buffer may be made ahead, frozen in a 50 ml polypropylene conical, thawed when needed, and refrozen. Once the dilution is made, 50 μl of the isotope dilutant is then aliquoted into 0.5 ml polypropylene microcentrifuge tubes and frozen in a −80 °C freezer. On the day of the assay, an appropriate number of microcentrifuge tubes are thawed and enough isotope is added to a volume of BB to produce 50,000 cpm (final concentration ∼ 50 pM) per test tube. As the isotope decays, more must be added to ensure that there is 50,000 cpm per test tube. The microfuge tubes containing any unused diluted isotope are not refrozen.

Membranes are prepared on the day of the assay as follows. An appropriate number of tubes of frozen cell pellets are removed from the −80 °C freezer and thawed on ice for 15 min. Each pellet typically provides enough protein for 100 assay tubes. One milliliter of ice-cold buffer A is then added to each tube. Using a 1 ml pipettor, along with vortexing, the pellet is resuspended into the buffer. The membrane suspension is then placed into 10 ml of ice-cold buffer A, homogenized using a polytron (Brinkman Instruments Co.) at setting 6 for 10 s and centrifuged at 10 °C at 30,000 × g for 10 min. After centrifugation, the pellet is resuspended into an appropriate amount of 25 °C TB and used in the binding assay. If using multiple tubes of frozen membrane, the final pellets are pooled. The reader should note that the actual ml of TB used to resuspend the pellets is determined by preliminary experiments in which the protein concentration is varied. We select a concentration that will provide $\sim$1000 cpm of "total" binding which, with the cells we use, is $\sim$50 μg/tube.

Each dose–response curve consists of 36 test tubes (12 conditions × 3 (triplicate samples) = 36). For agonist/inverse agonist curves, the curve setup is as follows: total binding (position 1), drug concentrations (position 2–10), standard (position 11), and NS binding (position 12). This design tests a nine-point dose–response curve. For shift experiments, the curve setup is as follows: total binding (position 1), 10 concentrations of the standard (position 2–11), and NS binding (position 12).

The assay components are pipetted, in five steps, into 12 × 75 mm polystyrene test tubes in the following order. (1) The total tubes (position 1) receive 50 μl of BB. For agonist/inverse agonist determinations, 50 μl of the compound to be tested is pipetted in increasing concentration into the next 9 positions, followed by 50 μl of a single concentration of the standard agonist into the position 11 tubes. The position 12 tubes receive 50 μl of NS solution. We typically test eight drugs in a single assay. (2) Each tube receives 50 μl of BB. (3) Every tube receives 50 μl of GDP. (4) Every tube receives 50 μl of [^{35}S]-GTP-γ-S. (5) The assay is initiated by the addition of 300 μl (50 μg of protein) of membrane/tube and then each test tube is vortexed. The final volume in each assay tube is 500 μl. After a 2–3 h incubation at 25 °C, bound and free [^{35}S]-GTP-γ-S are separated by vacuum filtration using a cell harvester (Brandel) using Whatman GF/B filters which are presoaked in WB. The test tubes are washed three times in ice-cold WB. The filters are then punched into 24-well plates, to which 0.6 ml of liquid scintillation cocktail is added. After an overnight extraction, the samples are counted in a Trilux liquid scintillation counter.

For shift experiments, the setup is similar, except that the standard compound is tested as 10-point dose–response curve, and that two curves comprise the experimental design. The first curve is the dose–response curve of the standard, and the second curve is the dose–response curve of the standard to which is added, in step 2, 50 μl of the test compound in BB. The concentration of the test compound must be chosen on the basis of preliminary experiments. We generally use a concentration equal to five times the Ki of the test compound as measured in an opioid binding assay.

Inverse agonist dose–response curves are setup as described for agonist curves, except that the standard compound is a compound known to produce a robust inverse agonist-mediated inhibition of [^{35}S]-GTP-γ-S binding.

Figures 21.1 and 21.2 provide examples of the use of these assays to measure constitutively active mu opioid receptors. For the experiments described in Fig. 21.1 (published in Xu et al., 2007), hMOR-CHO cells were incubated in fresh medium in the absence or presence of 10 μM drug (DAMGO [Tyr-D-Ala-Gly-N-Me-Phe-Gly-ol], or herkinorin [HERK, (2S,4aR,6aR,7R,9S,10aS,10bR)-9-(Benzoyloxy)-2-(3-furanyl)dodeca-hydro-6a,10b-dimethyl-4,10-dioxo-2H-naphtho-[2,1-c]pyran-7-carbox-ylic acid methyl ester]) for 20 h. DAMGO is a highly selective mu opioid agonist and HERK is a salvinorin A analog that is a selective mu agonist (Harding et al., 2005). Cells were washed three times with PBS and processed for the [^{35}S]-GTP-γ-S binding assay. The results (Fig. 21.1) showed that treatment with HERK increased basal [^{35}S]-GTP-γ-S binding by about 40%. Moreover, the addition of the mu opioid receptor antagonists CTAP and naloxone decreased the elevated basal [^{35}S]-GTP-γ-S binding to ~25% below baseline. These data demonstrate that treating cells with HERK generates constitutively active mu opioid receptors, and that, under these conditions, classic mu antagonists demonstrate inverse agonist activity. In contrast, naloxone and CTAP did not decrease basal

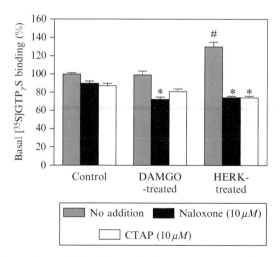

Figure 21.1 Comparison of the effects of naloxone (10 μM) and CTAP (10 μM) on basal [^{35}S]GTP-γ-S binding in the control or pretreated for (20 h) hMOR-CHO cells. Results are presented as mean ± SEM ($n = 3$). ★$p < 0.01$ when compared with no-addition group. #$p < 0.01$ when compared with no-addition condition of the control cells (two-tailed Students t-test). Taken from Xu et al. (2007).

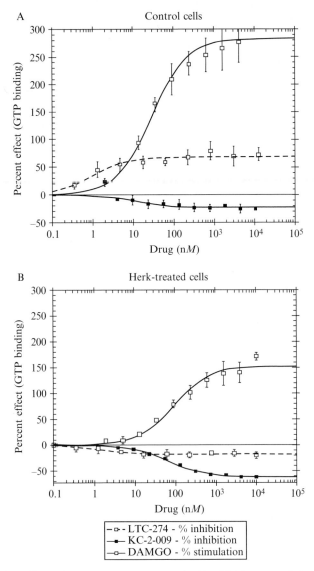

Figure 21.2 Effect of DAMGO, HERK, and LTC-274 in control (panel A) and HERK-treated cells (panel B). Each value is the mean ± SD ($n = 3$). The EC_{50} and E_{MAX} values are reported in the Section 2.2.3. Taken from Sally et al. (2009).

[^{35}S]-GTP-γ-S binding in control cells. Treating cells with DAMGO did not significantly increase basal [^{35}S]-GTP-γ-S binding, but naloxone exhibited inverse agonist activity, suggesting that DAMGO treatment does produces fewer constitutively active mu receptors than HERK.

Figure 21.2 (published in Sally et al., 2009) provides a useful example of agonist and inverse agonist dose–response curves. Panel A, which used control hMOR-CHO cells, shows that DAMGO produced a dose-dependent stimulation of [^{35}S]-GTP-γ-S binding with an EC$_{50}$ value of 28 $\pm$ 6 nM and an E_{MAX} of 283$\pm$ 10%. LTC-274 ((2)-3-cyclopropylmethyl-2,3,4,4aα,5,6,7,7aα-octahydro-1H-benzofuro[3,2-e]isoquinolin-9-ol) was a very low efficacy agonist, with an E_{MAX} value (34 $\pm$ 2%) 9.4% that of DAMGO. KC-2-009 ((1)-3-((1R,5S)-2-((Z)-3-Phenylallyl)-2-azabicyclo [3.3.1]nonan-5-yl)phenol hydrochloride) acted as an inverse agonist (EC$_{50}$ = 12 $\pm$ 4 nM, E_{MAX} = $-$24 $\pm$ 1.3%). As reported in Fig. 21.2B, treating the cells with 10 μM HERK for 20 h shifted the DAMGO dose–response curve to the right (EC$_{50}$ = 94 $\pm$ 17 nM) and decreased the E_{MAX} to 152 $\pm$ 6%, a finding consistent with the development of tolerance. HERK treatment substantially increased the inverse agonist efficacy of KC-2-009 but decreased the potency (EC$_{50}$ = 67 $\pm$ 5 nM, E_{MAX} = $-$63 $\pm$ 1.0%). HERK treatment also revealed inverse agonist activity for LTC-274, but the efficacy (E_{MAX} = 18 $\pm$ 1.0%) was much reduced as compared with KC-2-009.

Figure 21.3 (published in Sally et al., 2009) provides a useful example of shift experiments. These experiments used HERK-treated cells. Panel A reports dynorphin A(1–13) dose–response curves conducted in the absence and presence of 1 and 5 nM LTC-274. The dynorphin A(1–13) dose–response curves shifted progressively to the right with increasing concentration of LTC-274. This yielded a Ke value of $\sim$0.2 nM for LTC-274. Panel B reports CTAP (D-Phe-Cys-Tyr-D-Trp-Arg-Thr-Pen-Thr-NH$_2$) dose–response curves conducted in the absence and presence of 1 and 5 nM LTC-274. In this case, CTAP, a mu-selective antagonist, acts as an inverse agonist, inhibiting [^{35}S]-GTP-γ-S binding. The CTAP dose–response curves shifted progressively to the right with increasing concentration of LTC-274, yielding a Ke value similar to that determined with dynorphin A(1–13).

2.2.4. Data analysis used for the [^{35}S]-GTP-γ-S binding assay

As noted in Section 2.2.3, each data point is assayed in triplicate. The first step is to calculate the mean and SD of these points. The NS binding is then subtracted from the total binding and other data points to yield specific binding values. The percent stimulation of [^{35}S]-GTP-γ-S binding is calculated according to the following formula: $(S-B)/B \times 100$, where B is the basal level of [^{35}S]GTP-γ-S binding (total—NS) and S is the stimulated level of [^{35}S]GTP-γ-S binding observed in the presence of test agents. The data of several experiments, typically three, are pooled. In most cases, the stimulation produced by an agonist is expressed as a percent of the stimulation produced by the standard. Inverse agonist dose–response curves are similarly analyzed, except that percent inhibition of basal binding is used instead of percent stimulation. The EC$_{50}$ and E_{MAX} values are determined using the program MLAB-PC (Civilized Software, Bethesda, MD),

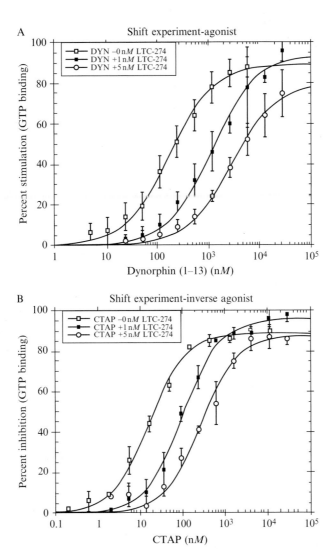

Figure 21.3 Representative "shift experiments" in membranes prepared from HERK-treated cells with an agonist (dynorphin A(1–13), panel A) and an inverse agonist (CTAP, panel B). Each value is the mean ± SD ($n = 4$ for panel A, $n = 3$ for panel B). Taken from Sally *et al.* (2009).

KaleidaGraph (Version 3.6.4, Synergy Software, Reading, PA), or Prism 4.0 (GraphPad Software, Inc., San Diego, CA), using a standard dose–response curve equation: $EC_{50} = E_{MAX} \times ([DRUG]/(EC_{50} + [DRUG]))$. For determination of antagonist Ke values, we used the "shift" experimental design. Agonist or inverse agonist dose–response curves are generated, in the absence and presence of an antagonist. The data of several experiments, typically three, are pooled, and the Ke values were calculated according to the equation: [antagonist]/(EC_{50-2}/EC_{50-1} − 1), where EC_{50-2} is the EC_{50} value in the presence of the antagonist and EC_{50-1} is the EC_{50} value in the absence of the antagonist.

3. cAMP Quantification Assay in CHO Cells Expressing Cloned Opioid Receptors

Opioid-induced adenylyl cyclase superactivation and agonist-mediated inhibition of forskolin-stimulated cAMP accumulation in intact CHO cells that express cloned opioid receptors can be measured in a variety of ways, such as cAMP cell-based assay kits based on competitive immunoassay methods (Xu et al., 2007) and original [^{3}H]cAMP competitive binding assay based on the competition between unlabelled cAMP and radiolabeled cAMP. In this part, we provide the optimal conditions and procedures for measuring the level of cellular cAMP using original [^{3}H]cAMP competitive binding assay. These assay monitored inhibition of [^{3}H]cAMP binding to cAMP-dependent protein kinase (cAMP binding protein). The procedures used followed published procedures with minor modifications (Gray et al., 2001; Nordstedt and Fredholm, 1990; Xu et al., 2004).

3.1. Materials and equipment required

- [^{3}H]cAMP–adenosine 3'5'-cyclic phosphate, [2,8-^{3}H] (SA = 43 Ci/mmol) was obtained from PerkinElmer Life and Analytical Sciences, Inc.
- Forskolin (F-6886), adenosine 3'5'-cyclic monophosphate sodium (cAMP, A-6885), 3-isobutyl-1-methylxanthine (IBMX, I-5879) and protein kinase, 3'5'-cyclic-AMP-dependent (P 5511) were obtained from Sigma Chemical (St. Louis, MO). 0.1 N HCl (80-0080) purchased from Assay Designs (Ann Arbor, Michigan)
- Prepare various growth medium for CHO cells expressing cloned μ, δ, and κ receptors, respectively
- For opioid dimer cells (cMyc-mδ-HμCHO cells), the following growth medium are used: F-12 nutrient mixture (HAM, GIBCO) containing 10% fetal bovine serum, 100 units/ml penicillin, 100 μg/ml streptomycin, 400 μg/ml hygromycin B (for δ receptor selection), and 400 μg/ml geneticin (G-418, for μ receptor selection)
- All medium and trypsin–EDTA need to be warmed up (at 37 °C water bath) before use
- T175 or T75 of cell culture flasks from USA Scientific and 24-well of cell culture plates from Costar (catalog no. 3524)

- Disposable polystyrene binding tubes (12 × 75 mm)
- Disposable 1.5 ml microcentrifuge tubes
- Disposable 50 ml of conical centrifuge tubes
- Vortex mixer
- Micro refrigerated centrifuge [TOMY, MTX-150]
- Scintillation cocktail (LSC-cocktail (CytoScint))
- 24-well sample plates suitable for counting in microplate format scintillation counter [1450 Microbeta Trilux Liquid scintillation and Luminescence Counter]

3.2. Stimulation

On the day of the experiment, cells in 24-well cell culture plates (0.5 ml/well), control or drug-treated, were washed by aspiration of the growth medium and the addition of 0.5 ml of serum-free medium [F-12 Nutrient Mixture (HAM, GIBCO)]. After two washes, cells were incubated with serum-free medium, containing IBMX (500 µM) for 20 min at 37 °C. After a 20-min incubation, medium was removed and then cells incubated with fresh serum-free medium containing IBMX (500 µM) and forskolin (100 µM) and appropriate agonist or antagonist for 15 min at 37 °C. The reaction was terminated at RT by aspiration of the medium and the addition of 0.5 ml of 0.1 N HCl. After chilling plates at 4 °C for at least 1 h, 0.4 ml was removed into 1.5 ml Eppendorf tubes, neutralized with 1 N NaOH, vortexed, and centrifuged in Micro refrigerated centrifuge [TOMY, MTX-150] at 13,000 rpm for 5 min (4 °C). Supernatants (i.e., neutralized HCl extracts, pH = 7.0) were used for cAMP assay.

3.3. [^{3}H]cAMP competition binding assay

cAMP binding buffer

- 50 mM Tris–HCl, pH 7.4, at RT
- 100 mM NaCl
- 5 mM EDTA (ethylenediaminetetraacetic acid disodium salt)

Steps for making buffer

1. Make serial dilutions in cAMP BB to prepare standards cAMP: 256, 128, 64, 32, 16, 8, 4, 2, 1, 0.5, 0.25, 0.125 pmol/50 µl. The highest concentration of standard cAMP (256 pmol/50 µl) was made by pipetting 12.8 µl of 10 mM cAMP into 25 ml of cAMP BB, mix well, and then make serial twofold dilutions with cAMP BB. A new pipette tip is used for each serial dilution.

2. Prepare [^{3}H]cAMP (NEN Life Sciences #NET275; 25–40 Ci/mmol; 1 µCi/µl). Dilute in cAMP BB to yield 30,000–40,000 cpm in 50 µL.
3. Add 50 µl of cold standard cAMP/or unknown sample to each binding tube. Each data point is determined in triplicate.
4. Add 50 µl of [^{3}H]cAMP to each binding tube.
5. Prepare diluted cAMP binding protein [from 50 µl of cAMP binding protein stock] using cAMP BB. The cAMP binding protein stock was made by adding 1.25 ml of distilled water into 10 mg of cAMP binding protein (Sigma P-5511), mix well, aliquot (50 µl) into Eppendorf tubes and store at $-80\ ^\circ$C for use.
6. Add 400 µl of diluted cAMP binding protein (~ 0.020–0.025 mg/ml) to each binding tube. Total volume in binding tube = 500 µl.
7. Shake tubes gently to mix and incubate 2–3 h at 4 $^\circ$C (protected from light). Two hours is usually sufficient.
8. Harvest with a Brandel cell harvester onto Whatman GF/B filters, 3× washes with ice-cold 10 mM Tris–HCl, pH 7.0. Place filters into 24-well sample plates and add 0.6 ml of scintillation cocktail. The plates are sealed and counted after an overnight extraction in a Trilux liquid scintillation counter.

3.4. Data analysis using GraphPad prism

The amount of cAMP in the samples is quantified from the cAMP standard curve ranging from 0.25 to 256 pmol of cAMP/assay tube. Forskolin (100 µM) stimulated cAMP formation in the absence of agonist is defined as 100%. The EC_{50} (the concentration of agonist that produces 50% inhibition of forskolin-stimulated cAMP formation) and E_{MAX} (percentage of maximal inhibition of forskolin-stimulated cAMP by agonists) were calculated using Prism 4 (GraphPad Software). A cAMP standard curve ranging from 0.25 to 256 pmol of cAMP/assay is generated by entering the standard cAMP concentrations (0.25–256 pmol) in the X column. The mean of the three binding cpm values for each data point is entered into the Y column. The X values are transformed using $X = \log [X]$. One then selects new analysis, nonlinear regression, and the dose–response equation. Figure 21.4 shows a typical cAMP standard curve.

3.5. Examples of using the cAMP assay to measure constitutively active mu opioid receptors

In the experiments described in Fig. 21.5, hMOR-CHO cells were pretreated for 20 h with no drug (control), DAMGO (10 µM), or HERK (10 µM) and the degree of forskolin-stimulated cAMP accumulation determined in the absence and presence of naloxone (10 µM) and CTAP

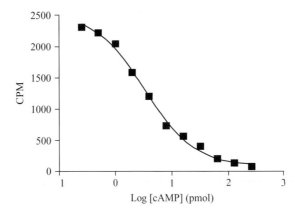

Figure 21.4 Sample cAMP standard curve.

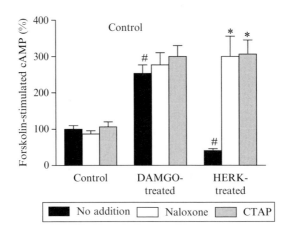

Figure 21.5 Comparison of the effects of naloxone (10 μM) and CTAP (10 μM) on forskolin-stimulated cAMP accumulation in the control or pretreated (20 h) hMOR-CHO cells. Results are presented as mean ± SEM (n=3). $^{\star}p < 0.01$ when compared with no-addition group. $^{\#}p < 0.01$ when compared with no-addition condition of the control cells (two-tailed Students t-test). Taken from Xu et al. (2007).

(10 μM). HERK treatment substantially reduced forskolin-stimulated cAMP accumulation, consistent with the generation of constitutively active mu receptors, which inhibit adenylyl cyclase activity in the absence of an agonist. The addition of either naloxone or CTAP inhibits the constitutively active mu receptors, relieving inhibition of adenylyl cyclase activity, thereby restoring the ability of forskolin to stimulate cAMP accumulation, and suggesting their inverse agonist effect at a basal (i.e., constitutively)

active mu receptor. The magnitude of forskolin-stimulated cAMP accumulation in the presence of naloxone is enhanced by both DAMGO and HERK treatment, an effect commonly called cAMP superactivation (Waldhoer et al., 2004).

Figure 21.6 provides examples of dose–response curves generated with the cAMP assay. Chronic DAMGO increased the EC_{50} values of both

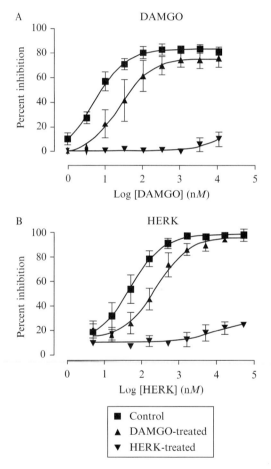

Figure 21.6 Drug-inhibition of forskolin-stimulated cAMP accumulation. Panel A: DAMGO dose–response curves were generated using membranes prepared from hMOR-CHO cells after 20 h treatment with DAMGO (10 μM), HERK (10 μM), or medium. The data were analyzed for the best-fit estimates of the E_{MAX} and ED_{50}. Each value is ±SEM ($n = 3$). Panel B: HERK dose–response curves were generated using membranes prepared from hMOR-CHO cells after 20 h treatment with DAMGO (10 μM), HERK (10 μM), or medium. The data were analyzed for the best-fit estimates of the E_{MAX} and ED_{50}. Each value is ±SEM ($n = 3$). Taken from Xu et al. (2007).

DAMGO (3.2-fold) and HERK (4.8-fold) for inhibiting forskolin-stimulated cAMP accumulation. Chronic HERK eliminated the ability of both DAMGO and HERK to inhibit forskolin-stimulated cAMP, reflecting the fact that chronic HERK reduced forskolin-stimulated cAMP accumulation by 60% (Fig. 21.4). The cAMP dose–response data suggest that chronic DAMGO produces auto- and cross-tolerance to opioid agonists.

4. *In Vivo* Assessment of Antagonist Potency in Opioid Naïve Subjects

As with characterization of drug effects in any *in vivo* system, there are a host of variables that can impact the behavior of interest and potentially confound interpretation of results. Opioid pharmacology has been extensively studied in rodents (primarily rats and mice) and in general these species serve as good models for predicting human pharmacology as it relates to analgesia, abuse potential, and many other classic opioid side effects. Standardization and strict adherence to protocols is essential for limiting variability, as is running positive and negative controls, completing full dose–response curves and having a reasonable handle on pharmacokinetic variables.

Our laboratories have completed most of the *in vivo* pharmacology of opioid constitutive activity and "neutral" antagonists using young–adult male ICR mice. This strain of mice is one of the more commonly used outbred strains and has a number of desirable characteristics related to measurement of MOR-mediated antinociception, physical dependence/withdrawal, and other opioid-mediated side effects. As described above, constitutive activity at the MOR is typically low in opioid naïve cell-based and *in vivo* test subjects. This likely explains the tolerability of naloxone and naltrexone in opioid-naïve humans and lack of gross behavioral effects of these antagonists in opioid-naïve rodents and nonhuman primates. Acute or chronic exposure to MOR agonists (e.g., morphine) significantly increases the level of constitutive activity and allows for enhanced sensitivity to detect inverse agonists and protean ligands such as naloxone and naltrexone (Sadée *et al.*, 2005). Either acute or chronic exposure paradigms can be used to increase constitutive activity, with higher dosing exposures typically being associated with greater increases in basal signaling and a stronger response to inverse agonists (Bilsky *et al.*, 1996; Raehal *et al.*, 2005). Variables such as strain, sex, age, and time of testing can impact the level of physical dependence and the expression and severity of withdrawal (e.g., Kest *et al.*, 2002).

With the MOR system, we were initially interested in modeling observations being made in cell-based systems where basal signaling could be detected in the absence of an agonist ligand, and in which naloxone and naltrexone exhibited negative intrinsic activity (Arden *et al.*, 1995; Wang *et al.*, 1994). Several predictions were made with respect to varying the level of opioid exposure (dose and time), the potency, and efficacy of the antagonist to precipitate withdrawal and the ability of a "neutral" antagonist to attenuate or prevent the actions of the inverse agonist. It was first necessary to generate full antagonist dose– and time–response curves in naïve subjects and to then compare these potencies to the potency of the ligand to precipitate withdrawal in opioid exposed subjects.

4.1. Materials and supplies required

A general list of reagents and supplies is provided below, with the understanding that variations on these protocols are possible.

- Mice: Male ICR mice (25–35 g) from Harlan (Indianapolis, IN); mice are allowed to acclimate to the vivarium for a minimum of 5 days before formal testing procedures are initiated. Mice, unlike rats, are typically not handled during this habituation process. Housing is under controlled conditions (12 h light:dark cycle, room temperature 20–22 °C, relative humidity of 30–70%)
- Syringes and needles (1-cm^3 syringes and 30-gauge needles for mice)
- Morphine sulfate (Mallinckrodt, St. Louis MO) or other suitable full MOR agonist
- Opioid ligands that vary in their intrinsic activity (inverse agonist, "neutral" antagonist, or partial agonist). Examples include naloxone, naltrexone, nalmefene, nalbuphine, various derivatives of naltrexol and naloxol, CTAP, and CTOP. Many of these compounds can be obtained through the NIDA Drug Supply Program (Bethesda, MD)
- Acute antinociceptive assays sensitive to MOR agonists (e.g., 55 °C tail-flick using a Neslab recirculating water bath); alternatives include the radiant heat tail-flick or the 55 °C hot-plate assay
- Additional behavioral assays that can detect MOR agonist activity (e.g., stereotypical circling behavior in an open field, inhibition of gastrointestinal transit, Straub tail, etc.)

4.2. Determination of antagonist potency and time of peak effect

MOR inverse agonists, "neutral" antagonists, and partial agonists can all precipitate withdrawal in a highly dependent animal (as can removal of the agonist from the system). It is important to accurately measure the potency of each ligand of interest with respect to blocking various opioid agonist effects (peripheral and/or central) in rodents not previously exposed to opioid agonists. Pharmacokinetic factors (time to peak effect, duration of action, etc.) vary widely among both the opioid agonists and antagonists. Determining these parameters is straightforward as the example with the 55 °C tail-flick illustrates below.

Sample protocol to determine antagonist potencies.

1. All mice should be weighed and marked (different colored permanent marker on base of the tail), with treatments assigned randomly and the investigator blinded to the actual treatments.
2. Baseline groups of mice (8–10/group) for tail-withdrawal latencies using the 55 °C tail-flick. Group baselines should average ~ 2 s.
3. Pretreat mice with doses of the test compound that are expected to produce between 20% and 80% inhibition of a CNS mediated effect. Most CNS penetrating opioid antagonists will have rapid onsets of action with peak effects reached between 10 and 30 min postinjection. The time to peak effect of the agonist also needs to be factored into the timing of drug administration (see below). Several different routes of injection can be explored with intravenous (i.v.) and oral (p.o.) administration being the most relevant to drug discovery/development and subcutaneous (s.c.) and intraperitoneal (i.p.) having some practical advantages. Direct injections into the CNS are also sometimes employed including intracerebroventricular (i.c.v.) and intrathecal (i.th.).
4. Administer an $\sim A_{90}$ dose of a MOR agonist and then test for antinociception at the time of peak agonist effect. Morphine is a good choice for an agonist given the extensive preclinical and clinical literature on the compound, its MOR dominant actions at doses typically administered for antinociceptive studies, and characteristics that make it an ideal agent for induction of opioid physical dependence (reasonable duration of action, general tolerability of high dose exposure, and availability of morphine implant pellets). The goal of the two injections is to have the times of antagonist and agonist peak effects line up at the time the tail-flick test is performed. There are always tradeoffs when working with two or more drugs. The administration of i.p. or s.c. derivatives of naloxone and naltrexone typically involve pretreatments in the 10–20 min ranged followed by morphine administration at $t = 0$ and the tail-flick test at $t = 20$ min.
5. Once an initial 1–2 groups have been run, additional doses of the antagonist can be adjusted up or down with the goal of generating a full antagonist dose–response curve (Fig. 21.7). Adjustments in ½ log or ¼ log increments work well, and the goal is to have 3–4 doses of antagonist that result in between 20% and 80% inhibition of the agonist effect.
6. The percentage inhibition of the morphine effect is determined for each mouse, and an ID_{50} value (and 95% confidence interval) is calculated using linear regression (e.g., FlashCalc software; Dr. Michael Ossipov, University of Arizona, Tucson, AZ).

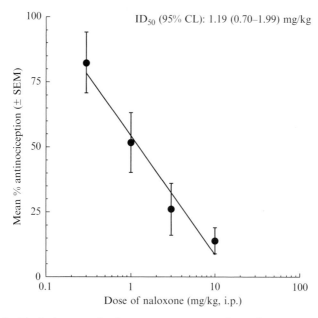

Figure 21.7 Typical antagonist dose–response curve of morphine antinociception in the 55 °C tail-flick assay. An A_{90} dose of morphine was administered to mice pretreated with i.p. doses of the general opioid antagonist naloxone. The ID_{50} value (inhibitory dose 50%) is calculated using linear regression software to provide an estimate of the potency of naloxone.

Variation #1 on the protocol to determine antagonist peak effect and duration of action:

1. Mice are marked, weighed, and baselined and then given a fixed dose of antagonist that produces ∼50% blockade of the agonist effect (determined in the initial dose-finding study above).
2. The antagonist pretreatment time is varied while the dose of antagonist and agonist are kept constant (as is the time of agonist injection and assessment of tail-withdrawal latency). The pretreatment intervals and number of groups run will depend on the antagonist being tested and the degree of precision desired.
3. Visual inspection of the data will typically indicate a pretreatment time that produces a maximum level of blockade. The choice of an initial intermediate dose of antagonist (i.e., ∼50% inhibition) provides the greatest sensitivity for detecting further increases or decreases in blockade. The data can also be analyzed to estimate the duration of blockade. In this case, a dose of antagonist that produces near maximal blockade can be administered (∼90%) and the level of blockade is tracked out to further time points by varying pretreatment time (keeping the agonist dose and tail-flick test point constant).

Variation #2 on the protocol to determine antagonist peak effect and duration of action:

The temporal resolution of the tail-flick test is limited in practice to ∼5 min intervals, and the assay also becomes labor and resource intensive if multiple drugs and time points are going to be tested. As an alternative, an automated open-field locomotor assay can be used to quantify morphine-induced hyperlocomotion and level of antagonist blockade. The Coulbourn Instruments (Allentown, PA) monitoring system with TruScan software allows data collection points every 100 ms, with 1-min summed bins of data providing reliable indications of agonist and antagonist drug effects (Osborn et al., 2010; Raehal et al., 2005). Importantly, the antagonist/agonist injections can be timed to determine potency to prevent or reverse the agonist effect (or be coadministered). Depending on the agonist used, the drug interactions can be tracked in the same animal over various periods of time. Morphine produces a more broad and longer duration effect whereas fentanyl has a very rapid onset and offset of effect.

1. Locomotor activity behaviors are measured using an activity monitoring system (Coulbourn Instruments) and TruScan software. Chambers consist of a 10-in. width × 10-in. length × 16-in. height arena surrounded by Plexiglas walls. Up to 10 chambers can be hooked up to a single card for high-throughput readouts. The floors consist of a removable plastic drop pan that is cleaned between sessions. A sensor ring surrounds the arena on the bottom outside edges of the four sides of the chamber and contains 16 infrared beams that transected both the length and the width of the chambers on all four sides (beam spacing = 0.6 in., resolution of 32 × 32 squares). Measurement of the animals' position is determined every 100 ms, and the software can calculate a number of parameters related to aspects of locomotor activity. For opioid agonists, the total distance traveled by each mouse is used as the primary measure of activity, with the data summed into evenly space intervals or "bins" (1- or 5-min intervals work well).

2. Mice are typically habituated to the chamber for 30 min, the program paused, and mice are then administered injections within a 1-min time span and placed immediately back into the chambers and monitored for an additional 30–90 min. Various routes of injections can be used, with i.p. and s.c. offering the quickest turnaround time for removal from the chamber, injection, and placement back into the chamber. Morphine (s.c.) produces a broad area of robust and stable hyperlocomotion. Full dose–response curves can be generated for the agonist alone and for a fixed agonist dose and varying antagonist doses (Fig. 21.8). A robust agonist effect is typically desired to provide plenty of separation between the agonist effect and the baseline level of activity following vehicle administration.

3. The data can be analyzed in a variety of ways including looking at time effects across bins and summing the data into area under the curve analyses. In one set of experiments (Raehal et al., 2005), the time period from 20 to 50 min after the injections was chosen for analysis, as this was a time when morphine was reaching its peak effect and when all antagonists were producing maximal blockade. For the calculation of ID_{50} values, data were expressed as a percentage of controls. The control-morphine effect was calculated by taking the mean distance traveled by the morphine control and subtracting the mean distance traveled by the vehicle control. The data for each individual mouse receiving a dose of the test compound was then expressed as a percentage of this control. ID_{50} values (and 95% confidence interval) were calculated using linear regression (FlashCalc software).
4. The onset of CNS blockade could, in theory, impact the severity of withdrawal in opioid-dependent subjects. A further variation of the locomotor protocol can be used to estimate the functional pharmacokinetic parameters of onset of antagonist effect (i.e., reversal of an established morphine effect). The total distance traveled is summed into 1-min intervals to give better temporal resolution of the drug effects. Following a 30-min habituation, mice receive an injection of morphine sulfate (30 mg/kg s.c.) and are immediately placed back into the chambers for an additional 30 min. Mice then received an i.p. injection of vehicle or equi-effective doses of various antagonists and placed back into the chambers for an additional 60 min of monitoring (Osborn et al., 2010; Raehal et al., 2005). Of the naloxone and naltrexone related compounds we have tested, the majority of them reduce the morphine effect by $\sim$50% within 3–5 min and produce near maximal blockade within 10 min.

More involved methods are available to estimate antagonist potency including *in vivo* apparent pA2 analysis (Walker et al., 1994; Woods et al., 1992). An example of this approach using naltrexone analogs is depicted in Fig. 21.9. Increasing doses of antagonist produces successive rightward shifts in the agonist dose–response curves. Antagonist potency estimates for the various compounds tested are in line with those obtained when the agonist dose is fixed ($\sim A_{90}$) and the antagonist dose is varied (unpublished results).

5. *In Vivo* Assessment of Antagonist Potency to Precipitate Withdrawal

Methods of Section 4. determined the potency of each test compound to block an acute CNS mediated effect of morphine (or similar MOR agonist). Additional *in vitro* and *in vivo* testing can be employed to rule out

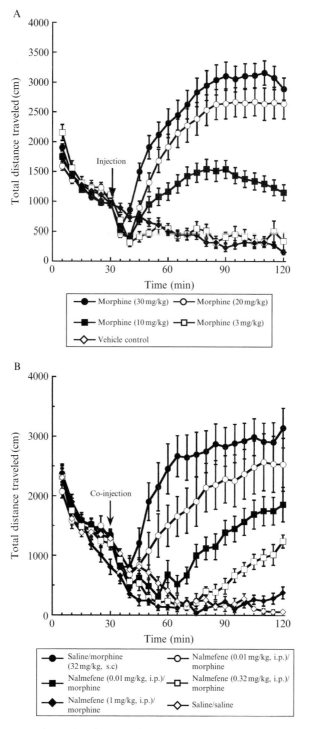

Figure 21.8 Dose-dependent hyperlocomotion with morphine alone (panel A) or in combination of increasing doses of the opioid antagonist nalmefene. Nalmefene produced a dose-related inhibition of the morphine effect (increase in distance traveled in 5-min bins). Injections were administered after a 30-min habituation to the chambers and animals were monitored for an additional 90 min postinjection.

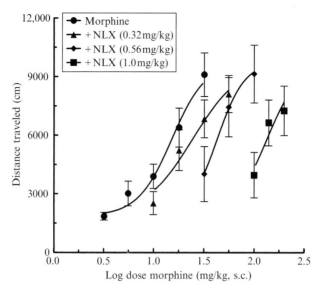

Figure 21.9 Sample data for generation of an *in vivo* apparent pA2 analysis of morphine interactions with naloxone. Note the parallel rightward shifts in the morphine dose–response curve with increasing doses of naloxone. Morphine-induced hyperlocomotion was used as the behavioral endpoint.

any partial agonist effects of the ligands if necessary. The next step in characterization is to determine the potency and efficacy of the drugs to precipitate withdrawal in opioid exposed (dependent) subjects. A "neutral" antagonist is expected to precipitate a less severe withdrawal syndrome compared to equi-effective antagonist doses of an inverse agonist (Raehal et al., 2005; Sadée et al., 2005). Furthermore, under certain conditions, the "neutral" antagonist is expected to attenuate/block the effects of the inverse agonist, assuming that the two drugs bind to similar binding domains and the process is competitive (Bilsky et al. 1996; Raehal et al., 2005; Sirohi et al., 2009). General methods for induction of opioid physical dependence and precipitation of withdrawal are described below.

5.1. Materials and supplies required

- Mice: Male ICR mice (25–35gs) from Harlan
- Syringes and needles (1-cm^3 syringes and 27–30-gauge needles for mice)
- Morphine sulfate (Mallinckrodt) or other suitable full MOR agonist
- Morphine implant pellets (25 or 75 mg, NIDA Drug Supply program) or osmotic mini-pumps for continuous and metered agonist drug delivery
- Opioid ligands that vary in their intrinsic activity (inverse agonist, "neutral" antagonist, or partial agonist). Examples include naloxone,

naltrexone, nalmefene, nalbuphine, naltrexol and naloxol derivatives, CTAP, and CTOP. Many of these compounds can be obtained through the NIDA Drug Supply Program
- Plexiglas observation tubes or similar setups that allow for unrestricted viewing/videotaping of behaviors
- Digital camcorders for recording and archiving of behaviors

5.2. Models of acute and chronic opioid physical dependence

The published literature has numerous protocols for establishing opioid dependence in rodents. They are typically categorized into acute or chronic exposures with no well-established guidelines on what constitutes acute versus chronic (Bilsky et al., 1996; Gold et al., 1994; Kest et al., 2002; Yano and Takemori, 1977). These models share a number of commonalities in terms of the initial adaptations of the MOR, its signaling pathways, neural adaptations and behavioral manifestations of withdrawal. A single administration of an opioid agonist can produce a dependent state that is characterized by a withdrawal or abstinence syndrome upon antagonist/inverse agonist administration in rodents and humans (Bilsky et al., 1996; Compton et al., 2004; Yano and Takemori, 1977). Chronic models of inducement include pellet or pump implantation or repeated dosing regimens (including escalating dose procedures) that treat the animals for days to weeks. It is often desirable to evaluate the test compounds in both acute (single dose) and chronic models that produce different levels of physical dependence. Traditional morphine pellets (75 mg) produces a very high level of dependence and a severe withdrawal that produces high levels of morbidity and mortality in mice. An alternative 25 mg pellet, in contrast, provides almost as high a level of physical dependence (as indexed by naloxone sensitivity/precipitated withdrawal) while being better tolerated. In other studies, it is desirable to precipitate withdrawal at times when the animals are still physically dependent, but blood levels of morphine are minimal. In these cases, repeated injections and precipitation 8–48 h after the last morphine injection can be used rather than the pellet regimen (removal of the pellet may leave residual drug at the site that can add to variability in response). Discussions of the data are contained in several published papers by our laboratories (Bilsky et al., 1996; Osborn et al., 2010; Raehal et al., 2005; Wang et al., 1994, 2001, 2004).

5.3. Acute physical dependence protocol

1. All mice should be weighed and marked (different colored permanent marker on base of the tail), with treatments assigned randomly and the investigator blinded to the actual treatments.

2. At $t = -4$ h, inject groups of mice (8–10 per group) with morphine sulfate (100 mg/kg, s.c.) and return to travel cages (food and water available *ad libitum*).
3. In preparation for precipitation of withdrawal, prepare Plexiglas observation cylinders with filter paper on bottoms. Filter paper can be weighed individually prior to and after precipitation to provide a rough guide on fecal and urine output. Prepare all syringes ahead of time to decrease time between injections.
4. At $t = 0$ min, inject animals with doses of test compound (typically i.p.) and immediately place individual animal into a Plexiglas chamber. Groups of five mice can be simultaneously videotaped with the same camera. Observe/videotape the animals for a minimum of 20 min (most animals receiving naloxone in the 1–3 mg/kg range will display withdrawal symptoms within 1 min with peak intensity seen around 10 min). As soon as the session has ended, mice should be removed from the chambers and weighed along with the individual filter papers. Filter papers can also be videotaped/photographed for additional assessment of urine and feces. Filter paper can be weighed and feces collected for quantification of output.
5. Mice should be immediately euthanized after the withdrawal session.
6. A blinded observer should review tapes and score the individual animals for vertical jumping, paw tremors, wet dog shakes, abnormal body posture/writhing, and other behaviors associated with opioid withdrawal. A number of classic reviews and book chapters on assessment of opioid withdrawal in rodents are available through PubMed or online used book retailers (Way *et al.*, 1973; Yano and Takemori, 1977).
7. Appropriate controls include vehicle/vehicle and opioid agonist/vehicle controls, along with a positive control (e.g., naloxone in the 1–10 mg/kg range, i.p.).

5.4. Chronic physical dependence protocol

1. Mice are prepared for surgery and then lightly anesthetized with isoflurane or other suitable short acting anesthetic.
2. A small (1–1.5 cm) incision is made along the nape of the neck under sterile conditions.
3. A morphine (25 or 75 mg) or placebo pellet is inserted through the incision and pushed caudally with a forceps so as to rest at the apex of the scapula.
4. The incision is closed using two surgical staples and the mice are returned to their home cages and allowed to recover.
5. Withdrawal is typically precipitated 72 h after pellet implantation with an i.p. injection of test compound.

6. Observation/videotaping of withdrawal is similar to that described in Section 5.3. The intensity of withdrawal is generally much higher than in the acute dependence assays, and the doses of antagonist needed to precipitate withdrawal are much lower. With neutral antagonists, higher doses may, however, be needed to precipitate equivalent withdrawal compared to an inverse agonist.
7. Alternative chronic dependence assays include repeated opioid agonist injections (e.g., two or three evenly spaced injections/day for 3–7 days). Doses can be escalated on each day of the procedure (e.g., start with an A_{90} dose and move up ½ log increments per day). Withdrawal is typically precipitated on the morning of the next day.

5.5. Competition studies

In theory, a "neutral" antagonist should be able to outcompete an inverse agonist that binds to a similar region of the MOR. The demonstration of these effects, in practice, is complicated by the effects of displacing the MOR agonist off the receptor (vs. removal of the MOR agonist and suppression of basal signaling). Put another way, if the level of physical dependence is high, the neutral antagonist will precipitate some withdrawal by itself. The assay and the doses selected for interaction studies need to be carefully chosen in order to best detect the potential blockade of the inverse agonist effects.

1. Acute or chronic physical dependence is established using a procedure similar to that described in Sections 5.3 and 5.4.
2. Ideally, as discussed above, a dose of the putative "neutral" antagonist should be chosen that blocks an acute effect of the agonist and precipitates minimal levels of withdrawal for a given level of physical dependence. The neutral antagonist is typically given 10–20 min prior to the inverse agonist.
3. A longer videotape/observation session may be employed to capture both initial effects of the neutral antagonist alone and then for a period of time after the inverse agonist is administered.
4. Complete dose–response curves should be run for lead compounds including varying the dose of the neutral antagonist and/or inverse agonist.

6. Summary

Constitutive activity of GPCRs has been demonstrated in many model systems (both *in vitro* and *in vivo*) and across a variety of receptors/tissues. Constitutive activity can be altered via mutations in the receptor/signaling pathways and changes in the micro-/macroenvironment of the receptor/cell, in response to disease and in response to chronic drug exposure. The clinical implications related to this activity are just recently being described for some receptor systems. Further information will emerge as the tools used to detect basal activity and inverse agonists, "neutral" antagonist and partial agonist activity become more refined. The implications for drug discovery/development are actively being discussed and debated and reclassification of ligands and drug candidates is underway.

With respect to opioid pharmacology, multiple groups have independently demonstrated constitutive activity at all three opioid receptors, and the activity is under dynamic regulation including changes in response to acute and chronic opioid exposure. This phenomenon has implications in the treatment of drug addiction, chronic pain patients receiving opioids, and other disease states that involve the endogenous opioid system (e.g., alcoholism). The chapter reviews some of the methods used to detect constitutive activity and to differentiate compounds based on their intrinsic activity at the MOR. The major advantages of the described assays include their ability to be conducted using standard laboratory techniques in addition to their capacity to screen compound libraries. The *in vivo* assays should be reserved for the most promising drug candidates. Other assays have now been described including some elegant operant conditioning paradigms (Hazel *et al.*, 2008) and methods that differentiate the reversal of opioid overdose versus propensity to elicit withdrawal (Sirohi *et al.*, 2009). Clinical trials are now also being conducted to extend the preclinical observations and help determine the importance of MOR constitutive activity to pharmacotherapy (http://clinicaltrials.gov/ct2/show/NCT00829777).

ACKNOWLEDGMENTS

This work was supported by the Intramural Research Program, National Institutes on Drug Abuse, NIH, DHHS. Dr. Bilsky holds several issued or pending patents on technology associated with opioid neutral antagonists and is a cofounder of Aiko Biotechnology, a start-up company that is testing 6β-naltrexol in clinical trials as a potential adjunct to reduce opioid-mediated side effects.

REFERENCES

Aloyo, V. J., Berg, K. A., Spampinato, U., Clarke, W. P., and Harvey, J. A. (2009). Current status of inverse agonism at serotonin2A (5-HT2A) and 5-HT2C receptors. *Pharmacol. Ther.* **121,** 160–173.

Arden, J. R., Segredo, V., Wang, Z., Lameh, J., and Sadee, W. (1995). Phosphorylation and agonist-specific intracellular trafficking of an epitope-tagged mu-opioid receptor expressed in HEK 293 cells. *J. Neurochem.* **65,** 1636–1645.

Bilsky, E. J., Bernstein, R. N., Wang, Z., Sadée, W., and Porreca, F. (1996). Effects of naloxone and D-Phe-Cys-Try-D-Trp-Arg-Thr-Pen-Thr-NH2 and the protein kinase inhibitors H7 and H8 on acute morphine dependence and antinociceptive tolerance in mice. *J. Pharmacol. Exp. Ther.* **277,** 484–490.

Breivogel, C. S., Selley, D. E., and Childers, S. R. (1997). Acute and chronic effects of opioids on delta and mu receptor activation of G proteins in NG108-15 and SK-N-SH cell membranes. *J. Neurochem.* **68,** 1462–1472.

Compton, P., Miotto, K., and Elashoff, D. (2004). Precipitated opioid withdrawal across acute physical dependence induction methods. *Pharmacol. Biochem. Behav.* **77,** 263–268.

Costa, T., and Herz, A. (1989). Antagonists with negative intrinsic activity at delta opioid receptors coupled to GTP-binding proteins. *Proc. Natl. Acad. Sci.* **86,** 7321–7325.

Covel, J. A., Santora, V. J., Smith, J. M., Hayashi, R., Gallardo, C., Weinhouse, M. I., Ibarra, J. B., Schultz, J. A., Park, D. M., Estrada, S. A., Hofilena, B. J., Pulley, M. D., et al. (2009). Design and evaluation of novel biphenyl sulfonamide derivatives with potent histamine H(3) receptor inverse agonist activity. *J. Med. Chem.* **52,** 5603–5611.

Gold, L. H., Stinus, L., Inturrisi, C. E., and Koob, G. F. (1994). Prolonged tolerance, dependence and abstinence following subcutaneous morphine pellet implantation in the rat. *Eur. J. Pharmacol.* **253,** 45–51.

Gray, J. A., Sheffler, D. J., Bhatnagar, A., Woods, J. A., Hufeisen, S. J., Benovic, J. L., and Roth, B. L. (2001). Cell-type specific effects of endocytosis inhibitors on 5-hydroxytryptamine(2A) receptor desensitization and resensitization reveal an arrestin-, GRK2-, and GRK5-independent mode of regulation in human embryonic kidney 293 cells. *Mol. Pharmacol.* **60,** 1020–1030.

Harding, W. W., Tidgewell, K., Byrd, N., Cobb, H., Dersch, C. M., Butelman, E. R., Rothman, R. B., and Prisinzano, T. E. (2005). Neoclerodane diterpenes as a novel scaffold for mu opioid receptor ligands. *J. Med. Chem.* **48,** 4765–4771.

Hazel, K. D., Fluck, J., Jones, N. T., Evola, M., and Young, A. M. (2008). Comparison of antagonist potency to evoke an acute withdrawal stimulus 4 h after pretreatment with morphine in rats. *FASEB J.* **22,** 712.

Jordan, B. A., and Devi, L. A. (1999). G-protein-coupled receptor heterodimerization modulates receptor function. *Nature* **399,** 697–700.

Kest, B., Palmese, C. A., Hopkins, E., Adler, M., Juni, A., and Mogil, J. S. (2002). Naloxone-precipitated withdrawal jumping in 11 inbred mouse strains: Evidence for common genetic mechanisms in acute and chronic morphine physical dependence. *Neuroscience* **115,** 463–469.

Klee, W. A., and Nirenberg, M. (1974). A neuroblastoma times glioma hybrid cell line with morphine receptors. *Proc. Natl. Acad. Sci. USA* **71,** 3474–3477.

Liu, J. G., and Prather, P. L. (2001). Chronic exposure to mu-opioid agonists produces constitutive activation of mu-opioid receptors in direct proportion to the efficacy of the agonist used for pretreatment. *Mol. Pharmacol.* **60,** 53–62.

Lunn, C. A. (2010). Updating the chemistry and biology of cannabinoid CB2 receptor-specific inverse agonists. *Curr. Top. Med. Chem.* **10,** 768–778.

Nordstedt, C., and Fredholm, B. B. (1990). A modification of a protein-binding method for rapid quantification of cAMP in cell-culture supernatants and body fluid. *Anal. Biochem.* **189,** 231–234.

Osborn, M. D., Lowery, J. J., Skorput, A. G., Giuvelis, D., and Bilsky, E. J. (2010). In vivo characterization of the opioid antagonist nalmefene in mice. *Life Sci.* **86,** 624–630.

Raehal, K. M., Lowery, J. J., Bhamidipati, C. M., Paolino, R. M., Blair, J. R., Wang, D., Sadee, W., and Bilsky, E. J. (2005). In vivo characterization of 6beta-naltrexol, an opioid ligand with less inverse agonist activity compared with naltrexone and naloxone in opioid-dependent mice. *J. Pharmacol. Exp. Ther.* **313,** 1150–1162.

Sadée, W., Wang, D., and Bilsky, E. J. (2005). Basal opioid receptor activity, neutral antagonists, and therapeutic opportunities. *Life Sci.* **76,** 1427–1437.

Sally, E. J., Xu, H., Dersch, C. M., Hsin, L. W., Chang, L. T., Prisinzano, T. E., Simpson, D. S., Giuvelis, D., Rice, K. C., Jacobson, A. E., et al. (2009). Identification of a novel "almost neutral" mu-opioid receptor antagonist in CHO cells expressing the cloned human mu-opioid receptor. *Synapse* **64**, 280–288.

Sirohi, S., Dighe, S. V., Madia, P. A., and Yoburn, B. C. (2009). The relative potency of inverse agonists and a neutral opioid antagonist in precipitated withdrawal. *J. Pharmacol. Exp. Ther.* **330**, 513–519.

Standifer, K. M., and Pasternak, G. W. (1997). G proteins and opioid receptor-mediated signaling. *Cell. Signal.* **9**, 237–248.

Toll, L., Berzetei-Gurske, I. P., Polgar, W. E., Brandt, S. R., Adapa, I. D., Rodriguez, L., Schwartz, R. W., Haggart, D., O'Brien, A., White, A., et al. (1998). Standard binding and functional assays related to medications development division testing for potential cocaine and opiate narcotic treatment medications. *NIDA Res. Monogr.* **178**, 440–466.

Traynor, J. R., and Nahorski, S. R. (1995). Modulation by mu-opioid agonists of guanosine-5′-O-(3-[35S]thio)triphosphate binding to membranes from human neuroblastoma SH-SY5Y cells. *Mol. Pharmacol.* **47**, 848–854.

Traynor, J. R., Clark, M. J., and Remmers, A. E. (2002). Relationship between rate and extent of G protein activation: Comparison between full and partial opioid agonists. *J. Pharmacol. Exp. Ther.* **300**, 157–161.

Waldhoer, M., Bartlett, S. E., and Whistler, J. L. (2004). Opioid receptors. *Annu. Rev. Biochem.* **73**, 953–990.

Walker, E. A., Makhay, M. M., House, J. D., and Young, A. M. (1994). In vivo pA2 analysis of naltrexone antagonism of discriminative stimulus and analgesic effects of opiate agonists in rats. *J. Pharmacol. Exp. Ther.* **271**, 959–968.

Wang, Z., Bilsky, E. J., Porreca, F., and Sadee, W. (1994). Constitutive mu opioid receptor activation as a regulatory mechanism underlying narcotic tolerance and dependence. *Life Sci.* **54**, PL339–PL350.

Wang, D., Raehal, K. M., Bilsky, E. J., and Sadee, W. (2001). Inverse agonists and neutral antagonists at mu opioid receptor (MOR): Possible role of basal receptor signaling in narcotic dependence. *J. Neurochem.* **77**, 1590–1600.

Wang, D., Raehal, K. M., Lin, E. T., Lowery, J. J., Kieffer, B. L., Bilsky, E. J., and Sadée, W. (2004). Basal signaling activity of mu opioid receptor in mouse brain: Role in narcotic dependence. *J. Pharmacol. Exp. Ther.* **308**, 512–520.

Wang, D., Sun, X., and Sadée, W. (2007). Different effects of opioid antagonists on mu, delta-, and kappa-opioid receptors with and without agonist pretreatment. *J. Pharmacol. Exp. Ther.* **321**, 544–552.

Way, E. L., Loh, H. H., Ho, I. K., Iwamoto, E. T., and Wei, E. (1973). Neuroanatomical and chemical correlates of naloxone-precipitated withdrawal. *Adv. Biochem. Psychopharmacol.* **8**, 455–470.

Williams, J. T., Christie, M. J., and Manzoni, O. (2001). Cellular and synaptic adaptations mediating opioid dependence. *Physiol. Rev.* **81**, 299–343.

Woods, J. H., Winger, G., and France, C. P. (1992). Use of in vivo apparent pA2 analysis in assessment of opioid abuse liability. *Trends Pharmacol. Sci.* **13**, 282–286.

Xu, H., Wang, X., Wang, J., and Rothman, R. B. (2004). Opioid peptide receptor studies. 17. Attenuation of chronic morphine effects after antisense oligodeoxynucleotide knockdown of RGS9 protein in cells expressing the cloned mu opioid receptor. *Synapse* **52**, 209–217.

Xu, H., Partilla, J. S., Wang, X., Rutherford, J. M., Tidgewell, K., Prisinzano, T. E., Bohn, L. M., and Rothman, R. B. (2007). A comparison of noninternalizing (herkinorin) and internalizing (DAMGO) mu-opioid agonists on cellular markers related to opioid tolerance and dependence. *Synapse* **61**, 166–175.

Yano, I., and Takemori, A. E. (1977). Inhibition by naloxone and dependence in mice treated acutely and chronically with morphine. *Res. Commun. Chem. Pathol. Pharmacol.* **16**, 721–734.

CHAPTER TWENTY-TWO

CONSTITUTIVELY ACTIVE μ-OPIOID RECEPTORS

Mark Connor* and John Traynor[†]

Contents

1. Introduction	446
1.1. Experimental questions	447
2. Methods for Measuring Constitutive Activity	447
2.1. General considerations	447
2.2. G protein measures: The [^{35}S]GTPγS/Eu-GTP assays	451
2.3. Native μ-opioid receptors	453
2.4. GTPγS binding after chronic agonist treatment	454
2.5. Constitutively active mutants	456
2.6. Adenylyl cyclase assays	458
2.7. Measuring constitutive activity using ion channels	460
2.8. Calcium channels	461
2.9. Potassium channels	464
3. Conclusions	465
Acknowledgments	465
References	466

Abstract

The μ-opioid receptor is the G protein coupled receptor (GPCR) responsible for the analgesic, rewarding and unwanted effects of morphine and similar drugs. Constitutive activity of GPCRs is a phenomenon that likely reflects receptors spontaneously adopting conformations that can activate G proteins, and is likely to be common to most if not all GPCRs. Basal constitutive activity has been observed in some systems with μ-opioid receptors, and constitutive activity is expressed by mutant μ-opioid receptors with amino acid substitutions in regions known to be important for signaling. However, μ-opioid receptors are unique in that a putative constitutively active state of the receptor, the μ*-state, has been suggested to be induced by prolonged agonist treatment. The μ*-state is thought to contribute to processes underlying adaptation to and

---

* Australian School of Advanced Medicine, Macquarie University, New South Wales, Australia
[†] Department of Pharmacology, Substance Abuse Research Centre, University of Michigan Medical School, Ann Arbor, Michigan, USA

withdrawal from opioid treatment, and may have a ligand sensitivity distinct from basal constitutive activity of the μ-opioid receptor or that exhibited by μ-opioid receptor mutants. In this chapter, we outline methods for measuring constitutively active μ-opioid receptors, including some that take advantage of the fairly direct coupling of the receptor to ion channels. We also briefly summarize the pharmacology of the different constitutively active μ-opioid receptor states, and highlight the areas where we need to know more. We hope that a better understanding of constitutive activity at the μ-opioid receptor may provide information useful in developing ligands that access subsets of receptor conformations, offering the potential to fine-tune opioid pharmacotherapy.

1. INTRODUCTION

The constitutive activity of G protein coupled receptors (GPCR) has been recognized for some time, and is not a surprising phenomenon given the stochastic principles that underlie signaling in biology. Antagonist-sensitive, agonist-independent GPCR activity was noted from the earliest studies of purified β-adrenergic receptors (Cerione et al., 1984), however, it was a comprehensive study of native δ-opioid receptors that showed G protein activity that could be stimulated by agonists and inhibited by "inverse agonists," with both effects being blocked by neutral antagonists (Costa and Herz, 1989). The constitutive activity of GPCR is an important idea in drug action and some diseases result from mutations in GPCR that enhance basal receptor activity (Smit et al., 2007). While there are varying degrees of constitutive activity measurable at different GPCR, the presumption that it arises as an inevitable consequence of continual movement of receptors through a series of different conformational states means that constitutive activity *per se* will be common to all receptors.

The μ-opioid receptor is the GPCR responsible for the analgesic and rewarding properties of morphine and similar drugs, as well as for their unwanted effects. Constitutive activity has been convincingly demonstrated for the μ-opioid receptor (e.g., Burford et al., 2000; Liu and Prather, 2001; Mahmoud et al., 2010), but a great deal of additional experimental interest has been generated by the proposition that chronic treatment with μ-opioid agonists induces a constitutively active form of the receptor known as the μ*-state (Sadee et al., 2005; Wang et al., 1994). This privileged state of the μ-opioid receptor is reported to be induced by agonist treatment *in vivo* or *in vitro* and is thought to reflect persistent phosphorylation of the receptor. The consequences of prolonged use of morphine for pain relief or other purposes are a major medical and societal challenge, and a significant research effort continues to be devoted to understanding the mechanisms underlying tolerance to the analgesic affects of morphine and the development of dependence on opioids. The suggestion that persistent constitutive activity of the μ-receptor is responsible for some or all of the adaptations that lead to tolerance to and dependence on morphine means that it is critically important to be able to measure μ-opioid receptor activity as directly as possible, and in as wide a variety of native tissues as possible. In this chapter, we will describe biochemical and electrophysiological techniques for detecting and quantifying the constitutive activity of μ-opioid receptors *in vitro* and endeavor to outline the strengths and potential shortcomings of each assay.

1.1. Experimental questions

The hypothesis that the constitutively active μ*-state of the μ-opioid receptor plays a pivotal role in driving cellular adaptation to prolonged opioid agonist treatment raises a number of important experimental questions specific to the μ-opioid receptor. These include whether the basal constitutive activity of μ-opioid receptors is operationally equivalent to that of the μ*-state and what functional consequences constitutive μ-opioid receptor activity has for neuronal excitability and synaptic transmission. It remains unresolved if constitutively active μ-receptors and μ*-state receptors activate the same suite of G proteins, whether their basal activity is sensitive to the same inverse agonists or even whether the constitutive activity observed with agonist naïve μ-opioid receptors simply reflects a low level of the phosphorylation event suggested to produce the μ*-state. It has also been extremely difficult to demonstrate μ-opioid receptor constitutive activity or any antagonist effects consistent with the reversal of constitutive μ-opioid receptor activity in isolated neurons or neurons in brain slices. It remains an open question whether techniques currently used in experiments on relatively intact neurons lack sufficient sensitivity to detect constitutively active μ-opioid receptors (μ* or otherwise) or whether there is simply nothing to see.

2. METHODS FOR MEASURING CONSTITUTIVE ACTIVITY

2.1. General considerations

The activity of GPCR can be quantified quite directly by measuring agonist-stimulated GTPase activity or the binding of stable GTP analogues to G proteins. One of the earliest events after receptor activation is the exchange of GDP on the Gα subunit for GTP (Fig. 22.1). Consequently, the [^{35}S]GTPγS assay, which measures binding of the radiolabeled, slowly hydrolysable, GTP analogue [^{35}S]GTPγS to activated Gα proteins, has been widely employed.

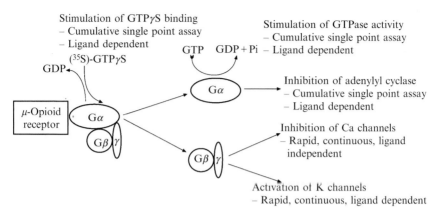

Figure 22.1 Common (and potential) assay points for constitutively active μ-opioid receptors. A diagram of where μ-opioid receptor activity can be assayed relatively straightforwardly, with respect to the G protein cycle. Brief comments about the nature of each assay are included.

The subsequent hydrolysis of GTP bound to the activated Gα proteins can also be measured, as can modulation of adenylyl cyclase (AC) downstream of GTP-bound Gα protein or ion channels downstream of released Gβγ subunits. It is, of course, important to define the receptor that is responsible for the activity, where possible to distinguish constitutively active receptors from constitutively active G proteins, and to ensure that the measured signal does not come from the activity of endogenous agonists remaining in the preparation. This latter consideration is less problematic for μ-opioid receptors, as the endogenous peptide agonists are likely to be scarce in the reduced preparations used for these studies, although lipophilic synthetic agonists used for pretreatment may be harder to remove.

The difficulty in measuring constitutive activity is of course that by its nature it is unstimulated. While it is possible to carefully measure basal GTPase activity while changing levels of receptor or G protein (Cerione et al., 1984), this approach is not practical for addressing most experimental questions. The voltage-dependent inhibition of calcium channels (I_{Ca}) by G protein βγ subunits provides an instantaneous and repeatable measurement of constitutive activity in single cells simply by varying membrane voltage (see below), but it is only good for isolated neurons. In general, constitutive activity is defined by the effects of ligands that stabilize poorly coupled state(s) of the μ-receptor. These compounds are called "inverse agonists" or described as ligands with "negative intrinsic efficacy" (Kenakin 2004). A list of readily available μ-opioid receptor ligands with reported negative intrinsic activity is found in Table 22.1. The degree to which these ligands inhibit the basal level of the output being measured is considered to reflect the amount of constitutive activity present in the system, although it is

Table 22.1 The activity of ligands for the μ-opioid receptor as modulators of [^{35}S]-GTPγS binding in basal and agonist pretreated conditions, and to constitutively active mutants of the μ-opioid receptor

Drug	Basal	Agonist pretreated ("μ*")	Constitutively active mutants	Comment
Naloxone	Positive (Burford et al., 2000; Liu and Prather, 2001; Wang et al., 1994, 2007); Neutral (Brillet et al., 2003; Divin et al., 2009; Li et al., 2001a; Traynor and Nahorski, 1995)	Neutral (Divin et al., 2009) Negative (Liu and Prather, 2001; Sally et al., 2010; Wang et al., 1994, 2001, 2007)	Negative, D164Q (Li et al., 2001a) Neutral C348/353A (Brillet et al., 2003)	
Naltrexone	Negative (Wang et al., 2001) Positive (Wang et al., 2007) Neutral (Brillet et al., 2003; Burford et al., 2000; Li et al., 2001a)	Neutral (Divin et al., 2009) Negative (Sally et al., 2010; Wang et al., 1994, 2001, 2007)	Negative, D164Q (Li et al., 2001a) Neutral C348/353A (Brillet et al., 2003)	
6β-Naltrexol	Negative (Wang et al., 2001) Weak positive (Wang et al., 2007) Neutral (Divin et al., 2009; Wang et al., 2001)	Neutral (Wang et al., 2001) Negative (Sally et al., 2010)		
CTAP	Neutral (Divin et al., 2009; Li et al., 2001a; Liu and Prather, 2001; Wang et al., 2001)	Neutral (Divin et al., 2009; Liu and Prather 2001; Wang et al., 1994, 2001) Negative (Sally et al., 2010)	Neutral, D164X (Li et al., 2001a)	CTAP had negative efficacy in both conditions in Hi K buffer in Divin et al. (2009)

(continued)

Table 22.1 (continued)

Drug	Basal	Agonist pretreated ("μ*")	Constitutively active mutants	Comment
CTOP	Positive (Brillet et al., 2003)	Neutral (Wang et al., 1994)	Negative, C348/353A (Brillet et al., 2003)	
β-Flunaltrexamine	Neutral (Wang et al., 2001)	Neutral (Wang et al., 2001)		Alkylating agent
β-Chlornaltrexamine	Neutral (Liu and Prather, 2001)	Negative (Liu and Prather, 2001; Wang et al., 2001)		Alkylating agent
	Negative (Burford et al., 2000; Wang et al., 2001)			
BNTX	Negative (Wang et al., 2001, 2007)	Negative (Sally et al., 2010; Wang et al., 2001, 2007)		Poorly correlated with μ-receptor affinity (Wang et al., 2001, 2007)
Naltrindole	Positive (Brillet et al., 2003)	Negative (Sally et al., 2010)	Neutral, C348/353A (Brillet et al., 2003)	
Diprenorphine	Neutral (Li et al., 2001a)	Negative (Sally et al., 2010; Wang et al., 1994)	Neutral, D164Q (Li et al., 20C1a)	
Cyprodime	Positive (Brillet et al., 2003)		Negative, C348/353A (Brillet et al., 2003)	

Ligand efficacy is reported as positive (stimulates binding), negative (inhibits binding), or neutral (no significant effect on binding) according to the results presented in the cited studies. Because of differing experimental conditions and methods of presenting data, we have assigned just direction to efficacy, not strength. Not all ligands tested in these studies are reported here; full references can be found at the end of the chapter.

difficult to prove that any ligand completely inhibits constitutive activity in the absence of a nonpharmacological method of defining this activity. Evidence that ligands with negative intrinsic activity are acting to change signaling via their cognate receptor is provided by reversal of the inhibition of signaling by a neutral antagonist (Costa and Herz, 1989). Putative neutral opioid antagonists are also listed in Table 22.1. An advantage of using ligands to define constitutive activity is that the receptor specificity of the drugs is usually well defined. However, different ligands have been reported to have different amounts of negative efficacy, which presumably reflects differences in their propensity to stabilize the G protein uncoupled states of the μ-receptor. Some laboratories report virtually all antagonists of μ-opioid receptor function have negative intrinsic activity (Sally et al., 2010), with essentially no compounds being completely devoid of positive or negative efficacy. Although somewhat at odds with the observed lack of effect of μ-opioid receptor antagonists in most assays, the notion that binding to a receptor stabilizes a subset of conformations suggests that it would be difficult for a high affinity ligand not to affect the distribution of receptor-coupled conformations one way or the other (Kenakin and Onaran, 2002).

2.2. G protein measures: The [^{35}S]GTPγS/Eu-GTP assays

Two methods are employed for measuring agonist activity at the level of the G protein. They respectively take advantage of the binding of GTP to the Gα subunit following agonist-induced dissociation of GDP and the subsequent hydrolysis of GTP bound to the Gα subunit to terminate the signal (Fig. 22.1). Both responses show an increase following agonist occupation of a GPCR and therefore should show a response to a receptor that constitutively activates Gα in the absence of agonist. In practice, the measure of GTP binding to the Gα subunit using labeled GTP analogues is the easiest and has been much more frequently employed in the study of constitutive activity.

The most common way to measure constitutive activity of μ-opioid receptors is to assess the nonagonist-stimulated binding of labeled GTP analogues to membranes from cells or tissues expressing the receptor. The analogues used are either ^{35}S-labeled GTP ([^{35}S]GTPγS) or europium-labeled GTP (Eu-GTP), although the commercial availability of the latter ligand is now uncertain. Assays for the two ligands are very similar except that the amount of [^{35}S]GTPγS bound is determined by scintillation counting and the binding of Eu-GTP by time-resolved fluorescence. In assays of ligand-stimulated receptor activity μ-opioid agonists increase the binding of the GTP analogues to appropriate Gα proteins. However, in the absence of agonist and even in the presence of a high concentration of unlabeled GTPγS to define nonspecific binding, there is some degree of "basal" binding of the labeled nucleotide. This basal binding of comprises binding of [^{35}S]GTPγS (or Eu-GTP) to unoccupied or constitutively active heterotrimeric G proteins, binding to small G proteins and binding caused by receptors activating Gα proteins in the absence of agonist, that is, binding reflecting constitutive activity.

To measure labeled GTP binding a low concentration of the labeled nucleotide (usually 0.1 nM for [^{35}S]GTPγS or 1 nM for Eu-GTP, which has lower affinity for Gα subunits) is incubated in membranes from cells or tissues expressing μ-opioid receptors for sufficient time to reach steady state (usually 60–120 min). Specific binding of the labeled nucleotide is defined in the presence of a saturating concentration of unlabelled GTPγS. GTP conjugate labeled Gα subunits remain bound to the cell membranes and the assay is terminated by filtration or immunoprecipitation, with signal from the retained proteins measured by traditional scintillation counting (Traynor and Nahorski, 1995) or scintillation proximity assay (DeLapp, 2004) for [^{35}S] GTPγS or time-resolved fluorescence for Eu-GTP (Labrecque et al., 2009).

To observe receptor-driven increases in binding it is necessary to include GDP, Na, and Mg in the incubation medium. GDP fills empty guanine nucleotide binding sites on the Gα protein which reduces the basal level of GTP conjugate binding and supplies GDP-bound Gα substrate for agonist-driven activation. The concentration of GDP needed to give an optimal signal-to-noise ratio must be determined experimentally for each system; it is usually between 3 and 30 μM for cells expressing cloned receptors, but may be higher for brain tissue. However, increasing the concentration of GDP increases the efficacy requirements of the system; conversely, lowering GDP will raise basal binding and has been used to identify inverse agonists in other systems (Roberts and Strange 2005). Mg is necessary to promote agonist-stimulated binding of [^{35}S]GTPγS, and the concentration can be critical for the identification of inverse agonists (Wang et al., 2001). Higher concentrations of Mg can inhibit binding, so again, the concentration should be optimized for each system. Assays are usually performed in the presence of 100 mM Na because Na selectively promotes agonist-stimulated binding over basal binding, improving the agonist signal. However, such a high concentration of Na tends to uncouple the receptor from G proteins and so reduces spontaneous coupling. Thus, the chances of observing constitutive activity can be improved by replacement of Na with K (Liu et al., 2001; Szekeres and Traynor, 1997) or with N-methyl-D-glucamine (Lin et al., 2006) to increase receptor–effector coupling and promote spontaneous activation of [^{35}S]GTPγS binding. This increased constitutive signal obviously allows for the easier determination of inverse agonist efficacy. Finally, endogenous adenosine can mask the signal arising from other GPCR, particularly in brain tissue, so membranes should be incubated with adenosine deaminase prior to experiments (Breivogel et al., 2004) or an adenosine receptor antagonist can be included in the assay (Horswill et al., 2007). A detailed review of the assay protocol can be found in Harrison and Traynor (2003), and specific experimental details in Kara and Strange (2010).

Several considerations need to be borne in mind when using GTP analogue binding assay to identify constitutive activity at μ-opioid receptors. First, the assay measures an event very close to the receptor itself and therefore the response is not amplified as seen with downstream assays. Consequently, it can be difficult to observe a small constitutively active signal unless receptors are over expressed or conditions such as Na concentration are altered. Further, because of the impermeable nature of the guanine nucleotides the assay is performed in isolated membranes, and this raises the possibility of disrupting the receptor–G protein coupling or scaffolding mechanisms potentially necessary to maintain constitutive activity. Finally, it is necessary to show that the constitutive effects derive from the receptor taking up active conformations and not some other process, such as changes in the spontaneous activity of the G proteins themselves. For Gαi/o-coupled receptors such as the μ-opioid receptor this can be achieved with pertussis toxin (PTX) pretreatment (usually 100 ng/ml overnight). PTX ADP ribosylates a Cys in the C-terminus of the Gα protein and prevents its coupling to receptor, thus inhibiting both agonist and constitutive activation of [^{35}S]GTPγS binding. Importantly, PTX does not alter other properties of the G protein, such as ability to bind guanine nucleotides. However, use of PTX does not identify which receptor in a system in responsible for the observed constitutive activity. For example, PTX reduces basal [^{35}S]GTPγS binding in SH-SY5Y cells (Traynor and Nahorski, 1995), although there is no specific evidence for μ-opioid constitutive activity in these cells. The observed effect could be due spontaneous activity of one of any number of Gαi/o-coupled GPCRs expressed in these cells, including 5HT1A, ORL1, and CB1 receptors. Nonetheless, PTX is an important tool to confirm the role of receptor G protein coupling, especially in heterologous systems overexpressing one receptor.

2.3. Native μ-opioid receptors

Unlike many other GPCRs that couple to Gαi/o proteins such as the δ-opioid receptor (Costa and Herz, 1989) and the CB1 receptor (Howlett, 2004), it has proved difficult to show basal, agonist-independent, activation of [^{35}S]GTPγS binding G protein by μ-opioid receptors in naïve native tissues (Raehal et al., 2005; Wang et al., 2004) or often even in heterologous expression systems (Neilan et al., 1999). This suggests that under normal physiological conditions there is insufficient receptor in active conformations to give a stimulation of [^{35}S]GTPγS binding that can be inhibited by ligands with negative efficacy. However, it has been demonstrated that the nonselective opioid antagonists 7-benzylidenenaltrexone (BNTX) and β-chlornaltrexamine (β-CNA) cause a small reduction in basal [^{35}S]GTPγS binding in homogenates of whole brain from wild-type mice but not their μ-opioid receptor knockout counterparts, suggesting constitutive activity of the native receptor (Fig. 22.2A; Wang et al., 2004). Moreover, changes in the levels of basal, nonagonist-stimulated [^{35}S] GTPγS binding have been reported in several cell lines in which the μ-opioid receptor has been overexpressed. For example, an approximately 25% increase in the basal [^{35}S]GTPγS signal is seen in HEK cells expressing a μ-opioid receptor (HEKμ) (Burford et al., 2000) and a 67% increase was reported in μ-opioid receptor expressing GH$_3$ cells (GH$_3$μ; Liu et al., 2001). By contrast, no change was reported in μ-receptor expressing C6 glioma cells (C6μ; Divin et al., 2009; Neilan et al., 1999). In general, the higher the μ-opioid receptor expression, the higher the level of constitutive coupling to Gα proteins that should be observed due to increased chances of fruitful collisions occurring. However, GH$_3$μ cells show higher than basal [^{35}S] GTPγS binding at receptor expression levels similar to those found endogenously (0.4 pmol/mg protein) and agonist-independent increases in [^{35}S] GTPγS binding have been reported in HEKμ cells expressing 1 pmol/mg receptor (Wang et al., 2001). As mentioned above, the constitutive increase in basal [^{35}S]GTPγS binding is increased by removal of Na ions (Liu et al., 2001), although in C6μ cells this still did not generate any constitutive activity that could be attributed to the μ-opioid receptor. The alkylating agent β-CNA inhibited the constitutive activation of [^{35}S]GTPγS binding in HEKμ cells, possibly by locking the receptor in an inactive conformation (Burford et al., 2000), although this was not seen at the μ-opioid receptor in naïve GH$_3$ cells (Liu and Prather, 2001).

2.4. GTPγS binding after chronic agonist treatment

Chronic opioid agonist treatment produces the putative constitutively activated μ*-state of the μ-opioid receptor both in heterologous systems and following *in vivo* administration of opioid ligands (Connor, 2009). Studies in SH-SY5Y cells suggest this process involves receptor phosphorylation (Wang et al., 1994). Thus, chronic treatment with agonists provides a strategy to increase the level of μ-opioid receptor constitutive activity from the rather low levels seen in naïve systems, or even to induce activity in silent systems. This makes it easier to identify compounds with inverse agonist activity. Many of these experiments have been performed by assaying the activity of AC (see below), but there are several examples that have used the [^{35}S]GTPγS binding assay.

Using brain membranes from mice, several studies found that 3-day treatment with morphine converts naltrexone and naloxone from ligands with a small positive efficacy to ones with modest negative efficacy, and this occurred in a region-specific manner (Raehal et al., 2005; Wang et al., 2004). The results suggest an induction of μ*-state receptors; however, the

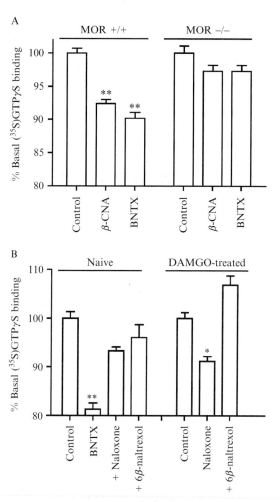

Figure 22.2 Detection of constitutively active μ-opioid receptors using the [^{35}S] GTPγS assay. (A) Basal [^{35}S]GTPγS binding in homogenates (10 μg) of whole brains from wild-type (MOR+/+) or μ-opioid receptor knockout (MOR−/−) mice in the absence or presence of the opioid antagonists β-CNA (1 μM) or BNTX (10 μM). β-CNA and BNTX reduced basal binding only in brain homogenates from the wild-type animals (**$p < 0.01$). The assays were performed using 0.1 nM [^{35}S]GTPγS in the presence of 10 μM GDP, 100 mM NaCl, and 4 mM MgCl$_2$ for 30 min at 30 °C as described in Wang et al. (2004). (B) In homogenates of untreated HEKμ cells (5 μg), BNTX (1 μM) acts as an inverse agonist to reduce basal [^{35}S]GTPγS binding; naloxone (10 μM) and 6β-naltrexol (10 μM) reverse the effect of BNTX. In DAMGO-treated HEKμ cells (5 μg) (1 μM, 24 h), naloxone is converted to an inverse agonist, but 6β-naltrexol is still a neutral antagonist (**$p < 0.01$; *$p < 0.0$). Assays were performed as above but in the presence of 150 mM KCl and 1 mM MgCl$_2$ (Wang et al., 2007). It should be noted from a practical standpoint that the differences observed are very small and the assay components are critical, emphasizing the low level of basal μ-opioid receptor constitutive activity and the difficulty of studying this phenomenon. The figure was redrawn from data in Wang et al. (2004) (A) and Wang et al. (2007), http://www.jpet.aspetjournals.org, with permission.

measured responses were small, with a less than 10% decrease in basal GTPγS binding produced by the inverse agonists. It should be noted that other studies have failed to see any increases in basal GTPγS activity after chronic morphine treatment of animals (e.g., Selley et al., 1997). By contrast, basal binding of [^{35}S]GTPγS in GH$_3$μ cells and CHO and HEK 293 cells expressing μ-opioid receptors is increased considerably following chronic opioid treatment (Liu et al., 2001; Wang et al., 2007; Xu et al., 2007), and there is a concomitant increase in the apparent negative efficacy of most tested antagonists (Fig. 22.2B). Interestingly, chronic treatment with the peptide agonist [D-Ala^2,N-MePhe^4,Gly-ol]-enkephalin (DAMGO) did not produce a constitutively active μ-opioid receptor in C6 glioma cells as defined by increases in [^{35}S]GTPγS binding (Divin et al., 2009), again highlighting the system dependence of μ-opioid receptor constitutive activity. The induction of constitutive activity is ligand dependent in some studies but not others, for example, Liu and Prather (2001) found that both DAMGO and morphine treatment induced constitutive activity in GH$_3$μ cells (although only DAMGO was reported to change basal [^{35}S]GTPγS binding), while Xu et al. (2007) found that herkinorin but not DAMGO pretreatment changed basal [^{35}S]GTPγS binding in CHO cells.

Of course, prolonged treatment with opioids can cause a number of changes in the cell including changes in receptor, Gα protein, and accessory protein expression, and these should be borne in mind when interpreting results (e.g., Liu and Prather, 2001; Xu et al., 2007). Also, it is vital to demonstrate that the opioid used for chronic treatment is washed out of the preparation since residual drug will stimulate [^{35}S]GTPγS, thus giving a false impression that constitutive activity is present.

2.5. Constitutively active mutants

Experimental mutagenesis of the μ-opioid receptor has helped define regions that are important for stabilization of inactive receptor conformations as well as those that contribute to the active conformations responsible for coupling to G proteins. Some mutations appear to produce constitutively active receptors and as consequence it has been possible to use such mutations in the search for ligands that stabilize inactive conformations.

2.5.1. The DRY (Asp-Lys-Tyr) motif

This amino acid sequence at the interface of TM3 and the second intracellular loop is highly conserved across GPCRs and is thought to be important for stabilization of inactive states of the receptor as well as to be playing a role in receptor activation (Rosenbaum et al., 2009). Mutation of amino acids in this region leads to constitutive activity in many GPCRs. In the μ-opioid receptor, the DRY motif is amino acids 164–166. Replacement of Asp^{164} with His produces a μ-opioid receptor that readily promotes a high degree of [^{35}S]GTPγS binding in the absence of agonist in both transiently transfected HEK293 cells and stably transfected CHO cells (Li et al., 2001b). Indeed, addition of the efficacious agonist DAMGO to the mutant receptor produces little further increase in [^{35}S]GTPγS binding. Qualitatively similar activity is obtained with when Asp^{164} is replaced by Tyr, Glu, or Met (Li et al., 2001b). The constitutive activity of these mutants is PTX-sensitive and is seen at receptor levels close to physiological expression levels. Similar to the results in some studies of prolonged agonist treatment of wild-type μ-opioid receptors, naltrexone and naloxone acted as inverse agonists at the constitutively active mutants, but CTAP (D-Phe-Cys-Tyr-D-Trp-Arg-Thr-Pen-Thr-NH$_2$) and diprenorphine were neutral (Li et al., 2001b). It is very important to note that expression of the constitutively active μ-opioid receptors was only detectable following prolonged incubation with high levels of naloxone (20 μM for 4 days). Naloxone stabilized the μ-opioid receptor protein and at the same time prevented constitutive internalization of the mutant receptor (Li et al., 2001a).

2.5.2. Junction of intracellular loop 3 and transmembrane domain 6

Another site implicated in the stabilization of inactive conformations of GPCRs is a conserved XBBXXB sequence (where B is a basic amino acid and X is nonbasic) at the junction of intracellular loop 3 and transmembrane domain 6. In the μ-opioid receptor this sequence is Leu^{275}-Arg-Arg-Iso-Thr-Arg^{280}. The exchange of Thr^{279} for Lys in this sequence produces a μ-opioid receptor that shows increased basal activation of [^{35}S]GTPγS binding when stably expressed in CHO cells, although in this case there is also an additional DAMGO-induced increase (Huang et al., 2001). This constitutively active μ-opioid receptor mutant behaves very similarly to the Asp^{164} mutants in that prolonged incubation with naloxone is needed for detectable protein expression and the constitutive activation of G protein is PTX sensitive.

2.5.3. C-terminal tail mutations

Replacement of both Cys^{348} and Cys^{353} in the C-terminal tail of the μ-opioid receptor also produces a receptor that shows increased basal activation of [^{35}S]GTPγS binding, 1.5-fold higher than wild-type receptors (Brillet et al., 2003). DAMGO further increased [^{35}S]GTPγS binding, in a PTX-sensitive manner. However, unlike the mutations at the junctions between transmembrane domains and intracellular loops, spontaneous activity of the C-terminal mutant was not completely reversed by PTX, suggesting coupling to non-Gαi/o receptors or other means of promoting [^{35}S]GTPγS binding. The C-terminal mutant receptor is stable and expressed at high levels even without prolonged naloxone treatment, presumably because the C-terminus is not critical for maintaining a stability of the transmembrane domain helices. In the Cys^{348}Cys^{353} double mutant, naltrexone and naloxone were neutral agonists, but cyprodime and the CTAP analogue CTOP (D-Phe-Cys-Tyr-D-Trp-Orn-Thr-Pen-Thr-NH$_2$) showed inverse agonist activity (Brillet *et al.*, 2003).

The advantage of working with chronically μ-opioid-treated systems or constitutively active mutant receptors is that they can express a high degree of constitutive activity with a large signal-to-noise ratio. On the other hand, the natural constitutively activated state(s) of the wild-type receptor may not be accurately or fully represented in these systems and so they may not identify compounds that might be inverse agonists at wild-type receptors. This problem is highlighted by the different inverse agonists identified using the different systems (Table 22.1). Although the molecular basis for these differences is unknown, a likely explanation for the distinct ligand efficacy profiles is that different receptor conformations are responsible for constitutive activity in the different systems, for example, the constitutively active mutants do not mimic the chronic morphine-dependent μ*-state, and the DAMGO-treated system may be different from the herkinorin-treated system. Moreover, photoaffinity labeling studies have suggested that there may be differences in the profile of agonist-stimulated versus constitutively stimulated Gα subunits (Liu *et al.*, 2001), implying that the agonist-stimulated and constitutively active states may not be equivalent. This idea is reinforced by the mutant μ-opioid receptors, where constitutive activity is obtained with very different mutations—some that involve amino acids that play a role in movement of the transmembrane helices and others that cause changes in the C-terminal tail downstream of helix movement.

2.6. Adenylyl cyclase assays

AC enzymes are a family of nine members (Sadana and Dessauer, 2009) that convert ATP to the second messenger cAMP. AC activity is stimulated by GTP-bound Gαs and inhibited by GTP-bound Gαi/o proteins. The AC assay is useful for studying constitutive activity and inverse agonism acutely, but since alterations are seen in the activity of the enzyme following chronic μ-opioid exposure, changes in AC have been widely used in the characterization of the μ*-opioid receptor state. Resting levels of AC in a cell are generally low, so to show effects of μ-opioids, the direct AC activator forskolin is often included at a concentration between 10 and 30 μM (an approximate EC$_{50}$ concentration) to stimulate AC activity. Alternatively, native Gαs-coupled receptors (e.g., PGE$_2$ or β-adrenergic receptors) can be activated by a suitable agonist to increase cAMP levels. Assays are run on adherent cells in culture medium (in the absence of serum) for between 5 and 15 min in the presence of a phosphodiesterase inhibitor such as 3-isobutyl-1-methylxanthine (IBMX, 1 mM). Accumulation of cAMP is measured using a variety of methods including ligand binding, ELIZA, or time-resolved fluorescence resonance energy transfer. An example of a detailed assay can be found in Clark and Traynor (2004). Additionally, there are newer methods of analysis of cAMP, such as cAMP-mediated activation of cyclic nucleotide gated ion channels (Reinsheid et al., 2003) and monitoring of cAMP signaling in neurons using genetically encoded FRET probes that allow for real-time monitoring (Vincent et al., 2008). An important advantage of cAMP measurements is that unlike G protein assays they can be made in intact cells, thereby avoiding alterations in membrane architecture and protein–protein interactions that might compromise constitutive activity. Moreover, activity at AC is downstream of G protein activation and thus benefits from amplification of the response.

In a system that expresses constitutively active Gαi/o-coupled receptors, AC activity should be reduced. Although measurements of cAMP levels would appear to be an ideal assay for the study of constitutive activity, it has seldom been employed to study basal constitutive activity or the constitutive activity of μ-opioid receptor mutants. In cells where the [^{35}S]GTPγS assay showed constitutive μ-opioid receptor signaling, the basal (unstimulated by forskolin) level of cAMP was about half of that seen with wild-type cells (Liu and Prather, 2001). The reduction in cAMP was inhibited by alkylation of the receptor with β-flunaltrexamine, this also reduced constitutive stimulation of [^{35}S]GTPγS binding (Liu and Prather, 2001). In striatal homogenates of the mouse, basal cAMP levels were increased by β-CNA and BNTX (Wang et al., 2004). Similarly, in HEKμ cells, forskolin-stimulated cAMP accumulation was increased by the irreversible antagonist β-CNA, the nonequilibrium antagonist clocinnamox, as well as BNTX, ligands that all showed negative efficacy in the [^{35}S]GTPγS assay in the same cells (Wang et al., 2001).

More use has been made of assays for AC activity to define changes occurring as a result of chronic opioid exposure. Prolonged agonist treatment of μ-opioid receptor expressing cells produces cellular homeostatic changes that result in a sensitization of AC. Following challenge with an antagonist there is a rebound increase in cAMP levels (cAMP overshoot) that is absent in naïve cells (Watts and Neve 2005). This cAMP overshoot response is also seen in brain tissue from animals following chronic opioid treatment (Williams et al., 2001) and has been linked to specific effectors mediating the opioid withdrawal syndrome (Bagley et al., 2005b). Measures of the cAMP overshoot response in chronic morphine-dependent systems have been used to describe the μ*-state and to differentiate neutral antagonists from inverse agonists. Indeed, the first description of the μ*-state was provided by Wang et al. (1994) using this methodology. To obtain a cAMP overshoot response, cells are treated chronically with μ-opioid agonist (usually 10 μM for 24–48 h), and then the agonist is removed and/or the antagonist added together with an agent to stimulate AC activity (either a Gαs receptor-coupled agonist or forskolin). In SH-SY5Y cells, washout of chronic morphine resulted in an increase in cAMP levels compared to naive cells, but addition of naloxone resulted in a greater overshoot than simply washing off the morphine. The difference was attributed to the ability of naloxone to reverse the μ*-state form of the receptor back to the basal μ-state (Wang et al., 1994). A similar degree of overshoot was seen with naltrexone and diprenorphine. In contrast, the peptidic antagonists CTAP and CTOP did not increase overshoot to any greater degree than simply removing the morphine, and they blocked the effect of naloxone. Thus, as defined by this experimental protocol, naloxone, naltrexone, and diprenorphine are inverse agonists that drive the μ*-state back to the resting receptor state whereas CTAP and CTOP are neutral antagonists. Similar findings have been shown in $GH_3μ$ cells (Liu and Prather, 2001), and in HEKμ cells (Wang et al., 2001). However, this effect of chronic agonists to produce a μ*-state of the receptor that is differentially sensitive to antagonists does not appear to be universal. In C6μ cells chronically exposed to morphine, all antagonists tested produced a degree of cAMP overshoot corresponding to that expected due to displacement of agonist from the receptor (Divin et al., 2009), while in herkinorin or DAMGO-treated CHOμ cells, CTAP and naloxone produced an equivalent degree of cAMP overshoot (Xu et al., 2007).

While these experimental protocols are straightforward and appear in certain systems to provide a constitutive form of the μ-receptor, care has to be taken with the interpretation of results. In particular, complete washout of agonist has to be confirmed to ensure that findings are not the result of antagonism of residual agonist. In addition, if neutral antagonists and inverse agonists are identified then they should compete. Further, cAMP overshoot is seen with prolonged activation of any Gαi/o-coupled receptor (Watts and Neve, 2005) and can be induced after only minutes of treatment (Levitt et al., 2009), so there is no guarantee that a rebound increase in cAMP levels actually represents a specific change in opioid receptor activity—dissociating these changes from those reflecting changes in the enzyme itself or other modulators of GPCR signaling is a challenge. Finally, as with the [^{35}S] GTPγS studies, whilst these experiments can inform about changes induced by chronic treatment of cells/tissues expressing μ-opioid receptors, they may not provide information on constitutively active receptors formed under other conditions.

2.7. Measuring constitutive activity using ion channels

The actions of Gα or Gβγ subunits on ion channels provide another readily measurable cellular effect that is one step removed from μ-opioid receptor activation of G protein heterotrimers. For Gi/Go-coupled receptors, including μ-opioid receptors, the most easily measured of these interactions are the inhibition of voltage-dependent calcium channels (I_{Ca}) and activation of G protein gated inwardly rectifying potassium channels (GIRK).

Both processes involve the direct binding of Gi/Go-derived βγ subunits to the channel proteins, and both processes have been used as sensitive probes for changes in μ-opioid receptor activity. The excess of Gi/Go proteins over μ-opioid receptors (Sternweis and Robishaw, 1984; Selley et al., 1998) and the relatively low abundance of ion channels in neurons mean that the amplification of receptor coupling to I_{Ca} and GIRK is significant, as demonstrated by the large receptor reserve for these processes (Christie et al., 1987; Connor et al., 1999).

2.8. Calcium channels

Although there are several pathways by which Gi/Go-coupled receptors can inhibit I_{Ca}, most attention has focused on the rapid, reversible, and voltage-dependent inhibition of N-type ($Ca_V2.2$) and P/Q-type ($Ca_V2.1$) I_{Ca} by G protein βγ subunits (Herlitze et al., 1996; Ikeda, 1996). This pathway is defined by a characteristic voltage dependence of the interaction between the Gβγ subunits and the channels—when the cell membrane is strongly depolarized, the Gβγ inhibition of I_{Ca} is transiently removed. This means that an index of channel inhibition can be obtained by comparing current amplitude before and after a strong depolarization (see Fig. 22.3). The ratio (S2:S1) of the current amplitude evoked by the control step (S1) and the step after the conditioning depolarization (S2) is often used as a measure of the amount of voltage-dependent inhibition; when inhibition is present, the I_{Ca} amplitude after the depolarization is larger and the ratio correspondingly greater. Importantly, the double pulse protocol (Fig. 22.3) can detect constitutive activation of the voltage-dependent inhibitory pathway; in this case, the basal ratio is greater than 1. Another feature of Gβγ inhibition of N- and P/Q-type I_{Ca} is an apparent slowing of channel activation when βγ subunits are bound (Herlitze et al., 1996; Ikeda, 1996). Although the mechanisms underlying this slowing of whole cell P/Q- and N-type channels are distinct (Colecraft et al., 2001), strong depolarization reverses the slowing, providing another measure of relief from Gβγ inhibition. The facilitation of N-type currents by a depolarizing conditioning step is usually more profound than that of P/Q-type channels, reflecting the larger inhibition of these channels by the Gβγ pathway.

The constitutive activity of several receptors has been studied using facilitation of I_{Ca} as the key assay (Beedle et al., 2004; Guo and Ikeda, 2004; Mahmoud et al., 2010; Pan et al., 1998). In these experiments, which have largely been performed in sympathetic neurons, microinjection of appropriate receptor mRNA increases the basal facilitation ratio and this increase is reversed by superfusion of appropriate antagonists, which also increase the absolute amplitude of the I_{Ca}. The constitutive activity of the opioid-related ORL1 receptor (NOP) has been explored in some detail in both tsA-201 HEK-293 cells and sympathetic neurons (Beedle et al., 2004,

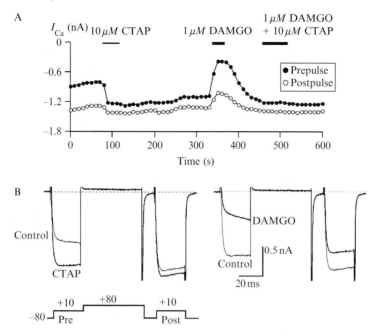

Figure 22.3 Detection of constitutively active μ-opioid receptors using voltage-gated calcium channels as a reporter. These patch clamp recordings were made from cultured sympathetic neurons injected with μ-opioid receptor mRNA. I_{Ca} were elicited using a double pulse protocol, neurons were stepped from -80 to $+10$ mV to elicit control currents (prepulse) and then βγ subunit-mediated inhibition of the channels was transiently relieved by a long step to $+80$ mV, the current amplitude was retested by another step to $+10$ mV following a brief rest at -80 mV (postpulse). (A) A representative time course of a typical experiment showing plotting pre- and postpulse amplitudes. (B) Representative current traces illustrating the effects of superfusion of the μ-opioid antagonist CTAP and the agonist DAMGO. Note that even before drug application, the amplitude of the postpulse is greater than that of the prepulse, indicating constitutive inhibition of I_{Ca}, which is reversed by the highly selective μ-opioid ligand CTAP and mimicked by a subsequent application of the agonist DAMGO. The figure is based on Fig. 8 in Mahmoud *et al.* (2010) (used with permission), with the traces kindly provided by Dr. Victor Ruiz-Velasco.

Mahmoud *et al.*, 2010), with the latter study also including a few experiments on neurons injected with μ-opioid receptor mRNA (Mahmoud *et al.*, 2010, see Figure 8). In these experiments, the μ-opioid receptor ligand CTAP increased the amplitude of the N-type I_{Ca} and reduced the facilitation ratio, while DAMGO produced a further inhibition of the I_{Ca}. This shows that high levels of receptor expression can induce constitutive activity detectable at a single cell level, without use of ligands.

The detection of the constitutive activity of native receptors using I_{Ca} assays is much more unusual (Pan *et al.*, 1998). In general, little or no facilitation of baseline I_{Ca} by a positive conditioning step is observed in native cells (e.g., Bagley *et al.*, 2005a; Borgland *et al*, 2001; Connor *et al.*, 1999), although a careful study reported modest facilitation in a subpopulation of small dorsal root ganglion neurons (Beedle *et al.*, 2004). In a small population of rat pelvic ganglion neurons expressing native cannabinoid CB1 receptors, the CB1 ligand SR 141716A was shown to inhibit tonic activation of the Gβγ pathway, probably by inhibiting constitutively active CB1 receptors (Pan *et al*, 1998). The only example of constitutive inhibition of I_{Ca} by native μ-opioid receptors was demonstrated in cultured dorsal root ganglion neurons from arrestin3 knockout mice (Walwyn *et al.*, 2007). In these neurons the basal facilitation ratio of I_{Ca} was around 1, but this ratio was reduced to about 0.8 by naltrexone. The effects of naltrexone were blocked by CTAP, indicating that they were mediated via the μ receptor. In neurons from wild-type mice, naltrexone had no effect on basal I_{Ca}. This study demonstrates the utility of the electrophysiological approach, even in cells with native levels of receptor expression.

The putative constitutively active μ*-state of the μ-opioid receptor should also produce a tonic inhibition of I_{Ca} in appropriate neurons, assuming that it activates a similar subset of G proteins as agonist-activated receptors. However, no differences in facilitation ratio were found between locus cocruleus neurons from chronically morphine-treated (CMT) and vehicle rats (Connor *et al.*, 1999), or in periaqueductal gray neurons from CMT and vehicle mice (Bagley *et al.*, 2005a). Both cell types exhibit an increased facilitation ratio and I_{Ca} activation when opioid agonists are superfused, indicating that the molecular machinery appropriate for expression of constitutive activity is present. Similarly, superfusion of naloxone and simply washing off morphine produced an equivalent increase in I_{Ca} in CMT SH-SY5Y cells, indicating that μ*-state receptors could not be detected using electrophysiology in this preparation (Kennedy and Henderson, 1992).

The Gβγ subunit modulation of I_{Ca} is ubiquitous and easy to measure using standard whole cell patch clamp techniques. There are, however, a few possible confounds that need to be kept in mind. First, many studies have shown that depolarizing steps can facilitate I_{Ca} independently of the Gβγ pathway, although the I_{Ca} affected is usually L-($Ca_V1.x$) or P/Q-type rather than N-type (Dolphin 1996). The facilitation can be associated with channel phosphorylation (Dolphin 1996) or be mediated by a direct effect of Ca (Chaudhuri *et al.*, 2007). Ca-dependent facilitation can be minimized without interfering with Gβγ-mediated channel inhibition by using Ba as the charge carrying cation (Chaudhuri *et al.*, 2007). Conversely, it can be difficult to obtain a basal facilitation ratio of around 1, owing to the propensity of I_{Ca} to desensitize. Ca-dependent desensitization can again be reduced by using Ba as a charge carrier while voltage-dependent processes can be attenuated by limiting the duration and amplitude of the test steps and the conditioning depolarization, and by allowing sufficient time between pulse sets to minimize the accumulation of channel desensitization. We and others have found that a conditioning depolarization to +80 mV for 50–80 ms is sufficient to produce a robust facilitation of μ-receptor inhibited I_{Ca} in a variety of central and peripheral neurons, without any significant channel desensitization (Bagley et al., 2005a; Connor et al., 1999; Mahmoud et al., 2010; Walwyn et al., 2007). Of course, it is also possible that a Gβγ-mediated inhibition of I_{Ca} results from constitutive activity of G proteins in a cell, quite independent of any altered G protein/receptor coupling. Given that Go-type G proteins are much more abundant than either μ-opioid receptors or I_{Ca}, small changes in basal G protein activity could produce significant effects on I_{Ca}, although in this situation the facilitation would not be sensitive to receptor antagonists.

Another possible pitfall is a change in the mix of I_{Ca} produced by a treatment. For example, in trigeminal ganglion neurons from morphine-treated mice there is a significant reduction in the amount of P/Q-type I_{Ca} (Johnson et al., 2006). If there were constitutively active receptors present in these cells, the basal facilitation ratio may actually increase, because the contribution of N-type I_{Ca} to the total current is greater, and N-type I_{Ca} are inhibited more effectively by the Gβγ pathway (Bourinet et al., 1996; Connor and Christie, 1998).

2.9. Potassium channels

GIRK (Kir 3.x) channels are directly gated by Gβγ subunits, and they also provide an easily measurable output very closely tied to G protein activation. Unlike measuring I_{Ca}, which is episodic and rarely sustainable for more than about 30 min, continuous recordings of current or membrane potential can be used to monitor GIRK activity for periods of hours. GIRK currents are also most readily measured in brain slices, preparation of which generally requires less tissue handling or exposure to unphysiological solutions than dissociation (and culturing) of neurons for recording I_{Ca}. There are a few reports of constitutively active GIRK in mammalian cells (Chen and Johnston, 2005; Dobrev et al., 2005), but we are not aware that constitutive activity associated with a specific GPCR has been detected in a mammalian cell using measurements of GIRK function. Several studies have examined coupling of GPCR to GIRK in sympathetic neurons and HEK 293 cells transfected with both receptors and channels, and constitutive activation of heterologously expressed GIRK was small and infrequently observed (Guo and Ikeda, 2004; Johnson et al., 2006; Ruiz-Velasco and Ikeda, 1998).

There are several reasons why constitutively active GIRK may be rarely reported, and activation of GIRK mediated by constitutively active receptors not at all. First, maximal Gβγ activation of the GIRK channels requires binding of multiple Gβγ subunits (Sadja et al., 2002), which is in contrast to the single βγ subunit required for inhibition of I_{Ca} (Dascal, 2001).

Thus, more constitutively active receptors may be required to produce a detectable Gβγ signal at GIRK than at I_{Ca}. Second, there is no obvious ligand-free strategy for detecting basally active GIRK currents; there are no specific GIRK channel blockers, and Gβγ modulation of the currents is not intrinsically voltage dependent (Doupnik et al., 1995).

Despite the necessity of using ligands to detect constitutive receptor activation of GIRK, μ*-state stimulation of GIRK should be apparent when opioids which inhibit μ*-state signaling are superfused onto neurons from CMT animals. Reversal of μ*-state activity would be seen as ligand-precipitated closing of a K conductance or a membrane depolarization. This has not been observed in studies where naloxone has been washed onto brain slices from morphine-treated animals (Christie et al., 1987), and studies that have reported opioid antagonist-mediated depolarization in PAG have attributed this to non-GIRK conductances (Bagley et al., 2005b; Chieng and Christie, 1996). Opioid receptor modulation of cAMP-regulated conductances or neurotransmitter release onto neurons in brain slices are a potential source of significant confounds in studies of constitutive activity in these preparations.

3. Conclusions

Constitutive activity reflects fundamental properties of GPCR. μ-Opioid ligands are extremely important therapeutic drugs, with significant unwanted effects, and a better understanding of constitutive activity at the μ-opioid receptor may provide information useful in developing ligands that access subsets of receptor conformations, offering the potential to fine-tune opioid pharmacotherapy. The unique role suggested for the putative μ*-state of the receptor also provides an intriguing possibility for therapeutic exploitation. While readily amenable to standard methods of measuring constitutive activity, μ-opioid receptors also couple to G proteins that modulate ion channels, providing the opportunity for studying receptor states in single neurons in real time. Together with emerging techniques for relatively direct studies of conformational states within GPCR (Yao et al., 2009), and access to purified μ-opioid receptors (Kuszak et al., 2009), the methods outlined in this chapter should provide the tools for many fruitful investigations of this intriguing property of the μ-opioid receptor.

ACKNOWLEDGMENTS

This work supported by NH&MRC of Australia Project Grant 512159 to M. C., and National Institutes of Health Grant DA04087 to J. T. We thank Dr. Victor Ruiz-Velasco for his generosity in sharing his data for Fig. 22.3 and Alisa Knapman for a careful reading of the manuscript.

REFERENCES

Bagley, E. E., Chieng, B. C. H., Christie, M. J., and Connor, M. (2005a). Opioid tolerance in periaqueductal gray neurons isolated from mice chronically treated with morphine. *Br. J. Pharmacol.* **146,** 68–76.

Bagley, E. E., Gerke, M. B., Vaughan, C. W., Hack, S. P., and Christie, M. J. (2005b). GABA transporter currents activated by protein kinase A excite midbrain neurons during opioid withdrawal. *Neuron* **45,** 433–445.

Beedle, A. M., McRory, J. E., Poirot, O., Doering, C. J., Altier, C., Barrere, C., Hamid, J., Nargeot, J., Bourinet, E., and Zamponi, G. W. (2004). Agonist-independent modulation of N-type calcium channels by ORL1 receptors. *Nat. Neurosci.* **7,** 118–125.

Borgland, S. L., Connor, M., and Christie, M. J. (2001). Nociceptin inhibits calcium channel currents in a subpopulation of small nociceptive trigeminal ganglion neurons in mouse. *J. Physiol.* **536,** 35–47.

Bourinet, E., Soong, T. W., Stea, A., and Snutch, T. P. (1996). Determinants of the G protein-dependent opioid modulation of neuronal calcium channels. *Proc. Natl. Acad. Sci. USA.* **93,** 1486–1491.

Breivogel, C. S., Walker, J. M., Huang, S. M., Roy, M. B., and Childers, S. R. (2004). Cannabinoid signaling in rat cerebellar granule cells: G-protein activation, inhibition of glutamate release and endogenous cannabinoids. *Neuropharmacology* **47,** 81–91.

Brillet, K., Kieffer, B. L., and Massotte, D. (2003). Enhanced spontaneous activity of the mu opioid receptor by cysteine mutations: Characterization of a tool for inverse agonist screening. *BMC Pharmacol.* **3,** 14.

Burford, N. T., Wang, D., and Sadee, W. (2000). G-protein coupling of μ-opioid receptors (OP$_3$): Elevated basal signalling activity. *Biochem. J.* **348,** 531–537.

Cerione, R. A., Codina, J., Benovic, J. L., Lefkowitz, R. J., Birnbaumer, L., and Caron, M. G. (1984). The mammalian β2-adrenergic receptor: Reconstitution of functional interactions between pure receptor and pure stimulatory nucleotide binding protein of the adenylate cyclase system. *Biochemistry* **23,** 4519–4525.

Chaudhuri, D., Issa, J. B., and Yue, D. T. (2007). Elementary mechanisms producing facilitation of Ca$_V$2.1 (P/Q-type) channels. *J. Gen. Physiol.* **129,** 385–401.

Chen, X., and Johnston, D. (2005). Constitutively active G-protein-gated inwardly rectifying K$^+$ channels in dendrites of hippocampal CA1 pyramidal neurons. *J. Neurosci.* **25,** 3787–3792.

Chieng, B., and Christie, M. J. (1996). Local opioid withdrawal in rat single periaqueductal gray neurons in vitro. *J. Neurosci.* **16,** 7128–7136.

Christie, M. J., Williams, J. T., and North, R. A. (1987). Cellular mechanisms of opioid tolerance: Studies in single brain neurons. *Mol. Pharmacol.* **32,** 633–638.

Clark, M. J., and Traynor, J. R. (2004). Assays for G-protein-coupled receptor signaling using RGS-insensitive Galpha subunits. *Meth. Enzymol.* **389,** 155–169.

Colecraft, H. M., Brody, D. L., and Yue, D. T. (2001). G-protein inhibition of N- and P/Q-type calcium channels: Distinctive elementary mechanisms and their functional impact. *J. Neurosci.* **21,** 1137–1147.

Connor, M. (2009). Shadows across μ-star? Constitutively active μ-opioid receptors revisited. *Br. J. Pharmacol.* **156,** 1041–1043.

Connor, M., Borgland, S. L., and Christie, M. J. (1999). Continued morphine modulation of calcium channel currents in acutely isolated locus coeruleus neurons from morphine-dependent rats. *Br. J. Pharmacol.* **128,** 1561–1569.

Connor, M., and Christie, M. J. (1998). Modulation of calcium channel currents of acutely dissociated rat periaqueductal grey neurons. *J. Physiol.* **509,** 47–58.

Costa, T., and Herz, A. (1989). Antagonists with negative intrinsic activity at δ opioid receptors coupled to GTP-binding proteins. *Proc. Natl. Acad. Sci. USA* **86,** 7321–7325.

Dascal, N. (2001). Ion-channel regulation by G proteins. *Trends Endocrinol. Metab.* **12**, 391–398.

DeLapp, N. W. (2004). The antibody-capture [^{35}S]GTPγS scintillation proximity assay: A powerful emerging technique for analysis of GPCR pharmacology. *Trends Pharmacol. Sci.* **25**, 400–401.

Divin, M. F., Bradbury, F. A., Carroll, F. I., and Traynor, J. R. (2009). Neutral antagonist activity of naltrexone and 6β-naltrexol in naive and opioid-dependent C6 cells expressing a μ-opioid receptor. *Br. J. Pharmacol.* **156**, 1044–1053.

Dobrev, D., Friedrich, A., Voight, N., Jost, N., Wettwer, E., Christ, T., Knaut, M., and Ravens, U. (2005). The G protein-gated potassium current IK, ACH is constitutively active in patients with chronic atrial fibrillation. *Circulation* **112**, 3697–3706.

Dolphin, A. C. (1996). Facilitation of Ca^{2+} current in excitable cells. *Trends Neurosci.* **19**, 35–43.

Doupnik, C. A., Lim, N. F., Kofuji, P., Davidson, N., and Lester, H. A. (1995). Intrinsic gating properties of a cloned G protein-activated inward rectifier K$^+$ channel. *J. Gen. Physiol.* **106**, 1–23.

Guo, J., and Ikeda, S. R. (2004). Endocannabinoids modulate N-type calcium channels and G-protein-coupled inwardly rectifying potassium channels via CB1 cannabinoid receptors heterologously expressed in mammalian neurons. *Mol. Pharmacol.* **65**, 665–674.

Harrison, C., and Traynor, J. R. (2003). The [^{35}S]GTPgammaS binding assay: Approaches and applications in pharmacology. *Life Sci.* **74**, 489–508.

Herlitze, S., Garcia, D. E., Mackie, K., Hille, B., Scheuer, T., and Catterall, W. A. (1996). Modulation of Ca^{2+} channels by G-protein βγ subunits. *Nature* **380**, 258–262.

Horswill, J. G., Bali, U., Shaaban, S., Kelly, J. F., Jeevaratnam, P., Babbs, A. J., Reynet, A. J., and In, P. W. K. (2007). PSNCBAM-1, a novel allosteric antagonist at cannabinoid CB1 receptors with hypophagic effects in rats. *Br. J. Pharmacol.* **152**, 805–814.

Howlett, A. C. (2004). Efficacy in CB1 receptor-mediated signal transduction. *Br. J. Pharmacol.* **142**, 1209–1218.

Huang, P., Li, J., Chen, C., Visiers, I., Weinstein, H., and Liu-Chen, L. Y. (2001). Functional role of a conserved motif in TM6 of the rat mu opioid receptor: Constitutively active and inactive receptors result from substitutions of Thr6.34(279) with Lys and Asp. *Biochemistry* **40**, 13501–13509.

Ikeda, S. R. (1996). Voltage-dependent modulation of N-type calcium channels by G-protein βγ subunits. *Nature* **380**, 255–258.

Johnson, E. A., Oldfield, S., Brakstor, E., Gonzalez-Cuello, A., Couch, D., Hall, K. J., Mundell, S. J., Bailey, C. P., Kelly, E., and Henderson, G. (2006). Agonist-selective mechanisms of μ-opioid receptor desensitization in human embryonic kidney 293 cells. *Mol. Pharmacol.* **70**, 676–685.

Kara, E., and Strange, P. (2010). Use of the [35S]GTPγS binding assay to determine ligand efficacy at G-protein coupled receptors. *In* "Protein-Coupled Receptors: Essential Methods," (D. Poyner and M. G. Wheatley, eds.), .Wiley-VCH.

Kenakin, T. (2004). Efficacy as a vector: The relative prevalence and paucity of inverse agonism. *Mol. Pharmacol.* **65**, 2–11.

Kenakin, T., and Onaran, O. (2002). The ligand paradox between affinity and efficacy: Can you be there and not make a difference? *Trends Pharmacol. Sci.* **23**, 275–280.

Kennedy, C., and Henderson, G. (1992). Chronic exposure to morphine does not induce dependence at the level of the calcium current in human SH-SY5Y cells. *Neuroscience* **49**, 937–944.

Kuszak, A. J., Pitchiaya, S., Anand, J. P., Mosberg, H. I., Walter, N. G., and Sunahara, R. K. (2009). Purification and functional reconstitution of monomeric mu-opioid receptors: Allosteric modulation of agonist binding by Gi2. *J. Biol. Chem.* **284**, 26732–26741.

Labrecque, J., Wong, R. S., and Fricker, S. P. (2009). A time-resolved fluorescent lanthanide (Eu)-GTP binding assay for chemokine receptors as targets in drug discovery. *Meth. Mol. Biol.* **552,** 153–169.

Levitt, E. S., Clark, M. J., Jenkins, P. M., Martens, J. R., and Traynor, J. R. (2009). Differential effect of membrane cholesterol removal on mu- and delta-opioid receptors: A parallel comparison of acute and chronic signaling to adenylyl cyclase. *J. Biol. Chem.* **284,** 22108–22122.

Li, J., Chen, C., Huang, P., and Liu-Chen, L.-Y. (2001a). Inverse agonist up-regulates the constitutively active D3.49(164)Q mutant of the rat μ-opioid receptor by stabilizing the structure and blocking constitutive internalization and down-regulation. *Mol. Pharmacol.* **60,** 1064–1075.

Li, J., Huang, P., Chen, C., de Reil, J. K., Weinstein, H., and Liu-Chen, L.-Y. (2001b). Constitutive activation of the mu opioid receptor by mutation of D3.49(164), but not D3.32(147): D3.49(164) is critical for stabilization of the inactive form of the receptor and for its expression. *Biochemistry* **40,** 12039–12050.

Lin, H., Saisch, S. G., and Strange, P. G. (2006). Assays for enhanced activity of low efficacy partial agonists at the D(2) dopamine receptor. *Br. J. Pharmacol.* **149,** 291–299.

Liu, J.-G., and Prather, P. L. (2001). Chronic exposure to μ-opioid agonists produces constitutive activation of the μ-opioid receptors in direct proportion to the efficacy of the agonist used for pretreatment. *Mol. Pharmacol.* **60,** 53–62.

Liu, J. G., Ruckle, M. B., and Prather, P. L. (2001). Constitutively active μ-opioid receptors inhibit adenylyl cyclase activity in intact cells and activate G-proteins differently than the agonist [D-Ala2, N-MePhe4, Gly-ol5]enkephalin. *J. Biol. Chem.* **276,** 37779–37786.

Mahmoud, S., Margas, W., Trapella, C., Calo, G., and Ruiz-Velasco, V. (2010). Modulation of silent and constitutively active nociceptin/orphanin FQ receptors by potent receptor antagonists and Na^+ ions in rat sympathetic neurons. *Mol. Pharmacol.* **77,** 804–817.

Neilan, C. L., Akil, H., Woods, J. H., and Traynor, J. R. (1999). Constitutive activity of the δ-opioid receptor expressed in C6 glioma cells: Identification of non-peptide δ-inverse agonists. *Br. J. Pharmacol.* **128,** 556–562.

Pan, X., Ikeda, S. R., and Lewis, D. L. (1998). SR 141716A acts as an inverse agonist to increase neuronal voltage-dependent Ca^{2+} currents by reversal of tonic CB1 cannabinoid receptor activity. *Mol. Pharmacol.* **54,** 1064–1072.

Raehal, K. M., Lowery, J. J., Bhamidipati, C. M., Paolino, R. M., Blair, J. R., Wang, D., Sadee, W., and Bilsky, E. J. (2005). In vivo characterization of 6beta-naltrexol, an opioid ligand with less inverse agonist activity compared with naltrexone and naloxone in opioid-dependent mice. *J. Pharmacol. Exp. Ther.* **313,** 1150–1162.

Reinsheid, R. K., Kim, J., Zeng, J., and Civelli, O. (2003). High-throughput real-time monitoring of Gs-coupled receptor activation in intact cells using cyclic-nucleotide gated channels. *Eur. J. Pharmacol.* **478,** 27–34.

Roberts, D. J., and Strange, P. G. (2005). Mechanisms of inverse agonist action at D2 dopamine receptors. *Br. J. Pharmacol.* **145,** 34–42.

Rosenbaum, D. M., Rasmussen, S. G., and Kobilka, B. K. (2009). The structure and function of G-protein-coupled receptors. *Nature* **459,** 356–363.

Ruiz-Velasco, V., and Ikeda, S. R. (1998). Heterologous expression and coupling of G protein-gated inwardly rectifying K^+ channels in adult rat sympathetic neurons. *J. Physiol.* **513,** 761–773.

Sadana, R., and Dessauer, C. W. (2009). Physiological roles for G protein-regulated adenylyl cyclase isoforms: Insights from knockout and overexpression studies. *Neurosignals* **17,** 5–22.

Sadee, W., Wang, D., and Bilsky, E. J. (2005). Basal opioid receptor activity, neutral antagonists, and therapeutic opportunities. *Life Sci.* **76,** 1427–1437.

Sadja, R., Alagem, N., and Reuveny, E. (2002). Graded contribution of the Gβγ binding domains to GIRK channel function. *Proc. Natl. Acad. Sci. USA* **99,** 10783–10788.

Sally, E. J., Xu, H., Dersch, C. M., Hsin, L.-W., Chang, L.-T., Prisinzano, T. E., Simpson, D. S., Giuvelis, D., Rice, K. C., Jacobson, A. E., Cheng, K., Bilsky, E. J., et al. (2010). Identification of a novel "almost neutral" μ-opioid receptor antagonist in CHO cells expressing the cloned μ-opioid receptor. *Synapse* **64**, 280–288.

Selley, D. E., Nestler, E. J., Brievogel, C. S., and Childers, S. R. (1997). Opioid receptor-coupled G proteins in rat locus coeruleus membranes: Decrease in activity after chronic morphine treatment. *Brain Res.* **746**, 10–18.

Selley, D. E., Liu, Q., and Childers, S. R. (1998). Signal transduction correlates of *mu*-opioid agonist intrinsic efficacy: Receptor-stimulated [^{35}S]-GTPγS binding in mMOR-CHO cells and rat thalamus. *J. Pharmacol. Exp. Ther.* **285**, 496–505.

Smit, M. J., Vischer, H. F., Bakker, R. A., Jongejan, A., Timmerman, H., Pardo, L., and Leurs, R. (2007). Pharmacogenomic and structural analysis of constitutive G protein-coupled receptor activity. *Annu. Rev. Pharmacol. Toxicol.* **47**, 53–87.

Sternweis, P. C., and Robishaw, J. D. (1984). Isolation of two proteins with high affinity for guanine nucleotides from membranes of bovine brain. *J. Biol. Chem.* **259**, 13806–13813.

Szekeres, P. G., and Traynor, J. R. (1997). Delta opioid modulation of the binding of guanosine-5′-O-(3-[35S]thio)triphosphate to NG108-15 cell membranes: Characterization of agonist and inverse agonist effects. *J. Pharmacol. Exp. Ther.* **283**, 1276–1284.

Traynor, J. R., and Nahorski, S. R. (1995). Modulation by μ-opioid agonists of guanosine-5′-O-(3-[^{35}S]thio)triphosphate binding to membranes from human neuroblastoma SH-SY5Y cells. *Mol. Pharmacol.* **47**, 848–854.

Vincent, P., Gervasi, N., and Zhang, J. (2008). Real-time monitoring of cyclic nucleotide signalling in neurons using genetically encoded FRET probes. *Brain Cell Biol.* **36**, 3–17.

Walwyn, W., Evans, C. J., and Hales, T. G. (2007). β-Arrestin and c-Src regulate the constitutive activity and recycling of μ-opioid receptors in dorsal root ganglion neurons. *J. Neurosci.* **27**, 5092–5104.

Wang, Z., Bilsky, E. J., Porreca, F., and Sadee, W. (1994). Constitutive μ opioid receptor activation as a regulatory mechanism underlying narcotic tolerance and dependence. *Life Sci.* **54**, 339–350.

Wang, D., Raehal, K. M., Bilsky, E. J., and Sadee, W. (2001). Inverse agonists and neutral antagonists at μ opioid receptor (MOR): Possible role of basal receptor signaling in narcotic tolerance. *J. Neurochem.* **77**, 1590–1600.

Wang, D., Raehal, K. M., Lin, E. T., Lowery, J. J., Keiffer, B. L., Bilsky, E. J., and Sadee, W. (2004). Basal signalling activity of μ opioid receptor in mouse brain: Role in narcotic dependence. *J. Pharmacol. Exp. Ther.* **308**, 512–520.

Wang, D., Sun, X., and Sadee, W. (2007). Different effects of opioid antagonists on μ-, δ-, and κ-opioid receptors with and without agonist pretreatment. *J. Pharmacol. Exp. Ther.* **321**, 544–552.

Watts, V. J., and Neve, K. A. (2005). Sensitization of adenylate cyclase by Gαi/o-coupled receptors. *Pharmacol. Ther.* **106**, 405–421.

Williams, J. T., Christie, M. J., and Manzoni, O. (2001). Cellular and synaptic adaptations mediating opioid withdrawal. *Physiol. Rev.* **81**, 299–343.

Xu, H., Partilla, J. S., Wang, X., Rutherford, J. M., Tidgewell, K., Prisinzano, T. E., Bohn, L. M., and Rothman, R. B. (2007). A comparison of noninternalizing (herkinorin) and internalizing (DAMGO) μ-opioid agonists on cellular markers related to opioid tolerance and dependence. *Synapse* **61**, 166–175.

Yao, X. J., Velez Ruiz, G., Whorton, M. R., Rasmussen, S. G. F., DeVree, B. T., Deupi, X., Sunahara, R. K., and Kobilka, B. (2009). The effect of ligand efficacy on the formation and stability of a GPCR-G protein complex. *Proc. Natl. Acad. Sci. USA* **106**, 9501–9506.

CHAPTER TWENTY-THREE

Protein Kinase CK2 Is a Constitutively Active Enzyme that Promotes Cell Survival: Strategies to Identify CK2 Substrates and Manipulate Its Activity in Mammalian Cells

Jacob P. Turowec, James S. Duncan, Ashley C. French, Laszlo Gyenis, Nicole A. St. Denis, Greg Vilk, *and* David W. Litchfield

Contents

1. Introduction	472
2. Purification of CK2 for *In Vitro* Studies	474
2.1. Purification of recombinant CK2 from bacteria	474
2.2. Isolation of CK2 by immunoprecipitation from mammalian cells	479
3. Assays for CK2 Activity	481
3.1. Kinase assays using GST-CK2 and the synthetic peptide RRRDDDSDDD	481
3.2. CK2 kinase assays with peptide arrays	483
3.3. CK2 kinase assays with protein substrates	484
4. Modulation of CK2 in Mammalian Cells	485
4.1. Transient expression of CK2 in mammalian cells	486
4.2. Inducible expression of CK2 in mammalian cells	487
4.3. Modulation of CK2 in mammalian cells: Additional considerations	490
5. Conclusions	491
Acknowledgments	491
References	492

Abstract

Protein kinase CK2 is a constitutively active protein serine/threonine kinase that is ubiquitously expressed and essential for the survival of eukaryotic cells.

Department of Biochemistry, Schulich School of Medicine & Dentistry, University of Western Ontario, London, Ontario, Canada

Methods in Enzymology, Volume 484   © 2010 Elsevier Inc.
ISSN 0076-6879, DOI: 10.1016/S0076-6879(10)84023-6   All rights reserved.

On the basis of its elevated expression in a number of human cancers and its ability to promote tumorigenesis in transgenic mice, CK2 has emerged as a promising candidate for molecular-targeted therapy. Accordingly, there has been considerable interest in identifying the cellular events that are regulated by CK2 and the cellular substrates of CK2 that are responsible for mediating its actions in cells. Large-scale phosphoproteomics studies are revealing extensive lists of candidate CK2 substrates on the basis that these proteins are phosphorylated at sites conforming to the consensus for phosphorylation by CK2. However, efforts to validate the vast majority of these candidates as *bona fide* physiological CK2 substrates have been hindered by the lack of systematic strategies to identify its direct substrates and manipulate its activity in intact cells. To overcome these limitations, we describe experimental procedures for isolating CK2 from bacteria and from mammalian cells to enable *in vitro* phosphorylation of candidate substrates. We also outline strategies for manipulating the levels and activity of CK2 in intact cells. Collectively, the methods that are presented in this chapter should enable the identification and characterization of CK2 substrates and CK2-regulated processes both *in vitro* and in living cells.

1. INTRODUCTION

Protein kinases catalyze the reversible phosphorylation of proteins to regulate the majority of cellular processes (Hunter, 2000). For example, protein kinases that are located at distinct nodes in complex signaling networks can mediate changes in transcription, protein structure and function, and protein–protein interactions. While many protein kinases are regulated by diverse stimuli that alter their activity via posttranslational modifications or second messengers, there are other protein kinases that are constitutively active. Due to the lack of a convenient, defined means for manipulating activity in cells, a detailed investigation of constitutively active protein kinases can be particularly challenging.

Protein kinase CK2 is a constitutively active kinase implicated in a broad range of cellular processes including the control of cellular proliferation and apoptosis (Litchfield, 2003). CK2 is a protein serine/threonine kinase that is ubiquitously distributed in eukaryotes and typically found in tetrameric complexes composed of two regulatory subunits (CK2β) and two catalytic subunits (CK2α and/or CK2α′). When expressed in bacteria, the catalytic subunits of CK2 display robust activity demonstrating that the enzyme is constitutively active (Grankowski *et al.*, 1991). Addition of the regulatory CK2β subunits to form the tetrameric CK2 holoenzyme does not serve strictly to activate or inactive the catalytic activity of CK2 (i.e., both free catalytic subunits and tetrameric CK2 complexes are catalytically competent). Instead, CK2β appears to modulate substrate specificity as the free catalytic CK2 subunits and tetrameric CK2 complexes do exhibit some differences in their abilities to phosphorylate individual substrates (Meggio et al., 1992). In this respect, while many substrates are phosphorylated to a similar extent by free catalytic subunits and tetrameric CK2, the presence of CK2β can either promote or inhibit the phosphorylation of other substrates.

Although CK2 is constitutively active, recent work has highlighted the potential for spatial and temporal regulation of CK2 within cells. For example, CKIP-1 is a CK2α interacting partner that targets a subpopulation of CK2α to the cell membrane where it appears to participate in modulating the actin cytoskeleton (Canton et al., 2005). In addition, the peptide prolyl-isomerase Pin1 interacts with CK2 in a phosphorylation-dependent manner in M-phase apparently altering the ability of CK2 to phosphorylate topo-isomerase II (Messenger et al., 2002). Overall, CK2 interacts with a large number of cellular proteins, many of which likely contribute to ensuring that CK2 phosphorylates its physiological substrates at the appropriate time and location within cells.

In recent years, CK2 has attracted considerable attention because it is frequently elevated in human cancers (Duncan and Litchfield, 2008; St-Denis and Litchfield, 2009). It is apparent that CK2 overexpression is not simply a result of transformation, but actually contributes to tumorigenesis. Notably, targeted overexpression of CK2 in T-cells and mammary glands of transgenic mice leads to lymphoma development and mammary tumorigenesis, respectively, though the precise mechanisms responsible for the tumorigenic potential of CK2 remain poorly defined (Landesman-Bollag et al., 2001; Seldin and Leder, 1995). However, in view of its constitutive activity, the ability of CK2 to protect caspase substrates from caspase-catalyzed cleavage is intriguing (Duncan et al., 2010). In this respect, increased phosphorylation of caspase substrates by the elevated levels of CK2 could promote cancer cell survival by preventing caspase cleavage to block apoptosis. Along these lines, there is a growing body of evidence supporting the notion that CK2 inhibition, via pharmacological inhibitors or antisense RNA, results in the induction of apoptosis in cancer cells (Duncan and Litchfield, 2008).

A detailed understanding of how CK2 contributes to tumorigenesis and other biological processes will ultimately require the comprehensive identification of its cellular substrates. While the identification of bona fide physiological substrates for CK2 has been hindered by a traditional lack of systematic strategies to manipulate its activity in experimental models, knowledge of its consensus phosphorylation motif (S/T-X-X-D/E/pS/pY) has promoted the identification of many candidate CK2 substrates (Meggio and Pinna, 2003). In fact, even before many genome sequences were available, recognition of sequences conforming to this consensus had prompted examination of the phosphorylation of many individual proteins by CK2. Together with striking technological advances, completion of genome sequencing projects has enabled the emergence of the field of proteomics and many large-scale projects directed at the identification and characterization of proteins on a systems level. Interrogation of phosphoproteomics databases is now revealing candidate CK2 substrates at unprecedented rates. Indeed, some predictive analyses attribute 20% of the phosphoproteome as being substrates of CK2 (Salvi et al., 2009). While phosphoproteomics databases contain a wealth of information, validation of candidate substrates as *bona fide* CK2 substrates remains challenging as do efforts to ascribe functional consequences to CK2-catalyzed phosphorylaton events. In this chapter, we describe a platform for systematically identifying and validating CK2 substrates and investigating the functional consequences of CK2-catalyzed phosphorylation. The first part of this chapter includes strategies for the isolation of CK2 followed by analysis of its catalytic activity using synthetic peptides, peptide arrays, or recombinant substrates. We subsequently present methods for modulating CK2 levels in mammalian cells that can enable examination of the functional consequences of CK2 phosphorylation in these cells.

2. Purification of CK2 for *In Vitro* Studies

In this section, we describe the isolation of CK2 primarily for the purpose of evaluating the phosphorylation of synthetic peptides or recombinant proteins to identify its direct substrates and/or sites of phosphorylation. Depending on the applications, obtaining catalytically competent CK2 from a given source may be preferred over another. For instance, purification of recombinant CK2 represents a simple means for attaining significant quantities of catalytically competent enzyme that requires only basic laboratory supplies necessary for handling recombinant protein. However, in instances where it is undesirable to generate a bacterial expression vector or where purification of such large quantities of active enzyme is unnecessary, we also provide alternative strategies for isolating CK2 from mammalian cells using epitope-tagged CK2. As well as being useful for the study of enzyme/substrate relationships, this latter method is also applicable for evaluating CK2 activity in response to cellular treatments.

2.1. Purification of recombinant CK2 from bacteria

The following sections describe the isolation of free catalytic subunits of CK2 (either CK2α or CK2α') as well as tetrameric CK2 complexes that comprise both catalytic CK2 subunits and regulatory CK2β subunits. To enable rapid purification of recombinant proteins by affinity chromatography, the catalytic CK2 subunits are expressed as GST-fusion proteins while a His-tag is incorporated at the N-terminus of CK2β.

2.1.1. Materials for purification of GST-fusion proteins

Plasmid DNA: pDB1 (pGEX-3X (GE Healthcare Life Sciences) with cDNA encoding human CK2α to express GST-CK2α), pDB6 (pGEX-3X with cDNA encoding human CK2α' to express GST-CK2α'), or pAB46 (pet28A (Novagen) with cDNA encoding human CK2β to express His-CK2β). For full list of plasmids described in this chapter, see Table 23.1.

Bacteria, plates, and media: electrocompetent BL21 *E. coli*, LB-agar plates with 100 μg/ml ampicillin, 2X YT media, isopropyl β-D-1-thiogalactopyranoside (IPTG).

Solutions and materials for purification and storage: phosphate-buffered saline (PBS), lysis buffer (PBS with 30 μg/ml aprotinin, 20 μg/ml leupeptin, and 1 mM phenylmethylsulfonyl fluoride), 10% Triton X-100 in ddH$_2$O, glutathione–agarose beads (Sigma), disposable plastic gravity flow columns, elution buffer 1 (10 mM reduced glutathione, 50 mM Tris–Cl (pH 8.0), 1 mM DTT), elution buffer 2 (30 mM reduced glutathione, 50 mM Tris–Cl (pH 8.0), 1 mM DTT), storage buffer (15 mM Mops (pH 7.0), 1.5 mM EDTA, 0.75 mM DTT, 300 mM (NH$_4$)$_2$SO$_4$, 25% glycerol).

2.1.2. Purification of GST-CK2α or GST-CK2α'

We routinely express GST-CK2α and GST-CK2α' in *E.Coli* (BL21). Beginning with freshly plated bacteria, individual colonies are used to initiate 5–30 ml starter cultures in 2X YT with 100 μg/ml ampicillin. After overnight (16–20 h) growth at 37 °C, a 1/100 dilution of starter culture is used to inoculate 50 ml–1 l of 2XYT with 0.1 mg/ml ampicillin. When an OD600 ~0.8 is reached, protein production is induced by 0.5 mM of IPTG for 3 h at 37 °C. Bacteria are isolated by centrifugation at 4500 × *g* for 15 min at 4 °C, resuspended in 100 ml cold PBS, and spun again at 4500 × *g* for 15 min at 4 °C to remove residual media. If necessary, at this point pellets can be stored at −20 °C. Next, pellets are resuspended in 15 ml of lysis buffer and passed two to three times through a French press at 10,000 psi. Alternatively, bacteria can be lysed by sonication on ice but our experience is that the French press is more efficient and leads to higher purification yields. To ensure complete lysis, we then tumble the lysate at 4 °C with Triton X-100 at a final concentration of 1% for 15 min. The lysate is then cleared by centrifugation at 27,000 × *g* for 20 min at 4 °C. Next, the supernatant is poured over a prepared, disposable gravity column containing 2 ml of a 1:1 slurry of glutathione–agarose:PBS beads and then rotated at 4 °C for 1 h. Note that this amount of slurry has an approximately 10 mg capacity, is used for 1 L of culture, and can be scaled to culture size. At this point, the beads are washed with 20–30 column volumes of cold lysis buffer and then eluted by capping the column, resuspending the agarose beads in 1 ml of cold elution buffer 1, incubating for 5 min at 4 °C, and letting the column empty by gravity flow after removing the cap. This step

Table 23.1 Plasmid constructs encoding CK2 described in this chapter

Plasmid	Vector	Insert(s)	Comments
pDB1	pGEX-3X	CK2α	Bacterial expression of GST-CK2α
pDB6	pGEX-3X	CK2α'	Bacterial expression of GST-CK2α'
pAB46	pET-28a	CK2β	Bacterial expression of His-CK2β
pZW6	pRC/CMV	CK2α	Mammalian expression of CK2α HA
pZW16	pRC/CMV	CK2α'	Mammalian expression of HA-CK2α'
pZW12	pRC/CMV	CK2β	Mammalian expression of Myc-CK2β
pGV15	pRC/CMV	CK2αK68M	Mammalian expression of CK2αK68M-HA (catalytically inactive mutant)
pGP18	pRC/CMV	CK2α'K69M	Mammalian expression of HA-CK2α'K69M (catalytically inactive mutant)
pAB07	pCDNA3.1	CK2β	Mammalian expression of Myc-CK2β
pRS3	pBI	CK2α, CK2β	Mammalian expression of tetracycline-regulated CK2α-HA and Myc-CK2β (used to generate RS 3.22 cells)
pRS2	pBI	CK2α', CK2β	Mammalian expression of tetracycline-regulated HA-CK2α' and Myc-CK2β (used to generate RS 2.31 cells)
pGV13	pBI	CK2αK68M CK2β	Mammalian expression of tetracycline-regulated CK2αK68M-HA & Myc-CK2β (used to generated GV 13.35 cells)
pGV7	pBI	CK2α'K69M CK2β	Mammalian expression of tetracycline-regulated HA-CK2α'K68M & Myc-CK2β (used to generated GV 7.21 cells)

The positions of fusions or epitope tags are designated by position in relation to insert. For example, an N-terminal fusion is indicated before the insert while a C-terminal epitope tag is indicated after the insert.

is repeated four more times with elution buffer 1 and two times using elution buffer 2. Fractions containing GST-CK2α or GST-CK2α′ are identified by SDS-PAGE. Finally, we dialyze GST-CK2 containing fractions into storage buffer containing 15 mM MOPS (pH 7.0), 1.5 mM EDTA, 0.75 mM DTT, 300 mM $(NH_4)_2SO_4$, 25% glycerol. We routinely obtain at least 2 mg of GST-CK2α or GST-CK2α′ with activities of at least 2 μmol/min/mg (when assayed with the synthetic peptide RRRDDDSDDD at 30 °C). Purity of GST-CK2α, and other forms of CK2 as described in Sections 2.1.3 and 2.1.4, as assessed by SDS-PAGE and Coomassie stain is shown in Fig. 23.1. Catalytic activity is maintained when stored at −20 °C, though we recommend reanalyzing activity every 6 months.

2.1.3. Reconstitution of CK2 holoenzyme with GST-CK2α or GST-CK2α′ and His-CK2β

Though GST-CK2α is catalytically competent in the absence of the regulatory CK2β subunit, documented examples of variations in substrate specificity between the holoenzyme and catalytic subunits highlight the importance of isolating both free catalytic subunits and tetrameric forms of

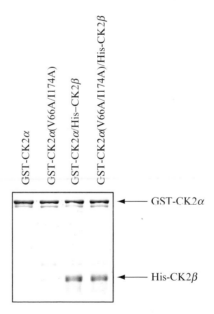

Figure 23.1 Purified recombinant CK2. Different forms of recombinant CK2 were purified as described in Section 2.1 and analyzed by SDS-polyacrylamide gel electrophoresis with Coomassie blue staining. Shown on this gel are two forms of free catalytic subunits, namely CK2α and an inhibitor-resistant form of CK2α (i.e., CK2αV66A/I174A described in Section 2.1.4) as well as each of these catalytic subunits in complexes with His-CK2β.

CK2 (Meggio et al., 1992). To enable reconstitution of the CK2 holoenzyme, we express His-CK2β. Although it can be purified with Nickel columns, our experience has been that isolation of recombinant His-CK2β has been hampered by solubility issues. Therefore, to enable the reconstitution of the CK2 holoenzyme, we circumvent this problem by mixing bacterial pellets expressing His-CK2β with the bacterial pellets expressing GST-CK2α or GST-CK2α' prior to lysis. A brief outline of this procedure is as follows. BL21 cells containing pAB46 are grown in LB media with 50 μg/ml kanamycin to an OD 600~0.800 and then induced with 0.5 mM IPTG for 24 h at 15 °C. The bacteria are then harvested by centrifugation at 4500 × g for 15 min at 4 °C as described and then mixed 2:1 (v/v) with bacterial pellets expressing GST-CK2α followed by further washing and then lysis and isolation as described in Section 2.1.2. Fractions that are eluted from glutathione agarose are examined by SDS-PAGE to ensure that holoenzyme has formed with GST-CK2α or GST-CK2α' and His-CK2β in equimolar ratio. Fractions containing CK2 holoenzyme are dialyzed for storage as described in Section 2.1.2.

2.1.4. Purification of recombinant CK2 from bacteria: Additional considerations

The procedures that have been described in the preceding sections represent the methods that we routinely employ to isolate recombinant forms of both the free catalytic CK2 subunits and the tetrameric CK2 holoenzyme. The purified enzymes can then be utilized to investigate the phosphorylation of candidate substrates in in vitro assays as described in Section 3. Depending on the application, there are modifications that may be considered. While detailed description of these modifications is beyond the scope of this chapter, a brief outline of two of our variations will be presented. For example, under some circumstances, the presence of a GST-tag on the catalytic subunits of CK2 may be undesirable. Therefore, we have generated constructs encoding GST-CK2α or GST-CK2α' where the linker between the GST and CK2α or CK2α' contains a site that can be cleaved by the TEV protease to enable removal of the GST. Another modification takes advantage of information that has emerged from structural studies of CK2 in complex with a number of inhibitors, particularly tetrabromobenzotriazole (TBB) and its derivatives (Sarno et al., 2005). Accordingly, we have generated constructs encoding variants of GST-CK2α or GST-CK2α', namely GST-CK2α(V66A/I174A) and GST-CK2α'(V67/I175A), that are catalytically competent but more than 20-fold less sensitive to treatment with TBB or some of its derivatives than wild-type CK2 (Duncan et al., 2008; Zien et al., 2005). These inhibitor-resistant mutants of CK2 can be employed in rescue experiments to validate inhibitor specificity by demonstrating that the inhibitor-resistant mutants of CK2 overcome inhibitor-dependent decreases in phosphorylation.

In addition to the modifications that are described above, we expect that other mutants of CK2α, CK2α′, or CK2β can be generated and isolated according to the strategies that we have described. Furthermore, while we have routinely used His-CK2β for our studies, we expect that untagged versions of CK2β will be amenable to purification using this methodology.

2.2. Isolation of CK2 by immunoprecipitation from mammalian cells

While recombinant proteins that are expressed in bacteria are suitable for many studies, it may be more desirable for particular applications to isolate CK2 from mammalian cells. Accordingly, in the following sections, we describe the methods that we use to isolate CK2 from mammalian cells by immunoprecipitation.

2.2.1. Materials for isolating CK2 by immunoprecipitation

Plasmid DNA: pZW6 (pRC/CMV (Invitrogen) with cDNA encoding human CK2α with a C-terminal HA tag), pZW16 (pRC/CMV with cDNA encoding human CK2α′ with an N-terminal HA tag), pEGFP-C2 (Clontech), and pZW12 (pRC/CMV with cDNA encoding human CK2β with an N-terminal Myc tag).

Cells and media: HeLa or U2OS cells, Dulbecco's modified Eagle's medium with 10% fetal bovine serum and antibiotic supplements (0.1 mg/ml streptomycin and 100 units/ml penicillin).

Solutions and materials for immunoprecipitation: PEI transfection reagents (Reed et al., 2006), PBS, NP-40 lysis buffer (1% NP-40, 150 mM NaCl, 50 mM Tris (pH 7.5)), protease inhibitors (30 μg/ml aprotinin, 20 μg/ml leupeptin, 10 μg/ml pepstatin, and 1 mM phenylmethylsulfonyl fluoride), 1 mM Na$_3$VO$_4$, protein A sepharose beads (Sigma), anti-HA 12CA5 and anti-Myc 9E10 antibodies (Roche Applied Science), and kinase assay buffer (50 mM Tris, pH 7.5, 150 mM NaCl, 10 mM MgCl$_2$, 1 mM dithiothreitol).

2.2.2. Procedure for isolation of epitope-tagged CK2 by immunoprecipitation

One method that we routinely employ to isolate CK2 from mammalian cells is to transfect cells with constructs encoding epitope-tagged versions of CK2α or CK2α′ in the presence or absence of epitope-tagged CK2β. Since the primary difference between CK2α and CK2α′ resides within their C-termini, our preference was to incorporate the HA tag at their N-termini. However, expression of CK2α with an N-terminal HA tag was consistently low, necessitating a switch to the use of a C-terminal HA tag (Penner et al., 1997). It should be noted here that exogenous overexpression of CK2 should be possible in most cell lines using any optimized transfection method. However, since we most frequently employ the PEI transfection method, this method will be described below. The day prior to transfection, we plate $\sim 1 \times 10^6$ HeLa cells on a 10 cm dish. The next morning, when cells are $\sim$60–80% confluent, cultures are fed with 5 ml of fresh media. In the afternoon, we transfect the cells with 8 μg of plasmid DNA and 2 μg of the reporter plasmid pEGFP-C2 using the PEI method (Reed et al., 2006). Cells can either be transfected to express free catalytic subunits or cotransfected to express catalytic CK2 subunits and Myc-CK2β. In the case of cotransfection, 4 μg of each plasmid are used. After allowing 16 h for expression, DNA precipitate is removed from the cells by washing twice in PBS. At this point, the cells can be harvested or fresh media added and the cells grown further. We harvest cells by removing residual media via 2 × 5 ml washes with PBS, followed by scraping the cells with ice-cold NP-40 lysis buffer. Following sonication on ice for 2 × 5 s, debris is cleared by centrifugation at 12,000 × g and 4 °C. Supernatants are then tumbled at 4 °C for 1 h with 30 μl of 50% protein A sepharose slurry and 2 μg of 12CA5 antibody/mg of lysate. Beads can then be isolated by centrifugation, washed once in lysis buffer and 3× in kinase assay buffer. The isolated CK2 can now be used in kinase assays as described in Sections 3.1.3 and 3.3. Furthermore, the presence of CK2α-HA, HA-CK2α′, and/or Myc-CK2β in immunoprecipitates can be examined by immunoblot analysis using 12CA5 antibodies to detect the HA tag or 9E10 antibodies to detect the Myc tag.

2.2.3. Isolation of CK2 by immunoprecipitation: Additional considerations

While the preceding section described the use of an epitope tag to immunoprecipitate active CK2 from mammalian cells, it should be possible to perform immunoprecipitations of endogenous CK2 using antibodies that are now available from a number of commercial sources. In this respect, our own studies with antibodies raised against C-terminal peptides from CK2α, CK2α′, or CK2β reveal that endogenous catalytically active CK2 can be isolated from mammalian cells by immunoprecipitation. Since there can be considerable variation between individual lots of commercial antibodies, the effectiveness of immunoprecipitation with a given antibody must first be verified.

Although immunoprecipitation offers a convenient strategy for isolating CK2 from cells, it may not be suitable for all applications. Under these circumstances, cell extracts can be fractionated by anion exchange chromatography (e.g., FPLC fractionation using Mono Q columns) to obtain column fractions enriched for the presence of CK2. Since a detailed description of this methodology is beyond the scope of this chapter, the reader is referred to other publications (Ahn et al., 1990).

3. Assays for CK2 Activity

This section presents methods that are employed to measure CK2 activity both for the purpose of assessing its catalytic activity and for the purpose of identifying its substrates. The first of these methods is a protocol that we routinely use for monitoring CK2 activity.

3.1. Kinase assays using GST-CK2 and the synthetic peptide RRRDDDSDDD

The method described in this section details how we measure the activity of purified CK2 using a synthetic peptide. The ability of CK2 to phosphorylate synthetic peptides is an important characteristic that allows for simple, specific measurements of CK2 activity using purified CK2 as well as crude or fractionated extracts of cells or tissues. The peptide RRREEETEEE was first described as being specific for CK2 when tested against a number of protein kinases (Kuenzel and Krebs, 1985). Subsequent studies revealed that RRRDDDSDDD has more favorable kinetic parameters (Km/Vmax 36.50 for RRRDDDSDDD vs. 0.94 for RRREEETEEE), so we typically use this peptide for routine assays of CK2 activity (Kuenzel et al., 1987). Importantly, the inclusion of three arginines on these peptides allows the use of P81 Whatman paper for separation of peptides from unincorporated γ-^{32}P-ATP, greatly reducing processing and analysis time when compared to the use of protein substrates that may require gel electrophoresis for separation. While the specific example that this is presented involves the measurement of GST-CK2 using the RRRDDDSDDD peptide, this assay can be readily adapted to monitor the activity of other forms of purified CK2.

3.1.1. Materials for performing kinase assays with GST-CK2 and RRRDDDSDDD

Enzymes and substrates: GST-CK2 (see Section 2.1), RRRDDDSDDD synthetic peptide (1–5 mM stock in 50 mM Tris, pH 7.5).

Reaction components and buffers: 5× CK2 assay buffer (250 mM Tris, pH 7.5, 750 mM NaCl, 50 mM $MgCl_2$, 500 μM ATP), CK2 dilution buffer (1 mg/ml bovine serum albumin, 5 mM MOPS, pH 7.0, 200 mM NaCl), γ-^{32}P-ATP (3000 Ci/mmol) (Perkin Elmer), 1% phosphoric acid, 95% ethanol. The 5× CK2 buffer and CK2 dilution buffer can be stored at $-20\ °C$.

Materials and equipment: shaking water bath set at 30 °C, P81 Whatman paper cut into 2 cm × 2 cm squares, PhosphorImager or X-ray film, and developer or scintillation counter.

3.1.2. Performing CK2 kinase assays using GST-CK2 and RRRDDDSDDD

First, prewarm the waterbath to 30 °C and thaw stock solutions on ice. Prepare 20 µl kinase reactions in microcentrifuge tubes by adding 100 µM RRRDDDSDDD (final concentration), 4 µl of 5× CK2 kinase reaction buffer with γ-^{32}P-ATP (typically 100–200 µCi/ml 5× kinase buffer) and ddH$_2$O to make a volume of 15 µl. To start the reaction, add 5 µl of freshly diluted GST-CK2. In our experience, our purified GST-CK2 preparations need to be diluted 1/500–1/2500 in dilution buffer before being added to the reaction. The diluted CK2 is not stable for long-term storage so is diluted immediately before use. Under the conditions described, reaction linearity is maintained for at least 10 min. To stop the reaction, spot 10 µl of the reaction on precut P81 paper. These papers are then placed in a wire mesh basket in a 500 ml beaker containing 1% phosphoric acid (typically at least 10 ml phosphoric acid/paper). The phosphoric acid is mixed by continuous stirring on a magnetic stir plate. After washing the papers 4 × 5 min in phosphoric acid and 5 min in 95% ethanol, they are dried under a heat lamp for 15 min or at room temperature for 1 h. To enable calculation of phosphate incorporation, 1 µl of 1/10, 1/50, and 1/250 dilutions of the 5× CK2 reaction buffer should also be spotted on filter paper and dried without washing. The dried P81 papers are exposed to a PhosphorImager screen or subjected to scintillation counting for quantification and determination of CK2 activity (the amount of phosphate transferred to the RRRDDDSDDD per minute per mg of CK2 at 30 °C).

3.1.3. Assays for CK2 activity with peptide substrates: Additional considerations

The method described in the preceding section can be readily adapted to monitor the activity of other recombinant forms of CK2 (e.g., free catalytic subunits and the CK2 holoenzyme) or CK2 from other sources (e.g., immunoprecipitated CK2 or CK2 in cell extracts or column fractions). In the case of immunoprecipitated CK2, after isolation as described in Section 2.2.2, the bead-immobilized immunoprecipitates are resuspended in the kinase assay buffer (we typically resuspend beads with 5–10 volumes of buffer). Resuspended beads are added to the kinase reactions as the source of enzyme. Since the CK2 will be immobilized on the beads, it may be necessary to cut the end of a pipette tip to ensure that beads are effectively transferred. Furthermore, to ensure appropriate mixing during the kinase reaction, it is necessary to ensure that kinase reactions are incubated with shaking to enable the contents to continue to mix during the reaction.

Another application where peptide substrates such as RRRDDDSDDD can be employed to monitor CK2 activity is for measurements of CK2 activity in cell extracts. For example, measurements of CK2 activity can be used to determine the extent to which CK2 is elevated when overexpressed. Since the overexpression of CK2 is described in Section 4, the preparation of cell extracts will be described in that section. To measure CK2 activity in cell extracts with RRRDDDSDDD as a substrate, the kinase reaction as described above is initiated by the addition of extract (typically 2–8 μg/20 μl reaction). When performing assays to measure CK2 activity in cell extracts, it is particularly important to verify that assays are linear with time and with the amount of extract that is added.

3.2. CK2 kinase assays with peptide arrays

Interrogation of phosphoproteomics databases with CK2 consensus motifs (i.e., S/T-X-X-D/E/pS/pY) reveal thousands of potential CK2 phosphorylation sites that are known to be phosphorylated in cells (Salvi et al., 2009). Not surprisingly, the vast majority of these candidate substrates remain unvalidated. Therefore, to streamline the identification of *bona fide* CK2 substrates, we outline a protocol for simultaneously testing hundreds of putative substrate sequences via peptide arrays and recombinant CK2 (see Section 2.1). As a means for validating substrates identified in this manner, we next present methods for performing kinase assays using recombinant proteins as substrates. Demonstrating direct phosphorylation by CK2 of substrates *in vitro* is generally required to substantiate conclusions made from *in vivo* experiments that attempt to corroborate a given phenotype with a phosphorylation event.

3.2.1. Materials for CK2 kinase assays on peptide arrays

Enzymes and arrays: GST-CK2 (see Section 2.1), peptide array constructed on Amino-PEG$_{500}$-UC540 using SPOT method

Reaction components and wash buffers: kinase buffer (50 mM Tris (pH 7.5), 30 mM MgCl$_2$, 50 mM KCl, 1 mM DTT), ATP stock (20–100 mM), γ-^{32}P-ATP (3000 Ci/mmol) (Perkin Elmer), CK2 dilution buffer (1 mg/ml bovine serum albumin, 5 mM MOPS, pH 7.0, 200 mM NaCl), 95% ethanol, 1 M NaCl, stripping solution (4 M guanidine hydrochloride, 1% SDS, 0.5% β-mercaptoethanol).

Equipment: shaking water bath set to 30 °C, PhosphorImager or X-ray film, and developer.

3.2.2. Procedure for CK2 kinase assays using peptide arrays as substrates

The following protocol for performing kinase assays on peptide arrays is adapted from a protocol previously described (Tegge and Frank, 1998; Vilk et al., 2008). Peptide arrays are moistened in 95% ethanol to rehydrate the peptides, followed by a wash in kinase assay buffer and incubation overnight on a rocker or shaker at room temperature in the same buffer. Depending on the number of peptide spots on the array, arrays can be upwards of 10 × 10 cm and, thus, may require over 5 ml buffer. Sufficient volume to permit complete coverage of the array during rocking should be used. The following morning, fresh kinase buffer is added to the array and incubated for 1 h with rocking or equivalent mixing at 30 °C. To perform CK2 assays with arrays, we use 10 μM ATP in the kinase assay buffer. For a 6 ml reaction, we use at least 10 μCi γ-^{32}P-ATP and 60 units of CK2 diluted in CK2 dilution buffer. The amount of γ-^{32}P-ATP and CK2 can be scaled to the volume of the kinase reaction. Reactions are performed for 20 min, but should also be performed for 10 and/or 40 min to verify linearity. To terminate the reaction, the buffer is decanted and the array washed 15× with 1 M NaCl and 10× with water. Washing in stripping solution (4 M guanidine hydrochloride, 1% SDS, 0.5% β-mercaptoethanol) for 1 h at 40 °C, followed by several water washes to remove precipitate left from the stripping solution, is effective in reducing background. Finally, the membrane is washed with ethanol and dried before being imaged with a PhosphorImager or autoradiography. It is important to incorporate both positive and negative controls. In this respect, peptides known to be phosphorylated by CK2 can be included on the array as a positive control. Similar peptides, without phosphorylatable residues, can serve as the negative control.

3.3. CK2 kinase assays with protein substrates

For many researchers, the desire to perform CK2 assays is to investigate the phosphorylation of a specific protein by CK2. For this purpose, the assays that are described in the preceding sections (particularly Section 3.1) can be readily adapted and will not be repeated in this section. A main adaptation is that the protein substrate is added in place of the synthetic peptide. When examining the phosphorylation of individual proteins by CK2, we typically analyze phosphorylation by subjecting the kinase reaction to SDS-PAGE followed by analysis with the PhosphorImager or by autoradiography to monitor ^{32}P incorporation. Therefore, to stop the reactions, we add 20 μl of 2× Laemmli buffer and heat to 95 °C prior to gel electrophoresis. Following electrophoresis, we dry the gel onto a filter paper and mark the molecular weight standards by spotting 1 μl of 1/50 dilutions of 5× CK2 reaction buffer with γ-^{32}P-ATP next to each marker. As well, 1 μl of 1/10, 1/50, and 1/250 dilutions of the 5× CK2 reaction buffer should be spotted on filter paper to allow for calculation of phosphate incorporation. β-casein (Sigma C8157) can be employed as a positive control for phosphorylation.

Monitoring the incorporation of radioactive phosphate enables determination of the stoichiometry of phosphorylation of a particular protein. However, an important consideration in examining the phosphorylation of a purified protein is to ensure that the phosphorylation under examination is physiologically relevant. In particular, it is critical to determine whether the site(s) that are phosphorylated in these *in vitro* assays are actually phosphorylated in intact cells. A full description of the methods that may be employed to identify phosphorylation sites and assess their physiological relevance is beyond the scope of this chapter.

Mass spectrometry has generally emerged as a preferred approach for the identification of phosphorylation sites, although there are some limitations. Generally, facilities for mass spectrometry do not accommodate radioactive samples. Consequently, it may be necessary to perform parallel phosphorylation assays in the absence of γ-^{32}P-ATP when samples will be destined for mass spectrometry. It is also possible that phosphorylation sites can be missed when performing analysis by mass spectrometry because phosphopeptides fail to ionize or because a protein lacks proteolytic cleavage sites that would be required to generate appropriately sized peptides. Before the widespread use of mass spectrometry for the identification of phosphorylation sites, it was common to perform traditional analyses such as phosphopeptide mapping and phosphoamino acid that are exceptionally sensitive to characterize the site(s) of phosphorylation (Boyle *et al.*, 1991). While these methods have been largely replaced by mass spectrometry, they may still have utility in the characterization of phosphorylation sites. Phosphopeptide mapping strategies are particularly useful when performing comparative analyses (e.g., when comparing sites that are phosphorylated *in vitro* with sites that are phosphorylated in cells). Overall, when evaluating the phosphorylation of a protein by CK2, the importance of ensuring that sites phosphorylated *in vitro* are indeed phosphorylated in cells cannot be overemphasized. Like other protein kinases, CK2 has robust enzymatic activity that can be rather indiscriminant *in vitro*.

4. Modulation of CK2 in Mammalian Cells

In this section, we describe various strategies that we use to increase or decrease CK2 activity in mammalian cells. Using transient overexpression or inducible expression of CK2 in stably transfected cell lines, we generally observe relatively modest changes in CK2 activity (two- to fivefold increases in CK2 activity are typically observed). In general, higher increases in CK2 activity are achieved when coexpressing CK2β with the catalytic CK2 subunits (Penner *et al.*, 1997). In view of its ubiquitous expression and constitutive activity, it can be readily envisaged that increased expression of CK2 could have limited effects on the phosphorylation of its physiological substrates since these proteins may be significantly phosphorylated by endogenous CK2 (Litchfield, 2003). Accordingly, under some circumstances, it could be expected that strategies to interfere with CK2 activity will have more dramatic effects on the phosphorylation of its substrates and on particular responses in cells. Indeed, in our own studies with cell lines that express catalytically component or catalytically inactive forms of CK2 under the control of tetracycline, the most dramatic effects were observed in those cells expressing catalytically inactive forms of CK2α' (Vilk et al., 1999).

4.1. Transient expression of CK2 in mammalian cells

The first method to be described in this section is one that we routinely employ to achieve transient expression of catalytically active CK2 or catalytically incompetent CK2 in mammalian cells. Since the method for the expression of epitope-tagged CK2 has already been presented in Section 2.2, many of the details will not be repeated here.

4.1.1. Materials for transient expression of CK2

Plasmid DNA: pZW6 (pRC/CMV (Invitrogen) with cDNA encoding human CK2α with C-terminal HA tag), pGV15 (CK2αK68M-HA in pRC/CMV), pZW16 (pRC/CMV with cDNA encoding human CK2α' with N-terminal HA tag), pGP18 (HA-CK2α'K69M), pZW12 (pRC/CMV with cDNA encoding CK2β with N-terminal Myc tag) or pAB07 (pCDNA3.1(+) (Invitrogen) with cDNA encoding human CK2β with N-terminal Myc tag), pEGFP-C2 (Clontech).

Cells and media: HeLa or U2OS cells, Dulbecco's modified Eagle's medium with 10% fetal bovine serum, and antibiotic supplements (0.1 mg/ml streptomycin and 100 units/ml penicillin).

Solutions and materials for overexpression: PEI transfection reagents (Reed et al., 2006), PBS.

Cell lysis solutions: PBS, extraction buffer (50 mM Tris–Cl (pH 7.5), 150 mM NaCl, 1 mM EDTA, and 1 mM dithiothreitol), and extraction buffer supplemented with protease and phosphatase inhibitors (30 µg/ml aprotinin, 20 µg/ml leupeptin, 1 mM Na$_3$VO$_4$, 10 µg/ml pepstatin, and 1 mM phenylmethylsulfonyl fluoride).

4.1.2. Procedure for transient expression of CK2 in mammalian cells

To overexpress CK2 in HeLa or U2OS cells, we follow the methods previously described in Section 2.2.2. As a companion to the expression of active CK2 in cells, we have also generated plasmids for the expression of catalytically inactive CK2α and CK2α'. All constructs incorporate an epitope tag to enable detection by immunoblotting or isolation by immunoprecipitation as described in Section 2.2.1. To assess the CK2 activity in transfected cells, the cells are washed twice with PBS (10 ml PBS for a 10 cm plate) and once in extraction buffer before being harvested with a cell lifter in extraction buffer supplemented with protease and phosphatase inhibitors

(0.5–1.0 ml/10 cm plate). All solutions are at 0–4 °C with procedures performed on ice. Cells are disrupted by sonication for 2 × 10 s before being cleared of debris by centrifugation at 140,000 × g for 20 min at 4 °C. Supernatants can be stored at −80 °C or used immediately for kinase assays with RRRDDDSDDD as described in Section 3.1. Modest increases in catalytic activity are generally observed when transfecting cells with catalytic CK2 subunits alone (two- to threefold), but greater increases (to approximately fivefold) observed when co-overexpressing the CK2β subunit (Penner et al., 1997). Increases in total CK2 activity, albeit less than is observed when transfecting cells to express active CK2 subunits, may also be observed when transfecting cells with constructs encoding catalytically incompetent mutants of CK2α or CK2α' (Vilk et al., 1999). In a similar respect, immunoprecipitates of catalytically incompetent CK2 do have measurable kinase activity (Vilk et al., 1999). One potential explanation for these observations is that exogenous CK2 subunits do form complexes with endogenous CK2 subunits that do exhibit kinase activity.

4.2. Inducible expression of CK2 in mammalian cells

Transient transfections have a number of significant limitations. For example, there can be considerable variation in expression between experiments and the expression of constructs is relatively short-lived. To overcome these limitations, we developed strategies to stably express CK2 in cells using a bidirectional plasmid that enables the coordinate expression of both catalytic (CK2α or CK2α') and regulatory (CK2β) subunits of CK2 under the control of tetracycline (Vilk et al., 1999). Furthermore, we generated constructs encoding both active and catalytically incompetent forms of CK2α and CK2α' (Vilk et al., 1999). Methods for the development and utilization of these cell lines are described below.

4.2.1. Materials for generating cell lines expressing tetracycline-regulated CK2

Plasmid DNA: pRS3 (pBI (Clontech) with cDNA encoding human CK2α with a C-terminal HA tag and cDNA encoding human CK2β with N-terminal Myc tag), pRS2 (pBI (Clontech) with cDNA encoding human CK2α' with a N-terminal HA tag and cDNA for human CK2β with N-terminal Myc tag), pGV13 (encoding CK2α(K68M)-HA and Myc-CK2β), pGV7 (encoding HA-CK2α'(K69M)-HA and Myc-CK2β), and pTk-Hyg (Clontech).

Cells, media, and antibiotics: UTA6 cells, Dulbecco's modified Eagle's medium with 10% fetal bovine serum, and antibiotic supplements (0.1 mg/ml streptomycin and 100 units/ml penicillin), G418 (Invitrogen), hygromycin (Roche), tetracycline (Sigma).

Other reagents: CaPO$_4$ transfection reagents or equivalent high-efficiency reagents, PBS, NP-40 lysis buffer (1% NP-40, 150 mM NaCl, 50 mM Tris (pH 7.5), 30 μg/ml aprotinin, 20 μg/ml leupeptin, 1 mM Na$_3$VO$_4$, 10 μg/ml pepstatin, and 1 mM phenylmethylsulfonyl fluoride), and antibodies for CK2.

4.2.2. Generating stable cell lines with tetracycline-regulated expression of CK2

The method that we have employed to generate cell lines with tetracycline-regulated expression of CK2 is as follows. The parental cells for these transfections are UTA6 cells that were derived from the human osteosarcoma U2-OS cells and express the tetracycline transactivator (tTA) to enable tetracycline-regulated expression of constructs (Englert *et al.*, 1995). We envisage that similar strategies can be employed to generate tetracycline-regulated expression of CK2 in other cell lines that harbor the tTA. For robust and coordinated expression of both catalytic and regulatory CK2 subunits, we used pBI (Clontech) that contains tetracycline responsive elements in between two partial CMV promoters. In the presence of tetracycline, expression from pBI is repressed, but upon removal of tetracycline from the growth media, protein expression begins within 6–12 h.

Before selecting for stable cell lines, a kill curve must first be performed by titrating the parental UTA6 cells with hygromycin (titrate from 0 to 1000 μg/ml) and G418 (0–1000 μg/ml). When selecting stable transfectants, we use the most dilute concentration of each drug that induces complete cell death in the parental UTA6 cells within 2 weeks. We begin the selection of stable cell lines by transfecting UTA6 cells, via the CaPO$_4$ method, at ~50% confluency with 5 μg pTk-Hyg (contains hygromycin resistance gene) and 35 μg pRS2 or pRS3 (contains G418 resistance gene) so that doubly stable clones can be selected. Note that once transfected, the cells should remain in media containing 1.5 μg/ml tetracycline to repress expression. After overnight incubation, DNA precipitates are removed and the cells allowed to recover for 48 h. At this point, cells are treated with the hygromycin and G418 at concentrations acquired from kill curves and continued incubation with 1.5 μg/ml tetracycline. Allow 1–2 weeks for cells to die off and colonies to form, and be sure to regularly change the media (every 2–3 days). Once colonies are well established and large enough to survive on 96-well plates, they can be isolated through the use of cloning cylinders and seeded (*Note:* cloning cylinders need to be prepared beforehand by using tweezers to dip clean cylinders into vacuum grease, placing them into pyrex Petri dishes and autoclaving). Cells should be monitored daily, as clonal expansion will occur at different rates for different clones. Once colonies are confluent on a 10 cm plate, they can be split 1:4 onto two 10 cm plates to test for tetracycline-regulated expression. The rest of the cells are plated onto a 15 cm plate to be frozen for storage.

To select for cells with robust and tightly regulated CK2 expression, cells are grown to 50–80% confluency in media containing 1.5 µg/ml tetracycline. Protein expression is initiated by washing tetracycline from the cells, with several thorough PBS washes, and refeeding the cells with tetracycline-free media. After allowing 24 h for expression, cells can be lysed as described earlier (Section 2.2.2) and analyzed by immunoblot analysis for the expression of HA-tagged CK2α or CK2α′ and Myc-CK2β in relation to endogenous CK2 subunits. Using this strategy, we have generated several cell lines with tetracycline-regulated expression of CK2 including: RS3.22 (UTA6 cells expressing CK2α-HA and Myc-CK2β), pRS2.31 (UTA6 cells expressing HA-CK2α′ and Myc-CK2β), GV13.35 (UTA6 cells expressing CK2αK68M-HA and Myc-CK2β), and GV7.21 (UTA6 cells expressing HA-CK2α′K69M and Myc-CK2β). In our experience, RS3.22 and GV13.35 cells express approximately double the amount of CK2α and CK2β as endogenous levels, whereas RS2.31 cells express about six times more CK2α′ and double the amount of CK2β. An example of the tight regulation achieved using GV13.35 cells is shown in Fig. 23.2. GV7.21 express the lowest levels of CK2α′ and CK2β but also exhibit a significant decrease in proliferation upon induction suggesting that the diminished expression could be a consequence of a loss of cell viability (Vilk et al., 1999). We have used similar strategies to generate cell lines with tetracycline-regulated expression of mutant forms of CK2α or CK2β in the absence of CK2α or CK2α′. As long as cells that express the tTA are available, it is envisaged that this strategy could be adapted to other cell lines.

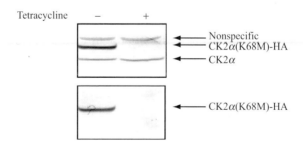

Figure 23.2 Tetracycline-regulated expression of CK2. As described in Section 4.2, we have developed vectors and cell lines that exhibit tetracycline-regulated expression of CK2. Immunoblot analysis of lysates from one representative cell line, GV13.35 cells, that exhibit coordinate expression of CK2αK68M-HA and Myc-CK2β under the control of tetracycline. Immunoblot analysis reveals that CK2αK68M-HA is expressed in the absence of tetracycline but repressed in the presence of tetracycline. The top panel was blotted using CK2α antibodies and the bottom panel using anti-HA. A nonspecific band on anti-CK2α immunoblots is also indicated.

4.3. Modulation of CK2 in mammalian cells: Additional considerations

As noted earlier, on the basis of the ubiquitous expression and constitutive activity of CK2, it can be expected that some of its physiological substrates will be phosphorylated to a significant extent even in the absence of overexpression. The expression of catalytically incompetent mutants of CK2 could be employed but this strategy does have limitations. For example, since CK2 is ubiquitously expressed and essential for viability, endogenous CK2 could compete with expressed CK2. Furthermore, the time course for expression in transfection experiments, and even with inducible expression, is on a time scale of hours. To address these concerns, strategies involving siRNA/shRNA or RNAi-mediated knockdown of endogenous CK2 or the use of pharmacological inhibitors of CK2 can be employed. While a detailed presentation of the methods for the use of these approaches is beyond the scope of our chapter, we will discuss some of the important considerations for the use of these strategies.

4.3.1. Knockdown of CK2

There is mounting evidence in the literature that knockdown strategies employing siRNA/shRNA or RNAi can be effective with CK2 (Di Maira et al., 2007; Faust et al., 2000). Furthermore, there are now reagents available from a number of commercial suppliers. Main advantages of these strategies include the fact that endogenous CK2 will be targeted and the prospect of selective knockdowns of individual CK2 subunits. However, in our own experience, there are a number of significant considerations. For example, it may be difficult to attain complete knockdown of CK2 due to its long half-life (>24 h for tetrameric CK2). There are also concerns of off-target effects with knockdown strategies. Consequently, it is important that observed effects be ascribed to changes in the levels of CK2. Validation strategies include the use of several different sequences to effect knockdown and rescue experiments performed with siRNA or RNAi-resistant constructs.

4.3.2. Pharmacological inhibition of CK2

Another alternative for inhibiting CK2 in cells involves the use of pharmacological inhibitors. There are now a number of potent and selective CK2 inhibitors that are available commercially, including TBB, TBBz, and DMAT (Duncan and Litchfield, 2008). Furthermore, there are many other inhibitors in various stages of development. In comparison to knockdown or expression strategies, pharmacological inhibition has appealing features, most notably the fact that effective inhibitors can result in almost immediate inhibition of kinase activity as compared to the lengthy times required to observe expression or knockdown. However, like other strategies, there are limitations that must be considered. For example, even the most specific pharmacological inhibitors of CK2 have off-target effects that are not fully understood (Duncan et al., 2008). Furthermore, there has not yet been a general biomarker for CK2 inhibition identified that can be used to monitor changes in CK2 activity. Structural insights obtained from complexes of CK2 with inhibitors have enabled the development of inhibitor-resistant forms of CK2 (described in Section 2.1) (Sarno et al., 2005). In principle, these forms of CK2 can be used to rescue the effects of inhibitors. However, in our own studies, we observed that these mutants did rescue the autophosphorylation of CK2 but failed to restore viability to cells treated with CK2 inhibitors (Duncan et al., 2008). We conclude from these observations that the inhibitor-resistant CK2 mutants may rescue some CK2-catalyzed phosphorylation events but there are other events that are likely unrelated to CK2 that cannot be restored. Collectively, these results suggest that while pharmacological inhibitors and the use of inhibitor-resistant forms of CK2 may enable the identification of some CK2 substrates and/or CK2-regulated cellular events, the use of these approaches may not be universally applicable.

5. CONCLUSIONS

Collectively, the experimental procedures in this chapter should enable the identification of *bona fide* physiological substrates for CK2 and characterization of cellular events regulated by CK2. In this respect, we have presented experimental procedures for isolating CK2 from bacteria and from mammalian cells. CK2, purified by these procedures, can be used to investigate the phosphorylation of candidate substrates *in vitro*. We have also outlined complementary strategies for manipulating CK2 levels in mammalian cells. A *bona fide* CK2 substrate is a protein that (i) can be phosphorylated directly at site(s) that are known to be phosphorylated in cells and (ii) exhibits changes in phosphorylation at these site(s) that correlate with changes in CK2 activity. By applying these criteria, it may ultimately be possible to capitalize on the vast amounts of information that are emerging from large-scale projects to determine how CK2 contributes to the regulation of cellular processes such as cell survival.

ACKNOWLEDGMENTS

Work on CK2 in our laboratory has been funded by Operating Grants from Canadian Cancer Society Research Institute and the Canadian Institutes of Health Research. We are grateful to Dr. Christoph Englert for the UTA6 cell line.

REFERENCES

Ahn, N. G., Weiel, J. E., Chan, C. P., and Krebs, E. G. (1990). Identification of multiple epidermal growth factor-stimulated protein serine/threonine kinases from Swiss 3T3 cells. *J. Biol. Chem.* **265,** 11487–11494.

Boyle, W. J., van der Geer, P., and Hunter, T. (1991). Phosphopeptide mapping and phosphoamino acid analysis by two-dimensional separation on thin-layer cellulose plates. *Methods Enzymol.* **201,** 100–149.

Canton, D. A., Olsten, M. E., Kim, K., Doherty-Kirby, A., Lajoie, G., Cooper, J. A., and Litchfield, D. W. (2005). The pleckstrin homology domain-containing protein CKIP-1 is involved in regulation of cell morphology and the actin cytoskeleton and interaction with actin capping protein. *Mol. Cell. Biol.* **25,** 3519–3534.

Di Maira, G., Brustolon, F., Bertacchini, J., Tosoni, K., Marmiroli, S., Pinna, L. A., and Ruzzene, M. (2007). Pharmacological inhibition of protein kinase CK2 reverts the multidrug resistance phenotype of a CEM cell line characterized by high CK2 level. *Oncogene* **26,** 6915–6926.

Duncan, J. S., and Litchfield, D. W. (2008). Too much of a good thing: The role of protein kinase CK2 in tumorigenesis and prospects for therapeutic inhibition of CK2. *Biochim. Biophys. Acta* **1784,** 33–47.

Duncan, J. S., Gyenis, L., Lenehan, J., Bretner, M., Graves, L. M., Haystead, T. A., and Litchfield, D. W. (2008). An unbiased evaluation of CK2 inhibitors by chemoproteomics: Characterization of inhibitor effects on CK2 and identification of novel inhibitor targets. *Mol. Cell. Proteomics* **7,** 1077–1088.

Duncan, J. S., Turowec, J. P., Vilk, G., Li, S. S., Gloor, G. B., and Litchfield, D. W. (2010). Regulation of cell proliferation and survival: Convergence of protein kinases and caspases. *Biochim. Biophys. Acta* **1804,** 505–510.

Englert, C., Hou, X., Maheswaran, S., Bennett, P., Ngwu, C., Re, G. G., Garvin, A. J., Rosner, M. R., and Haber, D. A. (1995). WT1 suppresses synthesis of the epidermal growth factor receptor and induces apoptosis. *EMBO J.* **14,** 4662–4675.

Faust, R. A., Tawfic, S., Davis, A. T., Bubash, L. A., and Ahmed, K. (2000). Antisense oligonucleotides against protein kinase CK2-alpha inhibit growth of squamous cell carcinoma of the head and neck in vitro. *Head Neck* **22,** 341–346.

Grankowski, N., Boldyreff, B., and Issinger, O. G. (1991). Isolation and characterization of recombinant human casein kinase II subunits alpha and beta from bacteria. *Eur. J. Biochem.* **198,** 25–30.

Hunter, T. (2000). Signaling—2000 and Beyond. *Cell* **100,** 113–127.

Kuenzel, E. A., and Krebs, E. G. (1985). A synthetic peptide substrate specific for casein kinase II. *Proc. Natl. Acad. Sci. USA* **82,** 737–741.

Kuenzel, E. A., Mulligan, J. A., Sommercorn, J., and Krebs, E. G. (1987). Substrate specificity determinants for casein kinase II as deduced from studies with synthetic peptides. *J. Biol. Chem.* **262,** 9136–9140.

Landesman-Bollag, E., Romieu-Mourez, R., Song, D. H., Sonenshein, G. E., Cardiff, R. D., and Seldin, D. C. (2001). Protein kinase CK2 in mammary gland tumorigenesis. *Oncogene* **20,** 3247–3257.

Litchfield, D. W. (2003). Protein kinase CK2: Structure, regulation and role in cellular decisions of life and death. *Biochem. J.* **369,** 1–15.

Meggio, F., and Pinna, L. A. (2003). One-thousand-and-one substrates of protein kinase CK2? *FASEB J.* **17,** 349–368.

Meggio, F., Boldyreff, B., Marin, O., Marchiori, F., Perich, J. W., Issinger, O. G., and Pinna, L. A. (1992). The effect of polylysine on casein-kinase-2 activity is influenced by both the structure of the protein/peptide substrates and the subunit composition of the enzyme. *Eur. J. Biochem.* **205,** 939–945.

Messenger, M. M., Saulnier, R. B., Gilchrist, A. D., Diamond, P., Gorbsky, G. J., and Litchfield, D. W. (2002). Interactions between protein kinase CK2 and Pin1. Evidence for phosphorylation-dependent interactions. *J. Biol. Chem.* **277**, 23054–23064.

Penner, C. G., Wang, Z., and Litchfield, D. W. (1997). Expression and localization of epitope-tagged protein kinase CK2. *J. Cell. Biochem.* **64**, 525–537.

Reed, S. E., Staley, E. M., Mayginnes, J. P., Pintel, D. J., and Tullis, G. E. (2006). Transfection of mammalian cells using linear polyethylenimine is a simple and effective means of producing recombinant adeno-associated virus vectors. *J. Virol. Methods* **138**, 85–98.

Salvi, M., Sarno, S., Cesaro, L., Nakamura, H., and Pinna, L. A. (2009). Extraordinary pleiotropy of protein kinase CK2 revealed by Weblogo phosphoproteome analysis. *Biochim. Biophys. Acta* **1793**, 847–859.

Sarno, S., Salvi, M., Battistutta, R., Zanotti, G., and Pinna, L. A. (2005). Features and potentials of ATP-site directed CK2 inhibitors. *Biochim. Biophys. Acta* **1754**, 263–270.

Seldin, D. C., and Leder, P. (1995). Casein kinase II alpha transgene-induced murine lymphoma: Relation to theileriosis in cattle. *Science* **267**, 894–897.

St-Denis, N. A., and Litchfield, D. W. (2009). Protein kinase CK2 in health and disease: From birth to death: The role of protein kinase CK2 in the regulation of cell proliferation and survival. *Cell. Mol. Life Sci.* **66**, 1817–1829.

Tegge, W. J., and Frank, R. (1998). Analysis of protein kinase substrate specificity by the use of peptide libraries on cellulose paper (SPOT-method). *Methods Mol. Biol.* **87**, 99–106.

Vilk, G., Saulnier, R. B., St Pierre, R., and Litchfield, D. W. (1999). Inducible expression of protein kinase CK2 in mammalian cells. Evidence for functional specialization of CK2 isoforms. *J. Biol. Chem.* **274**, 14406–14414.

Vilk, G., Weber, J. E., Turowec, J. P., Duncan, J. S., Wu, C., Derksen, D. R., Zien, P., Sarno, S., Donella-Deana, A., Lajoie, G., Pinna, L. A., Li, S. S., *et al.* (2008). Protein kinase CK2 catalyzes tyrosine phosphorylation in mammalian cells. *Cell. Signal.* **20**, 1942–1951.

Zien, P., Duncan, J. S., Skierski, J., Bretner, M., Litchfield, D. W., and Shugar, D. (2005). Tetrabromobenzotriazole (TBBt) and tetrabromobenzimidazole (TBBz) as selective inhibitors of protein kinase CK2: Evaluation of their effects on cells and different molecular forms of human CK2. *Biochim. Biophys. Acta* **1754**, 271–280.

CHAPTER TWENTY-FOUR

ASSESSMENT OF CK2 CONSTITUTIVE ACTIVITY IN CANCER CELLS

Maria Ruzzene, Giovanni Di Maira, Kendra Tosoni, *and* Lorenzo A. Pinna

Contents

1. Introduction	496
2. Assay of CK2 in Crude Biological Samples	499
2.1. Preparation of cell lysates	499
2.2. CK2 assay toward synthetic peptides	500
2.3. In-gel assay of CK2 catalytic subunit activity	502
3. In-Cell Assay of Endogenous CK2 Activity	504
3.1. Immunodetection of CK2-dependent phosphosites	505
3.2. Phosphorylation of a reporter substrate	506
4. Identification/Validation of *In Vivo* CK2 Targets with Specific Inhibitors	506
4.1. Cell treatment with specific CK2 inhibitors	507
4.2. Radioactive phosphorylation of endogenous proteins in treated and untreated cells	508
4.3. *In vivo* radioactive phosphorylation in treated and untreated cells	510
Acknowledgments	511
References	511

Abstract

At variance with the great majority of protein kinases that become active only in response to specific stimuli and whose implication in tumors is caused by genetic alterations conferring to them unscheduled activity, the highly pleiotropic Ser/Thr-specific protein kinase CK2 is constitutively active even under normal conditions and no gain-of-function CK2 mutants are known. Nevertheless, CK2 level is abnormally high in cancer cells where it is believed to generate an environment favorable to the development of malignancy, through a mechanism denoted as "non-oncogene addiction." This makes CK2 not only an appealing target to counteract different kinds of tumors but also a valuable

Department of Biological Chemistry, and VIMM (Venetian Institute of Molecular Medicine), University of Padova, Padova, Italy marker of cells predisposed to undergo neoplastic transformation owing to the presence in them of CK2 level exceeding a critical threshold. Such a prognostic exploitation of CK2 would imply the availability of methods suitable for the reliable, sensitive, and specific quantification of its activity in biological samples and in living cells. The aim of this chapter is to describe a number of procedures applicable to the quantitative determination of CK2 activity and to provide experimental details designed for rendering these assays as sensitive and selective as possible even in the presence of many other protein kinases. The procedures described roughly fall in three categories: (i) *in vitro* quantification of CK2 activity in crude biological samples and cell lysates; (ii) in-cell assay of endogenous CK2 activity based on the phosphorylation of reporter substrates; (iii) identification of CK2 targets in malignant and normal cells.

Abbreviations

CA	constitutive activity
DMAT	2-dimethylamino-4,5,6,7-tetrabromo-1H-benzimidazole
IQA	[5-oxo-5,6-dihydroindolo-(1,2-*a*)quinazolin-7-yl] acetic acid
K64	3,4,5,6,7-pentabromo-1H-indazole
TBB	4,5,6,7-tetrabromo-1H-benzotriazole
TBCA	4,5,6,7-tetrabromo-cinnamic acid
TBI	4,5,6,7-tetrabromo-1H-benzimidazole
WB	Western blot

1. Introduction

The inclusion of a protein kinase among proteins endowed with constitutive activity (CA) may sound quite surprising. Protein kinases are in fact signaling proteins *par excellence* whose activity is silent under basal conditions, being triggered only in response to specific stimuli. Also the implication of protein kinases in numerous pathologies, with special reference to cancer, reflects this general situation, considering that the products of many oncogenes are protein kinases with mutations conferring CA or, in general, gain of function to the kinase itself. This paradigm is exploited in so-called signal transduction therapy, a strategy based on the specific inhibition of oncogenic protein kinases, which is proving very successful to cure some kinds of tumors that are caused by the unscheduled activity of individual protein kinases (Cohen, 2002; Fabbro *et al.*, 2002; Sawyers, 2002).

In such a scenario, the Ser/Thr protein kinase denoted by the acronym CK2 (derived from the old misnomer "casein kinase"-2) represents a striking exception in that its catalytic subunits (α and/or α', very similar but encoded by two distinct genes) are active either combined or not with a dimer of its noncatalytic β-subunits and in the absence of any previous phosphorylation event (Litchfield, 2003; Pinna, 2003). Note that this latter feature not only excludes CK2 from "phosphorylation cascades" typical of most signal transduction pathways, but also it makes impossible to evaluate the state of activation of CK2 by immunodetection of its phosphorylation, as it is in the case of the great majority of protein kinases.

Also to note is that mutations causing gain of function of CK2—similar to those determining the oncogenic potential of many protein kinases—have never been reported. CA of CK2 under "basal" conditions is supposed to be instrumental to its terrific pleiotropy (Pinna, 2002): indeed repertoires of hundreds of *bona fide* physiological substrates of CK2 have been compiled (Meggio and Pinna, 2003), a still very reductive estimate considering that global proteomic analyses would indicate that CK2 alone could contribute to the generation of a substantial amount (possibly up to 20%) of the human phosphoproteome (Salvi et al., 2009).

The above observations would argue against the idea that unscheduled CK2 activity might be causative of neoplasia. In sharp contrast with this expectation, however, CK2 is clearly implicated in many of the cell biology phenomena associated with cancer (Duncan and Litchfield, 2008; Tawfic et al., 2001; Trembley et al., 2009) and its level is invariably elevated in malignant cells (Guerra and Issinger, 2008). Apparently, this is not due to further upregulation of its CA or overexpression of its catalytic subunits as revealed by mRNA analysis (Tawfic et al., 2001) but simply due to an excessive accumulation of normoactive CK2 protein. Seemingly such an accumulation of CK2 over a critical threshold predisposes cells to malignant transformation by generating a cellular environment especially favorable to the progression of the tumor phenotype: abnormally high CK2 level in fact permits cells to escape apoptosis, it enhances the multidrug resistance phenotype, it hyperactivates the chaperone machinery protecting the oncokinome, and it promotes neovascularization, as reviewed elsewhere (Sarno and Pinna, 2008). A rising concept to explain the implication of CK2 in neoplasia is that of "non-oncogene addiction," that is, overreliance of the perturbed signaling network responsible for malignancy on abnormally high CK2 level for its own maintenance (Ruzzene and Pinna, 2010). According to this concept, once oncogenic mutation(s) takes place in the genome, those cells where CK2 level reaches a critical threshold will be more prone to malignant transformation than cells where CK2 is normally represented, and will be "selected" by the tumor in such a way that at the end the predominant malignant phenotype, albeit not directly caused by CK2, is nevertheless relying on its abnormal accumulation for its own survival.

Given these premises, quantitative evaluation of CK2 may not only represent a criterion for "grading" malignancy, as outlined elsewhere (Unger et al., 2004), but also a powerful prognostic marker to evaluate the propensity of cells to undergo neoplastic transformation. Being abundant also in normal tissues however, and not susceptible either to activation by phosphorylation or to oncogenic mutations, the traditional immunological probes exploited for monitoring the appearance and the activity of oncogenic protein kinases are not applicable to CK2. However, a number of features make the enzymatic quantification of CK2 particularly handy and reliable. These are its high catalytic activity toward peptide substrates, its very peculiar site specificity, and its unusual ability to use also GTP, besides ATP, as phosphate donor in the kinase reaction.

Unlike the great majority of Ser/Thr protein kinases, whose targeting is specified by basic and/or prolyl residues in the proximity of the phosphoacceptor aminoacid, CK2 is extremely acidophilic in nature with a consensus sequence characterized by multiple carboxylic side chains mostly located downstream, the one at position $n + 3$ playing an especially crucial role (Meggio et al., 1994). It has been therefore a relatively straightforward task to develop excellent CK2 peptide substrates that not only are unaffected by all basophilic and proline-directed protein kinases but also are able to discriminate CK2 with respect to the other rare acidophilic Ser/Thr protein kinases.

Especially noteworthy are in this respect the peptide RRRADDSDDDDD and some derivatives of it which, owing to the lack of acidic residues at positions $n - 3/n - 4$ and to the replacement of aspartic for glutamic acid, are not phosphorylated by either CK1 ("casein kinase-1") or the Golgi apparatus casein kinase (GCK) (i.e., two other ubiquitous acidophilic Ser/Thr kinases) while being excellent CK2 substrates (Marin et al., 1994). Note that the presence of the N-terminal basic triplet in these peptides makes them amenable to phosphocellulose paper procedure to measure their phosphoradiolabeling by CK2 (Ruzzene and Pinna, 1999) without either hampering the catalytic efficiency of CK2 or rendering them susceptible to basophilic kinases which recognize basic residues only if they are closer to the phosphoacceptor aminoacid.

As discussed in Section 2.2, the CK2-tide RRRADDSDDDDD is readily phosphorylated by either CK2 holoenzyme or its isolated catalytic subunits, a behavior shared with the majority of CK2 protein substrates (so-called class I) (Pinna, 2002). A minority of CK2 protein substrates, however, do not follow this rule, their phosphorylation being either prevented by the β-subunit whenever the catalytic subunits are incorporated into the holoenzyme (class II) or, conversely, entirely dependent on the formation of the holoenzyme, their phosphorylation being critically relying on the β-subunit (class III). While peptide substrates belonging to class II are not presently available, a peptide reproducing the N-terminal segment of the eukaryotic translation initiation factor 2β (eiF2β) displays the properties of class III substrates and can be successfully exploited in combination with the CK2-tide for discriminating between constitutive activities displayed by either the holoenzyme or the uncombined catalytic subunits (Poletto et al., 2008; Salvi et al., 2006).

An invaluable advantage of these CK2 peptide substrates is that, at variance with the peptide substrates of most kinases, which are all more or less promiscuous, these are extremely specific and their phosphorylation by crude preparations containing numerous protein kinases can be definitely ascribed to CK2 alone (see below, Section 2.2). Thus, determining the phosphorylation rates of these peptides by crude extracts and biological preparations from different kinds of cells and tissues provides a first choice criterion to make a reliable comparative estimate of CK2 CA and to establish if the level of this kinase is "regular" or abnormally high. The selectivity of the assay can be further improved by replacing ATP with GTP which exceptionally is almost as good as ATP as phosphate donor if the phosphotransferase reaction is catalyzed by CK2. This strategy may prove particularly advantageous under circumstances where CK2 activity has to be assayed with a protein substrate (notably casein) which, unlike the CK2-tide, is also susceptible to phosphorylation by kinases other than CK2. This is the case, for example, of "in-gel assays" described in Section 2.3.

Whenever instead the CA of endogenous CK2 needs to be quantified in living cancer cells, eventually subjected to different treatments, advantage can be taken of methods based on the immunodetection of phosphosites specifically generated by CK2 in various reporters, notably Akt S129 (Di Maira et al., 2005). This will be dealt with in Section 3.

In addition to the protocols related to the evaluation of CK2 CA either *from* or *inside* cancer cells, a section will be devoted to illustrate the potential of selective cell permeable inhibitors to disclose and quantify the reliance of cancer cells on CK2 CA and to identify CK2 targets which are implicated in this kind of addiction. Section 4 will address this issue.

2. Assay of CK2 in Crude Biological Samples

2.1. Preparation of cell lysates

CK2 activity can be measured in total cell extracts or in specific cellular compartments, especially cytosol and nucleus where it is more abundant. Being CK2 quite soluble and not tightly associated with membrane fractions, no particular cell lysis buffer nor drastic conditions are required to solubilize it; treatments should be respectful of native conditions avoiding any step possibly detrimental to the kinase assay. Inhibitors of proteases and protein phosphatases should be always freshly added; temperature of lysates (and buffer before lysis) must be kept low (0–4 °C).

A suitable lysis buffer for performing exhaustive extraction of CK2 is the following: 20 mM Tris–HCl pH 7.5, 150 mM NaCl, 2 mM EDTA, 2 mM EGTA, 0.5% (v/v) Triton X-100, 2 mM DTT, protease inhibitor cocktail, 10 mM NaF, 1 mM Na$_3$VO$_4$, 1 µM okadaic acid. The lysis buffer volume should be calculated in order to obtain a protein concentration higher than 1 mg/ml; this allows enough dilution of buffer components in the kinase assay volume, to avoid any significant interference. Similarly, in case of subcellular fraction preparation, standard methods can be adopted, provided that denaturing conditions are avoided. Sonication can be applied (by short pulses to avoid overheating).

Usually, lysates can be stored at −20 °C and used several times for CK2 assays, without losing activity for several weeks.

2.2. CK2 assay toward synthetic peptides

Owing to high cellular CK2 CA and narrow selectivity of its peptide substrates (see Section 1), cell lysates can be directly assayed for CK2 activity without previous immunoprecipitation and/or purification steps. For a standard assay, incubate 1–2 µg of proteins at 30 °C for 10 min in a total volume of 20 µl, in the presence of 50 mM Tris–HCl pH 7.5, 100 mM MgCl$_2$, 100 mM NaCl, 10 µM ATP or, to increase the assay selectivity, GTP, [γ-^{33}P]ATP (GTP) (specific radioactivity ∼1000–5000 cpm/pmol), and 0.1 mM CK2-tide (whose sequence is shown in Table 24.1). ^{33}P-labeled nucleotides can be replaced by ^{32}P-labeled ones, if this is allowed by the lab safety rules. NaCl should be avoided in case CK2α (instead of α$_2$β$_2$) has to be monitored (see below, Section 2.2.1).

After incubation, reactions are stopped by spotting the whole volume on phosphocellulose filters (2 × 2 cm squares), then filters are washed three times in 0.5% (v/v) phosphoric acid (∼200–500 ml, in a beaker on a magnetic stirrer, 5 min for each washing) and counted in scintillation fluid.

When comparing different cell lysates, make sure to use not only equal amounts of proteins but also equal volumes of cell lysates (thus diluting more concentrate samples in the same lysis buffer; ideal volumes to be used are 1 and 2 µl for a 20-µl assay).

Replicates with 1 and 2 µg of proteins are usually appropriate in the case of cancer cells, where CK2 is highly expressed. However, sometimes higher amounts are required; be always sure to preserve linearity in activity, that is, increase of CK2 activity must be proportional to the amount of cellular proteins. Likewise, shorter or longer incubation times can be adopted, provided that linearity is maintained. To increase sensitivity and improve linearity, a higher CK2-tide concentration (up to 1 mM) can be used; usually this does not compromise the specificity of the assay, which can be assessed by means of CK2 inhibitors (see Fig. 24.1).

Table 24.1 Specific CK2 peptide substrates

Peptide	Sequence	CK2 $\alpha_2\beta_2$		CK2 α	
		K_m	V_{max}	K_m	V_{max}
CK2-tide	RRRADD$\underline{S}$DDDDD	14	56.5	25	18.9
eIF2β-tide	M$\underline{S}$GDEMIFDPTMSKKKKKKKKP	10	71.4	660	5.0

Residues phosphorylated by CK2 are underlined. K_m is expressed in μM, V_{max} in pmol/min. Reported values are from Poletto *et al.* (2008).

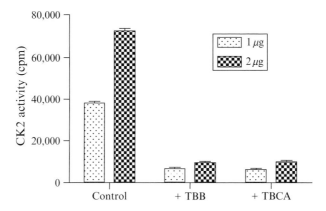

Figure 24.1 CK2 assay in cell lysate using a specific peptide substrate. One or two micrograms of protein from Jurkat cell lysates were incubated for 10 min with 0.1 mM CK2-tide (sequence shown in Table 24.1) and a radioactive phosphorylation mixture, as described in the text. Where indicated, cells were previously treated with 50 μM TBB or TBCA (CK2 inhibitors, see Section 4.1). Control blanks, performed without CK2-tide (all accounting to less than 2000 cpm), have been already subtracted from values showed.

Besides a blank with the peptide, without the addition of any cell lysate, blanks are also mandatory of each cell lysate sample, without the addition of the peptide substrate; in case these controls are significantly high, the possibility should be considered that they reflect real CK2 activity toward endogenous proteins able to bind to phosphocellulose papers; in this occurrence, more correct blanks can be obtained by incubation of samples in the presence of the peptide but with the addition of a CK2 inhibitor (see Section 4).

2.2.1. Discriminating between CK2 holoenzyme and isolated catalytic subunit activity

While usually the major CK2 form present in cell lysates is the holoenzyme $\alpha_2\beta_2$, the possibility exists that "free" catalytic α is also present. The detection of its activity can be particularly interesting, considering that it has been found in multidrug-resistant cancer cells (Di Maira et al., 2007). Both CK2 forms are catalytically competent, but some differences in their site specificity (see Section 1) can be exploited to discriminate between them. In particular, a recently developed CK2 peptide substrate (Poletto et al., 2008, Salvi et al., 2006) derived from the sequence of eiF2β (Table 24.1) is readily phosphorylated by $\alpha_2\beta_2$ but not by α. Therefore, assays can be performed under the same conditions described in Section 2.2, but replacing the CK2-tide with eiF2β-tide: comparing the results obtained with the two different substrates, it will be possible to quantify the contribution of free α to the whole CK2 activity of the lysates; in fact those lysates where free α exists are expected to display a ratio of CK2-tide/eiF2β-tide phosphorylation higher than those where only CK2 holoenzyme is present. This inference can be further corroborated by exploiting CK2 inhibitors with different efficacy toward either CK2 holoenzyme or its isolated catalytic subunits. This is the case of MNA whose IC_{50} values with CK2 holoenzyme and α are 0.3 and 2.8 μM, respectively (Salvi et al., 2006).

2.3. In-gel assay of CK2 catalytic subunit activity

This procedure allows the separation of cellular proteins in SDS-PAGE, according to their apparent molecular weight in denaturing conditions, and the detection of the kinase activity in the position where the catalytic subunit migrates. The applicability and sensitivity of this assay depend on the ability of the kinase to recover its active conformation once SDS is removed; in the case of CK2, we found a very good recovery of activity and a correlation with the amount of active kinase; therefore, this method provides an accurate assay whenever a measure of the activity of the catalytic subunit alone (separated from its regulatory subunit) is required. It also allows to appreciate the contribution of the two different CK2 catalytic subunits possibly present in a biological sample (α and α', whose M_w is slightly different), or the generation of truncated form of CK2 catalytic subunits that are still active (Sarno et al., 2002; Tapia et al., 2002) (Fig. 24.2).

To perform this kind of assay, a CK2 substrate has to be incorporated into a polyacrylamide–SDS gel prepared for protein separation; cellular proteins are normally run according to Laemmli (1970), then gel is washed and incubated in a renaturing buffer, followed by a radioactive phosphorylation mixture; gel is finally analyzed by autoradiography. Concerning the in-gel substrate, we are generally using casein, but theoretically any CK2 (protein) substrate would be preferable inasmuch it is more selective than casein.

The detailed protocol for this assay is the following:

1. Pour a normal 11% polyacrylamide gel according to Laemmli (1970), but add casein in both the running and the stacking gel, at a final concentration of 0.5 mg/ml.

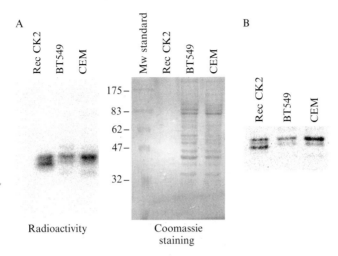

Figure 24.2 In-gel kinase assay of CK2 catalytic activity. Ten micrograms of proteins from cell lysates from BT549 (breast cancer) or CEM (lymphoblastoma) cells were analyzed by in-gel casein phosphorylation. Recombinant CK2 (10 ng) was also loaded, as a positive control, where indicated. Panel A shows the result of an experiment performed with [γ-^{32}P]ATP, according to the protocol described in Section 2.3; left panel corresponds to the autoradiography (16 h) with CyclonePlus (PerkinElmer), right panel to the Coomassie staining of the same gel. M_w standard migrations are also shown. Panel B shows the results of the electronic autoradiography (24 h) of a similar experiment performed according to the same protocol but with [γ-^{33}P]ATP.

2. Prepare samples by mixing proteins from cellular lysates with Laemmli sample buffer (Laemmli, 1970). For cancer cell lines, expressing high CK2 levels, 1–10 μg of total proteins are usually appropriate.
3. Load samples on the gel and run electrophoresis at 25 mA, with limiting voltage at 150 V.
4. After the electrophoresis, remove SDS by two 30-min washings of the gel in a buffer composed of 50 mM Tris–HCl pH 8.0, 20% (v/v) 2-propanol, at room temperature (RT).
5. Quickly rinse the gel in 50 mM Tris–HCl pH 8.0.
6. Incubate the gel for 1 h at RT in a buffer consisting of 50 mM Tris–HCl pH 8.0, 5 mM 2-mercaptoethanol (Buffer A).
7. Incubate the gel for 1 h at RT in Buffer A plus 6 M guanidine.
8. Incubate the gel overnight at 4 °C in Buffer A plus 0.04% (v/v) Tween-20.
9. Incubate the gel in a phosphorylation mixture containing 50 mM Tris–HCl pH 7.5, 100 mM MgCl$_2$, 10 μM ATP, [γ-^{32}P]ATP (specific radioactivity ∼1000–5000 cpm/pmol; it can be replaced by [γ-^{33}P]ATP). All incubation should be performed with gentle shaking in a volume ensuring total covering of the gel.

10. The gel can be normally stained and dried, then analyzed for the radioactivity by means of autoradiography (exposure to a photographic film or digital image obtained with instrument as Cyclone Plus, Perkin Elmer; the time of exposure ranges from few hours to few days).

Because casein is not a CK2-specific substrate, the selectivity of the assay is grounded on the colocalization of the CK2 catalytic subunit bands with the radioactive ones. It can be further improved, however, by replacing ATP with GTP; moreover, to make sure that the observed radioactive band(s) are really generated by CK2 activity, a parallel gel can be run and analyzed under the same conditions, but in the presence of a CK2 inhibitor in the phosphorylation mixture (see Section 2.2); radioactivity of the band(s) in this case should disappear or be significantly lower.

A representative experiment is shown in Fig. 24.2A; in the two cell lines analyzed (and also in many other tested, not shown), a major radioactive band corresponds to the size of the f.l. human CK2α, while a minor lower band is also present, which might correspond to CK2α′ or to a truncated product of CK2α; this is the case when recombinant CK2α is loaded. Both bands almost disappear when phosphorylation is performed in the presence of 2 μM TBB (not shown). Although other kinases able to phosphorylate casein are possibly present in cell lysates (Inglis et al., 2009), no other radioactive band is detectable under the described conditions. From Fig. 24.2B it can be appreciated that the use of ^{33}P-labeled ATP, although implying a lower sensitivity than that of ^{32}P-ATP, affords a neater resolution.

3. In-Cell Assay of Endogenous CK2 Activity

The in vitro assays described above are convenient whenever a rough quantitative estimate of constitutively active CK2 present in the cell, either normal or neoplastic, is needed; it fails, however, to provide reliable information about the actual occurrence of such an activity inside the intact living cell. This is a general problem with all protein kinases, and it is often addressed by immunochemical procedures designed for monitoring the phosphorylation state of either the kinase itself (which in many cases is symptomatic of an "active" conformation) or a reporter substrate of it.

The in-cell kinase activity can be also assessed following endogenous protein phosphorylation upon ^{32}P-radiolabeling of intact cells; in this case the employment of specific inhibitors is required to dissect proteins whose phosphorylation can be ascribed to CK2; this strategy will be considered under Section 4.

3.1. Immunodetection of CK2-dependent phosphosites

Since the great majority of protein kinases are activated by phosphorylation, generally occurring at the "activation loop," their activation state is frequently inferred by means of phospho-specific antibodies which recognize the phosphoresidue responsible for activation. This strategy, however, is entirely useless in the case of CK2: although in fact phosphorylation sites on CK2 are well known, and phospho-specific antibodies have been developed for these sites (St-Denis et al., 2009), they do not correlate with appearance of CK2 catalytic activity which, as mentioned in Section 1, is constitutively "on." To assess CK2 activity in cells, therefore, antibodies are required to recognize the phosphorylated form of known CK2 targets. Their reliability will depend on many factors, mainly related to specificity: on one hand, the antibody should be highly selective, that is, devoid of reactivity toward other phosphorylated sites and toward the unphosphorylated site itself; on the other hand, the recognized site should be affected only by CK2. These two crucial conditions must be carefully assessed, and several strategies should be adopted to validate the assay: whenever applicable, the absence of signal of a not-phosphorylatable mutant should be verified to test the specificity of the antibody, while the reduction of the signal by specific CK2 inhibitors or CK2 silencing should corroborate the concept that the site is indeed phosphorylated by CK2 also in living cells.

Many different phospho-specific antibodies have been developed for CK2 target sites; notably, those directed to pT117 of Bad (Klumpp et al., 2004), pS129 of Akt1 (Di Maira et al., 2005), pS17 of S6K1 (Panasyuk et al., 2006), pS13 of cdc37 (Miyata and Nishida, 2007), pS282 and pS559 of estrogen receptor α (Williams et al., 2009), pS201 of yeast Sic1 (Coccetti et al., 2006). The antibody toward pS392 of p53, although commercially available, is not recommendable for this purpose because this site, originally identified as a CK2 target (Meek et al., 1990), has been later reported to be a substrate for other kinases as well (Cox and Meek, 2010).

In all cases, the level of phosphorylation needs to be normalized to the amount of total protein present under the different conditions analyzed.

Of course, the value of the phospho-directed antibody tools will critically depend on the specific cellular environment: not only the protein target needs to be expressed in the analyzed cells, but also its amount should be quite high, and its phosphorylation level should correlate to the kinase activity which generally also implies a steady state equilibrium between its phosphorylated and dephosphorylated forms. To at least partially overcome these limits, we have set up a method based on the overexpression of a convenient target, as described in the following paragraph.

3.2. Phosphorylation of a reporter substrate

We found that Akt1 (PKBα) is phosphorylated at S129 by CK2, and that a phospho-specific antibody toward this site works very well (Di Maira et al., 2005). However, not always the Akt expression level is high enough to allow immunodetection with this antibody; therefore, we exploited the ectopic expression of Akt1 in the cells where CK2 activity needs to be quantified (Cozza et al., 2009). Cell transfection should be performed with the first choice technique applicable to that kind of cells. In case of HEK 293T cells, we use cells plated on to 60-mm-diameter dishes at ~80% confluence and we transiently transfect them with 4 μg of Akt cDNA using a standard calcium phosphate procedure. Cells are then lysed as described above, and analyzed by Western blot with antibody towards pS129 of Akt; usually 5 μg of total proteins are sufficient for immunodetection. Controls can be done by expressing the Akt mutant where S129 is replaced by alanine, which should fail to react with the antibody. The phospho-specific signal should be then normalized to the total amount of Akt expressed, detected by a general Akt antibody. A representative experiment is shown in Fig. 24.3.

4. IDENTIFICATION/VALIDATION OF *IN VIVO* CK2 TARGETS WITH SPECIFIC INHIBITORS

The assays of endogenous CK2 by means of model substrates, as described above, are satisfactory whenever the global activity of the kinase in a specific cell context has to be assessed; however, in signal transduction studies and especially in cancer cells, it is often necessary to establish if CK2 is actually responsible for the phosphorylation of a certain cellular substrate, in order to dissect the signaling pathways altered by its abnormally high

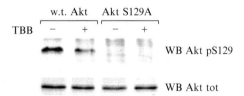

Figure 24.3 Assay of cellular CK2 activity by means of a reporter substrate. HEK 293T cells were transiently transfected (24 h) with wild-type Akt or S129A mutant Akt, and treated, where indicated, with the CK2 inhibitor TBB (20 μM) for the last 3 h of transfection. Ten micrograms of proteins from total cell lysates was analyzed by Western blot with anti-Akt antibodies either specific for the CK2-dependent phosphorylated site (pS129) or for total Akt protein.

activity. To this purpose, several strategies can be applied, mostly based on the usage of specific CK2 inhibitors. An ample repertoire of CK2 inhibitors are available, as reported elsewhere (Cozza et al., 2009; Pagano et al., 2008; Sarno and Pinna, 2008; Sarno et al., 2005a, in press). Of many of these, the selectivity has been analyzed by means of large panels of protein kinases, and the structural features underlying their potency and specificity have been deciphered by solving the structure of their complexes with CK2 catalytic subunits (e.g., Battistutta, 2009; Battistutta et al., 2007; Mazzorana et al., 2008; Sarno et al., in press). These studies have also led to the generation of CK2 mutants which are refractory to some of these inhibitors, thus providing a powerful tool to prove the implication of endogenous CK2 in a number of cellular events, including the targeting of individual protein substrates (Loizou et al., 2004). This approach is particularly helpful inasmuch the selectivity of commercially available CK2 inhibitors is turning out to be not as narrow as it was believed at the beginning, when their inhibitory activity was profiled toward relatively small panels of kinases, numbering up to 30 or so (Pagano et al., 2008). In general, therefore, it is highly recommendable to use more than one inhibitor, better if structurally unrelated between each other and displaying different inhibitory profiles on the same panel of protein kinases (e.g., TBB and quinalizarin) and/or to probe the actual implication of CK2 in the phosphorylation of the protein of interest by quenching the efficacy of the inhibitor by transfecting cells with a mutant displaying reduced sensitivity to the inhibitor itself. A very convenient mutant for this purpose is the one in which two hydrophobic residues nearly unique to CK2, V66 and I174, have been replaced by two alanines (V66, I174AA; Sarno et al., 2005b). This double mutant is normoactive but it displays one order of magnitude or more higher IC_{50} values with many CK2 inhibitors, including the commercially available ones, TBB (25-fold higher IC_{50}), DMAT (41-fold higher; Sarno et al., 2005b), emodin (55-fold higher; Cozza et al., 2009). Although for the time being this mutant is not commercially available, we are confident that it will soon become such. Presently, its cDNA can be obtained on request from the authors' lab.

The following section will be devoted to the technical details important to know when these chemical compounds are employed.

4.1. Cell treatment with specific CK2 inhibitors

The majority of the CK2 inhibitors described so far, with the possible partial exception of IQA (Sarno et al., 2005b), display their efficacy on cultured cells and are therefore suitable for this purpose. Of course their availability dictates the choice: DMAT, quinalizarin, TBB, TBCA, TBI, apigenin, emodin, ellagic acid are commercially available; however, their efficacy and, in particular, their selectivity are the main arguments to be considered when a cell-treatment experiment is planned. As a general rule, we can say that apigenin and emodin are quite promiscuous. This also applies to DMAT and TBI (Pagano et al., 2008), despite they were originally described as quite selective. In contrast, TBCA (Pagano et al., 2007) and quinalizarin (Cozza et al., 2009) are more specific. When different inhibitors are used, they are expected to induce similar effects, with a good correlation between their IC_{50} for CK2 in vitro and their efficacy in cells (see Meggio et al., 2004). In this connection, we remind that the concentrations required in vivo are higher than those effective in vitro; this is a general rule due to many factors, such as cell permeability, trapping and dilution in the medium and in cell compartments, high intracellular ATP concentration (being most inhibitors competitive with respect to ATP). As indicative figures, concentrations 50- to 100-fold higher than in vitro are frequently required to effectively inhibit the kinase in cells. A useful strategy can be to reduce fetal serum concentration in the cell culture medium, for example, from 10% to 1%; this allows a higher cell sensitivity, partly due to a lower trapping of the drugs, and is usually well tolerated for the short time of the treatment. Cell density should not be too high, usually lower than 1×10^6 cells/ml and 80% confluence for in-suspension cells or adherent cells, respectively. Concerning the time, it should be kept in mind that CK2 is essential for cell survival, thus prolonged treatments with CK2 inhibitors will invariably induce cell death; however, the length of time required to appreciate the reduction of a substrate phosphorylation usually ranges in few hours (2–6 h treatment can be initially considered), but it depends of course not only on the kinase inhibition, but also on the dephosphorylation rate.

Since the inhibition of CK2 is usually maintained even after cell disruption (see Fig. 24.1), the actual decrease of CK2 activity in response to the treatment with inhibitors can be checked by cell lysis and assay of the endogenous activity towards synthetic peptides, as described in Section 2.2.

4.2. Radioactive phosphorylation of endogenous proteins in treated and untreated cells

An easy and preliminary evaluation of the phosphorylation level of endogenous proteins can be achieved by incubating total cell lysates (2–10 μg proteins) with a radioactive phosphorylation mixture, as described in Section 2.2, but without the addiction of any artificial phosphoacceptor substrate. It could be convenient to avoid the addition of phosphatase inhibitors to the lysis buffer, to favor protein dephosphorylation, thus rendering substrates more prone to subsequent radioactive phosphorylation; in this case, phosphatase inhibitors should be added during the radioactive incubation. Proteins are then separated by SDS-PAGE and analyzed by autoradiography: each radioactive band corresponds to one (or more co-migrating) phosphorylated protein(s), which represents a target hit by an active kinase under these conditions. If CK2 inhibitors are added, those bands whose radioactivity is suppressed or reduced, as compared to untreated samples, represent putative CK2 substrates (Fig. 24.4). To further improve the specificity of the analysis with respect to CK2, GTP can replace ATP, as phosphate donor. CK2 inhibitors can be added during the phosphorylation reaction of untreated samples (at concentrations two- to fivefolds their *in vitro* IC_{50}), and they should induce effects similar to those observed if they are used for cell treatment. Another useful tool is provided by staurosporine, which drastically inhibits the majority of protein kinases with few exceptions including CK2 (Meggio *et al.*, 1995): when staurosporine (up to 2 μM) is added during the phosphorylation incubation, those bands due to CK2 activity should remain unchanged.

In our experience, CK2 is responsible for the generation of the majority of the radiolabeled bands detectable in lysates from unstimulated cells. Of course, this kind of assay does not necessarily correspond to real *in vivo* phosphorylation of substrates, since the accessibility of a target can be completely different in intact cells than it is in cell lysates. However, it provides an idea about the global kinase activity of CK2 in a certain cell type or condition, and can help to corroborate results obtained with peptide substrates. Moreover, the phosphorylation of a specific protein under these conditions can be evaluated; to this purpose, we suggest to immunoprecipitate the putative protein substrate after the phosphorylation reaction and to check its radioactivity in the immunoprecipitation pellet. A positive result

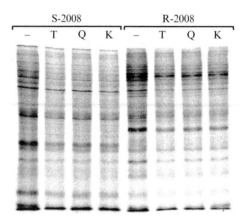

Figure 24.4 CK2-dependent phosphorylation of endogenous proteins. Five micrograms proteins of total lysate from human ovarian carcinoma 2008 cells (normal sensitive, S-2008, and their cisplatin-resistant variant, R-2008) were incubated with a radioactive phosphorylation mixture (see text). Where indicated, TBB (T, 1 μM) (Sarno *et al.*, 2001), quinalizarine (Q, 1 μM) (Cozza *et al.*, 2009), or K64 (K, 2 μM) (Pagano *et al.*, 2008) were added. Proteins were separated by 11% SDS-PAGE and the incorporated radioactivity was detected by Cyclone Plus.

will prompt to analyze the *in vivo* phosphorylation of the protein, as described in the following section.

4.3. *In vivo* radioactive phosphorylation in treated and untreated cells

Cell loading with [^{32}P]orthophosphate to produce an intracellular radioactive pool of ATP is a traditional strategy widely used to investigate protein phosphorylation in intact cells. However, the applicability of this assay to study CK2 activity derives from the possibility to detect CK2 targets based on their sensitivity to CK2-specific inhibitors and, eventually, their insensitivity to staurosporine (see above). ATP isotopic labeling should be carried out as appropriate for the cell line used; normally, a 2–6 h incubation is performed at 37 °C in atmosphere containing 5% CO_2, with 0.5–1 mCi/ml

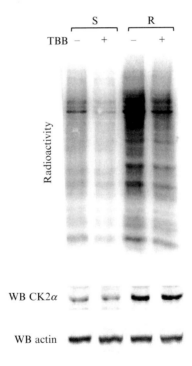

Figure 24.5 Detection of *in vivo* CK2 substrates by means of ^{32}P-metabolic labeling. CEM cells (normal sensitive, S, and their multidrug-resistant variant, R) were loaded with [^{32}P]orthophosphate for 4 h; where indicated, 20 μM TBB was added during metabolic labeling. Twenty micrograms of total proteins were analyzed by SDS-PAGE and blotting, followed by autoradiography with Cyclone Plus (radioactivity) and Western blot with anti-actin, as a loading control, or anti-CK2 α C-terminus (Di Maira *et al.*, 2007).

[^{32}P]orthophosphate in culture medium. Whenever available, a medium devoid of phosphate should be preferred, and, in case, fetal calf serum can be dialyzed before addition to the medium. To keep as low as possible radioactivity, cell density should be kept as higher as possible, compatible with viability. In treated cells, inhibitors are added from the beginning of the incubation with the radioactive phosphate. Cells are then lysed, and radioactivity associated to protein phosphorylation analyzed by SDS-PAGE or 2D electrophoresis, followed by autoradiography. A representative experiment is shown in Fig. 24.5, where it is evident that the number and intensity of radioactive CK2-dependent bands are proportional to the amount of CK2 present in the cells.

If a specific protein has to be considered, its radioactivity can be checked upon immunoprecipitation from total labeled lysate. Implication of CK2 in the phosphorylation of given protein bands can be further validated by quenching the efficacy of CK2 inhibitors upon transfection with a CK2 double mutant which is refractory to the inhibitor (see above). If the protein whose phosphorylation is inhibited by that inhibitor is really phosphorylated by CK2 rather than by other protein kinases sensitive to the same compound, its phosphorylation in the presence of the inhibitor is restored upon transfection with the mutant (Loizou et al., 2004).

ACKNOWLEDGMENTS

The research relevant to this paper was supported by the Associazione Italiana Ricerca sul Cancro (AIRC), the Italian Ministry of University and Research (PRIN-2007 to MR), and the Italian Cystic Fibrosis Research Foundation (Grant FFC#4/2007) with the contribution of "Banca Popolare di Verona e Novara" and "Fondazione Giorgio Zanotto."

REFERENCES

Battistutta, R. (2009). Protein kinase CK2 in health and disease: Structural bases of protein kinase CK2 inhibition. *Cell. Mol. Life Sci.* **66,** 1868–1889.

Battistutta, R., Mazzorana, M., Cendron, L., Bortolato, A., Sarno, S., Kazimierczuk, Z., Zanotti, G., Moro, S., and Pinna, L. A. (2007). The ATP-binding site of protein kinase CK2 holds a positive electrostatic area and conserved water molecules. *Chembiochem* **8,** 1804–1809.

Coccetti, P., Zinzalla, V., Tedeschi, G., Russo, G. L., Fantinato, S., Marin, O., Pinna, L. A., Vanoni, M., and Alberghina, L. (2006). Sic1 is phosphorylated by CK2 on Ser201 in budding yeast cells. *Biochem. Biophys. Res. Commun.* **346,** 786–793.

Cohen, P. (2002). Protein kinases—the major drug targets of the twenty-first century? *Nat. Rev. Drug Discov.* **1,** 309–315.

Cox, M. L., and Meek, D. W. (2010). Phosphorylation of serine 392 in p53 is a common and integral event during p53 induction by diverse stimuli. *Cell. Signal.* **22,** 564–571.

Cozza, G., Mazzorana, M., Papinutto, E., Bain, J., Elliott, M., Di Maira, G., Gianoncelli, A., Pagano, M. A., Sarno, S., Ruzzene, M., Battistutta, R., Meggio, F., *et al.* (2009).

Quinalizarin as a potent, selective and cell-permeable inhibitor of protein kinase CK2. *Biochem. J.* **421,** 387–395.

Di Maira, G., Salvi, M., Arrigoni, G., Marin, O., Sarno, S., Brustolon, F., Pinna, L. A., and Ruzzene, M. (2005). Protein kinase CK2 phosphorylates and upregulates Akt/PKB. *Cell Death Differ.* **12,** 668–677.

Di Maira, G., Brustolon, F., Bertacchini, J., Tosoni, K., Marmiroli, S., Pinna, L. A., and Ruzzene, M. (2007). Pharmacological inhibition of protein kinase CK2 reverts the multidrug resistance phenotype of a CEM cell line characterized by high CK2 level. *Oncogene* **26,** 6915–6926.

Duncan, J. S., and Litchfield, D. W. (2008). Too much of a good thing: The role of protein kinase CK2 in tumorigenesis and prospects for therapeutic inhibition of CK2. *Biochim. Biophys. Acta* **1784,** 33–47.

Fabbro, D., Ruetz, S., Buchdunger, E., Cowan-Jacob, S. W., Fendrich, G., Liebetanz, J., Mestan, J., O'Reilly, T., Traxler, P., Chaudhuri, B., Fretz, H., Zimmermann, J., et al. (2002). Protein kinases as targets for anticancer agents: From inhibitors to useful drugs. *Pharmacol. Ther.* **93,** 79–98.

Guerra, B., and Issinger, O. G. (2008). Protein kinase CK2 in human diseases. *Curr. Med. Chem.* **15,** 1870–1886.

Inglis, K. J., Chereau, D., Brigham, E. F., Chiou, S. S., Schöbel, S., Frigon, N. L., Yu, M., Caccavello, R. J., Nelson, S., Motter, R., Wright, S., Chian, D., et al. (2009). Polo-like kinase 2 (PLK2) phosphorylates alpha-synuclein at serine 129 in central nervous system. *J. Biol. Chem.* **284,** 2598–2602.

Klumpp, S., Mäurer, A., Zhu, Y., Aichele, D., Pinna, L. A., and Krieglstein, J. (2004). Protein kinase CK2 phosphorylates BAD at threonine-117. *Neurochem. Int.* **45,** 747–752.

Laemmli, U. K. (1970). Cleavage of structural proteins during the assembly of the head of bacteriophage T4. *Nature* **227,** 680–685.

Litchfield, D. W. (2003). Protein kinase CK2: Structure, regulation and role in cellular decisions of life and death. *Biochem. J.* **369,** 1–15.

Loizou, J. I., El-Khamisy, S. F., Zlatanou, A., Moore, D. J., Chan, D. W., Qin, J., Sarno, S., Meggio, F., Pinna, L. A., and Caldecott, K. W. (2004). The protein kinase CK2 facilitates repair of chromosomal DNA single-strand breaks. *Cell* **117,** 17–28.

Marin, O., Meggio, F., and Pinna, L. A. (1994). Design and synthesis of two new peptide substrates for the specific and sensitive monitoring of casein kinases-1 and -2. *Biochem. Biophys. Res. Commun.* **198,** 898–905.

Mazzorana, M., Pinna, L. A., and Battistutta, R. (2008). A structural insight into CK2 inhibition. *Mol. Cell. Biochem.* **316,** 57–62.

Meek, D. W., Simon, S., Kikkawa, U., and Eckhart, W. (1990). The p53 tumour suppressor protein is phosphorylated at serine 389 by casein kinase II. *EMBO J.* **9,** 3253–3260.

Meggio, F., and Pinna, L. A. (2003). One-thousand-and-one substrates of protein kinase CK2? *FASEB J.* **17,** 349–368.

Meggio, F., Marin, O., and Pinna, L. A. (1994). Substrate specificity of protein kinase CK2. *Cell. Mol. Biol. Res.* **40,** 401–409.

Meggio, F., Donella-Deana, A., Ruzzene, M., Brunati, A. M., Cesaro, L., Guerra, B., Meyer, T., Mett, H., Fabbro, D., Furet, P., Dobrowolska, G., and Pinna, L. A. (1995). Different susceptibility of protein kinases to staurosporine inhibition. Kinetic studies and molecular bases for the resistance of protein kinase CK2. *Eur. J. Biochem.* **234,** 317–322.

Meggio, F., Pagano, M. A., Moro, S., Zagotto, G., Ruzzene, M., Sarno, S., Cozza, G., Bain, J., Elliott, M., Deana, A. D., Brunati, A. M., and Pinna, L. A. (2004). Inhibition of protein kinase CK2 by condensed polyphenolic derivatives. An in vitro and in vivo study. *Biochemistry* **43,** 12931–12936.

Miyata, Y., and Nishida, E. (2007). Analysis of the CK2-dependent phosphorylation of serine 13 in Cdc37 using a phospho-specific antibody and phospho-affinity gel electrophoresis. *FEBS J.* **274,** 5690–5703.
Pagano, M. A., Poletto, G., Di Maira, G., Cozza, G., Ruzzene, M., Sarno, S., Bain, J., Elliott, M., Moro, S., Zagotto, G., Meggio, F., and Pinna, L. A. (2007). Tetrabromocinnamic acid (TBCA) and related compounds represent a new class of specific protein kinase CK2 inhibitors. *Chembiochem* **8,** 129–139.
Pagano, M. A., Bain, J., Kazimierczuk, Z., Sarno, S., Ruzzene, M., Di Maira, G., Elliott, M., Orzeszko, A., Cozza, G., Meggio, F., and Pinna, L. A. (2008). The selectivity of inhibitors of protein kinase CK2: An update. *Biochem. J.* **415,** 353–365.
Panasyuk, G., Nemazanyy, I., Zhyvoloup, A., Bretner, M., Litchfield, D. W., Filonenko, V., and Gout, I. T. (2006). Nuclear export of S6K1 II is regulated by protein kinase CK2 phosphorylation at Ser-17. *J. Biol. Chem.* **281,** 31188–31201.
Pinna, L. A. (2002). Protein kinase CK2: A challenge to canons. *J. Cell Sci.* **115,** 3873–3878.
Pinna, L. A. (2003). The raison d'être of constitutively active protein kinases: The lesson of CK2. *Acc. Chem. Res.* **36,** 378–384.
Poletto, G., Vilardell, J., Marin, O., Pagano, M. A., Cozza, G., Sarno, S., Falqués, A., Itarte, E., Pinna, L. A., and Meggio, F. (2008). The regulatory beta subunit of protein kinase CK2 contributes to the recognition of the substrate consensus sequence. A study with an eIF2 beta-derived peptide. *Biochemistry* **47,** 8317–8325.
Ruzzene, M., and Pinna, L. A. (1999). Assay of protein kinases and phosphatases using specific peptide substrates. *In* "Protein Phosphorylation. A Practical Approach," (D. G. Hardie, ed.) 2nd edn. pp. 221–253. Oxford University Press, Oxford, U.K.
Ruzzene, M., and Pinna, L. A. (2010). Addiction to protein kinase CK2: A common denominator of diverse cancer cells? *Biochim. Biophys. Acta* **1804,** 499–504.
Salvi, M., Sarno, S., Marin, O., Meggio, F., Itarte, E., and Pinna, L. A. (2006). Discrimination between the activity of protein kinase CK2 holoenzyme and its catalytic subunits. *FEBS Lett.* **580,** 3948–3952.
Salvi, M., Sarno, S., Cesaro, L., Nakamura, H., and Pinna, L. A. (2009). Extraordinary pleiotropy of protein kinase CK2 revealed by weblogo phosphoproteome analysis. *Biochim. Biophys. Acta* **1793,** 847–859.
Sarno, S., and Pinna, L. A. (2008). Protein kinase CK2 as a druggable target. *Mol. Biosyst.* **4,** 889–894.
Sarno, S., Reddy, H., Meggio, F., Ruzzene, M., Davies, S. P., Donella-Deana, A., Shugar, D., and Pinna, L. A. (2001). Selectivity of 4, 5, 6, 7-tetrabromobenzotriazole, an ATP site-directed inhibitor of protein kinase CK2 ('casein kinase-2'). *FEBS Lett.* **496,** 44–48.
Sarno, S., Moro, S., Meggio, F., Zagotto, G., Dal Ben, D., Ghisellini, P., Battistutta, R., Zanotti, G., and Pinna, L. A. (2002). Toward the rational design of protein kinase casein kinase-2 inhibitors. *Pharmacol. Ther.* **93,** 159–168.
Sarno, S., Salvi, M., Battistutta, R., Zanotti, G., and Pinna, L. A. (2005a). Features and potentials of ATP-site directed CK2 inhibitors. *Biochim. Biophys. Acta* **1754,** 263–270.
Sarno, S., Ruzzene, M., Frascella, P., Pagano, M. A., Meggio, F., Zambon, A., Mazzorana, M., Di Maira, G., Lucchini, V., and Pinna, L. A. (2005b). Development and exploitation of CK2 inhibitors. *Mol. Cell. Biochem.* **274,** 69–76.
Sarno, S., Papinutto, E., Franchin, C., Bain, J., Elliot, M., Meggio, F., Kazimierczuc, Z., Zanotti, G., Battistutta, R., and Pinna, L.A. (in press). ATP site-directed inhibitors of protein kinase CK2: An update. *Curr. Top. Med. Chem.*
Sawyers, C. L. (2002). Rational therapeutic intervention in cancer: Kinases as drug targets. *Curr. Opin. Genet. Dev.* **12,** 111–115.
St-Denis, N. A., Derksen, D. R., and Litchfield, D. W. (2009). Evidence for regulation of mitotic progression through temporal phosphorylation and dephosphorylation of CK2alpha. *Mol. Cell. Biol.* **29,** 2068–2081.

Tapia, J., Jacob, G., Allende, C. C., and Allende, J. E. (2002). Role of the carboxyl terminus on the catalytic activity of protein kinase CK2alpha subunit. *FEBS Lett.* **531,** 363–368.

Tawfic, S., Yu, S., Wang, H., Faust, R., Davis, A., and Ahmed, K. (2001). Protein kinase CK2 signal in neoplasia. *Histol. Histopathol.* **16,** 573–582.

Trembley, J. H., Wang, G., Unger, G., Slaton, J., and Ahmed, K. (2009). Protein kinase CK2 in health and disease: CK2: A key player in cancer biology. *Cell. Mol. Life Sci.* **66,** 1858–1867.

Unger, G. M., Davis, A. T., Slaton, J. W., and Ahmed, K. (2004). Protein kinase CK2 as regulator of cell survival: Implications for cancer therapy. *Curr. Cancer Drug Targets* **4,** 77–84.

Williams, C. C., Basu, A., El-Gharbawy, A., Carrier, L. M., Smith, C. L., and Rowan, B. G. (2009). Identification of four novel phosphorylation sites in estrogen receptor alpha: Impact on receptor-dependent gene expression and phosphorylation by protein kinase CK2. *BMC Biochem.* **10,** 36.

CHAPTER TWENTY-FIVE

Structural Basis of the Constitutive Activity of Protein Kinase CK2

Birgitte B. Olsen,* Barbara Guerra,* Karsten Niefind,[†] and Olaf-Georg Issinger*

Contents

1. Introduction 516
2. A Constitutively Active CK2α Structure and Its Stabilizing Elements 516
   2.1. Installation of COOT and basic introduction to the program 517
   2.2. Global superposition of CK2α structures—Fixation of the activation segment in a fully active conformation 518
   2.3. DWG rather than DFG: An additional H-bond stabilizes the magnesium-binding loop in its active conformation 520
3. Analyzing the Constitutive Activity of Protein Kinase CK2 521
   3.1. Analytical gel filtration chromatography to study aggregation 522
   3.2. Analysis of proteins by SDS-PAGE 524
   3.3. Autophosphorylation is not a requirement for aggregation 525
Acknowledgments 528
References 528

Abstract

Protein kinase CK2 (formerly referred to as casein kinase II) is an evolutionary conserved, ubiquitous protein kinase. In mammals, there are two paralog catalytic subunits, that is, CK2α (A1) and CK2α' (A2), and one CK2β dimer, which together form the heterotetrameric holoenzyme.

The presence of full functioning CK2α and CK2β subunits are absolutely mandatory for embryonic development. Total knockouts are lethal. The CK2α' paralog seems to be an exception inasmuch as a total knockout only leads to sterility in male mice.

The catalytic subunits are distantly related to the CMGC subfamily of protein kinases, such as the cyclin-dependent kinases (CDKs). There are some peculiarities associated with protein kinase CK2, which are not found with most of the other protein kinases: the enzyme is constitutively active, it can use ATP and GTP as phosphoryl donors, and it is found elevated in most tumors investigated

---

* Department of Biochemistry and Molecular Biology, University of Southern Denmark, Odense, Denmark
[†] Department of Chemistry, Institute of Biochemistry, University of Cologne, Cologne, Germany

Methods in Enzymology, Volume 484 © 2010 Elsevier Inc.
ISSN 0076-6879, DOI: 10.1016/S0076-6879(10)84025-X All rights reserved.

and rapidly proliferating tissues. In this review, we explain (i) its constitutive activity at the intramolecular level, and (ii) come forward with a model how this protein kinase could be regulated in cells by a mechanism involving intermolecular interactions.

1. INTRODUCTION

CK2 is a prosurvival kinase that regulates: signal transduction, cell cycle progression, apoptosis, angiogenesis, and ribosomal biogenesis (for reviews, see Guerra and Issinger, 2008; Trembley *et al.*, 2009). It induces mammary adenocarcinomas when overexpressed in the mammary glands of transgenic mice and causes acute lymphocytic leukemia in a transgenic mouse model when overexpressed in conjunction with c-myc (Kelliher *et al*, 1996; Landesman-Bollag *et al.*, 2001; Seldin and Leder, 1995). siRNA and antisense vectors, directed toward CK2 activity in xenograft models, induce potent apoptosis in cancer cells, but *minimal cell death in normal cells* (Slaton *et al.*, 2004). Especially the latter observation indicates that CK2 should have "disease-associated functions," which are different from its normal function and raises the possibility to pharmacologically target CK2 for induction of apoptosis in cancer cells under conditions that may spare normal cells.

Although the dogma that CK2 is a constitutively active kinase still persists, there is increasing evidence that it may be regulated not only by aggregation/disaggregation at the level of the holoenzyme structure but also by dissociation into its individual subunits and interaction with small cellular molecules (Niefind and Issinger, 2008, 2010). These modes of regulation are unique for CK2 and different from the common regulation found with other protein kinases, for example, activation by second messengers, phosphorylation, and association with regulatory molecules such as the cyclins in the case of CDKs (Pinna, 2002).

To specifically target CK2 in cancer, yet sparing "normal" cells, it is a prerequisite to understand its peculiar regulation for devising therapeutic approaches for optimal disease regimen.

2. A CONSTITUTIVELY ACTIVE CK2α STRUCTURE AND ITS STABILIZING ELEMENTS

A central goal of structural biology in the context of eukaryotic protein kinases (EPK) is the elucidation of the structural basis of control mechanisms. In the case of the CDKs and the MAP kinases—the closest relatives of CK2α within the CMGC family of EPKs—this approach was successful already in the 1990s: the activation of the CDKs by cyclin binding and phosphorylation, and of the MAP kinases by dual phosphorylation was convincingly rationalized on a structural level. Structural comparisons of various regulatory states revealed significant conformational variations, in particular, in the activation segment and the helix αC (see Huse and Kuriyan, 2002 for review).

In the case of CK2α, the approach was successful in a reversed sense: the "nonregulatory" CK2α, that is, the constitutive activity of CK2 on the level of the catalytic subunits was substantiated by more than 40 3D-structures of CK2α molecules (Niefind et al., 2009). These structures—more precisely the structural comparisons between pairs of these structures—demonstrate the extreme conformational rigidity of the canonical regulatory key elements.

Moreover, a detailed view reveals a number of structural elements stabilizing this constitutively active structure. In the following paragraphs, we will illustrate and discuss these elements. To perform such analyses with published structures and to generate detailed illustrations are typical tasks for structural biologists, but may also be useful for noncrystallographers. Therefore, we will describe in this context suitable procedures to do this.

Currently, the most popular program for structural illustrations is PyMOL (DeLano, 2002), which is now offered by Schrödinger, LLC. Nevertheless, we recommend here the "Crystallographic Object-Oriented Toolkit" (COOT) (Emsley et al., 2010), which is free for academic users. COOT is less versatile than PyMOL for illustrations; however, the program is under rapid development, provides very useful tools for validation and quality assessment, and is in our opinion easily comprehensible for beginners.

2.1. Installation of COOT and basic introduction to the program

COOT (Emsley et al., 2010) is a structural biology program, the principle tasks of which are "the visualization of macromolecular structures and data ... and the validation of existing models" (Emsley et al., 2010). It was designed with the primary goal "to make the software easy to learn in order to provide a low barrier for scientists who are beginning to work with X-ray data" (Emsley et al., 2010). The program runs on all common computer platforms, is available from the CCP4 (Collaborative Computational Project, Number 4, 1994) download site (www.ccp4.ac.uk/download.php), and can be easily installed. The installation comprises a tutorial and a manual suitable for nonexpert users.

- Users of a Windows operating system find an automatic installer file in the download area given above.
- After launching COOT, a graphical user interface (GUI) starts providing a number of tools for nonexpert users. These functions are clustered to various groups like "Edit," "Calculate," "Draw," "Measure," or "Validate" in the top menu.

- Provided the computer has access to the internet, any structure from the Protein Data Bank (PDB, Berman *et al.*, 2002) can be loaded via its four-character code. This code must be entered in a window that opens after clicking on "File" in the top command line and then on either "Get PDB using Accession code ..." or "Get PDB & Map using EDS ..." In the latter case, the structure is loaded together with the corresponding electron density map so that the fit between structure and electron density can be checked locally.

The program offers the possibility to load many structures in parallel and to compare them as described in the following chapter.

2.2. Global superposition of CK2α structures—Fixation of the activation segment in a fully active conformation

For illustration of the invariance of CK2α in the activation segment and the helix αC, a number of representative CK2α structures from different species, crystal packings, and molecular environments are superimposed and compared.

- Launch COOT and load the following PDB files: 1LP4 (maize CK2α in complex with an ATP-analog and magnesium ions), 1JAM (maize CK2α as an apoenzyme), 2PVR (human CK2α with an ATP-analog and two sulfate ions at the activation segment), 1JWH (human CK2 holoenzyme). There are now five CK2α protomers loaded, two of them belonging to the CK2 holoenzyme.
- Superimpose each of the CK2α chains on 1LP4 as a reference structure with the "Secondary Structure Matching" algorithm (Krissinel and Henrick, 2004) as implemented in COOT. For this purpose, click on "Calculate" in the top line of the COOT GUI and then on "SSM Superpose." In the upcoming window, select "1lp4" as the "Reference Structure" and one of the other CK2α chains as the "Moving Structure." Finish the 3D-fit by hitting "Apply," and repeat this procedure such that finally all other CK2α subunits lie on 1LP4. If the moving structure contains more than one protein chain (here the CK2 holoenzyme structure 1JWH), hit the box "Use Specific Chain" and select the desired protomer via the chain identifier. Further, choose the option "Move copy of Moving Structure" to ensure that after fitting the first holoenzyme-bound CK2α subunit to 1LP4 also the second one can be superimposed independently.
- Click on "Reset View" in the top menu of the COOT GUI to center the display on the 1LP4–CK2α subunit.
- Open the "Display Manager" of COOT and select the "C-alphas/Backbone" option for all five structures in the display window.

- Using the mouse in combination with the CTRL- and the SHIFT-key to rotate, translate, and zoom the molecules on the display such that the activation segment and the helix αC are optimally visible.
- For optimization, click "Edit" in the top menu of the COOT GUI. Select "Background Color ..." to change the background to white. Select "Bond Parameters" and "Bonds Color" to change either the line width or the color of the C_α-trace of any of the molecules in the display. Further optimization would be possible by selecting and highlighting certain zones that shall be particularly emphasized.
- By clicking on "Draw," "Screenshot," and then on "Simple," a basic figure showing an overlay of the five CK2α chains can be stored. The picture can be loaded in a presentation program like MS PowerPoint, cut to an appropriate size, and labeled. The result can be seen in Fig. 25.1.
- Before leaving COOT, it is advisable to save the novel coordinates of the superimposed structures via "File" in the top menu and then selecting the option "Save Coordinates." The resulting output coordinate files retain the pairwise 3D-fits and can be loaded to any other graphics program.

Figure 25.1—the result of the procedure outlined in the text—illustrates the perfect structural conservation of the ensemble of helix αC, activation segment (comprising among others both the activation loop and the P + 1 loop), and N terminal segment. The latter is a functional equivalent

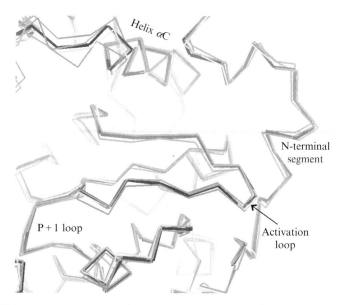

Figure 25.1 Illustration of the perfect structural conservation of the ensemble of helix αC, activation segment (comprising among others the activation loop and the P + 1 loop), and N-terminal segment of CK2α.

to the cyclin molecules in the case of the cyclin-dependent kinases because it constrains the activation segment and the helix αC in a fully active conformation. This arrangement is the most conspicuous basis of the constitutive activity of CK2α.

2.3. DWG rather than DFG: An additional H-bond stabilizes the magnesium-binding loop in its active conformation

The activation segment of EPKs begins with the magnesium-binding loop typically containing the highly conserved DFG-sequence motif. The plasticity of many protein kinases in the activation segment often starts in this region as illustrated in Fig. 25.2 for the inactive CDK2 in comparison to (partially) active CDK2 as part of the CDK2/cyclin A complex. In some other kinases, even larger structural changes can occur and the central phenylalanine residue can flip to a so-called DFG-out conformation. In CK2α, however, a tryptophan residue that replaces the canonical phenylalanine enables an additional hydrogen bond (Fig. 25.2) as a further

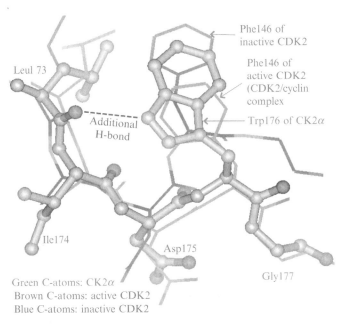

Figure 25.2 Superimposition of CK2α and inactive and (partially) active CDK2 created with COOT. In CDK2, the "DFG motif" is visible for both the inactive and active conformation. However, CK2α contains a "DWG motif," and here the tryptophan176 of CK2α makes an additional H-bond stabilizing the active conformation of CK2α. (See Color Insert.)

Structural Basis for Constitutive Active CK2   521

stabilizer of the fully active conformation. Figure 25.2 illustrates that these issues can be created in the following way:

- Launch COOT and load (via "File" → "Get PDB using Accession Code ...") 1LP4 as a typical CK2α structure, 1HCK as an inactive CDK2 structure, and 1FIN as a partially active CDK2 structure.
- Superimpose (via "Calculate" → "SSM Superpose ...") both 1HCK and chain A of 1FIN on 1LP4.
- Click on "Reset view" to center 1LP4 in the display.
- Copy the relevant fragments of the three structures to novel objects: click on "Extensions" in the top menu, then on "Modeling ..." and finally on "Copy Fragment ...." A small window opens in which you can define the mother structure and the fragment to be copied. For instance, for 1LP4 enter: "//A/173-177." In the same way, enter for 1HCK and for 1FIN: "//A/143-147." In general, relevant segments that shall be displayed isolated from the rest of the structure must be identified by clicking on the respective residues.
- Open "Display Manager"; deselect the complete structures, but select the three copied fragments for display.
- Set the background color to white ("Edit" → "Background Color ..." → "White").
- To allow a better distinction of the three overlaid fragments, change the colors of the C-atoms ("Edit" → "Bond Colors ..."; activate the box "Change color for Carbons only").
- Click on "Extensions," then on "Representations," and finally on "Ball & Stick ..." A small window opens in which "atom selection from 1lp4.pdb" must be selected and "//A/173-177" be entered.
- Translate and rotate the three fragments in the display via the mouse and the SHIFT- and the CTRL-key to a suitable view.
- Make a screen shot ("Draw" → "Screenshot" → "Simple").
- Load the resulting graphics file into a presentation program like MS PowerPoint. Set appropriate labels.

3. ANALYZING THE CONSTITUTIVE ACTIVITY OF PROTEIN KINASE CK2

In this part, we describe several methods for analyzing the constitutive activity of protein kinase CK2 *in vitro* considering both isozymes, that is, the CK2α- and the CK2α'-based holoenzymes in addition to a mutant form of the CK2α-based holoenzyme. The protocols include an analytical gel filtration chromatography approach to study the aggregation behavior at different salt concentrations because aggregations have been suggested as a mean of regulating the activity of CK2. Autophosphorylation of the regulatory β-subunit has also been linked to the supramolecular oligomerization of CK2 and as a prerequisite for high molecular mass aggregation observed for CK2 (Olsen et al., 2006; Pagano et al., 2005). A truncated version of CK2α has been used throughout the experiments; the deletion mutant (CK2α$^{1-335}$) is stable and undistinguishable from the full length CK2α with respect to all known characteristics (Guerra et al., 2001).

3.1. Analytical gel filtration chromatography to study aggregation

In this part, we provide procedures for studying the aggregation of recombinant human CK2 holoenzymes by analytical gel filtration chromatography. The protocol describes the analytical gel filtration chromatography at low and high monovalent salt concentrations as well as a SDS-PAGE analysis of the eluted peak fractions. In our experience, the Superdex200 chromatographic column offers the best separation when studying the aggregation of CK2 holoenzymes; however, also the Superose6 column can be used.

3.1.1. Required materials

Materials for analytical gel filtration chromatography of CK2 holoenzymes

- Cooling centrifuge
- Hamilton syringe
- Superdex200 column (PC3.2/30)
- Analytical chromatographic system (e.g., SMART system, GE Healthcare)
- Lyophilizer equipment
- 30 °C water bath

Additional materials and reagents

- Recombinant CK2α-based holoenzyme constituted of either wild-type CK2β or a mutant lacking the first four amino acids (β$^{\Delta 1-4}$) and CK2α'-based holoenzyme (purified from *E. coli* as described in Boldyreff et al., 1993a; Grankowski et al., 1991; Olsen et al., 2006)
- Gel filtration protein standards (Biorad)
- Low salt buffer (filtered (0.2 μ*M*) and degassed): 25 m*M* Tris–HCl pH 8.5, 1 m*M* DTT, 150 m*M* NaCl
- High salt buffer (filtered and degassed): 25 m*M* Tris–HCl pH 8.5, 1 m*M* DTT, 500 m*M* NaCl
- 100 μ*M* ATP (Sigma)
- 2 m*M* MgCl$_2$
- 20 m*M* (NH$_4$)HCO$_3$
- Bradford reagent (Biorad)

Disposables

0.5- and 1.5-ml (Eppendorf) tubes, dialysis tubing (6.3 mm, MWCO 12–14 kDa, MediCell International Ltd).

3.1.2. Analytical gel filtration chromatography in low salt buffer

The Superdex200 column is mounted on the SMART system and washed with 5 column volumes (i.e., 12 ml) ddH$_2$O prior to being equilibrated with 5 column volumes low salt buffer. When equilibrated, 7 µg of the CK2 holoenzyme is diluted to 0.14 µg/µl in low salt buffer supplemented with 100 µM ATP/2 mM MgCl$_2$ and incubated for 10 min at 30 °C in a total volume of 50 µl. Afterwards, the sample is centrifuged at 12,000 × g for 10 min at 4 °C and 50 µl injected into the injection loop using a Hamilton syringe. A flow of 40 µl/min is applied, and 100 µl fractions are collected. Generally, the void volume (< 670 kDa) elutes in fraction 9, and by fraction 20 all proteins are eluted.

An estimated molecular weight of the eluted peak can then be extrapolated based on a chromatography performed using a gel filtration protein standards (Biorad).

After completing the runs in low salt buffer, the column is equilibrated in high salt buffer and the CK2 holoenzymes are preincubated in the presence of ATP/MgCl$_2$ and injected onto the column as described above.

The gel filtration data performed in low salt buffer (Table 25.1) support the notion that only the CK2α-based holoenzyme is capable of aggregating (elutes at > 670 kDa) regardless of autophosphorylation because the CK2α-based holoenzyme containing a mutant form of CK2β ($\beta^{\Delta 1-4}$), which lacks the autophosphorylation sites, also aggregates at 150 mM NaCl, whereas the CK2α'-based holoenzyme does not aggregate.

The collected peak fractions from the chromatographies performed in high salt buffer (of the two wild-type holoenzymes, i.e., the CK2α- and α'-based holoenzymes) are dialyzed 3 × 1 h against 20 mM ammoniumbicarbonate prior to lyophilization overnight. The lyophilized samples are

Table 25.1 Analytical gel filtration analyses on Superdex200

	150 mM NaCl (kDa)	500 mM NaCl (kDa)
CK2α'-based holoenzyme	290	95
CK2α$_2$β$_2$	>670	145
CK2α$_2$β$^{\Delta 1-4}{}_2$	>670	–

The molecular weight indicates the elution of CK2 in 150 and 500 mM NaCl. Adapted from Olsen *et al.* (2008).

resuspended in 50 µl ddH$_2$O and the protein concentration determined according to the method of Bradford (1976).

3.2. Analysis of proteins by SDS-PAGE

Here we describe how to analyze proteins by SDS-PAGE followed by Coomassie staining and destaining. This protocol is not only used for analyzing the peak fractions eluted after analytical gel filtration chromatography but also prior to autoradiography following autophosphorylation assays.

3.2.1. Required materials

Materials for running and staining polyacrylamide gels

- 12.5% polyacrylamide gels
- 95 °C heating block
- Power supply
- Vertical gel chamber
- Trays for staining

Additional materials and reagents

- SDS sample buffer (0.5 M Tris–HCl pH 6.8, 10% SDS, 10 mM DTT, 14% glycerol, 0.025% bromophenol blue)
- Running buffer (100 mM Tris, 100 mM glycine, 3.5 mM SDS)
- Protein markers (Spectra Multicolor Broad Range Protein Ladder, Fermentas)
- Coomassie stain (0.25% Serva Blue R250, 45% ethanol, 10% acetic acid)
- Coomassie wash (45% ethanol, 10% acetic acid)

Disposables

1.5-ml (Eppendorf) tubes.

3.2.2. SDS-PAGE

One microgram from each dialyzed and lyophilized peak fraction is loaded onto a 12.5% SDS–polyacrylamide gel and the proteins are separated by SDS-PAGE. The gel is usually run at 60 mA for approximately 45 min until the bromophenol blue reaches the bottom of the gel. After carefully disassembling, the gel is placed in Coomassie stain for 10 min with gentle agitation prior to destaining in Coomassie wash for additional 10 min. As can be seen from Fig. 25.3, the loaded amount of CK2β derived from the peak fractions matches, whereas there is a clear difference with respect to the staining intensity of the two catalytic subunits (i.e., α and α′). This supports the elution pattern resulting from the gel filtration chromatography and suggests

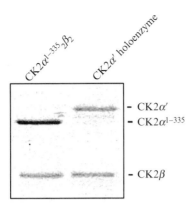

Figure 25.3 Coomassie stained SDS-PAGE of the peak fractions from SMART analyses at 500 mM NaCl of the wild-type CK2α- and CK2α′-holoenzymes. The loading is adjusted so that the intensity of the CK2β bands is the same. From Olsen *et al.* (2008).

that the CK2α-based holoenzyme is a heterotetramer (CK2$\alpha_2\beta_2$), whereas the CK2α′-based holoenzyme is a heterotrimer (CK2$\alpha'\beta_2$). The difference between the staining intensity of the two catalytic subunits cannot be solely explained by their amino acids composition because the two paralogs only differ for one amino acid with respect to amino acids able to react with the Coomassie stain (Bradford, 1976; Olsen *et al.*, 2008). Overall the gel filtration data presented in Table 25.1 and in Fig. 25.3 support a model, where only the CK2α-based holoenzyme is a heterotetramer, which at low salt concentrations forms high molecular mass aggregates. The CK2α′-based holoenzyme exists as a heterotrimer not capable of forming large aggregates (Fig. 25.4).

3.3. Autophosphorylation is not a requirement for aggregation

In the presence of ATP, the CK2β subunits of the CK2α-based holoenzymes are phosphorylated, by the CK2α subunits, at sites within the N-terminal domain (Boldyreff *et al.*, 1993b; Litchfield *et al.*, 1991; Pagano *et al.*, 2005). Interestingly, the CK2α′-based holoenzymes do not phosphorylate the CK2β subunits (Olsen *et al.*, 2006). Here, we describe methods to analyze the autophosphorylation of the CK2α- and α′-based holoenzymes.

3.3.1. Required materials

Materials for performing autophosphorylation studies of CK2α- and α′-based holoenzymes

- 30 °C water bath
- Microcentrifuge

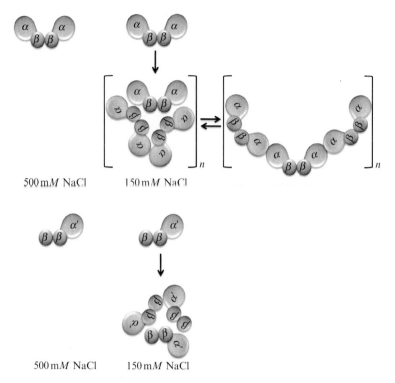

Figure 25.4 Model showing the CK2α-based holoenzyme, which at high salt concentrations (i.e., 500 mM NaCl) exists as a heterotetramer and at low salt concentrations (i.e., 150 mM) can form high molecular mass aggregates either in the form of large linear filaments or ring-like structures. The CK2α'-based holoenzyme exists as a heterotrimer at high salt concentrations, and at low salt concentrations these heterotrimers can form "trimer of heterotrimers" but not high molecular mass aggregates (Niefind and Issinger, 2005; Niefind et al., 2000; Olsen et al., 2006, 2008).

- Vortex
- X-ray films

Additional materials and reagents

- Recombinant CK2α-based holoenzyme constituted of either wild-type CK2β or a mutant lacking the first four amino acids ($\beta^{\Delta 1-4}$) and CK2α'-based holoenzyme (purified from *E. coli* as described in Boldyreff et al., 1993a; Grankowski et al., 1991; Olsen et al., 2006)
- Kinase buffer (25 mM Tris–HCl, pH 8.0, 150 mM NaCl, 5 mM MgCl$_2$, 1 mM DTT, 125 μM ATP)
- Dilution buffer (25 mM Tris–HCl, pH 8.0, 150 mM NaCl, 1 mM DTT)
- [γ-^{32}]ATP (specific activity of 3000 Ci/mmol) 10 μCi/μl

Disposables

1.5-ml (Eppendorf) tubes.

3.3.2. Autophosphorylation of the regulatory CK2β subunit

Both the CK2α-based holoenzymes (containing either wild-type CK2β or a mutant form lacking the autophosphorylation sites (CK2β$^{\Delta 1-4}$) and the α'-based holoenzyme (containing wild-type CK2β) are diluted in dilution buffer to 50 ng/μl and 2 μl are mixed with kinase buffer in Eppendorf tubes on ice in a final reaction volume of 20 μl. Finally 5 μCi [γ-^{32}P]-ATP per reaction are added and the tubes very briefly vortexed prior to a gentle spin to ensure the reaction is at the bottom of the tubes, and incubated in a 30 °C water bath for 30 min. The reactions are stopped by the addition of SDS sample buffer. Prior to loading onto a 12.5% polyacrylamide gel, the tubes are heated to 95 °C and the samples are loaded beside a protein markers mixture. After revealing the protein bands as described above (Section 3.2), the gel is placed in between plastic sheets and the edges are sealed. The gel is then placed in a cassette with an X-ray film. The films are exposed overnight at −80 °C using a cassette with an intensifying screen prior to development (Fig. 25.5).

As can be seen from the autoradiogram (Fig. 25.5), in the presence of ATP, the CK2α-based holoenzyme containing wild-type CK2β phosphorylates the CK2β-subunits. It is known that the N-terminal region of CK2β is a target for autophosphorylation and serines 2 and 3 have been shown to be phosphoacceptor sites (Boldyreff *et al.*, 1993b; Litchfield *et al.*, 1991; Pagano *et al.*, 2005), and indeed a CK2α-based holoenzyme containing

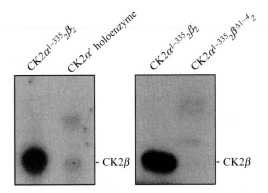

Figure 25.5 Autoradiograph showing the phosphorylated β-subunit from the CK2α- and CK2α'-based holoenzymes (left panel) and from CK2α-based holoenzymed comprising either wild-type CK2β or a mutant form (β$^{\Delta 1-4}$; right panel). From Olsen *et al.* (2008).

mutant CK2β ($\beta^{\Delta 1-4}$) lacking the aforementioned phosphorylation sites does not show any autophosphorylation. Furthermore, as can be seen, the CK2α'-based holoenzyme does not phosphorylate CK2β. Hence, it is not the intermolecular autophosphorylation that is responsible for aggregation, but rather the trimeric state of the CK2α'-based holoenzyme which is the reason why it cannot form high molecular mass aggregates.

Overall, the *in vitro* data suggest that it is only the CK2α'-based holoenzyme which is the truly constitutively active enzyme, since the CK2α-based holoenzyme can aggregate at physiological salt concentrations, and the higher ordered aggregates have been shown to be devoid of catalytic activity (Valero *et al.*, 1995).

ACKNOWLEDGMENTS

The research relevant to this chapter was supported by The Danish Cancer Society (grant no. DP08152 to B. G., DP06083 to B. B. O., DP07109 to O. G. I.), the Danish Natural Science Research Council (grant no. 272-07-0258 to B. G., 21-01-0511 to O. G. I.), and by the Deutsche Forschungsgemeinschaft (DFG) (grant no. NI 643/1-3 to K. N).

REFERENCES

Berman, H. M., Battistuz, T., Bhat, T. N., Bluhm, W. F., Bourne, P. E., Burkhardt, K., Feng, Z., Gilliland, G. L., Iype, L., Jain, S., Fagan, P., Marvin, J., *et al.* (2002). The Protein Data Bank. *Acta Crystallogr. D Biol. Crystallogr.* **58,** 899–907.

Boldyreff, B., Meggio, F., Pinna, L. A., and Issinger, O. G. (1993a). Reconstitution of normal and hyperactivated forms of casein kinase-2 by variably mutated beta-subunits. *Biochemistry* **32,** 12672–12677.

Boldyreff, B., James, P., Staudenmann, W., and Issinger, O. G. (1993b). Ser2 is the autophosphorylation site in the beta subunit from bicistronically expressed human casein kinase-2 and from native rat liver casein kinase-2 beta. *Eur. J. Biochem.* **218,** 515–521.

Bradford, M. M. (1976). A rapid and sensitive method for the quantitation of microgram quantities of protein utilizing the principle of protein-dye binding. *Anal. Biochem.* **72,** 248–254.

Collaborative Computational Project, Number 4 (1994). The CCP4 suite: Programs for protein crystallography. *Acta Crystallogr. D Biol. Crystallogr.* **50,** 760–763.

DeLano, W. L. (2002). The PyMOL molecular graphics system. http://www.pymol.org.

Emsley, P., Lohkamp, B., Scott, W. G., and Cowtan, K. (2010). Features and development of Coot. *Acta Crystallogr. D Biol. Crystallogr.* **66,** 486–501.

Grankowski, N., Boldyreff, B., and Issinger, O. G. (1991). Isolation and characterization of recombinant human casein kinase II subunits α and β from bacteria. *Eur. J. Biochem.* **198,** 25–30.

Guerra, B., and Issinger, O. G. (2008). Protein kinase CK2 in human diseases. *Curr. Med. Chem.* **15,** 1870–1886.

Guerra, B., Niefind, K., Ermakowa, I., and Issinger, O.G. (2001). Characterization of CK2 holoenzyme variants with regard to crystallization. *Mol. Cell. Biochem.* **227,** 3–11.

Huse, M., and Kuriyan, J. (2002). The conformational plasticity of protein kinases. *Cell* **109,** 275–282.

Kelliher, M. A., Seldin, D. C., and Leder, P. (1996). Tal-1 induces T cell acute lymphoblastic leukemia accelerated by CK2alpha. *EMBO J.* **15,** 5160–5166.

Krissinel, E., and Henrick, K. (2004). Secondary-structure matching (SSM), a new tool for fast protein structure alignment in three dimensions. *Acta Crystallogr. D Biol. Crystallogr.* **60,** 2256–2268.

Landesman-Bollag, E., Romieu-Mourez, R., Song, D. H., Sonenshein, G. E., Cardiff, R. D., and Seldin, D. C. (2001). Protein kinase CK2 in mammary gland tumorigenesis. *Oncogene* **20,** 3247–3257.

Litchfield, D. W., Lozeman, F. J., Cicirelle, M. F., Harrylock, M., Ericsson, L. H., Piening, C. J., and Krebs, E. G. (1991). Phosphorylation of the β subunit of casein kinase II in human A431 cells. Identification of the autophosphorylation site and a site phosphorylated by p34cdc2. *J. Biol. Chem.* **266,** 20380–20389.

Niefind, K., and Issinger, O. G. (2005). Neglected and overseen aspects of CK2 structure. *Mol. Cell. Biochem.* **274,** 3–14.

Niefind, K., and Issinger, O. G. (2008). Targeting protein/protein interactions: Protein kinase CK2 as a prime example for a novel concept. *Screening Trends Drug Discov.* **9,** 18–20.

Niefind, K., and Issinger, O. G. (2010). Conformational plasticity of the catalytic subunit of protein kinase CK2 and its consequences for regulation and drug design. *Biochim. Biophys. Acta* **1804,** 484–492.

Niefind, K., Guerra, B., Ermakowa, I., and Issinger, O. G. (2000). Crystallization and preliminary characterization of crystals of human protein kinase CK2. *Acta Crystallogr. D Biol. Crystallogr.* **56,** 1680–1684.

Niefind, K., Raaf, J., and Issinger, O. G. (2009). Protein kinase CK2: From structures to insights. *Cell. Mol. Life Sci.* **66,** 1800–1816.

Olsen, B. B., Boldyreff, B., Niefind, K., and Issinger, O. G. (2006). Purification and characterization of the CK2α′-based holoenzyme, an isozyme of CK2α: A comparative analysis. *Protein Expr. Purif.* **47,** 651–661.

Olsen, B. B., Rasmussen, T., Niefind, K., and Issinger, O. G. (2008). Biochemical characterization of CK2alpha and alpha′ paralogues and their derived holoenzymes: Evidence for the existence of a heterotrimeric CK2alpha′ holoenzyme forming trimeric complexes. *Mol. Cell. Biochem.* **316,** 37–47.

Pagano, M. A., Sarno, S., Poletto, G., Cozza, G., Pinna, L. A., and Meggio, F. (2005). Autophosphorylation at the regulatory beta subunit reflects the supramolecular organization of protein kinase CK2. *Mol. Cell. Biochem.* **274,** 23–29.

Pinna, L. A. (2002). A challenge to canons. *J. Cell Sci.* **115,** 3873–3878.

Seldin, D. C., and Leder, P. (1995). CK2 alpha transgene-induced murine lymphoma: Relation to theilerosis in cattle. *Science* **267,** 894–897.

Slaton, J. W., Unger, G. M., Sloper, D. T., Davis, A. T., and Ahmed, K. (2004). Induction of apoptosis by antisense CK2 in human prostate cancer xenograft model. *Mol. Cancer Res.* **2,** 712–721.

Trembley, J. H., Wang, G., Unger, G., Slaton, J., and Ahmed, K. (2009). Protein kinase CK2 in health and disease: CK2: A key player in cancer biology. *Cell. Mol. Life Sci.* **66,** 1858–1867.

Valero, E., De Bonis, S., Filhol, O., Wade, R. H., Langowski, J., Chambaz, E. M., and Cochet, C. (1995). Quaternary structure of casein kinase 2. Characterization of multiple oligomeric states and relation with its catalytic activity. *J. Biol. Chem.* **270,** 8345–8352.

CHAPTER TWENTY-SIX

MEASURING THE CONSTITUTIVE ACTIVATION OF C-JUN N-TERMINAL KINASE ISOFORMS

Ryan T. Nitta,* Shawn S. Badal,* *and* Albert J. Wong*,†

Contents

1. Introduction	532
2. Important Reagents for Studying JNK Activity	534
2.1. Immunoreagents	535
2.2. Important buffers	535
2.3. Kinase inactive JNK constructs	535
3. Protein Expression and Purification of JNK Proteins and c-JUN	535
3.1. Expression and purification of bacterially expressed 6× His-tagged JNK protein	536
3.2. Expression and purification of bacterially expressed GST-tagged c-JUN protein	537
4. Measuring the Autophosphorylation Ability of the JNK Isoforms	538
4.1. Western immunoblot *in vitro* kinase assay	539
4.2. Radioactive *in vitro* kinase assay	539
5. Determining the Kinase Activity of the JNK Isoforms	540
6. Monitoring the Formation of JNK Homodimers	540
6.1. Generating a standard curve	540
7. Measuring Nuclear Translocation of JNK Protein	541
7.1. Transfection of cells for studying constitutive activity of JNK Pathway	543
7.2. Immunofluorescence staining of cultured cells	543
8. Future Directions	544
8.1. Therapeutic potential of JNK inhibition in human disease	544
8.2. Methods to inhibit JNK	545
Acknowledgments	546
References	546

* Department of Neurosurgery, Cancer Biology Program, Stanford University Medical Center, Stanford, California, USA
† Cancer Biology Program, Stanford University School of Medicine, Stanford, California, USA

Methods in Enzymology, Volume 484
ISSN 0076-6879, DOI: 10.1016/S0076-6879(10)84026-1

© 2010 Elsevier Inc.
All rights reserved.

Abstract

The c-Jun N-terminal kinases (JNK) are important regulators of cell growth, proliferation, and apoptosis. JNKs are typically activated by a sequence of events that include phosphorylation of its T-P-Y motif by an upstream kinase, followed by homodimerization and translocation to the nucleus. Constitutive activation of JNK has been found in a variety of cancers including non-small cell lung carcinomas, gliomas, and mantle cell lymphoma. *In vitro* studies show that constitutive activation of JNK induces a transformed phenotype in fibroblasts and enhances tumorigenicity in a variety of cell lines. Interestingly, a subset of JNK isoforms was recently found to autoactivate rendering the proteins constitutively active. These constitutively active JNK proteins were found to play a pivotal role in activating transcription factors that increase cellular growth and tumor formation in mice. In this chapter, we describe techniques and methods that have been successfully used to study the three components of JNK activation. Use of these techniques may lead to a better understanding of the components of JNK pathways and how JNK is activated in cancer cells.

1. INTRODUCTION

The c-Jun N-terminal kinases (JNK) are a subgroup of the mitogen-activated protein kinases (MAPK) that translate extracellular stimuli into nuclear responses through the phosphorylation of transcription factors. The MAPK signaling modules have been conserved throughout evolution and adapted for various purposes from sporulation in yeast to regulation of the cell cycle in mammalian cells (Chen *et al.*, 2001b). The JNK pathway is characterized by a modular signaling cascade that comprises three phosphorylation events (Fig. 26.1A; Cuenda, 2000; Davis, 1999; Minden *et al.*, 1995; Olson *et al.*, 1995). Upon activation from growth factors or environmental stresses, the serine/threonine MAP kinase kinase kinases (MAPKKK) will phosphorylate and activate a dual specificity MAPKK, so named because they are capable of phosphorylating both threonine and tyrosine. Subsequently, these MAPKK will phosphorylate the threonine and tyrosine of the T-P-Y motif on JNK. After phosphorylation, a portion of the cytoplasmic JNK pool homodimerizes and translocates to the nucleus where it influences gene expression by regulating the activities of specific transcription factors such as c-JUN, ATF family members, c-MYC, p53, and NFAT4 (Chow *et al.*, 1997; Gupta *et al.*, 1995; Kallunki *et al.*, 1996; Khokhlatchev *et al.*, 1998). JNK signaling can initiate events of cell cycle progression or apoptosis, depending on the stimuli and cell type.

There are three distinct genes encoding JNK, *JNK1*, *JNK2*, and *JNK3*, and 10 different isoforms (Gupta *et al.*, 1996; Fig. 26.1B). The JNK1 and JNK2 proteins are expressed in all tissue, while JNK3 expression is restricted to the brain, heart, and testis (Bode and Dong, 2007; Cuevas *et al.*, 2007).

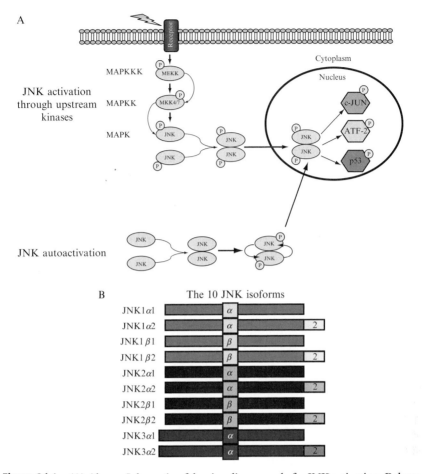

Figure 26.1 (A) Above: Schematic of the signaling cascade for JNK activation. Below: Model for constitutive JNK activation. (B) Diagram illustrating the 10 isoforms of JNK derived from *JNK1*, *JNK2*, and *JNK3* genes.

The 10 isoforms are produced by alternative splicing of the *JNK* genes. The α and β forms differ by an alternatively spliced exon in the middle of the transcript, while the −1 versus −2 forms differ by differential splicing at the 3′ end. There is now accumulating information that each of the *JNK* genes and isoforms mediates different functions. For example, JNK1 activity is associated with apoptosis and tumor suppression, while JNK2 activity can stimulate cell proliferation and tumor formation (Chen *et al.*, 2001a; Yang *et al.*, 2003). Consequently, each *JNK* gene and individual isoform most likely mediates different functions and there is compelling evidence indicating that JNK activity plays an important role in tumorigenesis (Bogoyevitch, 2006).

Recent research suggests that it is the constitutive activation of specific *JNK* genes that can enhance tumorigenesis within a variety of cancers. For instance, a constitutively active JNK mutant in which an upstream activator kinase was fused to JNK was shown to transform NIH-3T3 fibroblasts (Rennefahrt *et al.*, 2002). In addition, cellular transformation by viral or parasite infection was found to be dependent on constitutive activation of JNK (Wang *et al.*, 2009; Xu *et al.*, 1996). Immunohistochemical studies also demonstrated that JNK is constitutively active in the majority of gliomas (Li *et al.*, 2008), non-small cell lung (Khatlani *et al.*, 2007), mantle cell lymphoma (Wang *et al.*, 2009), and squamous cell carcinomas (Ke *et al.*, 2010). Together these findings indicate that constitutively active JNK proteins can induce tumorigenesis in a variety of cancers.

Additional studies analyzing the role of constitutive JNK activation in cancer revealed that the JNK2 isoforms have the unique ability to autophosphorylate and autoactivate (Tsuiki *et al.*, 2003). Several *in vitro* kinase studies have shown that JNK2 can phosphorylate itself on the T-P-Y motif in the absence of upstream kinases and phosphorylate downstream substrates such as c-JUN (Cui *et al.*, 2005; Nitta *et al.*, 2008; Pimienta *et al.*, 2007; Tsuiki *et al.*, 2003). A closer analysis of the mechanism of constitutive activation revealed that JNK2 isoforms initially dimerize, then the bound monomers phosphorylate each other in a *trans*-manner, and finally the phosphorylated homodimer translocates into the nucleus thereby regulating gene expression (Fig. 26.1A; Nitta *et al.*, 2008). Interestingly, these constitutively active JNK2 isoforms are highly expressed in a variety of cancers including brain and non-small cell lung carcinomas suggesting that these constitutively active genes directly lead to altered cell proliferation and tumor formation (Cui *et al.*, 2006; Khatlani *et al.*, 2007).

This chapter will focus on providing protocols and pointers for studying the constitutive activation of JNK isoforms. In order for JNK to be constitutively active, three main events must occur: (1) phosphorylation of the T-P-Y motif, (2) formation of a homodimer, and (3) translocation into the nucleus. Upon activation, JNK then regulates gene expression by phosphorylating transcription factors. To measure the constitutive activity of JNK, we describe several techniques that can be used to monitor the occurrence of each of these three events *in vitro* and *in vivo*.

2. Important Reagents for Studying JNK Activity

In this section, we describe the reagents that are necessary to accurately measure the constitutive activity of JNK.

2.1. Immunoreagents

The following materials have been especially useful for studying activation of the JNK pathway. Antibodies specific for phosphorylation on the T-P-Y motif of JNK (#9255), phosphorylated c-JUN (#9261), and JNK2 (#4672) are available from Cell Signaling Technology, Beverly, MA. Anti-JNK1 (#sc-1648), -JNK2 (#sc-827), and -JNK3 (#sc-130075) antibodies are available from Santa Cruz Biotechnologies, Santa Cruz, CA.

2.2. Important buffers

- Qiagen His Tag Lysis Buffer: 50 mM NaH$_2$PO$_4$, 300 mM NaCl, 10 mM imidazole, 2 μg/mL aprotinin, 2 μg/mL leupeptin, 100 μg/mL PMSF, 50 mM β-glycerolphosphate, pH 8.0
- Qiagen 6× His wash buffer: 50 mM NaH$_2$PO$_4$, 300 mM NaCl, 20 mM imidazole, pH 8.0
- Qiagen elution buffer: 50 mM NaH$_2$PO$_4$, 300 mM NaCl, 20 mM imidazole, pH 8.0
- Kinase buffer: 25 mM HEPES (pH 7.4), 25 mM MgCl$_2$, 2 mM dithiothreitol, 0.1 mM NaVO$_4$, and 25 mM β-glycerophosphate

2.3. Kinase inactive JNK constructs

To accurately measure the constitutive activity of JNK, a well-known kinase inactive mutant should be used as a negative control. Two common mutations that inactivate JNK are the mutation of the T-P-Y motif to A-P-F and the K55R mutation. The mutation of the threonine to alanine and tyrosine to phenylalanine prevents the phosphorylation of the activation loop and subsequently inhibits JNK kinase activity (Pimienta et al., 2007). The second mutation, K55R, mutates a lysine that is critical for phosphotransfer in the ATP-binding motif thereby inhibiting the kinase activity of JNK. Both mutations are commonly used as inactive JNK mutants and have also been used as dominant negatives for *in vitro* and *in vivo* assays, although exactly which mutant to be used depends on the context of the experiment. When overexpressed, these forms will compete with the corresponding wild type endogenous proteins for access to substrates and activating partners.

3. Protein Expression and Purification of JNK Proteins and c-JUN

In this part, we provide the optimal conditions to express and purify recombinant JNK proteins and a well-known JNK substrate, c-JUN, from *Escherichia coli* using 6× His or GST tags. The protocols are adapted from the

Qiagen 6× His tag expression handbook and GST fusion system from Amersham Biosciences (for additional details please consult the Qiagen and Amersham company webpages). Once the recombinant proteins are purified, they can be used for *in vitro* kinase assays, gel-filtration analysis, and binding assays.

3.1. Expression and purification of bacterially expressed 6× His-tagged JNK protein

Based on our experience, we recommend using the pET28 (Novagen) construct to express 6× His-tagged JNK isoform proteins in BL21(DE3) pLysS *E. coli* strain (Novagen). The following protocol will yield 4–10 mg of purified JNK (Fig. 26.2A).

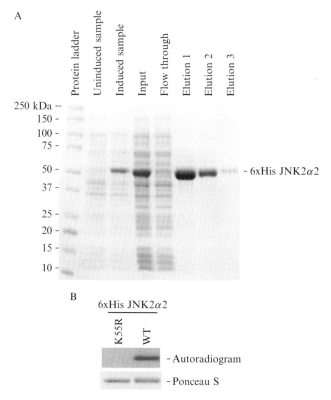

Figure 26.2 (A) Purification of recombinant JNK2α2 from *E. coli*. 6× His-tagged JNK2α2 was purified from cleared bacterial lysate and the samples were run on a SDS-PAGE followed by a Coomassie Blue stain. (B) Radioactive *in vitro* kinase assay using purified 6× His-JNK2α2 protein. (See Color Insert.)

1. A single colony of transformed *E. coli* is shaken overnight at 37 °C in 5 mL of LB medium containing suitable selection antibiotics.
2. The 5 mL starting culture is added to 200 mL of LB medium containing antibiotics and shaken for another 2–3 h until the OD600 reaches approximately 0.6–0.9. Collect a baseline sample (see below for additional information).[1]
3. The culture is chilled on ice for 10 min and isopropyl-β-D-thiogalactopyranoside (IPTG) is added to a final concentration of 0.5 mM and shaken for an additional 3–5 h at 30 °C. Collect an induced sample.[1]
4. The bacteria are pelleted, the supernatant removed, and then resuspended in 10 mL of Qiagen His Tag Lysis Buffer. The bacteria are then sonicated on ice for 20 s at the highest power with additional 20 s sonication intervals until the lysate clarifies. The lysate is then centrifuged at 15,000 × g for 20 min. The induced protein should be in the supernatant. Collect an input sample from the supernatant.[1]
5. Add 1 mL of the 50% Ni-NTA agarose bead slurry (Qiagen) to the 10 mL of cleared lysate and mix gently by shaking in an end-over-end manner at 4 °C for 1 h to overnight.
6. Load the lysate–Ni-NTA mixture onto a column (Bio-Rad, econo-Pac disposable chromatography column) and allow the lysate to flow through the column. Collect a flow-through sample as a control.[1]
7. Wash the column twice with 4 mL of Qiagen 6× His wash buffer.
8. Elute the protein from the column three times with 1.0 mL Qiagen elution buffer and collect into separate tubes.
9. Analyze all column samples by SDS-PAGE followed by staining with Coomassie blue.
10. Combine all the relevant eluates and concentrate the protein using Microcon Ultracel YM-10 (Millipore) and dilute with glycerol to obtain a final concentration of 20% glycerol at 1 mg/mL. Store recombinant protein at −80 °C.

3.2. Expression and purification of bacterially expressed GST-tagged c-JUN protein

The pGEX (Clontech) construct was used to efficiently express a GST-tagged domain of c-JUN in BL21(DE3)pLysS *E. coli* strain (Novagen). The following protocol will yield 5–10 mg of purified GST-c-JUN (Fig. 26.2B).

---

[1] To verify that protein expression and purification conditions are optimized, numerous samples should be obtained throughout the procedure. For the uninduced and induced samples, 0.5 mL of the bacteria solution is centrifuged for 30 s at 10 K rpm, after which the supernatant is removed. The remaining pellet is resuspended in protein sample buffer and boiled for 5 min. For the remaining samples, collect 200 μL, add sample buffer, and boil for 5 min.

Steps 1 and 2 are the same as previously described.

3. The culture was chilled on ice for 10 min and isopropyl-β-d-thiogalactopyranoside (IPTG) is added to a final concentration of 0.5 mM and shaken for additional 3–5 h at 37 °C. Collect 500 μL for an induced sample.[1]
4. The bacteria are centrifuged and resuspended in cold 10 mL of PBS containing 2 μg/mL aprotinin, 2 μg/mL leupeptin, 100 μg/mL PMSF, and 50 mM β-glycerolphosphate. Bacteria are sonicated on ice for 20 s at highest power with additional 20 s sonication intervals until the lysate clarifies. The lysate is then centrifuged at 15,000 × g for 20 min. The induced protein should be in the supernatant. Collect an input sample.[1]
5. Add 1 mL of 50% Glutathione Sepharose 4B slurry (GE Scientific) to the chromatography column (Bio-Rad, econo-Pac disposable chromatography column) and wash with 5 mL of cold PBS to remove residual ethanol.
6. Add the sonicated lysate to the column containing the washed Glutathione Sepharose and incubate for 1 h at 4 °C using an end-over-end rotation.
7. Allow the lysate to flow through the column. Collect a 500 μL flow-through sample.[1]
8. Wash the column by adding 10 mL of 1× PBS. Repeat twice more for a total of three washes.
9. Elute the fusion protein by adding 0.5 mL of elution buffer. Incubate the column at room temperature for 5 min and then collect the eluate. Repeat three to four times and collect samples in separate tubes.
10. Analyze all the samples by SDS-PAGE followed by staining with Coomassie blue.
11. Combine all the relevant eluates and concentrate the protein using a Microcon Ultracel YM-10 (Millipore) to a concentration of 1 mg/mL. Add glycerol to obtain a final concentration of 20% glycerol. Store recombinant protein at −80 °C for 6 months.

4. Measuring the Autophosphorylation Ability of the JNK Isoforms

One element of constitutive JNK activity is the ability of JNK proteins to autophosphorylate in the absence of upstream kinases. We have used two different *in vitro* kinase assays to analyze JNK autophosphorylation: a Western immunoblot assay and a radioactive assay. The Western analysis determines whether a JNK protein is being phosphorylated at the T-P-Y motif by using a specific phospho-JNK antibody. This technique is useful since phosphorylation of the T-P-Y motif has been widely shown to regulate the kinase activity of JNK. However, the use of ^{32}P ATP in the *in vitro* kinase assay is more sensitive than using Western immunoblotting; however, it cannot be used to identify which JNK residues are being phosphorylated. An example of the radioactive *in vitro* kinase is depicted in Fig. 26.2B.

4.1. Western immunoblot *in vitro* kinase assay

1. One microgram of the 6× His-JNK fusion protein was incubated in 25 μL of the kinase buffer containing 30 μM ATP at 30 °C for 30 min.
2. The reactions were terminated by adding protein loading buffer and boiling for 5 min.
3. The kinase solution was loaded onto a SDS-PAGE and subsequently transferred onto a nitrocellulose membrane.
4. After protein transfer, the membrane is incubated in Ponceau S solution for 1–2 min to visualize protein levels. Two washes with PBST are done to remove the Ponceau S stain from the membrane.
5. The membrane is then incubated in Blocking buffer (5% milk in PBST (PBS and 0.2% Tween 20)) for 1 h.
6. After washing the membrane with PBST, primary antibodies are diluted in 1% BSA and 0.01% thimerasol in PBST, and incubated at room temperature for 1–2 h or 4 °C for overnight incubations.
7. The membrane is washed three times with PBST, incubated with appropriate secondary antibody in blocking buffer for 30 min at room temperature followed by three additional washes.
8. The membranes are incubated with a chemiluminescence reagent and then developed using Kodak film.

4.2. Radioactive *in vitro* kinase assay

1. One microgram of the 6× His-JNK fusion protein is incubated in 25 μL of the kinase buffer containing 30 μM ATP and 5 μCi of γ-^{32}P ATP (3000 Ci/mmol) at 30 °C for 30 min.
2. The reaction is terminated by adding protein loading buffer and boiling for 5 min.
3. The kinase solution is then loaded onto a SDS-PAGE and subsequently transferred onto a nitrocellulose membrane.
4. After protein transfer, the membrane is incubated in Ponceau S solution for 1–2 min to visualize protein levels.
5. The nitrocellulose membrane is exposed to Kodak X-ray film and developed.

5. Determining the Kinase Activity of the JNK Isoforms

To verify that a JNK protein is constitutively active, we optimized the *in vitro* kinase assay to measure phosphorylation of a well-known JNK substrate c-Jun. This protocol can also be used to test other potential JNK substrates for their ability to interact with and be phosphorylated by a constitutively active JNK isoform. Previous research has shown that constitutively active JNK2 isoforms can phosphorylate oncoproteins such as Akt (Cui *et al.*, 2006), β-catenin (Wu *et al.*, 2008), Sirt1 (Ford *et al.*, 2008), and Ras (Nielsen *et al.*, 2007) and tumor suppressors such as p53 (Maeda and Karin, 2003; Oleinik *et al.*, 2007). Additional research is ongoing to identify additional JNK substrates that may be altered in cancer.

To measure the kinase activity of the purified 6× His-JNK fusion protein toward a new substrate, a similar *in vitro* kinase assay is conducted where 1 μg of the recombinant protein substrate (such as GST-c-Jun) is added with 1 μg of the 6× His-JNK isoform. As a control to verify that the phosphorylation is specific to the JNK isoform, a kinase dead JNK mutant (K55R) should also be tested. This radioactive kinase assay will quickly demonstrate the ability of the JNK protein to phosphorylate the potential downstream substrate, but in order to specifically identify the residue being phosphorylated, more elaborate methods, such as phosphoamino acid analysis or HPLC/mass spectrometric analysis of peptide fragments, would be required.

6. Monitoring the Formation of JNK Homodimers

Previous research has shown that MAPK homodimerization is an important element of MAPK activity. Extensive studies in *Xenopus* and human cell lines have shown that MAPK exists both as a dimer and a monomer, but upon phosphorylation the MAPK will dissociate from the MAPK kinase, form homodimers, and then translocate into the nucleus (Adachi *et al.*, 1999; Khokhlatchev *et al.*, 1998). We have previously shown that dimerization is necessary for JNK autophosphorylation and nuclear translocation (Nitta *et al.*, 2008). The following protocol will enable the identification of the JNK homodimers.

6.1. Generating a standard curve

Initially a standard curve is established to determine which protein fractions correspond to specific-sized proteins. The following protocol was adapted from the Sigma molecular weight gel-filtration markers protocol.

1. A prepacked gel-filtration column containing 90 mL of Sephacryl S-400 (GE Healthcare) is initially equilibrated by running PBS through the column at a flow rate of 0.2 mL/min at 4 °C. *Note:* The packing of a column is very critical. For additional instructions on packing your own column, refer to the protocol described by GE Healthcare.
2. 1 mL of 2 mg/mL of Blue Dextran (reconstituted in PBS and 5% glycerol) is loaded gently onto the column, and 1 mL fractions are collected.
3. The protein concentration of each fraction is determined by measuring the OD at 280 nm.
4. The fraction that has the highest concentration of the Blue Dextran is known as the void volume (Vo). For example, if the 40th fraction contains the highest concentration of Blue Dextran, then the Vo is 40. We collected 1 mL for each fraction; consequently the 40th fraction would mean the Vo is 40 mL. The Vo is used to normalize the molecular weight standards.
5. Load 1 mL of 2 mg/mL of a specific molecular weight standard and begin collecting 1 mL fractions.
6. Determine which fraction contains the molecular weight standard by measuring the OD at 280 nm. The fraction that contains the protein is known as the elution volume (Ve).
7. Repeat steps 6 and 7 for each molecular weight standard.
8. Plot the molecular weight versus Ve/Vo for each respective protein standard on semilog paper to derive the standard curve (Fig. 26.3A).

To determine if the 6× His-JNK fusion proteins exist as a homodimer or monomer, 1 mg of the 6× His-JNK fusion protein (at 1 mg/mL) is applied to the column and 1 mL fractions are collected as previously described. The protein concentrations are determined by measuring the OD280 for each fraction to determine the elution volume (Ve). To determine if the 6× His-JNK fusion protein is present in the fractions, a fixed amount of each fraction (usually 50 μl) is run on SDS-PAGE and then stained with Coomassie blue. The size of the protein is determined by plotting the Ve/Vo points on the standard curve. As seen in Fig. 26.3B, 6× His-JNK2α2 exists in two different states, a dimer and monomer. The 59-mL fraction contained proteins approximately 110–120 kDa in size, consistent with a JNK2α2 dimer, while the 67-mL fraction contained proteins 50–60 kDa in size, consistent with a JNK2α2 monomer.

7. MEASURING NUCLEAR TRANSLOCATION OF JNK PROTEIN

One hallmark of constitutive JNK activity is the ability to translocate to the nucleus. JNK does not contain an obvious nuclear localization sequence, but studies using the fission yeast *Schizosaccharomyces pombe*,

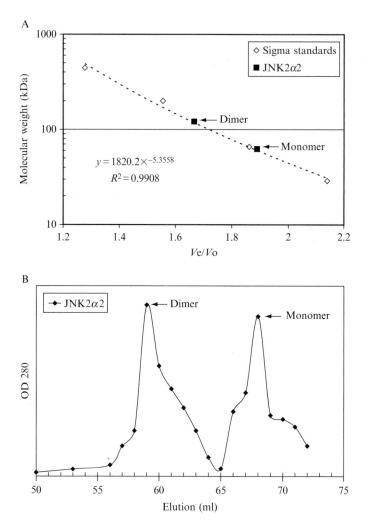

Figure 26.3 Measuring JNK homodimerization using size exclusion chromatography. (A) Standard curve using Sigma molecular weight standards. (B) Elution profile of 6× His-JNK2α2. The two peaks correspond to the monomeric and dimeric forms of JNK.

indicate that the yeast homologue of JNK, Spc1, is actively transported to the nucleus by Pim1, a homologue of the guanine nucleotide exchange factor RCC1 (Gaits and Russell, 1999). To determine if the constitutively JNK protein can readily translocate to the nucleus, we have transfected a GFP-tagged JNK protein into a variety of cancer cell lines. The following protocol will efficiently transfect the exogenous protein in cells and enable easy identification of nuclear localization.

7.1. Transfection of cells for studying constitutive activity of JNK Pathway

Certain biochemical studies involving the JNK pathway require high-efficiency transfections. To this end, we were able to obtain high-efficiency transfections using the Mirus TransIT®-LT1 Reagent for a variety of cell lines (HEK293, HCC-827, NCI-H2009, and several others).

1. Approximately 24 h prior to transfection, plate cells to obtain a cell density of 50–70% confluence the following day on a 100 mm plate with 10 mL of the appropriate media.
2. In a sterile plastic tube, add 1 mL of serum free media followed by 24 μL of TransIT®-LT1 Reagent. After mixing by gentle pipetting, the solution is incubated at room temperature for 5–10 min.
3. Add 4 μg of plasmid DNA to the diluted TransIT®-LT1 Reagent and mix by gentle pipetting and incubate at room temperature for 10–20 min.
4. Add the TransIT®-LT1 Reagent/DNA complex mixture, drop wise to the cells in complete growth medium. Incubate for 24–48 h.

7.2. Immunofluorescence staining of cultured cells

To monitor the nuclear localization of JNK, the following protocol is used. Constitutively active JNK isoforms have a strong nuclear localization compared to the kinase dead JNK mutant (Fig. 26.4).

1. Cells are cultured on coverslips (12 mm diameter grade) and washed with 1× PBS to remove residual media.
2. The cells are fixed with 4% paraformaldehyde for 10 min at room temperature, washed with PBS, and then permeabilized with 0.5% Triton X-100 in 1× PBS for 10 min and blocked with blocking buffer (1× PBS, 0.2% Tween 20, and 10% goat serum) for 30 min at 37 °C. If your cells are tagged with a fluorescent protein, then skip to step 5.
3. 200 μL of blocking buffer containing the primary antibody is incubated at 37 °C for 1 h, followed by 2–3 washes with the blocking buffer.
4. Fluorochrome-conjugated secondary antibody is applied in 200 μL of blocking buffer for 1 h at 37 °C.
5. Cells are then washed once with PBS before counter-staining the nuclei with 1 μg/mL 4′,6-diamidino-2-phenylindole (DAPI) dye in PBS, washed twice with PBS, and covered with mounting medium (Gel/Mount, Biomeda Corp.) and a cover glass.

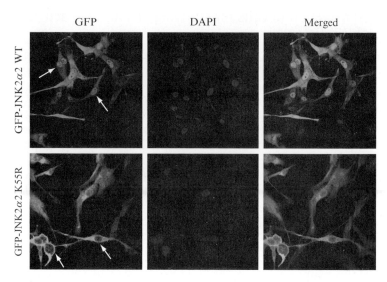

Figure 26.4 Direct immunofluorescence of GFP-tagged JNK constructs in U87-MG cells measuring nuclear translocation. Top: Nuclear localization of the constitutively active GFP-tagged wild type JNK2α2 construct. Bottom: Nuclear localization of the kinase dead GFP-tagged K55R JNK2α2 mutant. Arrows indicate cells with strong nuclear localization (top) or cytoplasmic stain (bottom). (See Color Insert.)

8. Future Directions

8.1. Therapeutic potential of JNK inhibition in human disease

The MAPKs are involved in a variety of diseases and are critical in cell-cycle regulation (Dhillon et al., 2007; McCubrey et al., 2006; Torii et al., 2006). Recently, constitutive activation of JNK has been found in a variety of cancers including non-small cell lung carcinomas, gliomas, mantle cell lymphoma, and squamous cell carcinomas (Ke et al., 2010; Khatlani et al., 2007; Li et al., 2008; Wang et al., 2009). Interestingly, a group of JNK isoforms possess the ability to autophosphorylate rendering them constitutively active. Overexpression of one constitutively active isoform, JNK2α2, was found to enhance cell proliferation, anchorage independent growth, and tumor formation in mice (Cui et al, 2006). While the *JNK* genes have been linked to tumorigenesis, they have been shown to influence other diseases as well. For example, JNK has been shown to be involved in Parkinson's, Alzheimer's, diabetes, and renal disease (Kim and Choi, 2010; Yang and Trevillyan, 2008). These findings suggest that the JNK family is an important target for therapeutic intervention.

8.2. Methods to inhibit JNK

Using a small molecule screen to inhibit JNK activity has been a popular method for elucidating potential JNK targets. Current JNK inhibitors, like SP600125, target the highly conserved ATP-binding sites of JNK (Ishii et al., 2004). Perhaps as a result of targeting this conserved binding site, SP600125 has been shown to inhibit 13 of 30 tested protein kinases, indicating a potential problem with specificity issues (Bain et al., 2003). Consequently, since recent reports demonstrate that each JNK gene and isoform can mediate different functions, the promiscuous nature of SP600125 decreases its potential as a successful therapeutic reagent to study JNK inhibition.

To enhance the specificity of JNK inhibition, non ATP-binding site inhibitors such as BI-78D3, a small molecule mimic of JNK-interacting protein 1 (JIP1), which is a scaffolding protein that binds JNK, have also been used (Stebbins et al., 2008). By targeting a specific binding site located preferentially within *JNK* genes, the specificity of inhibition is increased. However, by targeting these binding partners, only a small subset of the various pathways affected by JNK during disease progression will be successfully inhibited. Since JNK is found to regulate a wide variety of transcription factors in a cell type- or tissue-specific manner, it would be difficult to identify the specific JNK interaction that is the direct cause of tumorigenesis and disease.

Therefore, in lieu of using small molecule and binding partner inhibition, another exciting potential target for JNK inhibition is to block JNK homodimerization. As previously discussed, JNK homodimerization is critical for JNK activity and inhibition of dimer formation was shown to decrease JNK activity, and ameliorates JNK-induced tumorigenesis in glioma cell lines (Nitta et al., 2008; Wilsbacher et al., 2006). By targeting homodimerization of JNK, the need to inhibit effectors or binding partners of JNK to inhibit activity is reduced. Currently, there exists no defined method for blocking homodimerization, but a potential method is the utilization of peptide inhibitors.

Peptide inhibition of JNK is a more specific method to blocking JNK activity. By constructing short peptides against binding sites between JNK binding partners, such as a cell-permeable peptide against the sigma domain for c-JUN, JNK activity is preferentially inhibited (Holzberg et al., 2003). Since the crystal structure of JNK2 and JNK3 suggests that the JNK protein have different binding domains, peptide inhibition could be used to specifically target JNK isoforms thereby reducing cross-reactivity (Shaw et al., 2008; Xie et al., 1998).

The methods described above can be applied to the different JNK genes and isoforms, thanks in part to the similarity across the JNK family of genes. The variety of methods to inhibit JNK action enhances its widespread appeal as a therapeutic target, and its importance in the cellular milieu serves as a desirable target for scientists and clinicians.

ACKNOWLEDGMENTS

The research relevant to this chapter was supported by the Mark Linder/American Brain Tumor Association Fellowship, National Brain Tumor Foundation, and NIH grants CA69495, CA96539, and CA124832.

REFERENCES

Adachi, M., Fukuda, M., and Nishida, E. (1999). Two co-existing mechanisms for nuclear import of MAP kinase: Passive diffusion of a monomer and active transport of a dimer. *EMBO J.* **18,** 5347–5358.
Bain, J., McLauchlan, H., Elliott, M., and Cohen, P. (2003). The specificities of protein kinase inhibitors: An update. *Biochem. J.* **371,** 199–204.
Bode, A. M., and Dong, Z. (2007). The functional contrariety of JNK. *Mol. Carcinog.* **46,** 591–598.
Bogoyevitch, M. A. (2006). The isoform-specific functions of the c-Jun N-terminal Kinases (JNKs): Differences revealed by gene targeting. *Bioessays* **28,** 923–934.
Chen, N., Nomura, M., She, Q. B., Ma, W. Y., Bode, A. M., Wang, L., Flavell, R. A., and Dong, Z. (2001a). Suppression of skin tumorigenesis in c-Jun NH(2)-terminal kinase-2-deficient mice. *Cancer Res.* **61,** 3908–3912.
Chen, Z., Gibson, T. B., Robinson, F., Silvestro, L., Pearson, G., Xu, B., Wright, A., Vanderbilt, C., and Cobb, M. H. (2001b). MAP kinases. *Chem. Rev.* **101,** 2449–2476.
Chow, C. W., Rincon, M., Cavanagh, J., Dickens, M., and Davis, R. J. (1997). Nuclear accumulation of NFAT4 opposed by the JNK signal transduction pathway. *Science* **278,** 1638–1641.
Cuenda, A. (2000). Mitogen-activated protein kinase kinase 4 (MKK4). *Int. J. Biochem. Cell Biol.* **32,** 581–587.
Cuevas, B. D., Abell, A. N., and Johnson, G. L. (2007). Role of mitogen-activated protein kinase kinase kinases in signal integration. *Oncogene* **26,** 3159–3171.
Cui, J., Han, S. Y., Wang, C., Su, W., Harshyne, L., Holgado-Madruga, M., and Wong, A. J. (2006). c-Jun NH(2)-terminal kinase 2alpha2 promotes the tumorigenicity of human glioblastoma cells. *Cancer Res.* **66,** 10024–10031.
Cui, J., Holgado-Madruga, M., Su, W., Tsuiki, H., Wedegaertner, P., and Wong, A. J. (2005). Identification of a specific domain responsible for JNK2alpha2 autophosphorylation. *J. Biol. Chem.* **280,** 9913–9920.
Davis, R. J. (1999). Signal transduction by the c-Jun N-terminal kinase. *Biochem. Soc. Symp.* **64,** 1–12.
Dhillon, A. S., Hagan, S., Rath, O., and Kolch, W. (2007). MAP kinase signalling pathways in cancer. *Oncogene* **26,** 3279–3290.
Ford, J., Ahmed, S., Allison, S., Jiang, M., and Milner, J. (2008). JNK2-dependent regulation of SIRT1 protein stability. *Cell Cycle* **7,** 3091–3097.
Gaits, F., and Russell, P. (1999). Active nucleocytoplasmic shuttling required for function and regulation of stress-activated kinase Spc1/StyI in fission yeast. *Mol. Biol. Cell* **10,** 1395–1407.
Gupta, S., Barrett, T., Whitmarsh, A. J., Cavanagh, J., Sluss, H. K., Derijard, B., and Davis, R. J. (1996). Selective interaction of JNK protein kinase isoforms with transcription factors. *EMBO J.* **15,** 2760–2770.
Gupta, S., Campbell, D., Derijard, B., and Davis, R. J. (1995). Transcription factor ATF2 regulation by the JNK signal transduction pathway. *Science* **267,** 389–393.
Holzberg, D., Knight, C. G., Dittrich-Breiholz, O., Schneider, H., Dorrie, A., Hoffmann, E., Resch, K., and Kracht, M. (2003). Disruption of the c-JUN-JNK

complex by a cell-permeable peptide containing the c-JUN delta domain induces apoptosis and affects a distinct set of interleukin-1-induced inflammatory genes. *J. Biol. Chem.* **278**, 40213–40223.

Ishii, M., Suzuki, Y., Takeshita, K., Miyao, N., Kudo, H., Hiraoka, R., Nishio, K., Sato, N., Naoki, K., Aoki, T., and Yamaguchi, K. (2004). Inhibition of c-Jun NH2-terminal kinase activity improves ischemia/reperfusion injury in rat lungs. *J. Immunol.* **172**, 2569–2577.

Kallunki, T., Deng, T., Hibi, M., and Karin, M. (1996). c-Jun can recruit JNK to phosphorylate dimerization partners via specific docking interactions. *Cell* **87**, 929–939.

Ke, H., Harris, R., Coloff, J. L., Jin, J. Y., Leshin, B., de Marval, P. M., Tao, S., Rathmell, J. C., Hall, R. P., and Zhang, J. Y. (2010). The c-Jun NH2-terminal kinase 2 plays a dominant role in human epidermal neoplasia. *Cancer Res.* **70**(8):3080–3088.

Khatlani, T. S., Wislez, M., Sun, M., Srinivas, H., Iwanaga, K., Ma, L., Hanna, A. E., Liu, D., Girard, L., Kim, Y. H., Pollack, J. R., Minna, J. D., et al. (2007). c-Jun N-terminal kinase is activated in non-small-cell lung cancer and promotes neoplastic transformation in human bronchial epithelial cells. *Oncogene* **26**, 2658–2666.

Khokhlatchev, A. V., Canagarajah, B., Wilsbacher, J., Robinson, M., Atkinson, M., Goldsmith, E., and Cobb, M. H. (1998). Phosphorylation of the MAP kinase ERK2 promotes its homodimerization and nuclear translocation. *Cell* **93**, 605–615.

Kim, E. K., and Choi, E. J. (2010). Pathological roles of MAPK signaling pathways in human diseases. *Biochim. Biophys. Acta* **1802**, 396–405.

Li, J. Y., Wang, H., May, S., Song, X., Fueyo, J., Fuller, G. N., and Wang, H. (2008). Constitutive activation of c-Jun N-terminal kinase correlates with histologic grade and EGFR expression in diffuse gliomas. *J. Neurooncol.* **88**, 11–17.

Maeda, S., and Karin, M. (2003). Oncogene at last–c-Jun promotes liver cancer in mice. *Cancer Cell* **3**, 102–104.

McCubrey, J. A., Lahair, M. M., and Franklin, R. A. (2006). Reactive oxygen species-induced activation of the MAP kinase signaling pathways. *Antioxid. Redox Signal.* **8**, 1775–1789.

Minden, A., Lin, A., Claret, F. X., Abo, A., and Karin, M. (1995). Selective activation of the JNK signaling cascade and c-Jun transcriptional activity by the small GTPases Rac and Cdc42Hs. *Cell* **81**, 1147–1157.

Nielsen, C., Thastrup, J., Bottzauw, T., Jaattela, M., and Kallunki, T. (2007). c-Jun NH2-terminal kinase 2 is required for Ras transformation independently of activator protein 1. *Cancer Res.* **67**, 178–185.

Nitta, R. T., Chu, A. H., and Wong, A. J. (2008). Constitutive activity of JNK2 alpha2 is dependent on a unique mechanism of MAPK activation. *J. Biol. Chem.* **283**, 34935–34945.

Oleinik, N. V., Krupenko, N. I., and Krupenko, S. A. (2007). Cooperation between JNK1 and JNK2 in activation of p53 apoptotic pathway. *Oncogene* **26**(51):7222–7230.

Olson, M. F., Ashworth, A., and Hall, A. (1995). An essential role for Rho, Rac, and Cdc42 GTPases in cell cycle progression through G1. *Science* **269**, 1270–1272.

Pimienta, G., Ficarro, S. B., Gutierrez, G. J., Bhoumik, A., Peters, E. C., Ronai, Z., and Pascual, J. (2007). Autophosphorylation properties of inactive and active JNK2. *Cell Cycle* **6**, 1762–1771.

Rennefahrt, U. E., Illert, B., Kerkhoff, E., Troppmair, J., and Rapp, U. R. (2002). Constitutive JNK activation in NIH 3 T3 fibroblasts induces a partially transformed phenotype. *J. Biol. Chem.* **277**, 29510–29518.

Shaw, D., Wang, S., Villasenor, A. G., Tsing, S., Walter, D., Browner, M., Barnett, J., and Kuglstatter, A. (2008). The crystal structure of JNK2 reveals conformational flexibility in the MAP kinase insert and indicates its involvement in the regulation of catalytic activity. *J. Mol. Biol.* **20**(17), 5217–5220.

Stebbins, J. L., De, S. K., Machleidt, T., Becattini, B., Vazquez, J., Kuntzen, C., Chen, L. H., Cellitti, J. F., Riel-Mehan, M., Emdadi, A., Solinas, G., Karin, M., and Pellecchia, M. (2008). Identification of a new JNK inhibitor targeting the JNK-JIP interaction site. *Proc. Natl. Acad. Sci. USA* **105,** 16809–16813.

Torii, S., Yamamoto, T., Tsuchiya, Y., and Nishida, E. (2006). ERK MAP kinase in G cell cycle progression and cancer. *Cancer Sci.* **97,** 697–702.

Tsuiki, H., Tnani, M., Okamoto, I., Kenyon, L. C., Emlet, D. R., Holgado-Madruga, M., Lanham, I. S., Joynes, C. J., Vo, K. T., and Wong, A. J. (2003). Constitutively active forms of c-Jun NH2-terminal kinase are expressed in primary glial tumors. *Cancer Res.* **63,** 250–255.

Wang, M., Atayar, C., Rosati, S., Bosga-Bouwer, A., Kluin, P., and Visser, L. (2009). JNK is constitutively active in mantle cell lymphoma: cell cycle deregulation and polyploidy by JNK inhibitor SP600125. *J. Pathol.* **218,** 95–103.

Wilsbacher, J. L., Juang, Y. C., Khokhlatchev, A. V., Gallagher, E., Binns, D., Goldsmith, E. J., and Cobb, M. H. (2006). Characterization of mitogen-activated protein kinase (MAPK) dimers. *Biochemistry* **45,** 13175–13182.

Wu, X., Tu, X., Joeng, K. S., Hilton, M. J., Williams, D. A., and Long, F. (2008). Rac1 activation controls nuclear localization of beta-catenin during canonical Wnt signaling. *Cell* **133,** 340–353.

Xie, X., Gu, Y., Fox, T., Coll, J. T., Fleming, M. A., Markland, W., Caron, P. R., Wilson, K. P., and Su, M. S. (1998). Crystal structure of JNK3: A kinase implicated in neuronal apoptosis. *Structure* **6,** 983–991.

Xu, X., Heidenreich, O., Kitajima, I., McGuire, K., Li, Q., Su, B., and Nerenberg, M. (1996). Constitutively activated JNK is associated with HTLV-1 mediated tumorigenesis. *Oncogene* **13,** 135–142.

Yang, R., and Trevillyan, J. M. (2008). c-Jun N-terminal kinase pathways in diabetes. *Int. J. Biochem. Cell Biol.* **40,** 2702–2706.

Yang, Y. M., Bost, F., Charbono, W., Dean, N., McKay, R., Rhim, J. S., Depatie, C., and Mercola, D. (2003). C-Jun NH(2)-terminal kinase mediates proliferation and tumor growth of human prostate carcinoma. *Clin. Cancer Res.* **9,** 391–401.

CHAPTER TWENTY-SEVEN

MEASUREMENT OF CONSTITUTIVE MAPK AND PI3K/AKT SIGNALING ACTIVITY IN HUMAN CANCER CELL LINES

Kim H. T. Paraiso,[*,†] Kaisa Van Der Kooi,[‡] Jane L. Messina,[‡] and Keiran S. M. Smalley[*,†]

Contents

1. Introduction	550
2. Maintaining Melanoma Cell Lines	552
2.1. Culturing conditions	552
2.2. Plating cells	552
3. Western Blotting	553
3.1. Harvesting cells and protein extraction	554
3.2. Determine protein concentration	555
3.3. Sample preparation	555
3.4. Electrophoresis	556
3.5. Protein transfer	557
3.6. Antibody incubations and western blot development	558
4. Phospho-Flow Cytometry	559
4.1. Materials	561
5. Immunofluorescence	562
5.1. Reagents	563
5.2. Preparing the cells on coverslips	564
5.3. Fixing and staining the cells	564
6. Conclusions	565
Acknowledgments	566
References	566

[*] Department of Molecular Oncology, The Moffitt Cancer Center and Research Institute, Tampa, Florida, USA
[†] Department of Cutaneous Oncology, The Moffitt Cancer Center and Research Institute, Tampa, Florida, USA
[‡] Department of Pathology and Cell Biology, University of South Florida College of Medicine, Tampa, Florida, USA

Methods in Enzymology, Volume 484
ISSN 0076-6879, DOI: 10.1016/S0076-6879(10)84027-3

© 2010 Elsevier Inc.
All rights reserved.

Abstract

The growth and survival of cancer cells are often driven by constitutive activity in the mitogen-activated protein kinase (MAPK) and phospho-inositide 3-kinase (PI3K)/AKT signaling pathways. Activity in these signal transduction cascades is known to contribute to the uncontrolled growth and resistance to apoptosis that characterizes tumor progression. There is now a great deal of interest in therapeutically targeting these pathways in cancer using small molecule inhibitors. In this chapter, we describe methods to measure constitutive MAPK and AKT activity in melanoma cell lines, with a focus upon Western blotting, phospho-flow cytometry, and immunofluorescence staining techniques.

1. INTRODUCTION

Tumor cells are characterized by an escape from physiological mechanisms of growth control and cell survival. These processes can be driven through multiple mechanisms including the acquisition of activating mutations in receptor tyrosine kinases (RTKs) (such as *Her2/Neu* in breast cancer and Bcr-Abl in chronic myeloid leukemia), mutations in oncogenes (including *KRAS* in lung cancer and *BRAF* in melanoma), loss of expression/mutations in tumor suppressors (such as PTEN and p53), and autocrine growth factor loops (Cully and Downward, 2008; Salmena *et al.*, 2008; Smalley, 2003; Wong, 2010). Although diverse, these driving oncogenic events often rely upon the activation of a common set of signal transduction pathways to mediate their effects upon tumor behavior. The most highly studied intracellular signaling cascades in the context of cancer are the mitogen-activated protein kinase (MAPK) and phosphoinositide 3-kinase (PI3K)/AKT pathways.

Under physiological conditions, the MAPK pathway is primarily responsible for transducing extracellular growth signals, generated through the interaction of ligands with their respective RTKs, to the interior of the cell via the activation of the Ras-family GTPases. In its GTP-bound state, Ras activates a number of downstream signaling cascades involved in controlling cell growth and behavior (Cully and Downward, 2008). One such Ras-activated pathway is a family of serine/threonine protein kinases known as the MAPK cascade (Robinson and Cobb, 1997). Initially, Ras interacts with and activates the serine/threonine protein kinase Raf, which exists in three isoforms: ARAF, BRAF, and CRAF (Stokoe *et al.*, 1994; Wellbrock *et al.*, 2004). Once active, Raf serine phosphorylates MEK1 and MEK2, (Crews *et al.*, 1992; Dent *et al.*, 1992) which in turn tyrosine/threonine phosphorylates extracellular-signal-regulated kinase ERK1 and ERK2 (Kyriakis *et al.*, 1992). Upon activation, the ERKs either phosphorylate cytoplasmic targets or migrate to the nucleus (Lenormand *et al.*, 1993) where they phosphorylate and activate a number of transcription factors such as c-Fos and Elk-1 (Treisman, 1994).

The aberrant activation of the MAPK pathway is implicated in the growth and pathological behavior of many cancer types. In melanoma, constitutive MAPK signaling arises through activating mutations in *NRAS* (15–20% of cases), *BRAF* (50% of cases), and *c-KIT* (Curtin *et al.*, 2006; Davies *et al.*, 2002; Padua *et al.*, 1985; Smalley *et al.*, 2009). Activity in the MAPK pathway drives growth through the upregulation of cyclin D1 expression and the suppression of the cyclin-dependent kinase inhibitor p27^{KIP1} (Bhatt *et al.*, 2005; Smalley, 2003). Although much of the available evidence supports a role for the MAPK pathway in the uncontrolled proliferation of many cancer types, its potential role in the regulation of cell survival is less well characterized.

The PI3Ks are a family of lipid kinases that play a key role in regulating growth and survival (Cantley, 2002; Samuels and Velculescu, 2004; Wong, 2010). Structurally, PI3K forms a heterodimer consisting of a p85 regulatory and a p110 catalytic subunit (Cantley, 2002). It is recruited to the membrane following the activation of RTKs and associated adaptor proteins that bind to the SH2 domain of the p85 subunit (Cantley, 2002). The p110 domain can also be recruited and activated following the activation of Ras. Following membrane recruitment and activation, PI3K then phosphorylates the phosphatidylinositol-4,5-bisphosphate ring (PIP2) at the 3′ position, converting PIP2 to PIP3. Once generated, PIP3 recruits and activates the downstream serine/threonine kinases PDK1 and AKT (Cantley, 2002). The AKT family consists of three members, AKT1–3 (Brazil *et al.*, 2002), which exhibit different expression patterns depending upon cell type. AKT has a critical role in cancer development through its ability to regulate apoptosis via the direct phosphorylation of BAD, as well as effects upon many other pathways, including the stimulation of ribosomal S-6-kinase, the inhibition of Forkhead signaling, and the inhibition of glycogen synthase kinase-3 (Datta *et al.*, 1997; Robertson, 2005).

One of the most critical regulators of AKT is the phosphatase and tensin homologue (PTEN), which degrades the products of PI3K, therefore preventing AKT activation (Salmena *et al.*, 2008). Many cancers have constitutive activity in the PI3K/AKT signaling pathway, and this can result from loss/mutation of PTEN (which often occurs in prostate, brain, breast cancers, and melanoma), activating *Ras* mutations, mutations in PI3K, and activation/mutation of RTKs (such as EGFR, PDGFR, and HER2) (Cully and Downward, 2008; Salmena *et al.*, 2008; Samuels *et al.*, 2004; 2005; She *et al.*, 2008).

Recent studies have suggested that both the MAPK and AKT pathways may be good therapeutic targets in cancer (Carracedo *et al.*, 2008; Engelman *et al.*, 2008; Hoeflich *et al.*, 2009; Smalley *et al.*, 2006; Sos *et al.*, 2009). A number of pharmaceutical companies and academic institutions are now developing small molecule inhibitors for both the MAPK and PI3K/AKT pathways. As these compounds move from preclinical to clinical development, there is a need to accurately quantify the levels of constitutive MAPK and AKT signaling activity in both preclinical cancer models and in clinical specimens. Work in our lab is focused upon targeted therapy strategies for melanoma, a tumor that has been shown to rely upon both MAPK and PI3K/AKT signaling (Smalley, 2010). In this chapter, we describe methods to measure constitutive MAPK and AKT activity at both the single-cell level (microscopy and flow cytometry) as well as in populations of cells (Western blotting). As melanoma cell lines typically have high levels of constitutive MAPK and PI3K/AKT signaling, we have used this system to examine the cell signaling effects of the BRAF-specific kinase inhibitor PLX4720, which is currently being evaluated as a novel antimelanoma therapy (Flaherty et al., 2009; Tsai et al., 2008).

2. Maintaining Melanoma Cell Lines

2.1. Culturing conditions

Prior to working with the cell lines, it is recommended to expand and freeze down a large stock of cells at an early passage and to perform DNA fingerprinting analysis in order to authenticate genetic identity. To ensure stability of the cell lines between experiments, we recommend replacing the cultures every 2 months with freshly thawed, early passage cells. For the experiments described herein, cells were maintained in a 5% CO_2, 37 °C humidified incubator in growth media consisting of RPMI 1640 supplemented with 300 mg/L L-glutamine, 5% heat inactivated fetal calf serum, 100 U/mL penicillin, and 100 μg/mL streptomycin. WM164 and 1205Lu melanoma cell lines were a kind gift from Dr. Meenhard Herlyn (The Wistar Institute, Philadelphia, PA). Most of the cell lines described below are available from The Coriell Institute (Camden, NJ). Further details can be found at http://ccr.coriell.org/sections/Collections/Wistar.

2.2. Plating cells

ERK activity can be affected by cell density; therefore, it is important to plate the cells so that they remain subconfluent over the course of the experiment. At the same time, an adequate number of cells should be seeded so that there is enough material for a clear read-out of the assay. We recommend a cell density of 60–70% confluency on the day of experimentation.

3. Western Blotting

Signal transduction cascades are regulated through coordinated phosphorylation and dephosphorylation events (see Pratilas et al., 2009) for an overview of the phosphorylation events within the MAPK pathway). Constitutive activity in the MAPK and AKT signaling pathways can be easily assessed through the use of phosphorylation site-specific antibodies (aka phospho-specific antibodies). When used in traditional Western blotting, these antibodies allow for the sensitive detection of protein phosphorylation levels. In this section, we describe our technique for assessing ERK and AKT activity by Western blotting. Here we used Invitrogen's XCell SureLock Electrophoresis Cell for the separation of proteins and the XCell II Blot Module for the transfer of protein onto PVDF membrane; however, other equivalent apparatus can be used in place of the Invitrogen system. In the example shown in Fig. 27.1, BRAF V600E mutated 1205Lu melanoma cells are treated with increasing concentrations of the BRAF inhibitor PLX4720 for 1 and 24 h. These data show that although PLX4720 blocks constitutive pERK signaling activity, this leads in turn to a rebound increase in pAKT signaling (Fig. 27.1).

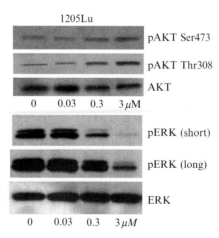

Figure 27.1 Measuring constitutive pERK and pAKT activity by Western blotting. 1205Lu melanoma cells were treated with either vehicle (DMSO, 0) or increasing concentrations of PLX4720 (0–3 μM, 1 h for pERK and 24 h for pAKT). Protein was then extracted, resolved by Western blotting and probed for expression of pAKT (Ser473 and Thr308) and pERK. Equal protein loading was demonstrated by stripping the original blot and reprobing for either total AKT or total ERK.

3.1. Harvesting cells and protein extraction
3.1.1. Materials
- RPMI growth media: RPMI 1640 supplemented with 300 mg/L L-glutamine, 5% heat inactivated fetal calf serum, 100 U/mL penicillin, and 100 μg/mL streptomycin
- 0.25% trypsin–EDTA
- 1 × PBS, (−) calcium chloride, (−) magnesium chloride, pH 7.4
- Roche Complete, Mini, EDTA-free protease inhibitors cocktail (# 11 836 170 001)
- RIPA lysis buffer: 50 mM Tris–HCl (pH 7.4), 1% Triton X-100, 24.1 mM Na-deoxycholate, 0.1% SDS, 154 mM NaCl, 1 mM EDTA Store RIPA buffer without protease inhibitors at 4 °C. To make the modified RIPA buffer, take a 10 mL aliquot of RIPA lysis buffer and add one tablet of Roche Complete, Mini, EDTA-free protease inhibitors
- 6-well tissue culture plates
- 15 mL Conical Falcon tubes
- 1.5 mL Eppendorf tubes.

1. 6-well tissue culture plates were seeded with 1 × 10^5 cells per 2 mL of growth media. Cells were grown overnight prior to treating with inhibitor.
2. Prior to protein extraction, melanoma cell lines were treated for 1 h with DMSO or 0.03–3 μM of the BRAF inhibitor, PLX4720 (Plexxikon, Inc., Berkley, CA). PLX4720 was dissolved in DMSO to make a 10 mM stock.
3. Prior to harvesting the adherent melanoma cells, growth media containing nonadherent cells was first collected and placed in a Falcon tube. The cells were then rinsed with 1 mL of 1 × PBS per well. The 1 × PBS from the rinse was placed into the same Falcon tube containing the nonadherent cells. To detach the cells, 0.5 mL of 0.25% Trypsin–EDTA was added to each well. After 5 min of incubation at room temperature (RT), plates were gently rocked back and forth to detach the cells followed by the addition of 1 mL of growth media to each well. Trypsinized cells were pipetted into the Falcon tube containing the nonadherent cells.
4. Falcon tubes containing the cells were centrifuged at 1500 rpm, 5 min, 4 °C. The supernatant was aspirated off and discarded, and the pellet was resuspended in 1 mL, cold 1 × PBS and transferred to a 1.5 mL Eppendorf tube.
5. Again, the cells were centrifuged at 1500 rpm, 5 min, 4 °C, and the supernatant was aspirated off and discarded. Cell pellets were resuspended in 35 μL modified RIPA buffer and pipetted up and down several times followed by vortexing to disrupt the pellet and lyse the cells.

6. The cells were centrifuged at maximum speed (13,200 rpm), 30 min, 4 °C. The soluble protein contained in the supernatant was collected and placed into new prechilled 1.5 mL Eppendorf tubes. Following protein extraction, cell lysates were stored at -20 °C.

3.2. Determine protein concentration

3.2.1. Materials

- Pierce BCA Protein Assay Kit and Instructions (Thermo Scientific Cat. #23,225)
- 96-well flat bottom microplates
- 37 °C Incubator
- 562 nm absorbance plate reader.

1. Prepare the BSA standards according to the standard test tube and microplate procedure using dH_2O as the diluent.
2. Dilute the protein extracts 1:10 in dH_2O in a total volume of 25 μL per replicate, and add 25 μL of standard or protein extract per well of a 96-well flat bottom microplate.
3. Prepare the BCA Working Reagent according to the instructions, and add 200 μL per well containing standard or diluted sample and incubate at 37 °C for 30 min. Following incubation, allow the plate to cool down to RT and measure the absorbance at 562 nm.

3.3. Sample preparation

3.3.1. Materials

- 1.5 mL Eppendorf tubes
- 2-mercaptoethanol
- NuPage (4×) LDS sample buffer (Invitrogen #NP0007)
- Heat block set to 95 °C or boiling waterbath

1. To detect phosphorylation of ERK and AKT, we run our proteins under denatured, reduced conditions. Denaturing unfolds the protein to reveal the epitope of interest and confers an overall negative charge to the sample. Therefore protein separation is based on molecular weight, not on conformation of the protein.
2. The samples are generally run in 1 mm × 10 or 12-well gels and are prepared in a total volume of 10–15 μL. However, the maximum loading volume for a 12-well gel is 30 μL. In our constitutively active cell lines, 5–30 μg of total protein is sufficient for the detection of

ERK phosphorylation. Detection of phospho-AKT usually requires 20–50 μg of total protein.
3. Samples are prepared in 1.5 mL Eppendorf tubes following the recipe below:

20 μg protein (10 μg/μL concentration)	2.0 μL
4 × NuPage (4×) LDS sample buffer	2.5 μL
2-mercaptoethanol (reducing agent)	0.1 μL
dH$_2$O	5.4 μL
Total volume	10.0 μL

4. Heat the samples for 5 min at 95 °C. Allow the samples to cool at RT for 5–10 min. Centrifuge the samples briefly to remove condensation from the tops of the tubes. Load the samples within an hour after adding the reducing agent.

3.4. Electrophoresis

3.4.1. Materials

- 10-well Novex 8–16% Tris–glycine gels (Invitrogen, #EC6045BOX)
- 12-well Novex 8–16% Tris–glycine gels (Invitrogen, #EC60452BOX)
- 10 × Tris–glycine–SDS buffer (Bio-Rad, #161-0732)
- XCell *SureLock*® Mini-Cell Kit (Invitrogen, #EI0001)[1]
- Precision Plus Protein Kaleidoscope Standard (Bio-Rad, #161-0375)

1. Wipe off the precast gel after removing it from the package and peel off the tape from the bottom of the cassette. You can also mark the wells with a sharpie marker to make them easier to see when you are loading the gel.
2. Place the gel in the chamber with the notched side facing the middle compartment. If only one gel is being run, place a gel dam on the other side of the electrode. Snap the tension wedge to secure the gels in place, and fill the middle chamber of the gel box with 500-mL 1× Tris–glycine running buffer allowing the buffer to flow to the outer compartment.
3. Remove the combs from the gels, and load each sample using gel loading tips. Alongside the samples, load 10 μL of Bio-Rad's Kaleidoscope standard or another standard with a molecular weight range within the range of phospho-ERK and phospho-AKT. Note that the molecular

---

[1] Please refer to Invitrogen's manual for detailed instructions.

weight for phospho-ERK(1/2) is 42 and 44 kDa while the molecular weight for phospho-AKT is 60 kDa.
4. Separate the proteins by electrophoresing at 125 V until the dye front reaches the bottom of the gel; this will usually take about 1.5 h.

3.5. Protein transfer

3.5.1. Materials

- Transfer buffer: 25 mM Tris base, 192 mM glycine, 20% methanol[2]
- Polyvinyldiene difluoride (PVDF) membranes
- Shaker
- Whatman filter paper
- Pipette or 15 mL conical tube for rolling out bubbles
- XCell II^{TM} Blot Module CE Mark (Invitrogen, #EI9051)[3]
- Sponge pads for blotting (Invitrogen, #EI9052)
- Power supply

1. Prewet the PVDF membrane for 30 s in methanol. Pour off the methanol and add 10–20 mL transfer buffer and shake for 5 min. Soak the filter paper briefly in transfer buffer immediately prior to use.
2. Open the gel cassette; the notched side of the cassette should face up. Carefully remove and discard the top plate; gel remains in the bottom slotted plate. Remove wells with the gel knife.
3. Place a piece of transfer buffer-soaked filter paper on top of the gel and lay just above the slot on the bottom of the cassette, leaving the "foot" of the gel uncovered. Keep the filter paper saturated with the transfer buffer and remove all trapped air bubbles by gently rolling over the surface using a pipette or 15 mL conical tube as a roller.
4. Turn the plate over so the gel and filter paper are facing downward over a flat surface. Use the gel knife to push the "foot" out of the slot allowing the gel to be released from the plate. Place the gel on a flat surface, and cut off the "foot" of the gel with a gel knife.
5. Wet the surface of the gel with transfer buffer, and place the presoaked transfer membrane on top of the gel. Use a pipette or conical tube to roll out all air bubbles. Place another presoaked filter paper on top of the membrane, and again ensure that all trapped air bubbles are removed.
6. Place two transfer buffer-soaked blotting pads into the deeper cathode (−) core of the blot module. Place the gel membrane assembly on top of the blotting pads in the same sequence making sure that the gel is closest to the cathode core.

---

[2] Store the transfer buffer at 4 °C.
[3] Please refer to Invitrogen's manual for further detailed instructions.

7. Add more presoaked blotting pads to the assembly so that the pads rise approximately 0.5 cm over the rim of cathode core. Place the anode (+) core on top of the pads.
8. Place the gel/membrane assembly core into the chamber and lock the core into place with the Gel Tension Wedge.
9. Fill the blot module with transfer buffer until the gel/membrane assembly is just covered in buffer. Do not fill all the way to the top, as this will only generate extra conductivity and heat.
10. Fill the outer buffer chamber with deionized water until it reaches approximately 2 cm from the top of the lower buffer chamber and place the lid on top of the unit. Transfer at 4 °C (cold room) at 35 V for 1 h or 12 V overnight.

3.6. Antibody incubations and western blot development

3.6.1. Materials

- 10 × TBST (recipe for 1 L, pH 7.6). Add 12.11 g Tris base and 87.66 g NaCl to 800 mL ddH$_2$O. Stir to completely dissolve the Tris and NaCl. Adjust the pH to 7.6 with 10 N HCl. Add 10 mL Tween 20 and stir to dissolve. Adjust the final volume to 1 L with ddH$_2$O.
- 1 × TBST: Dilute 10 × TBST to 1× with ddH$_2$O
- Blocking buffer: 5% nonfat milk in 1 × TBST
- Primary antibody dilution buffer: 5% BSA in 1 × TBST
- Phospho-p44/42 MAPK (ERK1/2) antibody (Cell Signaling Technology, #4370)
- p44/42 MAPK (ERK1/2) antibody (Cell Signaling Technology, #4695)
- Phospho-AKT (Ser473) antibody (Cell Signaling Technology, #4058)
- Phospho-AKT (Thr308) antibody (Cell Signaling Technology, #4056)
- AKT antibody (Cell Signaling Technology, #9272)
- Goat anti-rabbit IgG–HRP (GE Healthcare, #RPN4301)
- ECL Western Lightning (Perkin Elmer, #NEL100001EA)
- HyBlot CL Autoradiography Film (Denville Scientific Inc., #E3012)
- Autoradiography Cassette (Fisher Scientific, #FBCS 57)
- X-ray film developer
- Restore Western Blot Stripping Buffer (Thermo Scientific, #21059)

1. Rinse the PVDF membrane briefly with methanol and allow to air dry for ~15 min on top of a clean piece of filter paper.
2. Once dry, place the membrane into blocking buffer and shake at RT for 1 h.

3. Briefly rinse the membrane in 1 × TBST.
4. Make 1:1000 and 1:500 dilutions of the primary rabbit antiphospho-ERK and phospho-AKT antibodies, respectively.
5. Incubate in primary ab solution overnight at 4 °C (while shaking).
6. Wash three times, 10 min, RT on the shaker with 1 × TBST.
7. Prepare the secondary antibody (anti-rabbit IgG–HRP) in blocking buffer at a 1:2000 dilution.
8. Incubate in secondary ab, 1 h, RT. Wash three times 10 min, RT on the shaker w/1 × TBST.
9. Remove excess wash buffer by touching one corner of the membrane over a paper towel.
10. Place the membrane onto a plastic cover or saran wrap.
11. Wet the membrane with 1 mL of a 1:1 mix of Western Lightning ECL Reagents. Incubate at RT for 1 min.
12. Cover the membrane with a plastic cover or saran wrap and wipe off the excess ECL reagent.
13. Tape the encased membrane onto the inside of a cassette film holder.
14. Expose the film by placing it on top of the membrane and develop it in an X-ray developer. In order to increase the chances of detecting the protein of interest while avoiding overexposure of the film, it is best to carry out a range of exposure times.
15. To ensure even protein loading, the membrane should be stripped and reprobed for the total protein corresponding to the phosphorylated protein.
16. Briefly rinse the blot with 1 × TBST and cover the blot with Restore Western Stripping Buffer. Shake at RT for 15 min.
17. Repeat the wash with 1 × TBST and place the blot into a new container. Immerse the blot in blocking buffer and incubate at RT for 1 h. Briefly rinse in 1 × TBST.
18. For total ERK and total AKT, make 1:1000 dilutions in 5% BSA 1 × TBST. Repeat steps 5–14.

4. Phospho-Flow Cytometry

Phospho-flow cytometry allows for quantification of phosphorylation at the single-cell level. In addition, the flow cytometry platform through the combination of multiple antibodies allows for visualization of subpopulations of cells as well as multiple signaling events within a single cell or population. A sample experiment showing the ability of the BRAF inhibitor PLX4720 to inhibit constitutive pERK activity in two *BRAF* V600E mutated melanoma cell lines is shown in Fig. 27.2.

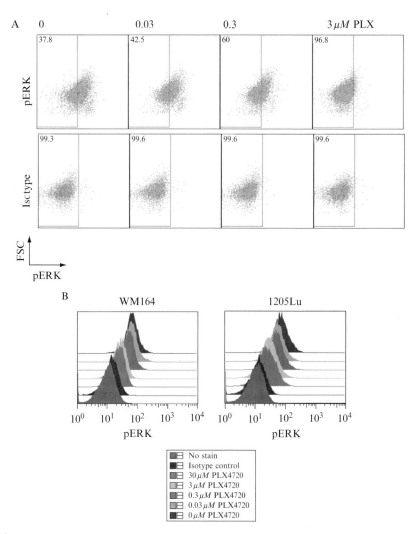

Figure 27.2 Using phospho-flow to measure constitutive pERK activity. (A) Demonstration of the specificity of response. WM164 melanoma cells were treated with PLX4720 (0–3 μM, 1 h), fixed, and stained for either pERK or a matching Rabbit IgG isotype control. Increasing concentrations of drug were found to inhibit pERK staining. The isotype control was used to gate for the analysis shown in Fig. 27.1B. (B) Quantification of levels of pERK signaling following PLX4720 treatment as measured by phospho-flow cytometry. Cell lines (WM164 and 1205Lu melanoma cell lines) were treated with PLX4720 (0–3 μM, 1 h), fixed, and stained for pERK. MFI (median fluorescence intensity) indicates the absolute level of pERK staining. (See Color Insert.)

4.1. Materials

- Anti-pERK1/2 (T202/Y204)-AF647 (BD Biosciences, #612593)
- Mouse IgG1, k-AF647 (BD Biosciences, #557732)
- BD Phosflow fix buffer I (BD Biosciences, #557870)
- BD Phosflow perm/wash buffer I (BD Biosciences, #557885)
- Staining buffer: 0.2% BSA, 1× PBS, pH 7.4
- FACS tubes (BD Falcon, #352235)
- BD FACSCalibur or other similar instrument equipped with a 635 or 638 nm laser capable of exciting AF647.

1. Melanoma cell lines were seeded at 2×10^6 cells per well in six-well tissue culture plates and allowed to attach overnight.
2. The following day, growth media (5% FBS, RPMI + L-glutamine) was replaced and individual wells containing cells at ~50% confluency were treated with 0.03, 0.3, 3, or 30 μM PLX4720. For vehicle controls, DMSO was added at an equivalent volume to the 30-μM PLX-treated wells.
3. The cells were treated for 1 h at which time nonadherent and adherent cells were collected as described in Section 3.
4. Following collection, cells were immediately fixed with an equivalent volume of phosflow fix buffer I for 10 min at 37 °C.
5. Fixed cells were pelleted and resuspended in Phosflow perm/wash buffer I and permeabilized for 30 min on ice.
6. Cells were washed with staining buffer (0.2% BSA, 1× PBS, pH 7.4) and counted.
7. Cells from each concentration of treatment were individually stained with pERK or the isotype control antibody. In addition, cells from each treatment were pooled and left unstained in order to set up the scatter and negative parameters of the flow cytometer.
8. A total of 1×10^5 cells in 100 μL staining buffer were stained with 3 μL pERK1/2-AF647 antibody or 1.25 μL pERK1/2 isotype control, mouse IgG1, k-AF647. Cells were incubated in the dark for 1 h at RT.
9. Following staining, cells were washed with 1 mL staining buffer and resuspended in 250 μL staining buffer.
10. At least 30,000 events were acquired on a FACSCalibur flow cytometer.
11. Analysis was carried out with FlowJo v8.7.1 software (Tree Star, Stanford, CA). A separate gate was generated for each treatment concentration by gating on approximately 99% of the corresponding isotype control. In order to delineate positive-stained cells, the isotype control gate, which represents the negative staining gate, was then copied to the pERK-stained sample.

5. Immunofluorescence

An alternate method for measuring constitutive MAPK and AKT activity in individual melanoma cells is to stain cell cultures with phospho-specific antibodies against either ERK or AKT and then use immunofluorescence microscopy to visualize and count the number of cells with pathway activity. Although the method requires the manual inspection of cultures, it allows rare events to be captured that may be missed by flow cytometry and is also well suited for examining nuclear versus cytoplasmic localization of the signals. There is evidence from immunohistochemical staining of melanoma samples that the intensity of pERK staining is heterogeneous within tumors and that levels of pERK expression can vary between the different cellular compartments (example shown in Fig. 27.3). In the method outlined below, melanoma cells are seeded subfluently onto glass coverslips and incubated overnight in the absence or presence of the BRAF inhibitor PLX4720. Cells are then fixed in paraformaldehyde, permeabilized, and stained for phospho-ERK. In the example shown in Fig. 27.4, increasing concentrations of the BRAF inhibitor PLX4720 reduces the phospho-ERK staining of the melanoma cell cultures.

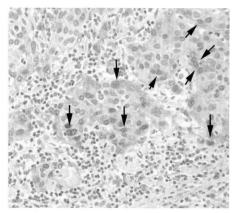

Metastatic melanoma

Figure 27.3 pERK is located in both the nuclear and cytoplasmic compartments of human melanoma metastases. Figure shows a representative clinical specimen of human melanoma stained for pERK by immunohistochemistry. Arrows indicate nuclear expression of pERK. Magnification: 200×.

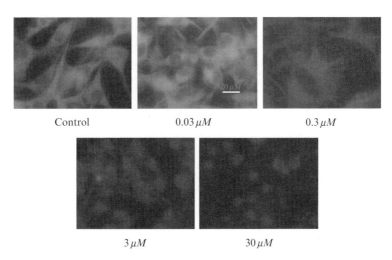

Figure 27.4 PLX4720 reduced the expression of pERK. WM164 cells were treated with increasing concentrations of PLX4720 (0.03–30 μM, 1 h) before being fixed, permeabilized, and stained for pERK. Scale bar: 20 μM.

5.1. Reagents

4% Paraformaldehyde solution

Dilute 10 mL of 16% paraformaldehyde solution (Electron Microscopy Systems, Hatfield, PA) in 30 mL of PBS in a 50-mL Falcon Tube. Cover the tube with foil and store in the dark. The diluted paraformaldehyde can be used for up to 14 days.

0.2% v/v Triton X-100 solution

Make a 10% v/v Triton solution in PBS by adding 2 mL of stock Triton X-100 to a 50 mL Falcon Tube. Make this up to a total final volume of 20 mL PBS. Vortex to ensure the Triton and PBS are thoroughly mixed. The resulting solution should be clear with no traces of undissolved Triton. Make a working stock of 0.2% v/v Triton by adding 0.2 mL of 10% v/v Triton solution to a Falcon tube and making to a final volume of 10 mL in PBS.

1% BSA–PBS blocking solution

Dissolve 0.2 g of bovine serum albumin (BSA, Fraction V) in 20 mL of PBS.

Primary antibodies

For dual pERK/pAKT staining, we use mouse antiphospho-ERK (Cell Signaling Technology #9106) and either rabbit Ser473 AKT

(Cell Signaling Technology, #4058) or anti-Thr308 AKT (Cell Signaling Technology, #2965).

Secondary antibodies
- Anti-mouse Alexa Fluor 488 (Invitrogen, #A21141)
- Anti-rabbit Alexa Fluor 594 (Invitrogen, #A11012).

5.2. Preparing the cells on coverslips

1. Place one glass coverslip into one well of a 6-well plate and immerse in 70% ethanol. Remove the ethanol using a vacuum line and allow the coverslips to air dry. Note: It is important for all the ethanol to evaporate and the coverslips to be completely dry before proceeding to the next stage.
2. Remove the cells from the flask by removing the media and adding 3 mL of Trypsin for 5 min. Once the cells have detached, add 7 mL of media and centrifuge at 1500 rpm for 5 min.
3. Remove the media and resuspend the cells in 10 mL of fresh media. Perform cell counts using a hemocytometer and resuspend the cells at a dilution of 50,000 cell/mL.
4. Add 100 µL of cell suspension to the center of each coverslip, and return the plates to the cell culture incubator and incubate at 37 °C for 15 min or until the cells have adhered.
5. Add 2 mL of media to each well of the plate and leave the cultures to equilibrate overnight.
6. Add either vehicle (DMSO) or increasing concentrations of the pharmacological inhibitor of choice (in Fig. 27.4, the BRAF inhibitor PLX4720 is used) for 1–24 h.

5.3. Fixing and staining the cells

1. Remove media using a vacuum line and add 1 mL of 4% paraformaldehyde per well (15 min). Once fixed, the coverslips can either be kept in the 6-well plates (wrapped in parafilm) and stored at 4 °C or used immediately.
2. Remove the formaldehyde and wash the cells twice with PBS.
3. Remove PBS and add 1 mL 0.2% v/v of Triton X-100 for 5 min to permeabilize the cells.
4. Remove Triton and wash the coverslips 2× with PBS.
5. Add 1 mL of PBS containing 1% BSA and leave to block for 15 min at RT.

6. Prepare the staining surface by laying the inverted lid of the six-well plate flat on the bench and placing parafilm on top of this.
7. Make up primary antibody at a concentration of 1:20–1:40 in BSA–PBS. Allow 50 μL of BSA–PBS per coverslip. Place 50 μL on top of the parafilm before carefully removing a coverslip using a pair of fine forceps. Gently blot the liquid off the coverslip before inverting the coverslip (so that the cells are facing downward onto the antibody) onto the 50 μL of BSA–PBS antibody solution. When you have performed this process for each of the coverslips, cover the lid with foil and place in a cell culture incubator (37 °C) for 1 h.
8. Return the coverslips to the 6-well plate and wash 2× in PBS over 30 min at RT.
9. Repeat steps 6–8 for the secondary antibody (1:200) in BSA–PBS and again incubate at 37 °C for 1 h.
10. Wash the coverslips 2× in PBS and once in distilled water (to remove salt deposits) and blot the coverslip gently on a Kim Wipe (or other tissue paper).
11. Put a single drop of antifade (we use Vectashield with DAPI) onto a glass slide and invert the coverslip gently on top of it (so the cells are in contact with the Vectashield). Seal the edges of the coverslip using nail polish.
12. Image the slides using an upright fluorescence microscope.

6. CONCLUSIONS

Measuring constitutive levels of MAPK and PI3K/AKT signaling is an essential part of assessing the activity of novel, targeted anticancer therapies. In this chapter, we have outlined methods for looking at MAPK/AKT activity in mass cell cultures using Western blotting, as well as techniques to quantify signaling events across cell populations (flow cytometry and immunofluorescence). In the examples given in Figs. 27.1–27.4, these methods were used to investigate the ability of the novel BRAF-specific kinase inhibitor PLX4720 to inhibit constitutive MAPK signaling in *BRAF* V600E-mutated human melanoma cell lines (Tsai *et al.*, 2008). Although Western blotting is regarded as being the gold standard for signal transduction studies, it is limited in giving only a "snap-shot" of the sum total of signaling activity within the whole cell population. There is a growing realization that Western blotting methods are not adequate at capturing the cell signaling heterogeneity that is commonly observed within populations of adherent cancer cells. Quantification of signaling at the individual cell level is critical to understand drug resistance and therapy escape, with recent studies suggesting that even minor differences in signaling dynamics within clonal populations of cancer cells lead to vastly different therapeutic outcomes (Cohen et al., 2008; Gascoigne and Taylor, 2008). Flow cytometry methods that utilize phospho-specific antibodies to measure signaling at the single-cell level are rapidly gaining popularity amongst those working upon nonadherent cell populations, such as immune cells and leukemia (Kotecha et al., 2008). In future, as the methods are refined we expect phospho-flow techniques to be more widely adopted by those working on adherent tumor cell lines. We further expect that enhanced methods to quantify signaling heterogeneity will bring important new insights into our understanding of constitutive MAPK and AKT signaling within cancer populations and that this will in turn lead to the development of optimized small molecule inhibitors for targeting signal transduction pathways in cancer.

ACKNOWLEDGMENTS

Work in the authors' lab is supported by The Melanoma Research Foundation, The Bankhead–Coley Research Program of the State of Florida (09BN-14), an Institutional Research Grant from the American Cancer Society #93-032-13, a Career Development Award from the Donald A Adam Comprehensive Melanoma Research Center (Moffitt Cancer Center), and the NIH/National Cancer Institute PSOC grant U54 CA143970-01. We would like to thank Gideon Bollag (Plexxikon, Inc.) for providing us with the PLX4720 used in these studies.

REFERENCES

Bhatt, K. V., et al. (2005). Adhesion control of cyclin D1 and p27Kip1 levels is deregulated in melanoma cells through BRAF-MEK-ERK signaling. Oncogene **24**, 3459–3471.
Brazil, D. P., et al. (2002). PKB binding proteins. Getting in on the Akt. Cell **111**, 293–303.
Cantley, L. C. (2002). The phosphoinositide 3-kinase pathway. Science **296**, 1655–1657.
Carracedo, A., et al. (2008). Inhibition of mTORC1 leads to MAPK pathway activation through a PI3K-dependent feedback loop in human cancer. J. Clin. Invest. **118**, 3065–3074.
Cohen, A. A., et al. (2008). Dynamic proteomics of individual cancer cells in response to a drug. Science **322**, 1511–1516.
Crews, C. M., et al. (1992). The primary structure of MEK, a protein kinase that phosphorylates the ERK gene product. Science **258**, 478–480.
Cully, M., and Downward, J. (2008). SnapShot: Ras signaling. Cell **133**(7), 1292–1292e1.
Curtin, J. A., et al. (2006). Somatic activation of KIT in distinct subtypes of melanoma. J. Clin. Oncol. **24**, 4340–4346.
Datta, S. R., et al. (1997). Akt phosphorylation of BAD couples survival signals to the cell-intrinsic death machinery. Cell **91**, 231–241.
Davies, H., et al. (2002). Mutations of the BRAF gene in human cancer. Nature **417**, 949–954.
Dent, P., et al. (1992). Activation of mitogen-activated protein kinase kinase by v-Raf in NIH 3T3 cells and in vitro. Science **257**, 1404–1407.
Engelman, J. A., et al. (2008). Effective use of PI3K and MEK inhibitors to treat mutant Kras G12D and PIK3CA H1047R murine lung cancers. Nat. Med. **14**, 1351–1356.
Flaherty, K. T., et al. (2009). Phase I study of PLX4032: Proof of concept for V600E BRAF mutation as a therapeutic target in human cancer. J. Clin. Oncol. **27**, Abstract 9000.

Gascoigne, K. E., and Taylor, S. S. (2008). Cancer cells display profound intra- and interline variation following prolonged exposure to antimitotic drugs. *Cancer Cell* **14**, 111–122.

Hoeflich, K. P., et al. (2009). In vivo antitumor activity of MEK and phosphatidylinositol 3-kinase inhibitors in basal-like breast cancer models. *Clin. Cancer Res.* **15**, 4649–4664.

Kotecha, N., et al. (2008). Single-cell profiling identifies aberrant STAT5 activation in myeloid malignancies with specific clinical and biologic correlates. *Cancer Cell* **14**, 335–343.

Kyriakis, J. M., et al. (1992). Raf-1 activates MAP kinase-kinase. *Nature* **358**, 417–421.

Lenormand, P., et al. (1993). Growth factors induce nuclear translocation of MAP kinases (p42mapk and p44mapk) but not of their activator MAP kinase kinase (p45mapkk) in fibroblasts. *J. Cell Biol.* **122**, 1079–1088.

Padua, R. A., et al. (1985). Activation of N-ras in a human melanoma cell line. *Mol. Cell. Biol.* **5**, 582–585.

Pratilas, C. A., et al. (2009). (V600E)BRAF is associated with disabled feedback inhibition of RAF-MEK signaling and elevated transcriptional output of the pathway. *Proc. Natl. Acad. Sci. USA* **106**, 4519–4524.

Robertson, G. P. (2005). Functional and therapeutic significance of Akt deregulation in malignant melanoma. *Cancer Metastasis Rev.* **24**, 273–285.

Robinson, M. J., and Cobb, M. H. (1997). Mitogen-activated protein kinase pathways. *Curr. Opin. Cell Biol.* **9**, 180–186.

Salmena, L., et al. (2008). Tenets of PTEN tumor suppression. *Cell* **133**, 403–414.

Samuels, Y., and Velculescu, V. E. (2004). Oncogenic mutations of PIK3CA in human cancers. *Cell Cycle* **3**, 1221–1224.

Samuels, Y., et al. (2004). High frequency of mutations of the PIK3CA gene in human cancers. *Science* **304**, 554.

Samuels, Y., et al. (2005). Mutant PIK3CA promotes cell growth and invasion of human cancer cells. *Cancer Cell* **7**, 561–573.

She, Q. B., et al. (2008). Breast tumor cells with PI3K mutation or HER2 amplification are selectively addicted to Akt signaling. *PLoS ONE* **3**, e3065.

Smalley, K. S. M. (2003). A pivotal role for ERK in the oncogenic behaviour of malignant melanoma? *Int. J. Cancer* **104**, 527–532.

Smalley, K. S. (2010). Understanding melanoma signaling networks as the basis for molecular targeted therapy. *J. Invest. Dermatol.* **130**, 28–37.

Smalley, K. S., et al. (2006). Multiple signaling pathways must be targeted to overcome drug resistance in cell lines derived from melanoma metastases. *Mol. Cancer Ther.* **5**, 1136–1144.

Smalley, K. S., et al. (2009). Genetic subgrouping of melanoma reveals new opportunities for targeted therapy. *Cancer Res.* **69**, 3241–3244.

Sos, M. L., et al. (2009). Identifying genotype-dependent efficacy of single and combined PI3K- and MAPK-pathway inhibition in cancer. *Proc. Natl. Acad. Sci. USA* **106**, 18351–18356.

Stokoe, D., et al. (1994). Activation of Raf as a result of recruitment to the plasma membrane. *Science* **264**, 1463–1467.

Treisman, R. (1994). Ternary complex factors: Growth factor regulated transcriptional activators. *Curr. Opin. Genet. Dev.* **4**, 96–101.

Tsai, J., et al. (2008). Discovery of a selective inhibitor of oncogenic B-Raf kinase with potent antimelanoma activity. *Proc. Natl. Acad. Sci. USA* **105**, 3041–3046.

Wellbrock, C., et al. (2004). The RAF proteins take centre stage. *Nat. Rev. Mol. Cell Biol.* **5**, 875–885.

Wong, K. K., et al. (2010). Targeting the PI3K signaling pathway in cancer. *Curr. Opin. Genet. Dev.* **20**, 87–90.

CHAPTER TWENTY-EIGHT

Constitutive Activity of GPR40/FFA1: Intrinsic or Assay Dependent?

Leigh A. Stoddart* *and* Graeme Milligan[†]

Contents

1. Introduction	570
2. Measuring FFA1-Mediated Calcium Mobilization	575
2.1. Required materials	577
2.2. Generation of Flp-In T-REx cell lines expressing FFA1	577
2.3. Cell preparation for calcium mobilization assay	579
2.4. Measuring changes in intracellular calcium	579
3. Measuring Direct Activation of G Proteins via FFA1	581
3.1. Required materials	583
3.2. Transient expression of FFA1 and membrane preparation	584
3.3. [^{35}S]GTPγS-binding assay with immunoprecipitation step	585
References	587

Abstract

Free fatty acid receptor 1 (FFA1; previously designated GPR40) is a potential therapeutic target for the treatment of diabetes and related metabolic disorders. Agonist-independent or constitutive activity is a feature associated with essentially all G protein-coupled receptors but the extent of this varies substantially between family members. In many situations, detection of such activity can be both assay- and context-dependent and may reflect the presence in the assay of an endogenous agonist. In studies on FFA1, experiments employing cell membrane preparations and the binding of [^{35}S]guanosine 5′-O-[γ-thio] triphosphate to G proteins produce data consistent with a high-level constitutive activity of this receptor. Herein, we detail these assays and discuss approaches to determine if this is a measure of intrinsic receptor constitutive activity or if such results reflect the presence of an endogenous agonist. FFA1 is coupled predominantly to G proteins of the Gα_q subfamily. Activation of the receptor results, therefore, in the transient elevation of intracellular

* Institute of Cell Signalling, School of Biomedical Science, Medical School, University of Nottingham, Nottingham, United Kingdom
[†] Molecular Pharmacology Group, College of Medical, Veterinary and Life Sciences, University of Glasgow, Glasgow, United Kingdom

Methods in Enzymology, Volume 484 © 2010 Elsevier Inc.
ISSN 0076-6879, DOI: 10.1016/S0076-6879(10)84028-5 All rights reserved.

569

[Ca^{2+}]. We also detail assays to measure such signals and consider whether they are appropriate to detect receptor constitutive activity.

Abbreviations

BSA	bovine serum albumin
FFA	free fatty acid
GDP	guanosine diphosphate
GSIS	glucose-stimulated insulin secretion
GPCR	G protein-coupled receptor
GTP	guanosine triphosphate
GTPγS	guanosine 5′-O-[γ-thio]triphosphate

1. Introduction

Free fatty acids (FFAs) were identified as the endogenous ligands for the orphan G protein-coupled receptor (GPCR) GPR40 in 2003 by three separate groups (Briscoe et al., 2003; Itoh et al., 2003; Kotarsky et al., 2003). At a similar time, short chain fatty acids were shown to activate the related receptors, GPR41 and GPR43 (Brown et al., 2003; Le Poul et al., 2003). To reflect their relatedness and the nature of their endogenous ligands, GPR40, GPR41, and GPR43 were classified as a family of fatty acid receptors by the International Union of Pharmacology in 2008 and the names of the receptors were changed to reflect this. GPR40 is, therefore, now referred to as the FFA receptor FFA1, GPR41 as FFA3, and GPR43 as FFA2 (Stoddart et al., 2008b).

Of these three FFA receptors, much of the published literature has focused on FFA1 to understand its role in pancreatic β cells and to attempt to validate it as a potential therapeutic target for the treatment of type 2 diabetes. Significant levels of FFA1 mRNA expression has been identified in a variety of β cell lines (Briscoe et al., 2003; Itoh et al., 2003; Shapiro et al., 2005) and in both human (Tomita et al., 2005) and rat (Feng et al., 2006; Salehi et al., 2005) pancreatic islets. Circulating long chain fatty acids play a key role in insulin release as they potentiate glucose-stimulated insulin secretion (GSIS). However, long-term exposure of islets to high levels of circulating fatty acids is associated with reduced insulin secretion capacity and β cell apoptosis (Nolan et al., 2006) and this has resulted in discussion over whether short-term agonism or longer term blockade of FFA1 might be the most effective therapeutic strategy. In cultured cells, it has been confirmed that the fatty acid potentiation of GSIS is mediated by FFA1 (Itoh et al., 2003; Salehi et al., 2005; Schnell et al., 2007; Shapiro et al., 2005). To further explore this, a number of FFA1 knock-out mouse lines have been generated and, although the results are not in complete agreement, it is now believed that FFA1 mediates the short-term positive effects of fatty acids on β cells but not the long-term, detrimental effects (Kebede et al., 2008; Lan et al., 2008; Latour et al., 2007; Steneberg et al., 2005; Tan et al., 2008). Expression of FFA1 has been detected in other tissues and cells, including immune cells (Briscoe et al., 2003), enteroendocrine cells in mice (Edfalk et al., 2008), and breast cancer cell lines (Hardy et al., 2005; Yonezawa et al., 2004), although comparatively little work has been carried out on the role of the receptor in these systems.

FFA1 is a predominately $G\alpha_{q/11}$-coupled receptor leading to activation of phospholipase C, which hydrolyses phosphatidylinositol-4,5-bisphosphate into inositol-1,4,5-trisphosphate (IP_3) and diacylglycerol. IP_3 activates IP_3 sensitive channels in the endoplasmic reticulum which release calcium from these stores and transiently increase the intracellular calcium concentration via multiple calcium release mechanisms (Briscoe et al., 2003; Fujiwara et al., 2005; Itoh et al., 2003). Some studies have also found that FFA1 can weakly activate $G\alpha_{i/o}$ (Itoh et al., 2003; Kotarsky et al., 2003; Yonezawa et al., 2004) and may, therefore, regulate other signaling pathways in particular cell types and tissues.

It is now accepted that the majority of GPCRs show some level of spontaneous activity in the absence of agonists (Milligan, 2003a). The exception to this is rhodopsin, as it has 11-cis-retinal covalently bound within the bundle of transmembrane helices, and this acts as a high-efficacy inverse agonist that suppresses potential basal activity (Bond and Ijzerman, 2006; Milligan, 2003a). To date, there is limited information focusing directly on the level of constitutive activity FFA1 displays. Determining the level of constitutive activity of any given GPCR can be challenging as levels of constitutive activity are often assay and recombinant expression systems dependent. This is even more problematic in native tissues and in vivo studies as it is often difficult to ensure washout or removal of endogenous agonists from the system by other means. As FFA1 is a $G\alpha_{q/11}$-coupled receptor, the majority of publications exploring the pharmacology and function of this GPCR have used calcium mobilization as a functional read-out. Calcium mobilization is, however, not well suited to examine receptor-constitutive activity, as the resting intracellular concentration of calcium is buffered within strict margins and alterations in concentration are generally transient. For example, the ghrelin receptor displays high levels of constitutive activity in many assays including inositol phosphate accumulation assays (Holst et al., 2003) and [^{35}S]guanosine 5′-O-[γ-thio]triphosphate ([^{35}S]GTPγS)-binding studies (Bennett et al., 2009) but such constitutive activity could not be measured when monitoring intracellular calcium (Holst et al., 2003) although cells with maintained expression of the ghrelin receptor frequently appear morphologically as if they are undergoing apoptosis driven by calcium-mediated toxicity (Ward et al., 2010). Therefore, it is not surprising that there are no reports of FFA1 constitutive activation in studies that have employed calcium mobilization based read-outs, even if the receptor may display significant constitutive activity in other assays. Another factor which is important in examining constitutive activity of a GPCR is the availability of selective antagonists/inverse agonists, as it is now appreciated that the majority of antagonists are actually inverse agonists that can reduce the basal activity of a receptor (Kenakin, 2001). Because FFA1 was only relatively recently deorphanized, there is a limited catalog of synthetic compounds that can be used to modulate the basal activity of this GPCR. To date, there are only two examples of FFA1 antagonists/inverse agonists published in the primary scientific literature (Briscoe et al., 2006; Hu et al., 2009) and neither class is available commercially. Furthermore, neither of the studies noted above used the identified FFA1 antagonists/inverse agonists to examine questions regarding constitutive activity of FFA1.

The most direct observations of FFA1 displaying apparent constitutive activity derive from studies carried out in our laboratory (Stoddart et al., 2007). We found that FFA1 displayed high levels of basal activity in a [^{35}S]GTPγS-binding assay performed on membranes of cells engineered to express human FFA1. Indeed, this activity was sufficiently high that known agonists of FFA1 were unable to increase substantially binding of [^{35}S]GTPγS above this basal level (Fig. 28.1) (Stoddart et al., 2007). This "activity" was reduced, in a concentration-dependent manner, by the addition of the FFA1 selective antagonist/inverse agonist GW1100 (Briscoe et al., 2006). Such results are potentially consistent with the FFA1 receptor displaying a high level of constitutive activity and this activity being diminished by the presence of the synthetic molecule which has the characteristics of an inverse agonist. However, indications that the situation might be more complex stemmed from studies showing that basal loading of [^{35}S]GTPγS in FFA1 expressing membranes could also be reduced substantially, and in a concentration-dependent manner, by the addition of fatty acid-free bovine serum albumin (BSA) (Fig. 28.1). This raised the question of whether we were measuring true, agonist-independent, constitutive activity or, rather, the capacity of an antagonist/inverse agonist (i.e., GW1100) to complete for the receptor-binding site with an unidentified and unappreciated endogenous agonist. Fatty acids are poorly soluble in water and are transported around the blood stream and body complexed to carrier proteins, predominantly serum albumin. [^{35}S]GTPγS-binding assays are performed routinely on cell membrane preparations, and we concluded that the high levels of basal [^{35}S]GTPγS binding probably reflected release of fatty acids from the cells during membrane preparation

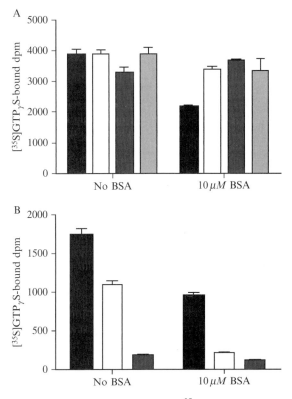

Figure 28.1 FFA1-Gα_q displays high levels of [^{35}S]GTPγS binding which can be reduced by the addition of BSA, an FFA1 antagonist or a Gα_q inhibitor. Membranes were prepared from HEK293 cells transiently transfected to express FFA1-Gα_q. [^{35}S] GTPγS-binding studies were performed in the absence or presence of 10 μM fatty acid-free BSA. In the absence of agonists (black bars), [^{35}S]GTPγS binding is reduced in the presence of BSA (A and B). (A) The addition of FFA1 agonists (white bars, 30 μM palmitic acid; dark gray bars, 30 μM troglitazone; light gray bars, 1 μM GSK250089A) in the absence of BSA gave no increase in [^{35}S]GTPγS binding but in the presence of BSA levels of [^{35}S]GTPγS binding were increased. (B) The FFA1 antagonist, GW1100 (10 μM, white bars) reduced levels of [^{35}S]GTPγS binding in the absence and presence of BSA. In the presence of the Gα_q inhibitor, YM254890 (100 nM, dark gray bars) levels of [^{35}S]GTPγS binding were substantially reduced and the addition of BSA could not reduce binding levels further.

and these fatty acids could fully activate FFA1 in the assay conditions employed. When BSA was added to the assay, this potentially sequestered these fatty acids and under these conditions, "basal" activity was reduced to an extent that allowed concentration–response curves to agonists to be performed (Stoddart et al., 2007). Further evidence in favor of this model was that at increasing BSA concentrations, as well as further reducing "basal" [^{35}S]GTPγS binding, the concentration–response curves of added fatty acids required higher concentrations to reach half-maximal effect (EC_{50}). The most obvious explanation was that the added BSA was now partially sequestering the added fatty acids as well as stripping the potential endogenous ligands from the receptor. Furthermore, when used in combination with BSA the FFA1 antagonist GW1100 further reduced the basal loading of [^{35}S]GTPγS to levels similar as seen in cell membranes that lacked expression of FFA1 (Fig. 28.1). In practice, addition of 10 μM BSA provided a compromise between reducing the "basal" loading of [^{35}S]GTPγS and allowing an increase in binding in the presence of agonist to be observed without the BSA binding large amounts of the added fatty acids. Considering these findings, it is hard to measure the true extent of FFA1 constitutive activity in a [^{35}S]GTPγS-binding assay, although the studies using combinations of GW1100 and BSA indicate that the basal [^{35}S]GTPγS binding in the absence of these agents may be a combination of true constitutive activity and the action of an unidentified endogenous agonist, most likely a fatty acid or a combination of different fatty acids.

When exploring potential FFA1 constitutive activity, it is important, therefore, to select, define, and optimize the assay(s) carefully. Even more than in many other situations, when studying FFA1, the use of a nonexpressing or nontransfected cell/membrane control is absolutely vital. Furthermore, careful examination of the direct data rather than reliance on replotting data as "% over basal activity" or other similar derivatives is essential to appreciate potential artifacts and to avoid masking key information. Assays centered on measures of calcium mobilization are unlikely to provide insight into basal signaling in the absence of ligand. [^{35}S]GTPγS-binding assays are often used to examine constitutive activity of GPCRs (Canals and Milligan, 2008), and this has been our assay of choice for FFA1. However, although in our studies on FFA1, these initially generated data suggesting this receptor to show very high levels of constitutive activity, careful pharmacological analysis indicated that much of this activity reflected the presence of an endogenous agonist. Clearly, other intact cell assays may provide insight into the level of constitutive activity of FFA1. Those based on the phosphorylation/activity status of kinases such as the ERK1/2 MAP kinases have suggested FFA1 to have low levels of constitutive activity (Smith et al., 2009), but it is important to remember that such studies are generally performed after a period, often quite sustained, of removal of most serum components from the cell growth medium and the phosphorylation of ERK1/2 is also a relatively transient event. It is common to use a cAMP-based assay to measure constitutive activity of GPCRs (Chen et al., 2001; Hall and Strange, 1997) but as FFA1 couples only weakly to G$α_{i/o}$, it would be difficult to gain a meaningful readout in a cAMP assay with this GPCR and, to date, there have been no reports of using a cAMP-based assay to measure FFA1 activity. Assays based on the production of inositol phosphates could also be used to study

FFA1-constitutive activity and have been used to study constitutive activity of other $G\alpha_q$-coupled GPCRs (e.g., Hein et al., 2001; Holst et al., 2003). Two previous studies on FFA1 have used inositol phosphate generation assays but neither explored the basal activity of FFA1 (Flodgren et al., 2007; Salehi et al., 2005). Finally, there is growing interest in the use of so-called "label-free" technologies to explore the pharmacology and function of GPCRs (Lee, 2009; Rocheville and Jerman, 2009) and these may offer useful additions to the range of approaches discussed above.

Due to the assay-dependent nature of constitutive activity, we will describe detailed methodology for a whole cell calcium mobilization assay that is unhampered by any constitutive activity FFA1 may show and for the $[^{35}S]GTP\gamma S$ method, which has been used to examine the extent of constitutive activity of FFA1, and whether this is an assay artifact.

2. MEASURING FFA1-MEDIATED CALCIUM MOBILIZATION

Measuring the ability of FFA1 to increase intracellular calcium has been utilized in multiple studies exploring the activity and function of this receptor (Briscoe et al., 2003; Fujiwara et al., 2005; Itoh et al., 2003; Schnell et al., 2007). There are a variety of different methodologies that are used to monitor calcium mobilization, but central to all are the use of calcium-sensitive dyes. As all $G\alpha_{q/11}$-coupled GPCRs are anticipated to increase intracellular calcium, at least transiently, there are a wide variety of such reagents available. Two of the most commonly used are the single wavelength dye, Fluo-4, and the ratiometric dye, Fura-2. The calcium-sensitive dyes are normally provided in the acetoxymethyl (AM) ester forms, which are cell permeable. Once inside the cell, the ester link is cleaved to yield the fluorescent dye. In certain cases, others may be used if the excitation and emission spectra of the commonly used dyes overlap with fluorescent proteins attached to the receptor under study or other fluorescent compounds present within the cells. Calcium mobilization may be studied via single cell imaging or in cell populations. Single cell studies have been used previously to measure the ability of FFA1 to signal (Stoddart et al., 2007; Yonezawa et al., 2004) but are time consuming and not well suited for the generation of concentration–response data. Generally, studies of ligand potency using calcium flux-based assays employ a multiwell flurometric imaging plate reader. The most widely used plate readers in this area are the FLIPR and the FlexStation, both supplied by Molecular Devices (http://www.moleculardevices.com). FLIPR plate readers are widely used in the pharmaceutical industry for high-throughput screening as they can read up to 1536 wells simultaneously with data generation in less than 5 min. The FlexStation is a more cost-sensitive solution that can analyze 96-well plates and read 8 wells at a time; thus, a 96-well plate requires around 45 min to read.

The method below describes in detail how to measure alterations in intracellular calcium in live cell populations using a FlexStation plate reader (Fig. 28.2). It also describes creating cells able to express FFA1 stably and how to use these for measuring calcium flux in a multiwell format. Measuring calcium flux using a multiwell platform requires a high proportion of the cells (around 80–90%) to respond in synchrony to allow the detection of the change in fluorescence that corresponds to the alterations in intracellular calcium concentrations. Transient transfection of GPCR cDNA frequently does not result in high-enough transfection efficiency; therefore, we have routinely used Flp-In T-REx 293 cells (Ward et al., 2010) with human FFA1 harbored at the Flp-In locus. Using Flp-In T-REx 293 cells ensures that the gene of interest is expressed from a single genomic location. Expression of the gene of interest is under the control of a Tet repressor, and expression only occurs in the presence of tetracycline or the related antibiotic doxycycline. This is useful when studying FFA1, as in the absence of doxycycline any pleiotropic, nonspecific effects of potential ligands can be assessed. We have generated multiple Flp-In T-REx cell lines able to express individual GPCRs on demand and found them a highly appropriate system for studying GPCRs (see Canals and Milligan, 2008; Ellis et al., 2006; Lopez-Gimenez et al., 2008; Stoddart et al., 2007; 2008a, see Ward et al., 2010 for review).

The method described below includes Brilliant Black BN in the assay buffer. Brilliant Black BN is a commonly used food dye and pharmaceutical excipient which is often used in calcium mobilization assays as it quenches the fluorescence from any of the calcium-sensitive dye that remains outside the cells. The use of Brilliant Black BN removes the need for a wash step

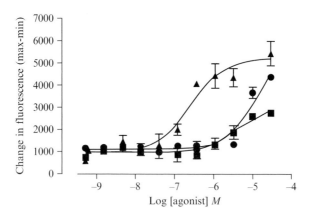

Figure 28.2 Examples of typical concentration–response curves generated using the described calcium mobilization assay. The graph shows concentration–response curves for three classes of FFA1 agonists on cells expressing FFA1. Triangles are GSK250089A, circles are linolenic acid and squares are rosiglitazone.

after loading the cells with the calcium-sensitive dye (May et al., 2010). This is useful when using cells derived from HEK293 cells as they are less adherent compared to other commonly used cell types and can detach easily from the plate during the wash steps.

2.1. Required materials

2.1.1. Fluorescence plate reader, plates, and materials required to run assay

To measure changes in fluorescence from the loaded Fluo-4 over time, you will need a benchtop multiwell plate reader with integrated fluid transfer and we use the FlexStation 3 Microplate reader from Molecular Devices (http://www.moleculardevices.com). The FFA1 expressing cells of interest must be grown in black 96-well, clear-bottomed plates and we use plates from Corning (3603). We also use preracked tips in the FlexStation (Molecular Devices 9000-0911). Fluo-4AM is purchased from Invitrogen (F-14201) in special 10 × 50 μg packaging.

2.1.2. Other materials and reagents

- Live cell buffer: we use a Hepes-buffered saline solution (HBSS) which is composed of the following: 25 mM HEPES, 146 mM NaCl, 5 mM KCl, 1 mM MgSO$_4$, 2 mM sodium pyruvate, and 1.3 mM CaCl$_2$, pH 7.4. The solution is autoclaved, and 10 mM glucose is added on day of assay.
- Probenecid is purchased from Sigma-Aldrich (P8761) and is used as it is an anion transporter inhibitor. This increases the concentration of Fluo-4 within the cell as it prevents the cleaved dye from being transported out from the cell.
- Pluronic F-127 (a 20% solution in DMSO) which is purchased from Invitrogen (P-300MP) and helps in the solubilization of the Fluo-4 AM dye.
- Brilliant black BN (211842 Sigma-Aldrich).
- Clear, round-bottomed 96-well plates (Corning 3799) for compounds.
- Materials for generation of Flp-In T-REx 293 cell lines: Flp-In T-REx 293 cell line, Optimem, Lipofectamine, FFA1 in the pcDNA5/FRT/TO vector, pOG44 vector. The cell line and transfection reagents are all purchased from Invitrogen.
- Poly-D-lysine (P6407 Sigma-Aldrich).

2.2. Generation of Flp-In T-REx cell lines expressing FFA1

To generate the Flp-In T-REx cell lines, FFA1 needs to be subcloned into the pcDNA5/FRT/TO vector. The multicloning site in pcDNA5/FRT/TO is similar to the multicloning site in pcDNA3.1; therefore with access to FFA1 in pcDNA3.1, it is likely that you will be able to generate FFA1 in pcDNA5/FRT/TO without having to introduce new restriction sites. The parental Flp-In T-REx cells are maintained as described by the manufacturer in high glucose D-MEM containing 10% fetal bovine serum, 2 mM L-glutamine, 1% penicillin–streptomycin antibiotic mixture, 100 μg/ml zeocin, and 15 μg/ml blasticidin. To transfect the cells with the construct of interest, the cells should be seeded at 60–80% confluency in 10 cm^2 tissue culture dishes. Normally, the cells are plated out the day before transfection from a confluent T75 flask to allow the cells to fully adhere to the plate. The following day, cells are transfected using Lipofectamine according to the manufacturers' instructions. The cells are transfected with a mixture containing nine parts of the pOG44 vector to one part FFA1 containing pcDNA5/FRT/TO vector. The pOG44 vector contains an Flp recombinase, which inserts the gene of interest into the FRT site within the genome of the Flp-In T-REx 293 cells. After the transfection, this mixture is removed and normal growth medium without zeocin is added to the cells. The following day, the cells are trypsinized from the plate and split 1:20 and 1:50 into fresh 10 cm^2 dishes. After 48 h, the medium is replaced with selection medium, which is the growth medium described above but with 200 μg/ml hygromycin in place of zeocin. This initiates selection of cells that have the gene of interest integrated into the FRT site. At this stage, all cells that do not have the gene of interest integrated into the FRT site should be hygromycin sensitive and die. The medium on the cells is changed every 2–3 days to ensure the cells are exposed to a consistently high concentration of hygromycin. Around 2–3 weeks after the initiation of selection, cell foci should be identifiable under a normal light microscope. As the gene of interest should be integrated into the same genomic location in all hygromycin-resistant cells, once foci can be identified the cells are trypsinized from the plate and transferred into a T75 tissue culture flask. Once the cells have grown to confluency, they should be tested for gene expression in the absence and presence of the inducer doxycycline. In our own studies, we routinely express GPCRs with enhanced yellow fluorescent protein (or an equivalent) fused in-frame with the intracellular C-terminal tail of the receptor. This allows rapid screening of positively expressing cells via fluorescence microscopy and more detailed subsequent analysis of the cellular localization of the receptor construct via confocal microscopy. We routinely use up to 1 μg/ml doxycycline for 24 h to induce expression of FFA1 maximally. If the cells express the gene of interest only in the presence of the inducer, cells are further characterized in doxycycline titration and time-course studies to determine the optimal time and doxycycline concentration required for induction of receptor expression.

2.3. Cell preparation for calcium mobilization assay

To prepare the cells for measuring calcium mobilization, 24–48 h prior to the day of assay, cells harboring FFA1 are seeded into poly-D-lysine-coated black-walled, clear-bottomed 96-well plates to be used in the assay. One T75 flask of cells is trypsinized to remove cells from the flask and these are centrifuged for 5 min at $288 \times g$. Medium is removed and the cell pellet resuspended in 5–6 ml of normal growth medium. Take 1 ml of resuspended cells and add 9 ml of normal growth media and mix well to ensure even distribution of the cells. To induce expression of the receptor, we include an appropriate concentration of doxycycline at this stage. Seed 100 µl of the diluted cells into each well of the assay plate using a repeater pipette; we use an Eppendorf Multipette repeater pipette. We do not routinely count cells but find a 90–100% confluent T75 flask to contain enough cells for 5–6 × 96-well plates. If you require more defined control of cell number, count cells with a hemocytometer and dilute cells to aliquot 30,000–50,000 cells per well and seed into plates as described above. After cells are seeded, the plates are incubated for 24–48 h in a humidified atmosphere at 37 °C/5% CO_2 to allow cells to attach and grow to 90–100% confluency. The assay works most effectively with a near confluent monolayer of cells.

2.4. Measuring changes in intracellular calcium

2.4.1. Loading cells with Fluo-4

For each 96-well plate, around 20 ml of HBSS is required which contains probenecid and Brilliant Black BN. This is prewarmed to 37 °C and is prepared as follows. A 250 mM probenecid solution is prepared fresh for each experimental day in a 1:1 ratio of 1 M NaOH:HBSS, and 200 µl of this is added to the 20 ml of HBSS. Brilliant Black BN can be prepared as a 500 mM solution, aliquoted, and stored at 20 °C. Brilliant Black BN is required at a 0.5 mM final concentration; therefore, 200 µl of the 500 mM solution is added to the HBSS containing probenecid. Each 50 µg tube of Fluo-4AM is reconstituted by adding 23 µl DMSO and 23 µl Pluronic F-127 to produce a 1 mM solution with 10% Pluronic F-127 in DMSO and 23 µl of this is added to 10 ml of HBSS plus probenecid plus Brilliant Black BN to give a final Fluo-4AM concentration of 2.3 µM.

To prepare the cells for loading with Fluo-4, remove normal growth medium carefully from cells by aspiration to avoid dislodging cells. To each well add 100 µl of the HBSS plus Fluo-4AM plus probenecid plus Brilliant Black BN using a Multipette, cover the plate in aluminum foil to prevent photobleaching of the dye and cells are incubated at 37 °C (no CO_2 is required) for 30–45 min.

2.4.2. Preparing compounds, setting up FlexStation protocol, and recording calcium mobilization in Fluo-4 loaded cells

Whilst cells are loading with Fluo-4, compounds are prepared at 6× final concentration in HBSS plus probenecid plus Brilliant Black BN and transferred into a clear, round-bottomed 96-well compound plate in required wells. Each well should always contain at least 30 μl more compound than required. We typically perform concentration–response curves that contain seven different concentrations of compound in triplicate and one basal control, also in triplicate. This allows four concentration–response studies to be performed per 96-well plate. Also, during the loading of the cells with Fluo-4, switch on the FlexStation and turn on the temperature control to allow the system to reach 37 °C. Also, place a new rack of tips in the "tip rack" chamber of the FlexStation and the compound plate in the "compound" chamber.

Once the cells have been loaded with Fluo-4 for the required time, the plate is transferred into the "reading chamber" of the FlexStation to allow cells to equilibrate to the change in temperatures for around 10 min. We normally run a single read, endpoint program prior to the full 5-min calcium mobilization read. We currently use SoftMax Pro version 4.8 to control the FlexStation. The endpoint program is set up with the wavelength set to 485 nm excitation, 520 nm emission with the auto cut-off set to 515 nm, and in the sensitivity settings, the readings are set to normal (6) and the PMT sensitivity is set to high. This endpoint read gives an idea of the loading of Fluo-4 across the plate and identifies any wells in which the fluorescence is significantly lower, allowing these to be eliminated during data analysis if necessary. The protocol we use as standard to measure calcium mobilization is set up as follows. The program is a Flex protocol with the wavelengths and sensitivity set as the endpoint read program. In the timing section, set the run time to 200 s with an interval of 1.52 s, which gives 152 reads. In the assay plate type section, there are preloaded settings of common plates and select "96 well Costar blk/clrbtm" and in the compound source section select "Costar 96 Ubtm clear.3 mL". For the compound transfer, set the initial volume in the assay plate to 100 μl and set the number of transfers to 1. In the transfer settings window, set the pipette height to 110 μl, the transfer volume to 20 μl, the transfer rate to 3, and the transfer time point to 15 s. You will also need to set up the compound and tip columns to ensure that the correct compound is transferred to the correct wells. Once the program is established and the cells have equilibrated with the change in temperature, the "read" button is pressed to start the program. A sharp rise in fluorescence, relating to a rise in intracellular calcium, should be observed about 5–10 reads (10–20 s) after the addition of the compounds and the fluorescence will diminish over time and may return to basal by 60 s after the addition of compound.

After the plate has read, the data needs to be extracted from the SoftMax Pro program. The data file will initially show both the fluorescence trace and a number underneath. This is the maximum fluorescence minus the minimum fluorescence (max − min) values. Click on the "display" button and choose "reduced" in the window that comes up. The plate map will now only show the max − min value. These values can be copied directly into a curve fitting software package such as GraphPad Prism where the max − min values are plotted as a function of concentration to generate concentration–response curves.

3. MEASURING DIRECT ACTIVATION OF G PROTEINS VIA FFA1

Below is a detailed description for carrying out a [^{35}S]GTPγS-binding assay on heterologously expressed FFA1-Gα_q fusion protein with an immunocapture step as used in our study into the pharmacology and potential constitutive activity of FFA1 (Stoddart et al., 2007). Conformational changes that are associated with agonist binding at GPCRs leads to activation of G proteins by promoting the dissociation of guanosine diphosphate (GDP) from the Gα subunit and this allows the binding of guanosine triphosphate (GTP). GTP-bound Gα is generally believed to dissociate from the corresponding G protein Gβγ complex and both the GTP-bound Gα and Gβγ dimer may regulate downstream pathways, including those that result in changes in the cAMP concentration within the cells and those that initiate calcium release. The intrinsic GTPase activity of the Gα subunit hydrolyses GTP to GDP and leads to a re-association of the Gα with Gβγ dimer which completes the G protein cycle. Poorly hydrolyzed analogues of GTP, such as GTPγS, can be used to trap Gα in the active form, and using a radiolabeled form of GTPγS, [^{35}S]GTPγS, allows the direct monitoring of G protein activation by a GPCR. As many GPCRs are expressed at low levels, it is often necessary to use heterologously expressed receptors to perform [^{35}S]GTPγS-binding assays effectively and to generate significant signal above background (Milligan, 2003b). Furthermore, historically it has been difficult to utilize [^{35}S]GTPγS-binding assays to explore the pharmacology of Gα_q- and Gα_s-coupled GPCRs as these G proteins have relatively low levels of basal guanine nucleotide exchange and to overcome this specific antibodies to Gα subunits have been employed to allow immunoprecipitation of the target G protein and improve signal to background in such studies (Milligan, 2003b). It has become relatively common to use GPCR-G protein α subunit fusion proteins in such [^{35}S]GTPγS-binding assays (Milligan et al., 2004, 2007). This ensures the close proximity of the receptor with the G protein of interest. To date, we are the only laboratory to have used a [^{35}S]GTPγS-binding assay to study the function of FFA1 (Stoddart et al., 2007). As mentioned in the introduction, in the absence of BSA FFA1-Gα_q displays high levels of GTPγS binding which could not be increased substantially by the addition of FFA1 agonist ligands. The presence of BSA reduced the basal loading of [^{35}S]GTPγS onto FFA1-Gα_q, and the addition of exogenous fatty acids or nonfatty acid agonist ligands in the presence of BSA allowed concentration–response curves to be generated for such ligands (Fig. 28.3). The method we describe here was optimized using a heterologously expressed FFA1-Gα_q fusion protein, although we have also performed studies on cells (such as INS-1E) that endogenously express FFA1. The method used for endogenously expressed FFA1 was essentially the same but the amount of membrane protein added per well was increased.

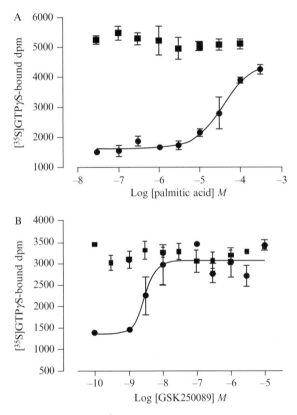

Figure 28.3 Examples of typical concentration–response curves generated using the described [^{35}S]GTPγS-binding assay. Membranes were prepared from HEK293 cells transiently transfected to express FFA1-Gα_q. [^{35}S]GTPγS-binding studies were performed in the absence (squares) or presence (circles) of 10 μM fatty acid-free BSA with increasing concentrations of (A) palmitic acid or (B) GSK250089A. No increase in [^{35}S]GTPγS can be observed in the absence of 10 μM BSA at any agonist concentration.

3.1. Required materials

3.1.1. Material and equipment required for transient transfection of an FFA1 fusion protein and membrane preparation

- A model cell system for transfections or cells endogenously expressing FFA1: we use HEK293 cells as our model cell system (ATCC CRL-1573). Cell growth medium and supplements are purchased from Invitrogen.
- Materials for transfection: Lipofectamine and Optimem are purchased from Invitrogen.
- DNA encoding FFA1 in an appropriate mammalian expression vector such as pcDNA3.1.
- Glass-on-teflon homogenizer (HOM3532 Scientific Laboratory Supplies).
- 15 ml centrifuge tubes.
- Centrifuge for harvesting cells: we use a refrigerated Eppendorf 5804 centrifuge.
- Ultracentrifuge and ultracentrifuge tubes: we use a Beckman Coulter optima LTX benchtop ultracentrifuge.
- Disposable cell scrapers.
- 25 gauge needles and 1 ml syringe.
- Materials for estimating protein concentration.

3.1.2. Equipment required for [^{35}S]GTPγS assay

A refrigerated microcentrifuge is required for the immunocapture step of the assay, we use an Eppendorf 5430R refrigerated centrifuge with a rotor to fit 1.5 ml microcentrifuge tubes. To incubate the samples, a water bath is required and a normal lab water bath that can be set to 30 °C will suffice. You will need a rotor fitted with a cradle for 1.5 ml microcentrifuge tubes, such as the Stuart Rotator (MIX2040) with the microcentrifuge tube cradle (MIX2062) from Scientific Laboratory Supplies. A liquid scintillation counter such as the Beckman LS 6500 Scintillation counter is required to estimate the levels of radioactivity in the samples.

3.1.3. [^{35}S]GTPγS and other required reagents and chemicals

- [^{35}S]GTPγS is purchased from Perkin Elmer Life and Analytical Sciences (NEG030H). Stock [^{35}S]GTPγS is diluted in a buffer composed of 10 mM Tris (pH 7.46), 10 mM DTT, and we store the diluted radiochemical in 100 µl aliquots at −80 °C.
- GDP (G7127) and GTPγS (G3776) both from Sigma-Aldrich.
- Fatty acid-free BSA (10775835001 Roche).

- The $G\alpha_{q/11}$ antisera that we use were produced in-house in New Zealand white rabbits using a decapeptide corresponding to the C-terminal of $G\alpha_{q/11}$ as the antigen (Mitchell et al., 1993). Commercial $G\alpha_{q/11}$ antisera and antibodies can be purchased from a range of sources including Santa Cruz Biotechnology (http://www.scbt.com/table-heterotrimeric_g_protein.html). The conditions for immunoprecipitation described below are optimized for our in-house antisera; therefore, the use of other antisera/antibodies will require optimization of amount of reagent required.
- Protein G (17-0618-01 GE Healthcare Life Sciences).
- Scintillation fluid: we use Ultima Gold (6013321) from Perkin Elmer.
- Pansorbin (507858 Merck Chemicals).
- Protease inhibitor cocktail tablets: we use Complete, EDTA-free tablets from Roche (11 873 580 001).
- 1.5 ml microcentrifuge tubes: care should be taken in selection of microcentrifuge tubes, as certain brands do not seal adequately and can leak during the immunoprecipitation step.
- 30 ml scintillation vials and caps.

3.1.4. Buffers required for assay

For harvesting cells, 1 × PBS is used: 13.7 mM NaCl, 0.27 mM KCl, 0.15 mM KH$_2$PO$_4$, 1.02 mM Na$_2$HPO$_4$, pH 7.4 which is stored at 4 °C. Crude cell membranes are made in a Tris-EDTA (TE) buffer: 10 mM Tris, 0.1 mM EDTA plus protease cocktail tablets, pH 7.4, which is stored at 4 °C.

The [^{35}S]GTPγS-binding assay requires three buffers at different steps, namely, assay buffer, solubilization buffer, and bead suspension buffer, and these are composed as follows:

- Assay buffer: 200 mM HEPES, 30 mM MgCl$_2$, 100 mM NaCl, 2 mM ascorbic acid. It is made up at 10× final assay concentration and is stored in 5 ml aliquots at −20 °C.
- Solubilization buffer: 100 mM Tris HCl, 200 mM NaCl, 1 mM EDTA, 1.25% (v/v) Nonidet P-40, protease cocktail tablets, pH 7.4, and stored at 4 °C.
- Bead suspension buffer: 2% BSA (w/v), 0.1% NaN$_3$ (w/v), protease cocktail tablet, and stored at 4 °C.

3.2. Transient expression of FFA1 and membrane preparation

On day prior to transfection, split HEK293 cells into 10 cm^2 dishes to be between 60% and 80% confluent the following day. Once cells have achieved the desired confluency, transfect the cells with DNA of interest using Lipofectamine according to the manufacturers' instructions. Cells are harvested 48 h posttransfection by discarding the growth medium and adding 10 ml ice-cold PBS. Cells are scraped from the flask using a disposable cell scraper and transferred to a 15-ml centrifuge tube. Cells are then pelleted by centrifugation for 5 min at 288× g at 4 °C. The supernatant is discarded and the pellets washed twice with fresh PBS with centrifugation steps in between washes to repellet the cells. We find that the yield for membranes is increased if the cell pellet is frozen to -80 °C prior to membrane preparation. If membrane preparation is not performed, immediately cell pellets can be stored at -80 °C until required.

[^{35}S]GTPγS-binding assays are carried out on crude cell membranes which are prepared as follows. The cell pellet containing cells expressing FFA1 is thawed on ice and resuspended in TE buffer; typically for one 10 cm^2 dish of cells add 1 ml of TE buffer. The resuspended cells are homogenized by 50 passes on a glass-on-teflon homogenizer. To remove the unbroken cells and nuclei, the resulting suspension is centrifuged at 288× g for 10 min. The supernatant is transferred into ultracentrifuge tubes; the tubes are balanced and centrifuged at 50,000× g for 30 min in an ultracentrifuge. This high-speed spin pellets the crude membranes from the supernatant and the pellet is resuspended in TE buffer (~200 μl per 10 cm^2 dish of cells). The membrane pellet is usually quite dense and to ensure even resuspension, the membranes are passed through a 25 gauge needle. The protein concentration of the preparation is then determined. We use the BCA (bincihoninc acid) method for protein concentration quantification, although any protein quantification method should suffice. Once protein concentration has been determined, membranes are diluted to 1 μg/μl in TE buffer and stored at -80 °C until required.

3.3. [^{35}S]GTPγS-binding assay with immunoprecipitation step

The [^{35}S]GTPγS-binding assay is carried out on membranes prepared as detailed in Section 3.2. The assay is set up in 1.5 ml microcentrifuge tubes and each agonist concentration is carried out in triplicate. The first thing to prepare is 2× assay buffer that contains 10× assay buffer, 2 μM GDP, and 50 nCi [^{35}S]GTPγS. This is maintained on ice until required. For each tube, you require 50 μl 2× assay buffer and, for example, if you require sufficient assay buffer for 45 tubes, the 2× assay buffer will contain 500 μl 10× assay buffer, 125 μl of 40 μM GDP stock, 50 μl [^{35}S]GTPγS (if at 100% original concentration), and 1825 μl H$_2$O. The amount of [^{35}S]GTPγS added is dependent on the specific activity of the radioisotope and can be calculated using one of the many radioactivity calculators found online, for example, the one provided by GraphPad (http://www.graphpad.com/quickcalcs/radcalcform.cfm). If fatty acid-free BSA is to be included, it should be added to the 2× assay buffer also at 2× final assay concentration.

The amount of cell membranes used per tube can vary vastly depending on the expression level of FFA1. For membranes from cells transiently expressing FFA1-Gα_q or from HEK293 Flp-In T-REx cells induced to express FFA1, we find that using between 2.5 μg and 10 μg of membranes per tube provides a robust signal. If 10 μg of cell membranes is to be used, dilute the membranes to 0.4 μg/μl in TE buffer because 25 μl of membranes is added per tube. Finally, agonists or antagonists are prepared at 10× the required final concentration in H_2O. Each of the components of the assay is kept on ice.

To each tube, 50 μl 2× assay buffer, 25 μl cell membranes, 10 μl agonist/antagonist/vehicle, and 15 μl H_2O are added. To determine nonspecific binding, an excess of unlabeled GTPγS is used. In tubes used to determine nonspecific binding, 10 μl of 1 mM GTPγS is added and the amount of water added is reduced correspondingly. The tubes are capped and briefly vortexed and then placed in a water bath set to 30 °C for 30 min. During the incubation time, the stop buffer, which is ice-cold 1× assay buffer plus a protease inhibitor tablet, must be prepared, and also 0.2% SDS must be added to the solubilization buffer; for 45 tubes, prepare 2.5 ml of solubilization buffer plus SDS. After the 30-min incubation time, tubes are removed from the water bath and immediately placed on ice and 500 μl ice-cold stop buffer is added per tube. The tubes are then centrifuged at 16,000× g at 4 °C in a microcentrifuge for 10 min to pellet the membranes. The tubes are then placed back on ice and the supernatant removed from the tubes to leave only the membrane pellet. Depending on the amount of membranes used, it may not be easy to see the membrane pellet. If the membrane pellet cannot be seen, the majority of the supernatant is carefully removed away from the section of the tube that the membranes would be expected to pellet to. The membranes are solubilized by the addition of 50 μl solubilization buffer plus SDS to each tube and rotating the tubes for 1 h at 4 °C. Meanwhile, the preclear solution is prepared, which for each tube is 20 μl of Pansorbin beads and 40 μl bead buffer. After 1 h, tubes are removed from the rotating wheel, 60 μl of the preclear solution is added, and the tubes are rotated again for an additional 1 h at 4 °C. This step removes a large portion of the nonspecific antigens that may interact with the G protein-specific antiserum/antibody. During this 1 h rotation, a new set of clean microcentrifuge tubes and the antiserum/antibody solution are prepared. For 45 tubes, the antiserum/antibody solution is composed of 500 μl protein G, 37 μl Gα_q-specific antibody, and 1963 μl bead buffer. The protein G beads are stored in ethanol, which must be removed before adding to the solubilized membranes. This is achieved by gently centrifuging the required amount of beads in a microcentrifuge for 1 min, removing the ethanol carefully, and adding the equivalent amount of bead buffer. This is repeated twice more. After the 1 h preclear step, the tubes are centrifuged at 4 °C at 16,000× g for 2 min. The supernatant is carefully transferred to the clean microcentrifuge tubes and 50 μl of the antiserum/antibody solution is added per tube. The tubes are again placed on a rotating wheel and incubated at 4° C overnight to allow the immunocapture of the $G\alpha_q$.

After the overnight incubation, the protein G beads are washed twice in solubilization buffer as follows to remove any nonprotein G-bound antibody. The tubes are removed from the rotating wheel and centrifuged for 5 min at 16,000× g at 4 °C. The supernatant is removed carefully as the protein G beads do not form a solid pellet. Then 500 μl of solubilization buffer is added per tube, each of which is gently vortexed to ensure the beads are washed. The tubes are centrifuged again for 5 min as above, the supernatant removed again, and another 500 μl solubilization buffer added. The tubes are centrifuged for a final time as above, and all the supernatant carefully removed. The routinely used scintillation counters are not equipped to count samples in microcentrifuge tubes and we find the easiest way to count the samples is to place each of the tubes into a 30 ml scintillation tube rather than transferring the beads. Therefore, to the resulting protein G pellet 1 ml of scintillation fluid is added per tube. The beads are required to be in suspension in the scintillation fluid so vigorous vortexing or "flicking" the bottom of the tubes is required to dislodge the beads from the bottom of the tubes. Normal microcentrifuge tubes do not fit easily into 30 ml scintillation tubes and to overcome this, the hinge region of the tubes needs to be removed. This can be done by cutting the hinge off with a pair of scissors or by using a scalpel blade. After the hinge has been removed, each tube is placed in a separate scintillation tube and the amount of radioactivity per sample is estimated by liquid scintillation counting.

The counts obtained from the liquid scintillation counter can be entered directly into graph fitting software such as GraphPad Prism where the counts in dpm are plotted as a function of concentration to gain concentration–response curves.

REFERENCES

Bennett, K. A., Langmead, C. J., Wise, A., and Milligan, G. (2009). Growth hormone secretagogues and growth hormone releasing peptides act as orthosteric super-agonists but not allosteric regulators for activation of the G protein $G\alpha_{o1}$ by the ghrelin receptor. *Mol. Pharmacol.* **76**(4), 802–811.

Bond, R. A., and Ijzerman, A. P. (2006). Recent developments in constitutive receptor activity and inverse agonism, and their potential for GPCR drug discovery. *Trends Pharmacol. Sci.* **27**(2), 92–96.

Briscoe, C. P., Tadayyon, M., Andrews, J. L., Benson, W. G., Chambers, J. K., Eilert, M. M., Ellis, C., Elshourbagy, N. A., Goetz, A. S., Minnick, D. T., Murdock, P. R., Sauls, H. R., et al. (2003). The orphan G protein-coupled receptor GPR40 is activated by medium and long chain fatty acids. *J. Biol. Chem.* **278**(13), 11303–11311.

Briscoe, C. P., Peat, A. J., McKeown, S. C., Corbett, D. F., Goetz, A. S., Littleton, T. R., McCoy, D. C., Kenakin, T. P., Andrews, J. L., Ammala, C., Fornwald, J. A., Ignar, D. M., et al. (2006). Pharmacological regulation of insulin secretion in MIN6 cells through the fatty acid receptor GPR40: Identification of agonist and antagonist small molecules. Br. J. Pharmacol. **148**(5), 619–628.

Brown, A. J., Goldsworthy, S. M., Barnes, A. A., Eilert, M. M., Tcheang, L., Daniels, D., Muir, A. I., Wigglesworth, M. J., Kinghorn, I., Fraser, N. J., Pike, N. B., Strum, J. C., et al. (2003). The orphan G protein-coupled receptors GPR41 and GPR43 are activated by propionate and other short chain carboxylic acids. J. Biol. Chem. **278**(13), 11312–11319.

Canals, M., and Milligan, G. (2008). Constitutive activity of the cannabinoid CB1 receptor regulates the function of co-expressed mu opioid receptors. J. Biol. Chem. **283**(17), 11424–11434.

Chen, A., Gao, Z. G., Barak, D., Liang, B. T., and Jacobson, K. A. (2001). Constitutive activation of A(3) adenosine receptors by site-directed mutagenesis. Biochem. Biophys. Res. Commun. **284**(3), 596–601.

Edfalk, S., Steneberg, P., and Edlund, H. (2008). Gpr40 is expressed in enteroendocrine cells and mediates free fatty acid stimulation of incretin secretion. Diabetes **57**(9), 2280–2287.

Ellis, J., Pediani, J. D., Canals, M., Milasta, S., and Milligan, G. (2006). Orexin-1 receptor-cannabinoid CB1 receptor heterodimerization results in both ligand-dependent and -independent coordinated alterations of receptor localization and function. J. Biol. Chem. **281**(50), 38812–38824.

Feng, D. D., Luo, Z. Q., Roh, S. G., Hernandez, M., Tawadros, N., Keating, D. J., and Chen, C. (2006). Reduction in voltage-gated K + currents in primary cultured rat pancreatic beta-cells by linoleic acids. Endocrinology **147**(2), 674–682.

Flodgren, E., Olde, B., Meidute-Abaraviciene, S., Winzell, M. S., Ahren, B., and Salehi, A. (2007). GPR40 is expressed in glucagon producing cells and affects glucagon secretion. Biochem. Biophys. Res. Commun. **354**(1), 240–245.

Fujiwara, K., Maekawa, F., and Yada, T. (2005). Oleic acid interacts with GPR40 to induce Ca2+ signaling in rat islet beta-cells: Mediation by PLC and L-type Ca2+ channel and link to insulin release. Am. J. Physiol. Endocrinol. Metab. **289**(4), E670–E677.

Hall, D. A., and Strange, P. G. (1997). Evidence that antipsychotic drugs are inverse agonists at D-2 dopamine receptors. Br. J. Pharmacol. **121**(4), 731–736.

Hardy, S., St-Onge, G. G., Joly, E., Langelier, Y., and Prentki, M. (2005). Oleate promotes the proliferation of breast cancer cells via the G protein-coupled receptor GPR40. J. Biol. Chem. **280**(14), 13285–13291.

Hein, P., Goepel, M., Cotecchia, S., and Michel, M. C. (2001). A quantitative analysis of antagonism and inverse agonism at wild-type and constitutively active hamster alpha (1B)-adrenoceptors. Naunyn Schmiedebergs Arch. Pharmacol. **363**(1), 34–39.

Holst, B., Cygankiewicz, A., Jensen, T. H., Ankersen, M., and Schwartz, T. W. (2003). High constitutive signaling of the ghrelin receptor: Identification of a potent inverse agonist. Mol. Endocrinol. **17**(11), 2201–2210.

Hu, H., He, L. Y., Gong, Z., Li, N., Lu, Y. N., Zhai, Q. W., Liu, H., Jiang, H. L., Zhu, W. L., and Wang, H. Y. (2009). A novel class of antagonists for the FFAs receptor GPR40. Biochem. Biophys. Res. Commun. **390**(3), 557–563.

Itoh, Y., Kawamata, Y., Harada, M., Kobayashi, M., Fujii, R., Fukusumi, S., Ogi, K., Hosoya, M., Tanaka, Y., Uejima, H., Tanaka, H., Maruyama, M., et al. (2003). Free fatty acids regulate insulin secretion from pancreatic beta cells through GPR40. Nature **422**(6928), 173–176.

Kebede, M., Alquier, T., Latour, M. G., Semache, M., Tremblay, C., and Poitout, V. (2008). The fatty acid receptor GPR40 plays a role in insulin secretion in vivo after high-fat feeding. Diabetes **57**(9), 2432–2437.

Kenakin, T. (2001). Inverse, protean, and ligand-selective agonism: Matters of receptor conformation. *FASEB J.* **15**(3), 598–611.

Kotarsky, K., Nilsson, N. E., Flodgren, E., Owman, C., and Olde, B. (2003). A human cell surface receptor activated by free fatty acids and thiazolidinedione drugs. *Biochem. Biophys. Res. Commun.* **301**(2), 406–410.

Lan, H., Hoos, L. M., Liu, L., Tetzloffl, G., Hu, W. W., Abbondanzo, S. J., Vassileva, G., Gustafson, E. L., Hedrick, J. A., and Davis, H. R. (2008). Lack of FFAR1/GPR40 does not protect mice from high-fat diet-induced metabolic disease. *Diabetes* **57**(11), 2999–3006.

Latour, M. G., Alquier, T., Oseid, E., Tremblay, C., Jetton, T. L., Luo, J., Lin, D. C. H., and Poitout, V. (2007). GPR40 is necessary but not sufficient for fatty acid stimulation of insulin secretion in vivo. *Diabetes* **56**(4), 1087–1094.

Le Poul, E., Loison, C., Struyf, S., Springael, J. Y., Lannoy, V., Decobecq, M. E., Brezillon, S., Dupriez, V., Vassart, G., Van Damme, J., Parmentier, M., and Detheux, M. (2003). Functional characterization of human receptors for short chain fatty acids and their role in polymorphonuclear cell activation. *J. Biol. Chem.* **278**(28), 25481–25489.

Lee, P. H. (2009). Label-free optical biosensor: a tool for G protein-coupled receptors pharmacology profiling and inverse agonists identification. *J. Recept. Signal Transduct. Res.* **29**(3–4), 146–153.

Lopez-Gimenez, J. F., Vilaro, M. T., and Milligan, G. (2008). Morphine desensitization, internalization, and down-regulation of the mu opioid receptor is facilitated by serotonin 5-hydroxytryptamine(2A) receptor coactivation. *Mol. Pharmacol.* **74**(5), 1278–1291.

May, L. T., Briddon, S. J., and Hill, S. J. (2010). Antagonist selective modulation of adenosine A(1) and A(3) receptor pharmacology by the food dye Brilliant Black BN: Evidence for allosteric interactions. *Mol. Pharmacol.* **77**(4), 678–686.

Milligan, G. (2003a). Constitutive activity and inverse agonists of G protein-coupled receptors: A current perspective. *Mol. Pharmacol.* **64**(6), 1271–1276.

Milligan, G. (2003b). Extending the utility of [^{35}S]GTPγS binding assays. *Trends Pharmacol. Sci.* **24**, 87–90.

Milligan, G., Feng, G.-J., Ward, R. J., Sartania, N., Ramsay, D., McLean, A. J., and Carrillo, J. J. (2004). G protein-coupled receptor fusion proteins in drug discovery. *Curr. Pharm. Des.* **10**, 1989–2001.

Milligan, G., Parenty, G., Stoddart, L. A., and Lane, J. R. (2007). Novel pharmacological applications of G-protein-coupled receptor-G protein fusions. *Curr. Opin. Pharmacol.* **7**, 521–526.

Mitchell, F. M., Buckley, N. J., and Milligan, G. (1993). Enhanced degradation of the phosphoinositidase c-linked guanine-nucleotide-binding protein g(q)alpha g(11)alpha following activation of the human m1-muscarinic acetylcholine-receptor expressed in Cho cells. *Biochem. J.* **293**, 495–499.

Nolan, C. J., Madiraju, M. S. R., Delghingaro-Augusto, V., Peyot, M. L., and Prentki, M. (2006). Fatty acid signaling in the beta-cell and insulin secretion. *Diabetes* **55**, S16–S23.

Rocheville, M., and Jerman, J. C. (2009). 7TM pharmacology measured by label-free: A holistic approach to cell signalling. *Curr. Opin. Pharmacol.* **9**(5), 643–649.

Salehi, A., Flodgren, E., Nilsson, N. E., Jimenez-Feltstrom, J., Miyazaki, J., Owman, C., and Olde, B. (2005). Free fatty acid receptor 1 (FFA(1)R/GPR40) and its involvement in fatty-acid-stimulated insulin secretion. *Cell Tissue Res.* **322**(2), 207–215.

Schnell, S., Schaefer, M., and Schofl, C. (2007). Free fatty acids increase cytosolic free calcium and stimulate insulin secretion from beta-cells through activation of GPR40. *Mol. Cell. Endocrinol.* **263**(1–2), 173–180.

Shapiro, H., Shachar, S., Sekler, I., Hershfinkel, M., and Walker, M. D. (2005). Role of GPR40 in fatty acid action on the beta cell line INS-1E. *Biochem. Biophys. Res. Commun.* **335**(1), 97–104.

Smith, N. J., Stoddart, L. A., Devine, N. M., Jenkins, L., and Milligan, G. (2009). The action and mode of binding of thiazolidinedione ligands at Free Fatty Acid receptor 1. *J. Biol. Chem.* **284**(26), 17527–17539.

Steneberg, R., Rubins, N., Bartoov-Shifman, R., Walker, M. D., and Edlund, H. (2005). The FFA receptor GPR40 links hyperinsulinemia, hepatic steatosis, and impaired glucose homeostasis in mouse. *Cell Metab.* **1**(4), 245–258.

Stoddart, L. A., Brown, A. J., and Milligan, G. (2007). Uncovering the pharmacology of the G protein-coupled receptor GPR40: High apparent constitutive activity in guanosine 5′-O-(3-[S-35] thio) triphosphate binding studies reflects binding of an endogenous agonist. *Mol. Pharmacol.* **71**(4), 994–1005.

Stoddart, L. A., Smith, N. J., Jenkins, L., Brown, A. J., and Milligan, G. (2008a). Conserved polar residues in transmembrane domains V, VI, and VII of free fatty acid receptor 2 and free fatty acid receptor 3 are required for the binding and function of short chain fatty acids. *J. Biol. Chem.* **283**(47), 32913–32924.

Stoddart, L. A., Smith, N. J., and Milligan, G. (2008b). International union of pharmacology. LXXI. Free fatty acid receptors FFA1,-2, and-3: Pharmacology and pathophysiological functions. *Pharmacol. Rev.* **60**(4), 405–417.

Tan, C. P., Feng, Y., Zhou, Y. P., Eiermann, G. J., Petrov, A., Zhou, C. Y., Lin, S. N., Salituro, G., Meinke, P., Mosley, R., Akiyama, T. E., Einstein, M., et al. (2008). Selective small-molecule agonists of G protein-coupled receptor 40 promote glucose-dependent insulin secretion and reduce blood glucose in mice. *Diabetes* **57**(8), 2211–2219.

Tomita, T., Masuzaki, H., Noguchi, M., Iwakura, H., Fujikura, J., Tanaka, T., Ebihara, K., Kawamura, J., Komoto, I., Kawaguchi, Y., Fujimoto, K., Doi, R., et al. (2005). GPR40 gene expression in human pancreas and insulinoma. *Biochem. Biophys. Res. Commun.* **338**(4), 1788–1790.

Ward, R. J., Alvarez-Curto, E., and Milligan, G. (2010). Using the Flp-In[TM] T-Rex[TM] system to regulate GPCR expression. *In* "Receptor Signal Transduction Protocols," (G. Willars and J. A. Challiss, eds.), 3rd edn. Methods in Molecular Biology. Humana Press.

Yonezawa, T., Katoh, K., and Obara, Y. (2004). Existence of GPR40 functioning in a human breast cancer cell line, MCF-7. *Biochem. Biophys. Res. Commun.* **314**(3), 805–809.

CHAPTER TWENTY-NINE

CONSTITUTIVE ACTIVITY OF TRP CHANNELS: METHODS FOR MEASURING THE ACTIVITY AND ITS OUTCOME

Shaya Lev *and* Baruch Minke

Contents

1. Introduction	592
2. TRP Channels and Cellular Degeneration	594
3. Constitutive TRP Channel Activity Which Does Not Lead to Cellular Degeneration	595
3.1. Examples of TRP channels displaying constitutive activity	596
3.2. Methods for determining constitutive activity	598
4. Constitutive TRP Channel Activity Which Leads to Cellular Degeneration	604
4.1. Examples of TRP channels which exert a degenerating effect	605
4.2. Methods for determination of the degenerating effect	606
4.3. Summary of the methods for viewing cellular degeneration	609
Acknowledgments	609
References	609

Abstract

TRP channels participate in many cellular processes including cell death. These channels mediate these effects mainly by changing the cellular concentration of Ca^{2+}, a prominent cellular second messenger. Measuring the current–voltage relationship and state of activation of TRP channels is of utmost importance for evaluating their contribution to a cellular process within a spatial and temporal context. The study of TRP channels and characterization of their mode of activation will benefit and progress our understanding of each channel's role in specific cellular mechanisms. Many TRP channels exhibit constitutive activity, which is mostly observed in cell-based expression systems. This constitutive activity can lead, in many cases, to cellular degeneration, which can be readily observed morphologically and by biochemical assays. This chapter describes in

Department of Medical Neurobiology and the Kühne Minerva Center for Studies of Visual Transduction, Institute of Medical Research Israel-Canada (IMRIC), The Edmond & Lily Safra Center for Brain Sciences (ELSC), The Hebrew University, Jerusalem, Israel brief different modes of TRP channel activity and their current–voltage relationships. The chapter outlines methods for visualizing this activity and methods to correlate between TRP channel activity and cell death, and it illustrates mechanisms that prevent cell death in spite of constitutive activity. Finally, it describes methods for qualitatively and quantitatively measuring the accompanied cellular degeneration.

1. INTRODUCTION

TRP channels constitute a large superfamily of channel proteins with diverse roles in many transduction and sensory mechanisms. The superfamily, which is conserved through evolution, consists of seven subfamilies and its members are expressed in many cell types, including excitable as well as nonexcitable cells (Damann et al., 2008). These channels participate in many sensory modalities, and they either open directly in response to ligands or physical stimuli (e.g., temperature, osmotic pressure, or noxious substance) or, indirectly, downstream of a signal transduction cascade (e.g., phototransduction; Katz and Minke, 2009; Ramsey et al., 2006). In many reports on various TRP channels, the channels reveal constitutive activity, mainly when expressed in tissue culture cells. Moreover, since usually there is no easy access to the channels where they natively reside (inaccessible membrane, tissue, or organelle), it is difficult to determine their actual activity under physiologically relevant conditions. A good example for a system in which the native signal can be accessed easily is the phototransduction cascade in the *Drosophila* eye, in which TRP and TRP-like (TRPL) channels are activated in response to light (Hardie and Minke, 1992; Niemeyer et al., 1996). In the photoreceptor cells, the channels are closed in the dark and open upon illumination. As will be discussed in brief below, we use electroretinogram (ERG) and whole-cell recordings from isolated grouped photoreceptor cells, in order to characterize the electrical activity of the cells. This activity changes in response to light, due to light-induced channel openings (Cosens and Manning, 1969; Hardie and Minke, 1992; Minke et al., 1975). Nevertheless, it must be pointed out that complete characterization, which entails single-channel analysis, is still difficult to obtain, due to inaccessible photoreceptor membrane which expresses the channels (Delgado and Bacigalupo, 2009). On the other hand, the TRPML1 channel protein is an example of a channel, which is in the most part nonaccessible for electrophysiological research in its native surroundings (Dong et al., 2008) due to exclusive native expression in intracellular vesicles, such as lysosomes (Vergarajauregui and Puertollano, 2006). Another difficulty arising from measurements of TRP channel activity in native surroundings is their spatial and temporal channel activity. In native systems, the complexity of the channel regulation (activation, trafficking, and posttranslational modifications) is for the most part unknown. There is evidence showing that some TRP channels are dynamic in their location and in the activity they exhibit within a time frame. Our ability to perform experiments on these channels might be affected by their physiological state because they reveal signal-dependent translocation between the surface membrane and intracellular compartments (Bahner et al., 2002; Cronin et al., 2006; Meyer et al., 2006; Stein et al., 2006). This could hinder our ability to interpret correctly the data obtained from experiments on native systems. For these reasons, we and others perform much of the research on TRP channels in cell culture expression systems with the aim of gaining insight into channel function. The use of expression systems for TRP channel research has many advantages: (1) The channels are for the most part easily expressed in a functional manner. (2) The dissection of biophysical properties such as conductance, mean open time, and permeability are readily obtained (Parnas et al., 2007). (3) In many cases, insight into the gating mechanisms can be achieved. (4) In many cases, TRP channels exhibit basal constitutive activity on the plasma membrane, which allows investigating TRP channels whose mode of activation is unknown. In spite of these advantages, care should be taken upon interpretation of results obtained from channels expressed in any expression system, which may show behavior differing from a native system. It is therefore advisable to compare between a specific TRP channel activity in a native system and that gained from an expression system. A good example for this methodology is the research on *Drosophila* TRPL channels conducted in our lab. Many of the properties were readily obtained in expression system and then verified with physiological relevance in the photoreceptor native system. In general, a major difficulty of TRP channel research is the limited pharmacological tools (e.g., activators, inhibitors), which in many cases are nonspecific. The fact that many channels have basal activity means that they are expressed functionally. This activity is helpful in cases where TRP channels' mode of activation and pharmacology are unknown. Not all TRP channels (e.g., *Drosophila* TRP channel) have been successfully expressed in heterologous expression systems (Minke and Parnas, 2006; but see Xu et al., 1997). In such cases, the inaccessibility of these channels to single-channel investigation in native cells puts a constraint on the full characterization of the channels. Without comparing the properties of a channel in both the native and expression systems, one cannot rule out that properties obtained in an expression system are different from those of the native system. Moreover, certain channels can display differing active states in different expression systems. For example, we have observed that the *Drosophila* TRPL channel reveals constitutive activity in the Schneider 2 (S2) and insect Spodoptera Frugiperda 9 (Sf9) expression systems (Chyb et al., 1999; Hu and Schilling, 1995), whereas when expressed in Human Embryonic Kidney (HEK) cells, these channels are closed, resembling the state of the channel in the photoreceptor cells, in the dark (Agam et al., 2000; Lev et al., 2010). This example demonstrates that different expression systems can affect channel properties differently. This point has to be taken into consideration when choosing an expression system. It would be advisable to compare activities of TRP channels in several expression systems in order to gain insight into channel regulation modes. Furthermore, differences in the basal activity can also be attributed to the expression system at hand and not necessarily indicate for inherent properties. Constitutive activity of TRP channels can be attributed to the direct activation of the channel (Grimm et al., 2007), or the channel can be activated by a known or unknown upstream element, which is constitutively active. This is mostly true for those TRP channels which are activated downstream of a transduction cascade (e.g., TRPC channels which are receptor mediated). In this respect, the *Drosophila* TRPL channel expressed in S2 cells might be affected by constitutive endogenous phospholipase C (PLC) activity. Constitutive activity of TRP channels may lead to detrimental cellular degeneration because of their Ca^{2+} permeability, or because of a change in the resting membrane potential. This topic is extensively discussed below and constitutes a main subject in this review. As noted above, there are channels that show plasma membrane expression but no constitutive activity even in expression systems, such as wild-type TRPML2 and TRPML3 channels expressed in HEK cells (Lev et al., 2010).

In conclusion, constitutive activity of a TRP channel can be beneficial in facilitating its characterization but can also lead to nonphysiological effects with pathological ramifications on both cellular and higher level functions. The scope of this chapter is to review TRP channels with and without constitutive activity, which lead to cellular degeneration in some but not all cases. We will also review several methods for determining the active state of TRP channels and describe methods for determining cellular degeneration and its quantification.

2. TRP Channels and Cellular Degeneration

Constitutive and nonconstitutive TRP channel activity can lead, in many cases, to Ca^{2+}-dependent cell death, via apoptotic and secondary necrotic processes (for review, see Dadon and Minke, 2010). The degeneration is induced by Ca^{2+} overload or by change in Ca^{2+} homeostasis (Orrenius et al., 2003). Other cations entering the cells via TRP channels besides Ca^{2+} can lead to cell death (Carini et al., 1999) by changing the resting potential level of the cells (van Aken et al., 2008). The fact that most TRP channels exhibit Ca^{2+} permeability constitutes a potential role for these channels to participate in processes of cell death under physiological or pathophysiological conditions. TRPM2 was shown to affect the susceptibility of cells to death, both in native and heterologous TRPM2 expressing cells, in response to stimulation by oxidative stress (Hara et al., 2002; Zhang et al., 2003). TRPM7 has also been shown to be involved in cell death under conditions of ischemia (Sun et al., 2009). In addition, TRP channels from the TRPC subfamily have been implicated in apoptosis-mediated cell death when activated, possibly due to Ca^{2+} influx leading to Ca^{2+} overload (for review, see Dadon and Minke, 2010). As stated above, in many cases, TRP channel activation has been linked to susceptibility to cellular degeneration. In addition, there are many examples of TRP channels that possess gain-of-function mutations, which also lead to cell death in a similar fashion. This is the case for the spontaneously occurring *varitint-waddler* mutation found in mouse TRPML3 expressing cells. In these cells, degeneration underlies cellular malfunction leading to behavioral pathology such as hearing loss and impaired vestibular function (Grimm et al., 2007; Nagata et al., 2008; Xu et al., 2007). We and others have shown that this mutation is conserved among other TRPML channels, leading to cellular degeneration as well (Lev et al., 2010).

To conclude, in some cases, TRP channels actively participate in programmed cell death. While in other cases, an increase in TRP channel activity (augmentation) is detrimental. Therefore, the activity level of TRP channels is crucial for maintaining cellular mechanisms at their physiological homeostatic point in native surroundings. As mentioned above, TRP channel activity in expression system can differ from the native system. Therefore, caution must be taken when drawing conclusions from such systems.

3. Constitutive TRP Channel Activity Which Does Not Lead to Cellular Degeneration

In this section, we briefly describe several TRP channels displaying activity which is not accompanied by cellular degeneration. We outline the current–voltage relationship (I–V curve) of different TRP channel types and point to those which would probably not be associated with cellular degeneration. We also describe the methods used to measure the constitutive activity, mainly the whole-cell patch–clamp recording technique. This technique is not a typical high-throughput technique, in which a vast population of channel-expressing cells can be evaluated, although there is advanced instrumentation in this direction by multiple electrodes performing simultaneous patch-clamp recordings. The patch-clamp technique gives insight into the I–V curve of a specific channel. Its power is in unfolding basic biophysical properties of a channel. This information is used to determine the characteristics and amplitude of the constitutive activity, which can be used to predict cell viability.

3.1. Examples of TRP channels displaying constitutive activity

Since most TRP channels are permeable to Ca^{2+}, it is plausible to assume that unregulated TRP channel activity, whether it is constitutive or not, would not occur under normal physiological conditions. Nevertheless, many TRP channels reveal constitutive activity (e.g., TRPL, TRPM8, TRPV1, TRPV5/6) when expressed in tissue culture cells. The open probability of these channels can be further enhanced by enzymatic or

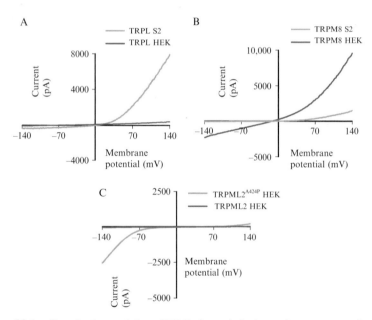

Figure 29.1 Constitutive activity of TRP channels in heterologous expression systems. (A) Representative I–V curves as measured from HEK and S2 cells expressing the *Drosophila* TRPL channel (red and green, respectively). Note that TRPL in HEK cells does not reveal constitutive channel activity, which can be initiated by linoleic acid (LA) application, whereas when expressed in S2 cells, it displays a pronounce outwardly rectifying current. (B) Representative I–V curves as measured from S2 and HEK cells expressing TRPM8 (green and red, respectively). Note that TRPM8 reveals constitutive activity in both expression systems albeit differences in current magnitude. (C) Representative I–V curves as measured from HEK cells expressing wild-type human TRPML2 (red) and TRPML2$_{A424P}$ (i.e., the Va mutant (green)). Note how the *varitint-waddler* paralogous mutation conveys constitutive activity to TRPML2 in HEK cells. Wild-type human TRPML2 has been shown to be surface membrane bound.

pharmacological means, which facilitates their whole-cell currents. Some TRP channels reveal constitutive activity in one expression system and not in another (Lev et al., 2010; Fig. 29.1A) and some reveal basal activity in all expression systems used (Fig. 29.1B). There are several examples of TRP channels which when mutated alter their activity and become constitutively active (Fig. 29.1C). This is observed for all TRPML channels possessing the alanine to proline (A–P) amino acid substitution (Va mutation) at the paralogous site in the transmembrane segment 5 (Dong et al., 2008; Grimm et al., 2007; Lev et al., 2010; Samie et al., 2009; Xu et al., 2007). This change in activity in a mutated channel was also observed for the *Drosophila* TRP channel with the phenylalanine to isoleucine (F550I) mutational substitution, causing photoreceptor degeneration (Yoon et al., 2000). However, the wild-type TRPML channels do not display any significant current (Xu et al., 2007; Fig. 29.1C), albeit their surface membrane expression. The same is true for the *Drosophila* TRP channel in the dark.

Constitutive activity of TRP channel can be investigated when upstream elements of its activating cascade display constitutive activity. This was shown for the *Drosophila* TRP channels in the *Drosophila* retinal degeneration A (*rdgA*) phototransduction mutant (Raghu et al., 2000). In *Drosophila* photoreceptor cells, mutations in diacylglycerol (DAG) kinase (RDGA), the enzyme which converts DAG into phosphatidic acid, induces constitutively active TRP and TRPL channels in the dark. This leads to photoreceptor degeneration. Similar constitutive activity of the TRP and TRPL channels was demonstrated upon anoxic conditions in the fly eye, which is thought to affect PLC activity (Agam et al., 2000).

It must be emphasized that only a fraction of the TRP channels displaying constitutive activity cause cellular degeneration (see below). A notable example for mammalian TRP channels showing constitutive activity of both native and heterologously expressed channels are the epithelial TRP channels TRPV5/6 (for review, see den et al., 2003). These are highly Ca^{2+} selective channels with distinctive physiological functions important for body Ca^{2+} homeostasis. Cells expressing these channels do not undergo degeneration in spite of their inwardly rectifying I–V curves (see Fig. 29.5). A possible mechanism which prevents Ca^{2+} overload in cells expressing these channels is their strong Ca^{2+}-dependent negative feedback regulation, leading to relatively fast Ca^{2+}-dependent inactivation of the channels (den et al., 2003). Several of the channels described above are not associated with cellular degeneration, presumably because the channels display an outwardly rectifying I–V curve (see below) and a minute current flow into the cells at resting membrane potentials, due to Ca^{2+}-dependent negative feedback mechanisms (i.e., TRPV1: Lukacs et al., 2007; Rosenbaum et al., 2004, TRPM8: Rohacs et al., 2005; TRPL and TRP: Parnas et al., 2007; Fig. 29.2 left, images, right, I–V curve with outward rectification).

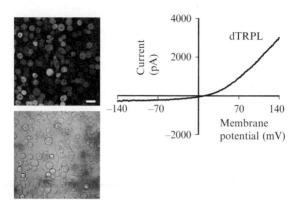

Figure 29.2 TRP channels with constitutive activity and no degeneration. Left: widefield confocal images showing representative cell morphology of S2 cells transfected with the indicated construct. *Drosophila* TRPL-GFP fluorescence (top) and DIC (bottom) images are presented. Scale bar: 20 µm. Right: representative *I–V* curve of whole-cell currents measured from S2 cells, expressing the indicated channel, displaying robust outward rectifying current.

However, when trying to predict a link between *I–V* curves and cell viability, one must consider the Ca^{2+} selectivity, resting potential, and other endogenous channel activities present.

3.2. Methods for determining constitutive activity

3.2.1. Whole-cell patch-clamp recordings

In this and the following sections, we do not describe standard protocols, which are readily available.

The most revealing technique for determination of channel activity is the whole-cell patch-clamp recording technique. This method is used for electrophysiological recordings of channel activity in native systems (such as isolated grouped photoreceptor cells) and in many expression systems (such as the *Drosophila* S2 cell line and the mammalian HEK cell line). In this method, a cell is approached with a borosilicate glass pipette. We suggest using capillaries containing glass filament inside them. The filament is important for easy back filling of the pipette with intracellular solution, due to capillary tension. Our experience leads us to recommend glass pipettes, which have a large inner radius and yet a large inner to outer segment, which allows the cell membrane attachment to the glass electrode with a tighter seal (higher resistance). We use glass pipettes with an outer 1.0 mm and inner 0.58 mm diameter for clamping photoreceptor and S2 cells. For clamping HEK cells, we use outer 1.5 mm and inner 1.1 mm diameter. In many cases, we also briefly (less than 1 s) heat the tip in order to melt and smoothen out the tip (heat polishing). This also greatly helps in creating a tight seal between the electrode and the cell membrane.

Under stereomicroscope visualization, using a micromanipulator, we approach the cell with the glass electrode applying positive pressure in the electrode in order to avoid particles from sticking to the tip upon entry into the extracellular solution. The pressure is applied via a "tygon"-type tubing connected to a 1 cc syringe. We approach the cell gently within the first third of the cell until one sees a dent in the cell (this is indicated by an optical black contour lining). At this point in time, we release the positive pressure and apply a minute negative pressure to form the gigaseal. Sometimes, the seal will form upon release of positive pressure. With some cells, applying negative pressure either with mouth or by way of the syringe will complete the seal. At this stage, it is possible to look at single-channel activity in the cell attached mode. In this mode, it is possible to determine the voltage at the outer membrane side of the patch. The membrane potential is the voltage difference between the inner and outer parts of the membrane patch, while the physiological resting membrane potential in most cells is between -20 and -90 mV (depending on other channel activities present in cells, such as potassium channels, and on specific cellular properties). In this part of the procedure, it is possible to pull and rupture the membrane patch from the cell (called inside-out patch). Our experience leads us to suggest a quick and strong pull upward with the micromanipulator (approximately half a turn: 500 μm). Sometimes it is advisable to lift the electrode outside the solution and back in, which causes a physical detachment of the membrane patch, due to air solution effect on membranes. If the inside-out technique is completed well, then a gigaseal is maintained. When doing so, any voltage clamping by the electrode is valid for the membrane potential of this patch. In this configuration, it is also possible to look at single-channel activity devoid of most of the cytosolic elements, which could affect channel activity. It has to be noted that in this configuration, the inner cytosolic part of the channel is facing toward the bath solution, which should be changed to intracellular solution, while the electrode solution should be changed to extracellular solution, in order to mimic physiological conditions. After performing a gigaseal, it is possible to rupture the membrane patch by applying pressure or by giving a short (0.2–1 ms duration) current pulse (zap) confined in duration. Rupturing the membrane connects chemically and electrically between the pipette content and the cell interior. In this mode (whole-cell recording), it is important to make sure that the only conductance seen is that of the expressed channels. One has to make sure that the resistance of the cell is not impaired by a low resistance seal between cell and electrode (i.e., less than 1 GΩ). In many cases (see below), the activity state of the channels can be large so that the input resistance is lower than 1 GΩ. The input resistance can be displayed by applying a seal test (this function is a built-in property of the amplifier).

The amplitude of the test, together with the resting potential of the cell and the channel's reversal potential (E_{rev}), have to be considered when performing the seal test and receiving a value for the input resistance. Most TRP channels have an E_{rev} around 0 mV in normal conditions. If the channels display constitutive activity, the seal test might show a lower resistance just because these channels are open. It is therefore important to apply pharmacological agents or enzymatic intervention to close the channels in order to reveal endogenous currents, or currents arising from an improper seal (leak current). At this point, we advise to perform two kinds of voltage clamping protocols. One should perform I–V curves by stepping the holding voltages from negative to positive voltages. Each step should be at least several hundreds of milliseconds in duration to allow observation of fast current changes and steady state. The other protocol is a voltage ramp, in which we change the voltage from negative to positive potentials continuously within 1 s. If both protocols give the same I–V curve, then it is easier to pursue all further experiments with the voltage ramp protocol. The advantage of this protocol is the ability to look at the full current–voltage relationship within a short time scale. It allows application of voltage ramps repeatedly every few seconds, giving a kinetics profile of the activity at positive as well as negative membrane potentials.

3.2.1.1. I–V curves and channel voltage dependence When measuring an I–V curve, there can be several possibilities related to the activity state of the channel: (1) the channel is integrated to the surface membrane but it is nonfunctional; (2) the channel is functional but it is not integrated to the plasma membrane; and (3) the functional channel is integrated to the plasma membrane but does not have constitutive activity. As mentioned, many TRP channels display basal activity, which indicates functional integration to the surface membrane. Different TRP channels exhibit differing I–V curves. For example, in the presence of divalent cations, TRPC3/4/5/6, TRPM4/5/6/7/8, TRPV1/2/3, and *Drosophila* TRP and TRPL channels all reveal outward rectifying I–V curves (Fig. 29.1A and B; Clapham, 2003; Hardie and Minke, 1994). TRPML1/2/3 (Xu *et al.*, 2007) and TRPV5/6 (Owsianik *et al.*, 2006) have an inward rectifying I–V curve (Fig. 29.1C) and TRPM2/3/4/5 have a linear I–V curve (Clapham, 2003).The voltage dependence (i.e., rectification) observed in most TRP channels can be either an intrinsic property of the channels or due to divalent cation open channel block (Parnas *et al.*, 2007). If a nonlinear I–V curve remains nonlinear even after removal of divalent cations, as in the case of TRPV1 or TRPM8 (Nilius *et al.*, 2005), then the voltage dependence is an intrinsic property of the channel. When removal of divalent cations leads to a linear I–V curve, as in the case of TRPL or TRPM7 (Nadler *et al.*, 2001; Parnas *et al.*, 2007), then the nonlinearity arises from open channel block. A change in extra- or intracellular ion concentrations can underlie and facilitate the open probability of the channels in native systems, leading to inward cationic currents, which could affect cell viability (Wei et al., 2007). Furthermore, there are TRP channels that produce an inward current when lowering the intra- or extracellular divalent ion concentration (e.g., Mg^{2+} or Ca^{2+}) as in the case of TRPV5 (Lee et al., 2005; Vennekens et al., 2000), TRPV6 (Bodding and Flockerzi, 2004), and *Drosophila* TRP (Hardie and Minke, 1994).

3.2.1.2. Endogenous currents in expression systems
It is important to identify the endogenous channel activity of the cells used for TRP channel expression (Lev et al., 2010). Many tissues natively express multiple types of TRP channels and there too, it is essential to recognize the specific TRP channel activity. There are several ways to separate between the various channel activities: (1) A pharmacological approach, in which one blocks a specific activity, unveiling the other (e.g., application of μM La^{3+} to *Drosophila* photoreceptor cells blocks the TRP but not the TRPL channel activity (Hardie and Minke, 1992)). (2) A genetic approach, in which TRP channels are eliminated (silenced using RNAi). (3) Clamping the membrane potential to values in which one channel activity is prominent and the other is negligible. In our research project with the TRPML channels, we meticulously characterized the endogenous current in the S2 expression system used to express TRPML2/3. In S2 cells, the endogenous current is small and almost negligible. At positive membrane potentials, it is mostly nonexistent, whereas at negative membrane potentials, it appears to have a small saturating inward rectifying voltage dependence with an E_{rev} at around -40 mV (Fig. 29.3A). In many cases, the endogenous current is

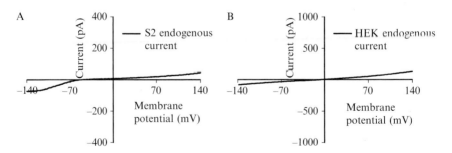

Figure 29.3 Endogenous currents in S2 and HEK expression systems. (A) Representative I–V curve of endogenous currents, as measured from S2 cells, by whole-cell patch-clamp recordings. Note the distinguishable inward rectifying, albeit small current, which is easily separated from currents in S2 cells expressing TRP channels (see Fig. 29.1A and B). (B) Same as A except representative I–V curve from HEK cells in control conditions. Note the lack of detectable current arising from absence in channel activity. The nonsignificant linear leak current is displayed.

nonexistent. In the HEK cell expression system that we also used for expression of TRPML1/2/3, the endogenous current is virtually absent when using intracellular solution containing Cs^+ instead of K^+ (Fig. 29.3B). In all cases, we perform extensive control experiments to identify the TRP channel activity. Other control experiments are performed using the large, TRP channel impermeable, organic cation N-methyl-D-glucamin (NMDG), in order to verify that the current is carried by cations, in a nonspecific manner (Lev et al., 2010). Several other experiments can be performed to identify the selectivity and ion permeability of a specific TRP channel (Kim et al., 2008, 2010; Xu et al., 2007).

3.2.2. Ca^{2+} imaging

Calcium imaging is widely used to determine activity of TRP channels which have been activated (Siemens et al., 2006), or show prominent constitutive activity, which is characterized by inward rectification (Xu et al., 2007). Many TRP channels, which reveal outward rectification, can be further activated (facilitated), resulting in opening of the channel at negative membrane potentials upon signaling, leading to increased intracellular Ca^{2+} (Varnai et al., 2006). The advantage of this technique is in its quick and high-throughput readout. However, a channel can be in an open state and yet exhibit low or no Ca^{2+} influx (e.g., as in the case of TRPM4/5, which are Ca^{2+} impermeable). This can be explained by the channel's I–V curve, open channel block, or low permeability to Ca^{2+}. Since most TRP channels are permeable to Ca^{2+}, we recommend using Ca^{2+} imaging as a start point and later on applying electrophysiological tools to further characterize the observed activity. Many channels that exhibit constitutive activity display outward rectifying I–V curve. In this paradigm, the Ca^{2+} imaging technique might not show a change in Ca^{2+} levels, because of the small inward current, and therefore will not give an indication for constitutive activity. However, inwardly rectifying TRP channels with constitutive activity and a certain extent of Ca^{2+} permeability will display this constitutive activity in terms of intracellular Ca^{2+} levels (Grimm et al., 2007; Xu et al., 2007). The Ca^{2+} indicator used should have a Kd that fits the calcium concentrations observed. In many cases, it is advisable to apply ionomycin (a nonspecific Ca^{2+} ionophore), which serves as an inner positive control in cases where the amplitude of the calcium signal is not known.

3.2.2.1. The Ca^{2+} imaging method Cells, which have been transfected with TRP channels, are then loaded with a Ca^{2+} indicator. Many indicators can be used, although the widely used reagents are the cell permeable Fura-2 AM and Fluo-4 AM. One should add Fluo-4 AM at a final concentration of ~1 μM (higher concentrations may act as Ca^{2+} buffers) to clean medium (no antibiotics, no fetal calf serum) and incubate for half an hour. It is suggested to add the Fluo-4 AM, which is dissolved in a DMSO stock solution containing 10% Pluronic F127, to medium and then add the cells to that medium so as to minimize DMSO toxicity. During the half-hour incubation, the indicator enters the cell where the AM hydrophobic component is cleaved by an endogenous esterase enzyme. There is a partitioning of the indicator between the inside and extracellular solution. Some of the indicator is in the plasma membrane. After this incubation, the cells are washed twice with clean medium to rid the extracellular solution from any additional indicator. This is followed by incubation in clean medium with 2% Pluronic F127 for half an hour. This incubation is helpful in releasing the indicator from the plasma membrane into the cell, in order to avoid a continuance inflow of indicator into the cell during the experiment. After the incubation time, cells are rinsed thoroughly but gently with clean medium. The cells are now ready for imaging. We use the Fluo-4 AM indicator which is excited by Green Fluorescence Protein (GFP) excitation spectrum. For imaging, we use confocal microscopy and a monochrome laser at a wave length of 488 nm.

Since the cells are at resting membrane potential, constitutive activity will be unmasked only if there is robust inward current, which is also carried by Ca^{2+}. Moreover, in order to measure the elevated Ca^{2+} levels, one should compare the indicator fluorescence to control (untransfected cells) or change the extracellular Ca^{2+} levels.

In a typical protocol, the cells are bathed in Ca^{2+}-free solution and basal Ca^{2+} levels are monitored. Then, cells are exposed to normal ($\sim$2 mM) or higher Ca^{2+} levels, while observing a change in fluorescence inside the transfected cells. Then, the extracellular solution can be changed back again to a low Ca^{2+} level to exhibit reversibility. One should be aware of the change in intracellular Ca^{2+} levels, which occurs due to the Na^+/Ca^{2+} exchanger (Fig. 29.4). This must be accounted for and separated from the signal arising from channel activity (Peretz et al., 1994).

3.2.3. Physiological mechanisms that are affected by constitutive channel activity

It is reasonable to assume that known physiological mechanisms can be altered when a constitutively active channel is present. In this case, the activity does not have to be confined to the plasma membrane, and electrophysiological techniques would "miss" the activity. In order to show constitutive channel activity, the TRP channel has to be directly linked to a known cellular mechanism and this mechanism must have a clear readout. One example of such a mechanism is the phototransduction cascade in the fly, which allows the fly to see. The *Drosophila* TRP and TRPL channels constitute the last element in this transduction cascade, leading to cation influx and subsequent depolarization when light is turned on. Without these channels, the fly is blind. A phototaxis bioassay (i.e., the attraction

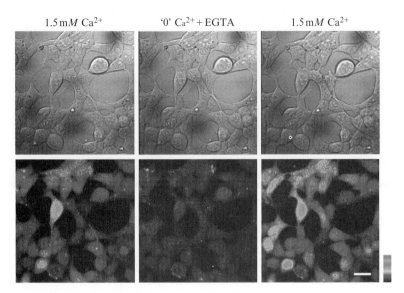

Figure 29.4 The intracellular basal Ca^{2+} concentration is affected by extracellular Ca^{2+}. Wide-field confocal images of native HEK cells loaded with Fluo4-AM indicator, displaying resting $[Ca^{2+}]_i$ in different levels of extracellular Ca^{2+}, as indicated above. Note how the intracellular Ca^{2+} concentration decreases significantly, upon lowering of the extracellular Ca^{2+} concentration, in a reversible manner (middle panel). This is most likely due to the activity of the Na/Ca^{2+} exchanger. Scale bar: 20 μm. (See Color Insert.)

of flies to light) can differentiate between loss of function, gain of function, and "normal" activity (Choe and Clandinin, 2005; Cosens and Manning, 1969). One could anticipate a change in phototaxis behavior in flies expressing constitutively active TRP and TRPL channels.

4. Constitutive TRP Channel Activity Which Leads to Cellular Degeneration

In this section, we describe TRP channels which have constitutive activity leading to cellular degeneration. The major causes of degeneration in cells expressing constitutively active TRP channels is Ca^{2+} influx leading to Ca^{2+} overload in the cells. The nature of the degeneration can be either necrotic or apoptotic, which can be differentiated easily by specific assays and also by the morphological changes, which accompany the degeneration process.

4.1. Examples of TRP channels which exert a degenerating effect

All three TRPML channels exhibit constitutive activity when possessing the Va mutation. The Va missense mutation (substitution of alanine or valine with helix breaking proline) is considered to introduce a kink in the transmembrane 5 segment (TM5 domain), thereby rendering the channel in the open state. This mutation was introduced into other TRP channels from other TRP channel subfamilies, reproducing the same effect as in TRPMLs (e.g., TRPV5 and TRPV6; Grimm et al., 2007). The channel activity is characterized by an inwardly rectifying I–V curve. The inward current is robust and is carried by cations including, but not limited to, Ca^{2+}. The E_{rev} is around 0 mV, which is typical for nonselective cation channels. In the case of all TRPMLs carrying the Va mutation, it is the Ca^{2+} overload which leads to cellular degeneration (Fig. 29.5 left, images of degeneration, right, I–V curve). Another TRP channel that leads to cellular degeneration is the founding member of the TRP superfamily, the *Drosophila* TRP channel. Photoreceptor degeneration is revealed in the constitutively active trp^{P365} mutant, because of unregulated channel activity leading to Ca^{2+} overload (Wang et al., 2005; Yoon et al., 2000). Although the mutated TRP channel displays mainly an outward current, a small but significant inward (mainly Ca^{2+}) current is observed in the dark, leading to photoreceptor degeneration (Yoon et al., 2000). Later, it was shown that a single missense mutation (Phe-550 to Ile) in TM5 is sufficient to account

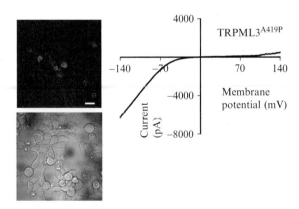

Figure 29.5 TRP channels with constitutive activity show degeneration. Left: widefield confocal images showing representative cell morphology of HEK cells transfected with the indicated construct. tdTomato-TRPML3^{A419P} fluorescence (top) and DIC (bottom) images are displayed. Scale bar: 20 μm. Note that most fluorescent cells (top) also show degeneration, as displayed in transmitted image (bottom). Right: representative I–V curve of whole-cell currents measured from HEK cells, expressing the indicated channel which displays robust inward rectifying currents.

for the degeneration observed in the original mutant, which harbors additional three mutations in TM4 and TM5 (Hong et al., 2002).

Strikingly, metabolic stress induces constitutively active TRP and TRPL channels in the dark, in a reversible manner, *in vivo* (Agam et al., 2000). A continuous uncontrolled and maximal activation of these channels most likely leads to photoreceptor cell death due to Ca^{2+} overload (Agam et al., 2004). This conclusion is strongly supported by the trp^{P365} mutation which makes the channel constitutively active with a magnitude similar to anoxia.

4.2. Methods for determination of the degenerating effect

4.2.1. Morphological changes associated with cellular degeneration

As mentioned in Section 2, it is easy to distinguish between normal viable cells and degenerating cells. It is always important to have a control which undergoes the same experimental manipulations with the tested cells expressing TRP channels to rule out possible degeneration effects originating from the experimental procedure.

We use S2 or HEK cell lines for transfection. The transfection procedure should be optimized to prevent a degenerating affect by virtue of transfection. A fluorescent marker or tag must be included in the transfection to allow separation between transfected and nontransfected cells. In our experience, we have used, with best results, TRP channels conjugated to the color probe, tdTomato, because of its high fluorescent yield and the fact that cells usually do not have autofluorescence in the long wavelength range. It is advisable to have the channels conjugated rather than coexpressed with a color probe to make sure that the TRP channel has been expressed. It is best to have several transfection plates to allow demonstration of the onset and kinetics of the degeneration process (start point and intensity). According to previous publications (Grimm et al., 2009) and our experience, degeneration starts around 10–15 h posttransfection with TRPML channels bearing the paralogous Va mutation in HEK cells, or wild-type TRPML2 in S2 cells. The first step is to look at a cover slip with cells under low microscope magnification. This gives a good estimate whether the cells have undergone degeneration. Next, the cover slip is taken to a fluorescent microscope for complete evaluation of the channel expression and its effect on cell viability. One should count the number of degenerating cells out of the cell population, which have undergone TRP channel expression. The morphological markers used to separate between degenerating and nondegenerating cells are rounding up of the cells when observing HEK cells (Xu et al., 2007; S2 cells are round under normal conditions), detachment of cells from the cover slip (this is to be considered when imaging those cells that are attached), swelling, total deterioration (loss of normal morphology), distorted contour of cells, and membrane roughness of cell surface. Not all cells will have all morphological changes because each cell can be at a different stage of the degenerating process. It is very important to recognize all features of the cell type used and how the cells look normally, before analyzing degeneration. The experiment should be carried out in a double blind fashion. Several areas of the plate are chosen randomly for scanning by confocal microscope, which gives a high spatial resolution (it is also possible to perform experiments with a regular fluorescent microscope, as long as morphology of the cells is readily seen). The cells should be imaged as differential interference contrast (DIC) and fluorescence (Fig. 29.5 left bottom). Once the images have been acquired, the fluorescent cells are counted and the percentage of degenerating cells is determined. This should formulate the degree of degeneration for a specific expressed channel. It is then possible to examine whether the degeneration results from Ca^{2+} overload by coexpressing the calcium extrusion pump, PMCA2, which rescues the cells from degeneration (Grimm et al., 2009; Lev et al., 2010). Another option is to grow transfected cells in minimal Ca^{2+} medium before imaging and immediately after transfection, in order to rescue cells from cellular degeneration. This experimental procedure was used to show a causal relationship between TRPML's constitutive activity and intracellular Ca^{2+} overload (Kim et al., 2008). According to unpublished observations made by us, when expressing TRPML2/3 harboring the paralogous Va mutation, we suggest reducing external Ca^{2+} from 1.5 mM Ca^{2+} (i.e., the physiological concentration in the extracellular solution) to ~ 50 µM. Lower Ca^{2+} concentrations usually lead to lower cell viability.

4.2.2. Annexin-based assay for determination of cellular degeneration

The Annexin-based assay allows determination of the extent of degeneration, and whether it is necrotic or apoptotic. One can use other assays such as PI (propidium iodide, a general marker for cellular deterioration), caspase 3 activity, nuclear staining, and DNA laddering to describe cellular degeneration and its characteristics. The Annexin assay is based on the translocation of phosphatidylserine (PS) to the outer leaflet from the inner leaflet of the plasma membrane at an early stage of the apoptotic process. This asymmetry, which is mediated by an inside-outside PS translocase, is necessary for cell recognition by macrophages, for their removal (Allen et al., 1997). The fact that this process is an inherent part of programmed cell death in all cells is utilized as an early biochemical marker, which can indicate for apoptosis before secondary degeneration proceeds.

To apply the Annexin method, one needs to grow cells on coverslips and transfect them with fluorescent protein constructs. Eighteen hours post-transfection, one should dilute Alexa Fluor 647 Annexin V (BioLegend) in Binding Buffer (BioLegend) and incubate with transfected cells for 15 min at room temperature in the dark. Other Annexin-conjugated probes are available depending on the fluorescent setup used and other fluorescent markers expressed in cells (conjugated fluorescent TRP channels). Following incubation with Annexin V, the cells are washed quickly with Binding Buffer and then fixed in the dark in 4% formaldehyde for 30 min at room temperature. Following three washes with 1% NH_4Cl (in PBS), coverslips are mounted onto glass slides with Antifade Solution (Vysis) and fluorescent images of Alexa Fluor 647 Annexin V staining, together with fluorescent protein emission, are acquired in a confocal microscope (e.g., Olympus Fluoview 300 IX70). The readout of such an experiment is the number of Annexin-positive cells from the total number of channel-expressing cells

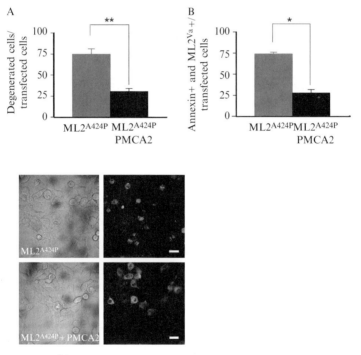

Figure 29.6 Ca^{2+}-induced degeneration caused by constitutive activity of h-TRPML2-^{A424P} is rescued by the calcium extrusion pump, PMCA2. (A) Top: histogram showing the percentage of degenerated HEK cells following cotransfection of h-TRPML2-^{A424P} (ML2-A424P) together with PMCA2 or empty vector (**, $P < 0.01$). Note the rescue of cell degeneration in cells cotransfected with h-TRPML2-^{A424P} and PMCA2; bottom: wide-field confocal images (left: DIC transmitted, right: fluorescent-conjugated channel-expressing cells) showing representative cell morphology in the two groups of cotransfected HEK cells described above (scale bar: 20 μm). Note that most cells expressing the channel without PMCA2 display degeneration (distinct changes in morphology). (B) Histogram showing the percentage of Annexin V-positive HeLa cells following cotransfection of h-TRPML2-^{A424P} with PMCA2 or with empty vector (*, $P < 0.05$). Note the rescue of cell degeneration in cells cotransfected with h-TRPML2-^{A424P} and PMCA2. (See Color Insert.)

(Fig. 29.6B). It is possible and also more accurate to use FACS analysis to quantify this degeneration. In combination with this assay, it is possible to look at various treatments such as coexpression with the Ca^{2+} extrusion pump to assess the contribution of Ca^{2+} overload to the degenerating effect.

4.3. Summary of the methods for viewing cellular degeneration

As stated above, it is advisable to use morphological alterations to distinguish between viable and nonviable cells when expressing a specific TRP channel in conjunction with other methods. Other methods use biochemical markers for determining degeneration, as we have done in our studies. Using several methods gives a more comprehensive description of the type of cell death, and verification of each method is achieved. It is easy to see that both methods result in the same degree (percentage-wise) of degeneration and that rescue of degeneration can also indicate the initial etiological effector which is Ca^{2+} overload in our case (Fig. 29.6A, top, histogram of morphological alterations, bottom, wide-field images of cells exhibiting degeneration and B, histogram of Annexin-positive TRP channel-expressing cells).

ACKNOWLEDGMENTS

We thank Ben Katz, Daniela Dadon, and Maximilian Peters for careful and critical reading of the chapter.
The experimental part of this review was supported by grants from the National Institute of Health (RO1-EY 03529), the German-Israel Foundation (GIF), the Israel Science Foundation (ISF), and the US-Israel Binational Science Foundation (BSF).

REFERENCES

Agam, K., von-Campenhausen, M., Levy, S., Ben-Ami, H. C., Cook, B., Kirschfeld, K., and Minke, B. (2000). Metabolic stress reversibly activates the *Drosophila* light-sensitive channels TRP and TRPL in vivo. *J. Neurosci.* **20,** 5748–5755.
Agam, K., Frechter, S., and Minke, B. (2004). Activation of the *Drosophila* TRP and TRPL channels requires both Ca^{2+} and protein dephosphorylation. *Cell Calcium* **35,** 87–105.
Allen, R. T., Hunter, W. J., III, and Agrawal, D. K. (1997). Morphological and biochemical characterization and analysis of apoptosis. *J. Pharmacol. Toxicol. Meth.* **37,** 215–228.
Bahner, M., Frechter, S., Da Silva, N., Minke, B., Paulsen, R., and Huber, A. (2002). Light-regulated subcellular translocation of Drosophila TRPL channels induces long-term adaptation and modifies the light-induced current. *Neuron* **34,** 83–93.
Bodding, M., and Flockerzi, V. (2004). Ca2+ dependence of the Ca2 + -selective TRPV6 channel. *J. Biol. Chem.* **279,** 36546–36552.
Carini, R., Autelli, R., Bellomo, G., and Albano, E. (1999). Alterations of cell volume regulation in the development of hepatocyte necrosis. *Exp. Cell Res.* **248,** 280–293.

Choe, K. M., and Clandinin, T. R. (2005). Thinking about visual behavior; learning about photoreceptor function. *Curr. Top. Dev. Biol.* **69**, 187–213.

Chyb, S., Raghu, P., and Hardie, R. C. (1999). Polyunsaturated fatty acids activate the *Drosophila* light-sensitive channels TRP and TRPL. *Nature* **397**, 255–259.

Clapham, D. E. (2003). TRP channels as cellular sensors. *Nature* **426**, 517–524.

Cosens, D. J., and Manning, A. (1969). Abnormal electroretinogram from a *Drosophila* mutant. *Nature* **224**, 285–287.

Cronin, M. A., Lieu, M. H., and Tsunoda, S. (2006). Two stages of light-dependent TRPL-channel translocation in Drosophila photoreceptors. *J. Cell Sci.* **119**, 2935–2944.

Dadon, D., and Minke, B. (2010). Cellular functions of transient receptor potential channels. *Int. J. Biochem. Cell Biol.* **42**(9):1430–1445.

Damann, N., Voets, T., and Nilius, B. (2008). TRPs in our senses. *Curr. Biol.* **18**, R880–R889.

Delgado, R., and Bacigalupo, J. (2009). Unitary recordings of TRP and TRPL channels from isolated Drosophila retinal photoreceptor rhabdomeres: Activation by light and lipids. *J. Neurophysiol.* **101**, 2372–2379.

den, D. E., Hoenderop, J. G., Nilius, B., and Bindels, R. J. (2003). The epithelial calcium channels, TRPV5 & TRPV6: From identification towards regulation. *Cell Calcium* **33**, 497–507.

Dong, X. P., Cheng, X., Mills, E., Delling, M., Wang, F., Kurz, T., and Xu, H. (2008). The type IV mucolipidosis-associated protein TRPML1 is an endolysosomal iron release channel. *Nature* **455**, 992–996.

Grimm, C., Cuajungco, M. P., van Aken, A. F., Schnee, M., Jors, S., Kros, C. J., Ricci, A. J., and Heller, S. (2007). A helix-breaking mutation in TRPML3 leads to constitutive activity underlying deafness in the varitint-waddler mouse. *Proc. Natl. Acad. Sci. USA* **104**, 19583–19588.

Grimm, C., Jors, S., and Heller, S. (2009). Life and death of sensory hair cells expressing constitutively active TRPML3. *J. Biol. Chem.* **284**, 13823–13831.

Hara, Y., Wakamori, M., Ishii, M., Maeno, E., Nishida, M., Yoshida, T., Yamada, H., Shimizu, S., Mori, E., Kudoh, J., et al. (2002). LTRPC2 Ca^{2+}-permeable channel activated by changes in redox status confers susceptibility to cell death. *Mol. Cell* **9**, 163–173.

Hardie, R. C., and Minke, B. (1992). The *trp* gene is essential for a light-activated Ca^{2+} channel in *Drosophila* photoreceptors. *Neuron* **8**, 643–651.

Hardie, R. C., and Minke, B. (1994). Calcium-dependent inactivation of light-sensitive channels in *Drosophila* photoreceptors. *J. Gen. Physiol.* **103**, 409–427.

Hong, Y. S., Park, S., Geng, C., Baek, K., Bowman, J. D., Yoon, J., and Pak, W. L. (2002). Single amino acid change in the fifth transmembrane segment of the TRP Ca2+ channel causes massive degeneration of photoreceptors. *J. Biol. Chem.* **277**, 33884–33889.

Hu, Y., and Schilling, W. P. (1995). Receptor-mediated activation of recombinant Trpl expressed in Sf9 insect cells. *Biochem. J.* **305**, 605–611.

Katz, B., and Minke, B. (2009). Drosophila photoreceptors and signaling mechanisms. *Front Cell Neurosci.* **3**, 2.

Kim, H. J., Li, Q., Tjon-Kon-Sang, S., So, I., Kiselyov, K., Soyombo, A. A., and Muallem, S. (2008). A novel mode of TRPML3 regulation by extracytosolic pH absent in the varitint-waddler phenotype. *EMBO J.* **27**, 1197–1205.

Kim, H. J., Yamaguchi, S., Li, Q., So, I., and Muallem, S. (2010). Properties of the TRPML3 pore and its stable expansion by the varitint-waddler causing mutation. *J. Biol. Chem.* **285**(22), 16513–16520.

Lee, J., Cha, S. K., Sun, T. J., and Huang, C. L. (2005). PIP2 activates TRPV5 and releases its inhibition by intracellular Mg^{2+}. *J. Gen. Physiol.* **126**, 439–451.

Lev, S., Zeevi, D. A., Frumkin, A., Offen-Glasner, V., Bach, G., and Minke, B. (2010). Constitutive activity of the human TRPML2 channel induces cell degeneration. *J. Biol. Chem.* **285**, 2771–2782.

Lukacs, V., Thyagarajan, B., Varnai, P., Balla, A., Balla, T., and Rohacs, T. (2007). Dual regulation of TRPV1 by phosphoinositides. *J. Neurosci.* **27**, 7070–7080.

Meyer, N. E., Joel-Almagor, T., Frechter, S., Minke, B., and Huber, A. (2006). Subcellular translocation of the eGFP-tagged TRPL channel in Drosophila photoreceptors requires activation of the phototransduction cascade. *J. Cell Sci.* **119**, 2592–2603.

Minke, B., and Parnas, M. (2006). Insights on TRP channels from in vivo studies in Drosophila. *Annu. Rev. Physiol.* **68**, 649–684.

Minke, B., Wu, C., and Pak, W. L. (1975). Induction of photoreceptor voltage noise in the dark in *Drosophila* mutant. *Nature* **258**, 84–87.

Nadler, M. J., Hermosura, M. C., Inabe, K., Perraud, A. L., Zhu, Q., Stokes, A. J., Kurosaki, T., Kinet, J. P., Penner, R., Scharenberg, A. M., *et al.* (2001). LTRPC7 is a Mg.ATP-regulated divalent cation channel required for cell viability. *Nature* **411**, 590–595.

Nagata, K., Zheng, L., Madathany, T., Castiglioni, A. J., Bartles, J. R., and Garcia-Anoveros, J. (2008). The varitint-waddler (Va) deafness mutation in TRPML3 generates constitutive, inward rectifying currents and causes cell degeneration. *Proc. Natl. Acad. Sci. USA* **105**, 353–358.

Niemeyer, B. A., Suzuki, E., Scott, K., Jalink, K., and Zuker, C. S. (1996). The *Drosophila* light-activated conductance is composed of the two channels TRP and TRPL. *Cell* **85**, 651–659.

Nilius, B., Talavera, K., Owsianik, G., Prenen, J., Droogmans, G., and Voets, T. (2005). Gating of TRP channels: A voltage connection? *J. Physiol.* **567**, 35–44.

Orrenius, S., Zhivotovsky, B., and Nicotera, P. (2003). Regulation of cell death: The calcium-apoptosis link. *Nat. Rev. Mol. Cell Biol.* **4**, 552–565.

Owsianik, G., Talavera, K., Voets, T., and Nilius, B. (2006). Permeation and selectivity of TRP channels. *Annu. Rev. Physiol.* **68**, 685–717.

Parnas, M., Katz, B., and Minke, B. (2007). Open channel block by Ca^{2+} underlies the voltage dependence of Drosophila TRPL channel. *J. Gen. Physiol.* **129**, 17–28.

Peretz, A., Suss-Toby, E., Rom-Glas, A., Arnon, A., Payne, R., and Minke, B. (1994). The light response of *Drosophila* photoreceptors is accompanied by an increase in cellular calcium: Effects of specific mutations. *Neuron* **12**, 1257–1267.

Raghu, P., Usher, K., Jonas, S., Chyb, S., Polyanovsky, A., and Hardie, R. C. (2000). Constitutive activity of the light-sensitive channels TRP and TRPL in the *Drosophila* diacylglycerol kinase mutant, *rdgA*. *Neuron* **26**, 169–179.

Ramsey, I. S., Delling, M., and Clapham, D. E. (2006). An introduction to TRP channels. *Annu. Rev. Physiol.* **68**, 619–647.

Rohacs, T., Lopes, C. M., Michailidis, I., and Logothetis, D. E. (2005). PI(4, 5)P2 regulates the activation and desensitization of TRPM8 channels through the TRP domain. *Nat. Neurosci.* **8**, 626–634.

Rosenbaum, T., Gordon-Shaag, A., Munari, M., and Gordon, S. E. (2004). Ca^{2+}/calmodulin modulates TRPV1 activation by capsaicin. *J. Gen. Physiol.* **123**, 53–62.

Samie, M. A., Grimm, C., Evans, J. A., Curcio-Morelli, C., Heller, S., Slaugenhaupt, S. A., and Cuajungco, M. P. (2009). The tissue-specific expression of TRPML2 (MCOLN-2) gene is influenced by the presence of TRPML1. *Pflugers Arch.* **459**, 79–91.

Siemens, J., Zhou, S., Piskorowski, R., Nikai, T., Lumpkin, E. A., Basbaum, A. I., King, D., and Julius, D. (2006). Spider toxins activate the capsaicin receptor to produce inflammatory pain. *Nature* **444**, 208–212.

Stein, A. T., Ufret-Vincenty, C. A., Hua, L., Santana, L. F., and Gordon, S. E. (2006). Phosphoinositide 3-kinase binds to TRPV1 and mediates NGF-stimulated TRPV1 trafficking to the plasma membrane. *J. Gen. Physiol.* **128,** 509–522.

Sun, H. S., Jackson, M. F., Martin, L. J., Jansen, K., Teves, L., Cui, H., Kiyonaka, S., Mori, Y., Jones, M., Forder, J. P., et al. (2009). Suppression of hippocampal TRPM7 protein prevents delayed neuronal death in brain ischemia. *Nat. Neurosci.* **12,** 1300–1307.

van Aken, A. F., Atiba-Davies, M., Marcotti, W., Goodyear, R. J., Bryant, J. E., Richardson, G. P., Noben-Trauth, K., and Kros, C. J. (2008). TRPML3 mutations cause impaired mechano-electrical transduction and depolarization by an inward-rectifier cation current in auditory hair cells of varitint-waddler mice. *J. Physiol.* **586,** 5403–5418.

Varnai, P., Thyagarajan, B., Rohacs, T., and Balla, T. (2006). Rapidly inducible changes in phosphatidylinositol 4, 5-bisphosphate levels influence multiple regulatory functions of the lipid in intact living cells. *J. Cell Biol.* **175,** 377–382.

Vennekens, R., Hoenderop, J. G., Prenen, J., Stuiver, M., Willems, P. H., Droogmans, G., Nilius, B., and Bindels, R. J. (2000). Permeation and gating properties of the novel epithelial Ca^{2+} channel. *J. Biol. Chem.* **275,** 3963–3969.

Vergarajauregui, S., and Puertollano, R. (2006). Two di-leucine motifs regulate trafficking of mucolipin-1 to lysosomes. *Traffic* **7,** 337–353.

Wang, T., Xu, H., Oberwinkler, J., Gu, Y., Hardie, R. C., and Montell, C. (2005). Light activation, adaptation, and cell survival functions of the Na^+/Ca^{2+} exchanger CalX. *Neuron* **45,** 367–378.

Wei, W. L., Sun, H. S., Olah, M. E., Sun, X., Czerwinska, E., Czerwinski, W., Mori, Y., Orser, B. A., Xiong, Z. G., Jackson, M. F., et al. (2007). TRPM7 channels in hippocampal neurons detect levels of extracellular divalent cations. *Proc. Natl. Acad. Sci. USA* **104,** 16323–16328.

Xu, X. Z. S., Li, H. S., Guggino, W. B., and Montell, C. (1997). Coassembly of TRP and TRPL produces a distinct store-operated conductance. *Cell* **89,** 1155–1164.

Xu, H., Delling, M., Li, L., Dong, X., and Clapham, D. E. (2007). Activating mutation in a mucolipin transient receptor potential channel leads to melanocyte loss in varitint-waddler mice. *Proc. Natl. Acad. Sci. USA* **104,** 18321–18326.

Yoon, J., Cohen Ben-Ami, H., Hong, Y. S., Park, S., Strong, L. L. R., Bowman, J., Geng, C., Baek, K., Minke, B., and Pak, W. L. (2000). Novel mechanism of massive photoreceptor degeneration caused by mutations in the *trp* gene of *Drosophila*. *J. Neurosci.* **20,** 649–659.

Zhang, W., Chu, X., Tong, Q., Cheung, J. Y., Conrad, K., Masker, K., and Miller, B. A. (2003). A novel TRPM2 isoform inhibits calcium influx and susceptibility to cell death. *J. Biol. Chem.* **278,** 16222–16229.

CHAPTER THIRTY

MEASUREMENT OF OREXIN (HYPOCRETIN) AND SUBSTANCE P EFFECTS ON CONSTITUTIVELY ACTIVE INWARD RECTIFIER K^+ CHANNELS IN BRAIN NEURONS

Yasuko Nakajima[*] *and* Shigehiro Nakajima[†]

Contents

1. Introduction	614
2. Dissociated Culture of Cholinergic Neurons in the Basal Forebrain	615
2.1. Culture materials	616
2.2. Obtaining slices of the basal forebrain	616
2.3. Dissociation of nucleus basalis neurons	616
2.4. Plating dissociated neurons in culture dishes	617
2.5. Culture medium and maintenance of culture	619
2.6. Selection of cholinergic neurons	619
3. Effects of Orexin (Hypocretin) and Substance P on Constitutively Active Inward Rectifier K^+ (KirNB) Channels	620
3.1. Equipment	620
3.2. Effects of orexin and substance P on KirNB channels: whole-cell recordings	620
3.3. Orexin and substance P effects on KirNB channels: single-channel recordings	622
4. Signal Transduction of Substance P and Orexin Effects on KirNB Channels	625
4.1. PKC inhibitor	626
4.2. Pseudosubstrate of PKC	627
4.3. Phosphatase inhibitor	628
Acknowledgments	628
References	629

[*] Department of Anatomy and Cell Biology, College of Medicine, University of Illinois at Chicago, Chicago, Illinois, USA
[†] Department of Pharmacology, College of Medicine, University of Illinois at Chicago, Chicago, Illinois, USA

Methods in Enzymology, Volume 484
ISSN 0076-6879, DOI: 10.1016/S0076-6879(10)84030-3

© 2010 Elsevier Inc.
All rights reserved.

Abstract

Electrophysiological experiments in our laboratory have led to the discovery that the cholinergic neurons in the nucleus basalis in the rat forebrain possess constitutively active inward rectifier K^+ channels. Unlike cloned inward rectifier K^+ channels, these constitutively active inward rectifier K^+ channels were found to have unique properties, and thus were named "KirNB" (inward rectifier K^+ channels in the nucleus basalis). We found that slow excitatory transmitters, such as orexin (hypocretin) and substance P, suppress the KirNB channel, resulting in neuronal excitation. Furthermore, it was discovered that suppression of KirNB channels by these transmitters is through protein kinase C (PKC).

This chapter describes detailed electrophysiological techniques for investigating the effects of orexin and substance P on constitutively active KirNB channels. For this purpose, we also present a method for culturing nucleus basalis cholinergic neurons in which KirNB channels exist. Then, we describe the procedures through which PKC has been determined to mediate inhibition of KirNB channels by orexin and substance P. There are probably many other transmitters which may produce effects on KirNB channels. This chapter will enable researchers to investigate the effects of such transmitters on KirNB channels and their roles in neuronal functions.

1. Introduction

In 1985, we discovered that unique constitutively active inward rectifier K^+ (Kir) currents exist in cholinergic neurons in the basal forebrain (Stanfield et al., 1985). These cholinergic neurons exist in the nucleus basalis, the diagonal band, and the medial septal nucleus. These neurons have been recognized to be important for memory and cognition (Deutsch, 1971; Perry et al., 1999; Riekkinen et al., 1990). It has also been found that these cholinergic neurons selectively degenerate in Alzheimer's disease in humans (Coyle et al., 1983).

We investigated the properties of this constitutively active inward rectifier K^+ channel in cholinergic neurons in the nucleus basalis by using the on-cell as well as the inside-out single-channel recordings. The constitutively active Kir channels in these neurons open and close spontaneously near the resting membrane potential. We designated such constitutively active Kir channels as "KirNB" channels (Bajic et al., 2002). When the recording method was changed from the on-cell recording to the inside-out recording, the activity of the KirNB channels remained intact with their constitutive activity unchanged. The mean open time of KirNB channels was ~ 1 ms and their unitary conductance was ~ 23 pS (155 mM [K^+]$_0$). These characteristic features of KirNB channels are different from those of cloned Kir channels (Bajic et al., 2002). At present, these KirNB channels have not been genetically determined.

Further characteristics of the KirNB channel were discovered. When a slow excitatory transmitter, such as substance P or neurotensin, is applied, KirNB channels are suppressed, and the cell produces neuronal excitation (Bajic et al., 2002; Farkas et al., 1994; Stanfield et al. 1985; Yamaguchi et al., 1990).

We then investigated the effects of orexin on KirNB channels. Orexins (also named "hypocretins") are recently discovered neuropeptides, consisting of orexin A and orexin B, and are implicated in the sleep disorder, narcolepsy (Chemelli et al., 1999; de Lecea et al., 1998; Nishino et al., 2000; Sakurai et al., 1998; Thannickal et al., 2000). In our laboratory, orexin A was found to inhibit KirNB channels, resulting in neuronal excitation (Hoang et al., 2004).

We further investigated the signal transduction mechanisms of substance P and orexin effects on KirNB channels (Hoang et al., 2004; Takano et al., 1995). It was observed through single-channel recordings that substance P effects are induced by a diffusible messenger. It was found that staurosporine (a protein kinase C (PKC) inhibitor) and PKC pseudosubstrate PKC (19–36) suppressed substance P effects on KirNB (Takano et al., 1995). In addition, it was noticed that substance P irreversibly suppressed KirNB channels in neurons treated with okadaic acid (a phosphatase inhibitor), suggesting that substance P effects on KirNB channels are mediated by PKC phosphorylation of KirNB channels (Takano et al., 1995).

We have also shown that the orexin A-induced suppression of KirNB channels is mediated by a pertussis toxin (PTX)-insensitive G protein (such as $G_{q/11}$). The recovery from this suppression is performed by dephosphorylation (Hoang et al., 2004), suggesting that the effects of orexin are mediated through PKC.

KirNB channels seem to always be active in the nucleus basalis cholinergic neurons. We have not encountered such a unique characteristics of KirNB channels in other types of neurons we have investigated, such as noradrenergic neurons in the locus coeruleus (Grigg et al., 1996; Nakajima et al., 1996). Thus, for KirNB channel investigations, we have been using the dissociated culture of cholinergic neurons (developed by Nakajima et al., 1985) from the basal forebrain.

In this chapter, we first explain our method of culturing cholinergic neurons. We then present electrophysiological techniques for investigating the orexin or substance P effects on the constitutively active KirNB channel.

2. Dissociated Culture of Cholinergic Neurons in the Basal Forebrain

Dissociated culture of cholinergic neurons in the basal forebrain was developed in our laboratory (Nakajima et al., 1985). In this culture, we discovered the existence of constitutively active Kir channels (Stanfield *et al.*, 1985); we later named them KirNB channels (Bajic *et al.*, 2002). This unique channel is so far encountered only in the basal forebrain cholinergic neurons. We used the cultured nucleus basalis cholinergic neurons for our studies on the effects of orexin or substance P on KirNB channels.

First, we describe the protocol of making dissociated culture of nucleus basalis cholinergic neurons. Since the initial reports of our culture methods (Nakajima and Masuko, 1996; Nakajima *et al.*, 1985), steady improvements have been made.

2.1. Culture materials

Culture neurons are obtained from young new born rats (2–5-day-old rats). After anesthesia with isoflurane, the forebrain region is rapidly removed, and the rats are quickly sacrificed by decapitation.

2.2. Obtaining slices of the basal forebrain

Four to six pieces of the forebrain are placed in a 3.5 cm petri dish with the anterior surface of the forebrain facing down. The forebrain is embedded in warm ($\sim$45 °C) 3.5% agar dissolved in a balanced salt solution; a buffer (pH 7.4) consisting of 130 mM NaCl, 4.5 mM KCl, 2 mM CaCl$_2$, 33 mM glucose, and 5 mM PIPES (piperazine-N,N'-bis[2-ethanesulfonic acid]). The petri dish is then immediately placed over ice-cold water for $\sim$5 min to solidify the agar.

To prepare for sectioning, the hardened agar block is trimmed and affixed to a cutting dish (a sterile disposable square petri dish, 10 × 10 × 1.5 cm) using instant glue (Alpha cyanoacrylate). The cutting dish is mounted on a Vibratome machine (Lancer 1000) (Fig. 30.1). The Vibratome is placed in a laminar flow cabinet. The cutting dish is then filled with oxygenated cold BBS solution (Fig. 30.1). The forebrain coronal slices of $\sim$400 µm thickness are aseptically sectioned by positioning the blade holder at a 16° angle with slow cutting speed and large amplitude vibration.

2.3. Dissociation of nucleus basalis neurons

From the forebrain sections, the area of the nucleus basalis is dissected under a microscope, using a pair of micro-knives (30 gauge hypodermic needles attached to 1 ml tuberculin syringes). Figure 30.2A shows an example of the nucleus basalis containing cholinergic neurons histochemically stained for acetylcholinestease (AChE). After training with stained materials, we successfully dissected the nucleus basalis without staining. The dissected pieces are collected in a conical 15 ml centrifuge tube and treated with oxygenated papain solution ($\sim$1.5 ml) for 15 min at 37 °C. Additional fresh oxygenated papain solution ($\sim$1 ml) is then added for another 15 min at 37 °C.

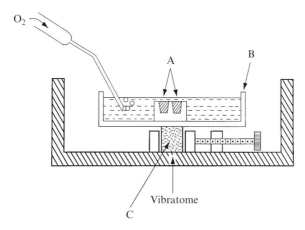

Figure 30.1 Schematic diagram showing brain slice sectioning. An agar block containing forebrains (A) is attached to the bottom of a square petri dish (10 × 10 × 1.5 cm) (B), and this cutting dish is mounted on a Vibratome by the plastic pedestal (C). The agar block is immersed in a balanced salt solution which is continuously oxygenated (Nakajima and Masuko, 1996).

After these treatments, the pieces of the nucleus basalis are rinsed three times with culture medium and then dissociated by gentle trituration using a fire-polished Pasteur glass pipette.

Papain solution is made by dissolving 12 units/ml papain (Worthington Biochemical Co.) in L-15 culture medium (~1.5 ml) containing 0.2 mg/ml DL-cysteine and 0.2 mg/ml bovine serum albumin (pH 7.3). When the papain solution is made, the solution is initially cloudy, but this turbidity disappears within 15 min. Then, the papain solution is sterilized by filtering.

2.4. Plating dissociated neurons in culture dishes

We plate dissociated nucleus basalis neurons in a central small well (~1.2 cm in diameter) inside a 3.5-cm culture dish (Fig. 30.3). The well was made by drilling a hole at the center of a culture dish. A plastic Aclar film or a cover glass is adhered to the bottom of the culture dish using inert glue (Dow Corning, 3140 RTV) (Fig. 30.3). This small well is necessary to efficiently utilize the limited number of cells obtained from small pieces of the nucleus basalis. The bottom of the well is coated with rat tail collagen and glial feeder layer cells (Fig. 30.3; Nakajima and Masuko, 1996).

Glial feeder layer cells are obtained by culturing cerebral cortex cells of newborn rats in a 10-cm culture dish. After a few weeks of culture, the cells start to divide. These cells are replated into another culture dish. As the procedure is repeated, eventually neurons die and glial cells remain. These glial feeder layer cells are plated in a well inside a culture dish 2–3 days

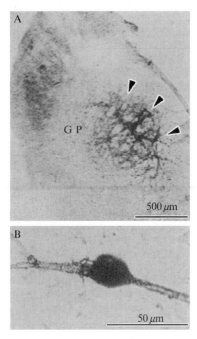

Figure 30.2 (A) Coronal vibratome slice of the forebrain of a 2-day-old newborn rat. The slice was stained for acetylcholinesterase. The dark areas are positive to acetylcholinesterase stain (arrowheads) and are located at the medial aspect of the globus pallidus (GP). They are neurons homologous to those in the nucleus basalis of Meynert in humans. (B) A cultured neuron from the nucleus basalis, showing intense staining for acetylcholinesterase, indicating that the neuron is cholinergic. Cultured for 21 days. Modified from Nakajima *et al.* (1985).

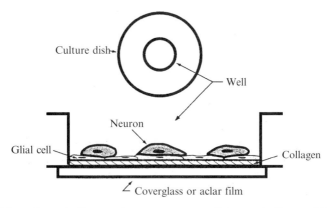

Figure 30.3 Schematic diagram showing a center well (1.2 cm in diameter) in a culture dish (3.5 cm in diameter). A well is made in the center of the culture dish. The bottom of the well is covered by a plastic Aclar film or a cover glass. The well is first covered by collagen and then by glial feeder layer cells. Dissociated neurons are cultured on the glial feeder layer which has been prepared beforehand (Nakajima and Masuko, 1996).

before culturing the neurons of the nucleus basalis. One day after the feeder layer cells are plated, antimitotic chemicals [5′fluoro-2′deoxyuridine (15 µg/ml) and uridine (35 µg/ml)] are applied to the feeder layer to suppress the overgrowth of feeder layer cells. This method of using the glial feeder layer promotes the survival of cultured neurons.

After the dissociation of dissected nucleus basalis pieces, ~0.15 ml of the cell suspension is placed in a feeder layer-covered well at a cell density of ~2–5 × 10^4 cm^{-2} (Fig. 30.3). The culture dishes are kept in an incubator with 10% CO_2 at 37 °C. After 2–3 h of incubation, 2.5 ml of culture medium is added to each culture dish.

2.5. Culture medium and maintenance of culture

The culture medium is a minimum essential medium with Earle's salts, modified with 0.292 mg/ml L-glutamine, 3.7 mg/ml $NaHCO_3$, 5 mg/ml D-glucose, 2% rat serum (prepared in our laboratory) and 10% horse serum, 10 µg/ml L-ascorbic acid, 50 units/ml penicillin, and 50 µg/ml streptomycin. A conditioned culture medium is always used, that is, a medium which has been kept in a culture dish with glial feeder layer cells overnight. With this procedure, toxic glutamate contained in fresh culture medium can be absorbed by glial cells (Baughman et al., 1991). The culture medium is usually exchanged with a new medium every 4 weeks.

Most of our electrophysiological experiments are performed on 2- to 3-week-old cultures. In some experiments, we use up to 2- to 3-month-old cultures.

2.6. Selection of cholinergic neurons

We found that cholinergic neurons in the basal forebrain tend to be quite large with a diameter range of 22–38 µm (Nakajima et al., 1985). After electrophysiological experiments, 13 cells were processed for AChE histocytochemistry. Twelve of these 13 (92%) cells produced AChE positivity (Nakajima et al., 1985), suggesting that they are cholinergic. According to Levey et al. (1983), all neurons with strong AChE positivity in the basal forebrain were choline acetyltransferase positive. Figure 30.2B shows a cultured cholinergic neuron stained with AChE histocytochemistry.

KirNB channels are a particular type of ion channels located in cholinergic neurons in the basal forebrain. To investigate KirNB channels of nucleus basalis neurons, it is important to select large neurons (about 30 µm in diameter), which are most likely to be cholinergic. It is also recommended that at an early stage of experiments, a certain number of cells be treated with AChE histochemistry or choline acetyltransferase immunochemistry after electrophysiological data are obtained, to confirm that experimented cells are cholinergic.

3. Effects of Orexin (Hypocretin) and Substance P on Constitutively Active Inward Rectifier K^+ (KirNB) Channels

3.1. Equipment

A simple oscilloscope for observation during experiments; an inverted phase-contrast microscope (with DC power supply for the light bulb to avoid AC interference); a binocular microscope (Axiovert 135; Zeiss); a manipulator (to insert a glass pipette into the cell); a patch–clamp amplifier (Axopatch 200B; Axon Instruments); a digital data recorder (Instrutech; VB-10B); P-clamp software (Molecular Devices); a videocassette recorder (VCR, Sony SLV-N51); a digidata 1440A converter (Axon Instruments); a drug application device (OCTAFLOW system; ALA Scientific Instruments), a pipette puller (Sutter Instrument, Co.); a stimulator (Medical Systems Corp; Greenvale, NY); a pipette tip polisher (Narishige); a Faraday cage (a home-made cage); an air-cushioned table (TMC; MICRO-g); a chart recorder (Astro-med; DASH II).

3.2. Effects of orexin and substance P on KirNB channels: whole-cell recordings

3.2.1. Introduction

Orexins are excitatory transmitters consisting of orexin A and orexin B and implicated in sleep disorders and narcolepsy (Chemelli et al., 1999; de Lecea et al., 1998; Sakurai et al., 1998). The orexin receptors (orexin receptor type 1 and type 2) are present in certain brain nuclei, including the ascending arousal system. It is reported that the nucleus basalis contains both orexin receptor type 1 and type 2 (Marcus et al., 2001; Trivedi et al., 1998). The nucleus basalis is located in the basal forebrain and contains large cholinergic neurons. Degeneration of the cholinergic neurons in the nucleus basalis, diagonal band, and medial septal nuclei would cause memory loss and Alzheimer's disease (Coyle et al., 1983).

In order to obtain neurons containing orexin receptors, neurons from the nucleus basalis are cultured for 2–4 weeks. Large neurons ($\sim$30 μm in diameter), which are mostly cholinergic (Nakajima et al., 1985), are used for the experiments. Application of orexin A or substance P inhibits a special type of channel, which we call KirNB channels, located in the basal forebrain cholinergic neurons. The KirNB channels seem to be one type of inward rectifying channels, causing neuronal inhibition (Bajic et al., 2002; Hoang et al., 2004; Stanfield et al., 1985; Yamaguchi et al., 1990).

Application of orexin A to a nucleus basalis neuron reduces cell conductance through a PTX-insensitive G protein (Hoang et al., 2004).

The orexin-suppressed currents are inwardly rectifying with a reversal potential around the K^+ equilibrium potential (E_K). Therefore, the orexin-induced inhibition of KirNB is mediated by a PTX-insensitive G protein ($G_{q/11}$). In many respects, the action of orexin is very similar to that of substance P.

3.2.2. Experiments using the whole-cell clamp

(a) External and internal solutions for KirNB channels: The external solution—141 mM NaCl, 10 mM KCl, 2.4 mM $CaCl_2$, 1.3 mM $MgCl_2$, 11 mM D-glucose, and 0.5 µM tetrodotoxin (pH 7.2). The patch pipette solution—141 mM K-D-gluconate, 10 mM NaCl, 5 mM HEPES–KOH, 0.5 mM EGTA–KOH, 0.1 mM $CaCl_2$, 4 mM $MgCl_2$, 3 mM Na_2ATP, 0.2 mM GTP (pH 7.4).

(b) Experimental procedures: Neurons, cultured for 2–4 weeks, were used. First, square-wave voltages are repetitively applied to the electrode. Then, let the electrode tip approach toward the cell surface. The square-wave current amplitude begins to decrease, indicating that the effective electrode resistance increases (because of the immediate presence of the cell membrane). Now, a negative pressure is applied to the electrode. This would suck up part of the cell membrane into the inside of the electrode tip. Addition of a larger negative pressure breaks a part of the membrane located inside the electrode. Thus, the electrode is now electrically continuous with the cell inside.

3.2.3. Effect of orexin A: experiments

Both orexin A and substance P suppress the inward rectifier K^+ channels in cultured nucleus basalis neurons. Figure 30.4 shows one of our experiments (Hoang et al., 2004). The patch pipette contained 0.2 mM GTPγS together with the routine ingredients. The holding potential was −84 mV.

In Fig. 30.4A, orexin A (3 µM) was applied. This resulted in a decrease of the membrane conductance. Each of the recurrent command pulses consisted of a square-wave depolarization (20 mV, 100 ms) and a hyperpolarization (50 mV, 100 ms).

Figure 30.4B shows the current–voltage (I–V) relation of the orexin A-suppressed current in a GTPγS-loaded NB neuron. First, a I–V relation was determined before (filled squares) and after (open circles) application of orexin. The difference between those two I–V curves would be the current amplitude that was suppressed by the orexin A application; this orexin A-suppressed current is plotted in Fig. 30.4C. It is clear from Fig. 30.4C that this current is inwardly rectifying K^+ channel since its reversal potential (−74 mV) is near E_K (−71 mV).

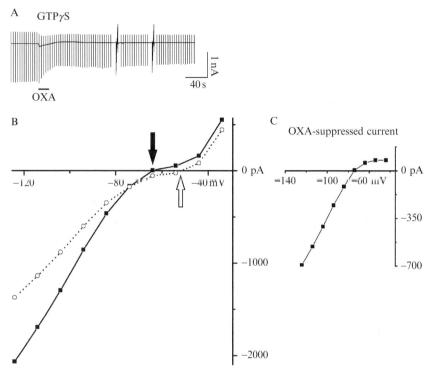

Figure 30.4 (A) Application of orexin A (3 μM) to a GTPγS-loaded nucleus basalis neuron resulted in an irreversible conductance decrease (closing of KirNB channels). (B) I–V relationship of whole-cell current before orexin A application (solid squares, solid line) and after orexin A application (hollow circles, dotted line). Comparison of the two curves indicates that the resting potential shifted from −64 mV (solid arrow) to −52 mV (hollow arrow) with about 12 mV depolarization by orexin A. (C) I–V relationship of the orexin A-suppressed (sensitive) current. Conductance showed an inward rectification and its reversal potential was close to E_K (−71 mV). Modified from Hoang et al. (2004).

3.3. Orexin and substance P effects on KirNB channels: single-channel recordings

Almost all large nucleus basalis neurons are cholinergic. These neurons show spontaneously active KirNB channels at the single channel level (Fig. 30.5A1). The unitary conductance of KirNB channels was ∼23 pS at 155 mM [K^+]$_0$ (Fig. 30.5C; Hoang et al., 2004). The mean open time of KirNB channels was 1.1–1.3 ms (Fig. 30.5D) (Hoang et al., 2004). These single-channel characteristics of KirNB channels are quite different from any of the known inward rectifier K^+ channels. (Nucleus basalis cholinergic neurons also possess Kir3 (G protein-gated inward rectifier potassium) channels. These Kir3 channels have larger unitary conductance, namely, ∼30–40 pS).

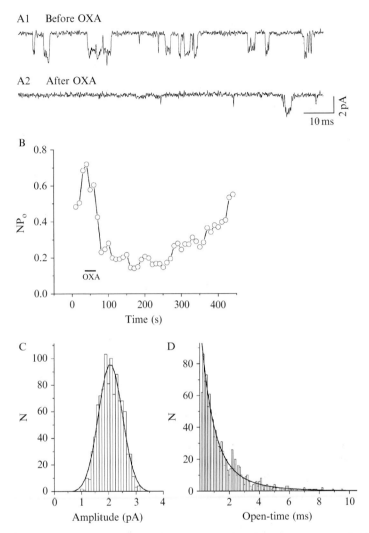

Figure 30.5 A1 and A2. On-cell single-channel events of KirNB were recorded from a nucleus basalis neuron before (A1) and after (A2) orexin A (OXA) (3 μM, 30 s) application. OXA closes channels in a nucleus basalis neuron. Patch pipette contained 155 mM KCl. Bath solution contained 5 mM KCl. Transmembrane potential was estimated to be −87 mV. (B) Application of OXA (3 μM, 30 s) resulted in a transient decline of NP_o. Each circle represents the average of a 10 s interval. (C) Amplitude histogram of KirNB channels. The unitary conductance was calculated to be 23 pS. (D) Open time histogram of KirNB channels. The mean open time was 1.33 ms. Modified from Hoang *et al.* (2004).

Previously, we noticed that orexin A and substance P suppressed KirNB channels through PKC (Bajic *et al.*, 2002; Hoang *et al.*, 2004; Takano *et al.*, 1995). In those experiments, we investigated the effects of orexin and substance P on KirNB channels using the cell-attached (on-cell) mode of single-channel recordings with the application of orexin A or substance P outside the patch pipette. The single-channel activity of KirNB channels was observed using both cell-attached mode as well as the inside-out mode of single-channel recordings as seen in Fig. 30.6A1 and A3 (Bajic *et al.*, 2002).

3.3.1. Solutions for the single-channel recordings with cell-attached mode

The external solution (5K Krebs solution): 5 mM KCl, 146 mM NaCl, 2.4 mM CaCl$_2$, 1.3 mM MgCl$_2$, 11 mM D-glucose, 0.5 μM tetrodotoxin, and 5 mM HEPES–NaOH (pH 7.4). The patch pipette solution: 155 mM KCl, 2.4 mM CaCl$_2$, 1.3 mM MgCl$_2$, 11 mM D-glucose, and 5 mM HEPES–NaOH (pH 7.4).

3.3.2. Experimental procedures

First, the culture medium is exchanged to the 5K$^+$ Krebs solution. Then, a rather large neuron (~30 μm in diameter) is selected. This is likely to be a cholinergic neuron. The cell is patched with a patch electrode with a resistance of ~5 MΩ. When the gigaohm seal is obtained, we do not break the cell membrane. With this procedure, spontaneously active

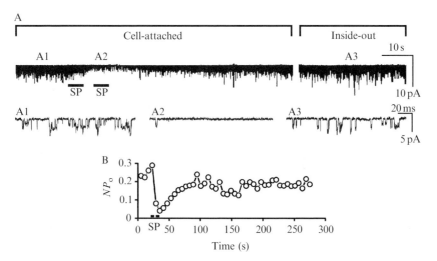

Figure 30.6 Single-channel activity of KirNB channels from a nucleus basalis neuron. (A) A1 and A2: Single-channel currents with cell-attached mode. The external solution was the 5K Krebs solution. 0.5 μM substance P (SP) application induced a transient decrease in channel activity. The membrane potential of the patched region was 27 mV more hyperpolarized than resting potential. A3: KirNB channels recorded with the inside-out configuration. (B) SP application produced a transient decline of NP_o. This graph was derived from the same patch as that in A. Modified from Bajic *et al.* (2002).

single-channel events (KirNB) will be observed as shown in Figs. 30.5A1 and 30.6A1. During the on-cell mode of single-channel recordings, the holding potential is about 27 mV more negative than the resting potential. The chance of the opening of KirNB channel in the patch is expressed as "NP_o" (N = the number of channels being recorded; P_o = probability of each channel to open) (Friedrich et al., 1988). The NP_o of the spontaneous activity of KirNB channels in the Fig. 30.5A1 was ~0.25 (Hoang et al., 2004). We also obtained NP_o of ~0.30 in another experiment (Bajic et al., 2002). After application of orexin A (3 μM) or substance P (0.3 μM) in the external solution, the opening of single-channel events drastically decreased as seen in Fig. 30.5A2 and B (Hoang et al., 2004) and in Figs. 30.6A2 and B (Bajic et al., 2002). Then, after washing out orexin or substance P, KirNB channel activity reappears as seen in Fig. 30.5B (Hoang et al., 2004) and in Fig. 30.6A3 and B (Bajic et al., 2002).

4. SIGNAL TRANSDUCTION OF SUBSTANCE P AND OREXIN EFFECTS ON KIRNB CHANNELS

Both orexin and substance P inhibit KirNB channels, leading to neuronal excitation. We have investigated the signal transduction mechanisms of substance P and orexin effects on KirNB channels (Hoang et al., 2004; Nakajima et al., 1988; Takano et al., 1995, 1996). First we observed that the substance P effect on KirNB was mediated through a PTX-insensitive G protein (Nakajima et al., 1988). Next, we identified this G protein to be a $G_{q/11}$(Takano et al., 1996). We also noticed that the substance P effect is mediated through phospholipase C-β1 (PLC-β1) (Takano et al., 1996). In addition, we found that staurosporine (a PKC inhibitor) as well as PKC pseudosubstrate PKC (19–36) suppressed the substance P effects on KirNB (Fig. 30.7; Takano et al., 1995), suggesting that PKC is a signal transducer. In addition, we observed that substance P irreversibly suppressed KirNB channels in neurons treated with okadaic acid (a phosphatase inhibitor), suggesting that the substance P effects on KirNB channels are mediated by phosphorylation of KirNB channels by PKC.

In orexin experiments, we observed that the orexin A-induced suppression of KirNB channels is mediated through a PTX-insensitive G protein (perhaps $G_{q/11}$) (Hoang et al., 2004). It was also found that the pretreatment with okadaic acid prevented the recovery of this orexin A effect. This fact also suggests that the orexin A effects on KirNB channels are mediated through phosphorylation of the channels by the signal transducer, PKC (Fig. 30.8) (Hoang et al., 2004). Figure 30.9 is a schematic diagram of our proposed signal transduction pathways, through which KirNB channels are inhibited by orexin or substance P.

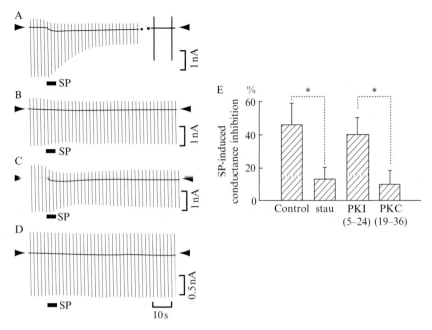

Figure 30.7 PKC inhibitors (staurosporine and PKC(19–36)) suppressed the substance P (SP) effects on KirNB currents in nucleus basalis neurons: this could indicate that PKC is a signal transducer. (A) SP effects on a control neuron. (B) SP effects on a staurosporine (100 nM)-treated neuron. (C) SP effects on a neuron loaded with protein kinase A inhibitor, PKI (5–24) (20 μM). (D) SP effects on a neuron loaded with PKC inhibitor, PKC (19–36). (E) SP-induced inhibition of neurons treated with staurosporine (stau) or loaded with protein kinase inhibitors PKI (5–24) or PKC (19–36). Modified from Takano et al. (1995).

Below, we describe three possible tests, by which the PKC involvement in orexin and substance P effects on KirNB channels could be determined. These strategies will use (1) PKC inhibitor, (2) pseudosubstrate of PKC, or (3) phosphatase inhibitor.

4.1. PKC inhibitor

Staurosporine, a broad spectrum protein kinase inhibitor (Tamaoki et al., 1986), was used in our experiments with the whole-cell clamp technique (Takano et al., 1995).

Cultured nucleus basalis neurons are first treated with staurosporine (100 nM in 0.01% DMSO for ∼40 min), followed by the application of SP (0.3 μM) or orexin A (3 μM). For controls, the external solution containing 0.01% DMSO is used. Figure 30.7 shows our experimental results indicating that staurosporine suppressed the substance P effect on KirNB channels (Takano et al., 1995).

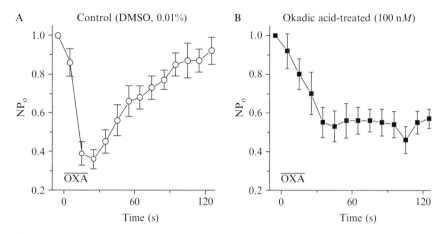

Figure 30.8 Effects of okadaic acid pretreatment on the orexin A-induced inhibition of KirNB channels. The cell-attached mode of single-channel recordings was used. Patch pipette contained 155 mM KCl. Bath contained 5 mM K^+. (A) Orexin A (OXA) (3 μM) application to control nucleus basalis cells (incubated in 0.01% DMSO in the external solution for 2–5 h) induced a transient decline in NP_o. (B) OXA application to cells pretreated with okadaic acid (0.1 μM in the external solution containing 0.01% DMSO, 2–3.5 h) induced a long-lasting decline in NP_o. Modified from Hoang et al. (2004).

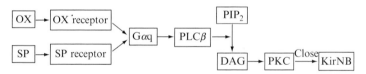

Figure 30.9 Proposed signal transduction pathways of orexin (OX) and substance P (SP) effects on KirNB channels. PLCβ (phospholipase Cβ); PIP_2 (phosphatidyl inositol-4,5-bisphosphate); DAG (diacylglycerol).

4.2. Pseudosubstrate of PKC

PKC (19–36), being a PKC pseudosubstrate, is a specific inhibitor of PKC since it competes with endogenous substrates of PKC (House and Kemp, 1987). PKC (19–36) is applied into the cell cytoplasm through a patch pipette. For this purpose, patch pipettes having a low resistance (∼2 MΩ) and filled with a patch pipette solution containing 30 μM PKC (19–36) are used. After patching a neuron and breaking the cell membrane, it is necessary to wait at least 5 min before starting to investigate the effects of orexin A (3 μM) or substance P (0.3 μM); waiting is necessary to ensure the introduction of PKC (19–36) into the cytoplasm. As a control for the PKC (19–36), the same concentration of a protein kinase A specific inhibitor,

PKI (5–24), is used. Figure 30.7C, D, and E (summary) show that the effect of substance P on KirNB was reduced with the intracellular application of PKC (19–36).

4.3. Phosphatase inhibitor

Inhibition of KirNB by orexin A and substance P recovers spontaneously (Hoang et al., 2004; Stanfield et al., 1985; Yamaguchi et al., 1990). We found that the recovery of these transmitters' effects was suppressed by the pretreatment with okadaic acid (Hoang et al., 2004; Takano et al., 1995). Okadaic acid is an inhibitor of protein phosphatase type 1 and type 2A, which belong to serine/threonine protein phosphatase (Bialojan and Takai, 1988; Nairn and Shenolikar, 1992). These experiments suggest that the suppression of KirNB channels by orexin A and substance P is due to the phosphorylation of KirNB channels by PKC and that the recovery would be caused by dephosphorylation of the channels (Bajic et al., 2002; Hoang et al., 2004; Takano et al., 1995).

In the experiments using okadaic acid, we employed on-cell single-channel recordings of KirNB channels. In this experiment, KirNB channel activities are expressed in NP_o (the possible numbers of channels multiplied by the open probability of the channel, namely, opening frequency of channels in the patch; see Section 3.3). For the okadaic acid experiments, 2–3-week-old nucleus basalis cultures are used. The culture medium is exchanged with 5K Krebs solution (5 mM KCl, 146 mM NaCl, 2.4 mM CaCl$_2$, 1.3 mM MgCl$_2$, 11 mM D-glucose, 0.5 μM tetrodotoxin, and 5 mM HEPES–NaOH, pH 7.4). Then, 5K Krebs solution is exchanged—5K Krebs solution containing 100 nM okadaic acid and 0.01% DMSO. The culture is incubated at 37 °C for about 2.5 h, after which on-cell single-channel experiments are performed, and orexin A (3 μM, 12 s) is applied. During the experiment, NP_o is obtained at a 10 second interval. Control experiments are performed in 5K Krebs solution containing 0.01% DMSO. Figure 30.8 is an example of such experiments (Hoang et al., 2004), showing an irreversible decrease of NP_o in okadaic acid treated neurons. In the control experiment, a decrease of NP_o after the application of orexin A is seen; however, as expected, its values spontaneously recovered.

The results of the above three types of experiments (use of staurosporine, use of PKC (19–36), and use of okadaic acid) all agree with the idea that PKC is involved in substance P- and orexin-induced inactivation of KirNB channels. Figure 30.9 is a schematic diagram showing this idea.

ACKNOWLEDGMENTS

We deeply thank the investigators who contributed greatly to the success of the experiments. The research relevant to this chapter was supported by the National Institute of Health grants NS043239 and AG06093.

REFERENCES

Bajic, D., Koike, M., Albsoul-Younes, A. M., Nakajima, S., and Nakajima, Y. (2002). Two different inward rectifier K^+ channels are effectors for transmitter-induced slow excitation in brain neurons. *Proc. Natl. Acad. Sci. USA* **99**, 14494–14499.

Baughman, R. W., Huettner, J. E., Jones, K. A., and Khan, A. A. (1991). Cell culture of neocortex and basal forebrain from postnatal rats. *In* "Culturing Nerve Cells," (Gary Banker and Kimberly Goslin, eds.), pp. 227–249. M.I.T. Press, Cambridge.

Bialojan, C., and Takai, A. (1988). Inhibitory effect of a marine-sponge toxin, okadaic acid, on protein phosphatases. *Biochem. J.* **256**, 283–290.

Chemelli, R. M., Willie, J. T., Sinton, C. M., Elmquist, J. K., Scammell, T., Lee, C., Richardson, J. A., Williams, S. C., Xiong, Y., Kisanuki, Y., Fitch, T. E., Nakazato, M., *et al.* (1999). Narcolepsy in orexin knockout mice: Molecular genetics of sleep regulation. *Cell* **98**, 437–451.

Coyle, J. T., Price, D. L., and DeLong, M. R. (1983). Alzheimer's disease: A disorder of cortical cholinergic innervation. *Science* **219**, 1184–1190.

de Lecea, L., Kilduff, T. S., Peyron, C., Gao, X.-B., Foye, P. E., Danielson, P. E., Fukuhara, C., Battenburg, E. L. F., Gautvik, V. T., Bartlett, F. S., II, Frankel, W. N., Van Den Pol, A. N., *et al.* (1998). The hypocretins: Hypothalamus-specific peptides with neuroexcitatory activity. *Proc. Natl. Acad. Sci. USA* **95**, 322–327.

Deutsch, J. A. (1971). The cholinergic synapse and the site of memory. *Science* **174**, 788–794.

Farkas, R. H., Nakajima, S., and Nakajima, Y. (1994). Neurotensin excites basal forebrain cholinergic neurons: Ionic and signal-transduction mechanisms. *Proc. Natl. Acad. Sci. USA* **91**, 2853–2857.

Friedrich, F., Paulmichl, M., Kolb, H. A., and Lang, F. (1988). Inward rectifier K channels in renal epithelioid cells (MDCK) activated by serotonin. *J. Membr. Biol.* **106**, 149–155.

Grigg, J. J., Kozasa, T., Nakajima, Y., and Nakajima, S. (1996). Single-channel properties of a G-protein-coupled inward rectifier potassium channel in brain neurons. *J. Neurophysiol.* **75**, 318–328.

Hoang, Q. V., Zhao, P., Nakajima, S., and Nakajima, Y. (2004). Orexin (hypocretin) effects on constitutively active inward rectifier K^+ channels in cultured nucleus basalis neurons. *J. Neurophysiol.* **92**, 3183–3191.

House, C., and Kemp, B. E. (1987). Protein kinase C contains a pseudosubstrate prototope in its regulatory domain. *Science* **238**, 1726–1728.

Levey, A. I., Wainer, B. H., Mufson, E. J., and Mesulam, M. M. (1983). Co-localization of acetylcholinesterase and choline acetyltransferase in the rat cerebrum. *Neuroscience* **9**, 9–22.

Marcus, J. N., Aschkenasi, C. J., Lee, C. E., Chemelli, R. M., Saper, C. B., Yanagisawa, M., and Elmquist, J. K. (2001). Differential expression of orexin receptors 1 and 2 in the rat brain. *J. Comp. Neurol.* **435**, 6–25.

Nairn, A. C., and Shenolikar, S. (1992). The role of protein phosphatases in synaptic transmission, plasticity and neuronal development. *Curr. Opin. Neurobiol.* **2**, 296–301.

Nakajima, Y., and Masuko, S. (1996). A technique for culturing brain nuclei from postnatal rats. *Neurosci. Res.* **26**, 195–203.

Nakajima, Y., Nakajima, S., Obata, K., Carlson, C. G., and Yamaguchi, K. (1985). Dissociated cell culture of cholinergic neurons from nucleus basalis of Meynert and other basal forebrain nuclei. *Proc. Natl. Acad. Sci. USA* **82**, 6325–6329.

Nakajima, Y., Nakajima, S., and Inoue, M. (1988). Pertussis toxin-insensitive G protein mediates substance P-induced inhibition of potassium channels in brain neurons. *Proc. Natl. Acad. Sci. USA* **85**, 3643–3647.

Nakajima, Y., Nakajima, S., and Kozasa, T. (1996). Activation of G protein-coupled inward rectifier K$^+$ channels in brain neurons requires association of G protein beta gamma subunits with cell membrane. *FEBS Lett.* **390**, 217–220.

Nishino, S., Ripley, B., Overeem, S., Lammers, G. J., and Mignot, E. (2000). Hypocretin (orexin) deficiency in human narcolepsy. *Lancet* **355**, 39–40.

Perry, E., Walker, M., Grace, J., and Perry, R. (1999). Acetylcholine in mind: A neurotransmitter correlate of consciousness? *Trends Neurosci.* **22**, 273–280.

Riekkinen, P., Jr., Sirvio, J., and Riekkinen, P. (1990). Similar memory impairments found in medial septal-vertical diagonal band of Broca and nucleus basalis lesioned rats: Are memory defects induced by nucleus basalis lesions related to the degree of non-specific subcortical cell loss? *Behav. Brain Res.* **37**, 81–88.

Sakurai, T., Amemiya, A., Ishii, M., Matsuzaki, I., Chemelli, R. M., Tanaka, H., Williams, S. C., Richardson, J. A., Kozlowski, G. P., Wilson, S., Arch, J. R. S., Buckingham, R. E., et al. (1998). Orexins and Orexin receptors: A family of hypothalamic neuropeptides and G protein-coupled receptors that regulate feeding behavior. *Cell* **92**, 573–585.

Stanfield, P. R., Nakajima, Y., and Yamaguchi, K. (1985). Substance P raises neuronal membrane excitability by reducing inward rectification. *Nature* **315**, 498–501.

Takano, K., Stanfield, P. R., Nakajima, S., and Nakajima, Y. (1995). Protein kinase C-mediated inhibition of an inward rectifier potassium channel by substance P in nucleus basalis neurons. *Neuron* **14**, 999–1008.

Takano, K., Yasufuku-Takano, J., Kozasa, T., Singer, W. D., Nakajima, S., and Nakajima, Y. (1996). Gq/11 and PLC-beta 1 mediate the substance P-induced inhibition of an inward rectifier K$^+$ channel in brain neurons. *J. Neurophysiol.* **76**, 2131–2136.

Tamaoki, T., Nomoto, H., Takahashi, I., Kato, Y., Morimoto, M., and Tomita, F. (1986). Staurosporine, a potent inhibitor of phospholipid/ Ca^{++} dependent protein kinase. *Biochem. Biophys. Res. Commun.* **135**, 397–402.

Thannickal, T. C., Moore, R. Y., Nienhuis, R., Ramanathan, L., Gulyani, S., Aldrich, M., Cornford, M., and Siegel, J. M. (2000). Reduced number of hypocretin neurons in human narcolepsy. *Neuron* **27**, 469–474.

Trivedi, P., Yu, H., MacNeil, D. J., Van der Ploeg, L. H., and Guan, X. M. (1998). Distribution of orexin receptor mRNA in the rat brain. *FEBS Lett.* **438**, 71–75.

Yamaguchi, K., Nakajima, Y., Nakajima, S., and Stanfield, P. R. (1990). Modulation of inwardly rectifying channels by substance P in cholinergic neurones from rat brain in culture. *J. Physiol.* **426**, 499–520.

CHAPTER THIRTY-ONE

CHARACTERIZATION OF G PROTEIN-COUPLED RECEPTOR KINASE 4 AND MEASURING ITS CONSTITUTIVE ACTIVITY *IN VIVO*

Bradley T. Andresen[*,†,‡]

Contents

1. Introduction	632
2. Selection of Cells/Model Systems to Study GRK4 Function	634
2.1. Identification of GRK4 splice variants	635
2.2. Identification of human GRK4 polymorphisms	636
3. Generation of Kinase Dead-GRK4	637
4. Functional Characterization of GRK4 Constitutive Activity	639
4.1. Receptor phosphorylation	639
4.2. Receptor internalization	642
4.3. β-Arrestin recruitment	646
5. Agonist-Mediated GRK4 Activity	648
6. Summary	649
Acknowledgments	649
References	650

Abstract

G protein-coupled receptor kinase 4 (GRK4) was originally identified in the brain and was initially thought to have a limited expression pattern and functionality; however, more recent studies have found that GRK4 is expressed in multiple tissues and cell types and that it contributes to cardiovascular disease. Additionally, human GRK4 exists as four splice variants and each variant can harbor at least three functionally relevant polymorphisms. The primary role of GRK4 is to phosphorylate G protein-coupled receptors (GPCR), which leads to desensitization of the G protein signaling mechanism while simultaneously recruiting β-arrestins and initializing the internalization of the receptor. Interestingly, GRK4 has been shown to be constitutively active in some, but not all, cases.

[*] Department of Internal Medicine, Division of Endocrinology, University of Missouri, Missouri, USA
[†] Department of Medical Pharmacology and Physiology, University of Missouri, Missouri, USA
[‡] Research Scientist, Harry S Truman VAMC, Columbia, Missouri, USA

Methods in Enzymology, Volume 484 © 2010 Elsevier Inc.
ISSN 0076-6879, DOI: 10.1016/S0076-6879(10)84031-5 All rights reserved.

A constitutive active GRK could lead to increased β-arrestin-mediated signaling while inhibiting traditional/canonical GPCR-mediated signaling mechanisms. Therefore, it is important to determine if GRK4 is constitutively active in a system. Measuring agonist-mediated activity of GRK4 is relatively straightforward since it inhibits second messenger signaling; however, only a few studies have directly examined the constitutive activity of GRK4 which requires techniques without an agonist. Since GRK4 has significant biological effects, identifying the mechanism underlying GRK4's constitutive activity and ligand-stimulated activity becomes increasingly important. Therefore, the methods provided here are designed to aid researchers in determining if GRK4 is expressed, and if so which GRK4 species is expressed, followed by procedures to identify if GRK4 is constitutively active in its model system. Last, procedures are explained for identifying if GRK4 is involved in its system in a nonconstitutive manner. The protocols described here are designed to be accessible to a wide range of scientists, which should allow for more laboratories to examine GRK4 constitutive activity as well as agonist-mediated activity.

1. Introduction

G protein receptor kinases (GRKs) are a family of seven serine/threonine kinases that phosphorylate GPCRs. All family members share a predominately amino-terminal regulator of G protein signaling (RGS) homology domain, also called a RH domain for RGS homology, followed by a serine/threonine kinase (STK) domain and require PIP_2 as a cofactor. A portion of the RH domain (two α-helices) follows the kinase domain of all GRKs, and the STK domain is related to the protein kinase C family. The GRK family is divided into three subfamilies: GRK1 and GRK7 are expressed in the eye and are involved in the signal transduction from rhodopsin; GRK2 and GRK3 are ubiquitously expressed and are involved in multiple signaling systems; and GRK4, GRK5, and GRK6 are expressed in a variety of tissues. Of all GRKs, GRK4 has high basal activity and can sequester receptors independent of ligand stimulation (Fig. 31.1) (Menard et al., 1996; Rankin et al., 2006); ergo GRK4 is constitutively active (see Section 2 for further discussion).

GRK4 was first named IT11 and was identified through positional cloning of the Huntington's disease locus (Ambrose et al., 1992). Shortly thereafter, the name GRK4 was given to the protein due to its similarity to the other GRK family members (Inglese et al., 1993). GRK4 exists as four splice variants: GRK4α is the full-length protein, GRK4β is missing exon 2 (32 amino acids) which contains the PIP_2 binding region (Pitcher et al., 1996), GRK4γ is missing exon 15 (46 amino acids), and GRK4δ is missing both exons 2 and 15 (Premont et al., 1996; Sallese et al., 1994). Both exon 2 and 15 encode for the beginning and end of the RH domain, suggesting

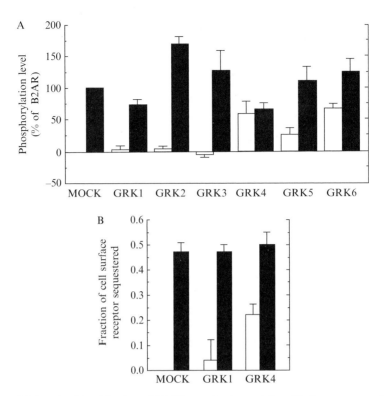

Figure 31.1 First identification of GRK4 constitutive activity; this figure was modified from Menard et al. with permission. Effect of the different GRKs on the phosphorylation of β_2AR (panel A) and effect of GRK1 and GRK4 on basal sequestration of β_2AR (panel B). All cells were transfected with β_2AR without (mock) or with 1 μg of the indicated GRK. In panel A, the cells were metabolically labeled with [$^{32}PO_4$], the receptors were immunoprecipitated and resolved on polyacrylamide gels, and the radioactivity migrating at the position of the glycosylated receptor (molecular mass 50–80 kDa) was quantitated using a PhosphorImager. Data for the β_2AR were normalized to the increase in radioactivity above basal in the mock sample [(3.6 ± 1.0)-fold above basal, 100%]. In panel B, the cells were prelabeled with 12CA5 antibody for 30 min on ice, washed, and stimulated with 100 μM ascorbate ± 10 μM isoproterenol (final concentrations) for 10 min at 37 °C. Cell surface receptor expression was measured by flow cytometry as described in Menard et al. (1996). Values were normalized to paired samples that were kept on ice. The results are the mean ± standard deviation of 3–5 experiments; open bars represent no isoproterenol and solid bars represent 10 μM isoproterenol.

that the splice variants could have significantly different biological properties. However, all variants are similarly localized at the plasma membrane, palmitoylated, and inhibit luteinizing hormone/choriogonadotropin receptor-mediated cAMP generation, albeit GRK4γ is the least effective at inhibiting cAMP generation (Premont et al., 1996). Additionally, GRK4 has a series of polymorphisms that have been identified: R65L, A142V, and A486V (Felder *et al.*, 2002; Premont *et al.*, 1996), which are associated with hypertension (Bengra *et al.*, 2002; Bhatnagar *et al.*, 2009; Felder *et al.*, 2002; Speirs *et al.*, 2004; Wang *et al.*, 2006; Williams *et al.*, 2004). Importantly, each polymorphism is present within all of the splice variants of GRK4 and increases the activity of GRK4 as measured by receptor phosphorylation (Felder *et al.*, 2002).

Original reports suggested that GRK4 is expressed predominately in the brain (Sallese *et al.*, 1994) and testes (Premont *et al.*, 1996); however, additional studies demonstrate that GRK4 is expressed in many other tissues and cell lines. Physiologically, the role of GRK4 has been defined primarily as phosphorylating $G\alpha_s$-coupled receptors leading to desensitization and internalization of the receptor. However, the GRK4 family contains a nuclear localization sequence (NLS); GRK4's NLS is located adjacent to the kinase domain and has been found in the nucleus (Johnson *et al.*, 2004). Additionally, GRK5, the most well studied of the GRK4 family, has recently been shown to act as a histone deacetylase kinase (Martini *et al.*, 2008) and phosphorylate p53 (Chen *et al.*, 2010). Since GRK4 and GRK5 are very similar and GRK4 is found in the nucleus, it is likely that GRK4 phosphorylates targets beyond GPCRs; however, since no targets are yet identified for GRK4, the methods detailed here will focus on GPCRs.

Constitutive activity of GRK4 is more difficult to measure than the function of GRK4. This is because, to deduce constitutive activity, the experiments must be conducted in the absence of ligand. However, it is also advisable to examine ligand-induced activation of GRK4 to ensure that GRK4 targets the receptor. Additionally, GRK4 does not target all receptors; therefore, it is also advisable to use GRK2, which targets most GPCRs, as a positive control for GRK activity.

2. Selection of Cells/Model Systems to Study GRK4 Function

There are three factors that should be taken into account when choosing a cellular system for analysis of GRK4 constitutive activity. The primary factor is whether to use a standard cell line, such as HEK293T or CHO cells; representative cell lines, such as PC-12 and MDCK cells; or primary cell lines, such as renal proximal tubule cells. Although this selection will be partially dictated by the larger approach of the project, it does impact the types of experiments that can be conducted and moreover can alter the results. For example, previous studies demonstrated that in CHO cells variants of GRK4γ have constitutive activity toward the D1 dopamine receptor (Felder *et al.*, 2002), yet in HEK293T cells GRK4α, but not

GRK4γ, has constitutive activity toward the D1 dopamine receptor (Rankin et al., 2006). Additionally, the initial in vitro data indicate that GRK4α requires an agonist to phosphorylate the β2-adrenergic receptor (AR) (Premont et al., 1996). Thus, the constitutive activity observed for GRK4 is likely dependent on a combination of the cellular environment and splice variant and polymorphism expressed. Consequently, it is advisable to either identify the GRK4 splice variant(s) in the system that is being modeled and use a primary cell line from that system or test each GRK4 splice variant in an appropriate model system. The second and third factors are to identify the GRK4 variants that are present in the chosen experimental system and to identify any polymorphisms, respectfully. For the remainder of this article, only human GRK4 will be discussed; consequently, any methods that are species specific, such as PCR, will have to be modified for each species examined.

2.1. Identification of GRK4 splice variants

The only way to precisely determine which splice variant of GRK4 is present in a sample is via reverse transcriptase (RT) PCR; however, since GRK4 has a limited range of expression, a quick test is to utilize Western blotting. As shown in Fig. 31.2, the Santa Cruz GRK4 antibody H-70 (sc-13079) is capable of detecting each human splice variant. When run on an 8% or 10% SDS-PAGE gel, it is possible to discriminate between GRK4α, which is 66.58 kDa; GRK4β/γ, which are 63.02 and 61.21 kDa, respectively; and GRK4δ, which is 57.65 kDa. In most systems, little to no GRK4 is present; thus a Western blot is a rapid and easy method to: (1) identify if GRK4 is present and (2) discriminate between three of the GRK4 splice variants. When running such an experiment, it is advisable to create positive controls.

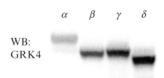

Figure 31.2 Western blot detection of GRK4 splice variants. HEK293T cells were transfected with 4 μg of the indicated GRK4 splice variant utilizing PEI, as described in Section 2.2, and run on a 10% SDS-PAGE gel. GRK4 was detected using the Santa Cruz GRK4 antibody H-70 (sc-13079) and horseradish peroxidase (HRP)-labeled Jackson ImmunoResearch Laboratories secondary antibodies that only recognize the light chain. The data were captured utilizing a Bio-Rad Chemidoc.

2.1.1. Generation of positive control samples

HEK293T cells, which express copious amounts of heterologously expressed protein, do not express GRK4 (Keever *et al.*, 2008); consequently, they can be transfected with plasmids harboring the GRK4 splice variants. The following protocol was used to transfect HEK293T cells to generate Fig. 31.2. First, the transfection agent, polyethylenimine (PEI) Sigma-Aldrich catalog number 408727, is prepared by making a 1 mg/mL solution in ultrapure water and bringing the pH to 7.4 with HCl, then filtering the solution with a 0.22 µm filter. The PEI transfection solution can be stored at $-70\,°C$ for at least 1 year. To transfect HEK293T cells in a 60-mm dish when they are $\sim$50–75% confluent:

1. Change the media before transfection (3 mL) so that the cells are in fresh growth media.
2. Mix 4 µg of DNA with 200–300 µL serum-free media (SFM) (both 200 and 300 µL SFM are effective).
3. Add 16 µL PEI and mix well.
a. If increasing the scale, keep the ratio of DNA to PEI at 1:4.
4. Let the solution incubate at room temperature for 10 min.
5. Add the solution to the HEK293T cells.

This procedure is nontoxic, inexpensive, and rapid; moreover, other easy-to-transfect cell lines can be transfected with PEI.

2.1.2. Primers to detect GRK4 splice variants

Utilizing RT-PCR, the expression of GRK4 splice variants within a sample can be partially discriminated via two separate approaches: selective amplification and size of the amplicon. Because GRK4δ is missing both exons 2 and 15, it is impossible to identify each GRK4 variant via selective amplification; therefore, the four primers listed in Table 31.1 can be used to discriminate between the splice variants. RT-PCR kits can be purchased from multiple vendors, and they have detailed methods on how to isolate RNA and perform the RT step. The primers in Table 31.1 will work with the following cycling program: 95 °C melt for 1 min, 59 °C anneal for 1 min, and 72 °C extension for 2 min. If using a species other than human, the primers in Table 31.1 can be used to determine the placement of species-specific primers.

2.2. Identification of human GRK4 polymorphisms

Currently, GRK4 polymorphisms can only be detected through sequencing, and the polymorphisms have only been found in humans. However, since they dramatically alter the activity of GRK4 (Felder *et al.*, 2002), when using human cells containing GRK4 the presence of polymorphic GRK4 should be examined. The most rapid method of identifying a polymorphism

Table 31.1 Primers to detect human splice variants in a cell or tissue of choice

# Sequence (5′–3′)	Direction	Location	Expected results
P1 ATGGAGCTCGAGAACATCGTG	Forward	1	P1 + P2: $\alpha/\gamma = 414$ bp and $\beta/\delta = 318$
P2 AGGGTTCTCCTCCTTCAGTCC	Reverse	414	
P3 CCGTGTTCAAGGACATCAACT	Forward	1340	P3 + P4: $\alpha/\beta = 387$ bp and $\gamma/\delta = 246$ bp
P4 TGGGTTCCACTTCCTTCTCAC	Reverse	1726	P1 + P4: $\alpha = 1726$ bp, $\beta = 1630$ bp, $\gamma = 1585$ bp, and $\delta = 1489$ bp

in a sample is to amplify, via PCR, the region around the polymorphism, gel purify the reaction, then submit the fragment for sequence analysis. The following protocol can be used to identify the polymorphisms in GRK4. First, utilize primers 1 and 4 from Table 31.1 with an error-checking polymerase, such as *pfu*, to generate near-full-length GRK4. Run the product on a 1% agarose gel and excise the band by cutting it out of the gel with a razor blade or X-ACTO knife. This gel and band slice can then be used in a commercially available agarose gel extraction kit, which is available from multiple vendors. Once purified, the DNA can be submitted for sequencing. Each sequencing center will have a slightly different set of instructions for how to prepare the sequencing reaction, so please see those instructions for further details. However, for this sequencing reaction, specific primers will have to be provided (Table 31.2).

3. Generation of Kinase Dead-GRK4

Kinase dead (KD)-GRK4α does not phosphorylate GPCRs and blocks receptor-mediated signaling (Sallese *et al.*, 2000). Additionally, KD-GRK4γ does not inhibit β-AR-mediated cAMP production (Keever *et al.*, 2008). To date, these are the only KD-GRK4 splice variants reported; however, KD-GRK4 (K216M, K217M GRK4α/γ or K182M, K183M GRK4β/δ) can be generated through site-directed mutagenesis. In both

Table 31.2 Primers to identify human GRK4 polymorphisms from cloned GRK4

Name[a]	Primer (5′–3′)	Position in GRK4α	Expected position in results	Sequence result (the polymorphic codon is in bold font)	
				Wild type	Polymorphic[b]
R65L-F	ATGGAGCTCGAGAACATCGT	1–20	[c]	AAGA**CGT**CTCT	AAGA**CTT**CTCT
R65L-R	GGTATTTCTGGTAAAGGGC	345–365	bp 151	AGAGA**CGT**CTT	AGAGA**AGT**CTT
A142V-F	TATGAAGTTGCCGATGATGA	267–287	bp 137	AAAA**GCC**TTTG	AAAA**GTC**TTTG
A142V-R	AATCCGCCTTTTCCTAGAAC	573–593	bp 148	CAAA**GCT**TTT	CAAA**GAC**TTTT
A486V-F	CAGAGGATGCCAAATCTATC	1240–1260	bp 197	CTCG**GCG**GTGA	CTCG**GTG**GTGA
A486V-R	TCAGCATTGCTTGGGTTCCA	1717–1737	[c]	TCAC**CGC**GAG	TCAC**CAC**CGAG

[a] F and R in the title refer to forward and reverse; it is recommended to use both primers to confirm that any result is not an artifact.
[b] The most likely polymorphic sequence.
[c] The position will alter depending on the splice variant being sequenced.

reported cases, Statagene's Quick Change Mutagenesis kit was used to generate KD-GRK4. The following primers can be used with their Quick Change II kit to generate human KD-GRK4 in any splice variant: 5′-gaaatgattcagggacattctccattcatgatgtacaaagagaaagtcaaatgggaggagg-3′ and 5′-cctcctcccatttgactttctctttgtacatcatgaatggagaatgtccctgaatcatttc-3′.

4. Functional Characterization of GRK4 Constitutive Activity

To date, GRK4 has only been shown to functionally regulate a small list of GPCRs; however, as more studies with GRK4 are conducted, this list of targets will likely expand to more GPCRs and other non-GPCR targets. In regard with other targets, Section 4.1 can be followed via replacing the receptor with the novel target protein. Throughout this section, GRK4 will be used to represent all the splice variants and polymorphisms.

4.1. Receptor phosphorylation

As their name indicates, the primary function of GRKs is to phosphorylate GPCRs. Therefore, the most direct examination of GRK4 activity is receptor phosphorylation. However, measuring receptor phosphorylation is not necessarily straightforward. For some receptors, there are phospho-specific antibodies that recognize GRK phosphorylation sites, but previous experience indicates that these antibodies are highly species specific (unpublished data). Additionally, it is notoriously difficult to generate a GPCR antibody and thus many of the antibodies are not considered of high quality and are not recommended for immunoprecipitation (IP), which makes it difficult to isolate GPCRs for analysis. Therefore, it is recommended that heterologously expressed GPCRs with an amino-terminal tag, such as Flag or AU1, are used in these studies. For these methods, it will be assumed that tagged receptors are used; however, untagged or endogenous receptors can be used in their place. Additionally, two methods will be provided for measuring GPCR phosphorylation by GRK4: a radioactive assay and, for those that do not want or do not have access to radioactive compounds, a nonradioactive assay.

4.1.1. Materials and solutions

- Inducible expression vector system (e.g., Tet-On) containing GRK4, KD-GRK4, and GRK2. (Note that these can be obtained from the laboratories that have published with these kinases, including this author.)
- Inducing agent (e.g., tetracycline or doxycycline).

- Expression vector containing a tagged GPCR of interest and the corresponding antibody.
- Transfection reagent (e.g., PEI—described in Section 2.1.1).
- RIPA lysis buffer (1% Triton X-100, 100 mM NaCl, 20 mM Tris pH 7.5, 2 mM EDTA, 10 mM MgCl$_2$, 10 mM NaF, and 40 mM β-glycerol phosphate can be made as a stock and stored at 4 °C for quick use at cold temperatures; before using, add phosphatase and protease inhibitors: 1 mM PMSF, 2 mM Na$_3$VO$_4$, 10 μg/mL aprotinin, and 10 μg/mL leupeptin).
- Protein A, or A/G agarose beads.
- GPCR agonist.
- ^{32}P-labeled sodium orthophosphate or phosphoserine antibodies (not all phosphoserine antibodies are identical; multiple vendors antibodies can be mixed together for this experiment).
- If looking at endogenous GRK4, then siRNA toward GRK4 will be needed to specifically knock down GRK4 and look for a reduction in effect. siRNA can be purchased through numerous commercial vendors, and each vendor will have detailed instructions for use of their product.

4.1.2. Methods: Radioactive assay

When beginning these assays, transfect cells so that each experiment has the following conditions: two negative controls (receptor + empty vector inducible system), two experimental groups (receptor + GRK4), two negative controls (receptor + KD-GRK4), and two positive controls (receptor + GRK2). To date, all receptors phosphorylated by GRK4 are also phosphorylated by GRK2, but GRK2 is not constitutively active and will require use of the GPCR agonist. After 2 days of transfection, add SFM (or media with 1% FBS depending on what cells are being used) containing 0.5 mCi/mL ^{32}P-labeled sodium orthophosphate to the cells. It is best to use phosphate-free medium, but not all media formulations are provided in phosphate-free versions; if this is the case, then ^{32}P orthophosphate can be added 2 h prior to the experiment. Six to eight hours before the experiment, treat half of the cells with the inducing agent (e.g., doxycycline) and half with vehicle. The 6–8 h time point should be optimized in the cellular system prior to these experiments; the goal is to do the experiments at the initial peak of GRK4 expression (please see the notes below for more details). For the GRK2 sample, add a maximal concentration (generally 1–10 μM) of agonist for 5 min.

Lyse the cells in 500 μL RIPA buffer, clarify the lysate by centrifuging at 10 k + rpm for 5 min at 4 °C, and transfer the supernatant to a new 1.5-mL tube; discard the pellet. Add the antibody used for IP to the samples and allow the samples to incubate while shaking at 4 °C for 1 h, add 25 μL of the protein A/G beads to the solution, and continue the incubation for 30 additional minutes. Spin the samples for roughly 7 s in a microcentrifuge at 4 °C using the quick spin function (if this is not available spin for 7 s at 9000 rpm), remove the supernatant, and wash the pelleted beads with ~750 µL PBS containing phosphatase and protease inhibitors at least three times by repeating the 7-s spin. Boil the pellet in 25 µL of 2 × Laemmli sample buffer for 5 min and run the samples on a SDS-PAGE gel and blot onto a PVDF or nitrocellulose membrane. This membrane can be exposed to film or a PhosphorImager and the bands that correspond to the GPCR analyzed by densitometry. Constitutive activity will be identifiable by an increase in phosphorylation of the GPCR in the induced state and not in the noninduced state or in the presence of KD-GRK4. Addition of an agonist to the GRK2 expressing cells will likely represent the maximal phosphorylation of the receptor under these conditions.

If examining endogenous GRK4 activity, then a similar assay can be done without the addition of the inducible GRK4; instead siRNA targeting GRK4 should be added to the cells. The same procedure can be followed; however, constitutive activity would be identified by a loss of receptor phosphorylation in the presence of the siRNA but not controls.

4.1.2.1. Methods: Nonradioactive assay A similar protocol will be followed to avoid the use of radioactive materials. Briefly, the same experimental setup should be followed except that when serum starving the cells no radioactive material should be added. Additionally, a Western blot with the phosphoserine antibodies should be performed on the PVDF or nitrocellulose membrane instead of exposing the blot directly to film or a PhosphorImager. A Zymend phosphoserine antibody (now a part of Invitrogen, Carlsbad, CA; catalog number 61-8100) has been shown to be effective for this assay (Felder *et al.*, 2002); however, as stated previously, multiple phosphoserine antibodies can be mixed together for this Western blot to ensure that all phosphorylation sites are covered.

4.1.3. Notes and tips

A phosphatase and protease pellet can also be used instead of the list given above. These are available from multiple vendors and are very easy alternatives to adding each component separately. Additionally, PMSF should be dissolved in EtOH.

An inducible system is recommended, but not necessary, for these experiments because when examining constitutive activity, the longer a protein is expressed the more likely there will be prolonged signaling mechanisms that are altered by the constitutively active protein. These prolonged alterations could impact the results through feedback loops or other unforeseen mechanisms such as β-arrestin-mediated signaling. When examining expression of GRK4 in an inducible system, it is advisable to test every 2 h for 8 h. Additionally, if the inducible system is chosen, it is advisable to make stably transfected cell lines. These lines can then be used in all the methods described here as well as for additional experiments; moreover, stable lines will reduce the variability between experiments as well as the cost of transfection reagents.

A simple and inexpensive method for scraping cells off a tissue culture dish is to use cut up credit cards. A credit card cut into three pieces makes a wonderful cell scraper that can be washed with distilled water after each experiment.

In an IP experiment, increasing the number of washes increases the clarity of the Western blot. If the Western blots have a lot of streaks or nonspecific bands, try increasing the number of washes, reducing the concentration of the antibody in the Western blot, or using less protein.

An IP of the GPCR and Western blot for the same GPCR should be conducted to identify at which molecular weight(s) the receptor resides. GPCRs are glycosylated; thus there is generally a large range of sizes that correspond to a single GPCR. For these experiments, it is advisable to use the Jackson ImmunoResearch Laboratories (West Grove, PA) secondary antibodies that only recognize the light chain, catalog numbers 115-035-174 and 211-032-171 for HRP (horseradish peroxidase)-conjugated anti-mouse and -rabbit antibodies, respectively. Generally, when conducting an IP and subsequent Western blot with the same antibody or an antibody from the same species, the heavy chain obscures most proteins around 50 kDa. This is near unglycosylated GPCRs, which can obscure the GPCR band; however, the heavy chain is not visible with the Jackson antibodies.

4.2. Receptor internalization

GRK phosphorylation of GPCRs leads to internalization of the receptor and in the case of constitutively active GRK4, basal receptor internalization is increased (Fig. 31.1B) (Menard et al., 1996). Therefore, a second measure of GRK4 activity is receptor internalization. Internalization of receptors can be measured via a variety of methods. As shown in Fig. 31.1, prelabeling the receptor with an antibody and then examining for the presence of the antibody via flow cytometry is one method. This assay is highly dependent on the quality of the antibody used; however, it works very well with Flag-tagged GPCRs. Additionally, this assay can be adapted to be performed in a fluorometer (see below) if a flow cytometer is not available. A second method is to conduct radiolabeled-ligand-binding assays. Although this assay requires the use of radioligands, it is the most sensitive and quantitative assay for receptors and can be used on endogenous receptors. Fortunately, there are a number of radiolabeled ligands for GPCRs, and many of these ligands are ^{3}H-labeled, which is a relatively harmless radionuclide.

4.2.1. Imaging methods for receptor internalization

The key material required for these assays is an antibody that recognizes the extracellular domains of a GPCR. Unfortunately, not all receptors have antibodies that target this area and are suitable for flow cytometry or immunohistochemistry. Consequently, this procedure may require a tagged GPCR, and thus transfection of the GPCR and the inducible GRK4 system. The same series of cells should be used as described in Section 4.1.2 with the following modifications: the cultured cells should be induced to express GRK4 for different times (i.e., 0 [control], 2, 4, 6, 8, and 12 h) where GRK4 is expected to be not expressed (controls) to expressed for a few hours. Additionally, a separate set of cells should be stimulated with agonist for 10 min as a positive control for receptor internalization; therefore, the cells expressing GRK2 are not required for these assays.

4.2.1.1. Flow cytometry measurements of membrane-bound GPCRs The stimulated cells should be washed with ice-cold PBS, then 1 mL flow buffer (1% BSA in PBS pH 7.4, 0.1% Azide, and 5 mM EDTA) is added to the cells, and they are gently scraped from the plate and collected in a 1.5-mL tube. Once the cells are in the tube, centrifuge the cells at 1500 rpm for 3 min to obtain a pellet of the live cells. Resuspend the pellet in the flow buffer, if using a 60-mm plate, 50 µL of buffer is sufficient, and add the primary antibody. The dilution/concentration of the antibody should be provided on the antibody spec sheet, but a 1:50 dilution is a good starting point if this information is not provided. Note that this dilution should be tested for each antibody used in pilot experiments. Allow the antibody to incubate for 30–60 min on ice. Wash the cells with 500 µL of the flow buffer three times, then resuspend in 200 µL of the flow buffer, and add the fluorescently labeled secondary antibody so that there is 1 µg of antibody per 10^6 cells. Allow this mixture to incubate on ice and in the dark for 30 min. Wash the cells with 500 µL flow buffer three times. The cells should be run through a flow cytometer immediately. Additionally, as negative controls, a sample treated with only the secondary antibody as well as a nontreated sample should be used to examine background fluorescence.

An alternate approach is to fix the cells with 3% paraformaldehyde (PFA) after the last wash step, and then run through a second series of washes to remove the PFA. This allows for the samples to be stored at 4 °C in the dark for days to weeks before flow cytometry analysis. This process is recommended if a flow cytometer is not immediately available.

Constitutive activity of GRK4 toward the GPCR examined would result in a decrease in the level of fluorescence, which in flow cytometry would be observed by a leftward shift in the fluorescent population compared to the control samples. This should not be seen with KD-GRK4, but should be seen with ligand stimulation.

4.2.1.2. Fluorometer measurements of membrane-bound GPCRs
Although flow cytometry is widely available, microplate fluorometers are more commonly available; therefore, the previous protocol can be modified as follows to measure GPCR internalization via a fluorometer. The cells should be grown in a multiwell plate; the plate type (number of wells) should be determined based on what the fluorometer can measure and the number of samples that will be examined. Once the cells are induced, the cells should be washed with ice-cold PBS, then fixed with cold 3% PFA for 30 min at 4 °C, and washed with PBS three times. The cells can then be treated like a Western blot: they should be blocked with 5% BSA in PBS followed by incubation of the primary antibody (targeting the GPCR) in 3% BSA in PBS for 1 h, and then washed six times with PBS. Following the washes, the fluorescently tagged secondary antibody and Hoechst 33342 should be added to the cells in 3% BSA in PBS. The antibody fluorophore should match the filter set on your fluorometer. The Hoechst 33342 is used as a control for the number of cells in the plate (Fig. 31.3) but should only be used if the fluorometer can excite around 350 nm and detect 480 nm. Last, the cells should be washed three times in PBS and then imaged. The results should be the same for this assay and the flow cytometry, and the data should be normalized to the Hoechst signal. Similar to the flow cytometry studies, the negative controls of secondary antibody only should be run to determine the nonspecific signal.

Note that the Hoechst 33342 signal is only linear if the cells are subconfluent to confluent; if the cells grow past the point of confluence, this assay will not necessarily provide a linear relationship. Therefore, these assays should be done at ∼95% confluence, and a standard curve should be generated similar to those presented in Fig. 31.3.

4.2.2. Radiolabeled-ligand-binding assays of receptor internalization
The gold standard in GPCR quantification is radiolabeled-ligand-binding assays. These assays can be adopted for internalization easily. Utilizing the same series of experiments described in the opening of this section as well as a radiolabeled, cell impermeant, antagonist, and a nonlabeled antagonist, the following procedure can be used to examine receptor internalization. After induction of the cells, wash the cells with ice-cold media on ice, and then add ice-cold media containing the radiolabeled ligand at ∼5 nM (this may change depending on the expression of the receptor) with Hoechst 33342 (as a normalization factor). One additional set of experiments must be added: cells treated with the radiolabeled ligand plus an excess of a nonlabeled "cold" antagonist; it would be best if the cold antagonist is not the same as the radiolabeled antagonist. An excess should be approximately two

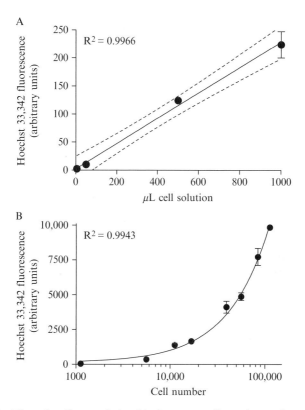

Figure 31.3 There is a linear relationship between cell number and Hoechst 33342 fluorescence. Mouse embryonic fibroblasts were seeded into a 12-well plate based on volume (panel A) and a 48-well plate based on cell number (panel B). Less than 16 h after seeding, the plates were treated with Hoechst 33342 for 1 h and imaged in a Bio-Tek FLx800. Each data point represents the mean ± standard error of the mean of three experiments. Dotted lines in panel A represent the 95% confidence intervals.

log orders greater; if using 5 nM, then 0.5 μM would be appropriate. This additional set controls for nonspecific binding of the radioligand. The samples should be incubated at 4 °C for 1 h. Incubation in the cold or on ice is important because it halts internalization of GPCRs and thus allows for quantification of the number of receptors on the cell. After the cold incubation, the radioactive solution is removed, the samples are washed twice with serum-free ice-cold media, and finally washed with ice-cold PBS. At this point, the samples are ready for analysis. If using the Hoechst as a normalizing agent (which is advisable since constitutively active GRK in the nucleus could conceivably induce cell growth in SFM), then the PBS should be removed and the samples imaged in a fluorometer. Following this, or if Hoechst is not used, the cells should be trypsinized and moved to a scintillation vial and analyzed via use of a scintillation counter. Note that after the washes the integrity of the cells is no longer a concern; the cells can be trypsinized for hours to ensure that all the material is transferred to the scintillation vials.

GRK4 constitutively inducing internalization of the GPCR would be observed by a decrease in the amount of ligand bound to the surface of the cells. Thus, if the increasing exposure to GRK4 decreases the scintillation counts, then GRK4 is causing the receptor to internalize independent of agonist and is constitutively active. As with the previous assays, this should not occur with KD-GRK4, but would in the cells treated with an agonist.

4.3. β-Arrestin recruitment

Phosphorylation of a GPCR by a GRK leads to β-arrestin recruitment and stable association of β-arrestin with the receptor (Gurevich et al., 1995); recently an excellent review has been written regarding β-arrestin activation and methodology for measuring β-arrestin activation (Rajagopal et al., 2010). Since activation of GRKs induce β-arrestin recruitment and binding to the receptor, this can also serve as a second measure of GRK4 constitutive activity for more β-arrestin will be bound to the receptor if GRK constitutively phosphorylates the GPCR. Two methods will be described to measure β-arrestin recruitment: a co-IP procedure to detect β-arrestin bound to the GPCR of interest and an imaging-based protocol to examine β-arrestin redistribution to the plasma membrane. The first protocol's strength is that it allows for the specific examination of a single GPCR, but, as in Section 4.1, it requires a receptor that can be IPed and thus may require expression of a tagged receptor and the inducible GRK4 system. The second protocol only requires transfection of the inducible GRK4 system, but it does not indicate which GPCR is phosphorylated by GRK4. Consequently, the second assay is a simple test to determine if GRK4 is constitutively active in a system, but to provide any specificity the first experiment would have to be conducted.

4.3.1. Coimmunoprecipitation of β-arrestin with a GPCR

Since this is an IP assay, the initial setup of these experiments is identical to what was described in Section 4.1.2 with the following exceptions. When conducting co-IP experiments, it is advisable to run two complementary experiments: IP the GPCR and conduct a Western blot for the β-arrestin, and IP the β-arrestin and conduct a Western blot for the GPCR of interest. The latter can be conducted much more readily on endogenous receptors; Cell Signaling (Danvers, MA) has a rabbit monoclonal antibody that recognizes β-arrestin 1 and 2, catalog number 4674, which can be used for IP and Western blot. Additionally, the Jackson light chain specific antibodies should be used since β-arrestin is 50 kDa.

Additionally, a second consideration for co-IP experiments is to save some sample to use as a total cell lysate. This is important when transfecting cells or inducting expression because it allows for comparison of the level of the proteins of interest across the samples. In the protocol identified in Section 4.1.2, a total cell lysate sample can be removed after the clarification step. These samples should be boiled in the presence of Laemmli sample buffer and stored at $-20\ °C$ until the Western blot is run. If using 60 mm dishes to grow the cells, alter the previous protocol as follows: lyse the cells with 250 µL, remove 25 µL (10%) for a total cell lysate sample, remove the remaining supernatant from the pellet and place in a clear 1.5-mL tube, add 275 µL lysis buffer to the sample, and then add the antibody for IP according to the manufacturer's recommended concentration. When the Western blot is run, the total cell lysate samples should be run with or in parallel to the IP samples.

If GRK4 is constitutively active with respect to the receptor, then an interaction should be seen between β-arrestin and the GPCR in samples expressing GRK4, but not KD-GRK4. A lack of an interaction would be determined by either no band appearing at the correct molecular weight or the band being in equal density as the cells not treated with the inducer and the controls (empty vector transfected). Cells expressing GRK2 and treated with an agonist serve as a positive control. The total cell lysate samples should show that GRK4 was induced in each case and that the receptor (if expressed) is also present. It is also advisable to show that β-arrestins are present in equal numbers; this will act as a loading control for the experiment.

4.3.2. Redistribution of β-arrestin

Redistribution of β-arrestins refers to the alteration in cellular localization (trafficking to the plasma membrane) after agonist stimulation and GRK phosphorylation of GPCRs. As with the previous assays, inducible GRK4 should be utilized, which if constitutively active will phosphorylate GPCRs resulting in recruitment of β-arrestins to the plasma membrane, but GRK2 is not needed. Two imaging procedures for examining the redistribution of β-arrestins will be briefly described; full procedures for conducting imaging experiments can be found elsewhere.

First, the easiest method is to induce GRK4 expression, as described in the previous sections, and then fix the cells with 3% PFA and prepare them for imaging. The following is an abbreviated protocol for preparing cells for confocal imaging. The cells should be grown on poly-D-lysine coated coverslips, GRK4 should be induced, and then the samples should be washed with cold PBS, fixed, and permeabilized with 15 ml 0.1% Triton X-100 in PBS for 2 min. Permeabilization is required for the antibodies to access the intracellular β-arrestins. The rest of the procedure is similar to a Western blot: the samples are blocked with 5% BSA, incubated with the primary antibody (in this case, a β-arrestin antibody) in 3% BSA, washed, and then a fluorescently labeled secondary antibody in 3% BSA is added. Finally, the samples are washed and the coverslips mounted onto slides and imaged via confocal microscopy. Confocal microscopy is ideal for these experiments because it will be easier to separate the cytosol from the membrane, but epifluorescence can also be used if a confocal microscope is not accessible. Additional controls for this experiment can include a membrane dye or CAAX-box tagged fluorescent protein (which will bind to the plasma membrane) as well as a nuclear stain (e.g., Hoechst 33342). If GRK4 is constitutively active, then β-arrestin should localize at the membrane when GRK4 is induced, but should not when KD-GRK4 is induced. Although GRK2 is not needed in this system, treatment with an agonist in a control set of cells is advisable to act as a positive control. Because all cells have GRKs, it should not be expected that KD-GRK4 would inhibit agonist-mediated β-arrestin redistribution unless the GPCR utilizes only GRK4, which has yet to be reported.

Second, a fluorescently tagged β-arrestin, such as enhanced green florescent protein (EGFP) tagged β-arrestin 1 or 2, can be used to monitor β-arrestin localization to the plasma membrane (Zhang et al., 1999). Utilizing chamber slides or tissue culture dishes with a glass coverslip on the bottom and live cell microscopy, the movement of β-arrestins can be measured in the same cell over time. Again an inducible system should be used in these experiments; however, first, a time course should be run to determine when GRK4 is first expressed. The imaging experiments should be conducted immediately before GRK4 is expressed. These studies will require a microscope with a heated stage to keep the cells at 37 °C and an environmental cabinet to maintain 5% CO_2. Similar to the experiments described previously, constitutively active GRK4, but not KD-GRK4, would be expected to induce β-arrestin redistribution.

5. Agonist-Mediated GRK4 Activity

Although this edition of *Methods in Enzymology* is focused on constitutive activity in receptors and other proteins, ligand-mediated activation of the GPCR, and consequently GRK4, should also be measured in all the discussed assays because any result of no constitutive activity in the assays described could also indicate that GRK4 does not target that specific GPCR in the cells chosen to study. In other words, a negative result could be a false negative. Therefore, the assays described previously should be repeated in the presence of an agonist for the GPCR being studied. The dose and duration of stimulation will vary slightly depending on the GPCR and cell type examined; these variables should be defined experimentally. When treating with agonists, cells expressing GRK4 should present increased phosphorylation of the receptor, an increased rate of receptor internalization (which entails a time course study), and increased association with β-arrestin; KD-GRK4 should ablate these effects. Thus, cells expressing KD-GRK4 and treated with agonist are likely to behave similar to control cells treated with agonist.

Additionally, functional studies examining canonical second messenger signaling is a classic method for examining GRK function. Furthermore, GRK function engages the β-arrestin signaling pathway which can also be functionally measured. Because both of these pathways can be different depending on the GPCR examined, these methods should be pursued in relation to the GPCR being studied. Such studies will allow for the physiological role of GRK4 activity to be more fully described.

6. Summary

Little is known about the molecular mechanism of GRK4 action; yet, much is known regarding the physiology of the species (combination of splice variants and polymorphisms) of GRK4 expressed. Previous studies have indicated that GRK4 is constitutively active and that its activity is dependent on ligand stimulation of a GPCR. Since some studies have utilized the same receptor system in different cells with different GRK4 splice variants (Felder et al., 2002; Rankin et al., 2006), it is likely that there is an unknown factor that is facilitating the constitutive activity of GRK4. Additionally, there are clear molecular and physiological alterations of function by the GRK4 polymorphisms (Felder et al., 2002). Consequently, there are a total of 32 different species of GRK4 to be studied to fully examine the effect of GRK4 in a system! This number can be narrowed down by determining which GRK4 species is present in the system that is being studied. The methods provided here are designed to be as simple and as inexpensive as possible so that scientists and students from undergraduate institutions to major medical schools can study GRK4 constitutive activity.

ACKNOWLEDGMENTS

This work was supported by a VA Heartland Network (VISN 15) grant. Additional thanks to Douglas R. Elliot II and Roger D. Tilmon for reading through the chapter and ensuring that this is accessible to undergraduates and beginning graduate students.

REFERENCES

Ambrose, C., James, M., Barnes, G., Lin, C., Bates, G., Altherr, M., Duyao, M., Groot, N., Church, D., Wasmuth, J. J., Lehrach, H., Housman, D., et al. (1992). A novel G protein-coupled receptor kinase gene cloned from 4p16.3. *Hum. Mol. Genet.* **1**, 697–703.
Bengra, C., Mifflin, T. E., Khripin, Y., Manunta, P., Williams, S. M., Jose, P. A., and Felder, R. A. (2002). Genotyping of essential hypertension single-nucleotide polymorphisms by a homogeneous PCR method with universal energy transfer primers. *Clin. Chem.* **48**, 2131–2140.
Bhatnagar, V., O'Connor, D. T., Brophy, V. H., Schork, N. J., Richard, E., Salem, R. M., Nievergelt, C. M., Bakris, G. L., Middleton, J. P., Norris, K. C., Wright, J., Hiremath, L., Contreras, G., Appel, L. J., and Lipkowitz, M. S. (2009). G-protein-coupled receptor kinase 4 polymorphisms and blood pressure response to metoprolol among African Americans: Sex-specificity and interactions. *Am. J. Hypertens.* **22**, 332–338.
Chen, X., Zhu, H., Yuan, M., Fu, J., Zhou, Y., and Ma, L. (2010). G-protein-coupled receptor kinase 5 phosphorylates p53 and inhibits DNA damage-induced apoptosis. *J. Biol. Chem.* **285**, 12823–12830.
Felder, R. A., Sanada, H., Xu, J., Yu, P. Y., Wang, Z., Watanabe, H., Asico, L. D., Wang, W., Zheng, S., Yamaguchi, I., Williams, S. M., Gainer, J., Brown, N. J., et al. (2002). G protein-coupled receptor kinase 4 gene variants in human essential hypertension. *Proc. Natl. Acad. Sci. USA* **99**, 3872–3877.
Gurevich, V. V., Dion, S. B., Onorato, J. J., Ptasienski, J., Kim, C. M., Sterne-Marr, R., Hosey, M. M., and Benovic, J. L. (1995). Arrestin interactions with G protein-coupled receptors. Direct binding studies of wild type and mutant arrestins with rhodopsin, beta 2-adrenergic, and m2 muscarinic cholinergic receptors. *J. Biol. Chem.* **270**, 720–731.
Inglese, J., Freedman, N. J., Koch, W. J., and Lefkowitz, R. J. (1993). Structure and mechanism of the G protein-coupled receptor kinases. *J. Biol. Chem.* **268**, 23735–23738.
Johnson, L. R., Scott, M. G., and Pitcher, J. A. (2004). G protein-coupled receptor kinase 5 contains a DNA-binding nuclear localization sequence. *Mol. Cell. Biol.* **24**, 10169–10179.
Keever, L. B., Jones, J. E., and Andresen, B. T. (2008). G protein-coupled receptor kinase 4gamma interacts with inactive Galpha(s) and Galpha13. *Biochem. Biophys. Res. Commun.* **367**, 649–655.
Martini, J. S., Raake, P., Vinge, L. E., DeGeorge, B., Jr., Chuprun, J. K., Harris, D. M., Gao, E., Eckhart, A. D., Pitcher, J. A., and Koch, W. J. (2008). Uncovering G protein-coupled receptor kinase-5 as a histone deacetylase kinase in the nucleus of cardiomyocytes. *Proc. Natl. Acad. Sci. USA* **105**, 12457–12462.
Menard, L., Ferguson, S. S., Barak, L. S., Bertrand, L., Premont, R. T., Colapietro, A. M., Lefkowitz, R. J., and Caron, M. G. (1996). Members of the G protein-coupled receptor kinase family that phosphorylate the beta2-adrenergic receptor facilitate sequestration. *Biochemistry* **35**, 4155–4160.
Pitcher, J. A., Fredericks, Z. L., Stone, W. C., Premont, R. T., Stoffel, R. H., Koch, W. J., and Lefkowitz, R. J. (1996). Phosphatidylinositol 4, 5-bisphosphate (PIP2)-enhanced G protein-coupled receptor kinase (GRK) activity. Location, structure, and regulation of the PIP2 binding site distinguishes the GRK subfamilies. *J. Biol. Chem.* **271**, 24907–24913.
Premont, R. T., Macrae, A. D., Stoffel, R. H., Chung, N., Pitcher, J. A., Ambrose, C., Inglese, J., MacDonald, M. E., and Lefkowitz, R. J. (1996). Characterization of the G protein-coupled receptor kinase GRK4. Identification of four splice variants. *J. Biol. Chem.* **271**, 6403–6410.
Rajagopal, S., Rajagopal, K., and Lefkowitz, R. J. (2010). Teaching old receptors new tricks: Biasing seven-transmembrane receptors. *Nat. Rev. Drug Discov.* **9**, 373–386.

Rankin, M. L., Marinec, P. S., Cabrera, D. M., Wang, Z., Jose, P. A., and Sibley, D. R. (2006). The D1 dopamine receptor is constitutively phosphorylated by G protein-coupled receptor kinase 4. *Mol. Pharmacol.* **69,** 759–769.

Sallese, M., Lombardi, M. S., and De Blasi, A. (1994). Two isoforms of G protein-coupled receptor kinase 4 identified by molecular cloning. *Biochem. Biophys. Res. Commun.* **199,** 848–854.

Sallese, M., Salvatore, L., D'Urbano, E., Sala, G., Storto, M., Launey, T., Nicoletti, F., Knopfel, T., and De Blasi, A. (2000). The G-protein-coupled receptor kinase GRK4 mediates homologous desensitization of metabotropic glutamate receptor 1. *FASEB J.* **14,** 2569–2580.

Speirs, H. J., Katyk, K., Kumar, N. N., Benjafield, A. V., Wang, W. Y., and Morris, B. J. (2004). Association of G-protein-coupled receptor kinase 4 haplotypes, but not HSD3B1 or PTP1B polymorphisms, with essential hypertension. *J. Hypertens.* **22,** 931–936.

Wang, Y., Li, B., Zhao, W., Liu, P., Zhao, Q., Chen, S., Li, H., and Gu, D. (2006). Association study of G protein-coupled receptor kinase 4 gene variants with essential hypertension in northern Han Chinese. *Ann. Hum. Genet.* **70,** 778–783.

Williams, S. M., Ritchie, M. D., Phillips, J. A., III, Dawson, E., Prince, M., Dzhura, E., Willis, A., Semenya, A., Summar, M., White, B. C., Addy, J. H., Kpodonu, J., *et al.* (2004). Multilocus analysis of hypertension: A hierarchical approach. *Hum. Hered.* **57,** 28–38.

Zhang, J., Barak, L. S., Anborgh, P. H., Laporte, S. A., Caron, M. G., and Ferguson, S. S. (1999). Cellular trafficking of G protein-coupled receptor/beta-arrestin endocytic complexes. *J. Biol. Chem.* **274,** 10999–11006.

CHAPTER THIRTY-TWO

VOLTAGE-CLAMP-BASED METHODS FOR THE DETECTION OF CONSTITUTIVELY ACTIVE ACETYLCHOLINE-GATED $I_{K,ACh}$ CHANNELS IN THE DISEASED HEART

Niels Voigt,* Samy Makary,[†] Stanley Nattel,[†] *and* Dobromir Dobrev*

Contents

1. Introduction	654
2. Recording of Constitutive $I_{K,ACh}$ Using Patch-Clamp Techniques	655
2.1. Basic principles of patch-clamp technique	655
2.2. Isolation of human atrial cardiomyocytes	657
2.3. Single-channel recordings of $I_{K,ACh}$ in cell-attached configuration	658
2.4. Whole-cell recording of constitutive $I_{K,ACh}$	664
2.5. Identification of the underlying mechanism(s) of constitutive $I_{K,ACh}$	669
3. Conclusions and Perspective	671
Acknowledgments	673
References	673

Abstract

Vagal nerve stimulation can promote atrial fibrillation (AF) that requires activation of the acetylcholine (ACh)-gated potassium current $I_{K,ACh}$. In chronic AF (cAF), $I_{K,ACh}$ shows strong activity despite the absence of ACh or analogous pharmacological stimulation. This receptor-independent, constitutive $I_{K,ACh}$ activity is suggested to represent an atrial-selective anti-AF therapeutic target, but the underlying molecular mechanisms are unknown. This chapter provides an overview of the voltage-clamp techniques that can be used to study

---

* Division of Experimental Cardiology, Department of Internal Medicine I—Cardiology, Angiology, Pneumology, Intensive Care and Hemostaseology, Medical Faculty Mannheim, University of Heidelberg, Theodor-Kutzer-Ufer, Mannheim, Germany
[†] Research Center, Montreal Heart Institute, Montreal, Quebec, Canada

constitutive $I_{K,ACh}$ activity in atrial myocytes and summarizes briefly the current knowledge about the potential underlying mechanism(s) of constitutive $I_{K,ACh}$ activity in diseased heart.

1. INTRODUCTION

Stimulation of the vagal nerve decelerates the heart rate due to release of acetylcholine (ACh). This was demonstrated for the first time by Otto Loewi in 1921 and the "Vagusstoff" (ACh) became the first neurotransmitter ever discovered (Loewi, 1921). However, it took more than 50 years until it was suggested that ACh activates a specific population of K^+ channels (ACh-gated $I_{K,ACh}$) leading to hyperpolarization of the cell membrane thereby decreasing pacemaker activity in sinoatrial node cells (Noma and Trautwein, 1978). In 1983, Sakmann et al. recorded single-channel openings of $I_{K,ACh}$ in rabbit atrial and sinus node cells and showed that $I_{K,ACh}$ channels exhibit kinetic properties clearly different from those of the background inwardly rectifying K^+ channel I_{K1} (Sakmann et al., 1983). Cardiac $I_{K,ACh}$ channels were identified as heterotetramers consisting of two Kir3.1 and two Kir3.4 channel subunits (Krapivinsky et al., 1995).

Today, it is well known that ACh binding to type 2 muscarinic receptors (M_2 receptors) leads to dissociation of inhibitory G_i proteins and to activation of $I_{K,ACh}$ due to direct interaction of G protein $\beta\gamma$-subunits with the channel (for review, see Hibino et al., 2010; Yamada et al., 1998). Apart from the canonical M_2-receptor-mediated activation of $I_{K,ACh}$, purinergic A_1 (Dobrev et al., 2000; Kurachi et al., 1986) and sphingolipid Edg-3 (Himmel et al., 2000) receptors, coupled to G_i proteins as well, can also activate cardiac $I_{K,ACh}$ channels. Taking account of this heterologous regulation, $I_{K,ACh}$ channels are also designated as G protein-activated inwardly rectifying K^+ channels (GIRK channels). In neurons and various endocrine cells, GIRK channels are also activated by neurotransmitters that target α_2-adrenoceptors, opioid (μ, δ, and κ), D_2 dopamine, $GABA_B$, m-Glu, 5-hydroxytryptamine-1A, and somatostatin receptors, thereby contributing to formation of slow inhibitory postsynaptic potentials and inhibition of hormone release (for review, see Hibino et al., 2010; Yamada et al., 1998). Because of strong activation of GIRK channels by neurotransmitters, tonic (constitutive) $I_{K,ACh}$ activity in the absence of receptor agonists is often regarded as negligible. However, it was early recognized that $I_{K,ACh}$ possesses an agonist-independent "resting" activity, which occurred with a much lower opening frequency (Sakmann et al., 1983). Physiologically constitutive $I_{K,ACh}$ channels may underlie a major part of basal K^+ conductance in sinoatrial cells which lack I_{K1}, playing an important role in regulating heart rate (Ito et al., 1994). In addition, constitutively active GIRK channels have been shown to contribute to the resting membrane potential of murine hippocampal pyramidal cells (Luscher et al., 1997) and rat locus coeruleus slices (Blanchet and Luscher, 2002).

In normal heart constitutive $I_{K,ACh}$ activity is very low but may increase substantially during cardiac diseases. For instance, agonist-independent constitutive $I_{K,ACh}$ activity increases in atrial myocytes from patients with chronic atrial fibrillation (cAF) and dogs with atrial tachycardia-induced remodeling (ATR; Cha et al., 2006; Dobrev et al., 2005; Voigt et al., 2008). Constitutive $I_{K,ACh}$ channels may contribute to the shortening of the atrial action potential duration (APD), which is a hallmark of the AF-associated changes in atrial electrical properties (electrical remodeling; Dobrev, 2006; Nattel et al., 2008). In vivo, vagal nerve stimulation abbreviates APD, increases APD heterogeneity, and enhances atrial vulnerability to tachyarrhythmias, which induce and perpetuate AF (Liu and Nattel, 1997). In knockout mice lacking the Kir3.4 channel subunit, M-receptor stimulation did not induce AF, clearly suggesting that the effects of vagal nerve activation are exclusively mediated by $I_{K,ACh}$ (Kovoor et al., 2001). Thus, agonist-independent constitutive $I_{K,ACh}$ channels during cAF are expected to increase atrial vulnerability to tachyarrhythmias and to promote persistence of AF. Accordingly, inhibition of constitutive $I_{K,ACh}$ channels with the highly selective $I_{K,ACh}$-blocker tertiapin reverses the APD abbreviation and AF promotion in dogs with ATR (Cha et al., 2006). Since $I_{K,ACh}$ is absent in ventricles, constitutively active $I_{K,ACh}$ is considered as a promising atrial-selective anti-AF target, without proarrhythmic side effects in the ventricles (Dobrev and Nattel, 2010).

Here, we describe the voltage-clamp techniques that can be used to study constitutive $I_{K,ACh}$ activity in atrial myocytes and we summarize briefly the current knowledge about the potential underlying mechanism of constitutive $I_{K,ACh}$ activity in diseased heart.

2. Recording of Constitutive $I_{K,ACh}$ Using Patch-Clamp Techniques

2.1. Basic principles of patch-clamp technique

The development of the patch-clamp technique was based on the ambition to record currents through individual ion channels in intact cells. Erwin Neher and Bert Sakmann developed the patch-clamp technique in the late 1970s and early 1980s (Hamill et al., 1981; Neher and Sakmann, 1976; Sigworth and Neher, 1980). The major breakthrough was the discovery that application of slight suction within a freshly prepared glass pipette used as an electrode (Fig. 32.1) results in formation of very high seal resistances (up to several giga-ohms) between the pipette and the cellular membrane (giga-seal). The giga-seal formation reduces the electrical background noise and enables to measure ion

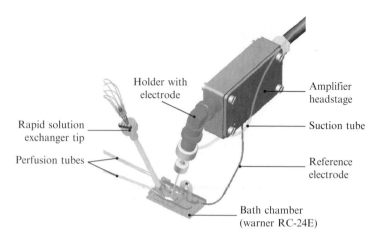

Figure 32.1 Bath chamber with associated devices for electrophysiological measurements, bath perfusion, and rapid solution exchange. The reference electrode is connected to the bath solution via an agar bridge (*), allowing separation of bath and reference electrode solution.

currents through single channels in a pA range (cell-attached configuration, Fig. 32.2A). Furthermore, it was recognized that giga-seal formation provides mechanical stability, so that the membrane patch can be excised from the intact cell (cell-free inside-out (I-O) configuration, Fig. 32.2A). However, breaking the patch by suction on the pipette provides electrical continuity between the patch pipette and the cell interior (whole-cell configuration, Fig. 32.2A), similar to conventional microelectrode recordings. Here, we do not discuss other advanced configurations like outside-out or perforated-patch and refer the interested readers to excellent descriptions of these techniques in previous issues of this book (Cahalan and Neher, 1992).

The application of correct voltages is challenging because depending on the patch-clamp configuration used (Fig. 32.2B) the potential applied through the pipette (E_P) may differ substantially from the potential "seen" by the patch membrane (V_C). In the cell-attached configuration, the potential across the membrane patch (V_C) equals the membrane potential of the cell (E_M) minus the pipette potential (E_P). Since the pipette potential applied by the patch-clamp amplifier (E_P) indicates the potential of the patch electrode with reference to the bath electrode (reference electrode; Fig. 32.1), the pipette potential (E_P) equals the membrane potential ($E_P = E_M = V_C$) only in whole-cell configuration, where the pipette is connected to the interior of the cell. In the case of I-O configuration, the electrodes are reversed with the patch pipette facing the exterior and the bath solution facing the interior of the membrane patch. Therefore, the potential across the membrane patch equals the reverse potential of the pipette ($V_C = -E_P$).

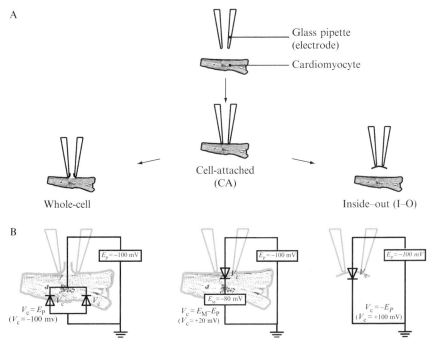

Figure 32.2 Different patch-clamp configurations that can be used to study $I_{K,ACh}$ (A) and corresponding considerations for voltage application (B). Depending on the configuration, the voltage seen by the studied ion channel (V_C) is set by variation of the pipette potential (E_P), which indicates the potential of the patch electrode with reference to the bath electrode (reference electrode, see Fig. 32.1). In inside-out configuration, V_C is also influenced by the membrane potential (E_M), which depends largely on the potassium concentration in the bath solution. See text for further details.

2.2. Isolation of human atrial cardiomyocytes

Studying electrophysiological properties of cardiac ion channels with the patch-clamp technique requires isolated single myocytes, which are usually obtained following the *in vitro* exposure of cardiac tissue samples to digestive enzymes (collagenase, hyaluronidase, peptidase, etc.). Since the first report of isolation of viable cardiac myocytes in 1955 (Margaret, 1955), a large quantity of protocols have been developed in order to harvest single atrial and ventricular myocytes from different species including mouse, rat, rabbit, dog, guinea pig, and human. Here, we describe in detail the isolation of human atrial cardiomyocytes, which generally derives from the procedure described by Bustamante *et al.* (1982). Regarding procedures used for isolation of atrial myocytes from other species, we refer to the "Worthington Tissue Dissociation Guide" provided by Worthington Biochemical Corp., USA (www.tissuedissociation.com).

During routine cannulation procedures in patients undergoing open-heart surgery for cardiopulmonary bypass grafting, the tip of the right atrial appendage is usually removed and can be used for isolation of atrial cardiomyocytes. After excision, the tissue sample is transferred immediately into sterile 2,3-butanedione monoxime (contractile inhibitor, preventing myocyte contracture) containing Ca^{2+}-free solution for transportation (see Table 32.1). At a general transport time of about 30 min, we did not recognize a clear advantage of cooling with respect to both number and quality of isolated atrial cardiomyocytes.

After removal of fat and connective tissue, the myocardial tissue, usually between 200 and 400 mg, is transferred to Ca^{2+}-free solution (without 2,3-butanedione monoxime; Table 32.1) and chopped into small chunks of ~ 1 mm^3 in size. The following steps should be executed at 37 °C and under continuous gassing with 100% O_2. Primarily the tissue chunks are washed three times for 3 min in Ca^{2+}-free solution. For the first digestive step, the tissue is stirred for 45 min in 20 ml Ca^{2+}-free solution containing collagenase I and protease XXIV (enzyme solution E1; Table 32.1) with $[Ca^{2+}]$ being raised to 0.02 mM (add 40 µl of 10 mM stock solution) after 10 min. The enzyme solution is then exchanged for 20 ml enzyme solution E2 (Table 32.1) containing collagenase I only and stirring is continued until rod-shaped, striated cells appear. Due to the high variability among tissue samples from different donors, the duration of the second enzyme exposure varies between 5 and 45 min. After settling down of the tissue chunks, the supernatant is replaced by 20 ml of storage solution (Table 32.1; Feng et al., 1998) and dissociation of the cells is achieved by gentle mechanical trituration using a Pasteur pipette. After separation from the debris by filtration through nylon gauze (200 µm mesh) and centrifugation for 10 min at 95 × g, the precipitate was redissolved in 4 ml of storage solution. The final Ca^{2+} concentration is set stepwise to 0.5 mM by adding a total amount of 0.2 ml 10 mM $CaCl_2$ within 20 min (80–40–80 µl). Usually, we obtain around 50 cells/10 µl and 30 cells/10 µl in samples from sinus rhythm (SR) and cAF patients, respectively. However, human cells are often more difficult to isolate than animal cells and cell yield varies largely, depending on degree of tissue fibrosis, age and cardiac diseases of the donor. The isolated atrial cardiomyocytes are kept at room temperature until use and should be used within 8 h.

2.3. Single-channel recordings of $I_{K,ACh}$ in cell-attached configuration

Since electrical properties of single-channel $I_{K,ACh}$ openings are clearly different compared to those of other ion channels (i.e., I_{K1}), single-channel registrations are suitable to provide direct evidence for increased constitutive $I_{K,ACh}$ activity in patients with cAF. Current amplitudes during

Table 32.1 Solutions for cell isolation

	Transport solution	Ca^{2+}-free solution	Enzyme solution E1	Enzyme solution E2	Storage solution
Albumin	–	–	–	–	1%
2,3-Butanedione monoxime	30	–	–	–	–
DL-β-Hydroxy-butyric acid	–	–	–	–	10
EGTA	–	–	–	–	10
Glucose	20	20	20	20	10
Glutamic acid	–	–	–	–	70
KCl	10	10	10	10	20
KH$_2$PO$_4$	1.2	1.2	1.2	1.2	10
MgSO$_4$	5	5	5	5	–
MOPS	5	5	5	5	–
NaCl	100	100	100	100	–
Taurin	50	50	50	50	10
Collagenase I (Worthington)	–	–	286 U/ml	286 U/ml	–
Protease XXIV	–	–	5 U/ml	–	–
pH	7.0	7.0	7.0	7.0	7.4
Adjusted with	1 M NaOH	1 M NaOH	1 M NaOH	1 M NaOH	1 M KOH

Concentrations in mM unless otherwise stated. All drugs were from Sigma–Aldrich (USA) unless otherwise stated.

single-channel measurements are lower (pA range) compared to whole-cell recordings (nA range) and therefore require higher efforts on reduction of electrical noise to improve the signal-to-noise ratio. Here we will describe briefly some steps that we found critical to reduce electrical noise and improve quality of single-channel recordings.

The first important step is to remove all electrical devices except the preamplifier from the Faraday cage including the light source of the microscope, which should be exchanged with a cold light source positioned outside the cage. We use a gravity-driven perfusion system to apply fresh bath solution (Table 32.2). The Faraday cage should be closed on all sides during the experiment. Furthermore, high efforts should be spend on pipette fabrication. We use sylgard (World Precision Instruments, USA)-coated pipettes fabricated from borosilicate glass capillaries (1.5 mm outer diameter, 0.87 mm inner diameter, 0.2 mm filament, Hilgenberg, Germany). Since the pipette acts as a capacitor with a capacitance proportional to 1/wall thickness, thereby causing electrical noise, coating of the pipette with a thick nonconducting layer reduces electrical noise. In addition, the hydrophobic nature of sylgard prevents the bath solution from creeping up the sides of the pipette, thereby reducing electrical contact between pipette and bath solution. Sylgard should be applied as near to the pipette tip as possible (10–50 μm) and extend until the pipette shoulder. It is cured by placing the pipette tip into a heated wire coil for a few seconds. After

Table 32.2 Bath solutions

	CA and whole-cell experiments	I-O experiments (phosphatases intact)	I-O experiments (phosphatases inhibited)
$CaCl_2$	2	–	–
EGTA	–	0.1	0.1
Glucose	10	–	–
HEPES	10	5	5
KCl	20	145	145
$MgCl_2$	1	1	1
NaCl	120	20	–
NaF	–	–	1
Na_3VO_4	–	–	0.1
$Na_2P_2O_7$	–	–	10
$ZnSO_4$	–	–	5
pH	7.4	7.4	7.4
Adjusted with	1 M NaOH	1 M Tris-acetat	1 M Tris-acetat

Concentrations in mM; CA, cell-attached configuration; I-O, inside-out configuration. All drugs were from Sigma-Aldrich (USA) unless otherwise stated.

Table 32.3 Pipette filling solutions

	CA and I-O experiments	Whole-cell experiments
$CaCl_2$	2	2
DL-Aspartic acid K^+ salt	–	80
EGTA	–	5
GTP-Tris	–	0.1
HEPES	5	10
KCl	145	40
Mg-ATP	–	5
$MgCl_2$	1	–
NaCl	–	8
pH	7.4	7.4
Adjusted with	1 M KOH	1 M KOH

Concentrations in mM; CA, cell-attached configuration; I-O, inside-out configuration. All drugs were from Sigma-Aldrich (USA) unless otherwise stated.

coating, pipettes should be fire polished using a microforge to clean and smooth the tip. Finally, pipettes should be cut to 2 cm total length. Because of the high $I_{K,ACh}$ channel density, pipettes should be relatively small, with tip resistances being ∼8–10 MΩ when filled with pipette solution (Table 32.3, CA Experiments). Small-tip pipettes increase the likelihood of having only one channel within the patch, which allows a much more reliable analysis of single-channel properties (see below).

Figure 32.3A shows representative recordings of $I_{K,ACh}$ in the absence and presence of an M-receptor agonist in atrial myocytes from patients with sinus rhythm (SR) and cAF. Constitutive $I_{K,ACh}$ activity was quite apparent in cAF, whereas it occurred only sporadically in myocytes from patients with SR. Inclusion of the nonselective M-receptor agonist carbachol (CCh, 10 μM) in the pipette solution strongly activated $I_{K,ACh}$ in both SR and cAF, causing frequent channel openings.

A detailed description of the single-channel analysis algorithms is clearly beyond the scope of this chapter. For more details regarding analysis of single-channel properties, see Colquhoun (1994), http://www.utdallas.edu/∼tres/microelectrode/me.html. An elementary property characterizing single-channel activity is the sojourn time at the open and closed level. To characterize closed and open times, transitions between the two current levels must be detected. We usually use the threshold-crossing method, which automatically detects transitions from the closed to the open state or vice versa every time the observed current crosses the 50% threshold (idealization). The opening times obtained in this way can be displayed in an open-time histogram as shown in Fig 32.3B and open-time constants (τ_{open}) can be calculated by mono-exponential fits. Open-time constants

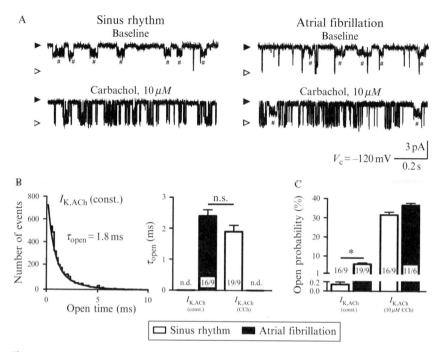

Figure 32.3 Single-channel properties of constitutively active $I_{K,ACh}$ channels in atrial myocytes from patients with chronic atrial fibrillation (cAF). (A) Representative cell-attached single-channel recordings of I_{K1}, constitutively active and cabachol (CCh)-activated $I_{K,ACh}$ in myocytes from sinus rhythm (SR, left) and cAF patients (right). Closed and open levels are indicated by filled and empty arrowheads, respectively. In the absence of M-receptor agonists (baseline), myocytes from the cAF patients exhibit both I_{K1} (#) and constitutive $I_{K,ACh}$ openings (*), whereas the latter is a rare event in myocytes from SR patients. (B) Representative histogram of open times of the constitutively active $I_{K,ACh}$ channel (left). Bin-width was 0.2 ms. Open-time constants were calculated for CCh-activated $I_{K,ACh}$ in SR patients and constitutively active $I_{K,ACh}$ in cAF patients by monoexponential fits and are expressed as mean ± S.E.M. (right). n.d., not determined. (C) Channel open probability of constitutively active and CCh-activated $I_{K,ACh}$ channels are expressed as mean ± S.E.M. *$P < 0.05$. (B, C) Numbers within the columns indicate number of myocytes/patients. (Reprinted with permission from Dobrev et al., 2005.)

from constitutive $I_{K,ACh}$ channels are comparable to $I_{K,ACh}$ channels activated with CCh, providing strong evidence for the identity of these spontaneous current openings as $I_{K,ACh}$. Since density of $I_{K,ACh}$ channels in atrial cardiomyocytes is high, patch membranes usually contain two or more channels under the patch pipette, indicated by two or more opening levels caused by simultaneous channel openings. However, since reliable determination of closed-time duration can only be performed with one channel in the patch, we usually refrain from analyzing closed times.

An additional parameter that can be obtained after calculation of open-time durations is the probability of a channel to open (open probability, P_{open}). P_{open} can be estimated by the total open time divided by the total length of the record. Again, a reliable estimate of P_{open} can be obtained only in the presence of only one individual $I_{K,ACh}$ channel in the membrane patch. Since presence of only one channel is a rare event precluding a reliable determination of the exact number of channels in the membrane patch, open probability is often expressed as nPo, where n indicates the usually unknown number of channels. Figure 32.3C shows higher nPo of constitutive $I_{K,ACh}$ in myocytes from patients with cAF compared to SR. Inclusion of the M-receptor agonist CCh further increases nPo without significant differences between cAF and SR.

The analysis of single-channel amplitudes in relation to the applied voltage (V_C) provides a further evidence for the identity of the studied channel (Fig 32.4A). To calculate $I_{K,ACh}$ amplitude, we generate point amplitude histograms by plotting all digitized current events according to their individual current values in a histogram (Fig. 32.4B). There is usually one peak at the

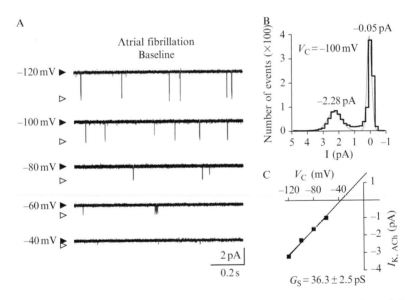

Figure 32.4 Single-channel conductance of constitutively active $I_{K,ACh}$ channels in a myocyte from one patient with chronic atrial fibrillation. (A) Representative single-channel recordings at various potentials (V_C; see Fig. 32.2). Closed and open levels are indicated by filled and empty arrowheads, respectively. (B) Corresponding amplitude histogram at -100 mV. Bin-width was 0.2 pA. After fitting two Gaussian distribution curves, single-channel $I_{K,ACh}$ amplitude equals the difference between the two peaks at -0.05 pA (closed level) and -2.28 pA (open level). (C) Current–voltage relation calculated from the corresponding amplitude histograms.

closed level and one peak at the open level. The single-channel $I_{K,ACh}$ amplitude equals the difference between both peaks. Since again any drift in baseline will bias the results, an appropriate baseline correction should be performed in advance. In addition, single-channel openings of constitutive $I_{K,ACh}$ are relatively rare; therefore, we recommend to confine the histogram to those points that correspond to periods when the channel is open. Plotting the amplitude values against the corresponding voltage applied on the patch (V_C) results in an almost linear relationship with the slope representing the single-channel conductance (Fig. 32.4C). The conductance obtained in atrial myocytes from cAF patients in absence of an M-receptor agonist (36.3 pS) is comparable to published values of agonist-activated $I_{K,ACh}$ and is clearly different from typical single-channel conductances of other inward-rectifier potassium currents like I_{K1} and ATP-dependent K^+ currents ($I_{K,ATP}$) (Heidbuchel et al., 1990b).

2.4. Whole-cell recording of constitutive $I_{K,ACh}$

Whereas single-channel measurements are valuable for the direct proof of constitutive $I_{K,ACh}$ activity, they are not feasible to evaluate the relative contribution of constitutive $I_{K,ACh}$ activity to total inward-rectifier K^+ current. In addition, in whole-cell configuration, the channels are not shielded by the patch pipette and the applied drugs reach the channel quickly without any access restrictions, which allows a reliable testing of drug effects on channel function (Kitamura et al., 2000). In whole-cell configuration, direct electrical access to the cell interior is obtained by disrupting the patch membrane after giga-seal formation. Successful disruption, achieved by applying gentle suction to the pipette, is indicated by the occurrence of capacitance artifacts in response to a distinct voltage step in the mV range. Since these capacitance artifacts are mainly due to cell membrane recharges, they are usually used for estimation of the cardiomyocyte surface area, assuming a membrane capacitance per unit area of about 1 µF/cm^2 (Gentet et al., 2000).

In whole-cell configuration, the patch-pipette filling solution is in direct contact with the cytosol and should therefore have a similar ion concentration (Table 32.3). Borosilicate glass microelectrodes (1.5 mm outer diameter, 0.87 mm inner diameter, 0.2 mm filament, Hilgenberg, Germany) with tip resistances of 2–5 MΩ are acceptable. The myocytes are superfused with bath solution (Table 32.2). We use high (20 mM) extracellular potassium concentration because this shifts the reversal potential to more positive values and allows to record much larger (more easily measurable/comparable) inwardly directed K^+ currents (Dobrev et al., 2000, 2001, 2005; Himmel et al., 2000; Voigt et al., 2007). We recorded the currents at room temperature because current amplitudes at room temperature are comparable to those at 37 °C and the success rate of experiments is much higher at room temperature (Voigt et al., 2010b), especially when using human atrial myocytes.

Drugs are generally applied via a rapid solution-exchange-system (ALA-Scientific-Instruments, USA). Agonist-inducible $I_{K,ACh}$ is stimulated with CCh (2 μM) and defined as the difference current between total current in the presence of CCh and basal current. Basal inward-rectifier currents and $I_{K,ACh}$ are specifically assessed as Ba^{2+} (1 mM)-sensitive currents. Since both I_{K1} and constitutive $I_{K,ACh}$ channels contribute to basal inward-rectifier K^+ current in atrial myocytes, the whole-cell configuration is not feasible for direct detection of constitutive $I_{K,ACh}$. However, selective $I_{K,ACh}$-channel blockers and specific voltage protocols targeting differences in time-dependent $I_{K,ACh}$ "relaxation" allow sufficient differentiation between constitutive $I_{K,ACh}$ currents and I_{K1} also in the whole-cell voltage-clamp configuration.

2.4.1. Use of selective $I_{K,ACh}$-channel blockers

Drugs like NIP-142, NIP-151, and AVE0118 are shown to block $I_{K,ACh}$ in atrial myocytes; however, their selectivity for $I_{K,ACh}$ is imperfect (Christ et al., 2008; Hashimoto et al., 2008; Tanaka and Hashimoto, 2007). In addition, inhibition of constitutively active $I_{K,ACh}$ may contribute to the efficacy of commonly used class I and III antiarrhythmic drugs (Voigt et al., 2010a). Tertiapin, a 21-amino acid containing peptide initially isolated from the venom of the European honey bee, has been characterized as a selective blocker of Kir3.1 and Kir3.4 heteromultimers, renally expressed Kir1.1a (ROMK1) channels and Ca^{2+}-dependent large-conductance K^+ channels (BK channels) in a nanomolar range (Jin and Lu, 1998; Kanjhan et al., 2005). Kir1.1b (ROMK2) and Kir2.1 (Jin and Lu, 1998) channels are inhibited in the micromolar range only (Jin and Lu, 1998; Sackin et al., 2003). Since ROMK channels are not expressed in cardiac myocytes and inhibition of BK channels occurs only after myocyte incubation for at least 15 min (Kanjhan et al., 2005), acute application of tertiapin in atrial myocytes can be considered as highly selective for $I_{K,ACh}$ inhibition. It is suggested that tertiapin blocks $I_{K,ACh}$ channels by interaction of its α-subunit C-terminus with the external side of the channel pore (Jin et al., 1999). This interaction with the extracellular side of the channel may explain the delayed inhibitory effect of tertiapin in cell-attached configuration, where the channel is largely isolated from the bath solution by the patch-pipette (Voigt et al., 2008). For experimental purposes, we recommend the usage of a non-air-oxidizable derivate, called tertiapin-Q, in which the methionine residue 13 is replaced by glutamine (Jin and Lu, 1999). Tertiapin-Q functionally resembles native tertiapin in both affinity and specificity, but is much more stable. On experimental days, we prepare fresh working solutions of tertiapin (0.1–100 nM) using a 100-μM stock solution stored at −15 °C. In addition, 1 mg/ml albumin should be added to the bath solution to reduce nonspecific protein binding of tertiapin.

At the whole-cell configuration, we activate the inward-rectifier currents using a ramp pulse from −100 to +40 mV (Fig 32.5A), which is much better tolerated by human atrial myocytes compared to clamp steps. A direct comparison between results obtained with ramp pulses versus clamp steps indicated their similarity and supported the validity of results obtained with ramp-pulse protocols (Voigt et al., 2010b). Figure 32.5B shows the time course of inward-rectifier current at −100 mV. Application of the M-receptor agonist CCh (2 μM) resulted in a rapid initial current increase (Peak), followed by a decrease to a quasi-steady state level (QSS). The agonist-dependent $I_{K,ACh}$ currents at Peak and QSS level are defined as CCh-sensitive current increases (Peak- and QSS-$I_{K,ACh}$). In the absence of M-receptor agonists, application of the highly selective $I_{K,ACh}$-blocker tertiapin (0.1–100 nM) to myocytes from SR patients resulted only in a slight reduction of basal inward-rectifier K$^+$ current, which was comparable to values in control experiments with application of bath solution only. In contrast, tertiapin reduced basal current in cAF patients in a concentration-dependent manner (Fig. 32.5C). In addition, tertiapin inhibited the CCh-mediated activation of $I_{K,ACh}$ in both SR and cAF, confirming the potency and efficacy of tertiapin as a selective $I_{K,ACh}$-channel blocker (Fig. 32.5D). Taken together, these data suggest the existence of substantial constitutive $I_{K,ACh}$ activity in cAF only.

2.4.2. Estimation of time-dependent $I_{K,ACh}$ "relaxation"

Apart from the use of selective $I_{K,ACh}$ blockers, differences in electrophysiological properties can be also utilized to differentiate I_{K1} and constitutive $I_{K,ACh}$ currents in whole-cell patch-clamp experiments. An elegant method is the analysis of activation kinetics upon a hyperpolarizing voltage step pulse. Whereas an instantaneous increase of current amplitude upon hyperpolarization is characteristic for almost all inward-rectifier K$^+$ currents, the time-dependent current increase, called "relaxation," is a hallmark of $I_{K,ACh}$ currents only (Hibino et al., 2010). Although the exact mechanism is largely unknown, regulators of G protein signaling (RGS-proteins), which accelerate GTPase activity of the Gα-subunit, appear essential for reconstituting the relaxation behavior of $I_{K,ACh}$ current in Xenopous oocytes expressing Kir3.1/Kir3.4 channels together with M$_2$ receptors (Fujita et al., 2000). As shown in Fig. 32.6, Cha et al. used this phenomenon to demonstrate increased time-dependent hyperpolarization-activated inward-rectifier current (I_{KH}) in atrial myocytes from dogs with ATR (Cha et al., 2006). I_{KH} is inhibited by tertiapin-Q, and single-channel recordings in atrial myocytes from dogs with ATR directly confirmed the identity of I_{KH} as constitutive $I_{K,ACh}$ (Voigt et al., 2008).

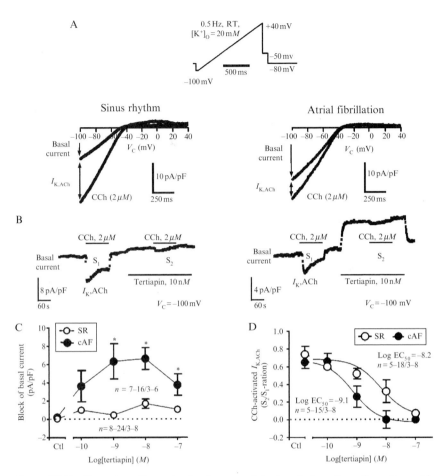

Figure 32.5 Detection of constitutive $I_{K,ACh}$ activity in whole-cell configuration with the highly selective $I_{K,ACh}$-blocker tertiapin. (A) Representative inward-rectifier current recordings under basal conditions (basal current) and in response to 2 μM carbachol (CCh) in myocytes from a patient with sinus rhythm (SR, left) and chronic atrial fibrillation (cAF, right), respectively. Top: Ramp protocol. (B) Corresponding time course of basal inward-rectifier current and CCh-activated $I_{K,ACh}$ at -100 mV. Tertiapin (10 nM) was applied before, during and after a second CCh application (S_2) with the first CCh application (S_1) serving as internal control. (C) Concentration-dependent block of basal current with tertiapin in SR and cAF patients unmasks increased constitutively active $I_{K,ACh}$ channels in cAF patients only. $*P < 0.05$ versus corresponding values in SR. (D) Concentration-dependent effects of tertiapin on the S_2/S_1 ratio of $I_{K,ACh}$. (C, D) Each point represents values (mean ± S.E.M., at -100 mV) from n independent experiments. Numbers within the figures indicate number of myocytes/patients. (Reprinted with permission from Dobrev et al., 2005.)

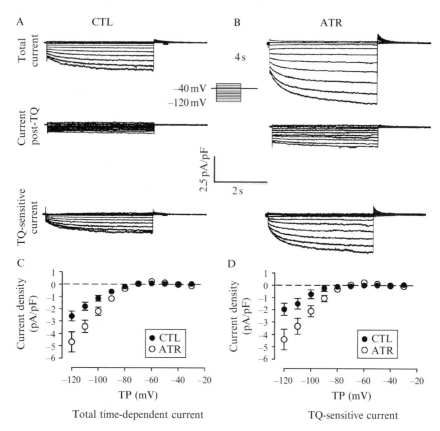

Figure 32.6 Estimation of constitutive $I_{K,ACh}$ activity in whole-cell configuration with a voltage-step protocol in canine atrial myocytes. (A and B, top) Representative current recordings in response to hyperpolarizing voltage steps (inset) in atrial myocytes from control (CTL) and atrial tachycardia-remodeled dogs (ATR). Middle: Currents from the same cells as in top panel in the presence of 100 nM tertiapin-Q, which selectively suppresses $I_{K,ACh}$, leaving essentially only I_{K1}. Bottom: Tertiapin-sensitive currents obtained in cells illustrated above by subtracting currents in presence of 100 nM tertiapin-Q from currents before superfusion. (C) Mean ± S.E.M. of constitutively active $I_{K,ACh}$ current density–voltage relations based on time-dependent activating current based on 18 cells from nine dogs (CTL) and 10 cells from four dogs (ATR), respectively. Currents in ATR dog cells were significantly greater than in CTL cells ($P = 0.02$). (D) Mean ± S.E.M. of constitutive $I_{K,ACh}$ density–voltage relations based on tertiapin-sensitive current densities based on nine cells from six dogs (CTL) and eight cells from three dogs (ATR), respectively. Currents in ATR dog cells were significantly greater than in CTL cells ($P = 0.02$). (C, D) TP indicates test potential. (Reprinted with permission from Cha *et al.*, 2006.)

2.5. Identification of the underlying mechanism(s) of constitutive $I_{K,ACh}$

Previous studies showed that the M-receptor antagonist atropine does not abolish the constitutive $I_{K,ACh}$ activity, suggesting that an agonist-independent mechanism might be implicated in the formation of constitutive $I_{K,ACh}$ channels (Dobrev et al., 2005). An increased receptor-independent dissociation of G_α- and $G_{\beta\gamma}$-subunits appears unlikely contributor because neither pertussis toxin nor the absence of GTP affected the $I_{K,ACh}$-like component of basal whole-cell current in dog atrial myocytes (Ehrlich et al., 2004). In addition, indirect evidence also suggests that upregulation and stronger membrane translocation of PKCε might cause hyperphosphorylation of $I_{K,ACh}$ channels in cAF patients, likely contributing to the increase in constitutive $I_{K,ACh}$ activity (Voigt et al., 2007). Together, these findings point to a membrane-delimited phosphorylation-mediated mechanism of constitutive $I_{K,ACh}$.

In the I–O configuration, a small, cell-free piece of cell membrane containing the channel of interest is attached to the pipette with the cytosolic side facing the bath solution. This configuration allows direct alterations of the intracellular milieu which makes it particularly suitable for the exploration of the intracellular mechanism of constitutive $I_{K,ACh}$ activity. Here we describe how we use this method to study increased constitutive $I_{K,ACh}$ activity in dogs with ATR (Makary et al., unpublished). For excision of the membrane patch, it is necessary that the myocytes are properly attached to the glass bottom of the chamber. Therefore, it is essential to allow enough time ($\sim$30 min) for myocytes to settle down before starting an experiment. Coating of the glass bottom with laminin may also improve the adhesion of the cells, but we did not find an advantage of using laminin coating. I-O patches are made based on the cell-attached configuration (see above). Very slow elevation of the pipette under continuous monitoring of single-channel activity should result in excision of the membrane patch out of the intact cell while preserving the giga-seal. Pulling off a membrane patch often results initially in the formation of a membrane vesicle in the pipette tip indicated by disappearance of the channel openings at preserved seal resistance. In order to reduce the chance of vesicle formation, we use pipettes with low seal resistance around 2–3 MΩ. If vesicle formation should still be a problem, the outer face of the vesicle can be opened by briefly taking the membrane through the bath solution/air interface, by positioning the tip in front of the outflow of the bath perfusion or by momentarily making contact with a droplet of paraffin or a piece of cured sylgard. However, these mechanical procedures are always critical because they may spoil the intact giga-seal.

After I-O formation, the cytosolic side of the channel faces the bath solution which is exchanged to a high K^+ bath solution (Table 32.2) in order to represent intracellular conditions and avoid K^+ gradients across the membrane patch. Furthermore, it has to be considered that after I-O formation the potential seen by the patch is directly controlled by the amplifier and that the influence of the membrane potential is eliminated ($E_P = +100$ mV results in $V_c = -100$ mV, see above and Fig. 32.2B). After excision of the patch and formation of the I-O configuration, the constitutive $I_{K,ACh}$ activity of ATR cardiomyocytes is strongly reduced (Fig. 32.7A) and the resulting low opening probability becomes comparable to that of control cardiomyocytes, pointing to the contribution of a cytosolic factor to ATR-associated increase of constitutive $I_{K,ACh}$. Previous studies of agonist-activated $I_{K,ACh}$ revealed that inhibition of phosphatases is necessary to prevent run-down in cell-free I-O patches (Huang et al., 1998; Kaibara et al., 1991; Shui et al., 1997), suggesting that normal $I_{K,ACh}$ activity requires channel phosphorylation (Heidbuchel et al., 1990a; Medina et al., 2000).

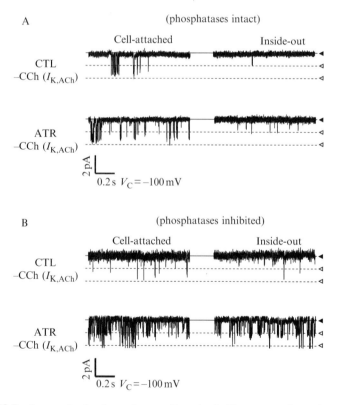

Figure 32.7 $I_{K,ACh}$ single-channel recordings in inside-out configuration. (A, B) Representative cell-attached (left) and inside-out (right) recordings of constitutively active $I_{K,ACh}$ in the absence (A) and presence (B) of protein phosphatase inhibitors. Closed and open levels are indicated by filled and empty arrowheads, respectively (Makary et al., unpublished observations).

To assess the contribution of channel phosphorylation to constitutive $I_{K,ACh}$ activity in ATR, we added phosphatase inhibitors (fluoride, vanadate, and pyrophosphate, Table 32.2) to the bath solution. Under these conditions, the open probability of constitutive $I_{K,ACh}$ channels in ATR cardiomyocytes was only slightly reduced after I-O formation (Fig. 32.7B) compared to cell-attached experiments and the differences between control and ATR were preserved. The results in Fig. 32.7 indicate a crucial role of phosphorylation in development of constitutive $I_{K,ACh}$ channel activity. These findings provide an example of how I-O recordings may be used to explore the molecular mechanisms of constitutive $I_{K,ACh}$. Further studies are necessary to elaborate whether and how specific phosphorylation processes contribute to the development of constitutive $I_{K,ACh}$ activity in cAF.

3. Conclusions and Perspective

Although impaired channel phosphorylation is a potential contributor to development of constitutive $I_{K,ACh}$ activity in cAF, the complex regulation of agonist-activated $I_{K,ACh}$ makes several underlying mechanisms possible (Fig. 32.8A). Physiological activation of $I_{K,ACh}$ requires binding of $G_{\beta\gamma}$-subunits to the $I_{K,ACh}$ channel, which strengthens the interaction between channel subunits and cell membrane-located phosphatidylinositol 4,5-bisphosphate (PIP$_2$; Huang et al., 1998). It is known that depletion of PIP$_2$ and removal of ATP both prevent activation of $I_{K,ACh}$ by $G_{\beta\gamma}$-subunits, suggesting tonic channel phosphorylation and sufficient PIP$_2$ levels in the cell membrane as absolute requirements for $G_{\beta\gamma}$-mediated activation of $I_{K,ACh}$ (Medina et al., 2000; Sui et al., 1998). In contrast, enhanced channel phosphorylation and/or increased amount of PIP$_2$ close to the channel may activate the channel even in the absence of $G_{\beta\gamma}$-subunits (Kaibara et al., 1991; Kim, 1993; Okabe et al., 1991). In addition, Na$^+$ enhances the interaction of PIP$_2$ with the $I_{K,ACh}$ channel indicating that AF-associated Na$^+$-overload might also contribute to constitutive $I_{K,ACh}$ (Mark and Herlitze, 2000). As mentioned above, activation of $I_{K,ACh}$ requires ATP which may modulate the channel through multiple mechanisms: (i) direct phosphorylation of the channel and/or channel regulatory proteins (Kim, 1993; Medina et al., 2000), (ii) generation of PIP$_2$ (Huang et al., 1998), and (iii) transphosphorylation between adenosine and guanosine nucleosides via nucleoside diphosphate kinase (NDPK; Heidbuchel et al., 1990a). In principle, each of these mechanisms may contribute to the development of cAF-associated constitutive $I_{K,ACh}$ activity. The here described patch-clamp techniques in combination with suitable biochemical and molecular biology methods (i.e., adenoviral transfection of cultured human atrial myocytes,

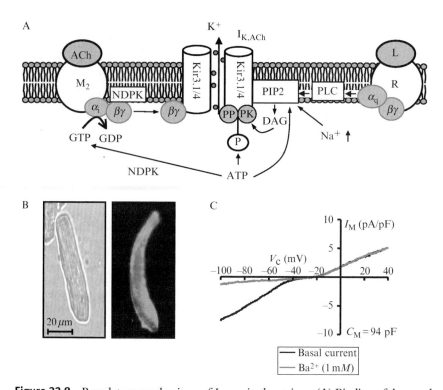

Figure 32.8 Regulatory mechanisms of $I_{K,ACh}$ in the atrium. (A) Binding of the vagal neurotransmitter acetylcholine (ACh) to muscarinic M_2 receptors leads to dissociation of G_i protein α- and $\beta\gamma$-subunits. In addition, transphosphorylation between adenosine and guanosine nucleosides via nucleoside diphosphate kinase (NDPK) may cause receptor-independent G_i protein dissociation. In both cases, the consecutive binding of G_i protein $\beta\gamma$-subunits to the $I_{K,ACh}$ channel subunits Kir3.1 and Kir3.4 strengthens their interaction with cell membrane-located phosphatidylinositol 4,5-bisphosphate (PIP_2), thereby increasing $I_{K,ACh}$ open probability. Similarly increased intracellular Na^+ also strengthens the interaction of PIP_2 with the channel leading to receptor-independent current activation. In contrast, stimulation of G_q-coupled receptors (R; i.e., AT_1-receptors or α_1-adrenoceptors) activates phospholipase-C (PLC), thereby lowering the PIP_2 membrane content, which results in $I_{K,ACh}$ inhibition. Furthermore, the PIP_2 metabolite diacylglycerol activates protein kinase C that phosphorylates $I_{K,ACh}$ channel subunits. In general, channel phosphorylation, which is dynamically regulated by protein kinases (PK) and phosphatases (PP), is an absolute requirement for normal $I_{K,ACh}$ function. Adenosine triphosphate (ATP) modulates the channel through multiple actions: (i) direct phosphorylation of the channel and/or channel regulatory proteins, (ii) generation of PIP_2, and (iii) transphosphorylation between adenosine and guanosine nucleosides via nucleoside diphosphate kinase (NDPK). (B) Adenoviral-mediated expression of "enhanced green fluorescence protein" and the $G\beta\gamma$-scavenger transducin-α. With bright-field microscopy, adenovirally transfected human right atrial myocytes are indistinguishable from nontransfected cells (left), whereas the fluorescence emission light at an excitation wavelength of 488 nm indicates successful transfection (right). (C) Basal inward-rectifier current (I_M) in response to a ramp pulse from -100 to $+40$ mV (see also Fig. 32.5A) measured in a human right atrial myocyte transfected with transducin-α. Identity as inward-rectifier current was specifically assessed by application of Ba^{2+} (1 mM). C_M, cell capacity. (See Color Insert.)

Fig. 32.8B) are powerful tools to discover the underlying mechanism(s) of constitutive $I_{K,ACh}$ activity in the diseased heart.

ACKNOWLEDGMENTS

The authors thank Dr Thomas Wieland, University of Heidelberg, for the kind provision of adenoviruses for atrial cardiomyocyte transfection.
The authors' research is supported by the Deutsche Forschungsgemeinschaft (Do769/1-1-3 to D. D.), the German Federal Ministry of Education and Research through the Atrial Fibrillation Competence Network (01Gi0204 to D. D.), the Canadian Institutes of Health Research (MOP44365 to S. N.), and the European–North American Atrial Fibrillation Research Alliance (ENAFRA) grant of Fondation Leducq (07CVD03 to D. D. and S. N.).

REFERENCES

Blanchet, C., and Luscher, C. (2002). Desensitization of mu-opioid receptor-evoked potassium currents: Initiation at the receptor, expression at the effector. *Proc. Natl. Acad. Sci. USA* **99**, 4674–4679.
Bustamante, J. O., et al. (1982). Isolation of single atrial and ventricular cells from the human heart. *Can. Med. Assoc. J.* **126**, 791–793.
Cahalan, M., and Neher, E. (1992). Patch clamp techniques: An overview. *Methods Enzymol.* **207**, 3–14.
Cha, T. J., et al. (2006). Kir3-based inward rectifier potassium current: Potential role in atrial tachycardia remodeling effects on atrial repolarization and arrhythmias. *Circulation* **113**, 1730–1737.
Christ, T., et al. (2008). Pathology-specific effects of the IKur/Ito/IK,ACh blocker AVE0118 on ion channels in human chronic atrial fibrillation. *Br. J. Pharmacol.* **154**, 1619–1630.
Colquhoun, D. (1994). Practical analysis of single channel records. In "Microelectrode Techniques. The Plymouth Workshop Handbook," (D. C. Ogden, ed.), pp. 101–139. The Company of Biologists, Cambridge, UK.
Dobrev, D. (2006). Electrical remodeling in atrial fibrillation. *Herz* **31**, 108–112, quiz 142–143.
Dobrev, D., and Nattel, S. (2010). New antiarrhythmic drugs for treatment of atrial fibrillation. *Lancet* **375**, 1212–1223.
Dobrev, D., et al. (2000). G-Protein beta(3)-subunit 825T allele is associated with enhanced human atrial inward rectifier potassium currents. *Circulation* **102**, 692–697.
Dobrev, D., et al. (2001). Molecular basis of downregulation of G-protein-coupled inward rectifying K(+) current (I(K, ACh)) in chronic human atrial fibrillation: Decrease in GIRK4 mRNA correlates with reduced I(K, ACh) and muscarinic receptor-mediated shortening of action potentials. *Circulation* **104**, 2551–2557.
Dobrev, D., et al. (2005). The G protein-gated potassium current I(K, ACh) is constitutively active in patients with chronic atrial fibrillation. *Circulation* **112**, 3697–3706.
Ehrlich, J. R., et al. (2004). Characterization of a hyperpolarization-activated time-dependent potassium current in canine cardiomyocytes from pulmonary vein myocardial sleeves and left atrium. *J. Physiol.* **557**, 583–597.
Feng, J., et al. (1998). Ultrarapid delayed rectifier current inactivation in human atrial myocytes: Properties and consequences. *Am. J. Physiol.* **275**, H1717–H1725.

Fujita, S., et al. (2000). A regulator of G protein signalling (RGS) protein confers agonist-dependent relaxation gating to a G protein-gated K + channel. *J. Physiol.* **526**(Pt 2), 341–347.

Gentet, L. J., et al. (2000). Direct measurement of specific membrane capacitance in neurons. *Biophys. J.* **79**, 314–320.

Hamill, O. P., et al. (1981). Improved patch-clamp techniques for high-resolution current recording from cells and cell-free membrane patches. *Pflugers Arch.* **391**, 85–100.

Hashimoto, N., et al. (2008). Characterization of in vivo and in vitro electrophysiological and antiarrhythmic effects of a novel IKACh blocker, NIP-151: A comparison with an IKr-blocker dofetilide. *J. Cardiovasc. Pharmacol.* **51**, 162–169.

Heidbuchel, H., et al. (1990a). ATP-dependent activation of atrial muscarinic K + channels in the absence of agonist and G-nucleotides. *Pflugers Arch.* **416**, 213–215.

Heidbuchel, H., et al. (1990b). Three different potassium channels in human atrium. Contribution to the basal potassium conductance. *Circ. Res.* **66**, 1277–1286.

Hibino, H., et al. (2010). Inwardly rectifying potassium channels: Their structure, function, and physiological roles. *Physiol. Rev.* **90**, 291–366.

Himmel, H. M., et al. (2000). Evidence for Edg-3 receptor-mediated activation of I(K.ACh) by sphingosine-1-phosphate in human atrial cardiomyocytes. *Mol. Pharmacol.* **58**, 449–454.

Huang, C. L., et al. (1998). Direct activation of inward rectifier potassium channels by PIP2 and its stabilization by Gbetagamma. *Nature* **391**, 803–806.

Ito, H., et al. (1994). Background conductance attributable to spontaneous opening of muscarinic K + channels in rabbit sino-atrial node cells. *J. Physiol.* **476**, 55–68.

Jin, W., and Lu, Z. (1998). A novel high-affinity inhibitor for inward-rectifier K + channels. *Biochemistry* **37**, 13291–13299.

Jin, W., and Lu, Z. (1999). Synthesis of a stable form of tertiapin: A high-affinity inhibitor for inward-rectifier K + channels. *Biochemistry* **38**, 14286–14293.

Jin, W., et al. (1999). Mechanisms of inward-rectifier K + channel inhibition by tertiapin-Q. *Biochemistry* **38**, 14294–14301.

Kaibara, M., et al. (1991). Regulation of spontaneous opening of muscarinic K + channels in rabbit atrium. *J. Physiol.* **433**, 589–613.

Kanjhan, R., et al. (2005). Tertiapin-Q blocks recombinant and native large conductance K + channels in a use-dependent manner. *J. Pharmacol. Exp. Ther.* **314**, 1353–1361.

Kim, D. (1993). Mechanism of rapid desensitization of muscarinic K + current in adult rat and guinea pig atrial cells. *Circ. Res.* **73**, 89–97.

Kitamura, H., et al. (2000). Tertiapin potently and selectively blocks muscarinic K(+) channels in rabbit cardiac myocytes. *J. Pharmacol. Exp. Ther.* **293**, 196–205.

Kovoor, P., et al. (2001). Evaluation of the role of I(KACh) in atrial fibrillation using a mouse knockout model. *J. Am. Coll. Cardiol.* **37**, 2136–2143.

Krapivinsky, G., et al. (1995). The G-protein-gated atrial K + channel IKACh is a heteromultimer of two inwardly rectifying K(+)-channel proteins. *Nature* **374**, 135–141.

Kurachi, Y., et al. (1986). On the mechanism of activation of muscarinic K + channels by adenosine in isolated atrial cells: Involvement of GTP-binding proteins. *Pflugers Arch.* **407**, 264–274.

Liu, L., and Nattel, S. (1997). Differing sympathetic and vagal effects on atrial fibrillation in dogs: Role of refractoriness heterogeneity. *Am. J. Physiol.* **273**, H805–H816.

Loewi, O. (1921). Über humorale Übertragbarkei der Herznervenwirkung. *Pflügers Arch.* **189**, 239–242.

Luscher, C., et al. (1997). G protein-coupled inwardly rectifying K + channels (GIRKs) mediate postsynaptic but not presynaptic transmitter actions in hippocampal neurons. *Neuron* **19**, 687–695.

Margaret, W. C. (1955). Pulsation, migration and division in dissociated chick embryo heart cells in vitro. *J. Exp. Zool.* **128**, 573–589.

Mark, M. D., and Herlitze, S. (2000). G-protein mediated gating of inward-rectifier K + channels. *Eur. J. Biochem.* **267**, 5830–5836.

Medina, I., et al. (2000). A switch mechanism for G beta gamma activation of I(KACh). *J. Biol. Chem.* **275**, 29709–29716.

Nattel, S., et al. (2008). Atrial remodeling and atrial fibrillation: Mechanisms and implications. *Circ. Arrhythm. Electrophysiol.* **1**, 62–73.

Neher, E., and Sakmann, B. (1976). Single-channel currents recorded from membrane of denervated frog muscle fibres. *Nature* **260**, 799–802.

Noma, A., and Trautwein, W. (1978). Relaxation of the ACh-induced potassium current in the rabbit sinoatrial node cell. *Pflügers Arch.* **377**, 193–200.

Okabe, K., et al. (1991). The nature and origin of spontaneous noise in G protein-gated ion channels. *J. Gen. Physiol.* **97**, 1279–1293.

Sackin, H., et al. (2003). Permeant cations and blockers modulate pH gating of ROMK channels. *Biophys. J.* **84**, 910–921.

Sakmann, B., et al. (1983). Acetylcholine activation of single muscarinic K + channels in isolated pacemaker cells of the mammalian heart. *Nature* **303**, 250–253.

Shui, Z., et al. (1997). ATP-dependent desensitization of the muscarinic K + channel in rat atrial cells. *J. Physiol.* **505**(Pt 1), 77–93.

Sigworth, F. J., and Neher, E. (1980). Single Na + channel currents observed in cultured rat muscle cells. *Nature* **287**, 447–449.

Sui, J. L., et al. (1998). Activation of the atrial KACh channel by the betagamma subunits of G proteins or intracellular Na + ions depends on the presence of phosphatidylinositol phosphates. *Proc. Natl. Acad. Sci. USA* **95**, 1307–1312.

Tanaka, H., and Hashimoto, N. (2007). A multiple ion channel blocker, NIP-142, for the treatment of atrial fibrillation. *Cardiovasc. Drug Rev.* **25**, 342–356.

Voigt, N., et al. (2007). Differential phosphorylation-dependent regulation of constitutively active and muscarinic receptor-activated IK,ACh channels in patients with chronic atrial fibrillation. *Cardiovasc. Res.* **74**, 426–437.

Voigt, N., et al. (2008). Changes in I K,ACh single-channel activity with atrial tachycardia remodelling in canine atrial cardiomyocytes. *Cardiovasc. Res.* **77**, 35–43.

Voigt, N., et al. (2010a). Inhibition of IK, ACh current may contribute to clinical efficacy of class I and class III antiarrhythmic drugs in patients with atrial fibrillation. *Naunyn Schmiedebergs Arch. Pharmacol.* **381**, 251–259.

Voigt, N., et al. (2010b). Left-to-right atrial inward-rectifier potassium current gradient in patients with paroxysmal versus chronic atrial fibrillation. *Circ. Arrhythm. Electrophysiol.* doi: 10.1161/CIRCEP.110.954636.

Yamada, M., et al. (1998). G protein regulation of potassium ion channels. *Pharmacol. Rev.* **50**, 723–760.

CHAPTER THIRTY-THREE

Assaying WAVE and WASH Complex Constitutive Activities Toward the Arp2/3 Complex

Emmanuel Derivery *and* Alexis Gautreau

Contents

1. Introduction	678
2. Establishment of Stable Cell Lines Expressing Tagged WAVE and WASH Complexes	680
2.1. Reagents	680
2.2. Procedure	681
3. Large-Scale Purification of WAVE and WASH Complexes	682
3.1. Reagents	682
3.2. Procedure	682
4. Aggregation Analysis of WAVE and WASH Multiprotein Complexes	684
4.1. Procedure	685
5. Activity Measurements using Pyrene Actin Polymerization Assays	687
5.1. Reagents	687
5.2. Procedure	688
6. Detection of an Endogenous Activity of the WAVE Complex	691
7. Concluding Remarks	693
Acknowledgments	693
References	694

Abstract

The Arp2/3 complex generates branched actin networks when activated by Nucleation Promoting Factors (NPFs). Among these, WAVE proteins are required for lamellipodia and ruffle formation, whereas WASH proteins are required for the fission of endosomes. Both WASH and WAVE NPFs are embedded into multiprotein complexes that provide additional functions and regulations. Understanding how these complexes regulate the activity of their NPF starts with the determination of the constitutive activity of the complex. In this chapter, we describe how to efficiently purify the WAVE and WASH complexes

Laboratoire d'Enzymologie et de Biochimie Structurales, Gif sur Yvette, France

Methods in Enzymology, Volume 484 © 2010 Elsevier Inc.
ISSN 0076-6879, DOI: 10.1016/S0076-6879(10)84033-9 All rights reserved.

from human stable cell lines. We also describe how to verify that these complexes are not aggregated, a prerequisite for activity assays. We then provide a protocol to measure their activity toward the Arp2/3 complex using the well-established pyrene actin assay. Finally, we show how our fast purification protocol can be modified to detect the endogenous activity of the WAVE complex, providing an easy readout for the level of WAVE activation in cells.

1. INTRODUCTION

The dynamic actin cytoskeleton is an essential player of cell morphogenesis. Polarized assembly of actin filaments often occurs on membranes, causing their deformation or movement. These actin-based processes are initiated by several classes of actin nucleators (Pollard, 2007). Among these, the Arp2/3 complex nucleates actin filaments from the side of an existing filament, thus generating dendritic arrays. The Arp2/3 complex is constitutively inactive and needs to be activated by Nucleation Promoting Factors (NPFs). The most active NPFs contain a C-terminal domain, called the VCA domain for Verprolin Homology (also known as WH2), Connector and Acidic. The VCA domain binds and activates the Arp2/3 complex (Pollard, 2007). To date, there are four families of VCA containing NPFs, namely WASP, WAVE, WHAMM/JMY, and WASH. WAVE proteins are required for the formation of lamellipodia and membrane ruffles (Takenawa and Suetsugu, 2007). WASP proteins are required for the internalization step of endocytosis, podosome formation, and endosome motility under certain conditions (Benesch et al., 2002, 2005; Jones et al., 2002). WHAMM is required to maintain Golgi morphology and for anterograde membrane transport (Campellone et al., 2008). JMY, despite significant homologies to WHAMM, is involved in transcription and cell migration (Zuchero et al., 2009). Finally, WASH is required for endosome fission (Derivery et al., 2009b).

An interesting property of NPFs is their incorporation into multiprotein complexes. Indeed, WASP proteins have been shown to be in complex with proteins of the WIP family (Derivery and Gautreau, 2010b), WAVE proteins have been shown to be embedded into a complex comprising five subunits namely Sra, Nap, Abi, WAVE, and Brick1 (Eden et al., 2002; Gautreau et al., 2004; Innocenti et al., 2004), and WASH has been recently shown to be part of a complex containing seven subunits namely VPEF, KIAA1033, Strumpellin, WASH, Ccdc53, and the heterodimer of Capping protein (Derivery and Gautreau, 2010a; Derivery et al., 2009b). Interestingly, WASH and WAVE complexes might be structurally analogous (Jia et al., 2010). The incorporation into a multiprotein complex usually provides additional functions and regulations to the NPF. For instance, the WAVE complex provides a molecular link between WAVE proteins and Rac signaling through the Sra subunit and the WASH complex combines a NPF with an actin barbed end capper. Since both WAVE and WASH proteins in isolation are active toward the Arp2/3 complex (Duleh and Welch, 2010; Eden et al., 2002; Innocenti et al., 2004; Machesky et al., 1999), one might wonder what the intrinsic activity of their respective complex is.

The major bottleneck in the assessment of the activity of multiprotein complexes is their purification. The purification of endogenous multiprotein complexes from cells or tissues usually requires many steps of chromatography, which leads to an overall low yield. For instance, between four and eight steps are required for the purification of WAVE1 and 2 complexes (Eden et al., 2002; Gautreau et al., 2004; Kim et al., 2006; Lebensohn and Kirschner, 2009). This procedure increases the likelihood of denaturation since it lasts several days and use physicochemical conditions that are not necessarily optimal for the stability of the complex.

An alternative to the purification of an endogenous complex is the reconstitution of this complex from recombinant proteins. This method has high yields due to overexpression and has the advantage to simplify the composition of the purified complex. Indeed, cells can express several isoforms for each subunit, giving rise to a combinatorial complexity in native preparations. But this approach is also a long and difficult process. Moreover, such a reconstitution is considered successful only when the reconstituted complex has the same activity as the native one, and thus cannot be used if this constitutive activity is not yet known. The reconstitution of fully recombinant complexes has been obtained for the Arp2/3 (Gournier et al., 2001), the WAVE (Ismail et al., 2009), and the WASH complexes (Jia et al., 2010).

We have developed an intermediate approach between native purifications and fully recombinant reconstitutions. A recombinant tagged subunit is stably expressed in a mammalian cell line and the endogenous complex assembled around this recombinant subunit is purified efficiently by affinity chromatography. Using this strategy, we have greatly improved the yield of the WAVE complex purification over native purification (Derivery et al., 2009a). Importantly, this method can also be used to quickly identify unknown complexes, as we have recently shown for the WASH complex (Derivery et al., 2009b). Using actin polymerization assays, we have shown that the WASH complex purified using this procedure was constitutively active toward the Arp2/3 complex, unlike the WAVE complex (Derivery et al., 2009a,b).

In this chapter, we describe how to quickly and efficiently purify the WASH and WAVE complexes from human stable cell lines. We also provide a protocol to verify that purified complexes are not aggregated, a prerequisite for activity assays. We then describe how to measure their activity toward the Arp2/3 complex using the quantitative pyrene actin assay. Finally, we show how the purification protocol described here can be adapted to detect an endogenous activity of the WAVE complex, which likely reflects its activation in cells.

2. ESTABLISHMENT OF STABLE CELL LINES EXPRESSING TAGGED WAVE AND WASH COMPLEXES

We purify WASH and WAVE complexes from cell lines stably expressing one subunit fused to purification tags. In stable cell lines, an interesting property occurs: the exogenous subunit is incorporated into the endogenous complex at the expense of the endogenous one. This property is the result of the degradation of unassembled subunits (Derivery and Gautreau, 2010b; Derivery et al., 2009a). The endogenous complex assembled around this tagged subunit can then be retrieved efficiently by affinity purification. Purification starting from cell lines expressing different subunits can be compared to find the one providing the best incorporation. Actually, we have demonstrated the feasibility of purifying WAVE complexes tagged on any subunit using this system, although with variable efficiency (Derivery et al., 2009a). We will only focus here on the purification of complexes assembled around tagged WAVE and WASH subunits.

To generate these stable cell lines, we use a commercial system in which the plasmid of interest is inserted at a unique well-expressed locus in the host HEK 293 cell line through site directed homologous recombination. As purification tags, we commonly use FLAG and Protein C that are small epitope tags, DYKDDDDK and EDQVDPRLIDGK, respectively. They are recognized by highly specific monoclonal antibodies. FLAG tagged proteins are eluted by competition with an excess of FLAG peptide, whereas PC tagged proteins are eluted by Ca^{2+} chelation since the antibodies requires Ca^{2+} for binding.

2.1. Reagents

pCDNA5 plasmid derived from pCDNA5/FRT/V5-His (Invitrogen) and containing the ORF of human WAVE2 fused to FLAG-HA (M-DYKDDDDK-YPYDVPDYA) or of murine WASH fused to $(His)_6$-PC (M-HHHHHH-EDQVDPRLIDGK) followed by a TEV binding and cleavage site (DYDIPTTENLYFQG). Both tags are fused at the N-terminus and the V5-His tags in the C-terminus are not translated because the natural stop codon of the ORF is kept.

pOG44 plasmid containing the recombinase (Invitrogen).
Flp-In^{TM} T-REx^{TM} 293 cells (Invitrogen).
Hygromycin (Hygrogold, Invivogen).
Growth medium: DMEM supplemented with 10% (v/v) Fetal Calf Serum (FCS) and Penicillin/Streptomycin (0.1 mg/ml each), all reagents from PAA Laboratories.

$2\times$ *HBS*: 274 mM NaCl, 10 mM KCl, 1.5 mM Na$_2$HPO$_4$, 11.1 mM D-glucose, 40 mM Na–HEPES, pH 7.05.
XB: 100 mM KCl, 20 mM HEPES, pH 7.7.

2.2. Procedure

250,000 cells are plated in a 3-cm dish. The next day, the medium is replaced by 1.2 ml of fresh medium and cells are transfected using calcium phosphate precipitation procedure 1 h later. The transfection mixture is prepared as follows:

12.4 µl CaCl$_2$ (stock solution at 2 M, filtered)
0.96 µg of pOG44 plasmid
0.2 µg of pCDNA5 plasmid
Sterile Milli-Q water up to 100 µl

This mix is added dropwise into 100 µl of $2\times$ HBS while vortexing at mid power. The resulting solution is then added to cells. Medium is changed after 6–8 h and supplemented with 200 µg/ml hygromycin 24 h later. On the next day, cells are passed into a 10-cm dish and stable transfectants obtained by homologous recombination at the Flp-In site are selected by the continuous presence of 200 µg/ml hygromycin. Clones usually appear after 2 weeks. Stable transfectants are subsequently pooled, amplified, and saved. This homologous recombination procedure is highly efficient and we usually generate many stable cell lines in parallel. A negative control without the pOG44 plasmid is systematically performed to ensure that all the clones obtained are the result of homologous recombination events. These cell lines are quite fragile and we had to use a special protocol to ensure a good viability upon freezing: cells are grown in a 15-cm dish until 80% confluency, trypsinized, and resuspended in 2 ml of 20% growth medium and 80% FCS followed by the addition of 2 ml of 20% DMSO Hybri-Max™ (Sigma, D2650) and 80% FCS. The resulting solution is aliquoted into four vials, stored at $-80\ ^\circ$C for a week, and then transferred into liquid nitrogen for long-term storage. Frozen vials are thawed at 37 $^\circ$C, washed once in growth medium to remove DMSO, and plated in a 10-cm dish. Hygromycin selection is added at 200 µg/ml 2 days later.

The large-scale purification of WAVE and WASH complexes requires a large number of cells, which are amplified in spinner bottles as follows. Cells are amplified to obtain eight 15-cm dishes at 90% confluency and then trypsinized and diluted into 1 l of growth medium supplemented with 100 µg/ml hygromycin in a 2 l spinner flask (Techne, 6027608). The culture is maintained under permanent gentle agitation (50 rpm) on a stirring table (Techne, MCS-104L). Three days later, the culture is split into four spinners and each volume is adjusted to 1 l with growth medium supplemented with

100 μg/ml hygromycin. Volume is adjusted to 1.5 l 2 days later and cells are harvested after two more days. Cells are washed in XB buffer. Cell pellets are quickly frozen in liquid nitrogen and stored at −80 °C. A 6 l culture represents roughly 10^{10} cells and 20 ml of pellet. Flp-In^{TM} T-REx^{TM} 293 cells have retained characteristics of epithelial cells and form spherical hollow structures resembling balloons when cultured in nonadherent conditions. Such structures are not suited for long-term culture and so we usually do not exceed 1 week of culture in spinner bottles. This property also implies that cells must be extensively trypsinized before counting.

3. LARGE-SCALE PURIFICATION OF WAVE AND WASH COMPLEXES

3.1. Reagents

Pellet of cells stably expressing FLAG-HA-WAVE2 of $(His)_6$-PC-WASH (volumes are given for a 10-ml pellet, corresponding to roughly 5×10^9 cells)

WAVE lysis buffer (WLB): 150 mM NaCl, 5 mM EDTA, 0.1% SDS, 1% NP-40, 0.5% DOC, 5% glycerol, 50 mM Na–HEPES, pH 7.7.

WAVE buffer 1 (WB1): 200 mM KCl, 20% glycerol, 50 mM Na–HEPES, pH 7.7.

WAVE buffer 2 (WB2): 400 mM KCl, 20% glycerol, 50 mM Na–HEPES, pH 7.7.

FLAG elution buffer: WB1 supplemented with 0.15 mg/ml 3× FLAG peptide (Sigma F4799). Stock solution at 5 mg/ml in 0.2 M NaCl, 0.1 M Tris–HCl, pH 7.5.

Methylcellulose (Sigma, M0512). Stock solution made at 2% (4000 cP) in Milli-Q water. Let stir overnight at 4 °C to dissolve.

WASH lysis buffer (WHLB): 200 mM NaCl, 1 mM CaCl$_2$, 1% Triton X-100, 5% glycerol, 50 mM Na–HEPES, pH 7.4.

WASH buffer 1 (WHB1): 200 mM NaCl, 1 mM CaCl$_2$, 0.02% methylcellulose, 50 mM Na–HEPES, pH 7.4.

WASH buffer 2 (WHB2): WHB1 without CaCl$_2$.

PC elution buffer: WHB2 supplemented with 5 mM EGTA.

3.2. Procedure

To minimize degradation, all purification steps are performed at 4 °C. Buffers are prepared the day before the experiment and allowed to equilibrate at 4 °C overnight, along with centrifuge rotors and bottles. The cell pellet is thawed and resuspended in 40 ml of WLB supplemented with protease inhibitor cocktail (1:500, Sigma, P8340). The solution is rocked for 1 h at 4 °C and cell debris are removed by centrifugation at 3300 × g for 10 min in conical tubes using a swinging bucket rotor (Heraeus, Megafuge 1.0). The supernatant is further clarified by ultracentrifugation at 200,000 × g for 1 h in a fixed-angle rotor (Beckman Type 70 Ti). A thin layer composed of lipids that do not pellet during the high-speed centrifugation is usually observed and carefully removed by pipetting. A small contamination is unavoidable and does not affect subsequent steps. The clarified supernatant is filtered through a 0.22 μm syringe filter unit (Whatman) to avoid subsequent column clogging. The extract is then transferred onto a 25 ml plastic column (Bio-Rad, Econo-Pac) containing 1 ml of FLAG-M2 agarose resin (Sigma, A2220) preequilibrated in WLB buffer. The extract is allowed to recirculate for 4 h at 4 °C with a flow rate of 0.5 ml/min using a peristaltic pump (GE Healthcare, P-1 pump). The column is then washed with 50 ml of WLB buffer, 50 ml of WB1, 50 ml of WB2, and 50 ml of WB1 by gravity flow. The washing step in WB2 constitutes a salt "bump" that enhances purity (WB2 buffer contains twice more KCl than WB1 buffer). FLAG-M2 beads are then transferred into a 2 ml centrifuge tube and several elutions are performed by mixing the resin with 1 ml of FLAG elution buffer. Each elution is performed by rocking the slurry (30 min at 17 °C then overnight at 4 °C for the first one, 10 min at 4 °C for the others) and centrifuging at 300 × g for 1 min. The WAVE complex usually elutes in the first two fractions which are pooled and allowed to pass through a disposable column to remove remaining beads. The WAVE complex is further concentrated using ultrafiltration (Amicon Ultra-4 3 K, Millipore). The yield of this one-step purification is about 100–200 μg complex starting from 5×10^9 cells. The WAVE complex can be flash frozen in liquid nitrogen and stored at −80 °C. We prefer, however, to use fresh material for activity assays. Figure 33.1 (left panel) shows the pattern of the WAVE complex analyzed by SDS page on 4–12% Bis–Tris gradient gels (Invitrogen).

The WASH complex is purified using the above procedure with the following modifications: WLB is replaced by WHLB, WB1 by WHB1, WB2 by WHB2, FLAG elution buffer by PC elution buffer and FLAG-M2 agarose resin by Protein C affinity resin (Roche, 11815024001), respectively. Since elution from the PC affinity resin (Ca^{2+} chelation) is more efficient than from the FLAG resin, elution is performed by gravity flow at 4 °C. We usually prefer EGTA over EDTA for Ca^{2+} chelation because it does not affect the polymerization of Mg-ATPactin, hence the activity of eluted proteins can be directly tested. The yield of WASH complex purification is usually 5–10 times lower than the one of the WAVE complex. Figure 33.1 (right panel) shows the pattern of the WASH complex analyzed by SDS page on 4–12% Bis–Tris gradient gels.

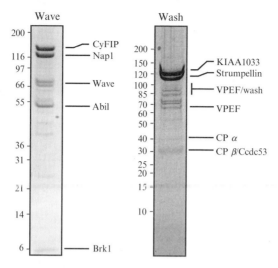

Figure 33.1 Patterns of WAVE and WASH complexes: 10 μg of purified complexes were analyzed by SDS-PAGE and Coomassie blue staining. Identity of the different subunits was determined by mass spectrometry. Molecular weight markers in kDa are indicated on the left of each gel. Left panel shows the pattern of the WAVE complex purified through a tagged Abi1 subunit. Reprinted with permission from Derivery et al. (2009a). The WAVE doublet observed here is likely due to phosphorylations and to the presence of two paralogous WAVE genes expressed in 293 cells, WAVE1, and WAVE2. Right panel shows the pattern of the WASH complex obtained using a tagged WASH subunit. Reprinted with permission from Derivery et al. (2009b).

4. Aggregation Analysis of WAVE and WASH Multiprotein Complexes

As with any protein, care must be taken concerning aggregation of multiprotein complexes when one wants to perform activity assays. For instance, an intramolecular inhibition can be released upon aggregation, leading to activity artifacts. This is especially the case for multiprotein complexes, which are bigger, so have larger contact interfaces, and whose purification yields are usually low, leading to diluted solutions. The WAVE complex offers a textbook example to illustrate this point. Depending on purifications, the native WAVE complex has been found active (Kim et al., 2006) or not (Derivery et al., 2009a; Eden et al., 2002; Ismail et al., 2009; Lebensohn and Kirschner, 2009). It was recently shown that denaturation by heat (Derivery et al., 2009a; Lebensohn and Kirschner, 2009) or freezing in the absence of cryoprotectant (Ismail et al., 2009) leads to the detection of some activity of the complex. This likely explains the results of Kim and colleagues and illustrates how important it is to find optimal buffer conditions for the stability of a new complex before starting activity assays.

Several methods are available to study the aggregation state of proteins, such as DLS and analytical gel filtration. These methods, however, require a large amount of material and thus cannot be practically used to study the aggregation of multiprotein complexes, whose purification yields are low. We describe here a quick and efficient method to analyze the aggregation state of WASH and WAVE complexes based on controlled ultracentrifugation experiments. This method requires only small amounts of starting material and thus can be performed on small-scale purifications, making it particularly suited for buffer screening. Unfortunately, there is no universal buffer and every complex is likely to require such buffer optimization. For instance, the WASH complex is not stable in the conditions optimized for the WAVE complex.

4.1. Procedure

The aggregation assay described here requires only limited amounts of starting material; so, it is performed with complexes retrieved in small-scale purifications. Small-scale purification of WASH or WAVE complexes is performed by scaling down the above described, large-scale protocol. We start from two 15-cm dishes of cells (6×10^7 cells) lysed in 1 ml of lysis buffer. Clarification of the lysate is performed by a single spin at $20,000 \times g$ for 10 min at 4 °C in a tabletop centrifuge (Eppendorf, model 5424). We use only 20 μl of affinity resin and all washes are performed with 1 ml of buffer. For elution, the resin is resuspended in 20 μl of elution buffer and allowed to sit overnight at 4 °C. The next day, the slurry is centrifuged at $300 \times g$ for 1 min and the supernatant carefully pipetted. 20 μl of fresh elution buffer is added to the beads and after centrifugation of the slurry at $300 \times g$ for 1 min, the supernatant carefully pipetted and pooled with the previous one. Remaining beads are removed by two sequential spins at $20,000 \times g$ for 30 s.

20 μl of eluate is spun in a Beckman TLA100 rotor at 40,000 rpm for 30 min at 4 °C in a 7×20 mm polycarbonate tube (Beckman, 343775). Apparent K factor for centrifuging 20 μl at 40,000 rpm in this rotor is 24.25. This K factor implies that 2.2 h are required to pellet the WASH complex, whose sedimentation coefficient is 11S (Gautreau et al., 2004). The WASH complex, with a sedimentation coefficient of 12.5S (Derivery et al., 2009b) requires 1.95 h to pellet. Thus the pellet after a 30 min spin contains only aggregated complexes. The supernatant is carefully pipetted and the pellet is resuspended in 20 μl of boiling 1× SDS-PAGE loading buffer (diluted with fresh elution buffer). Equal volumes of eluate, supernatant, and resuspended pellet are then analyzed by SDS-PAGE followed by Coomassie blue staining or Western blotting, depending on complex abundance.

Figure 33.2A illustrates how increasing the KCl concentration from 100 to 200 mM decreases the aggregation of the WAVE complex. Exact buffer

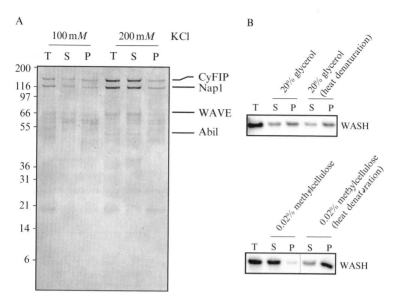

Figure 33.2 Screen for stability conditions of WAVE and WASH complexes. Small-scale purifications of WAVE and WASH complexes were performed in indicated buffer. WAVE and WASH complexes were purified using tagged WAVE and WASH subunits, respectively. Purified material was then subjected to ultracentrifugation and both supernatant and pellet were analyzed by SDS-PAGE. T: total, S: supernatant, P: pellet (aggregated material). Comparison of the amount of material between total and supernatant indicates the level of aggregation. (A) Coomassie blue stained gel showing that increasing the KCl concentration from 100 to 200 mM through the entire purification decreases the aggregation of the WAVE complex. (B) Western blot showing that replacement of 20% glycerol by 0.02% methylcellulose in wash buffers decreases the aggregation of the WASH complex. Heat denaturation (95 °C for 5 min) is used here as a positive control of aggregation.

compositions in this experiment are as follows (buffer names refer to the large-scale purification scheme):

Sample	100 mM KCl	200 mM KCl
WLB	150 mM NaCl, 5 mM EDTA, 0.1% SDS, 1% NP-40, 0.5% DOC, 5% glycerol, 50 mM Na–HEPES, pH 7.7.	
WB1 = WB2	20% glycerol, 50 mM Na–HEPES, pH 7.7	
	100 mM KCl	200 mM KCl

Figure 33.2B illustrates how the substitution of 20% glycerol with 0.02% methylcellulose decreases the aggregation of the WASH complex. Heat denaturation (95 °C for 5 min) is used as a positive control of aggregation. Exact buffer compositions in this experiment are as follows:

Sample		
WHLB	20% glycerol	0.02% methylcellulose
	200 mM NaCl, 1 mM CaCl$_2$, 1% Triton X-100, 5% Glycerol, 50 mM Na–HEPES, pH 7.4.	
WHB1	200 mM NaCl, 1 mM CaCl$_2$, 50 mM Na–HEPES, pH 7.4	
	20% glycerol	0.02% methylcellulose
WHB2	200 mM NaCl, 50 mM Na–HEPES, pH 7.4	
	20% glycerol	0.02% methylcellulose

To find optimal buffer conditions, we commonly vary salt concentration (NaCl or KCl: 50–400 mM) and additives (glycerol: 0–30%, sucrose 0–0.4 M or methylcellulose 0–0.1%). We usually keep the same salt concentration through the entire purification and add the additives only in the wash buffers. Using this assay, we found that the stability of the WAVE complex requires a KCl concentration between 200 and 400 mM in addition to glycerol (between 15% and 20%). On the other hand, the stability of the WASH complex requires a KCl concentration of 200 mM in addition to 0.02% methylcellulose.

5. Activity Measurements Using Pyrene Actin Polymerization Assays

Once complexes are purified and in an unaggregated state, their constitutive activity toward the Arp2/3 complex can be assayed. To monitor the kinetics of Arp2/3 mediated actin polymerization, we use the robust and quantitative pyrene actin assay, in which actin polymerization is followed in a fluorimeter through the incorporation of a fluorescent tracer, pyrene actin. The fluorescence intensity of pyrene actin increases by an ∼25-fold upon polymerization (Kouyama and Mihashi, 1981), allowing quantitative analysis of polymerization kinetics. Alternatively, a microscopy-based assay of Arp2/3 activity can be used (Blanchoin et al., 2000). This assay requires less material but its quantification is less convenient. Hence, this assay will not be described here.

5.1. Reagents

G buffer: 0.2 mM ATP, 0.1 mM CaCl$_2$, 0.1 mM DTT, 0.01% NaN$_3$, and 5 mM Tris–HCl, pH 8.0.
10× KMEI: 500 mM KCl, 10 mM MgCl$_2$, 10 mM EGTA, and 100 mM Imidazole–HCl, pH 7.0.
10× Exchange buffer: 2 mM EGTA and 0.2 mM MgCl$_2$.

Rabbit muscle actin, isolated from acetone powder as described (Spudich and Watt, 1971) and stored in G buffer. Alternatively, this reagent can be purchased from Cytoskeleton, Inc.

Pyrene actin, prepared as described in a previous volume of *Methods in Enzymology* (Zigmond, 2000) and stored in G buffer. Alternatively, this reagent can also be purchased from Cytoskeleton, Inc.

Arp2/3 complex, purified as described (Egile *et al.*, 1999) or purchased from Cytoskeleton, Inc.

5.2. Procedure

To minimize concentration errors due to pipetting, a common stock solution of actin/pyrene actin (90/10%) at 12 μM in G buffer is prepared and kept on ice. A typical actin polymerization assay is made as follows:

- MgATP-G-actin is first prepared by mixing 37.5 μl of the actin/pyrene actin stock (12 μM) with 4.5 μl of 10× Exchange buffer and 3 μl of G buffer. This solution is incubated on ice for 5 min and the resulting MgATP-G-actin (at 10 μM) is used within 1 h.
- The polymerization mixture is then prepared at room temperature by adding components in this order:

Volume added	Final concentration
16 μl of 10× KMEI	1×
87.5 μl G buffer	
1.5 μl Arp 2/3 complex (stock at 4.5 μM)	43 nM
15 μl WAVE or WASH complex	20–200 nM
40 μl of MgATP-G-actin (stock at 10 μM)	2.5 μM (10% pyrene)
160 μl total	

It is important to add the G buffer before fragile components like the Arp2/3, WASH, or WAVE complexes so that the high concentration of KCl of the 10× KMEI is diluted. We do not recommend increasing too much the volume of WAVE/WASH complex since additives such as glycerol and sucrose affect actin polymerization. Low activities are often hidden by the spontaneous nucleation of actin, which can be reduced by diminishing the actin concentration in the assay down to 1 μM.

- The reaction is mixed, transferred immediately to a quartz cuvette, and fluorescence is monitored in a SAFAS Xenius fluorimeter (Safas SA, Monaco), where up to 10 kinetics can be followed in parallel. Pyrene fluorescence is measured at 407 nm with excitation at 365 nm.

One must be extremely careful about controls when observing an activity with material retrieved with only one step of purification, since an increase of the actin polymerization rate can arise from many sources other than the constitutive activity of the complex assayed. We recommend the following controls:

- Assaying the activity of an equal amount of buffer without complex since additives can affect actin polymerization.
- Removing the Arp2/3 complex from the polymerization mix, since other actin nucleators, like formins, might copurify, as described for the WAVE complex (Beli *et al.*, 2008; Yang *et al.*, 2007). A "nucleation" effect might also be observed if F-actin stabilizing proteins or filament cappers are present in the preparation.
- Assaying the activity of material purified from a cell line expressing the empty vector, or changing the purification tags, to exclude a possible contamination from the affinity resin.
- Using blocking antibodies targeting the supposedly active part of the complex, if available, to ascribe the observed activity to a subunit in particular.
- Assaying the activity of complexes purified through tags present on different subunits. It is indeed reasonable to think that the measured activity might be due to a contamination by a small pool of unassembled, unregulated subunit, since the purification protocol described here only relies on the tag present on the exogenous subunit. This is especially true when the tagged subunit is the one harboring the activity, like WASH or WAVE. Fortunately, in the stable cell line context described here, we observed that the excess of unassembled exogenous subunit is usually degraded (Derivery *et al.*, 2009a). Along with this idea, we could not detect any differences in the constitutive activity of WAVE complexes tagged on either WAVE2 or Abi1 (Derivery *et al.*, 2009a).

Constitutive activities of WAVE and WASH complexes are displayed in Fig 33.3. Figure 33.3A shows that the WAVE complex is constitutively inactive since no increase of the polymerization rate is observed when it is added to Actin and the Arp2/3 complex. The WAVE complex has the counterintuitive property to reveal its activity upon heat denaturation (Derivery *et al.*, 2009a; Lebensohn and Kirschner, 2009), providing a useful internal positive control in the experiment. On the other hand, Fig. 33.3B shows that the WASH complex is constitutively active. Indeed, addition of WASH complex highly increases the polymerization rate of actin in the presence of the Arp2/3 complex. This phenomenon is specifically due to the activation of the Arp2/3 complex by the VCA domain of WASH, since a blocking antibody targeting the VCA domain can alleviate this increase of the polymerization rate. Interestingly, in the absence of Arp2/3 complex,

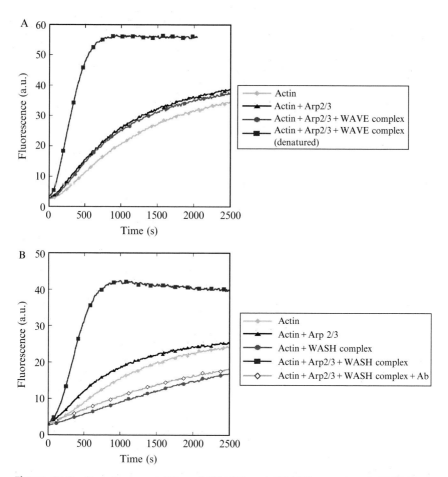

Figure 33.3 Constitutive activity of WAVE and WASH complexes. WAVE or WASH complex activities toward purified Arp2/3 complex was monitored by pyrene fluorescence of Mg-ATP actin. WAVE and WASH complexes were purified using tagged WAVE and WASH subunits, respectively. Conditions: 10 mM imidazole–HCl, pH 7.0, 50 mM KCl, 1 mM MgCl$_2$, 1 mM EGTA, 0.2 mM ATP, 0.2 mM CaCl$_2$, 0.5 mM DTT, and 3 mM NaN$_3$ at room temperature. $\sim$20 nM WAVE complex or 35 nM WASH complex was added to 43 nM Arp2/3 complex and 2.5 μM actin (10% pyrene labeled). WASH-VCA blocking antibody was added at 40 nM. (A) The WAVE complex is constitutively inactive. Heat denaturation (58 °C for 10 min), which reveals the activity of the WAVE complex, is shown as a positive control. The purified WAVE complex used in this experiment came from a small-scale purification, but no difference in terms of activity is observed in purification performed at small or large scale. (B) The WASH complex is constitutively active. Reprinted with permission from Derivery et al. (2009b). The complex used in this experiment was purified as described above, except that methylcellulose (0.02%) was substituted by sucrose (0.4 M). Substituting sucrose by methylcellulose does not affect the activity of the complex but prevents its denaturation upon flash freezing in liquid nitrogen.

the WASH complex slowed down the kinetics of spontaneous actin polymerization, a property attributed to a low barbed end capping activity of the complex (Derivery et al., 2009b).

6. Detection of an Endogenous Activity of the WAVE Complex

The WAVE complex is thought to alternate between a cytosolic, inactive conformation and a membrane-bound, active one (Suetsugu et al., 2006). Although we recover both pools with our purification method, we do not detect any activity of the WAVE complex. This suggests that the transient interactors responsible for the activity of the membrane pool are washed away during the purification. It is indeed intuitive to distinguish the weak interactions formed between the WAVE complex and its activator(s) in response to signaling from the strong interactions between the subunits maintaining complex cohesion. If one manages to preserve such weak interactions, the detection of the endogenous activity of the WAVE complex should be possible. Having a quantitative readout of the activity level of the WAVE complex in cells is useful to decipher the signaling pathways that activate the WAVE complex. For instance, this assay can be combined with siRNA or drug treatments. We will now describe how the purification protocol described here can be modified to detect the endogenous activity of the WAVE complex.

Transient interactors are lost during the purification mainly for two reasons. First, if handling times are long, the system can reach equilibrium during washing steps, leading to dissociation due to the law of mass action. The only way to overcome this phenomenon is to develop the fastest purification protocol possible. The second phenomenon is that the salts contained in washing buffers compete with the partners for binding to the complex of interest. This is especially true in case of weak interactors, which have a lower number of contacts. To overcome this phenomenon and preserve such interactions, one can think to screen milder buffer conditions. Buffer modifications must, of course, be performed according to stability conditions. The small-scale purification protocol of the WAVE complex described here fulfills both the speed requirement and the possibility to screen various buffers, since handling time is short and that it requires only a limited amount of starting material. In addition, it is compatible with activity measurements (Fig. 33.3A).

Figure 33.4 shows that an endogenous activity of the WAVE complex can be detected if the high salt wash (WB2 buffer) is omitted during the purification (compare "high salt" and "low salt" curves). This activity depends on the Arp2/3 complex. Thus, this activity cannot be accounted

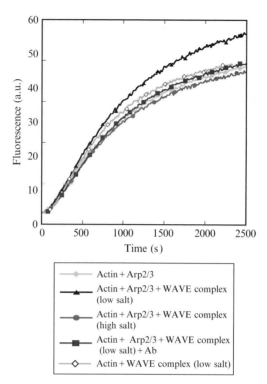

Figure 33.4 Detection of an endogenous activity of the WAVE complex. Two small-scale purifications of the WAVE complex were performed by tagging the WAVE subunit. The washing step at high concentration of KCl (WB2 buffer) was omitted in one sample, referred to as "low salt." WAVE complex activity in both samples was monitored by pyrene fluorescence using conditions identical to those of Fig. 33.3. Interestingly, an activity is detected in the "low salt" sample. This activity is specific of the WAVE-Arp2/3 pathway since removing the Arp2/3 complex or adding a WAVE-VCA blocking antibody (100 nM) can alleviate it. This activity is thus due to a transient activator of the WAVE complex, which is lost during the high salt wash.

for by the formin mDia1, another actin nucleator previously shown to interact with the Abi subunit of the WAVE complex (Beli *et al.*, 2008; Yang *et al.*, 2007). Moreover, this activity depends on WAVE, since it is lost when a blocking antibody specifically targeting the VCA domain of WAVE is added. This control establishes that this activity cannot be accounted for by N-WASP, another NPF previously shown to interact with Abi (Innocenti *et al.*, 2005). This experiment shows that an activity of the WAVE complex can be detected and that an uncharacterized activator weakly associated to the WAVE complex is responsible for it. This activity likely reflects the level of WAVE complex activation in cells.

7. Concluding Remarks

In this chapter, we have described a method to purify and address the constitutive activity of WAVE and WASH complexes. Importantly, this method can be used to identify novel multiprotein complexes and should thus be applied to other NPFs for which the embedment into multiprotein complex is still unknown, like JMY or WHAMM.

It is quite intriguing to find the WASH complex constitutively active. It is indeed unlikely that NPFs are unregulated in cells, given the major cytoskeleton rearrangements they induce. A WAVE-like model in which the NPF is maintained inactive through its incorporation within a regulatory complex that connects it to signaling pathways is more satisfactory. As illustrated for the case of the WAVE complex, it is possible that the one-step purification protocol described here preserved interactions with unknown activator(s). Alternatively, several posttranslational modifications essential for activity might have been preserved, like phosphorylations, as shown for the Arp2/3 and WAVE complexes (Lebensohn and Kirschner, 2009; LeClaire et al., 2008). This might explain why a longer, four-step purification of a recombinant WASH complex using coinfection with multiple baculoviruses yielded an inactive complex (Jia et al., 2010). To understand this discrepancy between purifications, one can notice that, like the WAVE complex, the WASH complex is present into two pools in cells. One pool is cytosolic (Duleh and Welch, 2010), whereas the other is bound to endosomal membranes, where it colocalizes with actin and the Arp2/3 complex (Derivery et al., 2009b). The cytoplasmic pool is likely recruited onto endosomes upon signaling and must reflect the true constitutive activity of the complex. It would be thus interesting to address the activity of the WASH complex purified from cytosolic extracts using the protocol described here. We expect it to be inactive, as shown for the WAVE complex (Suetsugu et al., 2006).

We believe that this purification method based on stable cell line provides a tremendous number of possibilities to dissect the activity of virtually any mammalian complexes. For instance, one can imagine addressing the activity of mutant complexes by introducing point mutations in the tagged subunit, provided that the mutation does not impair cell growth. Another exciting potentiality of the system would be to address the activity of complexes lacking regulatory subunits by purifying material from siRNA-treated cells.

ACKNOWLEDGMENTS

This work was supported by Fondation pour la Recherche Médicale (INE20071110919) and Agence Nationale pour la Recherche (ANR-08-BLAN-0012-03 and ANR-08-PCVI-0010-03).

REFERENCES

Beli, P., Mascheroni, D., Xu, D., and Innocenti, M. (2008). WAVE and Arp2/3 jointly inhibit filopodium formation by entering into a complex with mDia2. *Nat. Cell Biol.* **10**, 849–857.
Benesch, S., Lommel, S., Steffen, A., Stradal, T. E., Scaplehorn, N., Way, M., Wehland, J., and Rottner, K. (2002). Phosphatidylinositol 4, 5-biphosphate (PIP2)-induced vesicle movement depends on N-WASP and involves Nck, WIP, and Grb2. *J. Biol. Chem.* **277**, 37771–37776.
Benesch, S., Polo, S., Lai, F. P., Anderson, K. I., Stradal, T. E., Wehland, J., and Rottner, K. (2005). N-WASP deficiency impairs EGF internalization and actin assembly at clathrin-coated pits. *J. Cell Sci.* **118**, 3103–3115.
Blanchoin, L., Amann, K. J., Higgs, H. N., Marchand, J. B., Kaiser, D. A., and Pollard, T. D. (2000). Direct observation of dendritic actin filament networks nucleated by Arp2/3 complex and WASP/Scar proteins. *Nature* **404**, 1007–1011.
Campellone, K. G., Webb, N. J., Znameroski, E. A., and Welch, M. D. (2008). WHAMM is an Arp2/3 complex activator that binds microtubules and functions in ER to Golgi transport. *Cell* **134**, 148–161.
Derivery, E., and Gautreau, A. (2010a). Generation of branched actin networks: Assembly and regulation of the N-WASP and WAVE molecular machines. *Bioessays* **32**, 119–131.
Derivery, E., and Gautreau, A. (2010b). Evolutionary conservation of the WASH complex, an actin polymerization machine involved in endosomal fission. *Commun. Integr. Biol.* **3**, 227–230.
Derivery, E., Lombard, B., Loew, D., and Gautreau, A. (2009a). The wave complex is intrinsically inactive. *Cell Motil. Cytoskeleton* **66**, 777–790.
Derivery, E., Sousa, C., Gautier, J. J., Lombard, B., Loew, D., and Gautreau, A. (2009b). The Arp2/3 activator WASH controls the fission of endosomes through a large multi-protein complex. *Dev. Cell* **17**, 712–723.
Duleh, S. N., and Welch, M. D. (2010). WASH and the Arp2/3 complex regulate endosome shape and trafficking. *Cytoskeleton (Hoboken)* **67**, 193–206.
Eden, S., Rohatgi, R., Podtelejnikov, A. V., Mann, M., and Kirschner, M. W. (2002). Mechanism of regulation of WAVE1-induced actin nucleation by Rac1 and Nck. *Nature* **418**, 790–793.
Egile, C., Loisel, T. P., Laurent, V., Li, R., Pantaloni, D., Sansonetti, P. J., and Carlier, M. F. (1999). Activation of the CDC42 effector N-WASP by the Shigella flexneri IcsA protein promotes actin nucleation by Arp2/3 complex and bacterial actin-based motility. *J. Cell Biol.* **146**, 1319–1332.
Gautreau, A., Ho, H. Y., Li, J., Steen, H., Gygi, S. P., and Kirschner, M. W. (2004). Purification and architecture of the ubiquitous Wave complex. *Proc. Natl. Acad. Sci. USA* **101**, 4379–4383.
Gournier, H., Goley, E. D., Niederstrasser, H., Trinh, T., and Welch, M. D. (2001). Reconstitution of human Arp2/3 complex reveals critical roles of individual subunits in complex structure and activity. *Mol. Cell* **8**, 1041–1052.
Innocenti, M., Zucconi, A., Disanza, A., Frittoli, E., Areces, L. B., Steffen, A., Stradal, T. E., Di Fiore, P. P., Carlier, M. F., and Scita, G. (2004). Abi1 is essential for the formation and activation of a WAVE2 signalling complex. *Nat. Cell Biol.* **6**, 319–327.
Innocenti, M., Gerboth, S., Rottner, K., Lai, F. P., Hertzog, M., Stradal, T. E., Frittoli, E., Didry, D., Polo, S., Disanza, A., Benesch, S., Di Fiore, P. P., et al. (2005). Abi1 regulates the activity of N-WASP and WAVE in distinct actin-based processes. *Nat. Cell Biol.* **7**, 969–976.
Ismail, A. M., Padrick, S. B., Chen, B., Umetani, J., and Rosen, M. K. (2009). The WAVE regulatory complex is inhibited. *Nat. Struct. Mol. Biol.* **16**, 561–563.

Jia, D., Gomez, T. S., Metlagel, Z., Umetani, J., Otwinowski, Z., Rosen, M. K., and Billadeau, D. D. (2010). WASH and WAVE actin regulators of the Wiskott-Aldrich syndrome protein (WASP) family are controlled by analogous structurally related complexes. *Proc. Natl. Acad. Sci. USA* **107,** 10442–10447.

Jones, G. E., Zicha, D., Dunn, G. A., Blundell, M., and Thrasher, A. (2002). Restoration of podosomes and chemotaxis in Wiskott-Aldrich syndrome macrophages following induced expression of WASp. *Int. J. Biochem. Cell Biol.* **34,** 806–815.

Kim, Y., Sung, J. Y., Ceglia, I., Lee, K. W., Ahn, J. H., Halford, J. M., Kim, A. M., Kwak, S. P., Park, J. B., Ho Ryu, S., Schenck, A., Bardoni, B., *et al.* (2006). Phosphorylation of WAVE1 regulates actin polymerization and dendritic spine morphology. *Nature* **442,** 814–817.

Kouyama, T., and Mihashi, K. (1981). Fluorimetry study of N-(1-pyrenyl)iodoacetamide-labelled F-actin. Local structural change of actin protomer both on polymerization and on binding of heavy meromyosin. *Eur. J. Biochem.* **114,** 33–38.

Lebensohn, A. M., and Kirschner, M. W. (2009). Activation of the WAVE complex by coincident signals controls actin assembly. *Mol. Cell* **36,** 512–524.

LeClaire, L. L., 3rd, Baumgartner, M., Iwasa, J. H., Mullins, R. D., and Barber, D. L. (2008). Phosphorylation of the Arp2/3 complex is necessary to nucleate actin filaments. *J. Cell Biol.* **182,** 647–654.

Machesky, L. M., Mullins, R. D., Higgs, H. N., Kaiser, D. A., Blanchoin, L., May, R. C., Hall, M. E., and Pollard, T. D. (1999). Scar, a WASp-related protein, activates nucleation of actin filaments by the Arp2/3 complex. *Proc. Natl. Acad. Sci. USA* **96,** 3739–3744.

Pollard, T. D. (2007). Regulation of actin filament assembly by Arp2/3 complex and formins. *Annu. Rev. Biophys. Biomol. Struct.* **36,** 451–477.

Spudich, J. A., and Watt, S. (1971). The regulation of rabbit skeletal muscle contraction I. Biochemical studies of the interaction of the tropomyosin-troponin complex with actin and the proteolytic fragments of myosin. *J. Biol. Chem.* **246,** 4866–4871.

Suetsugu, S., Kurisu, S., Oikawa, T., Yamazaki, D., Oda, A., and Takenawa, T. (2006). Optimization of WAVE2 complex-induced actin polymerization by membrane-bound IRSp53, PIP(3), and Rac. *J. Cell Biol.* **173,** 571–585.

Takenawa, T., and Suetsugu, S. (2007). The WASP-WAVE protein network: Connecting the membrane to the cytoskeleton. *Nat. Rev. Mol. Cell Biol.* **8,** 37–48.

Yang, C., Czech, L., Gerboth, S., Kojima, S., Scita, G., and Svitkina, T. (2007). Novel roles of formin mDia2 in lamellipodia and filopodia formation in motile cells. *PLoS Biol.* **5,** e317.

Zigmond, S. H. (2000). In vitro actin polymerization using polymorphonuclear leukocyte extracts. *Methods Enzymol.* **325,** 237–254.

Zuchero, J. B., Coutts, A. S., Quinlan, M. E., Thangue, N. B., and Mullins, R. D. (2009). p53-cofactor JMY is a multifunctional actin nucleation factor. *Nat. Cell Biol.* **11,** 451–459.

Author Index

A

Aanonsen, L., 5
Abagyan, R., 59
Abassi, Y. A., 42
Abbondanzo, S. J., 571
AbdAlla, S., 120
Abdullah, N. A., 120
Abdulla, M. H., 120
Abell, A. N., 233–234, 236–238, 277, 432
Abel, P. W., 114
Abizaid, A., 55
Abo, A., 432
Abood, M. E., 8
Abu-gharbieh, E., 204
Abuin, L., 110
Abush, H., 5–6
Achiriloaie, M., 167
Acierno, J. S. Jr., 76
Adachi, M., 440
Adams, J., 8
Adams, M. D., 166
Adan, R. A., 271
Adapa, I. D., 416
Addy, J. H., 634
Adkisond, K. K., 213, 214
Adler, M., 430, 438
Adler-Wailes, D. C., 268
Advenier, C., 203
Agam, K., 594, 597, 606
Agrawal, D. K., 607
Aguilera, G., 268
Ahituv, N., 268
Ahmed, K., 490, 497–498, 516
Ahmed, S., 440
Ahn, J. H., 359, 362, 365, 679, 684
Ahn, N. G., 480
Ahonen, T. J., 341
Ahow, M., 76–77
Ahren, B., 575
Ahuja, S., 58–60
Aichele, D., 505
Aimi, T., 400, 403
Akahane, M., 209, 213
Akahane, S., 209, 213
Akakpo, J. P., 332, 340, 345, 347, 349, 352–353
Akamizu, T., 55, 379
Akerman, K. E., 33
Akil, H., 453–454

Akirav, I., 5–6
Akiyama, T. E., 571
Akopian, A. N., 4
Alagem, N., 464
Albanesi, J. P., 167
Albano, E., 594
Alberghina, L., 505
Alberti, L., 268, 379, 381
Albsoul-Younes, A. M., 614–616, 620, 623–625, 628
Aldrich, M., 615
Alewijnse, A. E., 319
Alexander, M., 57
Ali, S., 332, 341, 349
Allende, C. C., 502
Allende, J. E., 502
Allen, L. F., 110, 298
Allen, R. T., 607
Allgeier, A., 377, 379
Allison, S., 440
Almog, S., 8
Alonso, R., 33, 203
Aloyo, V. J., 414
Alpini, G., 4
Alquier, T., 571
Alspaugh, J. A., 400
Altenbach, C., 58
Altherr, M., 632
Altier, C., 461, 463
Altshuler, D., 332
Alvarez-Curto, E., 572, 576
Alvarez, R., 309
Amann, K. J., 687
Amberg, D. C., 404
Ambrose, C., 632–635
Amemiya, A., 615, 620
Aminian, S., 268
Ammala, C., 572
Anand, J. P., 465
Anborgh, P. H., 77, 648
Andersen, P. H., 61
Anderson, J., 54
Anderson, K. I., 678
Andersson, D. A., 4
Andersson, S., 299
Andresen, B. T., 631, 636–637
Andrews, J. L., 570–572, 575
Andrews, Z. B., 55
Andric, N., 233, 240

697

Andy, J., 57
Ankersen, M., 56, 61, 571–572, 575
Annunziata, M., 55
Anselmi, E., 119–120
Antosiewicz-Bourget, J., 282
Anyamale, V., 57
Aoki, H., 282, 292
Aoki, T., 445
Apaja, P. M., 234
Appel, L. J., 634
Applanat, M., 345
Arai, Y., 209
Arasaradnam, R. P., 55
Araya, R., 369–370
Archer, Z. A., 63
Arch, J. R. S., 198, 615, 620
Arcidiaco, M., 204
Ardecky, R. J., 166
Arden, J. R., 431
Areces, L. B., 678–679
Arena, J. P., 54
Arévalo-León, L. E., 119
Ariyasu, H., 55
Arkenau, C., 378, 381
Arlow, D. H., 59
Arnhold, I. J. P., 233–234, 236–238
Arnold, A., 198
Arnon, A., 603
Arora, K., 282
Arrais, C. A., 7
Arrigoni, G., 499, 505–506
Aschkenasi, C. J., 620
Ascoli, M., 233–234, 237, 240, 246
Ashby, B., 96, 98
Ashby, C. R. Jr., 5
Ashizawa, K., 378
Ashworth, A., 432
Asico, L. D., 634, 636, 641, 649
Ast, C., 359, 362–365, 367–370
Atayar, C., 434, 444
Atiba-Davies, M., 594
Atienza, J. M., 42
Atkinson, M., 432, 440
Attisano, L., 282
Aubin, N., 203
Aung, M. M., 8
Autelli, R., 594
Autschbach, F., 234
Avallone, R., 199, 202
Awakura, T., 365
Awatramani, R. B., 369–370

B

Babbs, A. J., 452
Babinet, C., 330
Babitt, J. L., 292, 369–370
Babwah, A. V., 75–77

Bach, A., 55, 57, 59, 61
Bachelot, A., 332, 340, 345, 347, 349–350, 352–353
Bach, G., 594–595, 597, 601–602, 607
Bach, M. A., 54
Bachvarova, M., 173
Bachvarov, D. R., 173
Bacigalupo, J., 592
Badachi, Y., 332, 340, 345, 347, 349, 352–353
Badal, S. S., 431
Badone, D., 215
Baekelandt, V., 180
Baek, K., 597, 605–606
Baetscher, M., 268
Bagdy, G., 4
Bagley, E. E., 459, 463–465
Bahner, M., 593
Bahn, Y. S., 398
Bailey, C. P., 464
Bain, J., 445, 506–509
Baixeras, E., 345
Bajic, D., 614–616, 620, 623–625, 628
Baker, D., 33
Baker, J. G., 8,199, 212
Bakker, R. A., 56, 76, 129–131, 133, 136, 139–140, 142, 144, 167, 446
Bakris, G. L., 634
Balani, S., 273
Baldwin, J. M., 406
Bali, U., 452
Balla, A., 597
Balla, T., 597, 602
Balleine, B. W., 180, 189, 194
Ballesteros, J. A., 58–60, 143, 167, 313
Balster, R. L., 8
Balthasar, N., 268
Baltz, J. M., 319
Bang, D. D., 369
Bangi, E., 369
Banks, W. A., 55
Ban, T., 381
Barakat, K. J., 54
Barak, D., 574
Barak, L. S., 167, 632–633, 642, 648
Baranski, T. J., 274
Baratti-Elbaz, C., 378
Barbara, G., 204
Barbe, P., 202
Barber, D. L., 693
Bardakci, F., 332
Bardin, S., 381
Bard, J. A., 110
Bardoni, B., 679, 684
Bardou, M., 203
Barnes, A. A., 570
Barnes, G., 632
Barnett, J., 445

Author Index 699

Barnhart, C. B., 171
Baroni, M., 215
Barra, J., 330
Barreiro-Morandeira, F., 379, 381
Barrere, C., 461, 463
Barrett, D. G., 214
Barrett, R. L., 8
Barrot, M., 203
Barsh, G. S., 271
Barshop, N. J., 151
Barth, F., 4–5, 8, 33
Bartles, J. R., 595
Bartlett, F. S., II., 615, 620
Bartlett, S. E., 429
Bartoov-Shifman, R., 571
Bartram, C. R., 332
Basbaum, A. I., 602
Bass, A. S., 201
Bastiaens, P. I. H., 83
Basu, A., 505
Bates, G., 632
Bates, R. W., 330
Battenburg, E. L. F., 615, 620
Battistutta, R., 478, 491, 502, 506–509
Battistuz, T., 518
Baufreton, J., 298
Baughman, R. W., 619
Baumgartner, M., 693
Bayne, M., 110
Bazan, F., 331
Beaudet, A., 173
Beaulieu, P., 6
Becattini, B., 445
Beck-Sickinger, A. G., 61–63, 64
Bednarz, M. S., 213
Beedle, A. M., 461, 463
Beisaw, N., 54
Bejsovec, A., 105
Beletskaya, I., 8
Beli, P., 689, 692
Bell, G. I., 151
Bellomo, G., 594
Belmaker, R. H., 57
Beltramo, M., 31, 33–34, 46
Ben-Ami, H. C., 594, 597, 606
Benesch, S., 678, 692
Bengra, C., 634
Benjafield, A. V., 634
Ben Jonathan, N., 330
Benko, G., 58
Bennett, G. J., 12–13
Bennett, K. A., 571
Bennett, P., 488
Benoit, S. C., 55
Benovic, J. L., 32, 167, 297, 425, 446, 448, 646
Ben-Shabat, S., 8
Ben-Shlomo, A., 149
Benson, W. G., 570–571, 575

Bentwood, J. L., 359, 365–366
Berger, D. M., 369–370
Bergis, O., 203
Berg, K. A., 133, 141, 414
Berglund, B. A., 4
Bergmann, M., 234
Bergsdorf, C., 136
Berkemeier, L. R., 268
Berlanga, J. J., 345
Berlan, M., 202
Berman, H. M., 518
Bernardi, G., 23
Bernardini, N., 204
Bernard, J., 42
Bernetiere, S., 273
Bernichtein, S., 331–333, 340, 343–345, 347, 349–350, 352–353
Bernstein, R. N., 430, 437–438
Bertacchini, J., 490, 502, 510
Bertin, B., 408
Bertrand, L., 632–633, 642
Bertuzzi, F., 55
Berzetei-Gurske, I. P., 416
Bessis, A. S., 77
Bettecken, T., 271
Bettendorf, M., 234
Bhamidipati, C. M., 430, 434–435, 437–438, 453–454
Bharatam, P. V., 206
Bharucha, A. E., 204
Bhatnagar, A., 425
Bhatnagar, V., 634
Bhattacharya, M., 76–77
Bhatt, K. V., 551
Bhat, T. N., 518
Bhoumik, A., 434–435
Bhuiyan, M. A., 165–166
Bialojan, C., 628
Bianchetti, A., 203–204
Biancone, L., 55
Bichet, D. G., 118
Biebermann, H., 270
Bikle, D. D., 255, 259–260, 262–263
Billadeau, D. D., 678–679, 693
Billerbeck, A. E., 233–234, 236–238
Billings, P. C., 359, 362, 365–366
Bilsky, E. J., 413–415, 430, 431, 434–435, 437–438, 446, 449–455, 459–460
Bilson, H. A., 96
Binart, N., 330, 341
Binda, E., 282
Bindels, R. J., 597, 601
Binns, D., 445
Bioulac, B., 298
Bircan, R., 379
Birnbaumer, L., 32, 233, 297, 446, 448
Birnbaumer, M., 233
Birnbaum, S., 55

Bisogno, T., 4
Blair, J. R., 430, 434–435, 437–438, 453–454
Blanchet, C., 655
Blanchoin, L., 679, 687
Blanc, S., 42
Blandizzi, C., 204
Blanpain, C., 76
Bleve, L., 209, 212–215
Blin, N., 201
Bloch, K. D., 292, 369–370
Bloch, W., 199, 202
Blond, O., 8
Bloom, J. D., 201
Bluhm, W. F., 518
Blundell, M., 678
Bluthe, R. M., 24
Bo-Abbas, Y., 76
Bobadilla-Lugo, R. A., 120
Bober, L., 33
Boccardi, G., 215
Bockaert, J., 398
Bodding, M., 601
Bode, A. M., 432–433
Bogani, D., 255–257
Bogorad, R. L., 329, 332, 340, 345, 347, 349–350, 352–353
Bogoyevitch, M. A., 433
Böhm, M., 212
Bohn, L. M., 421, 425, 428–429, 456, 460
Boime, I., 240
Bokoch, M. P., 59
Boldyreff, B., 472–473, 478, 522, 525–527
Bole-Feysot, C., 330, 341
Bo, M., 240
Bond, R. A., 32, 141, 210, 313, 571
Bonhaus, D. W., 5
Bonner, T. I., 5, 54, 180, 189
Bonomi, M., 234, 277
Borgland, S. L., 461, 463–464
Borhani, D. W., 59
Borkowski, D., 110
Bornstein, S. R., 332
Borok, E., 55
Bortolato, A., 507
Bosga-Bouwer, A., 434, 444
Boss, O., 202
Boss, V., 129
Bost, F., 433
Boston, B. A., 268
Bottzauw, T., 440
Bouaboula, M., 4–5, 8, 33
Bougault, I., 203
Bourinet, E., 461, 463–464
Bourne, P. E., 518
Bouthillier, J., 173
Boutin, J. M., 332
Bouvier, M., 118, 234, 297, 321, 377
Bouxsein, M. L., 292, 369–370

Bowers, C. Y., 54
Bowman, J. D., 597, 605–606
Boyle, T. P., 166
Boyle, W. J., 485
Bradbury, F. A., 449, 454, 456, 460
Bradford, M. M., 524–525
Braestrup, C., 32
Brakstor, E., 464
Branchek, T. A., 110
Brandish, P. E., 136
Brandt, E., 61, 62
Brandt, S. R., 416
Brann, M. R., 139–140, 258
Brath, F., 33
Brault, V., 273
Brauner-Osborne, H., 139–140, 256–258
Brazil, D. P., 551
Breivogel, C. S., 415, 452
Brelière, J. C., 33
Brem, H., 282
Brennan, T. J., 180
Bretner, M., 478, 491, 505
Breuer, A., 8
Breuiller-Fouché, M., 203
Brézillon, S., 76, 570
Briddon, S. J., 577
Briend-Sutren, M. M., 215
Brievogel, C. S., 456
Brigham, E. F., 504
Briley, E. M., 8
Brillet, K., 449–450, 457–458
Briones, A., 120
Briscoe, C. P., 570–572, 575
Brito, V. N., 233–234, 236–238
Brixius, K., 199, 202
Brockie, H. C., 4, 8, 33
Brockunier, L. L., 209
Brody, D. L., 461
Broggi, G., 282
Bromberg, Y., 268
Brophy, V. H., 634
Brousse, N., 330
Broutin, I., 331, 353
Brown, A. J., 570, 572–573, 575–576, 581–582
Brown, E. M., 254–257, 260, 263
Browner, M., 445
Brown, F., 21
Brown, K., 213
Brown, M. S., 55
Brown, N. J., 634, 636, 641, 649
Brownstein, M. J., 110, 180, 189
Brown, T. G. Jr., 198
Bruel, H., 256
Brumm, H., 270, 271
Brunani, A., 268
Brunati, A. M., 508–509
Brusa, R., 31, 33–34, 46
Brustolon, F., 490, 499, 502, 505–506, 510

Author Index

Bruysters, M., 167
Bryant, J. E., 594
Bubash, L. A., 490
Buchdunger, E., 496
Buckingham, R. E., 615, 620
Buckley, N. J., 584
Bukowiecki, L. J., 201
Bumpus, F. M., 166
Burford, N. T., 446, 449, 450, 454
Burkey, B. F., 207
Burkhardt, K., 518
Burkholder, A. C., 399
Burnat, G., 6
Burns, M. G., 201
Burstein, E. S., 133, 139, 258
Burtt, N. P., 332
Burwinkel, B., 332
Buscher, R., 111, 119
Busch, S., 203
Bustamante, J. O., 657
Buteau, H., 330, 345
Butelman, E. R., 421
Butler, A. A., 268
Butler, B., 54
Butters, R., 254–256, 260
Bylund, D. B., 110
Byrd, N., 421

C

Cabral, G., 5
Cabrera, D. M., 632, 635, 649
Cabrol, D., 203
Cabrol, S., 65
Caccavello, R. J., 504
Cahalan, M., 656
Calandra, B., 4–5, 8, 33
Caldecott, K. W., 507, 511
Caldwell, J. S., 180
Calebiro, D., 180, 381
Calo, G., 446, 461–462, 464
Calton, M. A., 268
Camara, J., 273
Cameselle-Teijeiro, J., 379, 381
Camilleri, M., 204
Campbell, C. A., 204
Campbell, D., 432
Campbell, S., 110
Campellone, K. G., 678
Campfield, L. A., 268
Camus, A., 330
Camuso, N., 76–77
Canaff, L., 257, 263
Canagarajah, B., 432, 440
Canalis, E., 292
Canals, M., 4, 6, 574, 576
Canat, X., 4–5
Canbay, E., 332

Candelore, M. R., 206, 209, 214
Cantley, L. C., 551
Canton, D. A., 473
Cao, L., 268
Cao, T., 167
Capel, W. D., 167
Caplan, M. J., 187
Caput, D., 8, 203
Caracoti, A., 273
Cardiff, R. D., 473, 516
Carel, J.-C., 76, 234
Carillon, C., 203
Carini, D. J., 166
Carini, R., 594
Carlier, M. F., 678–679, 688
Carlson, C. G., 615–616, 618–620
Carney, S. T., 5
Caron, M. G., 32, 110, 167, 297–299, 306–307, 313, 446, 448, 632–633, 642, 648
Caron, P. R., 445
Caron, R. J., 359, 365–366
Carracedo, A., 551
Carrasquer, A., 33
Carrier, E. J., 5
Carrier, L. M., 505
Carrillo, J. J., 581
Carroll, F. I., 449, 454, 456, 460
Carroll, L., 284, 359–360, 362
Caruso, M. G., 204, 205
Cascieri, M. A., 209
Casellas, P., 5, 8, 33
Castaneda, T. R., 55
Castiblanco, A., 256–257
Castiglioni, A. J., 595
Casti, P., 5, 8
Castro, I., 381
Catapano, F., 55
Catterall, W. A., 461
Catt, K. J., 166
Cavallo, R., 105
Cavanagh, J., 432
Cawthorne, M. A., 202
Cecchi, R., 215
Cegla, J., 679, 684
Celeste, A. J., 281, 369–370
Cellek, S., 204
Cellitti, J. F., 445
Cendron, L., 507
Centonze, D., 23
Cerchio, K., 54
Cerione, R. A., 32, 297, 446, 448
Cerrato, F., 76–77
Cesaro, L., 474, 483, 497, 509
Cettour-Rose, P., 55
Chaar, Z. Y., 299
Chabot, J. G., 173
Chai, B. X., 55, 271

Chakhtoura, Z., 332, 340, 345, 347, 349, 352–353
Chakkalakal, S. A., 371
Chalothorn, D., 111, 117
Chambaz, E. M., 528
Chambers, J. K., 401, 570–571, 575
Champion, T. M., 8
Chan, C. B., 57
Chan, C. P., 480
Chan, D. W., 507, 511
Chang, F., 401
Chang, L.-T., 418, 422–424, 449–451
Chang, W., 253, 255–256, 258–260, 262–263
Chan, J., 6
Chan, W. W., 54
Chan, Y. M., 76–77
Chao, E. Y., 213
Charbono, W., 433
Charlier, C., 8
Charpentier, S., 299
Cha, S. K., 601
Cha, T. J., 655–666, 668
Chatzidaki, E. E., 76
Chaudhuri, B., 496
Chaudhuri, D., 463
Chaussain, J. L., 76
Chazenbalk, G., 378
Cheeseman, M. T., 255–257
Cheetham, T., 268
Che, J. L., 273
Chemelli, R. M., 615, 620
Chen, A. S., 268, 574
Chen, B., 679, 684
Chen, C., 241, 269, 449–450, 457, 570
Cheng, C. H., 57
Cheng, I., 332
Cheng, K., 54, 449–451
Cheng, X., 5, 592, 597
Cheng, Y.-C., 175
Chen, J., 59, 376
Chen, L. H., 445
Chen, M. H., 54
Chen, N., 433
Chen, S., 634
Chen, T. H., 255–256, 258–260, 262–263, 402
Chen, X. B., 96, 464, 634
Chen, Y. A., 180, 203, 317
Chen, Y. G., 282
Chen, Z., 111, 117–119, 432
Chereau, D., 504
Cherezov, V., 59, 274
Chesnokova, V., 151, 157–158, 162
Cheung, B. M., 270
Cheung, G. C., 270
Cheung, J. Y., 595
Chhatwal, J. P., 6
Chian, D., 504
Chiba, Y., 133

Chico-Galdo, V., 377
Chidiac, P., 297, 321
Chieng, B. C. H., 463–465
Childers, S. R., 415, 452, 456, 461
Childs, S. R., 282
Chiou, S. S., 504
Chiu, A. T., 166
Chiu, G., 110
Choe, H. W., 58, 273
Choe, K. M., 604
Choi, E. J., 444
Choi, H. J., 59, 274
Choi, I. H., 282, 359–363
Cho, T.-J., 282, 359–363
Chou, C. L., 96
Chow, C. W., 432
Chow, K. B., 57
Christiansen, J. S., 54
Christiansen, L. M., 268
Christie, M. J., 415, 459, 461, 463–465
Christ, T., 464, 665
Chu, A. H., 180, 434, 440, 445
Chuang, H., 4
Chuang, J. C., 55
Chu, K. M., 57
Chung, J. M., 14
Chung, K. S., 408
Chung, N., 632–635
Chung, S., 400
Chuprun, J. K., 634
Church, D., 632
Chu, X., 595
Chyb, S., 593, 597
Cicirelle, M. F., 525, 527
Cirakoglu, B., 379
Civelli, O., 459
Claing, A., 167
Clandinin, T. R., 604
Clapham, D. E., 592, 595, 597, 600, 602, 606
Claret, F. X., 432
Clarke, D. E., 110
Clarke, W. P., 414
Clark, M. J., 419, 459, 460
Clauser, E., 381
Claus, M., 381
Clement, K., 268, 271
Clevenger, C. V., 331–332
Clifford, E. E., 211
Clohisy, D. R., 18
Cobb, H., 421
Cobb, M. H., 432, 440, 445, 550
Coccetti, P., 505
Cochaux, P., 34, 379
Cochet, C., 528
Cockcroft, S., 89
Cockcroft, V., 33
Codina, J., 32, 297, 446, 448
Cogolludo, A. L., 119

Author Index

Cohen, A. A., 180, 566
Cohen Ben-Ami, H., 597, 605
Cohen, C., 203
Cohen, P., 445, 496
Cohen, R. N., 381
Colabufo, N. A., 212–215
Colapietro, A. M., 632–633, 642
Colecraft, H. M., 461
Cole, D. E. C., 257, 263
Coleman, R. A., 96
Collins, F., 359
Collins, J., 190–191
Coll, J. T., 445
Coloff, J. L., 434, 444
Colpaert, F. C., 33
Colquhoun, D., 661
Colucci, R., 204
Combes, T., 33
Communi, D., 76
Compton, D. R., 8
Compton, P., 438
Cone, R. D., 268, 271
Congy, C., 8, 33
Conklin, B. R., 270
Connell, D. W., 63
Conner, D. A., 254–255
Connor, J. M., 54, 282, 284, 358–363
Connor, M., 359, 445
Conn, P. M., 118
Conrad, K., 595
Conrath, M., 6
Constantinescu, C. S., 23
Conti, M., 379, 381
Contreras, G., 634
Cook, B., 594, 597, 606
Coolen, H. K., 8
Cooper, J. A., 473
Copolov, D., 298
Coquerel, A., 203
Corbett, D. F., 572
Cornford, M., 615
Corvol, P., 381
Cosens, D. J., 592, 604
Coscntino, M., 204
Costa, A., 204
Costagliola, S., 234, 376–377, 381
Costa, T., 32–33, 59, 61, 199, 297–298, 306, 313, 414, 446, 451, 453
Cotecchia, S., 32–33, 59, 61, 110, 199, 297, 306, 313, 575
Cottney, J., 8
Couch, D., 464
Coulter, S., 167
Coun, F., 180
Courtillot, C., 332, 340, 345, 347, 349–350, 352–353
Coutts, A. A., 5
Coutts, A. S., 678

Covel, J. A., 414
Cowan, C., 213
Cowan-Jacob, S. W., 496
Cowell, S., 209
Cowsik, S. M., 4
Cowtan, K., 517
Cox, H. M., 55, 57, 59, 77
Cox, M. L., 505
Coyle, J. T., 614, 620
Cozza, G., 499, 502, 506–509, 522, 525, 527
Crema, A., 204
Crema, F., 204
Cremers, B., 212
Crews, C. M., 550
Crochet, S., 33
Croci, T., 199, 202–204, 215
Cronin, M. A., 593
Crook, E. D., 57
Croston, G. E., 138–139
Croxford, J. L., 5
Crozat, A., 171
Cuajungco, M. P., 594–595, 597, 602, 605
Cuenda, A., 432
Cuevas, B. D., 432
Cui, H., 595
Cui, J., 434, 440, 444
Cui, Y., 180, 189, 194
Cullen, M. J., 268
Culler, M. D., 150–152, 157–158, 162
Cully, D. F., 54
Cully, M., 550–551
Cuny, G. D., 292, 369–370
Curcio-Morelli, C., 597
Curtin, J. A., 551
Cutler, G. B. Jr., 232
Cygankiewicz, A., 56, 61, 571–572, 575
Czech, L., 689, 692
Czerwinska, E., 601
Czerwinski, W., 601
Czyzyk, T. A., 55

D

Dabew, E., 212
Dadon, D., 594–595
Dahlback, H., 299
Dai, M., 273
Dalal, P., 211
Dal Ben, D., 502
Dale, L. B., 77
Dalgaard, L. T., 268
Dallanoce, C., 213
Damann, N., 592
Damotte, D., 330
D'Angelo, L., 204
Daniel, K. Y., 168
Daniels, D., 570
Danielson, P. D., 282

Danielson, P. E., 615, 620
Dantzer, R., 24
D'Aoust, J.-P., 295, 299, 306, 321
Dart, M. J., 34
Das, A. K., 131
Dascal, N., 464
da Silva, I., 55
DaSilva-Jardine, P., 209, 213
Da Silva, N., 593
Date, Y., 54
Datta, R., 55
Datta, S. R., 551
Davenport, A. P., 54
Davidson, N., 465
Davies, H., 551
Davies, S. P., 509
Davies, T. F., 377
Davies, W. L., 33
Davis, A. T., 490, 497, 498, 516
Davis, D. L., 299
Davis, H. R., 571
Davison, J. S., 5
Davis, R. J., 190–191, 432
Dawson, E., 634
Daza, A. V., 34
De Amici, M., 213
Deana, A. D., 508
De Angelis, D., 182
Dean, N., 433
Deaton, D. N., 213–214
De Backer, M. D., 128
De Blasi, A., 87, 378, 632, 634, 637
de Boer, J., 282, 292, 365
De Bonis, S., 528
DeCamp, D., 190–191
Decobecq, M. E., 570
Decobert, M., 203
de Costa, B. R., 5
Decoster, C., 377
Decosterd, I., 12, 16
Decoulx, M., 379
Dees, C., 180
De Felice, M., 377
de Filippis, T., 180
De, F. M., 4, 6
DeGasparo, M., 166
DeGeorge, B. Jr., 634
Degerli, N., 332
Degiorgio, R., 204
de Gorter, D. J. J., 281–282, 292, 365
Deirmengian, G. K., 358–359
de Kerdanet, M., 65
Dekoning, J., 268
Delacourte, A., 180
Delai, P., 282, 284, 358–363
DeLano, W. L., 517
DeLapp, N. W., 452
Delavier-Klutchko, C., 215

de Lecea, L., 615, 620
Delgado, R., 592
Delghingaro-Augusto, V., 570
de Ligt, R. A. F., 211
Delisle, M. J., 379
Delling, M., 592, 595, 597, 600, 602, 606
DeLong, M. R., 614, 620
Del Tacca, M., 204
De Maertelaer, V., 377
de Marval, P. M., 434, 444
De Micheli, C., 213
DeMorrow, S., 4
De, M. P., 8
den, D. E., 597
Denef, J. F., 180, 190
Deng, C., 202
Deng, D. Y., 292, 369–370
Deng, H., 8
Deng, T., 432
Dennedy, M. C., 203, 209
Dennis, M., 297, 321
Dennis, S. G., 17
Dent, P., 550
Dent, R., 268
Deoliveira, R. M., 55
DePaepe, A., 369
Depatie, C., 433
De Petrocellis, L., 4
De, P. L., 4
De Ponti, F., 204
de Reil, J. K., 457
Derijard, B., 432
Derivery, E., 677–680, 684–685, 689, 691, 693
Derksen, D. R., 483, 505
Derocq, J. M., 8, 33
de Roux, N., 76–77
Dersch, C. M., 413, 418, 421–424, 449–451
Desbiens, K., 6
De, S. K., 445
de Souza, C. J., 207
Dessauer, C. W., 317, 458
Destefanis, S., 55
Desvignes, C., 203
Detheux, M., 76, 570
Deupi, X., 58–59, 465
Deutsch, J. A., 614
Devane, W. A., 5, 8
Devi, L. A., 6, 415
Devine, N. M., 574
Devogelaer, J. P., 54
DeVree, B. T., 58, 465
De Vries, T. J., 6
Dewey, W. L., 8
DeYoung, M. B., 111
Dhillon, A. S., 444
Diamond, P., 473
Diano, S., 55
Diaz, C., 54

Di Blasio, A. M., 268
Dickens, M., 432
Dickinson, C. J., 271
Dickinson, K. E. J., 213
Dickinson, R. E., 234
Dickson, S. L., 55, 63
Didry, D., 692
Dieguez, C., 63
Dietzel, C., 401
Di Fiore, P. P., 678–679, 692
Dighe, S. V., 441
Dijksman, J. A., 8
Dilauro, R., 377
DiLeone, R. J., 203
Di Maira, G., 490, 495, 499, 502, 505–510
Di Marzo, V., 4
Dimeco, F., 282
Di, M. V., 4
Dina, O. A., 140
Ding, Y., 166
Dinh, Q. T., 131
Dion, S. B., 646
Di, P. P., 8
Disanza, A., 678–679, 692
Dittrich-Breiholz, O., 445
Divin, M. F., 449, 454, 456, 460
Dixon, J., 76
Dixon, R. A., 206
Dixon, W. J., 20
Djiane, J., 332
Dobrev, D., 464, 653–655, 657, 664, 669
Dobrowolska, G., 509
D'Ocon, P., 119–120
Doering, C. J., 461, 463
Doherty-Kirby, A., 473
Doi, R., 570
Dolan, J. A., 201
Dolphin, A. C., 463
Donadeu, F. X., 233
Donahoe, P. K., 282
Donaldson, K. H., 214
Donaldson, L. F., 211
Donella-Deana, A., 483, 509
Dong, P., 298
Dong, X. P., 592, 595, 597, 600, 602, 606
Dong, Z., 432–433
Donlan, M., 33
Dooijes, D., 105
Doorn, J., 282, 292, 365
Dorsch, S., 213
Dorshkind, K., 330
Doupnik, C. A., 465
Downward, J., 550–551
Dow, R. L., 209, 213
Doyle, D. A., 234
Drabic, S. V., 273
Dressler, H., 199, 202
Drews, J., 398

Droogmans, G., 600–601
Dror, R. O., 59
Drumare, M. F., 201, 206
D'Souza-Li, L., 257, 263
Dubner, R., 21
Dubuisson, D., 17
Dubyak, G. R., 211
Ducruix, A., 331, 353
Duleh, S. N., 679, 693
Dulley, F. L., 7
Dumitrescu, A. M., 381
Dumont, J. E., 34, 376–377, 379, 381
Dumont, Y., 173
Duncan, J. S., 471, 473, 478, 483, 490–491, 497
Duncan, W. C., 234
Duncia, J. V., 166
Dunmore, J. H., 268
Dunn, G. A., 678
Duprez, L., 34, 379
Dupriez, V., 570
Durand, E., 271
Duranteau, L., 234
D'Urbano, E., 87, 637
Dutia, M. D., 201
Duyao, M., 632
Dvorak, M. M., 253, 259–260, 262–263
Dykshorn, S. W., 330
Dymeki, S. M., 369–370
Dzhura, E., 634

E

Ebihara, K., 570
Eckhart, A. D., 634
Eckhart, W., 505
Economides, A. N., 359, 369–370
Edelmann, S. E., 111, 115, 117
Edelstein, E., 268
Eden, S., 678–679, 684
Edery, M., 330, 332, 341, 345
Edfalk, S., 571
Edlund, H., 571
Egbuna, O. I., 254, 257, 263
Egecioglu, E., 55
Egerod, K. L., 63, 64
Eggerickx, D., 180, 190
Egile, C., 688
Ehrlich, J. R., 669
Eiermann, G. J., 571
Eikenburg, D. C., 110
Eilers, M., 58
Eilert, M. M., 570–571, 575
Einat, H., 57
Einstein, M., 571
Eisch, A. J., 203
El-Alfy, A. T., 5
Elashoff, D., 438
El-Gharbawy, A., 505

Elizondo, G., 120
El, J. A., 7
El-Khamisy, S. F., 507, 511
Elling, C. E., 54–55, 57–59
Elliott, M., 445, 506–509
Ellis, C., 570–571, 575
Ellis, J., 576
el Massiery, A., 120
Elmquist, J. K., 615, 620
Elshourbagy, N. A., 570–571, 575
Elsohly, M. A., 5
Elsworth, J. D., 55
Emdadi, A., 445
Emerson, S. G., 359, 370
Emlet, D. R., 434
Emorine, L. J., 215
Emsley, P., 517
Endo, K. I., 365
Endo, T., 378
Engelhardt, S., 212
Engel, J. A., 55
Engel, M. A., 6
Engelman, J. A., 551
Engelstoft, M. S., 59, 61
England, P., 331, 353
Englert, C., 488
Engle, S. J., 330
Engstrom, U., 284
Enomoto, S., 282, 292
Epelbaum, J., 65
Eppig, J. J., 180
Ericsson, L. H., 525, 527
Ermakowa, I., 526
Ernst, O. P., 58, 273
Ersoy, B. A., 268
Esbenshade, T. A., 111
Estrada, S. A., 414
Eszlinger, M., 378, 381
Etienne-Manneville, S., 140
Etinger, A., 8
Evans, B. A., 209, 213
Evans, C. J., 449–450, 455–456, 463–464
Evans, J. A., 597
Evola, M., 441
Evsikov, A. V., 180

F

Fabbro, D., 496, 509
Faccenda, E., 87
Fagan, P., 518
Fahrenholz, F., 173
Fairchild-Huntress, V., 268
Faison, W., 213
Falqués, A., 499, 502
Fandl, J. P., 359, 369–370
Fanelli, F., 110, 233–237, 239
Fang, Q., 268

Fan, P., 8
Fantinato, S., 505
Fan, Y., 34
Fan, Z. C., 270–272, 274–275
Farid, N. R., 381
Farkas, R. H., 615
Farooqi, I. S., 268
Farr, A. L., 168, 223
Farrer, C. A., 273
Farr, S. A., 55
Fastnacht-Urban, E., 284
Faust, R. A., 490, 497
Favor, J., 255–257
Fedorov, V. V., 202
Feighner, S. D., 54
Felder, C. C., 5, 8
Felder, R. A., 634, 636, 641, 649
Feldman, G. J., 282, 359–363
Fendrich, G., 496
Feng, D. D., 570
Feng, G.-J., 581
Feng, J., 658
Feng, N., 268
Feng, X., 233–237, 239, 246–249
Feng, Y. H., 166, 171, 268, 571
Feng, Z., 518
Fenstermacher, D. A., 282, 359–363
Ferguson, S. S., 77, 167, 185, 378, 632–633, 642, 648
Fernando, S. R., 8
Ferrara, P., 199, 202–203
Ferreira, D., 5
Ferreira, M., 268
Ficarro, S. B., 434–435
Fidock, M. D., 110
Figueroa Garcia, M. C., 120
Filetti, S., 379, 381
Filhol, O., 528
Filipek, S., 58
Filonenko, V., 505
Fine, J. S., 33
Finke, M. P., 18
Fioravanti, B., 4, 6
Fiori, J. L., 359, 362, 365–366
Fischer, D. F., 180
Fischer, J., 6
Fisher, L. G., 213
Fisher, L. W., 365–366
Fisher, M. H., 209, 214
Fitch, T. E., 615, 620
Fitzsimons, C. P., 133
Flaherty, K. T., 552
Flavell, R. A., 433
Fleming, M. A., 445
Flockerzi, V., 601
Flodgren, E., 570–571, 575
Flores, C., 21
Fluck, J., 441

Author Index

Foglia, C., 34, 46
Foord, S. M., 54
Forder, J. P., 595
Ford, J., 440
Forman, S. J., 284, 359, 370
Fornwald, J. A., 572
Forray, C., 110
Forsti, A., 332
Forsyth, N. E., 273
Fourgeaud, L., 77
Fournier, A., 173
Fowler, T. J., 401, 403
Fox, T., 445
Foye, P. E., 615, 620
France, C. P., 435
Franchetti, R., 378
Franchin, C., 507
Françon, D., 203
Frankel, W. N., 615, 620
Franklin, R. A., 444
Frank, R., 483
Franzoni, M. F., 6
Frascella, P., 507
Fraser, C. M., 201
Fraser, I., 190–191
Frazier, E. G., 268
Frechter, S., 593, 606
Fredericks, Z. L., 632
Fredholm, B. B., 425
Freedman, N. J., 167, 632
Freissmuth, M., 377
Fremont, V., 376
French, A. C., 471
Frenzel, R., 378
Fretz, H., 496
Fricker, S. P., 452
Fride, E., 8
Friedrich, A., 464
Friedrich, F., 625
Friel, A. M., 203, 209
Frigerio, F., 213
Frigo, G., 204
Frigon, N. L., 504
Frimurer, T. M., 54, 58–61, 62
Frittoli, E., 678–679, 692
Froguel, P., 268, 271
Frohlich, O., 341
Frumkin, A., 594–595, 597, 601–602, 607
Fu, D., 59
Fueyo, J., 434, 444
Fu, J., 634
Fujii, M., 282, 292
Fujii, R., 570–571, 575
Fujikura, J., 570
Fujimoto, K., 570
Fujimura, T., 203
Fujino, H., 95–97, 104–106
Fujino, M., 168

Fujita, S., 666
Fujiwara, K., 571, 575
Fujiwara, M., 24
Fukuda, M., 440
Fukuda, T., 292, 365, 369–370
Fukuhara, C., 615, 620
Fukui, H., 128
Fukusumi, S., 570–571, 575
Fuller, G. N., 434, 444
Funahashi, H., 268
Funa, K., 284
Fung, J. J., 59
Furet, P., 509
Furness, L. M., 110
Furse, K. E., 206
Furth, J., 152
Furth, P. A., 331–332
Furutani, Y., 201, 209, 213
Fusco, A., 377

G

Gadsen, E. L., 152
Gagliani, M. C., 180
Gainer, J., 634, 636, 641, 649
Gaits, F., 442
Galal, A. M., 5
Gales, C., 234, 377
Galet, C., 233–234, 237, 240
Galitzky, J., 202
Gallagher, E., 445
Gallardo, C., 414
Gallardo-Ortiz, I. A., 119–120
Gamba, G., 254–256, 260
Gamper, N., 4
Ganellin, C. R., 33, 128
Ganguli, S. C., 33
Ganguly, A., 359
Gannon, F. H., 359, 362, 365
Gan, R. T., 199, 202
Gantz, I., 271
Gao, E., 634
Gao, X.-B., 55, 615, 620
Gao, Z. G., 574
Garcia-Anoveros, J., 595
García-Cazarin, M. L., 111, 117
Garcia, D. E., 461
García-Sáinz, J. A., 109–113, 115–117, 119
Gardeil, F., 203, 209
Gardell, L. R., 4, 6
Garret, M., 298
Garrison, T. R., 34
Garvey, C., 259–260, 262–263
Garvin, A. J., 488
Gascoigne, K. E., 566
Gaskin, F. S., 55
Gassenhuber, H., 180
Gauthier, C., 202–203

Gautier, J. J., 678–679, 684–685, 691, 693
Gautreau, A., 677–680, 684–685, 689, 691, 693
Gautvik, V. T., 615, 620
Gavai, A. V., 213
Gazzerro, E., 292
Gbahou, F., 33
Genestie, C., 332, 340, 345, 347, 349, 352–353
Geng, C., 597, 605–606
Genin, E., 76
Gentet, L. J., 664
Gerboth, S., 689, 692
Gerke, M. B., 459, 465
Gerlach, L. O., 58
Gertz, B., 54
Gervasi, N., 459
Gervy, C., 34, 379
Gether, U., 56, 58–59, 313
Gezen, C., 379
Ghanouni, P., 59, 209
Ghe, C., 55
Ghinea, N., 378
Ghisellini, P., 502
Giacobino, J.-P., 202
Gianoncelli, A., 506–509
Gibelli, G., 204
Gibson, D., 8
Gibson, T. B., 432
Gifford, A. N., 5
Gilchrist, A. D., 473
Gillespie, J., 332
Gillessen-Kaesbach, G., 284
Gilliland, G. L., 518
Gilman, A. G., 317
Giorgio, E., 213
Giovannoni, G., 5
Girani, M., 204
Girard, L., 434, 444
Girotra, R. N., 213
Gisbert, R., 119–120
Giudice, A., 215
Giuvelis, D., 413, 418, 422–424, 434–435, 438, 449–451
Glaser, D. L., 282, 284, 358–363, 369–370
Glasow, A., 332
Gloor, G. B., 473
Glover, J. S., 390
Gluck, O., 54
Godin, C., 77
Godinez-Hernández, D., 120
Goepel, M., 575
Goetz, A. S., 570–572, 575
Goffin, V., 329–333, 340–341, 343–345, 347, 349–350, 352–353
Goldenberg, D., 8
Goldhamer, D. J., 359, 369–370
Gold, L. H., 438
Goldsby, R. E., 359
Gold, S. J., 203

Goldsmith, E. J., 432, 440, 445
Goldstein, J. L., 55
Goldstone, A. P., 56
Goldsworthy, S. M., 570
Goley, E. D., 679
Golstein, J., 377
Gomes, I., 6
Gomes, W. A., 369–370
Gomez, T. S., 678–679, 693
Gómez-Zamudio, J., 119
Gonalons, N., 203
Goncalves, J. A., 58
Gong, Z., 572
Gonzalez-Cuello, A., 464
González-Espinosa, C., 112–113
Gonzalez, F. J., 120
González-Hernández, M. L., 120
Goodfriend, T., 166
Goodyear, R. J., 594
Goodyear, S., 55
Gopher, A., 8
Gorbatyuk, O. S., 276
Gorbsky, G. J., 473
Gordon, S. E., 593, 597
Gordon-Shaag, A., 597
Goswami, S., 204
Gottardi, C. J., 187
Gottschalk, M. E., 379
Gould, D., 332
Goumans, M.-J., 282, 292, 365
Gournier, H., 679
Gout, I. T., 505
Govaerts, C., 270
Govoni, M., 129, 131–132, 141, 144
Gozu, H. I., 379
Graaf de, C., 59
Grace, J., 614
Graham, R. M., 110–111, 166
Granata, R., 55
Grankowski, N., 472, 522, 526
Graves, L. M., 478, 491
Gray, J. A., 425
Gray, K. M., 204
Grayson, G. K., 34
Greene, K., 57
Greenwood, F. C., 390
Griebel, G., 203
Griffin, D. E., 23
Griffin, G., 8
Grigg, J. J., 615
Grimm, C., 594–595, 597, 602, 605–607
Grimsby, S., 284
Grishin, N. V., 55
Gronthos, S., 365–366
Groot, N., 632
Groppe, J. C., 359, 363
Gros, J., 206
Grossman, A. B., 56

Author Index

Grouselle, D., 65
Grubeck-Loebenstein, B., 377
Grueter, R., 330
Grynkiewicz, G., 113
Gualillo, O., 345
Guan, R., 234–235, 246–249
Guan, X. M., 268, 620
Guarino, R. D., 119
Guarnieri, F., 60
Gu, D., 634
Gudermann, T., 233
Gueorguiev, M., 56
Guerra, B., 497, 509, 515–516, 522, 526
Gu, F., 282
Guggino, W. B., 593
Guillaume, J. L., 206
Guitard, J., 203
Gulluoglu, B. M., 332
Gul, W., 5
Gulyani, S., 615
Guo, J., 461, 464
Gupta, R., 358–359
Gupta, S., 432
Gurevich, V. V., 646
Gustafson, E. L., 571
Guthrie, C. R., 57
Gutierrez, G. J., 434–435
Gutierrez, J. A., 55
Gu, W., 268
Gu, Y., 445, 605
Guy-Grand, B., 268, 271
Guzzi, U., 215
Gyenis, L., 471, 478, 491
Gygi, S. P., 678–679, 685
Gyombolai, P., 4

H

Haaksma, E., 167
Haber, D. A., 488
Hack, S. P., 459, 465
Hadac, E. M., 33
Hadcock, J. R., 209, 213
Hagan, S., 444
Hagen, D. C., 399, 401
Haggart, D., 416
Hague, C., 111, 117–119
Hahn, E. G., 6
Haiman, C. A., 332
Hajaji, Y., 65
Hale, J. E., 55
Hales, T. G., 449–450, 455–456, 463–464
Halford, J. M., 679, 684
Hall, A., 140, 432
Hall, D. A., 574
Hall, I. P., 212
Hall, K. J., 464
Hall, M. E., 679

Hall, R. A., 111, 117–119
Hall, R. P., 434, 444
Halsall, D. J., 268
Hamashima, H., 209, 213
Hamelin, M., 54
Hamid, J., 461, 463
Hamill, O. P., 655
Hamza, M. S., 5, 23
Han, C., 114
Handberg, K. J., 369
Hankenson, K., 371
Hankinson, S. E., 331–332
Hanley, M. R., 211
Hanlon, K. E., 3
Hanna, A. E., 434, 444
Hansen, B. S., 61
Hansen, T. K., 61
Hanson, M. A., 59, 274
Hanson, R. B., 204
Han, S. Y., 434, 440, 444
Hanus, L., 8
Hanzawa, H., 168
Han, Z. H., 34
Hao, F., 129
Harach, H. R., 377
Harada, H., 209, 213
Harada, M., 570–571, 575
Haraguchi, K., 378
Harashima, T., 399
Hara, T., 378
Hara, Y., 595
Harden, T. K., 167
Hardie, R. C., 592–593, 597, 600–601, 605
Harding, W. W., 421
Hardy, S., 571
Hargreaves, K. M., 4–5, 21
Hargrove, D. M., 209, 213
Harmar, A. J., 54
Harris, D. M., 634
Harrison, C., 453
Harris, R., 434, 444
Harry, A., 317
Harrylock, M., 525, 527
Harshyne, L., 434, 440, 444
Hartig, P. R., 110
Hart, R., 214
Hartwell, L. H., 399
Harutyunyan, L., 347, 349–350, 353
Harvey, J. A., 414
Hasegawa, H., 33, 96–98, 101, 103
Hashimoto, K., 111
Hashimoto, N., 111, 665
Haskell-Luevano, C., 270–271, 276
Hauffa, B. P., 234
Haupt, J., 357, 359, 362–365, 367–370
Hawcroft, G., 97
Hayashi, M., 379

Hayashi, R., 414
Hayashi, Y., 180, 190, 379
Haystead, T. A., 478, 491
Hazel, K. D., 441
Héaulme, M., 8, 33
Hebert, R., 54
Hebert, S. C., 254–256, 260
Hebert, T. E., 234–235, 246–249, 297, 321
Hediger, M. A., 254–256, 260
Hedrick, J. A., 571
Heidbuchel, H., 664, 670–671
Heidenreich, O., 434
Heilig, M., 55
Heiman, M. L., 55
Hein, P., 575
Heitman, J., 399
Heldin, C. H., 284
Hellemans, J., 369
Heller, S., 594–595, 597, 602, 605–607
He, L. Y., 572
Hémar, A., 77
Hemminki, K., 332
Henderson, B. E., 332
Henderson, G., 463, 464
Hendrick, A. G., 76
Hendy, G. N., 257, 263
Henrick, K., 518
Herblin, W. F., 166
Hercberg, S., 271
Herkenham, M., 5
Herlitze, S., 461, 671
Hermann, H., 4
Hermann, M., 377
Hermosura, M. C., 600
Hernandez, M., 570
Herremans, A. H., 8
Hershfinkel, M., 570–571
Herskowitz, I., 401
Hertel, C., 167
Hertzog, M., 692
Herz, A., 297, 414, 446, 451, 453
Hewick, R. M., 281, 369–370
Hewson, A. K., 63
Hibi, M., 432
Hibino, H., 654, 666
Hickey, G., 54
Hidaka, A., 381
Hidaka, H., 379
Hieble, J. P., 110, 199
Hiepen, C., 282, 359
Higgs, H. N., 679, 687
Hilal, L., 65
Hildebrand, P. W., 58, 273
Hiley, C. R., 33
Hillárd, C. J., 5
Hill, D., 8

Hille, B., 461
Hill, S. J., 128–129, 131, 136, 212, 577
Hill, W. S., 206
Hilton, M. J., 440
Himmel, H. M., 664
Hinney, A., 271
Hipkin, R. W., 33
Hirabayashi, A., 209, 213
Hirakawa, T., 233–234, 237, 240
Hiraoka, R., 445
Hirasawa, A., 111
Hiremath, L., 634
Hirohei, Y., 49
Hirokawa, Y., 213
Hirschhorn, J. N., 332
Hirshfeld, A., 58
Hiyama, Y., 213
Hizaki, H., 97
Hoang, Q. V., 615, 620–623, 625, 627–628
Ho, C., 254–255
Hoeflich, K. P., 551
Hoenderop, J. G., 597, 601
Hofer, A. M., 255–256
Hoffman, G. E., 76–77
Hoffmann, C., 199, 212–214
Hoffmann, E., 445
Hoffmann, M., 180
Hofilena, B. J., 414
Hofmann, K. P., 58, 273
Hogenboom, F., 6
Hogestatt, E. D., 4
Ho, H. Y., 678–679, 685
Ho, I. K., 439
Holgado-Madruga, M., 434, 440, 444
Holliday, N. D., 55, 57, 59, 77
Hollins, B., 179
Holloway, W. R., 24
Holst, B., 53–65, 77, 571–572, 575
Holtmann, M. H., 33
Holzapfel, H. P., 378, 381
Holzberg, D., 445
Hong, A., 54
Hong, C. C., 292, 369–370
Hong, D. W., 292, 369–370
Hong, Y. S., 597, 605–606
Honore, P., 18
Hooker, B. A., 34
Hoos, L. M., 571
Hoos, S., 331, 353
Hoover-Fong, J., 284
Hope, T. J., 378
Hopkins, E., 430, 438
Horlick, R. A., 273
Hornak, V., 58
Horn, L. C., 332
Horowitz, M., 8

Horré, K., 180
Horseman, N. D., 330
Horswill, J. G., 452
Horvath, B., 55
Horvath, T. L., 55
Ho Ryu, S., 679, 684
Ho, S. C., 376, 381
Hosey, M. M., 646
Hosoda, H., 54–55
Hosoda, K., 55
Hosoya, M., 570–571, 575
Hossain, M., 166
Hough, T. A., 255–257
Houle, S., 173
House, C., 627
House, J. D., 435
Housman, D., 632
Hou, X., 488
Ho, W., 5
Howard, A. D., 54, 61
Howard, M., 87
Howlett, A. C., 4–5, 8, 12, 453
Hoyng, S. A., 292, 369–370
Hreniuk, D. L., 54
Hruby, V. J., 209
Hsin, L.-W., 418, 422–424, 449–451
Hsueh, A. J., 240
Hsueh, R., 190–191
Hsueh, Y. P., 399–401, 405–407
Hua, L., 593
Huang, C. L., 601, 670–671
Huang, H., 267
Huang, P., 449–450, 457
Huang, S. M., 452
Huang, Y., 256–257
Huang, Z., 259
Hubbell, C. M., 58
Hubbell, W. L., 58
Huber, A., 593
Huettner, J. E., 619
Hufeisen, S. J., 425
Huffman, J. W., 8
Hughes, R. G., 214
Hu, H., 572
Huhtaniemi, I. T., 233
Huifang, G., 33
Hull, C. M., 400
Hull, M. A., 97
Hu, M., 369–370
Hunter, T., 472, 485
Hunter, W. J., III., 607
Hunter, W. M., 390
Hunyady, L., 4–5, 173
Hursh, D., 105
Husain, A., 166, 171
Huse, M., 282, 517
Hussain, A., 166
Huszar, D., 268

Hutchins, J. E., 214
Hutchinson, D. S., 209, 213
Hu, W. W., 87, 571
Hu, Y., 593
Hyman, C. E., 214

I

Iacovelli, L., 378
Ibarra, J. B., 414
Ibarra, M., 119–120
Ibrahim, M. M., 4, 6, 8
Ibrahim, N. M., 341
Ichijo, H., 282, 292
Ichikawa, A., 33, 96–98, 101, 103
Ichikawa, I., 166
Ignar, D. M., 572
Ignatov, A., 180
Ijzerman, A. P., 210–211, 214, 571
Ikeda, S. R., 461–464
Ikuyama, S., 379
Illert, B., 434
Illig, T., 271
Imada, M., 363
Imaizumi, S., 168
Imamura, E., 209, 213
Imamura, T., 282, 292
Inaba, T., 33
Inabe, K., 600
Inglese, J., 632–635
Inglis, K. J., 504
Innocenti, M., 678–679, 689, 692
Inoue, M., 625
In, P. W. K., 452
Inturrisi, C. E., 438
Iriyoshi, N., 131
Isgaard, J., 55
Ishiguro, M., 166
Ishii, K., 180
Ishii, M., 445, 595, 615, 620
Ishikawa, T., 209, 213
Ismail, A. M., 679, 684
Issa, J. B., 463
Issinger, O. G., 472–473, 478, 497, 515–517, 522–523, 525–527
Itarte, E., 499, 502
Itkin, O., 57
Ito, H., 654
Itoh, S., 284
Itoh, Y., 570–571, 575
Ivorra, M. D., 119–120
Iwakiri, K., 365
Iwakura, H., 570
Iwamoto, E. T., 439
Iwamura, H., 33
Iwanaga, K., 434, 444
Iwasa, J. H., 693
Iwasaki, K., 24

Iwasiow, R. M., 298–299, 313, 315
Iyengar, R., 317
Iype, L., 518
Izumi, H., 284
Izzo, A. A., 5

J

Jaattela, M., 440
Jackson, A., 298–299, 313, 315, 324
Jackson, M. F., 595, 601
Jackson, P. J., 271
Jacks, T., 54
Jacob, G., 502
Jacobson, A. E., 418, 422–424, 449–451
Jacobson, K. A., 574
Jaeger, L. B., 55
Jaffe, L. A., 180
Jain, S., 518
Jalink, K., 180, 592
James, C., 324
James, M., 632
James, P., 525, 527
James, T. Y., 400
Jamur, M. C., 83
Janaud, G., 332, 340, 345, 347, 349, 352–353
Jandacek, R. J., 55
Janovick, J. A., 118
Jansen, K., 595
Jansson, C. C., 33
Jarvie, K. R., 299
Järvinen, T., 34
Javitch, J. A., 58–60
Jeay, S., 333, 343–345, 353
Jeevaratnam, P., 452
Jenkins, L., 574, 576
Jenkins, P. M., 460
Jennifer, N. H., 168
Jensen, A. A., 258
Jensen, A. D., 59
Jensen, M. O., 59
Jensen, T. H., 56, 61, 571–572, 575
Jerlhag, E., 55
Jerman, J. C., 575
Jeske, N. A., 4
Jetton, T. L., 571
Jia, D., 678–679, 693
Jiang, H. L., 572
Jiang, L. I., 190–191
Jiang, M., 440
Jiao, X., 359, 365–366
Jia, X. C., 240
Jimenez-Andrade, J. M., 22–23
Jimenez-Feltstrom, J., 570–571, 575
Jin, C., 63, 64
Jin, J. Y., 96, 98, 434, 444
Jin, W., 665
Ji, R., 96

Jockers, R., 212–215
Joel-Almagor, T., 593
Joeng, K. S., 440
Johansen, P. B., 61
John, H., 168
Johns, E. J., 120
Johnson, E. A., 464
Johnson, G. L., 432
Johnson, L. R., 634
Johnson, M. R., 5
Johnson, N. B., 234
Johnson, R. A., 309
Johnson, S. W., 298
Johnston, D. B., 54, 464
Johnston, G. I., 110
Jolicoeur, C., 332
Joly, E., 571
Jomain, J. B., 331, 347, 349–350, 353
Jonas, S., 597
Jones, G. E., 678
Jones, J. D., 5
Jones, J. E., 636–937
Jones, K. A., 619
Jones, M., 595
Jones, N. T., 441
Jones, T., 105
Jongejan, A., 56, 76, 142, 167, 446
Joost, H. G., 55
Jordan, B. A., 415
Jorgensen, P. H., 369
Jorgensen, R., 58
Joris, J., 21
Jornvall, H., 299
Jors, S., 594–595, 597, 602, 605–607
Jose, P. A., 632, 634–635, 649
Jost, N., 464
Joyce, K. E., 8
Joynes, C. J., 434
Juang, Y. C., 445
Julius, D., 4, 602
Juni, A., 430, 438
Junior, A. M., 7
Junqueira, P. L., 7
Jüppner, H., 34

K

Kadiri, A., 65
Kaibara, M., 670–671
Kaiser, D. A., 679, 687
Kallunki, T., 432, 440
Kamber, B., 166
Kamijo, R., 363
Kaminski, N. E., 8
Kamiya, N., 292
Kamizono, J., 365
Kane, J. P., 268
Kaneko, T., 6

Author Index

Kaneshige, M., 378
Kangawa, K., 54–55
Kang, Y. S., 402
Kanjhan, R., 665
Kan, L., 369–370
Kanomata, K., 365
Kaplan, F. S., 284, 357–360, 362–371
Kara, E., 453
Karin, M., 432, 440, 445
Karnik, S. S., 166, 171
Kasai, K., 180
Kashaw, S. K., 209
Kash, T. L., 6
Katagiri, T., 292, 359, 362–365, 367–370
Kato, H., 201
Katoh, H., 96–97, 103
Katoh, K., 571, 575
Kato, K., 201
Kato, M., 209, 213, 282, 292
Kato, S., 209, 213
Kato, Y., 626
Katritch, V., 59
Katsanis, E., 118
Katsumata, Y., 49
Katyk, K., 634
Katz, B., 592–593, 597, 600
Kawabata, M., 282, 292
Kawaguchi, Y., 570
Kawai, T., 119, 120
Kawamata, Y., 570–571, 575
Kawamura, J., 570
Kawanabe, Y., 111
Kawano, M., 369–370
Kawarada, Y., 379
Kawashima, H., 213
Kaya, H., 332
Kaya, T., 33
Kayakiri, H., 209, 213
Kazimierczuc, Z., 507
Kazimierczuk, Z., 507–509
Keane, P. E., 203
Keating, D. J., 570
Kebede, M., 571
Keel, C., 8
Keever, L. B., 636–937
Ke, H., 434, 444
Keiffer, B. L., 453–455
Keizer, H. G., 8
Kek, B. L., 268
Kellermann, C. A., 6
Kelliher, M. A., 516
Kelly, E., 77, 464
Kelly, J. F., 452
Kelly, P. A., 330–333, 341, 343–345, 347, 349–350, 353
Kemp, B. E., 627
Kenakin, T., 32–33, 121, 129, 209–211, 321, 448, 451, 572

Kenakin, T. P., 33, 121, 572
Kennedy, C., 463
Kenny, C. D., 268
Kenyon, L. C., 434
Keogh, J. M., 268
Kerkhoff, E., 434
Kessler, J. A., 369–370
Kessler, M., 234
Kest, B., 430, 438
Kesterson, R. A., 268
Khan, A. A., 619
Khan, M. A., 120
Khatlani, T. S., 434, 444
Khokhlatchev, A. V., 432, 440, 445
Khong, K., 268
Khorana, H. G., 58
Khripin, Y., 634
Kieffer, B. L., 438, 449–450, 457–458
Kiepe, D., 234
Kifor, O., 254–256, 260
Kikkawa, U., 505
Kilduff, T. S., 615, 620
Kilo, S., 5
Kim, A. M., 679, 684
Kimble, R. D., 359, 369–370
Kim, C. M., 646
Kim, D., 671
Kim, E. K., 444
Kim, H. J., 602, 607
Kim, J., 459
Kim, K., 473
Kim, S. H., 14
Kimura, H., 378
Kimura, T., 377
Kim, Y. H., 434, 444, 679, 684
Kim, Y. J., 58, 273
Kinet, J. P., 600
Kinet, S., 333
King, D., 602
King, K., 408
King, T., 4, 6, 22–23
Kinghorn, I., 570
Kinoshita, A., 166
Kinoshita, E., 49
Kipnis, D. M., 270
Kirchner, H., 55
Kirschfeld, K., 594, 597, 606
Kirsch, J., 173
Kirschner, M. W., 678–679, 684–685, 689, 693
Kisanuki, Y., 615, 620
Kiselyov, K., 602, 607
Kitajima, I., 434
Kitamura, H., 664
Kitterman, J. A., 358–359
Kiya, Y., 168
Kiyonaka, S., 595
Kjelsberg, M. A., 110
Klaes, R., 332

Klee, W. A., 416
Kleinau, G., 381
Kleuss, C., 317
Klotz, K.-N., 199, 212–214
Kluin, P., 434, 444
Klumpp, S., 505
Knaus, P., 282, 359
Knaut, M., 464
Knight, C. G., 445
Knöpfel, T., 87, 637
Knowles, B. B., 180
Kobayashi, J., 209, 213
Kobayashi, M., 203, 570–571, 575
Kobilka, B. K., 56, 58–59, 209, 274, 296–297, 313, 456, 465
Kobilka, T. S., 59, 274
Koch, W. J., 632, 634
Ko, C. W. S., 97
Kodama, H., 204
Kofman, O., 57
Kofuji, P., 465
Kohda, M., 365
Kohler, U., 332
Kohn, L. D., 379, 381
Kohout, T. A., 167
Koh, T., 55
Koibuchi, Y., 203
Koike, M., 614–616, 620, 623–625, 628
Kojima, M., 54–55
Kojima, S., 689, 692
Kokkola, T., 34, 401
Kolb, H. A., 625
Kolch, W., 444
Kolonel, L. N., 332
Komatsu, Y., 369–370
Komatsuzaki, K., 408
Komoto, I., 570
Kondo, T., 365
Kong, Y. H., 199, 202
Konopka, J. B., 406, 409
Konturek, P. C., 6
Koob, G. F., 24, 438
Kopf, J., 282, 359
Korbonits, M., 56
Korchynskyi, O., 287–288
Koshimizu, T., 119, 120
Koski, G., 399
Kostenis, E., 180, 408
Koster, B., 284
Kosugi, S., 232, 381
Kotani, M., 76, 96
Kotarsky, K., 570–571
Kotecha, N., 566
Kourounakis, A. P., 211
Kouyama, T., 687
Kovoor, P., 655
Kozasa, T., 615, 625
Kozlowski, G. P., 615, 620

Kozlowski, J. A., 33
Kpodonu, J., 634
Kraakman, L., 399
Kracht, M., 445
Kragelund, B. B., 331, 353
Krapivinsky, G., 654
Krause, G., 381
Krauss, N., 58, 273
Krebs, E. G., 480–481, 525, 527
Kreienkamp, H. J., 180
Krieglstein, J., 505
Krissinel, E., 518
Kriz, R. W., 281, 369–370
Krogsgaard-Larsen, P., 258
Krohn, K., 379, 381
Kros, C. J., 594–595, 597, 602, 605
Krupa, D., 54
Krupenko, N. I., 440
Krupenko, S. A., 440
Krupnick, J. G., 167
Kubajewska, I., 23
Kubina, E., 158
Kudo, H., 445
Kudoh, J., 595
Kuehne, F., 202
Kuenzel, E. A., 481
Kuglstatter, A., 445
Kuhn, P., 274
Kukkonen, J. P., 33
Kumar, N. N., 634
Kumar, P. S., 206
Kunos, G., 6
Kuntzen, C., 445
Kuohung, W., 76
Kupin, W., 34
Kurachi, Y., 654
Kuratani, K., 204
Kurihara, R., 49
Kurisu, S., 693
Kurita, Y., 203
Kuriyan, J., 282, 517
Kurjan, J., 401–402
Kurosaki, T., 600
Kurose, J., 365
Kursawe, R., 376–377
Kurz, T., 592, 597
Kuszak, A. J., 465
Kuttenn, F., 332, 340, 345, 347, 349–350, 352–353
Kwak, S. P., 679, 684
Kwan, J., 5
Kwatra, M. M., 111
Kyriakis, J. M., 550

L

Labbe, O., 180, 190
Labrecque, J., 452

Author Index

Ladd, D. J., 254–255
Ladds, G., 406, 409
Lado-Abeal, J., 375, 379, 381
Laemmli, U. K., 502
Lafontan, M., 202
Lage, R., 55
Lahair, M. M., 444
Lahuna, O., 378
Lai, C. S., 292, 369–370
Lai, F. P., 678, 692
Lai, J., 4, 6
Lai, Y. M., 8, 271
Laitinen, J. T., 34
Lajoie, G., 473, 483
Lalli, E., 379
Lalude, O., 204
Lambert, D. M., 8
Lambert, N. A., 179, 185
Lameh, J., 431
Lam, K. S., 270
Lammers, G. J., 615
Lamorte, G., 282
Lamppu, D., 273
Lancaster, M. E., 214
Landesman-Bollag, E., 473, 516
Landgren, S., 55
Landi, M., 215
Lands, A. M., 198
Lane, J. R., 581
Lange, J. H., 8
Langelier, Y., 571
Langer, S. Z., 110
Lang, F., 625
Langlois, X., 203
Lang, M., 61–64
Langmead, C. J., 571
Langowski, J., 528
Lan, H., 571
Lan, R., 8
Lanham, I. S., 434
Lania, A., 379, 381
Lank, E. J., 268
Lannoy, V., 570
Lapatto, R., 76–77
Lapensee, C. R., 330
Lapensee, E. W., 330
Laporte, S. A., 167, 648
Laquerre, V., 295
Largent-Milnes, T. M., 22–23
Largis, E. E., 201
Larrivee, J. F., 173
Latif, R., 377
Latour, M. G., 571
Latronico, A. C., 233–234, 236–238
Laue, L., 232
Laugwitz, K. L., 377
Lau, J., 61
Launey, T., 87, 637

Lau, P. N., 57
Laurent, V., 688
Lavey, B., 33
Law, S. F., 151
Lázaro-Suárez, M. L., 119
LeBaron, M. J., 341, 343
Lebensohn, A. M., 679, 684, 689, 693
Leblond, F., 6
Lebrethon, M. C., 381
Lebrun, J. J., 332, 341, 349
Lechleider, R. J., 282
LeClaire, L. L., 3rd, 693
Leclerc, G., 114
Leclere, J., 379
Ledent, C., 5
Leder, P., 473, 516
Lee, C. E., 268, 615, 620
Lee, D., 298
Lee, F., 214
Lee, J., 601
Lee, K. W., 679, 684
Lee, P. H., 575
Lee, S. A., 332
Lee, S. E., 118, 119
Lee, T. W., 59
Lee, Y. H., 409
Lee, Y. S., 268
Le, F. G., 4–5
Leed, F., 213
Lefkowitz, R. J., 32–33, 59, 110, 167, 297–298, 306, 313, 378, 446, 448, 632–635, 642, 646
Lefort, A., 381
Leftkowitz, R. J., 167
Le Fur, G., 33, 203, 215
Legendre, M., 65
Lehrach, H., 632
Leitz, M. R., 199, 212–214
Le Merrer, M., 282, 358–363
Le, M. M., 24
Lenehan, J., 478, 491
Lengeler, K. B., 399
Lenkei, Z., 4
Lenormand, P., 550
Leonardi, A., 110
Leonetti, M., 203
Leopoldt, D., 128
Lepor, H., 110
Le Poul, E., 76, 570
Leroy, C., 256
Leroy, M. J., 203
Leshin, B., 434, 444
Lester, H. A., 465
Lesueur, L., 332
Leung, P. K., 57
Leung, W. Y., 110
Leurs, R., 56, 76, 127, 319, 446
Le, V., 369
Levey, A. I., 619

Levitt, E. S., 460
Lev, S., 591, 594–595, 597, 601–602, 607
Levy, S., 594, 597, 606
Lewis, D. L., 4, 461–463
Lewitan, R. J., 57
Le, W. W., 76–77
Liang, B. T., 295, 574
Liang, G., 55
Liao, X. H., 381
Liapakis, G., 59–60
Liberator, P. A., 54
Libert, F., 180, 190
Li, B. Y., 282, 634
Li, H. S., 593, 634
Li, J. Y., 55, 271, 434, 444, 449–450, 457, 678–679, 685
Li, L., 400, 595, 597, 600, 602, 606
Li, N., 572
Li, P. W., 166, 273
Li, Q., 434, 602, 607
Li, R., 688
Li, S. S., 473, 483
Li, W. M., 180, 199, 202
Li, Y. P., 369
Lichtner, P., 271
Liebetanz, J., 496
Lieu, M. H., 593
Ligneau, X., 33
Ligumsky, M., 8
Lima, L., 381
Lim, A. T., 298
Lim, G., 6
Lim, N. F., 465
Lin, A., 432
Lin, C., 632
Lin, D. C. H., 571
Lin, E. T., 438, 453–455
Lin, H. Y., 292, 369–370, 452
Lin, J. S., 33
Lin, K.-M., 190–191
Lin, S. N., 8, 56, 571
Lins, T. S. S., 233–234, 236–238
Lintzel, J., 180
Lipkowitz, M. S., 634
Litchfield, D. W., 471–473, 478–479, 485–487, 489–491, 497, 505, 525, 527
Litherland, S. A., 270, 276
Little, M. D., 5
Little, S. C., 359, 362–365, 367–370
Littleton, T. R., 572
Liu-Chen, L.-Y., 449–450, 457
Liu, C. W., 59
Liu, D., 434, 444
Liu, F., 282
Liu, G., 234
Liu, H., 572
Liu, J.-G., 414, 446, 449–450, 452, 454, 456, 458–460

Liu, L., 571, 655
Liu, N. A., 151, 157–158, 162
Liu, P., 634
Liu, Q., 8, 461
Liu, X. J., 201, 233–234, 236–238, 277, 319, 359, 369–370
Liu, X. P., 166, 171
Liu, X. S., 319
Liu, Y., 209
Liu, Z. W., 55
Liuzzi, A., 268
Llovera, M., 333
Lloyd, M. L., 5
Lobo, M. K., 180, 189, 194
Lodowski, D. T., 167
Loew, D., 678–680, 684–685, 689, 691, 693
Loewi, O., 654
Logothetis, D. E., 597
Loh, H. H., 439
Lohkamp, B., 517
Lohse, M. J., 180, 190, 193, 199, 212–214, 381
Loisel, T. P., 688
Loison, C., 570
Loizou, J. I., 507, 511
Loke, K. Y., 268
Lomasney, J. W., 110
Lombard, B., 678–680, 684–685, 689, 691, 693
Lombardi, M. S., 254–256, 260, 632, 634
Lommel, S., 678
Long, F., 440
Loosfelt, H., 378
Lopes, C. M., 597
Lopez-Gimenez, J. F., 576
Lopez-Grancha, M., 203
Lopez-Guerrero, J. J., 119–120
Lopez, M., 55
López-Sánchez, P., 120
Lorenz, W., 110
Lother, H., 120
Lounev, V. Y., 357, 359, 369–370
Loureiro, J., 105
Loustalot, C., 203
Lowe, J., 8
Lowery, J. J., 430, 434–435, 437–438, 453–455
Lowry, O. H., 168, 223
Loy, T., 234, 377
Lozano-Ondoua, A. N., 22–23
Lozeman, F. J., 525, 527
Lozinsky, I. T., 202
Lozza, G., 33
Lu, B., 365–366
Lu, Y. N., 572
Lu, Z. L., 87, 665
Lubrano-Berthelier, C., 270
Lucas, B. K., 330
Lucchini, V., 507
Ludaena, F. P., 198
Ludwig, M., 270

Lukacs, G. L., 87
Lukacs, V., 597
Lumpkin, E. A., 602
Lundell, D. J., 33
Lunn, C. A., 33, 414
Luo, J., 571
Luo, M. C., 4, 6
Luo, Z. Q., 570
Lupker, J., 4–5
Luscher, C., 655
Lutter, M., 55
Lutz, B., 4, 6
Lutz, M. W., 211
Lybrand, T. P., 206
Lynch, C. A., 268
Lynn, A. B., 5
Lynn, A. M., 4
Lyon, M. F., 255–257
Lytton, J., 254–256, 260

M

Maack, C., 212
Ma, A. L., 8
MacDonald, M. E., 632–635
Machesky, L. M., 679
Machleidt, T., 445
Mackie, K., 5–6, 8, 461
MacKinnon, S. E., 5
MacLennan, S. J., 5
MacNeil, D. J., 620
Macrae, A. D., 632–635
Madathany, T., 595
Madia, P. A., 441
Madiraju, M. S. R., 570
Madsen, A. N., 63, 64
Maeda, S., 440
Maekawa, F., 571, 575
Maeno, E., 595
Maestrini, S., 268
Maffrand, J. P., 4–5
Magnusson, R., 317
Mahan, A., 76–77
Maher, L., 6
Maheswaran, S., 488
Mahmoud, S., 446, 461–462, 464
Maidment, A. D. A., 359, 369–370
Maier, J., 381
Makary, S., 653
Makhay, M. M., 435
Makriyannis, A., 5, 8, 23, 33
Ma, L., 434, 444, 634
Malan, T. P. Jr., 4–6, 8, 23
Mallet, E., 256
Malloy, M. J., 268
Malnic, B., 369
Manara, L., 204, 215
Mancini, I., 31, 33–34, 46

Mandelbaum, A., 8
Manganaro, T., 282
Manning, A., 592, 604
Manning, B., 206
Mann, M. K., 5, 678–679, 684
Mansouri, J., 8
Mantovani, G., 379, 381
Mantyh, P. W., 18, 22–23
Manunta, P., 634
Manzoni, O., 415, 459
Mao, G. F., 96, 98
Mao, J., 6
Marceau, F., 173
Marchand, J. B., 687
Marchese, G., 5, 8
Marchiori, F., 473, 478
Marcotti, W., 594
Marcus, J. N., 620
Maresca, M., 377
Maresz, K., 5
Margaret, W. C., 657
Margas, W., 446, 461–462, 464
Marians, R. C., 377
Marinec, P. S., 632, 635, 649
Marin, O., 473, 478, 498–499, 502–506
Marinovic, A. C., 341
Markland, W., 445
Mark, M. D., 671
Markoff, E., 330
Marmiroli, S., 490, 502, 510
Marsh, D. J., 268
Marsicano, G., 5
Martens, J. R., 460
Martial, J. A., 333
Martin, B. R., 5, 8
Martinetti, M., 55
Martinez, S., 8
Martini, J. S., 634
Martin, J., 282
Martin, K. A., 209, 211, 213
Martin, L. J., 595
Martin, P., 203
Maruani, J., 8
Marullo, S., 215
Maruyama, M., 570–571, 575
Maruyama, T., 204
Marvin, J., 518
Masaki, T., 111
Mascheroni, D., 689, 692
Masini, M., 378
Masker, K., 595
Massague, J., 282
Massart, C., 377
Massotte, D., 449–450, 457–458
Masuko, S., 616–618
Masuzaki, H., 570
Mata, H. P., 5, 8, 23

Mathur, A., 213
Mathvink, R. J., 214
Matsuda, F., 76
Matsui, T., 204
Matsumoto, R., 203
Matsumoto, Y., 24
Matsuo, H., 54–55
Matsuo, Y., 168
Matsusaka, S., 49
Matsusaka, T., 166
Matsushima, K., 379
Matsuzaki, I., 615, 620
Matthews, R. P., 57
Maudsley, S., 87
Mäurer, A., 505
Maurer, T. S., 209, 213
Ma, W. Y., 433
Mayginnes, J. P., 479–480, 486
May, L. T., 577
May, R. C., 679
May, S., 434, 444
Mazzorana, M., 506–509
McAuliff, J. P., 198
McCallion, D., 8
McCann, P. J., 213
McClung, M., 54
McCoy, D. C., 572
McCreary, A. C., 8
McCubrey, J. A., 444
McCune, D. F., 111, 115, 117
McDall, D. E., 166
McDonald, D., 378
McGovern, R. A., 268
McGrath, J. C., 120
McGuire, K., 434
McGuire, T. L., 369–370
McKay, R., 433
McKeown, S. C., 572
McKnight, G. S., 57
McLauchlan, H., 445
McLean, A. J., 581
McManus, P. M., 292, 369–370
McNay, E. C., 55
McPherson, R., 268
McRory, J. E., 461, 463
Means, A. R., 57
Mechoulam, R., 5, 8
Medina, L. C., 109, 112–113, 116, 670–671
Medler, K. A., 4, 6
Meek, D. W., 505
Meggio, F., 473, 478, 497–499, 502, 506–509, 511, 522, 525–527
Mehlmann, L. M., 180
Meidute-Abaraviciene, S., 575
Meindl, A., 332
Meinke, P., 571
Melck, D., 4
Mele, M., 166

Melmed, S., 149
Melvin, L. S., 5
Menard, L., 632–633, 642
Mencarelli, M., 268
Mendonca, B. B., 233–234, 236–238
Menken, U., 234
Mentesana, P. E., 400–401, 408
Mercola, D., 433
Merendino, J. J. Jr., 232
Messager, S., 76
Messenger, M. M., 473
Messina, J. L., 549
Mestan, J., 496
Mestayer, C., 347, 349–350, 353
Mesulam, M. M., 619
Methven, L., 120
Metlagel, Z., 678–679, 693
Mett, H., 509
Metzger, J. M., 268
Meyer, M. D., 34
Meyer, N. E., 593
Meyer, S., 378, 381
Meyer, T., 509
Michailidis, I., 597
Michel, A., 33
Michel, I. M., 213
Michel, M. C., 111, 119, 575
Middleton, J. P., 634
Mifflin, T. E., 634
Mignot, E., 615
Mihashi, K., 687
Mikkilineni, A. B., 213
Milano, C. A., 407
Milasta, S., 576
Milazzo, F. M., 212–215
Milgrom, E., 76, 378
Millan, J., 8, 33
Millard, W. J., 270, 276
Millar, R. P., 76–77, 87, 234
Miller, B. A., 595
Miller, L. J., 33
Millet, L., 33, 202
Millhauser, G. L., 271
Milligan, G., 4, 6, 32, 141, 313, 321, 569–576, 581–582, 584
Milligan, L., 4–5
Milliken, T., 213
Mills, E., 592, 597
Milner, J., 440
Minden, A., 432
Minegishi, T., 232, 234
Minerds, K., 324
Minke, B., 591–595, 597, 600–603, 605–607
Min, L., 246
Minna, J. D., 434, 444
Minneman, K. P., 110–111, 114–115, 117–119
Minnick, D. T., 570–571, 575
Miotto, K., 438

Author Index

Miquel, R., 119–120
Mirzadegan, T., 58
Miserey-Lenkei, S., 381
Mishina, Y., 369–370
Mishra, P., 209
Mitchell, D. L., 96
Mitchell, F. M., 584
Mitchell, R. L., 8
Mitchell, S. E., 63
Mithbaokar, P., 377
Mitra, S., 110
Mitsock, L. M., 281, 369–370
Miura, M., 365–366
Miura, S., 166, 168
Miura, T., 209, 213
Miwa, S., 111
Miyao, N., 445
Miyata, K., 204
Miyata, Y., 505
Miyazaki, J., 570–571, 575
Miyazono, K., 282, 292
Mizrachi, D., 235–236
Mizuno, K., 209
Mockel, J., 34, 379
Modrall, J. G., 171
Moechars, D., 55
Mogil, J. S., 430, 438
Moguilevsky, N., 128
Moinat, M., 202
Mokhtarian, F., 23
Mokrosinski, J., 53, 61, 62
Molitch, M. E., 330
Momany, F. A., 54
Monck, E. K., 271
Monici, M., 81
Monroe, J., 234
Montanelli, L., 234, 377
Montecino-Rodriguez, E., 330
Monteggia, L. M., 203
Montell, C., 593, 605
Montplaisir, V., 319
Moolenaar, W. H., 180
Moore, D. J., 507, 511
Moore, R. Y., 615
Moreira, F. A., 6
Morello, J. P., 118
Moreno, A., 120
Morgan, M. M., 6
Morgan, P. H., 211
Morhart, P., 358–359
Morhart, R., 282, 359–363
Mori, E., 595
Mori, K., 96
Morimoto, M., 626
Morisset, S., 33, 65, 129
Mori, Y., 595, 601
Morley, J. E., 55
Moro, S., 502, 507–508

Morris, B. J., 634
Morrison, J. J., 203, 209
Morris, R., 25
Morse, K. L., 129
Mortier, G., 369
Mortin, M., 105
Mosberg, H. I., 276, 465
Moses, H. L., 282
Mosley, M. J., 110
Mosley, R., 571
Motter, R., 504
Mottershead, M., 55
Muallem, S., 602, 607
Muccioli, G. G., 8
Muenkel, H. A., 201
Muenke, M., 359, 362, 365
Mufson, E. J., 619
Muir, A. I., 570
Mukaiyama, H., 209, 213
Mukherjee, S., 34
Mukhopadhyay, S., 4
Mulholland, M. W., 55
Muller, S., 379
Muller, T., 235–236
Mulligan, J. A., 481
Mullins, M. C., 357, 359, 362–365, 367–370
Mullins, R. D., 678–679, 693
Munari, M., 597
Mundell, S. J., 77, 464
Mundlos, S., 359, 362–365, 367–370
Mural, R. J., 166
Muranaka, H., 209, 213
Murase, S., 379
Murayama, T., 95
Murdock, P. R., 570–571, 575
Murphy, M. G., 54
Murphy, V. L., 8, 33
Mussini, J. M., 5, 8
Muzzin, P., 55, 202
Myers, A. G., 186–188, 190–191, 193
Myers, E. W., 166
Myers, M., 234

N

Nadler, M. J., 600
Nagasawa, H., 331
Nagase, I., 204
Nagata, K., 595
Nagataki, S., 378
Nagatomo, T., 165
Nagayama, Y., 378
Nahmias, C., 201
Nahorski, S. R., 415, 419, 449, 452–453
Nairn, A. C., 628
Nakajima, S., 613
Nakajima, Y., 209, 213, 613

Nakamura, A., 365
Nakamura, H., 474, 483, 497
Nakamura, K., 203, 234, 277
Nakamura, T., 166
Nakao, K., 96
Nakashima, K., 330
Nakazato, M., 54, 615, 620
Naline, E., 203
Namba, H., 378
Nanamori, M., 171
Nano, R., 55
Nantel, M. F., 298–299, 315
Naoki, K., 445
Naor, Z., 87
Napp, A., 199, 202
Narayan, P., 376
Nardone, N. A., 209, 213
Nargeot, J., 461, 463
Nargund, R. P., 54
Narumiya, S., 96
Naruse, H., 49
Nasa, Y., 119, 120
Nasman, J., 33
Natsui, K., 55
Nattel, S., 653, 655
Nebane, N. M., 33
Neef, G., 32
Negishi, M., 33, 96–98, 101, 103
Neher, E., 655–656
Neilan, C. L., 453–454
Neliat, G., 8
Nelson, M., 22–23
Nelson, S., 504
Nemazanyy, I., 505
Nemeth, E., 259
Nerenberg, M., 434
Neri, M., 59
Nesbit, M. A., 255–257
Nestler, E. J., 203, 456
Neubig, R. R., 54, 171, 271
Neumann, S., 381
Nevalainen, M. T., 341
Neve, K. A., 459–460
New, D. C., 33
Newton, D. C., 5
Nguyen, D. G., 180
Nguyen, T. T., 271
Ngwu, C., 488
Nicoletti, F., 87, 637
Nicotera, P., 594
Niederstrasser, H., 679
Niefind, K., 515–517, 522–523, 525–527
Nie, J., 4
Nielsen, C., 440
Nielsen, K. K., 61
Nielsen, M., 32
Niemeyer, B. A., 592
Nienhuis, R., 615

Nievergelt, C. M., 634
Nijenhuis, W. A., 271
Nijmeijer, S., 127
Nikai, T., 602
Nikiforovich, G. V., 267, 274
Nikolaev, V. O., 180, 190, 193, 381
Nilius, B., 592, 597, 600–601
Nilsson, N. E., 570–571, 575
Nirenberg, M., 416
Nishida, E., 440, 444, 505
Nishida, M., 595
Nishino, S., 615
Nishio, K., 445
Niso, M., 209, 213
Nitta, R. T., 431, 434, 440, 445
Nivot, S., 65
Niwa, M., 378
Noben-Trauth, K., 594
Nobuhiro, K., 369–370
Noda, K., 166, 171
Noguchi, M., 570
Noguchi, Y., 365
Nogueiras, R., 55, 63
Noguera, M. A., 119–120
Nojima, J., 365
Nolan, C. J., 570
Noma, A., 654
Nomoto, H., 626
Nomura, A., 201
Nomura, M., 433
Nonaka, N., 55
Nordstedt, C., 425
Norford, D. C., 5
Norris, K. C., 634
North, R. A., 461, 465
Northup, J. K., 167
Notarnicola, M., 204, 205
Notcovich, C., 129
Nussbaum, B., 359, 365–366
Nwokolo, C. U., 55
Nygaard, R., 58–61, 62, 142
Ny, T., 240

O

Obara, Y., 571, 575
Obata, K., 615–616, 618–620
Obendorf-Maass, S., 199, 212–214
Oberwinkler, J., 605
O'Brien, A., 416
Ochsner, J. L., 54
O'Connell, M. P., 359, 365–366
O'Connor, D. T., 634
O'Connor, M. B., 282
Oda, A., 693
Oeda, E., 282
Offen-Glasner, V., 594–595, 597, 601–602, 607
Offermanns, S., 377

Author Index

Ogawa, Y., 55, 96
Ogi, K., 570–571, 575
O'Hara, P. J., 258
Ohashi, H., 379
Ohtake, H., 209, 213
Ohue, M., 201
Oikawa, M., 240
Oikawa, T., 693
Okabe, K., 671
Okadome, T., 282
Okajima, F., 381
Okamoto, I., 434
Okayama, H., 241, 269
O'Keefe, R. J., 366
Oki, N., 379
Olah, M. E., 601
Oland, L. D., 5
Olde, B., 570–571, 575
Oldfield, S., 464
Oleinik, N. V., 440
Olesnicky, N. S., 409
Olges, J. R., 111, 115, 117
Oliver, C., 83
Olivi, A., 282
Olivo-Marin, J. C., 77
Olsen, B. B., 515, 522–523, 525–527
Olson, M. F., 432
Olsten, M. E., 473
Omori, Y., 234
Onaran, O., 451
Onaya, T., 378
Onorato, J. J., 646
Oosterom, J., 271
O'Rahilly, S., 268
O'Reilly, T., 496
Orgiazzi, J., 379
Ormandy, C. J., 330
Orrenius, S., 594
Orser, B. A., 601
Orwoll, B., 259–260, 262–263
Orzeszko, A., 507–509
Osborne-Lawrence, S., 55
Osborn, M. D., 413, 434–435, 438
Oseid, E., 571
Oshikawa, S., 119, 120
Ossipov, M. H., 4, 6
Ostlund, S. B., 180, 189, 194
Ostrowski, J., 110
Otero, F., 180
Otsuka, A., 203
Otwinowski, Z., 678–679, 693
Oue, M., 209, 213
Ouerstreet, D., 24
Oustric, D., 8, 33
Overeem, S., 615
Owman, C., 570–571, 575
Owsianik, G., 600

Ozaki, M., 166
Ozono, S., 203

P

Pacher, P., 6
Padmanabhan, S., 185–188, 190–191, 193
Padrick, S. B., 679, 684
Padua, R. A., 551
Pagano, M. A., 499, 502, 506–509, 522, 525, 527
Page, S. O., 110
Paight, E. S., 209, 213
Pak, W. L., 592, 597, 605–606
Palczewski, K., 58
Palejwala, V., 199, 202–203
Pallais, J. C., 76–77
Palmese, C. A., 430, 438
Palos-Paz, F., 379, 381
Palyha, O. C., 54
Pampillo, M., 75–77
Panasyuk, G., 505
Pan, G., 282
Pang, Z., 199, 202
Pani, L., 5, 8
Pantaloni, D., 688
Pantel, J., 65
Pan, X., 461–463
Paolino, R. M., 430, 434–435, 437–438, 453–454
Paoloni-Giacobino, A., 202
Papinutto, E., 506–509
Paraiso, K. H. T., 549
Pardo, L., 56, 76, 167, 446
Parenty, G., 581
Paress, P. S., 54
Parfitt, A. M., 34
Park, C. G., 33
Park, D. M., 414
Parker, C., 270
Park, J. B., 679, 684
Park, J. H., 58, 273
Park, S., 597, 605–606
Parlato, R., 377
Parlow, A. F., 343
Parma, J., 34, 379, 381
Parmee, E. R., 209
Parmentier, M., 180, 190, 570
Parmentier, R., 33
Parnas, M., 593, 597, 600
Parnot, C., 59, 381
Parry-Smith, D. J., 110
Parsons, M. E., 128
Partilla, J. S., 421, 425, 428–429, 456, 460
Paschke, R., 376–379, 381
Pascual, J., 434–435
Pasternak, G. W., 415
Patchett, A. A., 54
Patel, Y. C., 150, 153

Patey, G., 215
Patwardhan, A. M., 4
Paul, A., 204
Paulmichl, M., 625
Paulsen, R., 593
Pauwels, P. J., 33
Pawson, A. J., 87
Payne, R., 603
Pearson, G., 432
Peat, A. J., 572
Peavy, R. D., 111
Pecceu, F., 4–5, 33
Peck, R. M., 282
Pediani, J. D., 576
Peifer, M., 105
Pei, G., 297
Peleg, G., 209
Pellecchia, M., 445
Pellegrino, S. M., 201
Pelleymounter, M. A., 268
Pendola, F. L., 180
Penhoat, A., 171
Pennacchio, L. A., 268
Penner, C. G., 479, 485, 487
Penner, R., 600
Perello, M., 55
Peretz, A., 603
Perez, D. M., 110–111, 115, 117, 119
Perez-Guerra, O., 379, 381
Perez-Tilve, D., 55
Pérez-Vizcaino, F., 119
Perich, J. W., 473, 478
Perkins-Barrow, A., 110
Perkins, J. P., 167
Perkins, L., 214
Perrachon, S., 4–5
Perraud, A. L., 600
Perrissoud, D., 55
Perrone, M. G., 197, 204, 205, 209, 212–215
Perry, E., 614
Perry, R., 614
Perry, S. J., 167
Persani, L., 180, 379, 381
Persson, U., 284
Pertwee, R. G., 4–6, 8, 33
Pérusse, F., 201
Peschke, B., 61
Petaja-Repo, U. E., 118, 234
Peters, E. C., 434–435
Petersen, E. N., 32
Petersen, P. S., 59, 61, 63, 64
Peters, M. F., 33
Peterson, R. T., 292, 369–370
Petersson, J., 4
Petitet, F., 33
Petrel, C., 256
Petroni, M. L., 268
Petrò, R., 33

Petrov, A., 571
Peyot, M. L., 570
Peyron, C., 615, 620
Pfeiffer, R., 173
Pfluger, P. T., 55
Pheng, L. H., 173
Phillips, J. A. III., 634
Piana, S., 59
Piascik, M. T., 110–111, 115, 117, 119
Picard, F., 270
Picciotto, M. R., 55
Piccirillo, S. G., 282
Pichard, C., 333
Pichat, P., 203
Pichette, V., 6
Pichon, C., 370
Pichurin, O., 151, 157–158, 162
Pickel, V. M., 6
Pidasheva, V., 257, 263
Piedras, I. C., 375
Piening, C. J., 525, 527
Pierce, K. L., 104
Pietila, E. M., 234
Pietri, F., 201
Pietrì-Rouxel, F., 206
Pigg, J. J., 4
Pignolo, R. J., 359, 363, 370
Pike, M. C., 332
Pike, N. B., 570
Pimienta, G., 434–435
Pin, J. P., 54, 77, 398
Pinna, G. A., 5, 8
Pinna, L. A., 473–474, 478, 483, 490–491, 495, 497–499, 502, 505–511, 516, 522, 525–527
Pintel, D. J., 479–480, 486
Pinto, S., 55
Piskorowski, R., 602
Pitcher, J. A., 167, 632–635
Pitchiaya, S., 465
Pizza, T., 8
Plas, P., 273
Plouffe, B., 295
Podtelejnikov, A. V., 678–679, 684
Poenie, M., 113
Pogozheva, I. D., 276
Pohlenz, J., 381
Poh, L. K., 268
Poinot-Chazel, C., 33
Poirot, O., 461, 463
Poitout, L., 273
Poitout, V., 571, 575
Poletto, G., 499, 502, 508, 522, 525, 527
Polgar, W. E., 416
Polites, H. G., 199, 202
Pollack, J. R., 434, 444
Pollak, M. R., 254–255
Pollard, T. D., 678, 687
Pollet, D., 180

Polo, S., 678, 692
Polyanovsky, A., 597
Ponomarev, E. D., 5
Pooler, L. C., 332
Poppitz, W., 8
Porreca, F., 5, 8, 23, 430–431, 437–438, 446, 449–450, 454, 459–460
Portier, M., 4–5, 8, 33
Poso, A., 34
Poss, K. M., 213
Postel-Vinay, M. C., 345
Post, G. R., 111, 115, 117
Postiglione, M. P., 377
Pott, C., 199, 202
Poulter, M. O., 77
Poupaert, J. H., 8
Prabhakar, B. S., 378
Prasad, B. M., 179, 185–188, 190–191, 193
Prather, P. L., 414, 446, 449–450, 452, 454, 456, 458–460
Pratilas, C. A., 553
Pratt, S., 255–256, 258–259
Premont, R. T., 632–635, 642
Prenen, J., 600–601
Prentki, M., 570–571
Price, D. L., 614, 620
Price, L. A., 401
Price, S. R., 341
Price, T. J., 4
Prince, M., 634
Prinster, S. C., 118, 119
Prisinzano, T. E., 418, 421–425, 428–429, 449–451, 456, 460
Proekt, I., 150, 152
Proneth, B., 270, 276
Prusoff, W. H., 175
Pryce, G., 5, 33
Ptasienski, J., 646
Puertollano, R., 592
Puett, D., 376
Puglisi, J. D., 59
Pula, G., 77
Pulley, M. D., 414
Pullinger, C. R., 268
Pupo, A. S., 115, 117–119
Puricelli, E., 268
Puzas, J. E., 366

Q

Qi, H., 203
Qin, J., 507, 511
Quadrato, G., 34, 46
Quinlan, M. E., 678
Quirion, R., 173
Quisenberry, L. R., 375
Quitterer, U., 120
Quraishi, N., 55

R

Raabe, T., 371
Raaf, J., 517
Raake, P., 634
Raehal, K. M., 430, 434–435, 437–438, 449–450, 452–454, 459–460
Raghay, K., 55
Raghu, P., 593, 597
Rajagopal, K., 378, 646
Rajagopal, S., 378, 646
Rajaniemi, H. J., 234
Ramachandran, R., 359, 369–370
Ramanathan, L., 615
Ramnaraine, M. L., 18
Ramsay, D., 581
Ramsey, I. S., 592
Randall, R. J., 168, 223
Rankin, M. L., 632, 635, 649
Rao, D. S., 34
Rapoport, B., 378
Rapp, U. R., 434
Rasmussen, M., 63, 64
Rasmussen, S. G. F., 58–59, 274, 296–297, 456, 465
Rasmussen, T., 523, 525–527
Rathi, L., 209
Rathmell, J. C., 434, 444
Rath, O., 444
Rauly, I., 33
Raun, K., 61
Rau, T., 6
Ravens, U., 464
Ravitsky, V., 359
Ray, M. K., 369–370
Raynal, B., 331
Rayner, D. V., 63
Rebres, R. A., 190–191
Reddy, H., 509
Reed, S. E., 479–480, 486
Reeves, P. J., 58
Refetoff, S., 180, 190, 379, 381
Regan, J. W., 95–97, 104–106
Re, G. G., 488
Reggiani, A., 33–34, 46
Reich, E.-P., 33
Reichwald, K., 271
Reinsheid, R. K., 459
Reisine, T., 151
Reitter, B., 214
Remmers, A. E., 419
Rennefahrt, U. E., 434
Ren, S. G., 150, 152, 166
Resch, K., 445
Ressler, K. J., 6
Reuveny, E., 464
Revelli, J.-P., 202
Reyes-Cruz, G., 112, 113, 115–117, 119

Reynen, P. H., 5
Reynet, A. J., 452
Reynolds, B. A., 282
Reynolds, G. A., 54
Reynolds, K. A., 59
Rhim, J. S., 433
Riccardi, D., 254–256, 260
Ricci, A. J., 594–595, 597, 602, 605
Rice, K. C., 5, 418, 422–424, 449–451
Richa, J., 371
Richard, E., 634
Richardson, G. P., 594
Richardson, J. A., 615, 620
Richardson, J. D., 5
Richter-Unruh, A., 234
Riddle, O., 330
Rief, W., 271
Riekkinen, P. Jr., 614
Riel-Mehan, M., 445
Rinaldi-Carmona, M., 4–5, 8, 33
Rincon, M., 432
Rios, C., 6
Ripley, B., 615
Ritchie, M. D., 634
Rizzoli, R., 54
Roach, T., 190–191
Roberts, D. J., 452
Robertson, G. P., 551
Roberts, P. J., 77
Robey, P. G., 365–366
Robinson, F., 432
Robinson, M. J., 432, 440, 550
Robishaw, J. D., 461
Rocheville, M., 575
Rockwell, S., 282
Rockwood, K., 54
Rodien, P., 376, 381
Rodionova, E. A., 77
Rodriguez-Cuenca, S., 55
Rodriguez, J. J., 6
Rodriguez, L., 416
Rodriguez, M., 203
Rodriguez-Mallon, A., 377
Rodriguez Pena, M. S., 319
Rodríguez-Pérez, C. E., 111–113, 115–119
Rodriguez-Sosa, M., 120
Roger, P. P., 377
Rogers, J. G., 282, 358–363
Rogers, S. D., 18
Rognan, D., 59, 256
Rohacs, T., 597, 602
Rohatgi, R., 678–679, 684
Rohner-Jeanrenaud, F., 55
Roh, S. G., 570
Romero-Ávila, M. T., 109, 111–113, 115–117, 119

Rom-Glas, A., 603
Romieu-Mourez, R., 473, 516
Romoli, R., 379, 381
Ronai, Z., 434–435
Rong, R., 270
Ronken, E., 8
Root, S., 282
Rosati, S., 434, 444
Rosebrough, N. J., 168, 223
Rosenbaum, D. M., 59, 274, 296–297, 456
Rosenbaum, T., 597
Rosenblum, C. I., 54, 268
Rosenkilde, M. M., 54, 58–60, 399
Rosen, M. K., 678–679, 684, 693
Rosen, V., 281, 369–370
Rosica, A., 377
Rosier, R. N., 366
Rosner, M. R., 488
Ross, E. M., 190–191
Rossier, O., 110
Rossignol, F., 77
Rossi, S., 23
Ross, R. A., 4, 8, 33
Roth, B. L., 425
Roth, C. B., 59
Roth, C. L., 270
Rothlisberger, U., 59
Rothman, J. E., 182
Rothman, R. B., 413, 421, 425, 428–429, 456, 460
Roth, R. H., 55
Rottner, K., 678, 692
Roubert, P., 273
Rouget, C., 203
Rouleau, A., 33
Rouot, B., 114
Rouquier, L., 203
Rovinsky, S. A., 55
Rowan, B. G., 505
Roy, M. B., 452
Rozec, B., 202–203
Ruat, M., 256
Rubins, N., 571
Rubio, C., 120
Ruckle, M. B., 452, 454, 456, 458
Rudy, T. A., 7
Rueda-Chimeno, C., 379, 381
Ruetz, S., 496
Ruffolo, R. R. Jr., 110
Rui, H., 341, 343
Ruiu, S., 5, 8
Ruiz-Velasco, V., 446, 461–462, 464
Russell, D. W., 286, 299
Russell, P., 442
Russell, R. W., 24
Russo, G. L., 505

Author Index

Rutherford, J. M., 421, 425, 428–429, 456, 460
Ruzzene, M., 490, 495, 497–499, 502, 505–510
Ryan, G. L., 233–235, 237, 239
Ryan, T. A., 182
Ryder, A., 343

S

Saad, Y., 166, 171
Saba, P., 5, 8
Saboya, R., 7
Sachidanandan, C., 292, 369–370
Sackin, H., 665
Sackur, C., 273
Sadana, R., 458
Sadée, W., 414–415, 419, 430–431, 434–435, 437–438, 446, 449–450, 453–454, 459–460
Sadja, R., 464
Sadoshima, J., 171
Saeki, Y. J., 180
Saez, J. M., 171
Saha, A. K., 55
Saha, B., 8
Saisch, S. G., 452
Saitoh, M., 282
Saito, K., 378
Saji, M., 379
Sakata, I., 55
Sakmann, B., 654–655
Sakmar, T. P., 58
Sakurai, T., 615, 620
Sala, G., 87, 637
Salak-Johnson, J. L., 18
Salamon, Z., 209
Salehi, A., 570–571, 575
Salem, R. M., 634
Salio, C., 6
Salituro, G., 571
Sallese, M., 87, 632, 634, 637
Sally, E. J., 418, 422–424, 449–451
Salmena, L., 550–551
Salome, N., 55
Salomon, Y., 307–309
Salo, O. M. H., 34
Salvatore, L., 87, 637
Salvi, M., 474, 478, 483, 491, 497, 499, 502, 505–507
Samama, P., 32–33, 59, 297, 306, 313
Sambrook, J., 286
Samie, M. A., 597
Sampath, T. K., 282, 292
Sampsell-Barron, T. L., 282
Samuels, Y., 551
Sanada, H., 634, 636, 641, 649
Sanders-Bush, E., 59, 313
Sanger, G. J., 204
Sangiao Alvarellos, S., 55
Sankaranarayanan, S., 182

Sansonetti, P. J., 688
Santana, L. F., 593
Santandrea, E., 209, 212–215
Santani, D., 204
Santiago, P., 270
Santora, V. J., 414
Santucci, V., 203
Saper, C. B., 620
Sargin, H., 379
Sargin, M., 379
Sarno, S., 474, 478, 483, 491, 497, 499, 502, 505–509, 511, 522, 525, 527
Sarran, M., 8, 33
Sartania, N., 581
Sartin, J. L., 270–272, 274–275
Sasamata, M., 204
Sassone-Corsi, P., 379
Sato, N., 445
Sato, S., 204
Sattar, M. A., 120
Saulnier, R. B., 473, 486–487, 489
Sauls, H. R., 570–571, 575
Saussy, D. L. Jr., 119
Savinainen, J. R., 34
Savola, J. M., 33
Sawa, M., 209
Sawyers, C. L., 496
Saxena, A. K., 209
Scammell, T., 615, 620
Scandroglio, P., 31, 33–34, 46
Scaplehorn, N., 678
Schaefer, M., 571, 575
Schaller, H. C., 180
Schaper, J., 234
Scharenberg, A. M., 600
Schatz, A. R., 8
Scheerer, P., 58, 142, 273
Scheld, W. M., 7
Schemann, M., 204
Schenck, A., 679, 684
Schenk, J. P., 234
Scherag, A., 271
Scheuer, T., 461
Schiano, M. A., 4
Schiferer, A., 377
Schiffmann, S. N., 76
Schilling, W. P., 593
Schipani, E., 34
Schleim, K., 54
Schlumberger, P., 271
Schmid, K., 158
Schmiechen, R., 32
Schmittmann-Ohters, K., 234
Schmutzler, R. K., 332
Schnabel, P., 212
Schnee, M., 594–595, 597, 602, 605
Schneider, H., 445
Schneider, S. R., 209, 213

Schnell, S., 571, 575
Schnitzer, T., 54
Schöbel, S., 504
Schoffelmeer, A. N., 6
Schofl, C., 571, 575
Schork, N. J., 634
Schuler, L. A., 331–332
Schulte, N. A., 117–119
Schultz, G., 377
Schultz, J. A., 414
Schunack, W., 33
Schurmann, A., 55
Schvartz, C., 379
Schwart, T. W., 66
Schwartz, J. C., 33, 114
Schwartz, R. W., 416
Schwartz, T. W., 54–62, 65, 77, 571–572, 575
Schwarz, E. M., 366
Schwei, M. J., 18
Schwinger, R. H. G., 199, 202
Schwinn, D. A., 110–111
Schwinof, K. M., 76
Sciarroni, A. F., 212–215
Scilimati, A., 197, 204–205, 209, 212–215
Scita, G., 678–679, 689, 692
Scott, C. W., 33
Scott, G., 369–370
Scott, K., 592
Scott, M. G., 634
Scott, W. G., 517
Scriba, G. K., 8
Sebring, N. G., 268
Sedaghat, K., 324
Seemann, P., 284, 357, 359–360, 362–365, 367–370
Segaloff, D. L., 231, 233–239, 246–249, 268–270, 277
Segredo, V., 431
Seidman, C. E., 254–255
Seifert, R., 4, 59, 152, 180, 313, 399
Seigman, J. G., 254–255
Seiler, S. M., 213
Sekler, I., 570–571
Seldin, D. C., 473, 516
Selley, D. E., 415, 456, 461
Selzer, E., 377
Semache, M., 571
Semenya, A., 634
Seminara, S. B., 76–77
Senitzer, D., 359, 370
Sen, M., 332
Seoane, R., 381
Serrano de la Peña, L., 359, 362, 365
Settanni, F., 55
Sevak, R., 204
Severynse, D. M., 299
Shaaban, S., 452
Shachar, S., 570–571

Shafritz, A. B., 359, 362, 365
Shagoury, J. K., 76
Shah, P. C., 282
Shanabrough, M., 55
Shapiro, E., 110
Shapiro, H., 570–571
Sharkey, K. A., 5
Shaw, A. M., 270, 276
Shaw, D. E., 59, 445
Shearer, B. G., 213
Shearman, L. P., 5
Shears, S. B., 88–89
Sheffler, D. J., 425
Shelton, F., 180
Sheng, Y., 319
Shen, J. X., 199, 202
Shenker, A., 232–234, 381
Shenolikar, S., 628
Shen, Q., 359, 362–365, 367–370
Sheppard, P. O., 258
She, Q. B., 433, 551
Sherman, B. W., 213
Sher, P. M., 213
Sheves, M., 58
Shibata, K., 111
Shi, L., 58, 60
Shiloah, S., 8
Shimizu, S., 595
Shinbo, H., 203
Shinoura, H., 111, 119–120
Shiraishi, K., 240
Shirakami, G., 55
Shire, D., 8, 33, 203
Shirota, M., 332
Shi, S., 365–366
Shiverick, K. T., 333
Shoback, D. M., 253, 255–256, 258–260, 262–263
Shore, E. M., 282, 357–371
Showalter, V. M., 8
Shriver, L. P., 5
Shugar, D., 478, 509
Shui, Z., 670
Sibley, D. R., 632, 635, 649
Sieber, C., 282, 359
Siegel, E. M., 5, 23
Siegel, J. M., 615
Siemens, J., 602
Sigal, I. S., 206
Sigworth, F. J., 655
Sillence, D., 359
Silve, C., 256
Silverman, L. S., 34, 46
Silvestro, L., 432
Simiand, J., 203
Simon, A., 4
Simoneau, I. I., 5, 23
Simon, S., 505

Simpson, D. S., 418, 422–424, 449–451
Simpson, P. C., 120
Sim-Selley, L. J., 8
Singer, M. A., 369
Singer, W. D., 625
Singh, L. P., 57
Singh, S. P., 378
Sinton, C. M., 615, 620
Sirohi, S., 441
Sirvio, J., 614
Skierski, J., 478
Skippen, A., 89
Skorput, A. G., 434–435, 438
Skwish, G. S., 213
Slade, D., 5
Slaton, J. W., 497, 498, 516
Slaugenhaupt, S. A., 597
Sleeman, M. W., 55
Sloper, D. T., 516
Smalley, K. S. M., 549–552
Smiley, D. L., 55
Smith, C. L., 505
Smith, D. F., 118
Smith, F. J., 268, 330
Smith, F. L., 8
Smith, J. M., 414
Smith, M. S., 119
Smith, N. J., 570, 574, 576
Smith, R., 358–359
Smith, S. O., 58–60
Smith, W. L., 96
Smit, M. J., 56, 76, 129, 131, 133, 136, 139–140, 142, 144, 319, 446
Smits, G., 234, 377
Snutch, T. P., 464
So, I., 602, 607
Solenberg, P. J., 55
Solinas, G., 445
Solly, K., 140
Soltis, E. E., 119
Sommercorn, J., 481
Sommers, C. M., 409
Sonenshein, G. E., 345, 473, 516
Song, C., 4
Song, D. H., 473, 516
Song, X., 434, 444
Song, Y., 377
Song, Z. H., 33, 180, 189
Soong, T. W., 464
Sorensen, N. B., 270
Sorgard, M., 4
Sos, M. L., 551
Soubrié, P., 33, 203
Souchelnytskyi, S., 284
Soudijn, W., 214
Sousa, C., 678–679, 691, 693, 884–685
Soyombo, A. A., 602, 607

Spada, A., 379, 381
Spalding, T. A., 258
Spampinato, U., 414
Spedding, M., 54
Speirs, H. J., 634
Speleman, F., 369
Spicher, K., 377
Spittaels, K., 180
Springael, J. Y., 570
Spudich, J. A., 688
Srinivasan, D., 104
Srinivasan, S., 270
Srinivas, H., 434, 444
Staes, M., 180
Stahl, N., 359, 369–370
Stahl, S. M., 203
Staley, E. M., 479–480, 486
Standifer, K. M., 415
Stanfield, P. R., 614–616, 620, 623, 625–626, 628
Stanhope, R. G., 268
Stanley, J. C., 171
Stark, H., 33
Staudenmann, W., 525, 527
St-Denis, N. A., 471, 473, 505
Stea, A., 464
Stebbins, J. L., 445
Steen, H., 678–679, 685
Steenhuis, J., 59
Stefan, C. J., 406
Steffen, A., 678–679
Stein, A. T., 593
Steiner, A. L., 270
Stemmelin, J., 203
Steneberg, P., 571
Steneberg, R., 571
Sterne-Marr, R., 646
Sternweis, P. C., 190–191, 461
Steven, J. F., 168
Stevenson, L. A., 4, 8, 33
Stevens, R. C., 274
Stewart, A. J., 234
Stewart, D. J., 5
Stewart, R., 282
Stinus, L., 438
Stoddart, L. A., 569–570, 572–576, 581–582
Stoffel, R. H., 632–635
Stokes, A. J., 600
Stokoe, D., 550
Stone, W. C., 632
St-Onge, G. G., 571
Stork, B., 8
Storto, M., 87, 637
St Pierre, R., 486–487, 489
Strack, A. M., 268
Stradal, T. E., 678–679, 692
Strader, C. D., 110, 206, 214

Stram, D. O., 332
Strange, P. G., 199, 211, 452, 453, 574
Stratakis, C. A., 332
Stretton, J. L., 204
Stricker, P., 330
Stringer, B. M., 377
Strong, L. L. R., 597, 605
Strosberg, A. D., 199, 201–202, 206, 215
Strum, J. C., 570
Struyf, S., 570
Stucky, C. L., 4, 6
Stuiver, M., 601
Suarez-Huerta, N., 76
Su, B., 434
Suda, M., 55
Suda, T., 363
Südkamp, M., 212
Suetsugu, S., 693
Sugg, E. E., 214
Sugimoto, Y., 96
Sui, J. L., 671
Sullivan, J. P., 34
Summar, M., 634
Summers, R. J., 209, 213
Su, M. S., 445
Sun, A., 254–256, 260
Sun, C. Q., 213
Sun, H. S., 595, 601
Sun, M., 434, 444
Sun, T. J., 601
Sun, X., 415, 419, 601
Sunada, S., 119, 120
Sunahara, R. K., 58, 317, 465
Sung, B., 6
Sung, J. Y., 679, 684
Sunthornthepvarakui, T., 379
Suss-Toby, E., 603
Sutton, G. G., 166
Su, W., 434, 440, 444
Suzuki, E., 592
Suzuki, H., 33
Suzuki, K., 213
Suzuki, S., 209
Suzuki, Y., 445
Svitkina, T., 689, 692
Swaminath, G., 59
Swigart, P., 89
Swillens, S., 234, 377, 381
Syrett, N., 58
Szanto, I., 55
Szekeres, P. G., 452
Szereszewski, J. M., 76–77
Szidonya, L., 4
Szkudlinski, M. W., 376, 381

T

Tacchetti, C., 180
Tadayyon, M., 570–571, 575
Tagami, T., 55
Tagliaferri, M., 268
Takahashi, I., 626
Takahashi, N., 363
Takahashi, S. I., 379
Takai, A., 628
Takano, K., 615, 623, 625–626, 628
Takasu, T., 204
Takaya, K., 55
Takeda, K., 282, 292
Takemori, A. E., 438–439
Takenawa, T., 693
Takeo, S., 119, 120
Takeshita, A., 378
Takeshita, K., 445
Talavera, K., 600
Tallet, E., 331, 353
Tamai, T., 209, 213
Tamaoki, T., 626
Tamargo, J., 119
Tambaro, S., 5, 8
Tamura, N., 96
Tamura, T., 203
Tanaka, H., 570–571, 575, 615, 620, 665
Tanaka, I., 96
Tanaka, K., 378
Tanaka, M., 330
Tanaka, N., 209, 213
Tanaka, S., 180
Tanaka, T., 240, 379, 570
Tanaka, Y., 570–571, 575
Tan, C. P., 110, 571
Tang, R., 110
Tang, V., 268
Tanikella, T. K., 201
Taniyama, K., 378
Tanoue, A., 111, 119–120
Taoka, I., 209
Tao, S., 434, 444
Tao, Y. X., 233–235, 237, 239, 267–268, 270–275, 277, 298
Tao, Y. x., 268, 270
Tapia, J., 502
Tappeser, S., 234
Tardivel-Lacombe, J., 33
Tarnow, P., 271
Tata, J. R., 54
Tateishi, H., 209
Tate, K., 215
Taub, R., 359, 362, 365
Taupignon, A. I., 298
Taussig, R., 190–191

Author Index

Tavernelli, I., 59
Tawadros, N., 570
Tawfic, S., 490, 497
Taylor, J. E., 76–77, 150–152
Taylor, S. S., 566
Taymans, S. E., 332
Tcheang, L., 570
Tchilibon, S., 8
Tedeschi, G., 505
Tegge, W. J., 483
Templeton, F., 8, 33
Ten Dijke, P., 281–282, 284, 287–288, 292, 365
Terranova, J.-P., 203
Terro'n, J. A., 120
Tesmer, J. J., 167
Tess, D. A., 209, 213
Tetzloffl, G., 571
Teves, L., 595
Thakker, R. V., 255–257
Thamsborg, G., 54
Thangiah, R., 204
Thangue, N. B., 678
Thannickal, T. C., 615
Thastrup, J., 440
Thathiah, A., 180
Theander-Carrillo, C., 55
Themmen, A. P. N., 233
Theroux, T. L., 111
Thian, F. S., 59, 274
Thomas, R., 211
Thompson, A. L., 8
Thompson, D. A., 271
Thomson, J. A., 282
Thor, D. H., 24
Thrasher, A., 678
Thresher, R. R., 76
Thyagarajan, B., 597, 602
Tian, S., 282
Tiberi, M., 295, 298–299, 306–307, 313, 315, 319, 321, 324
Tidgewell, K., 421, 425, 428–429, 456, 460
Timmerman, H., 56, 76, 319, 446
Tipker, K., 8
Tisdale, M. J., 202
Tjon-Kon-Sang, S., 602, 607
Tnani, M., 434
Tocci, M. J., 199, 202–203
Toews, M. L., 117–119
Tohyama, Y., 49
Tollin, G., 209
Toll, L., 416
Tomishima, Y., 209, 213
Tomita, F., 626
Tomita, T., 570
Tonacchera, M., 381

Tong, K. P., 376
Tong, Q., 595
Tookman, L., 63
Torii, S., 444
Torres-Padilla, M. E., 115, 116
Tortorella, P., 213
Tosoni, K., 490, 495, 502, 510
Tota, L., 209, 214
Touraine, P., 329, 331–333, 340, 345, 347, 349–350, 352–353
Tovar, S., 55, 63
Toyooka, S., 55
Toyosawa, K., 201
Tran, H., 282
Tran, T. H., 343
Trapella, C., 446, 461–462, 464
Trautwein, W., 654
Traxler, P., 496
Traynor, J. R., 415, 419, 445, 449, 452–454, 459–460
Treisman, R., 551
Tremblay, C., 571
Trembley, J. H., 497, 516
Trevillyan, J. M., 444
Trinh, T., 679
Trivedi, P., 110, 620
Troppmair, J., 434
Trovato, L., 55
Trumbauer, M. E., 268
Tsai, J., 552, 565
Tsao, P., 167
Tschop, M. H., 55
Tsien, R. Y., 113
Tsing, S., 445
Tsuchiya, Y., 444
Tsuiki, H., 434
Tsujimoto, G., 111, 117, 119–120
Tsujioka, T., 49
Tsujiuchi, H., 209
Tsunoda, H., 379
Tsunoda, S., 593
Tsutsumi, T., 203
Tu, C., 255
Tullis, G. E., 479–480, 486
Tulshian, D., 34, 46
Tumova, K., 298–299, 315
Tung, L. Y., 63
Turowec, J. P., 471, 473, 483
Turu, G., 4–5
Tutino, V., 204, 205
Tuusa, J. T., 234
Tu, X., 440
Twombly, V., 369
Tyldesley, R., 76
Tyroller, S., 212

U

Uberti, M. A., 111, 115, 117–119
Uchida, H., 204
Ueda, Y., 33
Uehling, D. E., 213–214
Uejima, H., 570–571, 575
Ufret-Vincenty, C. A., 593
Uhlenbrock, K., 180
Ukai, M., 204
Ulloa-Aguirre, A., 118
Ullrich, A., 332, 341, 349
Umetani, J., 678–679, 684, 693
Unger, G., 497, 516
Unger, G, M., 498, 516
Untch, M., 332
Upton, A. C., 152
Urist, M. R., 281, 369–370
Urizar, E., 234, 377
Urquiza-Marin, H., 119
Usher, K., 597
Usui, T., 55, 96
Utama, F. E., 343
Utiger, R., 270

V

Vaclavicek, A., 332
Vaisse, C., 268, 270–271
Valentin-Hansen, L., 59, 61
Valero, E., 528
Valiquette, M., 297, 321
Van Agthoven, J., 331, 353
van Aken, A. F., 594–595, 597, 602, 605
van Beest, W., 105
Vanbrabant, M., 180
Van Damme, J., 570
Vandenbogaerde, A., 76
Van Den Pol, A. N., 615, 620
Vandeput, F., 76
Vanderah, T. W., 3, 5, 8, 22–23
Vanderbilt, C., 432
van der Geer, P., 485
Van Der Kooi, K., 549
Van der Ploeg, L. H., 620
Vanderwinden, J. M., 76
Vandesompele, J., 369
Van de Wetering, M., 105
van Dinther, M., 281–282, 292, 365
van Es, J., 105
Van Keymeulen, A., 377
Vanni, S., 59
Vanoni, M., 505
Van Sande, J., 34, 377, 379, 381, 399
Van Sickle, M. D., 5
van Stuivenberg, H. H., 8
van Wijngaarden, I., 214
Vardanyan, A., 22–23
Varela, L., 55

Vargiu, R., 5, 8
Varnai, P., 597, 602
Vasina, V., 204
Vasquez, C., 4
Vassart, G., 34, 180, 190, 234, 376–377, 379, 381, 570
Vassileva, G., 571
Vaudry, R., 333, 343–345, 353
Vaughan, C. W., 459, 465
Vazquez-Cuevas, F. G., 120
Vázquez-Cuevas, F. G., 111, 116
Vazquez, J., 445
Vazquez, M. J., 55
Vázquez-Prado, J., 112–113, 115, 116
Veerman, W., 8
Velasques, R. D., 7
Velculescu, V. E., 551
Velez, R. G., 58
Velez Ruiz, G., 465
Velly, J., 114
Vennekens, R., 601
Venter, J. C., 166
Verfurth, F., 111, 119
Vergarajauregui, S., 592
Verloes, A., 381
Verti, B., 268
Verzijl, D., 319
Vescovi, A. L., 282
Vie Luton, M. P., 65
Vila, E., 120
Vilardell, J., 499, 502
Vilaro, M. T., 576
Vilk, G., 471, 473, 483, 486–487, 489
Villablanca, A. C., 211
Villalobos-Molina, R., 110, 111, 115, 119–120
Villasenor, A. G., 445
Vincent, P., 459
Vinge, L. E., 634
Virtanen, R., 33
Vischer, H. F., 56, 76, 127, 446
Visiers, I., 457
Visser, L., 434, 444
Visser, N., 282, 292, 365
Vitale, P., 209, 212–215
Vivekanandan, S., 204
Vlaeminck-Guillem, V., 376, 381
Voets, T., 592, 600
Voight, N., 464
Voigt, C., 378, 381
Voigt, N., 653, 655, 664–666, 669
Vo, K. T., 434
Vollmert, C., 271
von-Campenhausen, M., 594, 597, 606
Vonderhaar, B. K., 332
Vongs, A., 268
Von Zastrow, M., 167
Vos, T. J., 273
Vural, S., 379

W

Wade, R. H., 528
Wagner, K., 332
Wailes, L. M., 57
Wainer, B. H., 619
Wainwright, K. L., 24
Wakamori, M., 595
Walczak, J. S., 6
Waldhoer, M., 429
Waldo, G. L., 167
Waldrop, B. A., 111, 115, 117
Walker, E. A., 435
Walker, G. E., 268
Walker, J. K., 167
Walker, J. M., 452
Walker, M. D., 570–571, 614
Wals, H. C., 8
Walter, D., 445
Walter, N. G., 465
Walwyn, W., 449–450, 455–456, 463–464
Wang, B., 180
Wang, C., 434, 440, 444
Wang, D., 414–415, 419, 430, 434–435, 437–438, 446, 449–450, 452–454, 459–460
Wang, E. A., 281, 369–370
Wang, F., 592, 597
Wang, G., 497, 516
Wang, H. Y., 55, 434, 444, 572
Wang, J., 425
Wang, L., 319, 359, 369–370, 433
Wang, M., 434, 444
Wang, S. X., 6, 273, 445
Wang, T. C., 213, 282, 605
Wang, W. Y., 634, 636, 641, 649
Wang, X., 42, 199, 202, 421, 425, 428–429, 456, 460
Wang, Y., 397, 634
Wang, Z. Q., 267, 430–431, 437–438, 446, 449–450, 454, 459–460, 479, 485, 487, 632, 634–636, 641, 649
Wappenschmidt, B., 332
Wardeh, G., 6
Ward, R. J., 572, 576, 581
Warren, H. B., 254–255
Washizuka, K., 209, 213
Wasmuth, J. J., 632
Watanabe, H., 634, 636, 641, 649
Watt, S., 688
Watts, V. J., 459–460
Waugh, D. J., 111, 115, 117
Wawrowsky, K. A., 149
Way, E. L., 439
Way, M., 678
Webb, N. J., 678
Weber, A. E., 209, 214
Weber, J. E., 483
Wedegaertner, P., 434

Wehland, J., 678
Wei, E., 439
Weiel, J. E., 480
Weinberg, D. H., 110
Weiner, D. M., 138, 140
Weinhouse, M. I., 414
Weinshank, R. L., 110
Weinstein, H., 59, 143, 313, 457
Weintraub, B. D., 376
Weiss, J. M., 133, 211
Weiss, R. E., 381
Weiss, S., 54
Weis, W. I., 59, 274
Wei, W. L., 601
Welch, M. D., 678–679, 693
Wellbrock, C., 550
Wellendroph, P., 256–257
Welsch, C. W., 331
Welsh, W. J., 4
Weng, G., 317
Wenzel-Seifert, K., 4, 59, 180, 399
Wermuth, C. G., 114
Wessels, H. T., 234
West, K. A., 97, 105–106
Wettwer, E., 464
Wetzel, J. M., 110
Wharton, K. A., 369
Wheeldon, A., 204
Whisnant, R. E., 317
Whistler, J. L., 429
White, A., 416
White, B. C., 634
Whitebread, S., 166
Whitesell, L., 118
Whitters, M. J., 281, 369–370
Whitworth, J. A., 120
Whorton, M. R., 58, 465
Wiedmer, P., 55
Wieland, K., 129
Wieland, T., 152
Wieser, R., 282
Wigglesworth, M. J., 570
Wiley, J. L., 8
Wilfing, A., 377
Wilken, G. H., 4, 8
Willems, P. H., 601
Willency, J. A., 55
Williams, C. C., 505
Williams, D. A., 440
Williams, E. D., 377
Williams, J. N., 267
Williams, J. T., 415, 459, 461, 465
Williams, L. M., 63
Williams, S. C., 615, 620
Williams, S. M., 634, 636, 641, 649
Williams, T., 268
Williams, W. M., 33
Willie, J. T., 615, 620

Willis, A., 634
Wilsbacher, J. L., 432, 440, 445
Wilson, A. R., 6
Wilson, J. M., 359, 369–370
Wilson, K. H., 110
Wilson, K. P., 445
Wilson, S., 615, 620
Winchester, W. J., 204
Winger, G., 435
Winzell, M. S., 575
Wise, A., 408, 571
Wise, H., 57
Wislez, M., 434, 444
Woelfle, J., 270
Woldhye, D. P., 63, 64
Wolkenfeld, N. M., 150, 152
Woloszyn, J., 55
Wong, A. J., 431, 434, 440, 444–445
Wong, K. K., 550–551
Wong, R. S., 452
Wong, Y. H., 33
Wood, M. S., 270
Woods, J. A., 425
Woods, J. H., 435, 453–454
Woodward, D. F., 96
Woolf, C. J., 12, 16
Worman, N. P., 110
Wosczyna, M. N., 359, 369–370
Wouters, F. S., 83
Wouters, J., 8
Wozney, J. M., 281, 369–370
Wrana, J. L., 282
Wright, A., 432
Wright, C., 22–23
Wright, J., 634
Wright, S., 504
Wu, C., 483, 592
Wu, G., 213
Wu, M., 8
Wu, N., 199, 202
Wu, S., 199, 202
Wu, X., 234, 440
Wurch, T., 33
Wurster, S., 33
Wynford-Thomas, D., 377
Wyvratt, M. J., 209, 214

X

Xiang, Y., 59
Xiang, Z., 270, 276
Xi, B., 42
Xie, X., 445
Xie, Y. K., 12–13
Xiong, Y., 615, 620
Xiong, Z. G., 601
Xiu, C. H., 199, 202
Xu, A., 270

Xu, B., 432
Xu, C., 166
Xu, D., 689, 692
Xu, H., 413, 418, 421–425, 428–429, 449–451,
    456, 460, 592, 595, 597, 600, 602, 605–606
Xu, J., 634, 636, 641, 649
Xu, M., 282, 284, 357, 359–365, 367–370
Xu, R. H., 60, 282
Xu, W., 96
Xu, X. Z. S., 42, 434, 593
Xue, C., 397, 400, 408

Y

Yada, T., 571, 575
Yaksh, T. L., 7
Yalcin, I., 203
Yamada, H., 595
Yamada, M., 369–370, 654
Yamaguchi, I., 204, 634, 636, 641, 649
Yamaguchi, K., 445, 614–616, 618–620, 628
Yamaguchi, O., 203–204
Yamaguchi, S., 602
Yamamoto, M., 359, 369–370
Yamamoto, T., 203, 444
Yamamura, H. I., 4, 6
Yamashita, H., 209
Yamashita, S., 234, 378
Yamazaki, D., 693
Yanagisawa, M., 55, 620
Yanai, T., 363
Yan, E. C., 58
Yang, B., 8
Yang, C., 689, 692
Yang, F., 267
Yang Feng, T. L., 110
Yang, J. J., 55, 256–257
Yang, L., 6
Yang, M., 111, 119
Yang, R., 444
Yang, W., 256–257
Yang, X. W., 180, 189, 194
Yang, Y. M., 433
Yano, I., 438–439
Yanovski, J. A., 268
Yao, B. B., 34
Yao, X. J., 58, 465
Yasuda, K., 151
Yasufuku-Takano, J., 625
Yavuzer, D., 379
Yeager, M., 59
Yeo, G. S., 268
Yin, D., 408
Yin, H., 180
Yoburn, B. C., 441
Yokoyama, N., 378
Yonezawa, T., 571, 575
Yoon, J., 597, 605–606

Author Index

Yoon, S. O., 180
Yoshida, N., 213
Yoshida, T., 595
Yoshimi, T., 96
Young, A. M., 435, 441
Young, L. G., 341
Young, S. F., 268
Young, W. S., 180, 189
Yowe, D. L., 273
Ypma, A., 105
Yuan, M., 634
Yue, D. T., 461, 463
Yu, H., 268, 620
Yu, J., 268, 282
Yu, L., 54
Yu, M., 504
Yu, N., 42, 140
Yu, P. B., 292, 369–370
Yu, P. Y., 634, 636, 641, 649
Yu, S., 497

Z

Zackai, E., 359
Zagotto, G., 502, 508
Zahn, D., 76
Zajac, M., 76–77
Zambon, A., 507
Zamponi, G. W., 461, 463
Zanetti, N., 282
Zannini, M. S., 377
Zanotti, G., 478, 491, 502, 507
Zare, R. N., 209
Zasloff, M. A., 359, 362, 365
Zeevi, D. A., 594–595, 597, 601–602, 607
Zeng, J., 459
Zerkowski, J., 8
Zetterberg, C., 54
Zhai, Q. W., 572
Zhang, D., 76–77, 298–299, 366, 371
Zhang, J. Y., 167, 434, 444, 459, 648
Zhang, M. F., 233–235, 237, 239, 246–249, 369, 376

Zhang, S., 268
Zhang, W., 55, 595
Zhang, X., 199, 202, 234
Zhang, Y., 359, 370
Zhao, M., 365–366
Zhao, P., 615, 620–623, 625, 627–628
Zhao, Q., 634
Zhao, W., 330, 634
Zhao, X., 57
Zheng, L., 595
Zheng, S., 634, 636, 641, 649
Zhivotovsky, B., 594
Zhou, C. Y., 151, 157, 162, 571
Zhou, D., 5
Zhou, L., 166
Zhou, S., 602
Zhou, Y. P., 256–257, 571, 634
Zhu, H., 634
Zhu, J., 42
Zhu, Q., 600
Zhu, W. L., 59, 572
Zhu, Y., 505
Zhu, Z. T., 298
Zhyvoloup, A., 505
Ziani, K., 119–120
Zicha, D., 678
Zien, P., 478, 483
Zigman, J. M., 55
Zigmond, S. H., 688
Ziliox, M., 58
Zimmermann, J., 496
Zinsmeister, A. R., 204
Zinzalla, V., 505
Ziskoven, C., 199, 202
Zlatanou, A., 507, 511
Znameroski, E. A., 678
Zou, Y., 59
Zucconi, A., 678–679
Zuchero, J. B., 678
Zuker, C. S., 592
Zuscik, M. J., 366
Zygmunt, P. M., 4

Subject Index

A

Acetylcholine-gated $I_{K,ACh}$ channels
  giga-seal formation, 655–657
  human atrial cardiomyocytes isolation, 657–659
  identification, 669–671
  I-O configuration, 656–657
  patch membrane, 656–657
  regulatory mechanisms, 672
  single-channel recordings
    bath solution, 660
    chronic atrial fibrillation, 662
    open probability, 663
    open-time constants, 661–662
    pipette filling solutions, 661
    threshold-crossing method, 661
  whole-cell recording
    direct electrical access, 664
    $I_{K,ACh}$-channel blockers, 665–667
    time-dependent current, 666, 668
AcroWell™ filter plate, 36
ACTH. See Adrenocorticotropin hormone
Action potential duration (APD), 655
Adenylyl cyclase (AC)
  experimental protocol, 460
  FRET probes, 459
  phosphodiesterase inhibitor, 458
  TSHR, 377
α_{1D}-Adrenergic receptors
  adrenergic agents, 115–116
  BMY 7378, 110, 115, 117
  CEC, 114
  cellular model systems, 111
  ERK, 115
  methods
    cell culture and transfection, 111–112
    intracellular calcium concentration, 113–114
    radioligand binding assays, 112–113
  photoaffinity labeling, 110–111
  physiological implications
    hypertension, 120
    intracellular calcium, 119–120
    inverse agonism, 121
    phosphoinositide turnover/calcium signaling, 119
  plasma membrane
    amino terminus-truncation, 119
  GPCRs, 117
  HEK293 cells, 118
  heterodimers, 119
  pharmacological chaperones, 118
  protein kinase C, 116
  rat-1 fibroblasts, 114, 116
  toxin-insensitive G proteins, 111
  TPA, 116, 118
  WB4101, 115, 117
β_3-Adrenoceptor
  agonists, 199–200
  antagonists, 200–201
  β_1-and β_2-AR binding experiments, 223–224
  cell culture and membrane preparation, 223
  chemistry
    3a–d preparation, 216–217
    5a–d preparation, 216–217
    alkyl 2-[4-(bromoacetyl)phenoxy] propanoates (4c–d), 219–220
    alkyl 2-[4-(2-bromoethyl)phenoxy] propanoates (3c–d), 220–221
    alkyl 2-propanoates (2e and 2g), 221–222
    1-(alkyloxycarbonyl)ethyl 4-methylbenzenesulfonates (6c–d), 217–218
    alkyl 2-phenoxypropanoates (5c–d), 218–219
    5c–d preparation, 216–217
    2-propanoic acid (SP-1e, SP-1g), 222–223
    SP-1a–h preparation, 214, 216
  clinical use, 199
  DELFIA cAMP-Eu assay, 224
  inverse agonism and blockage, 210–211
  inverse agonists
    adenylyl cyclase stimulation, 212
    cAMP accumulation, 214–215
    enantiomers, 214
    metoprolol, 212
    propranolol, 213
    salmeterol, 213
    stereoisomers, 214
  ligand selectivity
    alcohol group, 207
    alkyl substitution, 209
    amine, 207, 209
    phenyl substituents, 209
    pyridine/m-chlorophenyl ring, 207
    SAR, 207–208

β_3-Adrenoceptor (cont.)
  properties, 198
  stereospecific interactions and biological
    activity, 209
  7TD aminoacid-ligand specific interactions,
    205–206
  therapeutic target
    amibegron, 203
    brown and white adipose, 201–202
    cachexia control, 202
    gastrointestinal tract, 204
    gene expression, 205
    human pregnant and nonpregnant
      myometrium, 203
    OAB, 204
    pathophysiological process, 201
    visceral adipose tissue, 202
  WT/CAM, 200
Adrenocorticotropin hormone (ACTH), 162
Aggregation analysis
  buffer compositions, 686–687
  intramolecular inhibition, 684
  small-scale purifications, 686
Agouti-related protein (AgRP), 273
Alkaline phosphatase (ALP), 366
ALK2 constitutive activity
  alkaline phosphatase induction, 289–291
  transcriptional activity
    determination, 289
    luciferase assay, 287–288
    materials, 287
  western blot analysis, 284–287
Amaxa®, 340
American Type Culture Collection
  (ATCC), 416
Angiotensin II receptor blockers (ARBs)
  inositol phosphate accumulation assay, 171
  internalization assay, 171–174
  radioligand binding assay, 169
Anion exchange chromatography, 480
ARBs. See Angiotensin II receptor blockers
Asp-Lys-Tyr motif, 456–457
AT$_1$ receptor. See Constitutively active mutants
Autophosphorylation, 538–539
Autosomal dominant hypocalcemia (ADH), 257

B

Ba/F3 cells, 340
BigDye®, 386
Bioassay
  FACS analysis, 344–345
  luciferase reporter assay, 342–343
  signaling, 341–342
  survival/proliferation, 343–344
Bioluminescence resonance energy transfer
  (BRET), 190–193
Bio-Safe II™, 310

BMP. See Bone morphogenetic protein
Bone mineral density (BMD), 263
Bone morphogenetic protein (BMP)
  ACVR1/ALK2 model, 370–371
  adenovirus production, 291
  alcian blue staining, 368
  alkaline phosphatase assay, 291
  ALK2 constitutive activity
    transcriptional activity, 287–289
    western blot analysis, 284–287
  ALP staining, 368
  antagonist, 292
  cell culture methods, 363
  cell transfections and luciferase assays, 364
  chicken assay, 366
  chondrogenic marker genes, qPCR, 368–369
  connective tissue progenitor cells, 365
  constructs and virus production, 367
  ectopic skeletogenesis, 358
  expression plasmid construction, 363
  immunoblot analysis, 364
  immunoprecipitation, 364–365
  isolation and cultivation, micromass culture,
    367–368
  knock-in mouse, 371
  micromass culture system, 367
  overactivity and overexpression, 370
  patient methodology
    ACVR1/ALK2 mutation, 360–362
    FOP definition, 359–360
    FOP mutational analysis, 360
  pleiotropic cytokines, 282
  SHED cell, 365–366
  signaling, FOP cells, 362–363
  Smad signaling pathway, 282–283
  zebrafish embryo, 369–370
Bovine serum albumin (BSA)
  Bio-Rad assay kit, 305
  GPR40/FFA1, 572–574, 582
BRET. See Bioluminescence resonance energy
  transfer
BSA. See Bovine serum albumin

C

Calcium channels
  desensitization, 464
  ganglion neurons, 463
  receptor expression, 462
  voltage-dependent inhibition, 461–462
Calcium (Ca^{2+}) imaging
  Fluo-4 AM, 602–603
  intracellular basal Ca^{2+} concentration,
    603–604
Calcium-sensing receptors (CaSRs)
  bone remodeling, 264
  materials and methods
    cell transfection, 258

Subject Index

immunoblotting, 259–261
InsP assays, 258–259
in vivo expression, 259–260
mutant construction, 258
post hoc analysis, 261
skeletal phenotypes assessment, 261
NSHPT, 257
osteoblasts, transgenic expression, 260, 263
PTH, 254
topology, 255–256
transgenic Act-CaSR mice, skeletal phenotyping, 263–264
Wt-CaSR *vs.* Act-CaSR, 264
cAMP response element binding (CREB) protein, 57
cAMP-responsive element modulator (CREM), 379
CAMs. *See* Constitutively active mutants
Cannabinoid-2 receptor
advantages and disadvantages, 48–49
cAMP assay
cell preparation, 41
concentration-dependent augmentation, 39
data analysis, 41
instrumentations, 40
reagents and reagent preparation, 40
CHO cells, 34
GPCR, 32
GPR55, 32–33
GTPγS assay
AcroWell^{TM} filter plate, 36
basal activity, nonspecific binding, and stimulated binding, 37–38
data analysis, 38–39
instrumentations, 35
membrane preparation, 37
Prewet filter plate, 38
reagents and reagent preparation, 36
label-free technologies, 33
RT-CES
cell impedance assay, 43–45
data analysis, 45–46
E Plate^{TM}, 43
instrumentations, 42
reagents and reagent preparation, 42–43
signal transmission, 34
SR141716A and SR144528, 33
WIN55212-3, 34
Cannabinoid (CB$_1$) receptor
cancer pain, 18–19
ECS, 4, 6
emesis, 23
inflammatory bowel disease, 6
inflammatory pain models, 17–18
learning and memory, 24–25
Marinol$^©$ (dronabinol), 5
MOPs, 6

multiple sclerosis, 23–24
neuropathic pain models
CCI, sciatic nerve, 13
L5/L6 SNL, 13–16
SNI, 16–17
spontaneous pain behaviors, 12
pain-related behaviors
Hargreaves method, 21–22
movement-evoked pain, 22–23
rotarod test, 19–20
spontaneous pain, 22
tail flick model, 20
Von Frey filament up-down method, 20–21
pharmacological manipulations
intrathecal catheter, 7, 9–10
lumbar puncture, 10
oral gavage, 10
parenteral injections, mice, 10–11
parenteral injections, rats, 11–12
solubility, 7–8
vehicle preparation, 7
selective and neutral antagonists, 5
sequester $G_{i/o}$ proteins, 4
Carrageenan, 18
Catalytic subunits isolation
bacterial purification, 478–479
GST-CK2α/GST-CK2α', 475–477
materials, 475
reconstitution, 477–478
CDKs. *See* Cyclin-dependent kinases
Cell model
Ba/F3 cells, 340
HEK-293 cells
cell line, 333–334
culture conditions, 334
generation and selection, 335
MCF-7 cells
cell line, 335
generation and selection, 340
transfection conditions, 336–337, 340
transient *vs.* stable transfections, 340–341
Cellular degeneration
Annexin-based assay, 607–609
E_{rev}, 605
morphological changes
differential interference contrast, 605, 607
transfection procedure, 606
TRPM2 and TRPM7, 595
Va missense mutation, 605
CFA. *See* Complete Freund's adjuvant
Chinese hamster ovary (CHO), 415
Chloroethylclonidine (CEC), 114
Cholinergic neurons, basal forebrain
acetylcholinestease, 616, 618
culture materials, 616
culture medium, 619
papain solution, 617

Cholinergic neurons, basal forebrain (cont.)
  plating, culture dishes, 617–619
  selection, 619
  slice sectioning, 616–617
Chronic atrial fibrillation (cAF), 662
Chronic constriction injury (CCI), 13
c-Jun N-terminal kinase (JNK) isoforms
  homodimer formation, 540–542
  immunoreagents, 535
  inhibit JNK, 545
  JNK genes, 533–534
  kinase activity, 540
  kinase inactive JNK constructs, 535
  MAPK, 532, 540
  NIH-3T3 fibroblasts, 534
  nuclear translocation measurement
    cell transfection, 543
    exogenous protein, 542
    immunofluorescence staining, cultured cells, 543–544
    Schizosaccharomyces pombe, 541
  protein expression and purification
    bacterially expressed GST-tagged c-JUN protein, 537–538
    bacterially expressed 6× His-tagged JNK protein, 536–537
  radioactive *in vitro* kinase assay, 539
  signaling cascade, 532–533
  therapeutic potential, human disease, 544
  T-P-Y motif, 532, 538
  tumorigenesis, 533–534
  western immunoblot *in vitro* kinase assay, 539
CK2α′-based holoenzyme, 521
CK2α structure
  canonical regulatory key elements, 517
  CDKs and MAP kinases, 516–517
  COOT installation, 517–518
  DWG, 520–521
  EPK, 516
  global superposition, 518–520
CK2 constitutive activity
  cell lysate preparation, 499–500
  identification/validation
    cell treatment, specific CK2 inhibitors, 507–508
    endogenous proteins, radioactive phosphorylation, 508–510
    hydrophobic residues, 507
    in vivo radioactive phosphorylation, 510–511
    signaling pathways, 506
  in-cell assay, endogenous CK2 activity
    immunodetection, CK2-dependent phosphosites, 505
    ^{32}P-radiolabeling, 504
    reporter substrate phosphorylation, 506
  in-gel assay, 502–504
  neovascularization, 497
  oncogenic mutations, 498
  phosphorylation cascades, 497
  phosphotransferase reaction, 499
  protein kinase, 496–497
  RRRADDSDDDDD peptide, 498
  synthetic peptides
    cell lysates, 500–501
    CK2 holoenzyme and isolated catalytic subunit activity, 501–502
Complete Freund's adjuvant (CFA), 17–18
Constitutively active μ-opioid receptor
  analgesic/rewarding properties, 446
  antagonist potency, precipitate withdrawal acute and chronic opioid model, 438
  acute physical dependence, 438–439
  chronic physical dependence, 439–440
  inverse agonist effect, 440
  materials, 437–438
  cAMP quantification assay
    data analysis, 427
    examples, 427–430
    [^{3}H]cAMP competition binding, 426–427
    materials and equipments, 425–426
    stimulation, 426
  cell culture
    disposables, 416–417
    growth medium, 416
    tissue sources, 415–416
  forskolin stimulation, 415
  hypothesis, 447
  in vivo assessment
    materials, 431
    neutral antagonist, 430
    peak effect, 431–435
  measurement methods
    active mutants, 456–458
    adenylyl cyclase assay, 458–460
    agonist treatment, 454–456
    calcium channels, 461–464
    GDP exchange, 447–448
    G protein measures, 451–453
    inverse agonists, 448
    ion channels measurement, 460–461
    native receptor, 453–455
    negative intrinsic efficacy, 448–451
    potassium channels, 464–465
    receptor specificity, 451
  neutral antagonist, 414
  receptor/signaling pathways, 440–441
  [^{35}S]-GTP-γ-S
    binding assay, 417–423
    data analysis, 423–425
    disposables, 417
    reagents, 417
  therapeutic drug, 465
Constitutively active mutants (CAMs)
  dose response, 321, 323
  epitope-tagged receptor, 319

Subject Index

Leu^{222} identification, 406
ligand binding properties, 320–321
MC4R, 272
N111G
  β-arrestins, 167
  cell culture, transfection, and membrane preparation, 168
  data analysis, 174–175
  GPCRs, 166–167
  inositol phosphate accumulation assay, 170–172
  internalization assay (see Internalization assay)
  radioligand binding assay, 168–170
  site-directed mutagenesis and plasmid preparation, 167
  TMD III and TMD V, 166
  western blot analysis, 173–174
Constitutive MAPK and PI3K/AKT signaling activity
  immunofluorescence
    cell fixing and staining, 564–565
    cells on coverslips, 564
    nuclear vs. cytoplasmic localization, 562
    pERK, 562–563
    reagents, 563–564
  melanoma cell lines, 552
  phospho-flow cytometry
    BRAF inhibitor, 559
    constitutive pERK activity, 559–560
    materials, 561
  p27^{KIP1}, 551
  RTKs, 550
  serine/threonine protein kinases, 550
  western blotting
    antibody incubations and western blot development, 558–559
    electrophoresis, 556–557
    harvesting cells and protein extraction, 554–555
    Invitrogen's XCell SureLock Electrophoresis Cell, 553
    protein concentration, 555
    protein transfer, 557–558
    sample preparation, 555–556
    signal transduction cascades, 553
Constitutive receptor internalization assay
  flow cytometry
    HBSS, 86
    HEK 293 cells, 86
    monoclonal anti-FLAG® M2 antibody, 86
    NT cells, 84
    sample calculation, 86–87
  immunofluorescence and confocal imaging
    autofluorescence, 81
    epitope-tagged empty vector transfected cells, 82
    fixing and permeabilizing solution, 83

FLAG-GPR54, 81
HEK 293 cells, 84–85
intracellular/intraorganellar antigens, 83
negative control, 81–82
Shandon Immu-Mount^{TM}, 84
indirect receptor radiolabeling, 87–88
COOT. See Crystallographic Object-Oriented Toolkit
COS-7 cells, 168, 170
COULTER®, 344
Cryptococcus neoformans, 399
Crystallographic Object-Oriented Toolkit (COOT), 517–518
Cyclic adenosine monophosphate (cAMP) assay
  cannabinoid-2 receptor
    cell preparation, 41
    concentration-dependent augmentation, 39
    data analysis, 41
    instrumentations, 40
    reagents and reagent preparation, 40
  data analysis, 427
  drug inhibition, 429–430
  EP receptors
    forskolin, 97
    [^{3}H]cAMP, 98
    IBMX, 98
    materials, 99
    PTX, 98–99
  [^{3}H]cAMP competition binding, 426–427
  materials and equipments, 425–426
  naloxone effects, 427–428
  protean agonism
    abrogation, 48
    AM630 pretreatment, 46–47
    R(+)AM1241 and L768242, 46
  stimulation, 426
Cyclin-dependent kinases (CDKs), 516–517
Cytoplasmic tail (CT), 298

D

Differential interference contrast (DIC), 605, 607
Dissociation enhanced lanthanide fluoroimmuno assay (DELFIA), 224
DMEM. See Dulbecco's Modified Eagle's medium
Dopamine D1 and D5 receptors
  agonist-independent/dependent activation, 312
  BSA, 305
  CAM phenotype
    constitutive activity, 321–322
    dose response, 321, 323
    epitope-tagged receptor, 319
    ligand binding properties, 320–321
  conformational changes, 297
  constructs and cloning strategy
    oligonucleotide sequences, 300

Dopamine D1 and D5 receptors (cont.)
  PCR procedure, 300–301
  recombinant DNAs, 299
  cytoplasmic tail (CT), 298
  diverse methods, 326
  forskolin stimulation
    inverse agonist, 317–318
    molecular mechanism, 319
  HEK293 cells
    isolation, 301
    preparation, 302
    transfection, 302–303
  intramolecular interactions, 296
  pharmacological properties, 299
  phenotypic expression
    allosteric mechanism, 321
    PMA-mediated regulation, 321, 325
    thermodynamic equilibrium, 324
  physiological functions, 298
  quantification
    cAMP assays, 315–316
    WT and mutant forms, 313–315
  radioligand-binding assay
    crude membrane preparation, 304
    saturation and competition, 304–305
    serum impacts, 313–314
    whole cell cAMP assays, 305–312
Dual-Luciferase®, 343
Dulbecco's Modified Eagle's medium (DMEM), 168, 171

E

ECS. See Endocannabinoid system
EIAs. See Enzyme immunoassays
ELISA. See Enzyme-linked immunosorbent assay
Endocannabinoid system (ECS), 4, 6
Enhanced green florescent protein (EGFP), 648
Enzyme immunoassays (EIAs), 190, 192–193
Enzyme-linked immunosorbent assay (ELISA), 68–69
E-type prostanoid (EP) receptors
  actin stress fiber formation, 103–105
  cAMP assay
    forskolin, 97
    [^{3}H]cAMP, 98
    IBMX, 98–99
    materials, 99
    PTX, 98–99
  EP3α and EP3γ receptor, 96–97
  EP3β isoform, 97
  Gi inhibitor pertussis toxin, 102–103
  GTPase activity, 99–101
  GTPγS binding assays, 100, 102
  prostanoid metabolites, 96

Tcf/β-catenin luciferase reporter plasmid assay, 105–106
Eukaryotic protein kinases (EPK), 516
Extracellular signal regulated kinase (ERK), 115

F

Familial benign hypocalciuric hypercalcemia (FBHH), 257
Fibrodysplasia ossificans progressiva (FOP)
  chicken assay, 366
  definition, 359–360
  knock-in mouse chimera, 371
  mutation analysis, 360
FlexStation protocol, 580–581
Fluorescence-activated cell sorting (FACS), 344
Fluorescence-resonance energy transfer (FRET), 190, 193
FOP. See Fibrodysplasia ossificans progressiva
Formalin, 17
Forskolin stimulation
  inverse agonist, 317–318
  molecular mechanism, 319
FRET. See Fluorescence-resonance energy transfer
Fungi constitutive activity, 409
FUS1–lacZ reporter gene, 406–407

G

Gαβγ heterotrimers, 166
Gene Pulser®, 340
Geneticin®, 335
Ghrelin receptor
  CREB, 57
  experimental procedures
    ELISA assay, 68–69
    inositol phosphate turnover assay, 67
    SRE reporter gene assay, 68
    T-REx™, 65
  GHS, 54
  GI tract, 55
  hypothalamus, 55
  inverse agonism and efficacy swap
    AwFwLL vs. KwFwLL agonism, 62–63
    definition, 60
    fQwFwLL heptapeptide, 62
    ligand binding, 62
    mutational mapping, 61
    single-residue substitutions, 63
    wFwLL peptide, 61
  IP, 57
  phylogenic tree, 55–56
  physiological relevance, 63–65
  signal transduction, 56
  structural basis
    aromatic cluster, 59
    basal and ligand-induced receptor, 61
    The Global Toggle Switch Model, 58

Subject Index

molecular model, 59–60
mutational analysis, 60
opsin, 58
rhodopsin vs. β2-adrenergic receptor, 59
7TM receptors, 57–58
Glucose-stimulated insulin secretion (GSIS), 570–571
Glutamate receptor 1a (mGluR1a), 77
GPCRs. See G-protein-coupled receptors
GPR6
  cell surface expression detection
    antibody labeling assay, 185–187
    cell surface biotinylation, 181–183
    chymotrypsin treatment, susceptibility, 183–184
    imaging based assays, 184–185
  cell surface protein detection
    bioluminescence resonance energy transfer, 190–191
    cAMP enzyme immunoassay, 190
    cell surface biotinylation assay, 187
    chymotrypsin, 187, 189
    HEK293 cells, 187–188
    pHD1, 187
    pHluorin fluorescence, 188–189
    phospho-CREB activity, 191–192
  constitutive Gs-activity detection, 192–194
  GFP, 180
  sphingosine-1-phosphate, 180
  WGA, 181
GPR40/FFA1
  calcium-mediated toxicity, 572
  calcium mobilization measurement
    Brilliant Black BN, 576
    calcium flux-based assays, 575–576
    calcium-sensitive dyes, 575
    cell preparation, 579
    FlexStation protocol, 580–581
    Flp-In T-REx cell line generation, 577–578
    Fluo-4 loaded cells, 579–580
    HEK293 cells, 577
    materials, 577
    Tet repressor, 576
    transient transfection, 576
  cAMP-based assay, 574
  $G\alpha_{q/11}$-coupled receptor, 571
  GSIS, 570–571
  IP_3, 571
  label-free technologies, 575
  [^{35}S] GTPγS-binding assay
    BSA, 572–574, 582
    buffers, 584
    concentration-response curves, 573, 582
    equipments, 583
    GDP and GTP, 581
    immunoprecipitation, 585–587
    reagents and chemicals, 583–584
    transient expression and membrane preparation, 584–585
    transient transfection, 583
G protein-coupled receptor kinase 4 (GRK4)
  β-arrestin recruitment
    coimmunoprecipitation, 646–647
    redistribution, 647–648
  GRK4β, 632
  kinase dead-GRK4, 637, 639
  polymorphisms, 636–638
  positive control samples, 636
  receptor internalization
    flow cytometry measurements, 643–644
    fluorometer measurements, 644–645
    radiolabeled-ligand-binding assays, 644–646
  receptor phosphorylation
    inducible system, 641–642
    materials and solutions, 639–640
    nonradioactive assay, 641
    radioactive assay, 640–641
  splice variants
    positive control samples, 636
    primers, 636–637
    western blot detection, 635
G-protein-coupled receptor 54 (GPR54), KISS1-R
  cell preparation and transfection, 79–80
  constitutive receptor internalization assay
    flow cytometry, 84, 86–87
    immunofluorescence and confocal imaging, 81–84
    indirect receptor radiolabeling, 87–88
  GRK2, 76
  IP turnover, intact cells
    anion exchange chromatography, 90
    DMEM, 89
    vs. FLAG-GPR54-expressing HEK 293 cells, 91
    HBSS, 89–90
    NT control, 88
  kisspeptin, 76–77
  materials
    flow cytometry assay, 78–79
    HEK 293 cells, 77–78
    immunofluorescence assay, 78
    inositol phosphate formation assay, 79
    receptor radiolabeling assay, 79
  mGluR1a, 77
  serum starvation, 80
  trypsinization and reseeding, 80
G-protein-coupled receptors (GPCRs)
  α_{1D}-Adrenergic receptors, 117
  CAMs N111G, 166–167
  cannabinoid-2 receptor, 32
  CaSRs, 255

G-protein-coupled receptors (GPCRs) (cont.)
coimmunoprecipitation of β−arrestin, 646–647
constitutive GPCR signaling
histabudifen and histapendifen, 129, 131
histamine-and mepyramine-induced modulation, 129–130
inactive (R) and active (R★) conformations, 129
Cpr2 protein identification, 400–401
GPCR-induced G_q activation, 138
G protein signaling, *C. neoformans*, 399–400
immunoprecipitation, 639, 642
internalization of receptors
flow cytometry measurements, 643–644
fluorometer measurements, 644
radiolabeled-ligand-binding assays, 644–646
ligand activation, 398
redistribution β−arrestin, 647–648
7-transmembrane domain GPCR, 150
yeast expression, 401–408
G protein gated inwardly rectifying potassium channels (GIRK), 460
GraphPad prism, 427
Green fluorescent protein (GFP), 180
GRK4. *See* G protein-coupled receptor kinase 4
Growth hormone secretagogue (GHS), 54
GSIS. *See* Glucose-stimulated insulin secretion
[^{35}S] GTPγS
binding assay
agonist-induced stimulation, 419
assay components, 420
BSA, 572–574, 582
buffers, 584
concentration-response curves, 573, 582
DAMGO, HERK effects, 421–423
equipments, 583
GDP and GTP, 581
immunoprecipitation, 585–587
naloxone effects, 421
reagents and chemicals, 583–584
shift experiment, 423–424
shift paradigm, 418
transient expression and membrane preparation, 584–585
transient transfection, 583
data analysis, 423–425
disposables, 417
reagents, 417
GTPγS assay
AcroWell^{TM} filter plate, 36
basal activity, nonspecific binding, and stimulated binding, 37–38
data analysis, 38–39
instrumentations, 35
membrane preparation, 37
Prewet filter plate, 38
reagents and reagent preparation, 36

Guanine nucleotide exchange factors (GEFs), 298
Guanosine diphosphate (GDP), 100, 581
Guanosine triphosphatase-activating proteins (GAP), 378
Guanosine triphosphate (GTP), 581

H

Hargreaves method, 21–22
HEK293 cells
cell growth, 240
cell model, 333–335
GPR54, 76
GPR40/FFA1, 577
isolation, 301
preparation, 302
transfection, 302–303
transient transfections, 241
HEK293T cells, 268–269
Histamine H_1 receptor
constitutive active mutants, 142–143
constitutive GPCR signaling
histabudifen and histapendifen, 129, 131
histamine-and mepyramine-induced modulation, 129–130
inactive (R) and active (R★) conformations, 129
expression level
agonist-independent signaling, 130
d-chlorpheniramine, 131
H_2R, 131
R★ conformation, 129–130
up-regulation, 131–132
$G\alpha_i$-coupled signaling, 142, 144
G proteins, 133
inflammatory H_1Rs, 128–129
inositol phosphate accumulation assay
Dowex columns, 135
[^{3}H]-InsP, 135–136
negative intrinsic activity, 136
SPA, 136
special materials and reagents, 134–135
label-free impedance biosensors, 133
prolonged inverse agonist treatment, 140–141
reporter gene assays
β-galactosidase readout, 138
luciferase readout, 138
special materials and reagents, 136–137
white-and black-bottom plates, 138
R-SAT
β-galactosidase marker gene, 139
contact-inhibited cells lines, 140
DMEM, 139
GPCR-induced G_q activation, 138
special materials and reagents, 139
three parameter equation, 140
RT-CES system, 140
signal transduction, 133–134

Subject Index

Histamine H_2 receptor (H_2R), 131
hSSTR. *See* Human somatostatin receptor
Human chorionic gonadotropin (hCG), 232
Human embryonic kidney (HEK) cells
  endogenous currents, 601
  HEK293 cells
    GPR54, 76
    isolation, 301
    preparation, 302
    transfection, 302–303
  $I-V$ curves, 596
  transfection procedure, 606
Human prolactin receptor
  bioassay
    luciferase reporter assay, 342–343
    signaling, 341–342
    survival/proliferation, 343–344
  breast tumorogenesis, 353
  cell model
    Ba/F3 cells, 340
    HEK-293 cells, 333–335
    MCF-7 cells, 335–340
    transient *vs.* stable transfections, 340–341
  constitutive activity, identification, 345–353
  experimental models, 333
  missense polymorphisms, 331–332
  pathophysiology, 330
  rodent models, 331
Human somatostatin receptor (hSSTR)
  immunocytochemistry
    AtT20 stable transfectants, 157
    binding affinities, 155–156
    illumination intensity, 157
    invitrogen, 155
    line cell receptor internalization, 158
    PBS, 154
    temporal and spatial resolution, 157
  mRNA expression, 154
Hyperparathyroidism (HPT), 254
Hypocretin. *See* Orexin

I

IBMX. *See* Isobutylmethylxanthine
Immunofluorescence
  confocal imaging
    autofluorescence, 81
    epitope-tagged empty vector transfected cells, 82
    fixing and permeabilizing solution, 83
    FLAG-GPR54, 81
    HEK 293 cells, 84–85
    intracellular/intraorganellar antigens, 83
    negative control, 81–82
    Shandon Immu-Mount™, 84
  constitutive MAPK and PI3K/AKT signaling activity
    cell fixing and staining, 564–565
    cells on coverslips, 564
    nuclear *vs.* cytoplasmic localization, 562
    pERK, 562–563
    reagents, 563–564
    staining, cultured cells, 543–544
In-cell assay, endogenous CK2 activity
  immunodetection, CK2-dependent phosphosites, 505
  ^{32}P-radiolabeling, 504
  reporter substrate phosphorylation, 506
Inducible cAMP early repressor (ICER), 379
In-gel assay, 502–504
Inositol monophosphate (IP), 57
Inositol phosphate (InsP) assays, 258–259
Internalization assay
  vs. AT_1 wild-type mutant, 172–174
  cell-associated radioactivity, 171
  GR231118, 173
  inverse agonists, 172
  serum-free DMEM, 171
Intracellular calcium concentration ($[Ca^{2+}]_i$), 113–114
Inverse agonism
  α_{1D}-adrenergic receptors, 121
  β_3-Adrenoceptor, 210–211
  AgRP
    MC3R, 273, 275
    MC4R, 271–272
  efficacy swap
    AwFwLL *vs.* KwFwLL agonism, 62–63
    definition, 61
    fQwFwLL heptapeptide, 62
    ligand binding, 62
    mutational mapping, 61
    single-residue substitutions, 63
    wFwLL peptide, 61
Invitrogen's XCell SureLock Electrophoresis Cell, 553
In vivo assessment, antagonist potency
  materials, 431
  neutral antagonist, 430
  peak effect
    hyperlocomotion, 434, 436
    morphine antinociception, 432–433
    naltrexone analog, 435, 437
    pharmacokinetic factors, 431–432
    sample protocol, 432
    variation, 433–434
Isobutylmethylxanthine (IBMX), 98–99, 191

J

JNK genes, 533–534

L

Label-free impedance biosensors, 133
Lactogenic hormone response elements (LHRE), 342

Large-scale purification, 682–684
LHR. *See* Lutropin receptor
Luciferase, 269
Lutropin receptor (LHR)
　BRET, 234–235
　cAMP production, 245–246
　cell surface hLHR expression
　　cell surface epitope-tagged hLHR, 244–245
　　maximal ^{125}I-hCG binding, 243–244
　dimers and oligomers, 234
　FSHR and TSHR, 233
　hCG, 232
　hormone binding affinity, 235
　principles, quantification
　　basal cAMP levels, 235–236
　　cell surface receptor densities, 235
　　dose-response curves, 237–238
　　hCG-stimulated cAMP, 236–237
　　HEK293 cells, 236
　　^{125}I-hCG binding, 238
　　L457R cells, 236–237
　　wt hLHR and hLHR CAMs, 239
　recombinant hLHR
　　cDNA encoding LHR, 240–241
　　gonadal cells, 239
　　HEK293 cell growth, 240
　　heterologous cells, 239–240
　　transient transfections, 241–243
　reproductive physiology, 232
　serpentine domain, 233
　signaling inactive hLHR
　　BRET saturation curves, 249
　　coexpression, 246, 248–249
　　dose-response curves, 246
　　HA-hLHR, 248
　　MC3R, 250
　　myc-tagged hLHR, 249
　　partial attenuation, 246–247
　　TM3 and TM6, 234

M

Mammalian cells
　immunoprecipitation, 479–480
　modulation, 490–491
MAPK. *See* Mitogenactivated protein kinases
Marinol© (dronabinol), 5
MCF-7 cells
　cell line, 335
　generation and selection, 340
　transfection conditions, 336–337, 340
MC4R molecule, 272–274
Microcomputed tomography (μCT), 261
Microswitches, 58
Mitogen-activated protein kinases (MAPK), 516–517, 532, 540
μ-opioid receptors (MOPs), 6. *See also*
　Constitutively active μ-opioid receptor

Morris water maze, 25
Multiple sclerosis (MS), 23–24

N

Neonatal severe hyperparathyroidism (NSHPT), 257
Neural melanocortin receptors
　β$_2$-adrenoreceptor, 276
　computational modeling
　　rhodopsin, 274
　　side chain orientation, 275–276
　HEK293T cells, culture and transfection, 268–269
　naturally occurring mutants, 270–271
　signaling pathway, 269–270
Nontransfected (NT) cells, 84
Nucleation promoting factors (NPFs), 678–679
Nucleofector®, 340

O

Orexin A (OXA), 623, 627
Orexin and substance P effects, KirNB channels
　equipment, 620
　signal transduction
　　phosphatase inhibitor, 627–628
　　PKC inhibitor, 626
　　PKC pseudosubstrate, 627–628
　single-channel recordings
　　A1 and A2, 623–624
　　cell-attached mode, 624
　　experimental procedures, 624–625
　whole-cell recordings
　　application, 620
　　experimental procedures, 621
　　I–V relation, 621–622
OXA. *See* Orexin A

P

Parathyroid hormone (PTH), 254
Patch-clamp techniques, $I_{K,ACh}$ channels
　basic principles, 655–657
　human atrial cardiomyocytes isolation, 657–659
　single-channel recordings
　　bath solution, 660
　　chronic atrial fibrillation, 662
　　open probability, 663
　　open-time constants, 661–662
　　pipette filling solutions, 661
　　threshold-crossing method, 661
　whole-cell recording
　　direct electrical access, 664
　　$I_{K,ACh}$-channel blockers, 665–667
　　time-dependent current, 666, 668
Pertussis toxin (PTX), 98–99, 102, 453

Subject Index

Phenotypic expression
  allosteric mechanism, 321
  PMA-mediated regulation, 321, 325
  thermodynamic equilibrium, 324
pHGPR6 cells, 190, 192
Phorbol-12-myristate-13-acetate (PMA), 324
Phosphatase and tensin homologue (PTEN), 551
Phosphate-buffered saline (PBS), 154
Phospho-cyclic AMP response element-binding (pCREB) assay, 193
Phospholipase C (PLC), 377
Physical dependence protocol
  acute, 438–439
  chronic, 439–440
PKC. *See* Protein kinase C
Polyacrylamide-SDS gel, 502
Polymerase chain reaction (PCR) method, 167
Prolactin (PRL), 330–331
Protean agonism
  abrogation, 48
  AM630 pretreatment, 46–47
  R(+)AM1241 and L768242, 46
Protein kinase C (PKC)
  inhibitors, 626
  pseudosubstrate, 627–628
Protein kinase CK2
  analytical gel filtration chromatography
    low salt buffer, 523–524
    materials, 522–523
  autophosphorylation
    materials, 525–527
    regulatory CK2b subunit, 527–528
  bona fide substrate, 491
  catalytic subunits isolation
    bacterial purification, 478–479
    GST-CK2α/GST-CK2α', 475–477
    materials, 475
    reconstitution, 477–478
  cellular processes, 472
  CK2α'-based holoenzyme, 521
  CK2α structure
    canonical regulatory key elements, 517
    CDKs and MAP kinases, 516–517
    COOT installation, 517–518
    DWG, 520–521
    EPK, 516
    global superposition, 518–520
  disease-associated functions, 516
  inducible expression
    materials, 487–488
    tetracycline-regulation, 488–489
  kinase assay
    GST-CK2 and RRRDDDSDDD, 482
    materials, 481
    peptide substrates, 482–483
  mammalian cells, immunoprecipitation
    anion exchange chromatography, 480
    isolation procedure, 479–480
    materials, 479
  mammalian cells, modulation
    knockdown strategies, 490
    pharmacological inhibition, 490–491
  peptide array, 483–484
  phosphoproteomics, 474
  protein analysis, SDS-PAGE, 524–525
  protein substrates, 484–485
  supramolecular oligomerization, 522
  transient expression, 486–487
  tumorigenesis, 473
PTX. *See* Pertussis toxin
Pyrene actin polymerization assays
  procedure, 688–691
  reagents, 687–688

Q

Quantitative PCR (qPCR), 260–261
Quasi-steady state level (QSS), 666

R

Radial maze, 24
Radioactive *in vitro* kinase assay, 539
Radioimmunoassay, 270
Radio-iodination, 390
Radioligand-binding assay
  crude membrane preparation, 304
  saturation and competition, 304–305
Real-time cell electronic sensing (RT-CES) system
  cell impedance assay, 43–45
  constitutive H_1R signaling, 140
  data analysis, 45–46
  E-Plate^{TM}, 43
  instrumentations, 42
  reagents and reagent preparation, 42–43
Receptor internalization
  imaging methods
    flow cytometry measurements, 643–644
    fluorometer measurements, 644–645
    radiolabeled-ligand-binding assays, 644–646
Receptor phosphorylation
  inducible system, 641–642
  materials and solutions, 639–640
  nonradioactive assay, 641
  radioactive assay, 640–641
Receptor selection and amplification technology (R-SAT)
  β-galactosidase marker gene, 139
  contact-inhibited cells lines, 140
  DMEM, 139
  GPCR-induced G_q activation, 138
  special materials and reagents, 139
  three parameter equation, 140
Receptor tyrosine kinases (RTKs), 550

Regulators of G protein signaling (RGS), 378
Reversal potential (E_{rev}), 600
R-SAT. See Receptor selection and amplification technology
RT-CES system. See Real-time cell electronic sensing system

S

SAR. See Structure-activity relationships
Schizophyllum commune, 401
Schizosaccharomyces pombe, 406
Scintillation proximity assay (SPA), 136
Sequester $G_{i/o}$ proteins, 4
Ser/Thr protein kinase, 497–498, 550
Signal transduction, KirNB channels
  phosphatase inhibitor, 627–628
  PKC inhibitor, 626
  PKC pseudosubstrate, 627–628
Simian virus 40 (SV40), 268
Single-channel recordings
  acetylcholine-gated $I_{K,ACh}$ channels
    bath solution, 660
    chronic atrial fibrillation, 662
    open probability, 663
    open-time constants, 661–662
    pipette filling solutions, 661
    threshold-crossing method, 661
  orexin and substance P effects, KirNB channels
    A1 and A2, 623–624
    cell-attached mode, 624
    experimental procedures, 624–625
Single nucleotide polymorphisms (SNPs), 332
SNI. See Spared nerve injury
SNL. See Spinal nerve ligation
Social recognition model, 24–25
Somatostatin receptor (SSTR)
  cell selection, 151–152
  cortistatins, 150
  growth hormone, 150–151
  knockdown
    AtT20 cells, 160
    mRNA, intracellular cAMP, and ACTH levels, 162–163
    siRNA, 160, 162
  overexpression
    cAMP assays, 159–160
    intracellular cAMP measurements, 158–159
    plasmids, 153
    stable transfectants, 153–154
    system validation (*see* Human somatostatin receptor)
  7-transmembrane domain GPCR, 150
Spared nerve injury (SNI), 16–17
Spinal nerve ligation (SNL), 13–16
SSTR. See Somatostatin receptor

Stem cells from human exfoliated deciduous teeth (SHED) cell, 365–366
Structure-activity relationships (SAR), 207–208

T

T-cell factor (Tcf)/β-catenin signaling, 97
Tetracycline-regulated expression system (T-REx^{TM}), 65
Tetradecanoyl phorbol acetate (TPA), 116, 118
Tet repressor, 576
Thyroid peroxidase (TPO), 377
Thyroid stimulating hormone receptor (TSHR)
  β arrestin proteins, 378
  binding assay, 389–390
  cell surface expression, 388–389
  constitutive activity *in vitro*
    materials, 386–387
    transfection and luciferase assay, 387–388
  constitutively active mutations, 379–380
  desensitization, 377–378
  DNA extraction, 383
  gene amplification
    materials, 384
    PCR and gene sequencing, 384–386
  G_s and G_q protein interaction, 380–382
  in silico analysis, 379
  phosphorylation analysis, 391–392
  posttranslational processes, 376
  receptor shedding, 377
  replication slippage, 381
7TM receptors, 55–58
TPA. See Tetradecanoyl phorbol acetate
Transient receptor potential (TRP) channels
  advantages, 593
  calcium imaging
    Fluo-4 AM, 602–603
    intracellular basal Ca^{2+} concentration, 603–604
  cellular degeneration
    Annexin-based assay, 607–609
    E_{rev}, 605
    morphological changes, 606–607
    TRPM2 and TRPM7, 595
    Va missense mutation, 605
  heterologous expression systems, 596
  outward rectification, 597–598
  physiological mechanisms, 603–604
  whole-cell patch-clamp recordings
    channel voltage dependence, 600–601
    electrophysiological recordings, 598
    endogenous currents, 601–602
    E_{rev}, 600
    I–V curves, 600–601
    stereomicroscope visualization, 599

Subject Index

Transient transfections
 HEK293 cells, 241
 wt and mutant LHR
  cell surface densities, 241–242
  cell surface levels, 242–243
Transmembrane domain (TMD), 166
TRITC-phalloidin, 103–105
TRP channels. See Transient receptor potential channels
TSHR. See Thyroid stimulating hormone receptor

U

UltraCULTURE™, 313

V

Venus flytrap (VFT), 256
Von Frey filament up-down method, 20–21

W

WAVE and WASH proteins, Arp2/3 complex
 aggregation analysis
  buffer compositions, 686–687
  intramolecular inhibition, 684
  small-scale purifications, 686
 endogenous activity, 691–692
 large-scale purification
  procedure, 682–684
  reagents, 682
 pyrene actin polymerization assays
  procedure, 688–691
  reagents, 687–688
 stable cell lines
  procedure, 681–682
  reagents, 680–681
WHAMM, 678
Western blotting
 antibody incubations and western blot development, 558–559
 electrophoresis, 556–557
 harvesting cells and protein extraction, 554–555
 Invitrogen's XCell SureLock Electrophoresis Cell, 553
 materials, 285–286
 Ponceau staining, 286
 protein concentration, 555
 protein transfer, 557–558
 sample preparation, 555–556
 signal transduction cascades, 553
 Smad1 phosphorylation, 284–285
Western immunoblot *in vitro* kinase assay, 539
Wheat germ agglutinin (WGA), 181
Whole cell cAMP assays
 adenylyl cyclase activity, 305
 constitutive activity, 306
 [^{3}H]-ATP conversion
  chromatography columns, 309–311
  medium, 307–308
  sample preparation, 308
 metabolic labeling [^{3}H]-adenine, 307
 quantification, 312
 well plates, preparation, 306–307
Whole-cell patch-clamp recordings
 channel voltage dependence, 600–601
 electrophysiological recordings, 598
 endogenous currents, 601–602
 endogenous currents, expression systems, 601–602
 E_{rev}, 600
 I-V curves, 600–601
 stereomicroscope visualization, 599
Whole-cell recordings
 acetylcholine-gated $I_{K,ACh}$ channels
  direct electrical access, 664
  $I_{K,ACh}$-channel blockers, 665–667
  time-dependent current, 666, 668
 orexin and substance P effects, KirNB channels
  application, 620
  experimental procedures, 621
  I-V relation, 621–622
 WT *vs.* mutant forms, 313–316

Y

Yeast expression
 constitutive activity
  β-galactosidase, 405–406
  CPR2 expression, 405
  Leu^{222}, 406–407
 limitations, 408
 materials
  β-galactosidase measurement, 405
  transformation, protocol, 404–405
 pheromone-responsive pathway, 401–402

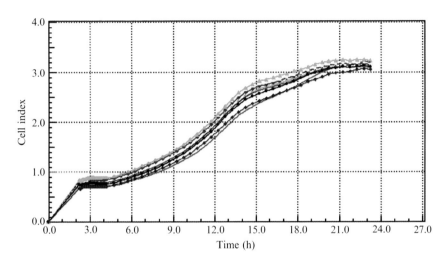

Massimiliano Beltramo et al., Figure 2.3 Time course of impedance variations during cell adhesion and growth. Cells were seeded at time zero and variation in cell impedance monitored over a period of 24 h. The cell index, an index related to cell impedance, is increased for several hours before reaching a plateau at around 24 h after cell seeding. Different curves represent different wells and each dot represents a recording point. Reported curves are representative of results obtained in several experiments.

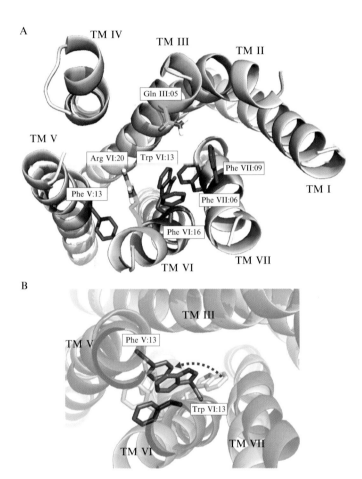

Jacek Mokrosiński and Birgitte Holst, Figure 3.2 Molecular model of the residues proposed to be part of the structural basis for the high constitutive activity in the ghrelin receptor. (*Panel A*) Molecular model of the ghrelin receptor built over the inactive structure of the β2-adrenergic receptor. The seven helical bundles are displayed without the loops as viewed from the extracellular side. Only the residues on the inner faces of TMs III, VI, and VII, which in mutational analysis have been identified to be involved in the constitutive activity are shown. The residues forming the aromatic cluster in between TMs VI and VII are indicated *in purple* and ionic/hydrophilic interaction in TMs III and VI are indicated in *green*. (*Panel B*) Presumed interaction between Trp VI:13 and Phe V:13 seen in Monte Carlo simulations of the activation process where inactive and active conformations are shown in *white* and *blue ribbons*, respectively.

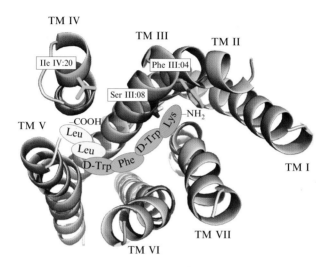

Jacek Mokrosiński and Birgitte Holst, Figure 3.3 Molecular model of the ghrelin receptor α-helical transmembrane domains viewed from the extracellular side. The residues identified to drive the efficacy swap from agonism toward inverse agonism as well as from inverse agonism toward agonism are shown *in red* and *in green* sticks, respectively. A schematic model of the KwFwLL inverse agonist has been "docked" into the model in a configuration where the N-terminal molecular switch region of the ligand is placed in proximity to the efficacy switch region of the receptor—positions III:04 and III:08. It is proposed that the alpha NH_2-group of the N-terminal residue (Lys in this case Ala in the agonist version) makes an interaction with the region in between TM III and TM II and that the central aromatic cluster makes key interactions with the aromatic cluster presented at the interface between TMs VI and VII (including a potential with Arg VI:20) and that the C-terminal double Leu sequence is located in the pocket between TMs III, IV, and V.

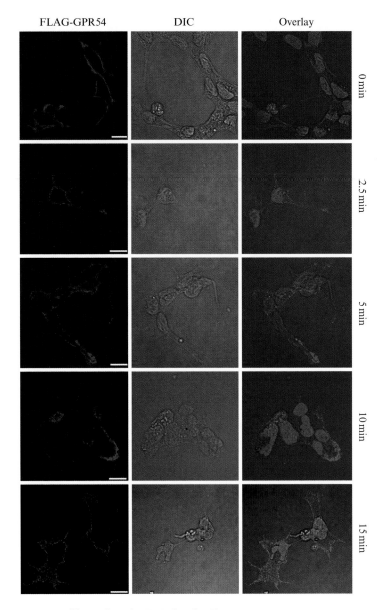

Macarena Pampillo and Andy V. Babwah, Figure 4.1 HEK 293 cells transiently expressing FLAG-GPR54 were surface labeled at 4 °C (0 min) using rabbit anti-FLAG antibody and then were incubated at 37 °C for 2.5, 5, 10, and 15 min. After fixation, surface GPR54 was detected by Alexa Fluor 568-conjugated anti-rabbit IgG. Images are representative of four independent experiments. Scale bars: 10 μm. DIC, differential interference contrast.

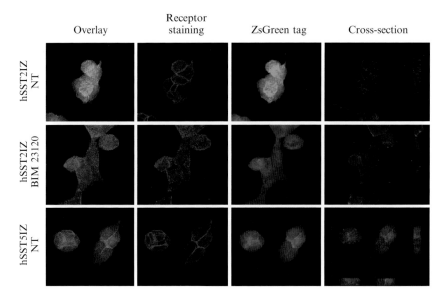

Anat Ben-Shlomo et al., Figure 8.1 hSSTR protein expression and localization in AtT20 stable transfectants. Human SSTR2 and hSSTR5 localize to the cell membrane in AhSSTR2IZ and AhSSTR5IZ cells, respectively. One hour treatment with selective SSTR2 agonist, BIM-23120 (100 nM), causes hSSTR2 internalization. Cells were plated on coverslips, treated or not for 1 h at 37 °C, fixed in 4% paraformaldehyde, stained with the receptor-directed antibody and visualized with confocal microscopy. DNA, blue; ZsGreen, green; and receptors, red. Adapted from Ben-Shlomo et al. (2009a).

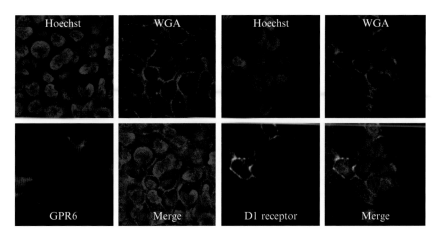

Balakrishna M. Prasad et al., Figure 10.1 GPR6 appears to be predominantly located in the intracellular compartments. HEK293 cells expressing GFP-tagged GPR6 or D1 receptors were labeled with WGA (plasma membrane marker) and Hoechst 33342 (nuclear stain). These images show that D1 receptor is primarily located along the plasma membrane, while GPR6 fluorescence is mostly located between plasma membrane and nuclei.

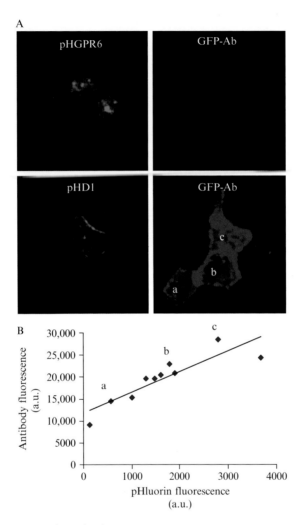

Balakrishna M. Prasad et al., Figure 10.5 Surface expression of GPR6 in HEK293 cells was not detected by antibody labeling. (A) Compressed images constructed from Z-stacks of pHluorin fluorescence pictures of pHGPR6 or pHD1 are shown in green. Staining of the same cells with GFP antibody is shown in red. (B) pHluorin fluorescence intensities of individual cells expressing pHD1 are plotted against corresponding GFP antibody fluorescence values. Fluorescence data points corresponding to the three cells labeled in (A) are indicated by a, b, and c. The slope (4.7) and intercept (11,906) of linear regression line provide a measure of the sensitivity of this assay ($R^2 = 0.78$). None of the pHGPR6 expressing cells have any detectable antibody staining and are not shown in this graph. Reproduced with permission from Padmanabhan et al. (2009).

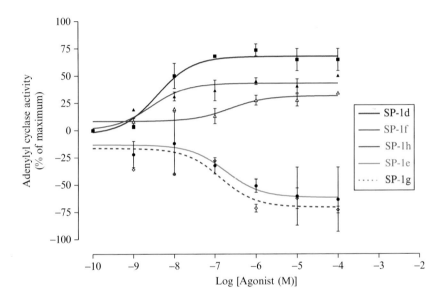

Maria Grazia Perrone and Antonio Scilimati, Figure 11.8 Representative curves of adenylyl cyclase activity evaluation in CHO cells stably expressing human β_3-AR subtype. Adenylyl cyclase stimulation was calculated as percentage (%) of the maximum effect by isoproterenol (ISO, Table 11.3).

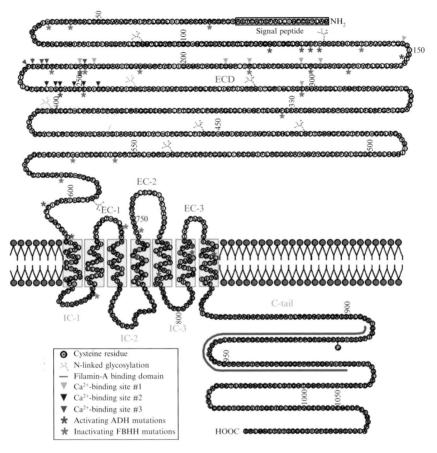

Wenhan Chang *et al.*, Figure 13.1 Topology of the bovine CaSR. The legend (lower left) shows symbols representing naturally occurring activating (gain-of-function) mutations causing ADH and inactivating (loss-of-function) mutations responsible for FBHH. Modified, with permission, from Brown *et al.* (1993).

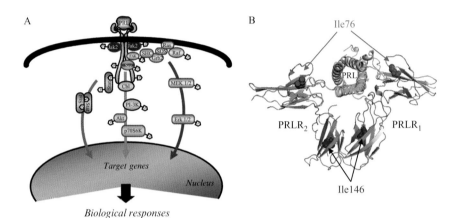

Vincent Goffin et al., Figure 17.1 Molecular features of the PRLR and its constitutively active variants. (A) This figure represents a simplified view of PRLR signaling. The main cascades involve the tyrosine kinase Jak2, which in turn activates three members of the STAT (signal transducers and activators of transcription) family, Stat1, Stat3 and, mainly, Stat5. The MAP Kinase pathway is another important cascade activated by the PRLR. It involves the Shc/Grb2/Sos/Ras/Raf intermediaries upstream to MEK and Erk-1/-2 kinases. Activation of other signaling pathways have been reported, among which Src and Akt kinases are known to play important roles in some responses such as cell proliferation. Yellow boxes symbolize phosphorylated tyrosine. Reproduced with permission from Goffin et al. (2005) (Copyright 2010, The Endocrine Society). (B) The crystal structure of human PRL (green) bound to two moieties of the rat PRLR extracellular domain (orange) has been solved recently (Broutin et al., 2010). Isoleucine 76 (red) and Isoleucine 146 (blue) are represented on the receptor.

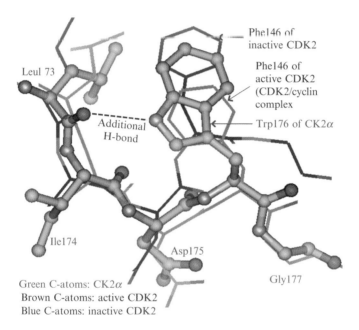

Birgitte B. Olsen *et al.*, Figure 25.2 Superimposition of CK2α and inactive and (partially) active CDK2 created with COOT. In CDK2, the "DFG motif" is visible for both the inactive and active conformation. However, CK2α contains a "DWG motif," and here the tryptophan176 of CK2α makes an additional H-bond stabilizing the active conformation of CK2α.

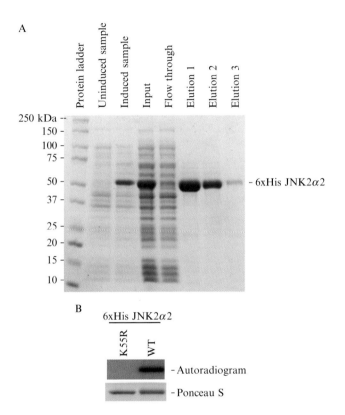

Ryan T. Nitta et al., Figure 26.2 (A) Purification of recombinant JNK2α2 from *E. coli*. 6× His-tagged JNK2α2 was purified from cleared bacterial lysate and the samples were run on a SDS-PAGE followed by a Coomassie Blue stain. (B) Radioactive *in vitro* kinase assay using purified 6× His-JNK2α2 protein.

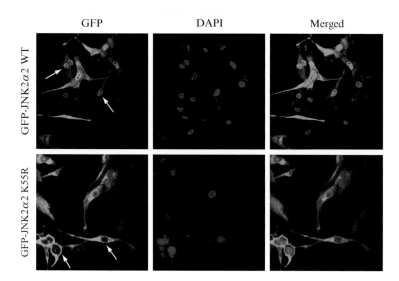

Ryan T. Nitta et al., Figure 26.4 Direct immunofluorescence of GFP-tagged JNK constructs in U87-MG cells measuring nuclear translocation. Top: Nuclear localization of the constitutively active GFP-tagged wild type JNK2α2 construct. Bottom: Nuclear localization of the kinase dead GFP-tagged K55R JNK2α2 mutant. Arrows indicate cells with strong nuclear localization (top) or cytoplasmic stain (bottom).

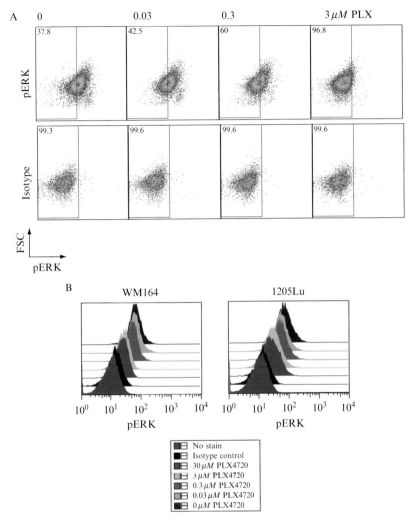

Kim H. T. Paraiso et al., Figure 27.2 Using phospho-flow to measure constitutive pERK activity. (A) Demonstration of the specificity of response. WM164 melanoma cells were treated with PLX4720 (0–3 μM, 1 h), fixed, and stained for either pERK or a matching Rabbit IgG isotype control. Increasing concentrations of drug were found to inhibit pERK staining. The isotype control was used to gate for the analysis shown in Fig. 27.1B. (B) Quantification of levels of pERK signaling following PLX4720 treatment as measured by phospho-flow cytometry. Cell lines (WM164 and 1205Lu melanoma cell lines) were treated with PLX4720 (0–3 μM, 1 h), fixed, and stained for pERK. MFI (median fluorescence intensity) indicates the absolute level of pERK staining.

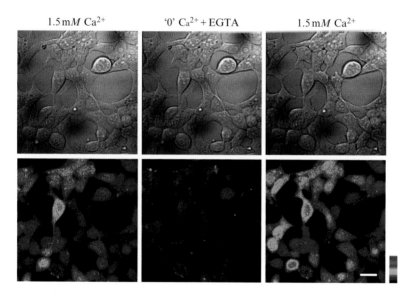

Shaya Lev and Baruch Minke, Figure 29.4 The intracellular basal Ca^{2+} concentration is affected by extracellular Ca^{2+}. Wide-field confocal images of native HEK cells loaded with Fluo4-AM indicator, displaying resting $[Ca^{2+}]_i$ in different levels of extracellular Ca^{2+}, as indicated above. Note how the intracellular Ca^{2+} concentration decreases significantly, upon lowering of the extracellular Ca^{2+} concentration, in a reversible manner (middle panel). This is most likely due to the activity of the Na/Ca^{2+} exchanger. Scale bar: 20 μm.

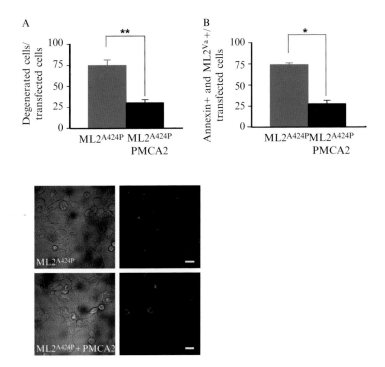

Shaya Lev and Baruch Minke, Figure 29.6 Ca^{2+}-induced degeneration caused by constitutive activity of h-TRPML2-^{A424P} is rescued by the calcium extrusion pump, PMCA2. (A) Top: histogram showing the percentage of degenerated HEK cells following cotransfection of h-TRPML2-^{A424P} (ML2-A424P) together with PMCA2 or empty vector (**, $P < 0.01$). Note the rescue of cell degeneration in cells cotransfected with h-TRPML2-^{A424P} and PMCA2; bottom: wide-field confocal images (left: DIC transmitted, right: fluorescent-conjugated channel-expressing cells) showing representative cell morphology in the two groups of cotransfected HEK cells described above (scale bar: 20 μm). Note that most cells expressing the channel without PMCA2 display degeneration (distinct changes in morphology). (B) Histogram showing the percentage of Annexin V-positive HeLa cells following cotransfection of h-TRPML2-^{A424P} with PMCA2 or with empty vector (*, $P < 0.05$). Note the rescue of cell degeneration in cells cotransfected with h-TRPML2-^{A424P} and PMCA2.

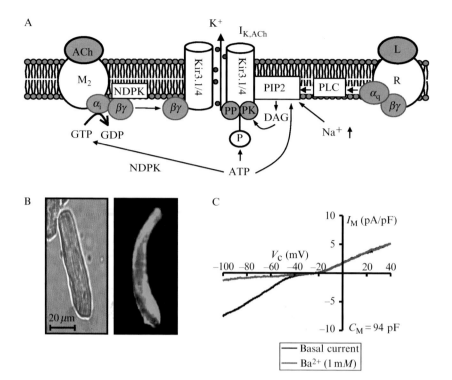

Niels Voigt et al., Figure 32.8 Regulatory mechanisms of $I_{K,ACh}$ in the atrium. (A) Binding of the vagal neurotransmitter acetylcholine (ACh) to muscarinic M_2 receptors leads to dissociation of G_i protein α- and $\beta\gamma$-subunits. In addition, transphosphorylation between adenosine and guanosine nucleosides via nucleoside diphosphate kinase (NDPK) may cause receptor-independent G_i protein dissociation. In both cases, the consecutive binding of G_i protein $\beta\gamma$-subunits to the $I_{K,ACh}$ channel subunits Kir3.1 and Kir3.4 strengthens their interaction with cell membrane-located phosphatidylinositol 4,5-bisphosphate (PIP_2), thereby increasing $I_{K,ACh}$ open probability. Similarly increased intracellular Na^+ also strengthens the interaction of PIP_2 with the channel leading to receptor-independent current activation. In contrast, stimulation of G_q-coupled receptors (R; i.e., AT_1-receptors or α_1-adrenoceptors) activates phospholipase-C (PLC), thereby lowering the PIP_2 membrane content, which results in $I_{K,ACh}$ inhibition. Furthermore, the PIP_2 metabolite diacylglycerol activates protein kinase C that phosphorylates $I_{K,ACh}$ channel subunits. In general, channel phosphorylation, which is dynamically regulated by protein kinases (PK) and phosphatases (PP), is an absolute requirement for normal $I_{K,ACh}$ function. Adenosine triphosphate (ATP) modulates the channel through multiple actions: (i) direct phosphorylation of the channel and/or channel regulatory proteins, (ii) generation of PIP_2, and (iii) transphosphorylation between adenosine and guanosine nucleosides via nucleoside diphosphate kinase (NDPK). (B) Adenoviral-mediated expression of "enhanced green fluorescence protein" and the $G\beta\gamma$-scavenger transducin-α. With bright-field microscopy, adenovirally transfected human right atrial myocytes are indistinguishable from nontransfected cells (left), whereas the fluorescence emission light at an excitation wavelength of 488 nm indicates successful transfection (right). (C) Basal inward-rectifier current (I_M) in response to a ramp pulse from -100 to $+40$ mV (see also Fig. 32.5A) measured in a human right atrial myocyte transfected with transducin-α. Identity as inward-rectifier current was specifically assessed by application of Ba^{2+} (1 mM). C_M, cell capacity.